Robert

Box 83

Austin, TX

233-72-4853

# PROJECT ENGINEERING
## of Process Plants

RASE, H. F. AND BARROW, M. H.

Project Engineering of Process Plants

RASE, H. F.

Piping Design for Process Plants

# PROJECT ENGINEERING
## of Process Plants

**HOWARD F. RASE**
*Associate Professor of Chemical Engineering*
*The University of Texas*

*and*

**M. H. BARROW**
*Project Engineer*
*Foster Wheeler Corporation*

*Original Illustrations*
*by*

JAMES R. HOLMES
*Assistant Professor of Drawing*
*The University of Texas*

**NEW YORK · JOHN WILEY & SONS, INC.**
**London · Sydney**

10 9



ISBN 0 471 70917 4

Library of Congress Catalog Card Number: 57-5929

Printed in the United States of America

To our wives

for their patience and help

# PREFACE

In the early days of the chemical industry, new plants were designed by a chemist and a mechanical engineer with the assistance of one or two draftsmen. The design work progressed slowly, but since the processes were simple, the job could be completed in a reasonable amount of time.

The simplicity of design methods used in industry's early days contrasts sharply with the complexity of modern plant design. The joint efforts of specialists in engineering, construction, and management and a close interrelationship between them are necessary for the design and construction of a modern plant.

The supervision and coordination of these specialists has become the job of a specialist. He is called the project engineer.

In many cases the project engineer has a background in chemical engineering. In addition to chemical engineering he must understand mechanical, electrical, and civil engineering problems. He must keep informed on procurement practices and be well schooled in the ways of office management. In the past this vast amount of know-how not included in a chemical engineering education has been gained by trial and error or from experienced project engineers.

It is the purpose of this book to record information and techniques used in plant design and to discuss ideas that many project engineers have developed through years of practice. It is written from the point of view of a project engineer, but is not intended only for him or for the individual aspiring to become a project engineer. It should be useful to any engineer who designs process equipment, since he must know the theories and techniques of mechanical design, procurement, and construction, in addition to basic chemical engineering principles.

*Project Engineering of Process Plants* is also intended as a text for chemical engineering plant design courses. Most chemical engineering curricula in the United States have courses in process plant design. The student finds such courses invaluable for three reasons. He is able to integrate the compartmentalized knowledge obtained from basic courses. He has greater opportunity for original thinking. And, third, he gains the healthy self-assurance that comes with being able to apply his knowledge to a comprehensive practical problem.

The principles of chemical engineering have been ably recorded in such texts as Hougen and Watson's *Chemical Process Principles* and Brown's *Unit Operations,* and the student will have such books for reference in making process design calculations.

The conversion of these process calculations to a workable plant design is project engineering. *Project Engineering of Process Plants* has been planned to acquaint the student with the methods, organization, and philosophy of project engineering.

The book is divided into four sections. The major steps in plant design are outlined in the first section, and the business and legal phases of plant design are discussed in part two. Part three gives principles of equipment design and selection and design of structure foundations and piping. Equipment common to all process plants is considered in detail whereas other equipment is treated briefly in a single chapter, although references to more complete sources of information are given. No attempt has been made to provide a handbook or guide to the selection of all types of equipment. Technological progress is too rapid to make such an undertaking worthwhile. Instead design principles and reasons for current practices are emphasized so that the reader can develop a sound basis for his own decisions.

The last part of the book is a description of construction operations, and any engineer engaged in plant design is advised to accept the first opportunity to observe the construction of a plant from start to finish. After that experience the engineer usually becomes a much better designer.

To aid in making the book useful to both student and practical engineer, an effort is made to adhere to the following principles:

1. The relationship between each design phase can be better understood if the sequence of chapters follows the usual steps in plant design.

2. Visualization of equipment is facilitated by isometric and pictorial views or cross-sections supplemented by other views.

3. Design equations, unless completely empirical, should be developed so that their limitations will be understood.

4. Only commonly used design methods should be presented.

5. Examples of actual design calculations, specifications, and contracts are at times preferable to detailed explanations.

6. Compilations of design data and sources of information should be given throughout the text.

7. Economics should be discussed in relation to each phase of plant design and procurement, rather than isolated in a separate chapter.

8. The young engineer will profit from the definition and use of terminology common to the construction industry.

The authors are grateful to the many companies which furnished illustrations. Specific acknowledgments are made in the text. Thanks are also due Professor M. Van Winkle for his many helpful suggestions.

*Austin, Texas*
*February, 1957*

HOWARD F. RASE
M. H. BARROW

# CONTENTS


1 The Project Engineer 1

**Part 1 MAJOR STEPS IN PLANT DESIGN 5**

2 Plant Location 7
3 Preliminary Data for Construction Projects 26
4 Process Engineering 43
5 Flow Diagrams 64
6 Plot Plans 80
7 Scheduling the Project 85
8 Engineering Design and Drafting 101

**Part 2 BUSINESS AND LEGAL PROCEDURES 118**

9 Procurement Operations 121
10 Office Procedure 136
11 Contracts and Contractors 156

**Part 3 DETAILS OF ENGINEERING DESIGN AND EQUIPMENT SELECTION 184**

12 Vessels 187
13 Heat Exchangers 214

14 Process Pumps 248
15 Compressors and Vacuum Pumps 297
16 Motors and Turbines 348
17 Other Process Equipment 373
18 Piping Design 394
19 Thermal Insulation 473
20 Process Instruments 497
21 Plant Utilities 545
22 Foundations 584
23 Structures and Buildings 618
24 Safety in Plant Design 632

**Part 4 CONSTRUCTION OF THE PLANT** 650

25 Construction 653

# CHAPTER 1

# THE PROJECT ENGINEER

The design and erection of a process plant could never be completed solely by members of one branch of engineering. Instead such an undertaking must result from coordinated efforts of chemical, mechanical, electrical, and civil engineers, chemists, and specialists from many other fields. This combined effort, however, must be directed by a single individual who can guide the engineering, anticipate the routine problems, and schedule the various phases of work. It has, therefore, become the practice in process industries to assign this overall responsibility for the entire design and erection of a process plant to a single individual called the project engineer or manager. The duties of the project engineer for a process plant, other than requiring a sound chemical engineering background,* demand knowledge of other engineering fields, business administration, and economics. Although it is not necessary for him to be an expert in any of these, he must have sufficient knowledge to coordinate the activities in all of them.

The larger operating companies in the process industries, such as chemical and petroleum, maintain extensive staffs of technologists in all branches of engineering for research, development, and plant maintenance. Such a firm will ordinarily explore any new process under consideration both technically and economically. However, when an

* Some practicing project engineers have mechanical engineering training and additional study in chemistry.

operating company decides to build a complete plant or a single process unit, an engineering and construction firm specializing in such work is usually retained.

Numerous variations are possible in the division of work between the engineering and construction firm (contractor) and the operating company (customer).

1. Process development, process design, engineering and construction by contractor. Examples: (*a*) Foreign customers desiring plants of American design. (*b*) Newly formed or young process industry firms. (*c*) Established firms not having the personnel available for active participation in any phase of the design or firms which prefer to have contractor handle all phases. (*d*) Contractor furnishing patented process or developing a suitable process.
2. Process developed by customer, process design a joint effort, engineering and construction by contractor. This is a common arrangement when the customer has, through bench scale and pilot plant work, developed a process and wishes to build a full-scale plant. The contractor uses the customer's basic data.
3. All process design done by customer. When the customer has a large process design group that has been closely associated with a process, it is often advisable for this group to supply a complete process design to the contractor. It is not uncommon, however, for the customer to request the contractor to check the design.

Other variations of the division of responsibility are possible. It is important to note, however, that the project engineer's role is extremely important in any arrangement. When both the contractor and customer are participating, the customer's project engineer and the contractor's project engineer have duties essentially parallel; and the chart in Fig. 1-1 depicts the duties of both.

The customer's project engineer must supply the contractor's project engineer with all the information concerning the customer's requirements and preferences. He must check and approve all designs, and obtain comments from various design and operating groups in his own organization. The contractor's project engineer must be responsible for transmitting information to the various groups in his own organization, for guarding both the contractor's and customer's interest in all engineering decisions, for contractual obligations, and above all for seeing that the plant is completed on time and that it operates according to specifications.

Referring again to Fig. 1-1, the duties of a project engineer can be briefly traced in a clockwise manner around the chart starting at the top. Assume that a process plant, Firm *A*, has, after economic studies

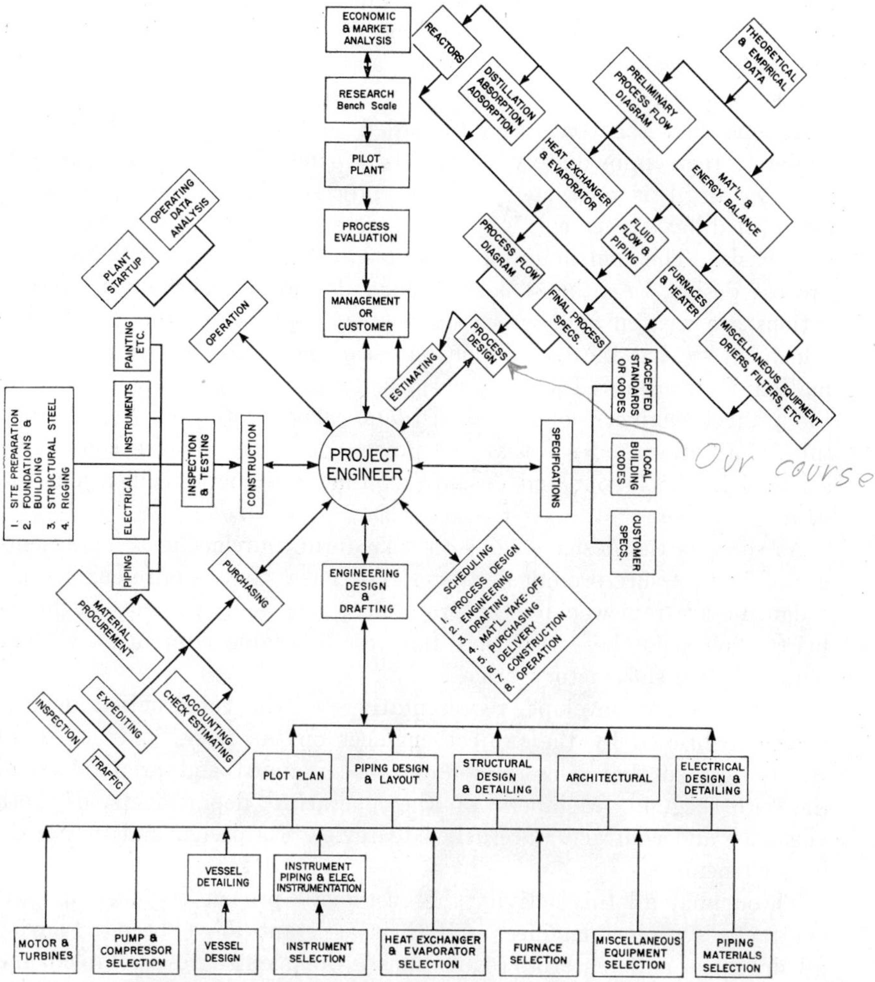


Fig. 1-1. Project organization.

and basic research, decided to construct a plant. Firm *A* sends out inquiries to various contractors which are prepared by *A*'s process department and project engineer. These include a description of the proposed location of the plant, the scope of the project, the production required, and any other information not available to the contractor. The contractor assigns a project engineer to the job, who immediately becomes the liaison between the customer and all departments of the contractor. The project engineer works with his process design and estimating groups in preparing an estimate for the cost of designing and constructing the proposed plant. The estimate is presented to the customer in bound form. It carefully outlines all duties and obligations of customer and contractor, describes the process units and equipment, gives the design specifications, operational and mechanical guarantees, and sets a completion date.

Assume that the contract is awarded to Contractor *B*, *B*'s project engineer then immediately starts organizing the job. The project manager of *A* is contacted and the process requirements are firmly established by agreement with *A*'s process group, using the proposal as a basis. The contractor's process group then proceeds with the final process design. As the design is being developed, mechanical specifications are written and agreed upon and all phases of the project are scheduled so that the completion date can be met. The process information is transmitted to the various design and drafting groups by the project engineer and his assistants. They interpret and develop this information and see that the process requirements are completely satisfied in the layout and design and that customer specifications are met.

As soon as the design begins to take form, purchasing commitments are made to assure prompt delivery. Purchases that require engineering judgment are reviewed by the project engineer and specialty engineers before being purchased so that the best available equipment will be obtained at a satisfactory price.

As the design develops, representatives of the construction department are briefed by the project engineer on the scope and details of the project so that construction may be planned and preparation of the site begun. Members of the operating departments of both customer and contractor are asked to review the design and to suggest improvements.

Throughout all this activity the customer's project engineer and the contractor's project engineer and their assistants have been reviewing all drawings and material requisitions. Comments are assembled and returned to the design groups so that the necessary changes can be made. The contractor's project engineer must see that these changes are properly executed and that both the customer's and contractor's interests are guarded. Changes must not alter the original process requirements, scope of work,* or specifications. In fact, the project engineer's responsibilities do not end until the operating department has shown that the plant meets the design specifications and process and mechanical guarantees.

* If changes in the scope of work are desired by the customer, he must ask for these changes in accordance with the terms of the contract. The contractor can ask for additional fees for these changes. (See Chapter 11.)

# PART I

# MAJOR STEPS IN PLANT DESIGN

THE project engineer coordinates and directs the entire design project. His active participation increases after the completion of process engineering, but in many instances he also takes part in the early stages of job planning. He may assist management in the selection of a site and will also direct the gathering of preliminary data. His most intensive work, however, comes during the period of detailed engineering design and drafting, for it is this work that culminates in the selection of equipment and production of drawings for use in construction.

# CHAPTER 2

# PLANT LOCATION

The proper location of a plant is as important to its success as the selection of a process. Not only must many tangible factors such as labor supply and raw material sources be carefully considered, but also a number of intangible factors which are more difficult to evaluate.

The selection of a plant site must be based upon a very detailed study in which all factors are weighed as well as possible. Such a study often requires a substantial outlay of money, but false economies at this point may lead to great losses in the future.

There is considerable literature on the subject of plant location ranging from practical considerations to the theory of plant location. The theoretical aspects are extremely interesting, and in general stem from the classical work of Weber[13] published in **1909** in the German and translated in **1928** in English. Many excellent outlines or check lists have been published which aid in plant location studies by listing factors that must be considered. An abbreviated listing of some of the literature in this field appears at the end of this chapter. These references will inevitably lead the interested reader to the entire literature on plant location.

## GENERAL PROCEDURE

Before proceeding with any plant location study, one must have a complete knowledge of the existing company organization, history, cost data, and marketing techniques. Just as a family must know its own requirements when purchasing a new house so a company must also know itself before locating a new plant. Acquiring such knowledge, however, is not always easy. In the fast growing process industry, many organizations have grown from small family-owned concerns into large corporate enterprises in a short span of years. Clear-cut organization and policy often fail to be defined during such phenomenal periods of growth. It is well worth while for a company in this situation to attempt definition of policy and organization before proceeding with expansion in new areas.

In devising a workable procedure for a plant location study, some compromise must be made between two extremes. One, an exhaustive statistical study requiring a decade for completion and the other, a short-cut study heavily weighted with preconceived ideas and lacking in objectivity. A complete study of thousands of possible locations is, of course, out of the question. Some means of decreasing the number of possibilities must be found. Aires[1] has suggested the method of apparent association. A survey of existing plants of the type being considered will often indicate that certain types of locations have been preferred in the past. Questioning of the management of these plants may indicate special considerations which dictated such locations. In this manner, it may be possible to restrict the study to several general areas.

Many processes have one or more predominant factors which serve to minimize the number of possibilities for plant location. Raw material and transportation costs may be such that a plant must be located near a source. This will automatically reduce the sites studied to those near sources of raw material, which may be only several in number. Labor requirements may be such that cities below a certain size must be excluded, because only a larger city can supply the large labor requirements. These and other factors serve as effective screening agents which save both time and money.

Having decided upon several possible regions for plant location, the next step is a detailed study of the possible locations. This study includes a survey of all the important factors involved, and a cost study. It is not possible to assign dollar values to all the factors that must be considered. Such intangible factors, however, can often be compared by assigning each a numerical value (100, 200, 300, etc) the size of which depends upon the importance attached to the particular item being considered. The assigning of the numerical values is based on opinion, but such opinion can be agreed upon by the management at the outset of the study so as to insure an objective approach to the accumulation and presentation of data.

After the detailed study of several areas indicates a particular region or community as being ideally suited, then a specific site in that region must be selected. Yaseen[15] has emphasized the importance of determining the specific site after the community has been selected. If during the study of several communities a particular site in one is found that has unusual appeal, the entire study may be influenced by it. Such influence in effect gives undue weight to the site itself, because it is so often possible for one or more individuals to favor the appearance of a given site and place excessive emphasis on its importance.

## PLANT LOCATION FACTORS

A discussion of the important factors to be considered in the study of areas and sites for plant location is given below. These factors are: raw materials, transportation, process water, waste disposal, fuel and power, labor, weather, and community factors.

### Raw Materials

The selection of the source of raw materials, although it may not be at the plant site, is an extremely important factor in the ultimate location of the plant. In many cases, the study of the raw material situation may precede the analysis of the other plant locational factors, since pilot plant work on a process requires at least a knowledge of the ultimate raw material supply. Process development work and economic studies will indicate the minimum standards for raw material selection. Once these standards are determined, then all possible sources of acceptable raw materials can be located and more detailed analysis of the sources can proceed.

The extent of each raw material source must be determined in the light of existing and estimated future requirements. An attempt to estimate the life of the raw material source based on future requirements must be made. Alternate sources or substitutes in the area should also be located and evaluated. The cost of raw material delivered to the plant site can then be determined for all those sources meeting the quality and quantity specifications required by the process.

Accurate freight rates must be used in determining these costs. Because of the complex freight rate structure that exists in the United States, it is not possible to use some simple dollar per mile figure (see below). A classical rule states that if the raw material loses considerable weight during processing to the final product, it is cheaper to locate the plant at the source of the raw material. As is true, however, with all such general rules, there are a number of notable exceptions. A delivered raw material cost analysis should always be made for each plant location being considered.

## Transportation

As stated previously, no simple dollar per mile cost rule can be applied to freight rates. Railroad freight rates in particular are extremely complex. An example given by Yaseen[15] illustrates this complexity most effectively. He describes a manufacturer who is considering a central Ohio location for a new plant and who has distribution in surrounding territory and also in the Los Angeles area. A major competitor is located in Cleveland, Ohio. Considering Marysville as a location, it is found that the freight rate per hundred pounds will be $5.56 to Los Angeles. The rate from Cleveland is the same. A location in Peoria, Ohio, only seven miles further west of Marysville will enjoy a rate of $4.83 per hundred pounds producing an advantage over the Cleveland competitor.

It is not possible to present a complete discussion of freight rates in this chapter. The engineer in charge of obtaining information related to plant location need only realize that transportation experts must be consulted in establishing freight charges and optimum location with respect to transportation. Such expert information is available from the freight agents and traffic experts of the railroads and other transportation facilities. In addition, the traffic manager of one's own company can be of great assistance in obtaining the necessary information and aiding in its interpretation.

The effect of transportation facilities and rates on plant location can be a controlling factor. Industries such as the plastic industry, for instance, which must deliver many small shipments to various users in the minimum of time, find location near the majority of users mandatory. Less-than-car-load-lot (LCL) rates are very high; and, therefore, the distance that the material must be shipped should be kept to a minimum.

### *Railroad Freight*

Although truck transportation has increased markedly in the past ten or fifteen years, it is safe to say that few process plants could exist without access to at least one railroad. The railroads continue to carry the bulk of industrial products. Discussions with railroad freight agents concerning freight rates and methods of handling in the various communities under consideration are necessary if a true picture is to be obtained for comparing the relative advantages of several locations.

A study of the railroad freight structure will reveal several fundamental principles. Because of the relatively high cost of transferring shipments from one railroad to another, it is desirable to locate plants along major railroad routes so that shipments will follow the major lines between the shipping point and ultimate destination. Where possible it is desirable to have more than one railroad servicing a community because of the greater flexibility such service provides.

Many larger communities served by several railroads have terminal companies or so-called belt-line railroads which are often jointly owned by the several railroads in the region. These terminal railroads provide service between the tracks of the major railroads entering the community and all plant sites. Such service is rapid and produces cheaper transfer from road to road and ultimately to the plant. Areas served by terminal companies, therefore, are to be desired.

If large amounts of LCL shipments are anticipated, the existence of a car-loading company in the area is advantageous. These car loaders offer door-to-door LCL service. They collect a number of LCL shipments from an area having the same destination, and are able to assemble complete carload shipments for delivery to the railroad. Some railroads provide similar service in large communities.

At first thought it may seem that a plant should either be located at the raw material source or at the market for its products. Considering transportation alone, however, an intermediate location may be just as advantageous. Railroads offer fabrication-in-transit privileges for certain commodities. This privilege permits the material to be shipped to a plant, processed, and then sent on its way to the ultimate destination for essentially the same rate that would have been charged for a non-stop trip. It is well to check the existence of such privileges for the commodities to be produced. Many other rate peculiarities have evolved as the railroads developed and competition became keen.

### *Trucking*

Interstate truck movements, as in the case of the interstate railroad transportation, are now regulated by the Interstate Commerce Commission. Truck lines are classed as scheduled or non-scheduled carriers, depending upon whether the line offers a regularly scheduled service between two points or operates on a contract basis at times convenient to the shipper. Truck lines have been able to compete vigorously with the railroads on small shipments and in particular on short hauls. Because of the competitive advantage in this phase of freight shipment, the truck lines have concentrated on the development of efficient means of handling less-than-truck-load-lots (LTL). As with railroad freight, excessive transfer should be avoided. Location of a plant on the route of a major carrier providing direct connections to various shipping points eliminates the need for transfer and cuts costs.

In considering truck transportation in a given community, the franchise of each line passing through the area should be consulted to determine if the the truck company is permitted to serve the particular community.

### *Water Transportation*

Water transportation is and will probably always be the cheapest means of transportation, especially for long hauls of great bulk. It is certainly an advantage for a process plant to be located adjacent to navigable water even though initial operations do not require water shipments. Many companies have been able to compete effectively in distant markets even though these markets are served by plants in the market area. A California producer of ammonia can ship liquid ammonia in sea-going barges to the Pacific Northwest and compete with plants located in the Pacific Northwest territory. A large Gulf Coast caustic producer ships 73% caustic by sea-going tanker to east coast terminals from whence it is forwarded by rail and truck to east coast users. Here again low cost water freight enables the producer to compete with producers in the market area.

In addition to ocean transportation, the United States is favored with many miles of navigable streams and canals. Perhaps the most spectacular canal system is the Intercoastal Canal which provides protected shipping lanes for barge traffic from Brownsville, Texas to New England. The Great Lakes and large rivers such as the Mississippi, Ohio, Hudson, and Tennessee provide additional thousands of miles of navigable waterways passing through the population centers of the United States.

Many process plants have found it advantageous to operate their own ships and barges, although the services of shipping companies are also available. To stimulate water transportation the federal government has for a number of years operated a large barge line called The Federal Barge Lines.

### *Other Means of Transportation*

The services offered by the air lines, express companies, the Post Office Department, and pipelines vary in importance depending upon the particular plant in question. Because of the increase in airline traffic and the tempo of present day industry, it is advantageous to be located conveniently to an airport if for no other reason than for the convenience of company personnel traveling on plant business. Although few process plants will require continuous air freight shipment, it is not uncommon to call for an emergency air shipment of parts for plant equipment. Here again the convenient location of an airport is an advantage.

Air and rail express and postal service are necessities for certain phases of any industrial plant operation. Availability and quality of such service in a given community should be checked thoroughly.

In any discussion of transportation for process plants, pipelines should not be overlooked. The petroleum industry is the largest user of pipelines for transporting both raw materials and products. The increasing network of natural gas pipelines has in some cases influenced the users of natural gas in chemical processing. Instead of locating at the source of natural gas supply, several companies have found it advantageous to locate their plants along pipelines and near the eventual market. Such a decision, however, must be carefully considered since the pipeline supply is less reliable than an underground supply at or near the plant.

### Process Water

The process industries rank above all others as users of water. No process plant could operate without water as a cooling medium and as a direct raw material in certain phases of a process. The water supply of an area, therefore, must be studied before that area can even be considered as a possible site. Advice should be sought from competent water experts, including geologists and chemical engineers specializing in water problems.

Prior to the study, a detailed estimate of water requirements both for the present and the future must be made. This should then be followed by careful study of the water available in the region being considered. If well water is to be used a complete study of the past history of the underground water table is necessary. Because of their lower temperature, underground water supplies are preferred if adequate.

Powell and Von Lossberg[9] have described a method for predicting the future availability and reliability of underground water supply by testing existing wells. Water levels in observation wells are measured during the pumping of nearby active wells. The formula of Theis[9] is then used to predict the drawdown anticipated over a long period of pumping.

Surface waters from streams or lakes also require careful consideration since they are often affected severely by seasonal variations. Fresh water streams discharging into the ocean during times of low run-off can become salty due to sea water back-up. Under these conditions, it may be necessary for the plant to provide large storage facilities for fresh water to be collected during the period of high run-off and used during the season when the stream becomes salty.

It is necessary to check the history of river flow or lake level for as many years back as possible so that assurance of an adequate supply can at least be predicted from historical data. Industries moving into relatively new areas often fail to consider the possibility of other plants following suit. The quantity of water supply should not only be adequate for future needs of the proposed plant in question, but also adequate for supplying the anticipated needs of other industries that might move into the area. It is also desirable to consider alternate sources of supply that may be required as the preferred water sources become deleted.

In addition to having the proper quantity of water available, the quality of water must also be studied. Chemical and bacteriological examination of the water will indicate the extent of treatment required and aid in the development of water cost figures for comparison with other locations. Possibility of the contamination of the water source by other industries in the area should be anticipated. This contamination may consist of only raising the temperature of the water to such a level that its use as a cooling medium will be impossible.

The chemical engineer is usually well equipped to make cost comparisons of various water supplies, but he rarely has the background necessary to make intelligent conclusions concerning the extent and reliability of a particular supply. Such studies deserve the attention of a competent water consultant trained in geology and meteorology.

### Waste Disposal and Noise Abatement

The forward-looking engineer will consider waste disposal and noise abatement just as important in thinly populated areas as in a heavily populated city having special ordinances related to these problems. Nothing is so unprincipled or injudicious as dumping waste in the atmosphere or in nearby streams. In addition to the moral or ethical considerations, it is not even good economics. Eventually the community will rise in indignation and impose laws which may be so burdensome that profitable operation would be impossible. It behooves the engineer, therefore, to study the waste disposal and noise problems and to consider the methods and costs of an effective control program for each region being studied.

### Fuel and Power

All process plants require both steam and electric power in their operations. Power is either purchased from local utility companies or generated at the plant site. Even if power is generated by the process plant, arrangements for stand-by power from the local utility must be made for emergency purposes. Steam is rarely purchased but is generated at the plant for use in the process and as a driving medium for pumps and compressors.

A detailed knowledge of the quantity of power and steam required for the operation of the projected plant must be obtained before the study can proceed. The cost of all fuels available in the area should be carefully analyzed. Natural gas pipelines are making a cheap gas available in many parts of the country. Industrial rates for natural gas appear attractive, but they must be examined carefully. Industries located remotely from the natural gas supply will be the first to suffer should an interruption in pipeline service occur or an unduly cold winter cause a taxing of the facilities. Further, cheap rates for industries in cities served by pipelines often only apply to summer conditions when the demand is low. In such cases, alternate fuel supply must be available for the winter months.

Evaluation of power costs is greatly aided by the ready cooperation of local utility companies. Most of these organizations maintain staffs that specialize in industrial power. The cost of producing power at the process plant which will include writing-off the construction of a power plant, is perhaps more difficult to determine. It will be based on the fuel available in the area, the company's experience with other power generating operations, and particularly the experience of other plants in the area.

It is advantageous if a progressive utility company serves the region being considered. Even process plants generating their own power find it necessary because of rapid expansion to purchase large amounts of power from time to time. Only a dependable and rapidly growing utility system can supply such demands.

## Labor

A large portion of the costs of any manufactured item is represented by labor costs. Although labor rates are more and more becoming similar in most parts of the country, factors such as skill, labor relations, and the general welfare of the laboring force affect labor productivity and efficiency materially. Each region being considered for plant location must be surveyed to determine the availability and the skills of the labor market. The skills need not exactly compare with those required by the process plant. This fact was most effectively demonstrated during the speculator influx of industry on the Gulf Coast of the United States. Chemical plant operating labor was recruited from former refinery operating men and oil field workers, both groups of which had experience which required the same degree of responsibility as that in the chemical plant.

Maintenance men make up a large portion of any process plant labor force, and it is necessary that the community have available at least a nucleus of such forces as well as training facilities for providing new men.

The wage rates existing in a given community are important, but it would be a definite mistake to plan on the continuation of lower rates indefinitely. The trend in the United States has been toward a gradual elimination of differences in wage rates between various parts of the country. Perhaps, a more important factor is the stability of such rates in a community. Some communities have been notorious for wide fluctuations, a condition which makes company planning extremely difficult.

A further consideration is the history of labor unrest in the region. Some cities in the United States, particularly the older industrial areas, have a history of labor unrest which began back in the days of early union organization and apparently has continued periodically to the present time. Conferences with other plant managers in the area and local union officials will aid in evaluating the general labor picture. A conference with local union representatives will also serve to begin the establishment of amiable relations. Talks with the management of existing plants will yield data on the important problem of labor turnover. A stable labor force is valuable in successful plant operation. A look at the homes of the laboring force in a community and knowledge of the percentage of home ownerships should give some clue to the contentment of the laboring group.

In addition to federal labor laws which apply throughout the United States, various states and communities have enacted laws affecting labor relations and workmen's compensation, minimum age, and other laws related to labor practices. These laws vary from region to region and competent legal advice is necessary for the purpose of interpretation.

### Weather

Weather data for a number of years should be assembled for each community being studied. Particular attention should be given to severe conditions of weather such as hurricanes, earthquakes, and floods. These catastrophic events, which must be assumed as probable, increase construction cost. Extremely cold weather often hampers process plant operation and requires special construction features to protect equipment from freezing. Predominantly warm weather permits cheaper construction, but classically it is said to reduce the efficiency of a labor force. Such a conclusion must be viewed with scepticism when the production records of process plants on the hot and humid Texas and Louisiana Gulf Coast are compared with those in other parts of the country.

Costs of heating or air conditioning may be estimated using weather data from former years and experiences of existing plants. Air conditioning in hot sections of the country can no longer be considered a luxury. Indeed, companies endeavoring to operate their offices and control rooms without air conditioning will find it difficult to obtain or keep employees.

### Community Factors

One of the most important aspects of plant location is often overlooked or only regarded lightly. This aspect is the effect of the character and facilities of the community being studied. It is most difficult to obtain objective information on a community because the usual sources of information, such as Chambers of Commerce or Industrial Commissions tend to present somewhat biased opinions because of their natural enthusiasm for their own home town.

A study of the community should begin with a look into its historical development. From such a study, the character of a city will emerge including its general attitude toward industrial developments.

Assuming that the historical study yields a satisfactory report, the general facilities offered by the community should be reviewed. A contented group of people require a certain minimum number of facilities for satisfactory living. If these do not exist, it often becomes a burden for the plant to subsidize such facilities.

Banks must be reliable and staffed with competent officials and of such a size that they will be capable of handling plant as well as employee accounts. The value of adequate shopping facilities, especially for the female members of family units, cannot be overemphasized.

Larger cities offer the advantage of factory warehouse facilities so that replacement parts for plant equipment can be readily obtained.

Adequate and gracious hotel facilities are always welcome, especially when it becomes the job of plant officials to entertain visiting dignitaries. At least one or more hospitals fully accredited by the American Hospital Association should be located in the area. In larger cities, transportation becomes a problem. Many workers prefer to use public transportation because of the difficulty of parking in certain areas. Such public transportation must be efficient and economical.

The cultural facilities of the community are important to sound growth. Churches, libraries, schools, civic theaters, concert associations, and other similar groups, if active and dynamic, do much to make a community progressive. Forward-looking plant managers will realize that the young people in a community will be the employees of tomorrow, and a good school system and adequate cultural opportunities will in the final analysis make better employees. Because the process industries depend on technical manpower, it is desirable that a university of the first class be located in the state under consideration.

Its engineering and science departments should be accredited by the appropriate agencies (Engineering Council for Professional Development, American Chemical Society, etc.).

A study of the population trends in a community is often revealing since it indicates the growth and tenor of a community. A breakdown of population statistics into age groups is helpful in predicting the possible vitality of a region.

The problem of recreation deserves special consideration. Recreation will include, of course, the cultural entertainment already mentioned, and, in addition, outdoor activities and other events ranging from country club dances to small gatherings in homes. The disadvantage of the extremely small communities or company towns is that the company must subsidize the construction of parks, golf courses, and clubs. Further, the so-called company town is predominantly occupied by employees and their families. Many people find such a situation narrowing since most of their friends are other employees and their families and all too often after-hours conversations develop into "shop talk." This generalization, however, is dangerous since it depends so much on the individuals involved. It might be said, nevertheless, that it requires people with better than average imagination to avoid these pitfalls.

Extremely small communities may fail to offer the opportunities for recreation required by young people and discontent often develops. Plants located near large cities have avoided such situations, while some remotely located plants have overcome these difficulties by the establishment of active recreation departments and clubs. These human factors are all rather nebulous, but none the less important. They should not be overlooked in any analysis of plant sites.

The efficiency, character, and history of both state and local government should be evaluated. Localities which have suffered long years of corrupt or inefficient government are poor risks. State and local laws and regulations which will affect plant operation and living conditions require the interpretation of a competent attorney. These include all tax ordinances, especially those applying directly to the industrial community. The existence of low taxes is not in itself a favorable situation unless the community is already well developed and is not in great debt. It is certain that a community having undeveloped sewer systems, roads and other similar facilities must soon increase taxes.

## FINAL SITE SELECTION

After an area or region has been selected for plant location, the next job is the selection of a specific site. A considerable amount of the data necessary for the information discussed above can be obtained without actually visiting the areas. The final site selection, however, requires careful scrutiny by a corps of experts. It is preferable to avoid publicizing such visits since real estate values invariably rise upon rumor of new plant construction. This necessary secretiveness is often best accomplished by dealing with only one reliable real estate firm. Topography and soil conditions of each site must be evaluated. Foundation costs are greatly increased on sites having soil of low bearing quality. Good natural drainage is another desirable feature; and, if the site is located near a stream or other body of water, flood history should be carefully checked. Often seemingly excellent sites have suffered from periodic flooding that would never be predicted on inspection of the area during normal conditions. In addition to the advice of competent soil experts and construction engineers regarding the feasibility of the site, officials of the neighboring plants can give helpful hints concerning the nature of various locations in the area.

## FINDING INFORMATION

The foregoing brief summary of factors that must be determined in a plant location study indicates the need for a vast amount of information both quantitative (statistical) and qualitative. Because of the large number of agencies, public and private, which publish useful information of this type little actual original gathering of data must be done. All sources of printed information should be exhausted before making a private survey on a given phase of the study. To aid in the search for printed information, a suggested list of sources together with types of information offered is given in outline form below. The works of Riggleman and Frisbee[10] and Coman[3] aided greatly in preparing this outline.

I. Indices and Aids to Searching

A. U.S. Government Publications*

1. Hirshberg, H. S., and Melinat, C. H., *Subject Guide to United States Government Publications,* American Library Association, Chicago, Ill. (1947).—Describes government publications and gives sources for a number of selected topics.

2. Schmeckebeir, L. F., *Government Publications and Their Use,* The Brookings Institution, Washington, D.C. (1936).—Another descriptive guide to government publications, although somewhat outdated.

3. *Monthly Catalog of the United States Government Publications*—Major index of government publications. A year-end issue contains annual subject index. Lists practically all publications both congressional and departmental, and starting in 1936 has included many processed (mimeographed, etc.) publications deemed of interest to the public. Most large universities and public libraries serve as

* All U.S. government publications listed unless otherwise noted are available from the Superintendent of Documents, Washington, D.C.

depositories for government publications. Hence much of the material listed is available from nearby library sources.

4. *Document Catalog* (discontinued in 1940)—Contains listing of all documents submitted to Congress and also publications of executive departments, bureaus, and offices of the government. Each volume covers a two-year period.

5. *Superintendent of Documents Price Lists*—Not a complete list of all publications but a handy list of those currently popular. Bi-weekly list of selected publications also available.

6. *Numerical Lists and Schedules of Volumes of Reports and Documents of each Session of Congress*—Guide to Congressional literature.

7. *Congressional Directory*—Lists official staff, responsibilities and activities of each department, bureau, and agency of federal government. Government organization changes from year to year and the Directory is helpful in the proper routing of letters of inquiry.

8. *U.S. Government Manual*—Describes organization and duties of government agencies. Helpful in locating services formerly handled by abolished agencies.

9. *Official Register of the United States* (yearly)—Lists persons occupying administrative, judicial and legislative positions in the Federal Government.

10. Departmental and Bureau Indices—A number of government departments and bureaus publish indices of their own publications. These include the following summaries:

a. *List of U.S. Bureau of Mines Publications, 1910–1949.*

b. *Index to Publications of the U.S. Dept. of Agriculture, 1936–1940.*

c. *Catalog of U.S. Census Bureau Publications, 1790–1945.*

d. *U.S. Dept. of Commerce Publications Catalog (1952)*—Monthly list of publications by these and other agencies are available.

B. State Publications

1. *U.S. Library of Congress Monthly Check-List of State Publications,* Division of Documents, Library of Congress, Washington, D.C.—Does not list all state publications.

2. Lists of state publications can be obtained from the various states through the office of Secretary of State for the individual state and from the state universities. Most states publish manuals describing governmental organization and listing current heads of departments and agencies.

C. Other Indices

1. Standard indices familiar to any library users are, of course, valuable in any search for information. These include *Industrial Arts Index, Chemical Abstracts, Engineering Index,* etc.

2. Certain organizations such as the Engineering Societies' Library in New York supply free bibliographical service on selected subjects.

3. Coman, E. T., Jr., *Sources of Business Information,* Prentice-Hall, Inc., New York, N.Y. (1949).

4. *National Trade and Professional Associations*—Listing published periodically by Government Printing Office.

II. Specific Sources of Information by Subject

A. General Business and Financial Information of the Industry

1. Moody's Investor Service, New York, N.Y.
*Industrial Securities*
*Manuals of Security Rating*

2. Standard & Poor's Corporation, New York, N.Y.
*Standard & Poor's Industry Surveys*

3. Trade Publications.
*Chemical Engineering*
*Chemical & Engineering News*
*Chemical Industries*
*Petroleum Refiner*
*Oil & Gas Journal*
*Petroleum Engineering*
*Oil, Paint & Drug Reporter*
*Oil World*

4. Trade and Professional Associations.
American Chemical Society
American Institute of Chemical Engineers
American Petroleum Institute
Natural Gasoline Association of America
National Petroleum Association

5. *World Almanac and Book of Facts, New York World-Telegram & Sun,* New York, N.Y. (yearly).

6. State Almanacs—Published by some newspaper in the state.

7. *Chemical Engineering Catalog,* Reinhold, New York, N.Y. (yearly).

8. *The Refinery Catalog,* Gulf Publishing Company, Houston, Texas (yearly).

9, *Thomas' Register of American Manufacturers,* Thomas Publishing Co., New York, N.Y. (yearly).

10. *Survey of Current Business,* Supt. of Documents, Government Printing Office, Washington, D.C. (monthly).

11. *Statistical Abstract of the United States,* Government Printing Office, Washington, D.C. (yearly).—Helps locate sources of the tremendous amount of statistical data available. Aids in locating statistical publications and sources both public and private.

12. U.S. Department of Commerce—Business statistics, maps, and advisory services.

13. Chamber of Commerce of the U.S., Washington, D.C.—Many publications and services. List available.

*14. Local Chambers of Commerce and State Industrial Commissions.

15. Association of American Railroads, Public Relations Dept., Washington, D.C.—List of railroads having Industrial Research Departments which make area surveys. Other services.

16. *Census of Manufacturers* and *Census of Business,* Bureau of Census, Washington, D.C.—Costs of material and value or products given in form of sales and retail trade in latter.

*17. Industrial Development Divisions of railroads and utility companies.

18. Bureaus of Business Research of state universities and other colleges. These bureaus publish results of business surveys of many areas in a state and represent an excellent source of information.

B. Raw Materials, Water, Fuel, and Power

1. *Minerals Yearbook,* U.S. Dept. of the Interior, Bureau of Mines (yearly).

2. Department of Interior, Bureau of Mines, Geological Survey—Many publi-

* Most organizations of this type supply honest and accurate statistical information. Their own enthusiasm for the area in which they are located may however understandably influence the conclusions based on these statistics. In addition to printed information, chambers of commerce and industrial agencies provide staffs of experts to aid companies in gathering information.

cations on resources are available from these departments. The Geological Survey can supply information on mineral and water resources, and is a source of information on all maps available from the government. The Bureau of Mines can furnish data on fuels and minerals.

3. Bureau of Plant Industry, Soils, and Agricultural Engineering, U.S. Dept. of Agriculture—Information on soils throughout the country.

4. Forest Service, U.S. Dept. of Agriculture—Forest resources.

5. Federal Power Commission, Washington, D.C.—Power surveys rate data, directory of utilities, natural gas data available.

6. Professional, Trade Publications, and Almanacs. (See A-3, 4 and 5.) The reader will be familiar with the literature in his own technical field.

7. American Gas Association, New York, N.Y.—Gas rates. Publishes *Gas Facts.*

8. Chambers of Commerce, Industrial Commissions, Railroad and Utility Industrial Divisions, etc. (See A-13, 14 and 17.)

9. Bureaus of Engineering, and Economic Geology of state universities collect and publish considerable useful data about the resources of a state.

10. *Petroleum Data Book,* Petroleum Engineering Publishing Company, Dallas, Texas.

C. Transportation

1. Association of American Railroads.

2. Local Railroad Representatives.

3. Local carloading companies and warehouses.

4. *Official Guide of Railway & Steam Navigation Lines,* National Railway Publishing Co., New York, N.Y.

5. *Handy Railway Atlas of the United States,* Rand McNally and Company, Chicago, Ill.

6. *Official Steamship and Airway Guide,* Transportation Guides, Inc., New York, N.Y.

7. Interstate Commerce Commission, Washington, D.C.—Statistics on railroads, trucks, and buses.

8. Civil Aeronautics Administration, U.S. Dept. of Commerce, Washington, D.C.—Statistical data on airlines available on request.

9. *American Aviation Directory,* Aviation Associates, Inc., Washington, D.C.

10. *Russel's Official National Motor Coach Guide,* Russel's Guides, Inc., Cedar Rapids, Iowa.

11. *Motor Carrier Directory* (trucks), Motor Carrier Directory Co., Atlanta, Georgia.

12. American Trucking Association.

13. State Highway Departments—Maps, highway regulation.

14. *Maps of Principal Waterways of the U.S.,* Corps of Engineers, Dept. of Army, Washington, D.C.—Also information on ports of entry and harbors available.

D. Labor

1. Bureau of Labor Statistics, U.S. Dept. of Labor, Washington, D.C.

*a.* Monthly employment report by states.

*b.* Cost of Living Indices.

*c. Handbook of Labor Statistics.* Published irregularly since 1926—Wages, cost of living, etc.

*d. Monthly Labor Review*—Wage surveys, statistical data, bibliography of labor literature.

2. Bureau of Labor Standards, U.S. Dept. of Labor, Washington, D.C.—Summaries of labor laws; will answer inquiries concerning local labor situations.

3. *The Conference Board Management Record,* National Industrial Conference Board, New York, N.Y. (monthly).—Discusses labor trends and presents statistics.

4. Labor Union Publications.

*a. Economic Outlook*—CIO, Dept. of Education and Research, Washington, D.C.

*b. Labor's Monthly Survey,* AFL, Washington, D.C.

5. *Handbook of Labor Unions,* American Council on Public Affairs, Washington, D.C.

6. *Who's Who in Labor,* Dryden Press, New York, N.Y.

7. Prentice-Hall Labor Publications, Prentice-Hall, New York, N.Y.

*a. Labor Report*—Weekly report in labor relations.

*b. Labor Guide*—Interprets existing laws.

*c. Labor Service*—Weekly supplements, complete texts or digests of all new labor laws.

8. Bureau of National Affairs, Washington, D.C., and the Commerce Clearing House, Chicago, Ill.—Publish considerable labor information.

9. Local union representatives and labor lawyers.

10. State employment offices—Local offices can supply information on area employment conditions.

11. *Index to Labor Articles,* Rand School of Social Science, New York, N.Y. (monthly).

E. Weather

1. Weather Bureau, U.S. Dept. of Commerce, Washington, D.C.—Price list of publications available.

2. Local Weather Bureau Offices are also helpful in supplying weather data for previous years.

F. Community Factors

1. Bureau of Census.

*a.* Digests of state laws relating to taxes (Gov. Div.).

*b.* City finances and employment (Gov. Div.).

*c.* Census of population.

2. *Statistics of Income,* Bureau of Internal Revenue, U.S. Treasury Dept., Washington, D.C.

3. *Municipal Yearbook,* International City Managers Association, Chicago, Ill.—Lists officials and form of city government.

4. *Management Almanac,* National Industrial Conferences Board, New York, N.Y.—Gives consumers price indices in 60 cities.

*5. *Automobile Facts & Figures,* Automobile Manufacturers' Association, Detroit, Mich.—Gives registration data.

*6. American Transit Assoc., New York, N.Y.—Write for information on local transportation systems.

*7. American Taxicab Association, Chicago, Ill.—Write for data on taxi statistics.

*8. American Telephone & Telegraph Co., New York, N.Y.—Contact for statistical data on phones.

*9. U.S. Post Office Dept., Washington, D.C.—Contact for receipts for individual post offices.

10. Department of Commerce—Statistical data on business operations of all kinds.

11. Banker's Directories.

*a.* Rand-McNally, Chicago, Ill.

*b.* R. L. Polk Company, Nashville, Tenn.

* Valuable in comparing communities.

12. State Banking Commissions.

13. *Moody's Bank and Financial Manual,* Moody's Investor Service, New York, N.Y.

14. Statistical Services.

*a.* Dun & Bradstreet

*b.* Standard & Poor's

15. Federal Reserve Bank.

16. *Federal Reserve Bulletin,* Board of Governors Federal Reserve System, Washington, D.C. (periodical)—Gives department store sales and consumer credit data.

17. Local banks.

18. Chamber of Commerce: city, county, or state Industrial Development Boards; state universities.

19. *Biennial Survey of Education,* Federal Security Agency, Washington, D.C.—Statistics on higher education and city schools.

20. Local PTA groups.

21. *Annual Directory of Hospitals,* American Hospital Association, Chicago, Ill.

22. *American Medical Directory* (annual), American Medical Association, Chicago, Ill.—Lists physicians.

23. State and city health departments.

24. National Office of Vital Statistics, U.S. Dept. of Public Health, Washington, D.C.—Births and deaths in cities.

25. *Official Hotel Redbook & Directory,* American Hotel Association Directory Corp., New York, N.Y.

26. *Directory of Newspapers & Periodicals,* N. W. Ayer & Sons, Philadelphia, Pa.

27. *American Guide Series,* Work Projects Administration—A series of guide books on states and cities in the U.S.A. written during the WPA days and published in 1940. Although outdated, much of the historical and general descriptive information is valuable in aiding in the understanding of a region. Hastings House, New York (1940).

The above listing is by no means complete. Heavy emphasis has been placed on the nontechnical data sources since it is assumed that one trained in engineering will have a working knowledge of the sources of scientific and technical literature.

The factual information available from industrial divisions of chambers of commerce, utility companies, and railroads deserve special emphasis here. Although, as already mentioned, the interpretation by these organizations will probably be filled with the same enthusiasm for an area as that held by any area booster, the factual data is usually reliable. It is arranged in a very useful manner and represents the most rapid means of obtaining a compilation of data, particularly for a preliminary survey. Certainly for detailed surveys additional facts which may have been omitted must be obtained and all other data checked.

As an example of the type of material obtainable from a chamber of commerce, it is interesting to take a brief look at the brochure issued by the Chamber of Commerce of Houston,[6] a rapidly growing industrial area, particularly in the process industries. Briefly, the information assembled includes the following:

1. Market—buying income of surrounding area by counties.

2. Transportation—description of all types and names.

3. Raw Materials—qualitative information.

4. Fuel, Water, and Power—rates and addresses of utility officials.

5. Labor—typical wage rates, consumer price index.

6. Taxes—rates. Addresses of city and state officials to contact for additional information—Explanation of tax structure.

7. Weather—charts.

8. Industries in Area—oil and gas, chemicals, and metals—Description of existing facilities, production and list of all plants on Gulf Coast with names of general managers, lists of products, and quantities available.

9. Miscellaneous Statistics over Period of Years—

*a.* Population.

*b.* Retail Sales.

*c.* New Construction.

*d.* Bank Accounts.

*e.* Employment.

*f.* Post Office Receipts.

*g.* Chamber of Commerce Statistical Summary (issued monthly)—includes statistics on Port of Houston tonnages, department store sales, telephones in service, and twenty-two other points which indicate city vitality.

When it is realized that all this helpful information is available from one organization at no cost, the value of private and public industrial development departments becomes more apparent.

## REFERENCES

1. Aires, R. S., and D. F. Othmer, *Chem. Eng. Progr., 45,* 285 (1949).

2. Atkins, R. M., *Harvard Business Review, 30,* No. 6, 113–121 (1952).

3. Coman, E. T., *Sources of Business Information,* Prentice-Hall, Inc., New York, 1949.

4. Faith, W. L., *Chem. Eng. Progr., 45,* 304 (1949).

5. Hoover, E. M., *Location of Economic Activity,* 1st Ed., McGraw-Hill Book Co., New York, 1948.

6. *Houston,* Houston Chamber of Commerce Brochure, Houston, Texas, 1952.

7. Hoyer, C. O., *Chem. Eng. Progr., 45,* 314 (1949).

8. Perry, John H., *Chemical Engineer's Handbook,* Section 26, McGraw-Hill Book Co., New York, 1950.

9. Powell, S. T., and L. G. Von Lossberg, *Chem. Eng. Progr., 45,* 289 (1949).

10. Riggleman, J. R., and I. N. Frisbee, *Business Statistics,* 3rd Ed., McGraw-Hill Book Co., New York, 1951.

11. Sittenfield, M., *Chem. Eng. Progr., 45,* 316 (1949).

12. Sommers, H. A., *Chem. Eng. Progr., 45,* 301 (1949).

13. Weber, A., *Über den Standort der Industrien,* Tübingen (1909). Translated by C. J. Friedrich, University of Chicago Press, Chicago, 1928.

14. *When Planning Plant Expansion,* Brochure of the Walter Kidde Constructors, Inc., New York.

15. Yaseen, L. C., *Plant Location,* Business Reports, Inc., New York, 1952.

# CHAPTER 3

# PRELIMINARY DATA FOR CONSTRUCTION PROJECTS

After a site has been selected for the construction of a process plant, a large amount of information must be accumulated before design work can begin. Engineers, purchasing agents, and construction supervisors must have detailed information about the site and surrounding area.

Several experienced men with backgrounds in both engineering and construction management can obtain the needed preliminary data on a visit to the construction site. The information they gather includes data on engineering for use in design, procurement practices in the area, and information necessary for planning construction. These preliminary data plus process requirements set the basis for design and construction of the plant.

## TYPICAL PRELIMINARY DATA

Preliminary data outlined below are typical of that gathered by the engineer on his survey visit to the job site.

### Engineering Data

Essential data necessary for the engineering of a process plant which should be obtained by the survey include:

#### *Weather Data*

For domestic projects (U.S.) information on weather may be obtained from government agencies such as the U.S. Weather Bureau, U.S. Department of Agriculture, and the U.S. Coast and Geodetic Survey. The American Society of Heating and Ventilating Engineers have also compiled considerable weather data, which may be useful in preliminary work.

PREVAILING WIND DIRECTION

Useful in initial plant arrangement. As an illustration assume that the plant is to be located in an area where the wind is predominantly from the Southeast. It would be undesirable to locate a process which frequently discharges noxious fumes on the Southeast side of the plant where they would be constantly blown through the plant. Again, assume that a process discharges particularly volatile and flammable vapors that are heavier than air. It would be unwise to locate an open fired heater in the path that the fumes would ordinarily follow. General wind direction must be considered in locating any process which is extraordinarily hazardous or subject to fires. The effect of wind direction on storage areas for flammable products should also be studied.

MAXIMUM WIND VELOCITY AND HURRICANE OR TORNADO FREQUENCY

Used in setting proper design conditions (pounds per square foot) for wind loads on exposed structures, equipment, or buildings. Certain U. S. Gulf Coast areas, for instance, use 40 pounds per square foot because of possibility of hurricanes. The final value has considerable influence on structural and foundation design.

MAXIMUM AND MINIMUM ATMOSPHERIC TEMPERATURE

Used in process design, in the selection of insulation, in determining plant-building types for process equipment protection, in building design for personnel comfort, and in selecting maintenance procedures. The design of the cooling water system and selection of equipment are also influenced by atmospheric temperature ranges. Temperature ranges should be studied over a two-year period, or longer, to provide the best estimates of average values.

RELATIVE HUMIDITY (range by day, month, and seasonal average)

Required for design of cooling towers, air driers for process air, and for determining personnel comfort requirements for buildings.

RAINFALL (maximum, yearly, and seasonal average)

Required for plant drainage design, building design, and possibly in process design.

SNOW CONDITIONS

Required for structure design, maintenance procedures, and building design.

Frost Line

Required for setting minimum depth of foundation or for designing concrete work below grade.

Local Flood or Tide Conditions

Useful when a plant is located near a major stream subject to flooding, or in a coastal area, which may be subject to flooding.

Weather conditions also affect construction operations. Extreme cold, heat, or rainfall may influence construction scheduling, and reduce construction personnel efficiency. Extremely low temperatures affect construction operations, particularly excavation and concrete work. Such work must be stopped or extraordinary precautions must be taken during very cold weather.

### Plant Terrain

*a.* Soils must be explored for foundation or load-carrying characteristics.

*b.* Contour maps showing ground elevations which are useful in planning plant drainage and plant roadway systems should be prepared.

*c.* General slope or pecularities of the site are also necessary in estimating preliminary grading which in turn indicates the type of construction grading equipment.

### Water Requirements and Supply

Sources of water for the following uses must be developed:

*a.* Process water (water used in the process).

*b.* Process cooling water.

*c.* Sanitary water (this must be potable water).

*d.* Utility water (water used for general cleaning and miscellaneous services).

*e.* Steam generation (makeup water for boilers).

*f.* Fire protection water.

*g.* Water for construction use (quantities of construction water depend on type of project, number of construction employees, and the length of the construction period).

Water for process cooling, sanitary use, and steam generation must be chemically pure or treated to the extent required for the process. The source of water supply may be from stream beds or other natural water supplies, provided turbidity and sedimentation is not excessive. Brackish or salt water is often used for cooling, but careful selection of the materials of construction is necessary in order to prevent corrosion and deposits.

Available water sources must be studied and water analyses obtained in order to decide upon the degree and type of water treatment. Water taken from streams or above ground sources will require a certain amount of screening to remove foreign matter. The quantities of water to be used, will determine the design and selection of this equipment.

Water wells are a common source of supply in many areas. Data on the particular area may sometimes be obtained from firms specializing in water well installation. A certain amount of potable water will be required during construction. Small temporary wells driven into the water table may be satisfactory for this service. Bacteriological analysis of all such natural ground water should be made to insure safe drinking water. Water to be used in concrete operations should be chemically analyzed. Local water unsatisfactory for personnel use may be chlorinated during construction operations to reduce bacteria count, or sufficient water for drinking purposes may be hauled to the plant site by tank truck.

### *Electric Power and Fuel Services*

#### Electric Power

When electric power is to be purchased, the location of incoming power feeders, and the plant substation must be determined. Primary voltages must be determined early, in order to start engineering and procurement of the main plant transformer and switchgear. The preliminary survey should determine the direction from which the lines will come; whether power from the property limit to the substation is to be overhead or underground, whether a single feeder system will be provided, or whether two or more sources are to be connected. A single supply feeder system is usually all that will be provided. In certain areas subject to severe atmospheric electrical disturbances, however, power failure frequency is a serious problem. The power supplier may be able to reduce frequency or probability of failure by adding one or more additional feeders providing power from a different direction. Past history of power failures may be determined from the supplier's records or from other local users.

The supplier must also be given, as early as possible, the date when the plant will start operation or when power will be required. If temporary power is to be provided for construction uses, the necessary arrangements for installation should be made early. Poor electric utility service or a record of high frequency of power failures must be considered when parts of the process become hazardous under power failure conditions. In certain processes many prime movers must be provided with spare steam drivers for safety reasons. In a major installation the duplication of drivers may be costly.

Small package-type steam generators are usually more economical than a central generating plant and distribution system. In nearly all U.S. industrial areas operations with purchased electric power are more economical than plant generation of power. These questions must be carefully analyzed as early as possible, since they will have much to do with initial design and design costs as well as operating economics.

### Fuel Sources

The selection of the type of fuel depends on heating value and selling price and is closely related to the economics of purchased power. Some of the sources are: (1) Natural gas; (2) fuel oils and liquefied petroleum gas; (3) coal or other solid fuels; (4) plant wastes.

*Natural gas* is often used as a raw material in certain processes. When the process requires natural gas as a raw material, gas is also used as plant fuel. Some of the plant uses of gas are as fuel for prime movers such as large internal combustion engine-driven compressors or gas turbines, fuel for steam generation, engine fuel for electrical generators, and for production of inert gases for purging or blanketing services. When natural gas is to be purchased it is important to obtain data on the supply pressure and an analysis of the gas as well as a guarantee of its composition or the limits of variation. Natural gas in some areas is wet or carries a certain amount of condensate. If such is the case, a trap should be provided near its entrance to the property. It must be decided and clearly pointed out in the contract for supply, which party shall furnish the trap and where the responsibility for drainage of the trap will lie, i.e., with the supplier or the purchaser. While certain of the liquid materials will immediately evaporate upon discharge, other components may remain in a liquid state producing a hazard. In congested industrial areas the drainage of trapped condensate from natural gas can become a considerable problem. Unless there will be some process use of these liquids it is better to insist on a dry gas, furnished at a certain pressure at the user's meter.

The purchaser must also ascertain precisely where the natural gas pipeline will enter the property and decide which party shall supply the meter, the primary device, the meter house, valving arrangement, and the piping accessories. The above suggestions are for supply services in the order of millions of cubic feet. They are not of importance for small installations with service connections perhaps of 3″-pipe size or less. However, it should always be clearly understood which party supplies and installs the equipment.

When the final contract is entered into with the supplier a drawing should be prepared locating the connecting point of the owner's piping and showing the type flange, valve, joint, etc. to which plant piping will be connected. These data should be given the purchaser by the supplier.

Natural gas supplied by pipeline will not ordinarily require any auxiliary pressure vessel or capacity tank since the system should provide the required capacity. Pressures above 30 to 40 psig for plant fuel systems are unnecessary. Usually gas supply pipeline pressures are much above this range. Ordinarily the problem is more one of reducing pipeline pressures to ranges suitable for plant services. For this reason the question of supply pressure should be discussed as early as possible. If the gas is to be used primarily for fuel, a supply line pressure range from 50 to 75 psig should be sufficient to allow for plant surges with adequate upstream pressure for satisfactory control and capacity in the plant fuel system. Pressures at the point of usage are ordinarily in the range of 20 to 30 psig, and are lower for burner headers.

*Fuel oils* are usually delivered by railroad tank car or motor tank truck. Facilities for storage of the liquid fuel must be provided. Usually heating must be provided for unloading. This is more fully discussed in Chapter 21. In the preliminary survey, thought should be given to proper location of railroad or truck unloading point so that the initial plant arrangement studies will include this facility.

*Liquefied petroleum gases* or gases held in a liquid state are becoming an important source of fuel for domestic heating in some areas, however, their use as industrial fuels is still somewhat limited. If the plant is small, this type fuel may be used to an advantage for isolated building heating.

*Coal* is not ordinarily used as a plant fuel for the process plant; however, coal may be used as a power-generating fuel if the project is located in a coal producing area or if other fuels prove uneconomical. Necessary facilities would be required for unloading, storage, and handling. A number of solid fuels other than coal exist; however, their use would probably require facilities similar to that for coal handling.

*Plant wastes* are often an important source of fuel. Their availability, of course, depends upon the process. If plant wastes may be burned as fuel, the problem of waste disposal may be simplified. Plant wastes that are to be intermittently used as fuel may produce a storage problem.

### Product and Raw Material Transportation and Storage

The preliminary data concerning plant terrain may have considerable effect on the location of such facilities as railroad unloading and loading points, and packaging and storage facilities. Certain plants located on sloping sites have the liquid product storage arranged so that all loading is by gravity. The location of storage points for hazardous products must be carefully considered since in order to meet safety requirements additional acreage may be required.

It is very important to determine the degree of interplant transportation and the storage requirements because of their effect on plant power requirements.

Raw materials sometimes enter plants and products leave the plant by pipeline. Wherever any piping enters or leaves the plant the exact points should be determined as early as possible so that interference with plant facilities can be avoided.

### Procurement Data

Preliminary data affecting procurement are:

1. Earliest date when materials may be received at jobsite.
2. Basic materials of construction available in area.
3. Connecting railroads and nearest switch point, connecting or serving motor truck lines.
4. Main highway locations.
5. Location of nearest metropolitan area.
6. Types of industrial equipment useable in plant erection available in the immediate area.
7. Types of industrial shops in the immediate area suitable for repair or maintenance or parts supply for construction equipment.
8. Storage space and storage area for plant equipment and construction material.

Procurement, as discussed in Chapter 9, is intimately related to project scheduling and planning (Chapter 7). The procurement organization will be most concerned with plant equipment and construction material transportation. The methods of transportation, time required, freight rates, construction material, storage facilities (at the jobsite), field fabrication of certain parts of the equipment, and delivery data all affect procurement. Transportation and jobsite storage most seriously affect preliminary procurement planning or scheduling.

The most vital question in procurement is always delivery or delivery dates. A project at a remote location with a possible transportation time of one month will require all scheduling to include this travel time. Availability of common materials in the area of construction together with price data must be ascertained. Price differentials may warrant the procuring of a large percentage of materials locally and thus justify a local procurement operation. Certain materials such as ready-mixed concrete or the materials required for concrete are always purchased locally on domestic projects. Sources of supply and prices for these materials must be ascertained as a part of preliminary survey data. Local sales taxes may affect price and, therefore, must be considered in the survey.

### Construction Data

All construction phases are affected by local conditions since the construction organization must become a part of the locale for the whole period of erection. The following data must be accumulated as early as possible.

### *Labor*

*a*. Area wage rates, trade union contracts in force and expiration dates, and availability of local skilled craftsmen.

*b*. Trade union representation, names, addresses of all local or area union business agents.

*c*. Other projects planned in area and times of peak employment.

*d*. Union permits—if construction personnel are to be brought into area, union permits must be obtained for their work by agreement (These are usually never completely satisfied but should be attempted).

*e*. Availability of personnel housing—this will be resolved if job is near a large metropolitan area.

*f*. Food and recreation facilities available.

*g*. Medical facilities and special climatic conditions which may affect personnel (most important on foreign project work).

*h*. Area practices—these questions may never be completely answered since labor practices vary considerably. See Chapter 25.

*i*. Transportation required by certain personnel—some crafts require that transportation be furnished or require an allowance for travel time somewhat similar to portal-to-portal pay.

*j*. Construction personnel parking facilities—if erection is to take place within an operating plant and construction employees must leave their own transportation at the plant gate, it is sometimes necessary to reach an agreement with the owner on parking facilities. The question may sometimes be resolved by providing a parking area for construction personnel within the plant. These points must be completely understood previous to start of work.

*k*. Change and sanitary facilities for construction personnel—it is becoming common practice for construction personnel to have specific space allotted for them to change clothing. This may require construction of a temporary building. Sanitary facilities of course must be provided early in the project. If plant facilities exist, the owner should make facilities available.

*l*. Construction personnel identification—on certain work, photographs and finger-prints may be required. Badges are usually required. When work is to be done in an existing plant, all owner's requirements must be determined.

*m*. Location of construction material and plant material and equipment storage area—the location often seriously affects jurisdictional problems in unloading, transporting, and the handling of materials. The selection of the storage area, therefore, aside from the best location for economic handling must also be considered on the basis of labor requirements.

*n*. Operating-plant labor unions—when the constructor uses union labor, local business agents must be contacted and any possible problems anticipated.

### Construction Interference

These questions are not usually as important for a complete project as for work within an operating plant.

Overhead or Underground Electrical Power Lines. Location, height above ground, depth below ground, voltages, source of power, area supplied, owner or operator, nearest switchpoint for opening circuit must be determined. If guy derrick is used, will guys clear cables? Location of deadmen.

Overhead or Underground Piping and Sewers. Determine location, size, etc. Existing piping in the operating plant is often a source of considerable extra cost to the constructor of the new facilities. Many plants do not have exact or reliable records of the locations of underground lines. The responsibility for removal of such lines should be established.

Plant Hazards. In many operating plants where work is being performed obnoxious or hazardous vapors or fumes may be released. In some cases work may be completely stopped until the condition is cleared. The responsibility for costs of unproductive labor during such periods should be understood. Often serious restrictions exist on flame cutting or welding or the use of any open fire on the premises. It must be determined if any such restrictions will be in force. In many hydrocarbon processing plants and other plants processing hazardous materials all matches and pocket lighters must be left at the plant gate. Smoking areas may be designated or smoking may be permitted in construction personnel change houses.

Receiving and Handling of Plant Equipment. If the work is in an operating plant, the owner may require certain permits or receiving procedures for material or equipment delivered by rail or truck. Permits may be required for trucks or cars leaving the plant. If all materials are received by the owner, the responsibility for the shipment, proper accounting, handling, and delivery at the construction site must be determined and clearly understood.

Railway and Highway Connections. Exact locations should be determined for possible personnel transportation use. The handling of plant and construction materials is greatly simplified by laying a spur railroad track directly into the plant from the nearest railroad.

## SPECIAL REQUIREMENTS FOR FOREIGN PROJECTS

The data discussed above are required as essential preliminary data for any project. Considerable additional preliminary data and decisions are required for foreign jobs.

### Permits

The erection of plants in foreign countries, where most of the equipment or materials will be manufactured in the United States, requires certain overall permits or licenses in order to export the various items from the United States and usually import permits or licenses in order to move the equipment into the country where the plant is to be erected. If the country is landlocked and shipments must move through an adjacent country, certain permits for in-transit shipments will be required.

A number of materials have been classified by U.S. Governmental agencies as critical to the national economy. Explicit permits must be obtained to move these materials out of the U.S.A. The destination and use of such materials affect the granting of export permits. Complete lists of all critical materials, the applicable restrictions on destination, and explanations of the necessary procedures are available from the U.S. Government Printing Office.

An application to the U.S. Government for an export license must usually be accompanied by a complete listing of all materials and equipment components used in such equipment, together with weight or volume (as required) and dollar value for each item. A numerical and alphabetical designation has been developed for various items, which designation must be used in preparing the lists. The shipping dates by yearly quarters must also be included. Later, as shipments proceed the specific items must again be listed for each shipment and checked off the master list.

The preparation of the detailed lists is somewhat simplified by the intent of the governmental control, which is more concerned with maximums than with precise or exact evaluations. Care must be taken, however, to account accurately for items as they are shipped. Since export license restrictions change frequently, it is mandatory that each issue of all applicable governmental publications offered by the various agencies be carefully studied.

The data necessary to meet the export license requirements will only tentatively permit shipment. All items shipped must be accompanied in detail by the standard export documentation to meet U.S. customs and shipping requirements. Movement of shipments to dockside, warehousing, lighterage, and handling and preparation of shipping documents are each specialized functions requiring considerable experience. These are best handled by firms known as freight forwarders or forwarding agents. Such firms will undertake the complete handling of all shipments, including shipping notifications to vendors, completion of traffic agreement with domestic carriers, and the preparation of standard documents for a very small fee. This fee may be as little as ½ of 1% of the declared export value of the shipment. As a matter of fact, in many of the large U.S. ports it is practically impossible or at least completely impractical, to handle export shipments in any other way. Furthermore, the quantities of materials or the number of shipments which would be necessary for a project will require a fairly large organization of experienced personnel to handle the innumerable shipping problems. Thus, it is usually justifiable to retain a firm experienced in this work.

### Planning Export Shipping

The major cost in export shipping is ocean freight, which is rated by the ton or by cubic measurement, usually 40 cu ft, whichever is greater. The ton used may be the short ton (2000 lb), the long ton, 2240 lb or the metric ton (2204.6 lb). Weights may be based on gross, packaged weight, or net weight, depending upon destination. Measured packaged dimensions are overall measurements taken at right angles only. Requirements for weights and measurement vary with the international area served, and may also vary with the carrier, the commodity, or the export agreement.

The forwarding agent can usually save the shipper his fee many times over, by timely advice on the peculiarities of the assessment of ocean freight charges. While ocean freight rates for all areas have been set in certain international agreements, the area into which shipment is to be made should be investigated thoroughly. The standard rates, commodity limits, the steamship lines serving the area, and particular requirements of the area should be considered.

Marine schedules are subject to seasonal changes produced by cargo requirements. The marine sailing date, except for certain passenger liners, is usually a tentative one. Here also the forwarding agent can save the shipper expense by holding back heavy pieces of equipment and avoiding extraordinary dock or lighterage charges caused by early receipt of the equipment.

As a simple illustration of possible ocean freight charges, assume that a section of 24-in. pipe 30 ft in length, weighing 3000 lb is to be shipped. Rates are to be assessed on weight or cubic measurement, whichever is greater. Assume the base rate for this commodity to be \$30.00 per ton (2000 lb) or per 40 cu ft (whichever is greater). The freight will be $1\frac{1}{2} \times \$30.00$ or \$45.00 on a weight basis. The cubage will be $2 \times 2 \times 30$

or 120 cu ft, however, and on the cubic basis the freight will be (120/40) × \$30.00 or \$90.00. Shape also usually affects the freight assessment. Assume that the above pipe is fabricated as an elbow with two 15-ft legs. Cubic measurement will then be 15 × 15 × 2 or 450 cu ft with freight assessed at \$337.50. This is admittedly an extreme condition. Close attention would have to be given to shapes and weights, if this rating basis happened to be in effect.

It is often possible to nest materials, or package materials to maximize the weight per unit volume when cubic measurement governs cost. The above 24-in. pipe, for instance, could be made to carry considerable additional weight by placing smaller diameter pipe inside the large pipe. If materials are to be crated or packaged, by packing relatively light materials, such as insulation, together with heavy materials, such as loose pipe fittings, the cubage can be made to approach its freight weight rating. Commodity Classification Rates must be carefully studied in such cases. Certain commodities require a higher rate because of peculiarities of storage or handling. A package or crate containing a higher rate commodity may cause an increase in basic freight rate even though the major portion of the contents would ordinarily travel at a lower rate.

Additional dock and lighterage charges called heavy lift charges are applied to single shipments if the weight exceeds a certain limit. These charges can become a considerable portion of the cost required for equipment transportation.

All charges should be carefully ascertained both for the primary U.S. port (or dock) and for the receiving dock at the destination. A function of the forwarding agent is to warn the shipper of these costs and problems and where possible to effect savings. The forwarding agent, however, can do little with a piece of equipment after it is fabricated. Thus these limitations should be considered in planning initial design.

Travel time is a most important factor in foreign operations, since this time must be added to the delivery time promised by vendors or fabricators so that equipment will arrive on schedule. A complete study should be made of marine freight carriers serving the entry-port, and average travel times developed. It will ordinarily be advantageous to utilize a single steamship line. Freight agreements may be reached and more preferential handling may be obtained.

Marine freight carriers usually do not follow a definite fixed schedule throughout the year. The schedule may vary from port to port according to cargo and loading. If a single carrier is selected it may be possible to work out definite schedules so that warehousing and storage costs are reduced. Forewarning of the shipment of pieces of equipment with extraordinary weight or dimensions should be given to the carrier as early as possible so that space can be provided for the shipment.

### Export Packing

Precautions should be taken to protect equipment being shipped to tropical countries against exposure in shipment, or during its storage prior to erection. A number of plastic (non-bonding) materials which are easily removed are available for coating mechanical working parts, or machined surfaces during shipment.

There is considerable argument for and against extensive export preparation or packaging. If the shipment is to be unloaded essentially at the jobsite and would thus be transferred immediately from the original carrier (ship) to the owner or erector, there would be little need for any preparation other than the normal shipping precautions. If, however, the jobsite or erection point is inland and subject to additional carriers, lay-overs in foreign freight yards, or open sidings, the matter of pilferage may become serious. This is particularly true in some of the more primitive countries. It would appear impractical to assume that domestic preparation, which is usually the absolute minimum, can suffice in all cases. Certainly small loose parts should always be consolidated into larger crates or packages. For this reason it is often desirable to retain a warehousing or export packaging company at a consolidation point. These firms also should assist the owner in avoiding heavy lift charges and multiple consolidations of commodities where a higher tariff due to one of the commodities will apply.

### Planning Freight Movement within Foreign Country

In order to develop limitations in equipment size or weight, the preliminary survey should include an investigation of all the foreign railroad and seaport facilities. The installations at the initial port-of-entry of first foreign destination should be thoroughly inspected to determine docking capacity, the capacities of cranes or lifting equipment, storage or warehouse space, and the degree of protection afforded. Dock, storage, and handling charges should be ascertained. It is often necessary to provide personnel at the port-of-entry to process documentation and to insure that the shipments will be properly handled and expedited to the jobsite.

Direct rail connections at the entry-port will usually eliminate additional handling and storage costs since heavy shipments can be unloaded by ship's tackle directly onto railroad cars. Railroad freight rates, routes, schedules, and freight travel-time should be obtained from the serving carriers. If shipments are handled by several carriers or moved through distribution points, or across the borders of other countries it may be necessary to provide personnel at these points to expedite their movement.

The capacities of the serving rail systems, including the type of rolling stock and operating schedules should be investigated. In some of the smaller countries, shipments of heavy industrial equipment may be seriously delayed due to the lack of proper cars to handle particular loads.

Many foreign railroad systems are narrow-gage. Rail car sizes, and all limitations on size or weight over the particular route must be obtained. If the service is over a narrow-gage carrier, dimensional limitations will seriously affect the size of fabricated equipment. This information must be available early in the project so that manufacturers and fabricators can be instructed by specification and drawings. Size limitations often require partial fabrication of major equipment with final fabrication at the jobsite by the construction forces. When this procedure is required, the necessary tools and the particular type of skilled personnel will have to be provided.

Each time a piece of heavy equipment is loaded or unloaded labor and lifting equipment time is expended. Rail connections directly into the project site provide the least expensive method of transportation. The preliminary survey should determine rail connections nearest the jobsite and obtain necessary permits for extension of a spur track into the plant. Obtaining permits to connect to a railroad in a foreign country can become a time-consuming operation. Negotiations should be started as early as possible. Track equipment in the more primitive countries is often scarce. The erector will probably save time and costs by furnishing the necessary switch, frog rail, and other appurtenances for the connection (rail past the switch or turn-out may be second-hand). Turn-out specifications and rail weights must be obtained since the frog, lead-rail, and switch-block are often closely specified. Railroad specifications and construction methods are more fully discussed in Chapters 21 and 25.

If rail facilities do not exist, the erector will have to provide the necessary trucking and lifting equipment and perhaps prepare roadways outside the plant. The so-called low-boy, a multiple-wheeled low-bodied dray, may be obtained for almost unlimited capacities; however, studies of the heavier pieces of permanent plant equipment should be made. Reduction in fabricated sizes will enable reduction in transportation equipment size.

### Large Equipment Planning

All large or heavy items of equipment should be listed early in the project so that limitations in shipping may be discovered and the proper steps taken to avoid high costs in handling or transportation.

Heavy machinery such as large internal combustion engines or compressors, weighing as much as 100 tons, are rarely shipped completely assembled. In foreign operations, shipping limitations may further reduce the amount of assembly possible. Personnel with special skills may be required to complete assembly in the field. The manufacturer must be notified of any particular limitations in size or weight, preferably during the initial inquiry so that extra charges will not be assessed for dismantling or additional export packaging.

Manufacturers of large mechanical equipment sometimes require an inspection of their equipment by their own service engineers, usually just before the machine is grouted on its foundation. They may also require that their engineer be present when equipment is being initially operated, in order to meet their guarantee. If equipment is being erected in foreign countries, such requirements should be fully understood and agreed upon, or waived, in the purchase contract, so that guarantee clauses are not affected.

Large pressure vessels; diameters in excess of 12 or 14 ft, or vessels of extraordinary length, more than 60 to 80 ft, are often only partially fabricated due to railroad clearance requirements. On foreign projects the dimensions or weights may be even further reduced. Thus the preliminary survey information on size limitations must be available for the fabricators of this type of equipment as well as for mechanical equipment. In addition all the necessary tools or equipment and qualified personnel must be provided for completion of large vessels by the construction forces.

### Transmitting Correspondence and Drawings to Jobsite

Airmail should be used for transmitting specifications, and correspondence for the foreign project. Drawings are sent by air express or air parcel-post since they do not fall into the classification of correspondence. The preliminary survey should include an investigation of the customs requirements at the destination. In certain countries all such packages are opened for inspection, and it is desirable to be familiar with the inspection procedure and facilities. In a number of the smaller countries packages containing nothing but ordinary papers and drawing prints may be held by the local authorities without notice to either party, whether dutiable or not.

It is sometimes advantageous to provide a direct printing reproduction machine at the erection site. Data can then be mailed in some reproducible form, and prints made as required. The printing process should be checked for operation in the climate, and the effect of the climate on the sensitized paper determined. Many such papers quickly fail under high temperature or humidity. The cost of shipping replacement paper could easily exceed the costs of shipping prints.

**Construction Personnel**

Construction operations in foreign countries by U.S. contractors will usually require a number of experienced key U.S. personnel in the labor craftsman categories. The percentage of U.S. expatriates to nationals used on the project will vary with the degree of industrialization of the foreign country. In certain European countries labor in the skilled crafts is easily obtainable. The availability of skilled labor should be investigated since savings may be effected by a reduction of U.S. expatriates.

Labor regulations, laws, and income taxes or other labor taxes should be investigated and fully understood. In certain countries labor regulations may also affect expatriate personnel. Any such laws or regulations must be explained to expatriate personnel at the time of hiring. On large projects it is often common practice to institute training programs so that the nationals can become skilled in the various crafts.

Construction personnel hiring for foreign operations is usually an involved process. All personnel so hired should be employed under a contract for a certain number of months. A bonus is frequently offered as an incentive to complete the project. All personnel should be subjected to a physical examination and all necessary vaccinations taken to meet the requirements of the particular area.

If the project is remote from a metropolitan area, medical facilities should be provided. It is sometimes possible to erect some of the permanent buildings early in the project. In such instances facilities for a certain amount of hospitalization can be provided during the construction period. In the primitive countries medical supplies are often unobtainable. Provisions must be made to provide adequate medical supplies for all employed personnel, both nationals as well as expatriates since many of the countries require that an employer (particularly foreign) must provide medical attention for all employees regardless of their national origin. Furthermore, the costs are negligible compared to the possible ill-will generated if medical care is not supplied.

U.S. expatriates must obtain a passport and the proper visa from the U.S. representatives of the country to be visited. To obtain a U.S. passport the person must prove his U.S. citizenship by birth certificate or affidavit; he must have a local police release and usually a letter from the employer stating the reason for the visit.

Certain countries restrict and limit the amount of personal effects which may be carried by personnel. Many countries require payment of duties on personal effects such as shoes and clothing if they are new and un-used. These requirements should be ascertained and put into a printed form so that they can be understood completely.

Many employers buy travel or term insurance of one kind or another without cost to the employee, payable to the employee's beneficiary in case of death or injury. This procedure is a good morale builder and is not an important cost item.

It is important that personnel housing be adequate. All furnishings should be of a good durable quality and food should meet U.S. standards. Good housing and food is an important asset to any foreign project. Nothing can affect morale so seriously as uncomfortable living conditions or poor food. Savings in food costs cannot possibly approach personnel transportation costs caused by a high labor turnover.

A frequent problem with expatriate personnel on extended foreign projects is the question of single-status vs. family-status. If housing must be provided, the number of employees permitted to take their families may become an important cost item. It is the practice at times on foreign projects to allow key construction personnel to move their families to the locality. At times it is also common practice to permit employment only on a single-status basis. These procedures and practices vary considerably with owners and construction firms and are also governed by the local conditions. All such points must be developed and should be included in the printed form or the contract with the employee.

## REFERENCE

1. Leeming, J., *Modern Export Packing,* Trade Promotional Series No. 207, Bureau of Foreign and Domestic Commerce, Dept. of Commerce, Washington, D.C., 1940.

CHAPTER 4

# PROCESS ENGINEERING

Process engineering is concerned with the development, evaluation, and design of chemical processes. Perhaps in no other phase of chemical engineering can the engineer better apply the technical knowledge and skill gained in his formal college education. Although many operating companies may departmentalize process engineering into development, economic analysis, and design, it is not uncommon for a single process engineer to originate a new process and follow it through the design stages to a completed plant. Indeed, many process engineers with considerable experience are called upon to direct the design of a new plant as project engineers on the basis of their intimate knowledge of the process.

The complete discussion of process engineering and particularly of process design would require a book of considerable length. Certain brief comments, however, concerning the various phases of process engineering will aid in integrating it into the overall plant design picture.

## FUNDAMENTAL ENGINEERING RESEARCH

Fundamental research in chemistry and chemical engineering is performed to a large degree in the laboratories of universities and privately endowed research institutes. Many of the process industries, however, and a number of design firms have found it necessary to conduct certain types of fundamental research. It is the industrial-type fundamental research that pertains to this discussion.

New products are often born in the beakers and test tubes of chemists in industrial laboratories when studying certain types of reactions to gain fundamental knowledge in much the same manner as the chemists at a university or research institute. The difference is that the industrial chemist chooses reactions that, if successful, have some immediate commercial significance.

The chemical engineer also is engaged in research fundamental to his profession. Many companies carry out continuous experimental programs on unit operations, reaction kinetics, and thermodynamics. The purpose of such experimentation is to gain a greater knowledge of these operations so that design procedures and plant operation can be improved.

Chemical industry cannot progress without a continuous supply of new fundamental knowledge and information. The early American chemical industry obtained much fundamental knowledge from Europe, particularly in the field of pure chemistry. More and more, however, American universities, research institutes, and industrial laboratories are furnishing fundamental research necessary for the growth of the process industries. Congress established The National Science Foundation to encourage the development of fundamental research activities in the United States.

A competent process engineer must maintain intimate contact with developments of fundamental knowledge in his particular field. He does this by periodical review of the literature and by close association with work progressing in his own organization.

### Process Development

New design principles, novel reactions and new compounds usually suggest applications of industrial importance to alert technical management. Such applications may be a new process or the improvement of an existing one, both of which require the design and construction of new facilities and may involve expenditures of several million dollars. Since so much of engineering design is of an empirical nature, it is seldom possible to jump from the beaker to the finished plant. A considerable amount of what might be called applied developmental research must be done. The information obtained permits the design of process equipment with a reasonable amount of assurance. The process engineer who will be concerned with the design and perhaps the ultimate operation of a process becomes intimately associated with the study at this point. He should be supplied with all data and interpretations of data obtained in the laboratory. His background of plant design and operation make his counsel extremely valuable to the developmental groups. The process engineer can be assured of having all possible data necessary for design calculations only if he maintains a close liaison with the process development department.

Process development often begins with a bench-scale laboratory study. Small-scale glass or metal equipment is employed to obtain data of an exploratory nature. Such experimentation is valuable since it is often an inexpensive means of proving the feasibility of a process and of directing the way to further research efforts. The extent of chemical engineering knowledge at the present time seldom permits the design of a new plant from bench scale laboratory data alone. There are so many unknowns and intangibles which have defied theoretical analysis that it is necessary to evaluate these factors empirically on larger scale equipment, so-called pilot plant or semi-plant equipment. In pilot plant and semi-plant operation a procedure as nearly like that which will be employed in the commercial plant is followed so that the data obtained can be used with confidence in the scaling up of the pilot plant to commercial size.

Pilot plants are costly to build and operate. It is not uncommon for a single pilot plant to cost close to a million dollars. It is, therefore, imperative that all theoretical knowledge available to the chemical engineer be used to reduce the amount of pilot plant work required and, if possible, eliminate it entirely. It is possible, for instance, to obtain reaction rate data for a new reaction on simple bench-scale equipment. Using the principles of applied kinetics, rate equations can then be developed which will permit the calculation of reactor size for a number of different conditions of operation enabling the choice of the most economical design with the minimum amount of experimental data. It may still be considered wise in such a case to operate a pilot plant for the purpose of obtaining corrosion data, heat transfer information, and general operating characteristics of the process. The knowledge of kinetics gained in the bench-scale operation would aid greatly in the design of the pilot plant and would indicate very definitely the minimum amount of data required. The application of theoretical knowledge in this fashion saves a great deal of time and expense and is far superior to the completely empirical approach of "let's try this today."

Although the empirical method of research has been successful in the development of the process industries, a continuous effort must be made to strengthen the theoretical background of the various unit operations so that more and more results may be calculated, thereby reducing or eliminating completely much of the pilot plant and semi-plant work now required.

## PROCESS EVALUATION

Process evaluation consists of both an engineering and economic analysis of a process and in the broadest sense is a continuing operation. It should be started even before any laboratory work has been done. A careful analysis of cost at an early stage will indicate if the process proposed has promise and may also suggest the direction of experimentation. Process evaluation should continue at regular intervals as significant and new data are uncovered.

This continuous reevaluation of the engineering and economic factors in the proposed process aids in the overall planning of bench-scale and pilot plant work. There are often many ways of accomplishing a given unit operation, but these possibilities can be restricted to only a few when economics is considered. Dybdal[2] has presented the relation of project evaluation to well-planned experimental work, particularly the planning that occurs between laboratory and pilot plant work. One of the less tangible factors in favor of process evaluation at this stage is that making a cost estimate requires knowledge of the equipment to be employed and, therefore, automatically forces the chemical engineer to think objectively. Such thinking inevitably suggests many items of information that should be obtained from the pilot plant operation.

## COST ESTIMATES AND PROCESS ENGINEERING

The type of cost estimate required for the studies thus far mentioned is termed the predesign cost estimate since it is based only on information that is available prior to the actual design of plant equipment. The predesign estimate is not only valuable in that it directs research projects into the most profitable paths, but it is also a very real aid in design calculations. As the process design of a plant proceeds, alternate means of operation for portions of the plant often become apparent. Various arrangements of heat exchangers, pressure or temperature of reactors, and operating pressures of absorbers are typical of the factors to be considered in alternate designs.

Cost estimating does not end when the decision to build a plant is made. Inquiries are sent to various contractors, and they must make preliminary design calculations and obtain quotations on those items of equipment for which they have little or no recent cost data. From this information contractors develop preliminary cost estimates upon which their bids are based. If the contractor makes a firm quotation, his cost analysis must be exceedingly meticulous and contingencies must be kept as low as possible to insure the winning of the contract. So-called cost plus contracts, which are usually popular during accelerated building programs typical of war-time periods, protect the contractor from the many incalculable contingencies which always arise in rush projects that entail considerable risks. As the design and construction work on projects proceeds, the customer usually requests new estimates of the ultimate cost each month or for some other convenient interval. Each successive estimate becomes more accurate since more information is available as the job approaches completion. These data always become of great value to the contractor for cost studies of his own operations.

Often the "running-over" of preliminary estimates can be traced to additions or to improvements in the original design. Design drawings are usually circulated through the various departments of the customer's organization and ideas for additions to the plant often creep in during this process. The customer's project engineer can better control cost, therefore, by requesting frequent reestimates of the final plant cost.

Economic evaluation of processes should continue even after a plant goes into operation. Plans for increasing production or for obtaining lower production costs must all be evaluated on a dollar basis. Process evaluation and its primary tool cost estimation permeate every phase of the development, design, construction, and operation of a plant.

Both operating companies and construction firms maintain competent estimating departments. These groups are divisions of the central engineering organization and are usually staffed by men of considerable experience in estimating work. Most frequently each man specializes in one phase of estimating, such as piping or vessels. The primary duty of estimating departments is to develop accurate construction cost estimates based on reasonably firm design information and the latest equipment costs available. Such estimates are time consuming and costly, but are necessary in the development of quotations by contractors or appropriation estimates by operating companies, which are presented to management prior to the allocation of funds for a project.

Chemical engineers acting as process engineers or project engineers often, however, require more rapid and less costly estimates to aid in design decisions and in the direction of research and development. The process or project engineer can always request detailed estimates from the estimating department of his firm. It should be realized, though, that such estimates are costly, and for many purposes are not necessary. Nichols[6] has given an excellent commentary on methods of capital cost estimating varying from the very approximate method to a detailed estimate. Fig. 4-1, which was presented by Nichols, gives nine different estimate types with varying degrees of precision based on the information available. From this figure the approximate cost required to prepare an estimate of each type can be determined. The precision of this estimate and a recommended contingency for each type of estimate is given. This chart which is based on certain actual observations of one large firm is an illustration of the type of chart that proves helpful in selecting the type of estimate most suitable for a given purpose.* Nichols illustrates its

* Every group wishing to use such a chart should construct one of its own based on its own experience.

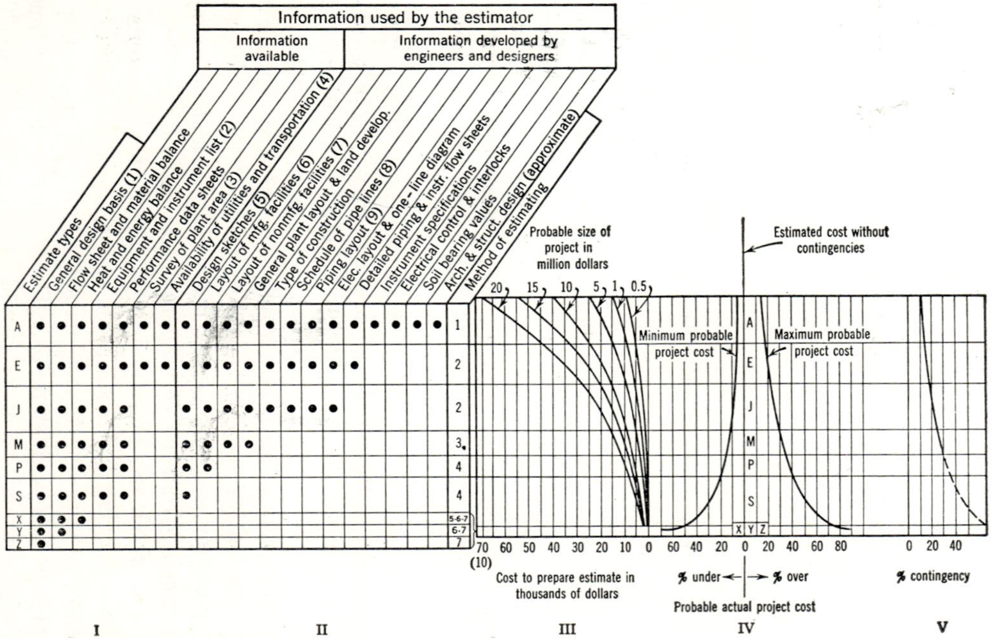


1. Product to be made, specifications and storage provisions; plant capacity; operating stream time; provisions for expansion; raw materials, source and storage
2. Showing number of each item required, capacity and materials of construction
3. Land values and land development
4. Description of laboratory and service facilities required
5. Unusual items of equipment
6. Showing equipment in plant and elevation
7. Utilities and service facilities, roads, sewers, etc.
8. Sizes, materials of construction
9. Superimposed on equipment and plant layout
10. Cost for preparing estimates contemplates no alternate studies and will not result in any construction drawings or specifications. Does not include cost of preparing the information available or the cost of sub-soil investigations to determine soil bearing values

Fig. 4-1. Comparison of cost estimating methods. (Reprinted by permission, W. T. Nichols, *Ind. Eng. Chem.*, *43*, 2297 (1951), based on a technique developed by E. H. Buford, M. Welhoelter, and R. E. Wright of Monsanto Chemical Company.)

**Estimating Method 1**

| | | |
|---|---|---|
| *A.* | Quoted equipment delivered cost | $ 0,000 |
| *B.* | Estimated installation cost for each equipment item | 0,000 |
| *C.* | Material cost take-offs for pipe, electrical, architectural, structural, instruments, etc. | 0,000 |
| *D.* | Labor estimated from practice factors for items under *C* | 0,000 |
| | Subtotal | $00,000 |
| *E.* | Engineering and contingency (% subtotal) | 0,000 |
| *F.* | Construction facilities (% subtotal) | 0,000 |
| *G.* | Contractor's fee or profit (% subtotal) | 0,000 |
| | Total | $00,000 |

**Estimating Method 2**

| | | |
|---|---|---|
| *A.* | Installed cost equipment = $\frac{\text{delivered cost (quoted and estimated)}}{0.70}$ | $ 0,000 |
| *B.* | Buildings, pipe, sewers, electrical, instruments, structural, foundations, insulation, painting and miscellaneous based on unit cost per item, length, area, or volume from preliminary design and take-off | 000 |
| | Subtotal | $00,000 |
| *C.* | Engineering and contingency (% subtotal) | 0,000 |
| *D.* | Construction facilities (% subtotal) | 0,000 |
| *E.* | Contractor's fee or profit (% subtotal) | 0,000 |
| | Total | $00,000 |

**Estimating Method 3**

| | | |
|---|---|---|
| *A.* | Equipment cost from established or published data and quotations | $ 0,000 |
| *B.* | Equipment installation, buildings, pipe, sewers, electrical, instruments, structural, foundations, insulation, painting, and miscellaneous, based on per cent of equipment cost using experience factors | 000 |
| | Subtotal | $ 0,000 |
| *C.* | Engineering and contingency (% subtotal) | 000 |
| *D.* | Construction facilities (% subtotal) | 0,000 |
| *E.* | Contractor's fee or profit (% subtotal) | 0,000 |
| | Total | $00,000 |

**Estimating Method 4**

Capital cost = factor × total delivered cost of equipment determined by quotations on unusual items, and with other items from established or published data

| Factors | |
|---|---|
| For solids process plant | 3.10 |
| For solid-fluid process plant | 3.63 |
| For fluid process plant | 4.74 |

**Estimating Method 5**

Capital cost = factor × total delivered cost of equipment determined by established or published data (factors same as in method 4)

**Estimating Method 6**

Capital cost estimated on basis of dollars per ton per year from published figures such as those by Wilcoxon multiplied by factor to bring figures up to date

Example: Wilcoxon states 1942 additions to large ammonia plant cost $86 per ton per year. Factor to bring figure up to 1951 = 1.84 [Marshall and Stevens]. 30,000 ton per year addition will cost: 30,000 × $86 × 1.84 = $4,800,000

**Estimating Method 7**

Capital cost = factor × annual capacity × unit price

Factor varies from about 2 for low profit margin operations to less than 0.5 for higher profit margin operations. Overall average for chemical industry is about 1.0

application by assuming a five-million-dollar project and comparing a type *A* estimate, the most accurate, with a less accurate type *J* estimate. The cost of the type *A* estimate would be approximately $25,000 while that for the type *J* would be $9,000. The best type *J* estimate would give a spread between minimum and maximum actual cost of about 30%, while the type *A* spread would be approximately 20%. Will the increase in precision from 30% to 20% spread justify the additional cost of $16,000 required for the type *A* estimate? In many instances, particularly design studies, such added cost would certainly not be justifiable. All too frequently time and money are wasted in attempting to estimate projects with greater precision than is possible with the technical data available.

## Estimates for Economic Evaluation

Economic evaluation is an indispensable guide to process decisions. These include decisions required for the guidance of research, process design, capital expenditures (appropriation requests to management), and optimum plant operating conditions. A complete evaluation of a new project requires the estimation of the new fixed capital required together with working capital. Assuming an annual production rate and a selling price of the product, the annual sales can be calculated. From the total value of annual sales, the cost of manufacturing, selling, administration, and taxes can be deducted to yield the income after taxes to be expected from the project. This income can then be expressed as a percentage of the capital investment which is a useful figure for study by management. Many studies, of course, do not require complete economic evaluation. In selecting alternate designs or arrangements of equipment, for instance, only a comparison of the equipment cost in the several cases need be made.

Short-cut procedures for estimating the fixed capital, the manufacturing costs, and the annual return are most helpful when used with the proper judgment. Many methods have been published. For convenience several of those currently regarded with favor by practicing engineers will be presented below.

### *Fixed Capital Cost Estimating*

When the design of new facilities has progressed to the mechanical design stage, the most accurate means of estimating the final costs of the plant is by a detailed cost estimate typical of that produced by an estimating department staffed with men who devote full time to cost estimation. The estimates required by process engineers, however, invariably must be made before a large amount of information is available. In such cases, a detailed estimate is seldom warranted or accurate.

Fig. 4-2 presents two short-cut methods, which have proved popular.

The first due to Lang[5] is best used for so-called "quickie" estimates and is also useful for checking the work of other engineers. The total estimated equipment cost is multiplied by a factor to obtain the total plant cost.

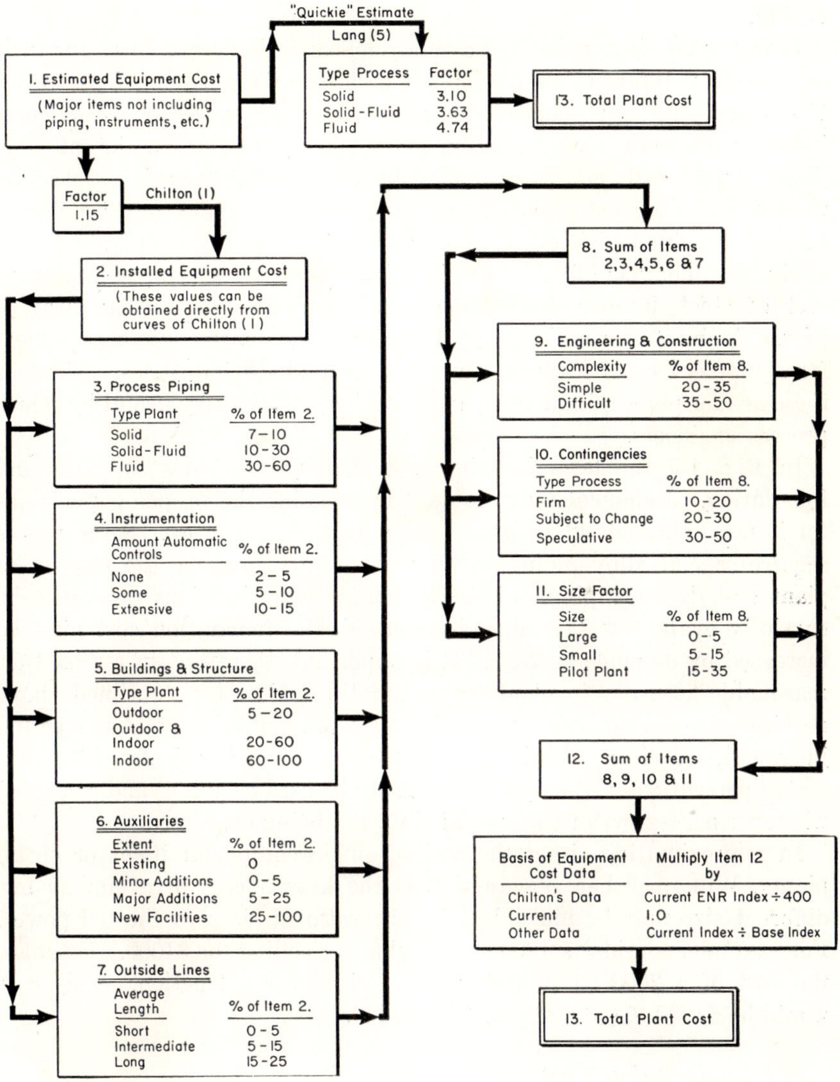


Fig. 4-2. Estimation of total plant costs.

Chilton's method[1] presents a more detailed estimate and permits the variation of important portions of plant cost in accordance with the estimator's judgement and knowledge of the particular situation. Chilton presents a series of curves giving the installed cost of a number of different equipment items based on an *Engineering and News Record* index of 400. The final plant cost is then corrected to current cost by multiplying by ratio of the current *Engineering News Record* index to the base index of 400.

Many practicing engineers have developed similar procedures for their own use from information available in a company purchasing department. Cost curves for equipment items most frequently used in a particular industry can be developed readily. To avoid continuous revision of these data, current costs can be estimated by multiplying the original cost by the ratio of a current value of some published cost index to the value of the index upon which the charts are based. Several such indexes are available. That compiled by *Engineering News Record* is based largely on cost of steel, lumber, foundation materials, and labor. The firm of Marshall and Stevens of Chicago, Illinois[8] issues quarterly index numbers for equipment costs covering a number of different industries. Eight of these are for process industries, and have been found very useful by many process engineers.

In Fig. 4-2 the term building and structure includes supports and foundation. Judgment must be used in selecting the proper percentages for items 3 through 11 to insure the maximum accuracy. Some plants require special alloy piping which may greatly increase the portion of plant cost due to piping. Similarly foundation conditions must also be carefully evaluated. If piling is required, the foundation cost may be increased by as much as 25%. It is important, therefore, to realize that reasonable accuracy in plant cost estimating of the type outlined above requires the development of considerable amount of sound engineering judgment, based on experience with many previous projects. The suggested percentages given by Chilton are to be used as guides and corrected wherever necessary for the special situation being evaluated.

In approximating either the cost of single equipment items or entire plants, Williams[9] has suggested that the known cost of a similar but different size unit be multiplied by the ratio of sizes to the 0.6 power. For example, consider a 1200 sq ft exchanger selling for \$4000. Estimate the cost of a 2000 sq ft exchanger of similar construction. This cost would be $\$4000 \times (2000/1200)^{0.6} = \$5440$.

### Manufacturing Cost

The total manufacturing cost can be estimated as suggested by Dybdal[2] and as outlined in Fig. 4-3. It is the sum of labor, overhead, fixed charges and maintenance, raw materials, utilities, laboratory expenses, and royalties. Multipliers for obtaining the annual fixed charges plus maintenance and the general overhead are given in Fig. 4-3. Where possible however,

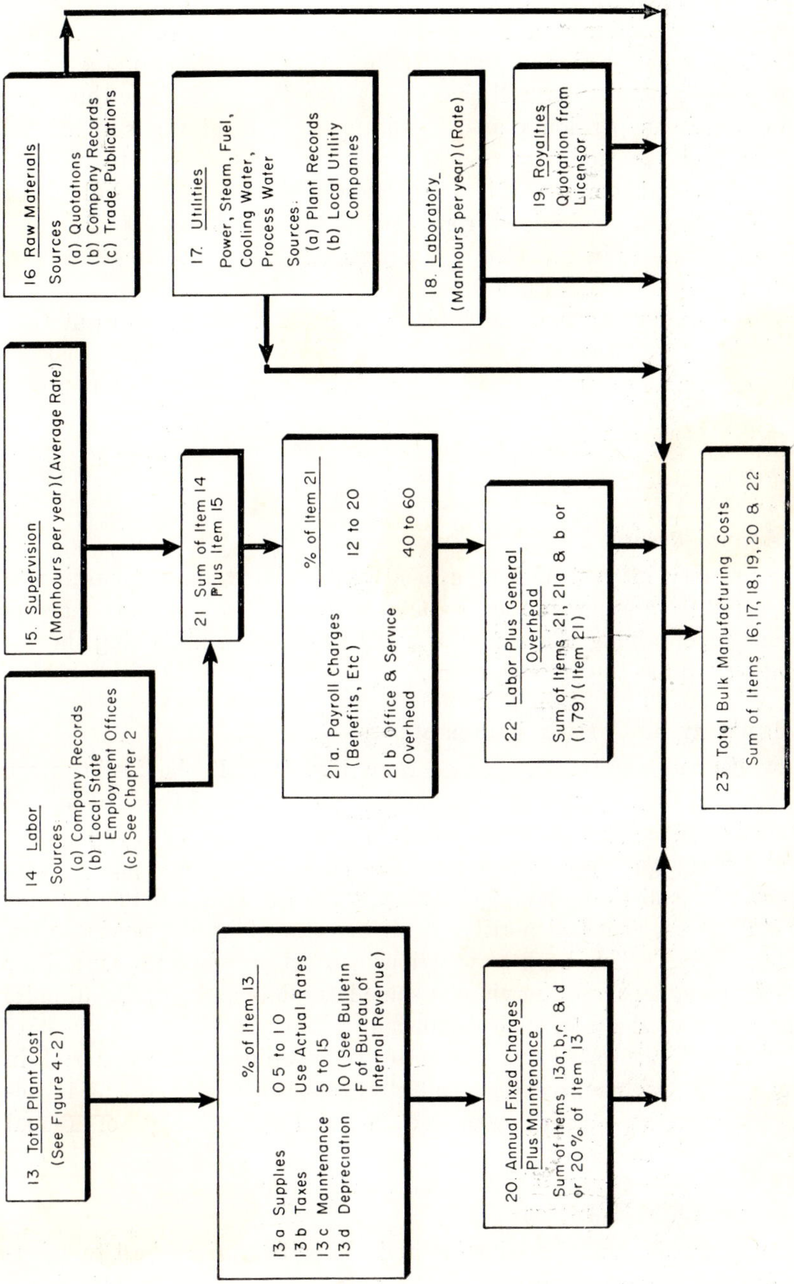


Fig. 4-3. Estimation of manufacturing costs. (Annual basis. Bulk cost only. Based on work of E. C. Dybdal.2)

these values should be calculated in a step-wise procedure. Taxes, for example, represent such a large portion of fixed charges and are subject to such extreme variations that it would be wise to use actual rates currently in effect. A large portion of plant overhead these days is represented by the item designated as payroll charges. These charges include the many benefits, such as pension plans, hospitalization, employee recreational facilities, and vacation pay. Such benefits have continued to increase year by year and represent a larger portion of the total labor bill. Raw material costs can be obtained from trade publications, but actual quotations of the costs to be expected at the particular plant site are, of course, superior.

There are many agencies of both federal and state government that supply wage rates in various localities, and it is wise to use these accurate values rather than some suggested handbook wage rate. Labor represents such a large portion of manufacturing cost that significant errors in estimation of labor cost will destroy the accuracy of the overall economic evaluation.

### *Annual Return*

The primary interest of management is, of course, the net earning to be expected from a new process or plant. The following outline suggests a means of obtaining this value and is helpful in summarizing the information required by management.

Summary of Capital Requirements

New Fixed Capital. Estimated from Fig. 4-1. If the facilities are to be located in an existing plant and thereby use existing steam and electrical generation equipment, part of the value of such equipment should be charged to the new installation. Many large plants have adopted the realistic policy of charging to the fixed capital required for new facilities, a prorated portion of all investment utility services even though existing. This book transaction in effect accumulates capital for the ultimate expansion of services at some later date and permits all units in the plant to share in such expansion.

Working Capital. (1) Raw material inventory; (2) material in process; (3) product inventory and stores (spare parts, etc.); (4) accounts receivable, working cash account, and others; 15% of annual sales.[2]

Summary of Earnings

Determine (*a*) annual production and (*b*) selling price—best to calculate on basis of several possible selling prices; then calculate (*c*) annual sales ($a \times b$); (*d*) gross profit—Item *c* minus sum of manufacturing costs and packaging and shipping costs; (*e*) net profit before taxes —Item *d* minus sum of selling, administration, and research costs (approximately 10% of annual sales); (*f*) net profit after taxes—Item *e* minus taxes. Always check latest tax rates. In recent years corporate income taxes have varied considerably.

RETURN ON INVESTMENT

This value is most conveniently expressed as a percentage of the fixed capital or fixed plus working capital required. Executives find it helpful in making their judgments to have this return calculated both on the basis of net profit before and after taxes.

## PROCESS DESIGN PROCEDURE

### Basis of Design

Before proceeding with any design calculations, it is essential that a complete basis of design be written. This should include the quantity and quality of products desired, the raw materials and their characteristics, utilities and their temperatures and pressures, and other factors such as the probable disposition of by-products. Two other important decisions must be made at this time. These are the selection of safety factors to be used in design and the determination of a completion date for the process design.

The first, safety factors, is an important but often neglected item. Frequently, numerous hidden safety factors creep into the design of a single item of equipment as various individuals work on different phases of its development. This procedure may often lead to serious overdesign and unnecessary added cost. It is preferable to use one safety factor, and factors for each type of equipment should be selected at the start of the job.

The second item, the time for completion, is more important than might be realized from a casual observation. The time available for a process design dictates the entire method of approach. If only a short time is available, it is impossible to attempt any rigorous calculations which are time consuming. Instead, it is necessary to make a number of assumptions based on previous jobs, and one must be prepared to accept the risks that go with such a procedure. It is certainly wise to make these decisions before proceeding with the actual design since the philosophy of the safety factors should be influenced accordingly.

### Schematic Flow Diagram and Material and Energy Balance

Experience usually dictates the most desirable arrangement of equipment for an old process. If the process is new, experimental work and economic evaluation should supply the information essential for choosing the unit operations required. Experimental work is not always complete, and the process design engineer must recognize any deficiencies which exist so that additional data can be requested. Plants have been designed and built based on experimental data for production of a new product, but without data for the purification of this product. The purification operations which were designed without the benefit of laboratory data proved inadequate and caused the entire operation to fail economically.

Assuming there is adequate information the process engineer should construct a schematic diagram indicating all the operations required for the production of the desired quantity and quality of finished product. Each of the unit operations necessary may offer several alternates from an economic standpoint. Material and energy balances around each unit should then be made and the results recorded in an orderly fashion so that they may be used for the many design calculations of individual equipment items and for the writing of specifications. Orderliness and neatness are valuable virtues in this work. Having an easily understood flow diagram and an accurate material energy balance early in the process design of a plant eliminates error and permits a number of people to work efficiently on various phases of the design.

## Design Procedures

The usual process design procedure might be separated into two major divisions. The first involves the actual design of equipment to be built to order. The second consists of specifying the operating conditions for standard manufactured equipment. In the first category, there is equipment, such as distillation towers, absorption towers, adsorption and extraction units and reactors. Those items involving mass transfer are designed by calculating the ideal number of stages or the number of equilibrium stages (in distillation this would be the number of trays). The actual number of stages is then determined by the application of appropriate empirically determined efficiencies.

In some cases it has been found more desirable to employ a differential stage instead of an equilibrium stage and determine the number of transfer units by an integration. The height of a transfer unit can be estimated by certain generalized correlations. The design of packed columns is done in this manner.

Reactors are designed by the application of chemical kinetics as described by Hougen and Watson.[4] Rate equations are determined from experimental data, and empirical constants for the equations are evaluated. If the effect of temperature on these constants has been established, it is possible to calculate reactor sizes for a number of assumed operating conditions. This technique enables complete economic analysis of the reactor design without excessive experimentation. In many cases, however, the reactions are so complex that rational analysis by the principles of applied kinetics is too difficult at our present state of knowledge. In this case, the empirical pilot plant procedure must be used, in which the optimum operating conditions are determined by a planned experimentation program requiring many hundreds of runs; and, therefore, considerable expense. The application of kinetics and mass transfer theory in pilot plant work should be attempted wherever possible since it will greatly aid in limiting the amount of Edisonian-type investigation required.

A large portion of the other unit operations equipment commonly employed in process plants is in the second category. It is not actually designed by the process engineer although knowledge of the theory related to each item of equipment is extremely helpful. Items such as screens, crushers and grinders, classifiers, settlers and thickeners, pumps, compressors, heat exchangers, filters, centrifuges, crystallizers, agitators and evaporators are all purchased from companies that specialize in one or more of these items. Theory for the design of most of this equipment is very sparse and empiricism is the only guide. The proper selection of an agitator, for instance, must be entrusted to a manufacturer who has specialized in the building of agitators and has a considerable backlog of experimental and actual operating data. The manufacturer can, by application of the principles of dimensional similitude, specify a large agitator installation on the basis of tests made on small laboratory equipment. Similar procedures are followed by the manufacturers of filters, dryers, centrifugals and other equipment of this type.

The process engineer must be familiar with the methods of selection used by the manufacturer so that the operating specifications required for each item of equipment can be intelligently compiled. Then it will not be necessary for the manufacturer to carry on voluminous correspondence in order to obtain additional facts required for the selection of the desired item. In subsequent chapters information is given that will aid in the selection of different types of equipment.

In specifying operating conditions for equipment or in the original design of process equipment, it is essential that the design engineer select the materials of construction. No one is more familiar with the process than the process engineer; and he, therefore, is in a favorable position to aid the mechanical designer in the intelligent specification of materials which will resist corrosion and erosion. There are many handy tables which have been printed to aid in selection of a material of construction that will resist the corrosion of a given substance. These tables are intended only as preliminary guides and extreme caution must be employed in their use. Often a slight change in temperature conditions can cause a metal to corrode which would ordinarily withstand the attack of the fluid in question. The study of corrosion is still largely empirical, and it is not possible for the average engineer to know the latest developments since the combination of materials and corroding agents is almost limitless. This suggests that for all but the simplest cases a competent corrosion engineer should be consulted. The excellent periodicals in the field will also prove helpful.

As the design and selection of the process equipment is completed, specifications for each item should be summarized in an orderly fashion. Many organizations have developed forms for the various equipment items. These insure neatness and also prevent the engineer from overlooking some important item of information. The completed forms can be reproduced and distributed to other departments in the engineering organization for integration into the complete plant design. In addition to these specifications, it is desirable to produce a final process flow diagram. This diagram shows all major items of equipment, gives temperatures, pressures, and flow rates throughout all parts of the process. The diagram then becomes an invaluable aid to specialists who must work on the mechanical design of the plant.

The process engineer should prepare a brief but complete written description of the operation of the plant. Such a description aids both management and all other engineers and draftsmen who must contribute to the final design. Distribution of this information, however, should be made with caution. Workers with incomplete background for understanding of technical process data may misinterpret the information with possible disastrous results.

## THE PROCESS ENGINEER AND THE PROJECT ENGINEER

Although the project engineer is in complete charge of the plant design and construction, it is necessary that the process engineer be ready to assist the project engineer at any time. Normally, once the process design is completed the process engineer will move on to other assignments, but questions involving the process continue to arise as mechanical design proceeds. The project engineer, even though having a background of process design, should always refer these questions to the process engineer who directed the original design. Only he will know the background of the design calculations.

It is of great value for the process engineer to accompany the project engineer during visits to the plant site as construction proceeds so that he might be better able to make decisions regarding changes in the design or arrangement of equipment. Companies interested in developing their design know-how should permit process engineers to follow the startup operations of new plants, and the process engineer should spend his time accumulating operating data. Such data should not just be placed in the company files, but it should be analyzed while still "fresh" and made a definite part of the organization's operating experience.

## SOURCES OF BASIC DATA

In addition to the experimental data or operating data which are used in the design of equipment, a large amount of basic physical and chemical data are also required. Many companies maintain very excellent data books in which such data are accumulated and periodically revised. A single department is often charged with the full-time duty of maintaining the data book. Those engineers who do not have such information available can accumulate data by requesting reprints of basic information appearing in the literature and by obtaining compilations of data that have appeared through the years. The following list represents some recommended compilations of data which have proved useful and popular with practicing engineers. No attempt has been made to list the vast amount of data that has appeared in the periodical literature or the excellent data that is published by manufacturers of various chemical products. The list of references is not to be considered complete, and the omission of certain material does not necessarily indicate that it is not recommended. Certain of these references are desirable as part of a personal library and are so indicated by an asterisk, but many are expensive and are best made a part of a company library for general technical use.

### A. Bibliographies and Reviews

1. Katz, D. L., and M. J. Rzasa, *Bibliography for Physical Behavior of Hydrocarbons under Pressure and Related Phenomena,* J. W. Edwards, Inc., Ann Arbor, Mich., 1946.
2. Rollefson, G. K., and R. E. Powell, *Ann. Rev. Phys. Chem.,* Annual Reviews, Inc., Stanford, Calif. (annual). Reviews literature on various subjects in physical chemistry.
3. Rose, A., and E. Rose, *Distillation Literature & Abstracts* (1946–52), Applied Science Laboratories, State College, Pa. (1953).
4. Unit Operations Reviews, *Ind. Eng. Chem.* (yearly feature). Reviews work in each unit operation. Many other bibliographies available on special subjects.

### B. Handbooks & General References

1. *Beilsteins Handbuch der Organischen Chemie,* Springer-Verlag, Berlin, (4th edition begun in 1914, supplements currently appearing). Latest editions cover literature up to 1929 on well-established organic compounds and reactions; however, it is still primary source of information.
2. Bennett, H., *The Chemical Formulary,* Chemical Publishing Co., Brooklyn, N.Y. (1943). Formulas of many industrial products.
3. *Chemical Economics Handbook,* Stanford Research Institute, Calif. (loose-

leaf; additional sheets four times per year). Gives financial data on chemical companies, chemical market prices, production and consumption figures.

4. Davis, D. S., *Chemical Engineering Nomographs,* McGraw-Hill Book Co., New York (1944). Collection of various nomographs found useful by Mr. Davis in solving chemical engineering problems.

5. Egloff, Gustav, *The Reactions of Pure Hydrocarbons,* Reinhold Publishing Co., New York (1937).

6. *Gmelins Handbuch der Anorganischen Chemie,* Verlag Chemie, Weinheim (additional volumes appearing currently). Source book of behavior of inorganic compounds. Some sections more recent than Mellor (see below).

*7. Hodgman, C. D., *Handbook of Chemistry & Physics,* 31st Ed., Chemical Rubber Pub. Co., Cleveland (1949).

8. Jacobson, C. A., *Encyclopedia of Chemical Reactions,* Reinhold Publishing Co., New York (1951). Arranged according to elements, four volumes.

9. Josephy, E., and F. Radt, *Elsevier's Encyclopedia of Organic Chemistry,* Elsevier Pub. Co., Houston (1946). Many volumes. They have started with more recent information with the goal of filling in the gap left by Beilstein.

10. Kirk, R. E., and O. F. Othmer, *Encyclopedia of Chemical Technology,* The Interscience Encyclopedia, Inc., New York (1947—).

*11. Lange, N. A., *Handbook of Chemistry,* 8th Ed., Handbook Publishers, Inc., Sandusky, Ohio (1952).

12. *McGraw-Hill Directory of Chemicals & Producers,* McGraw-Hill Publishing Co., Inc., New York (Annual). Lists chemical producers.

13. Mellor, J. W., *A Comprehensive Treatise on Inorganic & Theoretical Chemistry,* Longman's, Green & Company, London. 16 volumes. Handy reference to inorganic chemistry.

14. *Merck Index to Chemicals & Drugs,* 6th Ed., Merck & Co., Inc., Rahway, N.J. (1952). Eight thousand chemicals described.

*15. Perry, John H., *Chemical Engineer's Handbook,* 3rd Ed., McGraw-Hill Book Co., New York (1950).

16. Rodd, E. H., *Chemistry of Carbon Compounds,* Elsevier Press, Houston (1953).

*17. Furnas, C. C., *Rogers Manual of Industrial Chemistry,* 6th Ed., Van Nostrand Co., New York (1942). Good descriptive information on well-established processes.

*18. Sax, N. I., *Handbook of Dangerous Materials,* Reinhold Publishing Co., New York (1951).

## C. Compilations of Thermodynamic, Physical and Chemical Data

1. Azeotropic, Distillation, Equilibrium, and PVT

Chu, Ju Chin, R. J. Getty, L. F. Brennecke, and R. Paul, *Distillation Equilibrium Data,* Reinhold Publishing Co., New York (1950).

Dreisbach, R. R., *Pressure—Volume—Temperature Relationships of Organic Compounds,* 3rd Ed., Handbook Publishers Inc., Sandusky, Ohio (1952).

Horsley, L. H., *Azeotropic Data,* American Chemical Society, Washington, D.C. (1952).

2. Fuels

*Fuel—Flue Gases,* American Gas Association, New York (1941).

*Gaseous Fuels,* American Gas Association, New York (1948).

Spiers, H. M., *Technical Data on Fuel,* 5th Ed., British National Comm., World Power Conference, London (1950). Calorific values, handy charts and tables for many types of fuels.

3. Hydrocarbons

Brooks, B. T., and A. E. Dunstan, *The Science of Petroleum,* Volume 5, Oxford Univ. Press, London (1950). Crude oils: chemical and physical properties.

Brown, G. G., D. L. Katz, G. G. Oberfell, and R. C. Alden, *Natural Gasoline and the Volatile Hydrocarbons,* Natural Gasoline Association of America, Tulsa, Oklahoma (1948).

Egloff, G., *Physical Constants of Hydrocarbons,* Reinhold Publishing Co., New York (1947).

Doss, M. P., *Physical Constants of the Principal Hydrocarbons,* 4th Ed., The Texas Company, New York (1943).

*Kobe, K. A., *Thermochemistry for the Petrochemical Industry, Petroleum Refiner,* Reprint, Gulf Publishing Co., Houston (1951).

*Maxwell, J. B., *Data Book on Hydrocarbons,* D. Van Nostrand Co., Inc., New York (1950).

**Natural Gasoline Supply Men's Association Data Book,* 6th Ed., Tulsa, Okla. (1951). Data on light hydrocarbons.

*Rossini, F. D., et al, *Selected Values of Physical and Thermodynamic Properties of Hydrocarbons and Related Compounds,* Carnegie Press, Pittsburgh, Pa. (1953).

Sage, B. H., and W. N. Lacey, *Thermodynamic Properties of the Lighter Hydrocarbons and Nitrogen,* American Petroleum Institute, New York (1950).

4. Chemical Substances (General)

*Dreisbach, R. R., *Physical Properties of Chemical Substances,* Dow Chemical Co., Midland, Michigan (1952).

*International Critical Tables of Numerical Data,* National Research Council, McGraw-Hill Book Co., New York (1930).

Kobe, K. A., and R. E. Lynn, *The Critical Properties of Elements and Compounds,* Chemical Reviews *52,* Feb. 1953 (Reprint).

*Landolf—Bornstein Physikalisch—Chemische Tabellen,* Springer-Verlag, Berlin (1931 and supplements).

*Rossini, F. D., et al, *Selected Values of Chemical Thermodynamic Properties,* National Bureau of Standards Circular 500, Washington, D.C. (1952).

Other Sources: Many other general sources are available. Literature can be obtained on request from the various chemical manufacturers on the products they manufacture. The Monographs on specific elements and compounds published by the American Chemical Society ranging from various metals to styrene, glycerol, and glass provide valuable data. In addition, chemical engineering texts such as that by Hougen, O. A., and Watson, K. M., *Chemical Process Principles,* John Wiley & Sons, Inc., New York (1950) provide generalized methods for estimating properties not otherwise obtainable.

5. Kinetics

Berkman, S., J. C. Morrell, and G. Egloff, *Catalysis,* Reinhold Publishing Co., New York (1940). Good source of data on industrial catalysts of all types.

**Tables of Chemical Kinetics.* Homogeneous Reactions, Circular No. 510 of the National Bureau of Standards, Washington, D.C. (1951).

6. Solubilities

Seidell, A., *Solubilities of Inorganic & Organic Compounds,* 3rd Ed., Vol. I, II, & Supplements, Van Nostrand Co., New York (1941–52).

Hildebrand, J. H., and R. L. Scott, *The Solubility of Nonelectrolytes,* 3rd Ed., Reinhold Publishing Co., New York (1950).

7. Solvents

Mellan, I., *Industrial Solvents,* Reinhold Publishing Co., New York (1950).

8. Thermal Conductivity

Sakiadis, B. C., and J. Coates, *A Literature Survey of the Thermal Conductivity of Liquids,* Eng. Exp. Station Bulletin No. 34, Louisiana State Univ., Baton Rouge, La. (1952).

*Standards of the Tubular Exchangers Manufacturers Association,* 3rd Ed., TEMA, Inc., New York (1952).

9. Miscellaneous

Kennan, J. H., and J. Kaye, *Gas Tables,* John Wiley & Sons, Inc., New York (1945).

Kennan, J. H., and J. Kaye, *Thermodynamic Properties of Air,* John Wiley & Sons, Inc., New York (1945).

*Kennan, J. H., and F. G. Keyes, *Thermodynamic Properties of Steam,* John Wiley & Sons, Inc., New York (1936).

*Properties of Commonly Used Refrigerants,* Air Conditioning and Refrigerating Machinery Association, Washington, D.C. (1946).

Zimmerman, D. T., and I. Lavine, *Psychrometric Tables & Charts,* Industrial Research Service, Dover, New Hampshire (1945).

## D. Materials

1. Everhart, J. L., et al., *Mechanical Properties of Metals & Alloys,* Bureau of Standards Circ. C447, Washington, D.C. (1943).
2. DuMond, T. C., *Engineering Materials Manual,* Reinhold Publishing Co., New York (1951). Generalized and in simple terms.
3. Rabald, E., *Corrosion Guide,* Elsevier Publishing Co., Houston (1951). Handy tables of corrosion resistant material recommendations.
4. Simonds, H. R., A. J. Weith, and M. H. Bigelow, *Handbook of Plastics,* 2nd Ed., Van Nostrand, New York (1949).
5. *Technical Data on Plastics,* Manufacturing Chemists Assoc., Washington, D.C. (1952). Properties and Manufacturers.

The possession of the best technical date and theoretical knowledge for design of all types of process equipment represents only a portion of the requirements for successful process engineering. The process engineer must also have a working knowledge of equipment operation and mechanical design gained by actual operating experience or at least by observing operation. An engineer without this background will often size equipment in such a way that it is impractical to build and operate. Those engaged in mechanical design of equipment are reluctant to question or alter the process design since it is not uncommon for unusual arrangements or designs to be required due to peculiarities of the process conditions. An impractical design, therefore, initiated by the process engineer might be used in the final plant design without question. So it is necessary for the project engineer to be on the look-out for all impractical process designs. His experience in both mechanical and process phases of design makes him unusually qualified for such work, and as director of the project it most certainly is his duty.

## REFERENCES

1. Chilton, C. H., *Chem. Eng., 56,* No. 6, 97 (1949).
2. Dybdal, E. C., *Chem. Eng. Progr., 46,* 57 (1950).

3. Harris, J. McA., Jr., *Chem. Eng. Progr., 44,* 333 (1948).

4. Hougen, O. A., and K. M. Watson, *Chemical Process Principles,* Vol. III, John Wiley & Sons, New York, 1947.

5. Long, H. J., *Chem. Eng., 55,* No. 6, 112 (1948).

6. Nichols, W. T., *Ind. Eng. Chem., 43,* 2295 (1951).

7. "Process Equipment Estimation," Report *Chem. Eng., 54,* No. 5, 103 (1947).

8. Stevens, R. Q., *Chem. Eng., 54,* No. 10, 124 (1947).

9. Williams, R., Jr., *Chem. Eng., 54,* No. 12, 124 (1947).

10. Zimmerman, O. T., and I. Lavine, *Chemical Engineering Equipment Costs,* Industrial Research Service, Dover, New Hampshire, 1950.

# CHAPTER 5

# FLOW DIAGRAMS

It is important for the chemical engineer to be able to use the material and energy balance and the flow diagram in solving problems in design and operation. Just as the balance sheet and income account enable the accountant to solve complex financial problems, the material and energy balance, together with the flow diagram of a process, aid the chemical engineer to visualize and solve his problems in a rapid and straightforward manner.

Sketching a flow diagram is almost an instinctive manipulation for the experienced chemical engineer. It is part of his thinking process as well as his explanatory technique. In this form it is usually a rough sketch or schematic flow diagram. In the design and construction of a plant a more complete flow diagram called the engineering flow diagram becomes the vital cog in the transmission of process information to all departments of an engineering design group.

## SCHEMATIC FLOW DIAGRAMS

It seems desirable to suggest the least number of rules to follow in constructing schematic diagrams. By their very nature they should be individualistic, since they are used for varied purposes. When used to explain a process or mode of operation to a colleague, superior, or customer, originality often provides the needed emphasis. Clarity, however, must never be sacrificed, and, to assure this clarity, the audience must be considered.

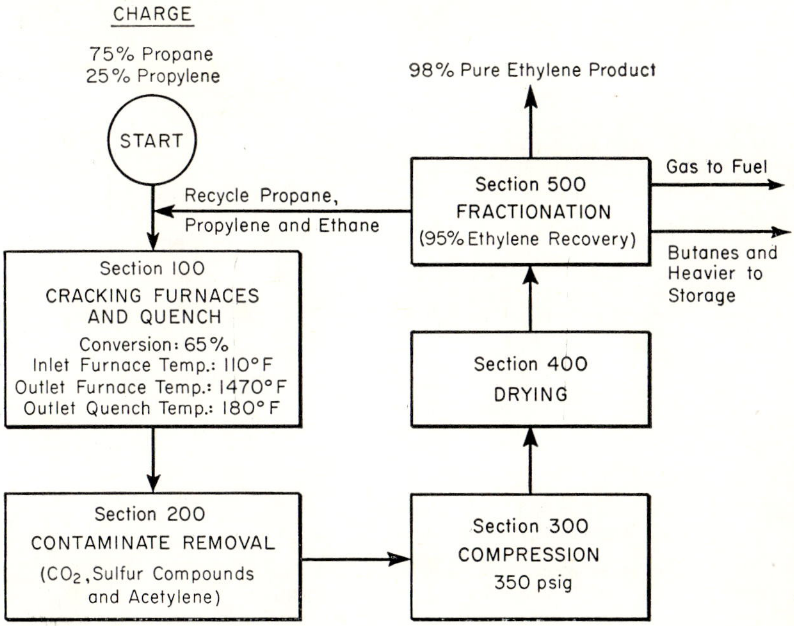


Fig. 5-1. Block diagram for an ethylene plant.

There are three types of schematic diagrams in general use:

1. Block diagram.
2. Process flow diagram.
3. Graphic flow diagram.

## Block Diagram

The block diagram is the simplest but least descriptive of the schematic diagrams. As the name implies, it consists of rectangular blocks which usually represent a single unit operation in a plant or an entire section of the plant. These blocks are connected by arrows indicating the flow sequence. One such diagram is shown in Fig. 5-1.

The block diagram is extremely useful in the early stages of a process study and is particularly valuable in presenting the results of economic or operating studies since the significant data can be placed within the blocks.

## Process Flow Diagram

The process flow diagram as illustrated in Fig. 5-2 is used most frequently by the process engineer in design work and in process studies. It must be drawn so that the process flow and operations are immediately

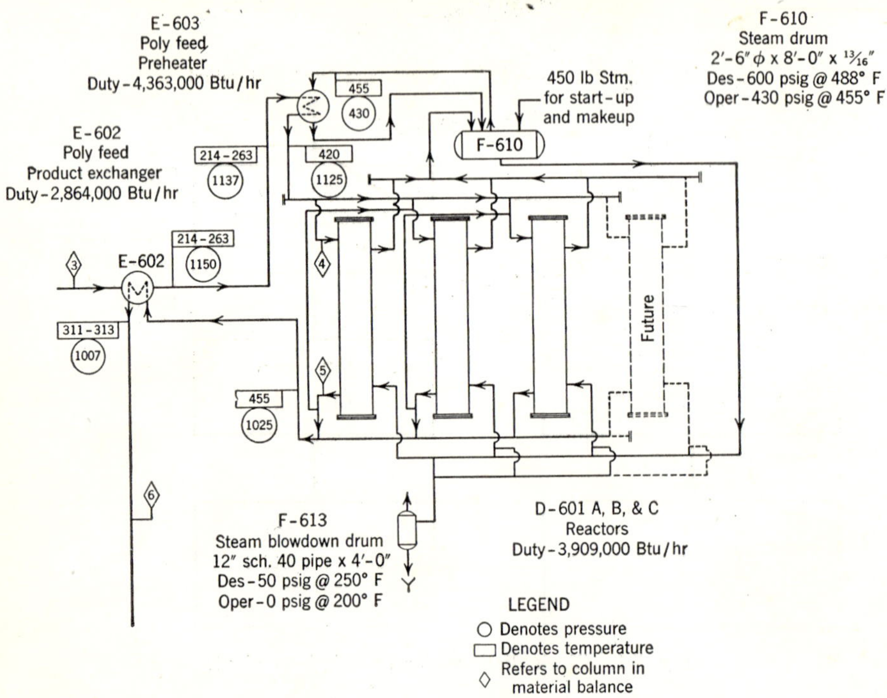


Fig. 5-2. Process flow diagram: polymerization unit reactor section. (Courtesy of Foster Wheeler Corporation.)

apparent. This is accomplished by omitting all but the essential detail using frequent arrows to indicate direction of flow, employing heavy lines for major flow lines, and indicating temperatures, pressures, and flow quantities at various significant points in the diagram. Pertinent process design data are shown, such as heat exchanger duty, vessel design information, and special requirements such as required elevations of certain equipment.

Convenient symbols for standard items, such as pumps and exchangers, are often used as shown in Fig. 5-2. Valves, utility lines, and spare items of equipment are omitted except where needed to clarify the process. Only instruments essential to the control of the process are shown.

Clarity, accuracy, and utility are the principal criteria for a good process flow diagram. It is used throughout the early phases of plant design and many engineers must see and understand it. It is from this diagram that the more complete engineering flow diagram is developed. The importance of a careful workmanlike job on the process flow diagram cannot be stressed too strongly.

## Graphic Diagram

The graphic flow diagram (Fig. 5-3) is used most frequently in advertising, company financial reports, and technical reports in which certain features of the flow diagram require extra emphasis. Few rules can be suggested for this type of diagram, since often unusual originality accomplishes the desired purpose most effectively. It should present the desired information clearly and in an eye-catching fashion that is both novel and informative.

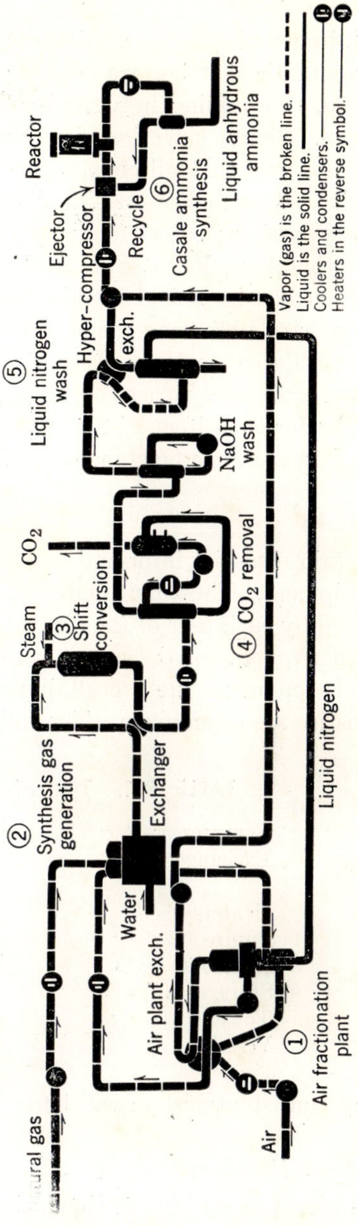


Fig. 5-3. Graphic flow diagram: ammonia plant. (Courtesy of Foster Wheeler Corporation.)

## ENGINEERING FLOW DIAGRAMS

Through the early stages of process study and development the various schematic flow diagrams serve very well. When the process calculations are completed and the construction contract has been made, the more comprehensive engineering or mechanical flow diagram is required (Figs. 5-4 and 5-5).

To understand what is needed on this diagram one must understand its use. In any engineering office the flow diagram is the central source of information for all design groups. These groups include electrical, piping, structural, instrument, and pump and compressor designers. It serves as a common meeting ground for design engineers and executives. Since so many men of varied backgrounds must refer to it, standardization is essential in order to avoid needless confusion. No important information should be omitted nor should useless information be included.

### Process Equipment

All items of process equipment are shown, including spares. The diagrammatic sketches are made to resemble the equipment as nearly as possible, thus aiding in visualization of the items by all users of the diagram. Simple elevation views are generally used, but they are altered in such a way that all connections appear on the plane of the diagram. This procedure makes for simplicity. Nozzles are located in the proper relative position; i.e., at the proper tray in a fractionating column. It is impractical to make these diagrams to scale, since smaller items would be unduly reduced. It is desirable, however, to preserve a suggestion of the relative size differences; and this can sometimes be accomplished by using a different scale for the horizontal than for the vertical.

Since equipment varies considerably, it is not possible to have standard sketches for each item. The representations of pumps, motors, and turbines, however, can be standardized as in Fig. 5-5.

### Notations

Certain essential information must be provided adjacent to each item of equipment. These notations must supply in a concise manner the data required by each user of the diagram. Experience has dictated the information required for the common items such as pumps and vessels. For special equipment, the overall dimensions and significant operating characteristics are often given. The following notations are suggested:

**TABLE 5-1. Typical Equipment Designations**

| | |
|---|---|
| Compressors | K, C |
| Exchangers | E, C (for condensers) and RB (for reboiler) |
| Heaters | H |
| Pumps | P or PU |
| Reactors | R |
| Storage tanks | ST |
| Towers | T |
| Vessels | V |

Any consistent system will be satisfactory.

#### *Item Numbers*

All pieces of equipment should be given an item number which should appear on or next to the equipment. If the item numbering system is planned with care, it aids immeasurably in the use of the flow diagram and in the execution of all phases of the design and construction work. A typical system which has been used with success, in cases where the process unit is divided into areas or sections, employs a number such as P-101. The letter indicates the type of equipment, in this case a pump; the 101 indicates that the pump is the number one pump in area 100. If this pump has an identical spare, it can be labeled P-101A. Common letter designations for equipment are given in Table 5-1.

### *Process Information*

Important process information should also be listed. The following are suggestions for several common items.

*Vessels*

Service
Diameter, height, thickness
Special features (lining, etc.)
Design conditions
Operating conditions

*Heat Exchangers*

Service
Differential pressure across shell and tubes
Heat transfer area
Duty in Btu/hr
Design conditions
Temperature and pressure at inlet and outlet

*Pumps*

(This information may be given on the flow diagram or on accompanying attachments.)

Service
Size and type
Fluid
Pump temperature
Density at pump temperature
Design GPM at pump temperature
Design differential
NPSH (Net positive suction head)

*Compressors*

(This information may be given on the flow diagram or on accompanying attachments.)

Service
Stages
Suction conditions
First stage suction
Second stage suction
Second stage discharge

BHP (Brake horsepower) first stage
BHP second stage
Capacity SCFM

## Piping and Valves

All process lines and utility lines serving process equipment are shown on engineering flow diagrams. Often the main process lines are drawn heavy so that the primary flow sequence will be more apparent. Utility

**TABLE 5-2. Typical Valve and Pipe Symbols for Flow Diagrams**

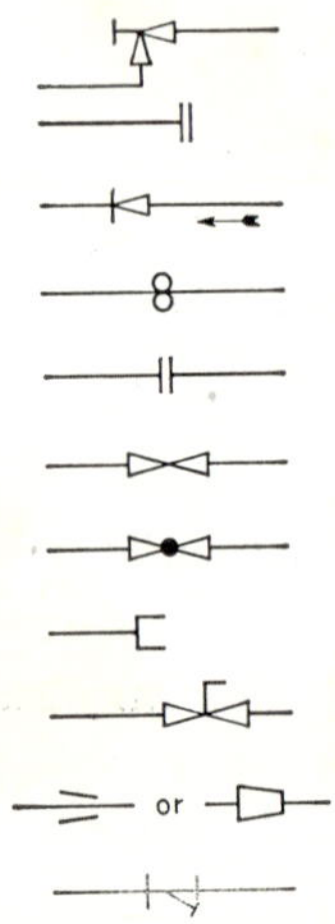

| ITEM | SYMBOL |
|---|---|
| Valves & Fitting | |
| Angle Valve | |
| Blind Flange | |
| Check Valve | |
| Figure 8 Flange | |
| Flange | |
| Gate Valve | |
| Globe Valve | |
| Hose Connection | |
| Plug Cock | |
| Reducer | or |
| Strainer | |

| Line Designations | |
|---|---|
| Main Process | ——————— (heavy) |
| Secondary | ——————— |
| Air | —//——//— |
| Condensate | ——————— |
| Sewer | — — — — — — |
| Steam | ——·——·— |
| Steam Traced | ======= |
| Water | ——··——··— |

Note: Usually no attempt is made to distinguish flanged and screwed fittings on flow diagrams. The piping specification as noted by the line number designation provides this information

lines are drawn using various symbols (Table 5-2) to permit ready differentiation of the several services. Arrows are used wherever necessary to aid in following the flow. All required valves and check valves are indicated, but flanges and fittings are not shown except at equipment. Common symbols for the valves in general use are given in Table 5-2. Sizes of valves for those other than line size are shown adjacent to the valve.

Since the piping designer uses the flow diagram as his main source of information, much data must be supplied on the diagram. Temperatures and pressures, flow rates, and a description of the fluid flowing must be provided. In addition, the specifications for the lines must be given so that the size, material, and thickness of pipe, class of flanges, and rating of fittings and valves will be known for each line. This information can be conveniently provided by means of a line numbering system in which each line extending from one piece of equipment to another is given a separate number.

These numbers can be recorded on forms together with the flow rate, pressure, temperature, and fluid flowing in the line. The line number changes whenever the line size or line specification changes. To avoid confusion, a line number is eliminated when a line is deleted. The line number can include the area designation. One typical system employs a number such as 3″06403E (Fig. 5-4).

| 3″ | 0 | 6 | 403 | E |
|---|---|---|---|---|
| Line size | Fluid class | Area number | Line number | Piping spec. |

Since, as explained in Chapter 18, piping specifications are written in detail for each class of service, the piping designer can obtain all the information required to layout and design the piping from the flow diagram, the line list, and the piping specification. These line numbers further serve in the marking and identifying of shop-fabricated pipe. A poorly numbered flow diagram, therefore, can cause considerable confusion throughout the project.

Valves are identified on the flow diagram, both by number and by symbol. The numbers refer to detailed valve specifications and are very helpful to both designers and material procurement men. Although some organizations do not number the valves on the flow diagram, almost all use numbers on the piping drawings.

## Instrumentation

All instrumentation for the controlling, recording, and indicating of plant operation must be shown on the flow diagram. Usually, the process engineer indicates in a schematic fashion the essential instrumentation on the process flow diagram. This is further engineered and added to the engineering flow diagram with all other required instrumentation.

The Instrument Society of America has prepared a set of standard instrument symbols which should be followed to avoid confusion. These are reproduced in Table 5-3. The instrument number is generally included in the symbol, as shown in the tabulation. Control valve and relief valve sizes should be shown adjacent to these items to aid in piping design.

## Arrangement of Diagram

A well-planned arrangement of the engineering flow diagram can do much to assure efficient and high quality workmanship throughout all stages of the job. If some idea of the actual plot plan is known when the flow diagram is drawn, it is very helpful to have the diagram arranged in a similar manner insofar as it is practical. Such an arrangement aids

TABLE 5-3 (1). Instrumentation Flow Plan Symbols.

(Reprinted by permission, Instrument Society of America, *Tentative Recommended Practice*)

**BASIC INSTRUMENTATION SYMBOLS**

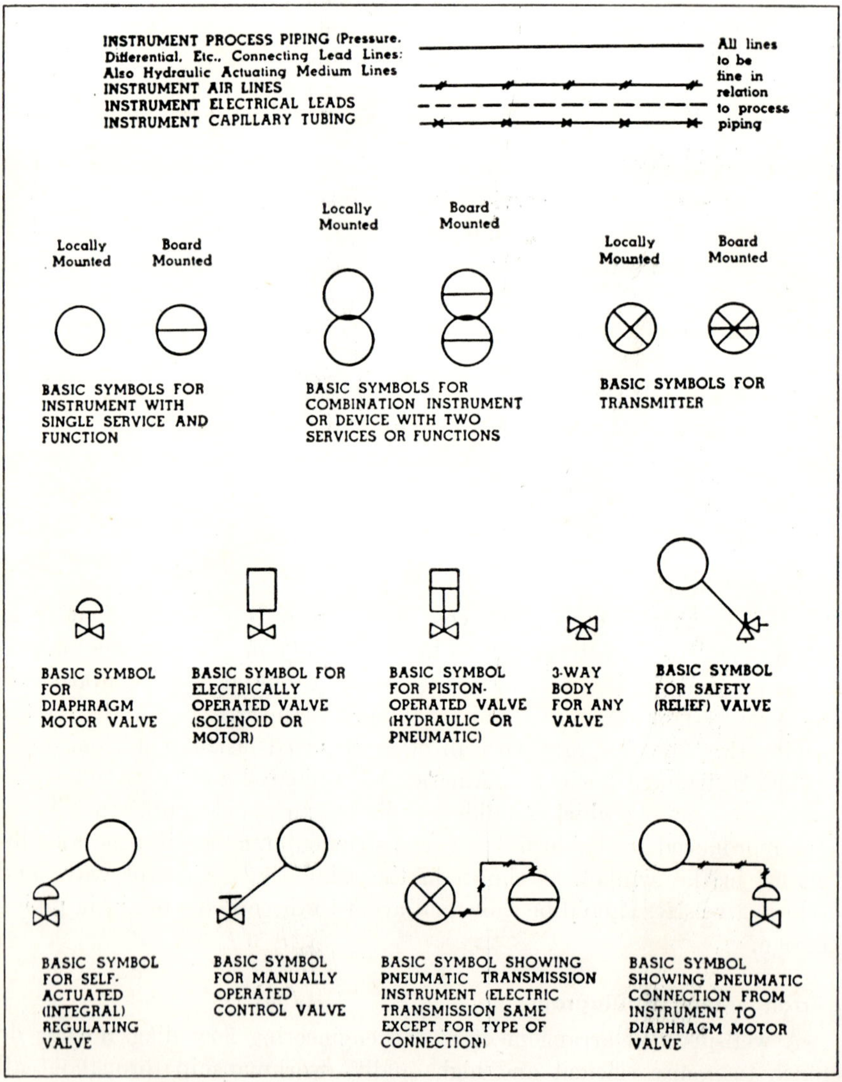

## TABLE 5-3 (2). Instrumentation Flow Plan Symbols.

(Reprinted by permission, Instrument Society of America, *Tentative Recommended Practice*)

TYPICAL INSTRUMENTATION
SYMBOLS FOR FLOW

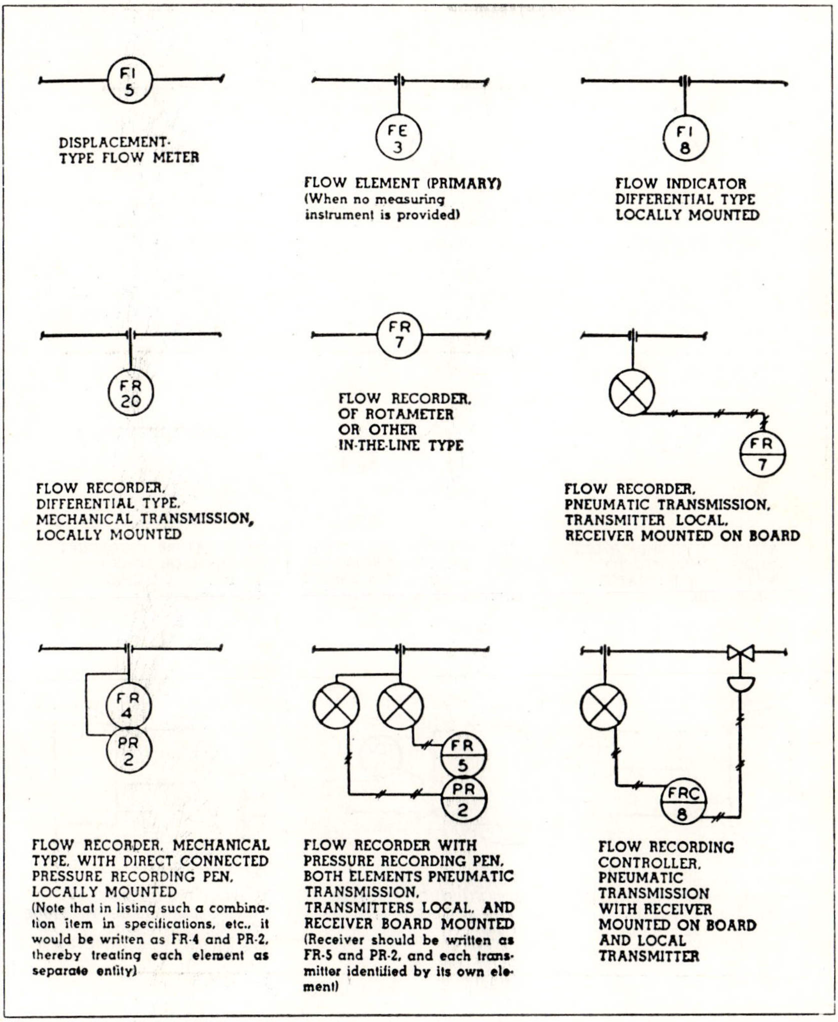

TABLE 5-3 (3). Instrumentation Flow Plan Symbols.

(Reprinted by permission, Instrument Society of America, *Tentative Recommended Practice*)

**TYPICAL INSTRUMENTATION SYMBOLS FOR TEMPERATURE**

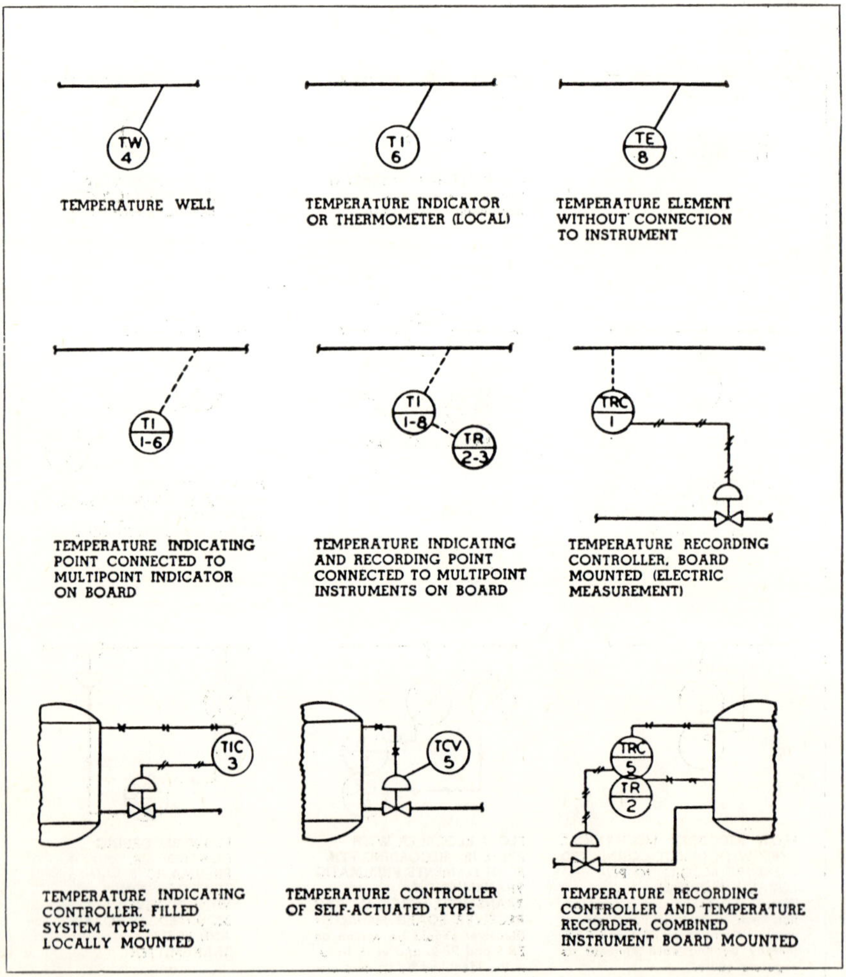

TABLE 5-3 (4). Instrumentation Flow Plan Symbols.

(Reprinted by permission, Instrument Society of America, *Tentative Recommended Practice*)

**TYPICAL INSTRUMENTATION SYMBOLS FOR LEVEL**

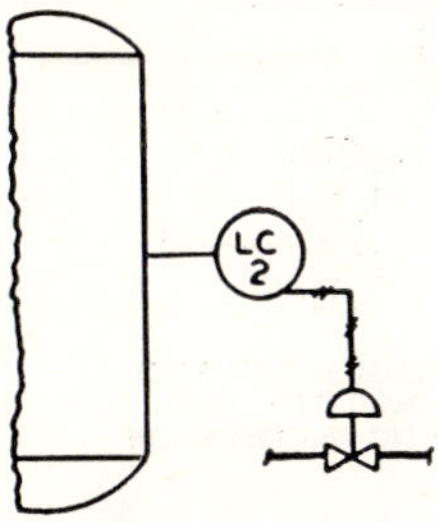


BLIND LEVEL CONTROLLER, INTERNAL TYPE

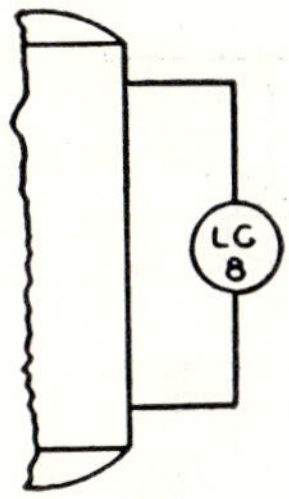


GAGE GLASS

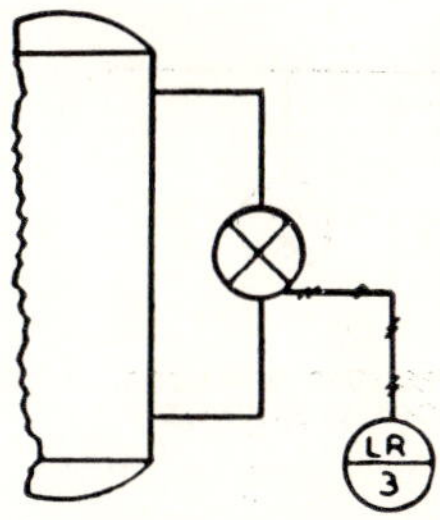


LEVEL RECORDER, PNEUMATIC TRANSMISSION, WITH BOARD MOUNTED RECEIVER. EXTERNAL TYPE TRANSMITTER

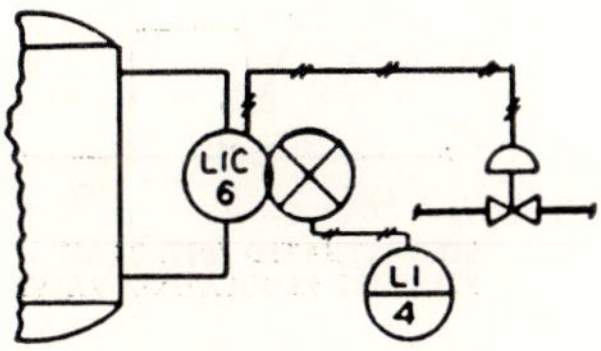


LEVEL INDICATING CONTROLLER AND TRANSMITTER COMBINED WITH BOARD MOUNTED LEVEL INDICATING RECEIVER

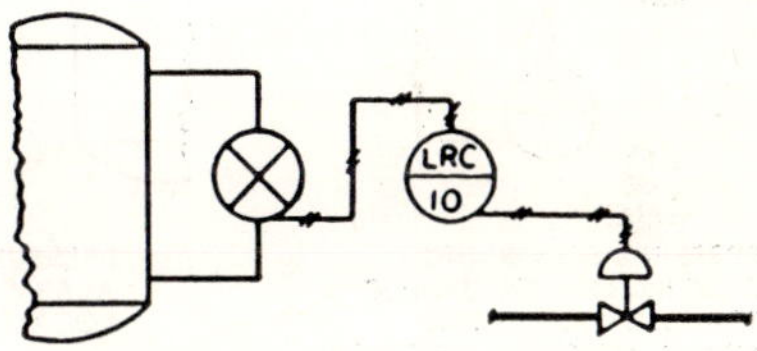


LEVEL RECORDING CONTROLLER. EXTERNAL TYPE. PNEUMATIC TRANSMISSION

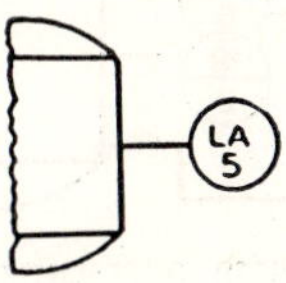


LEVEL ALARM. INTERNAL TYPE

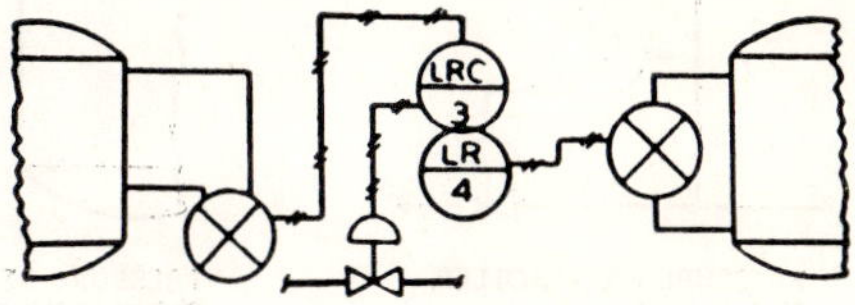


LEVEL RECORDING CONTROLLER AND LEVEL RECORDER, PNEUMATIC TRANSMISSION COMBINED RECEIVER BOARD MOUNTED

TABLE 5-3 (5). Instrumentation Flow Plan Symbols.

(Reprinted by permission, Instrument Society of America, *Tentative Recommended Practice*)

**TYPICAL INSTRUMENTATION SYMBOLS FOR PRESSURE**

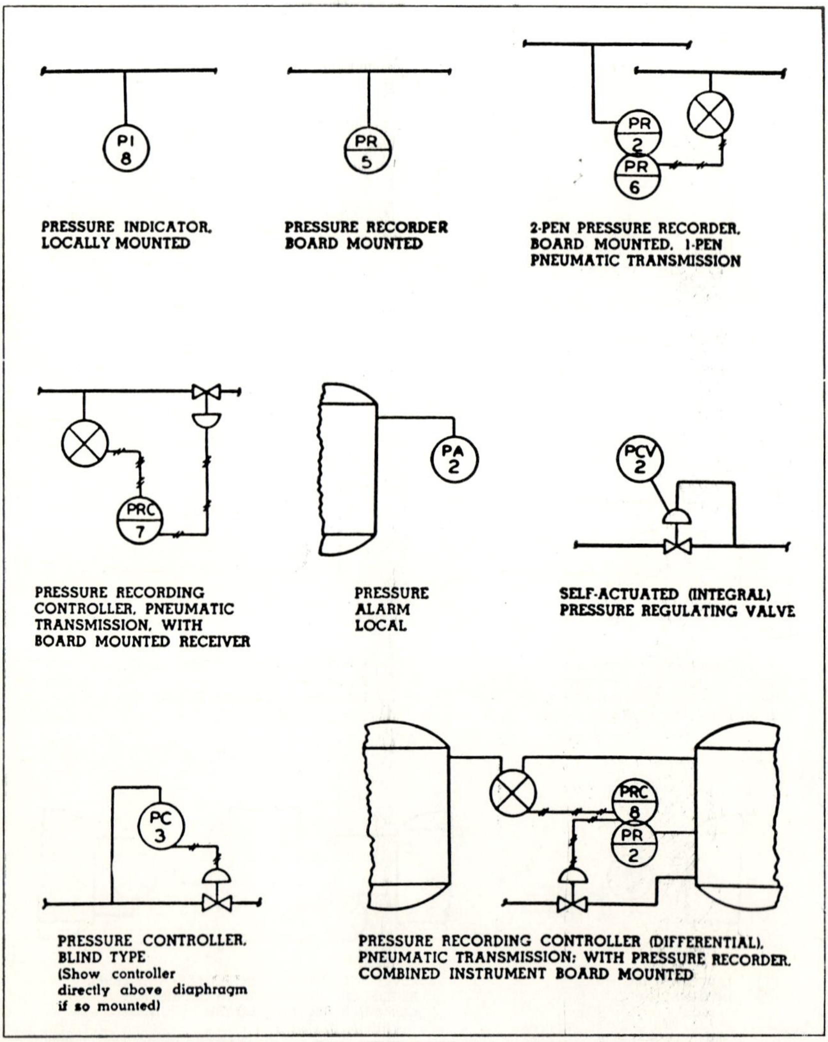

## TABLE 5-3 (6). Instrumentation Flow Plan Symbols.

(Reprinted by permission, Instrument Society of America, *Tentative Recommended Practice*)

TYPICAL INSTRUMENTATION SYMBOLS (MISCELLANEOUS)

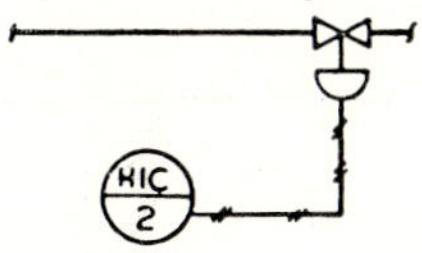

HAND ACTUATED PNEUMATIC CONTROLLER, BOARD MOUNTED, WITH INDICATION

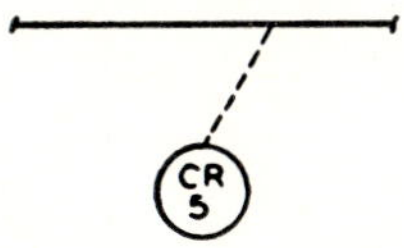

CONDUCTIVITY RECORDER, LOCALLY MOUNTED

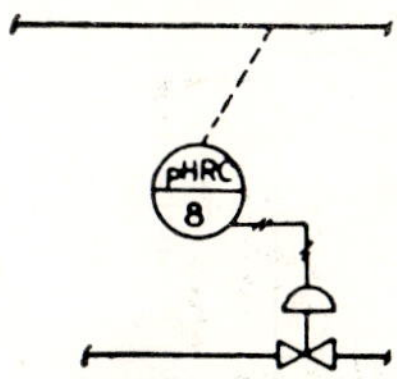

pH RECORDING CONTROLLER, BOARD MOUNTED

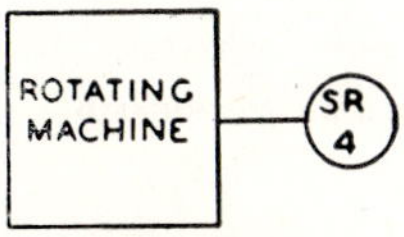

SPEED RECORDER, LOCALLY MOUNTED

WEIGHT RECORDER, LOCALLY MOUNTED

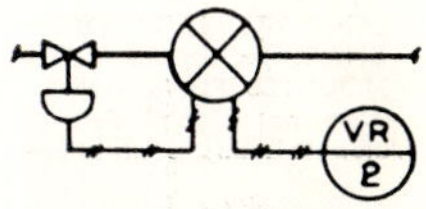

VISCOSITY RECORDER, PNEUMATIC TRANSMISSION, BOARD MOUNTED (Element in sample flow line)

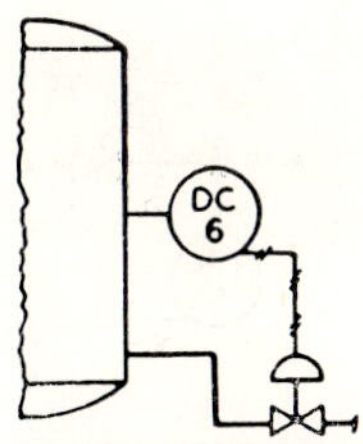

DENSITY CONTROLLER BLIND, INTERNAL ELEMENT TYPE

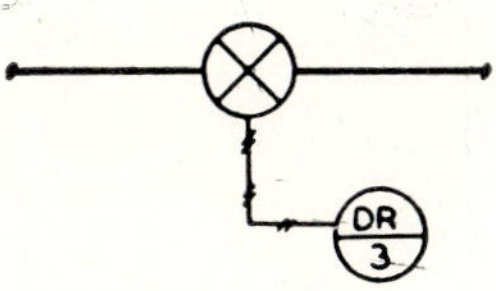

DENSITY RECORDER, PNEUMATIC TRANSMISSION BOARD MOUNTED (Element in sample flow line)

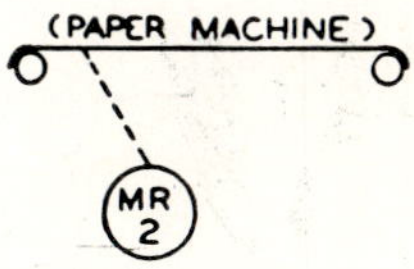

MOISTURE RECORDER, LOCALLY MOUNTED

TABLE 5-3 (7). Instrumentation Flow Plan Symbols.

(Reprinted by permission, Instrument Society of America, *Tentative Recommended Practice*)

**TYPICAL INSTRUMENTATION SYMBOLS FOR COMBINED INSTRUMENTS**

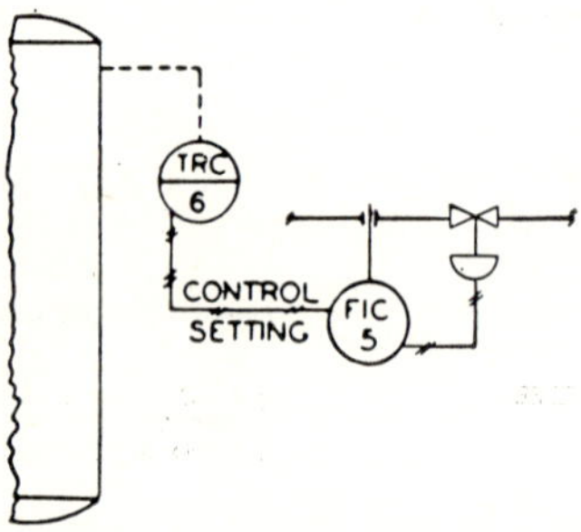


TEMPERATURE RECORDING CONTROLLER, BOARD MOUNTED, RESETTING LOCALLY MOUNTED FLOW INDICATING CONTROLLER (Note that "Control Setting" should be shown alongside air line to indicate cascade control)

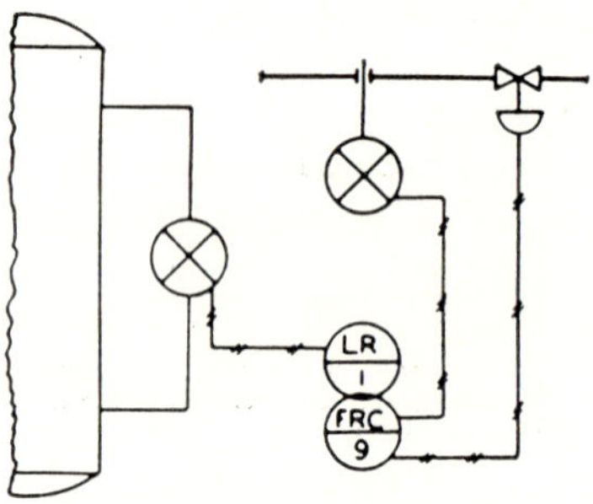


FLOW RECORDING CONTROLLER WITH LEVEL RECORD. BOTH ELEMENTS PNEUMATIC TRANSMISSION. LEVEL TRANSMITTER EXTERNAL TYPE. COMBINED RECEIVER BOARD MOUNTED

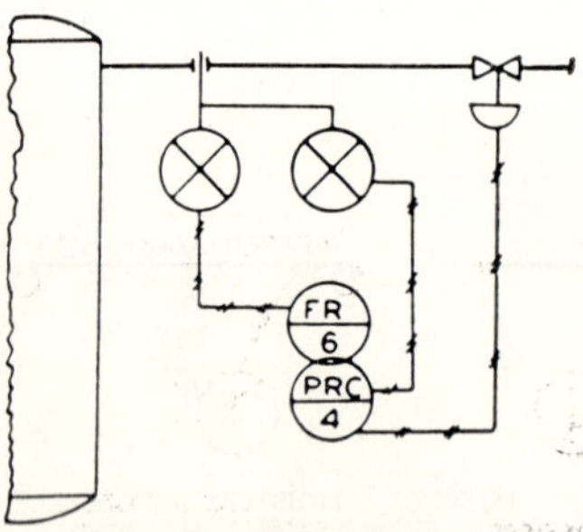


PRESSURE RECORDING CONTROLLER WITH FLOW RECORD. BOTH ELEMENTS PNEUMATIC TRANSMISSION. COMBINED RECEIVER BOARD MOUNTED

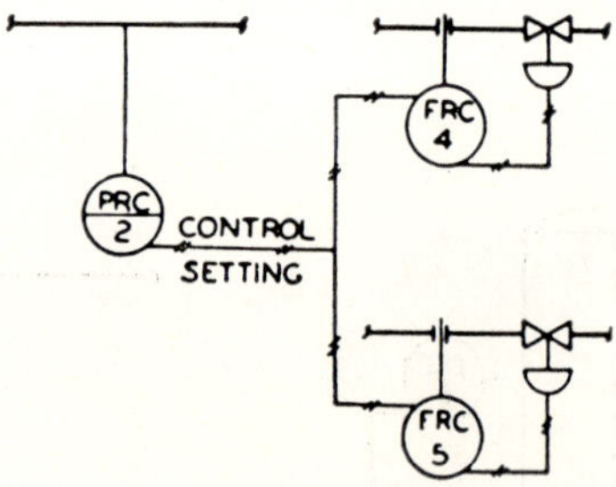


PRESSURE RECORDING CONTROLLER, BOARD MOUNTED, RESETTING LOCALLY MOUNTED FLOW RECORDING CONTROLLERS

the layout men and piping designers to visualize the plant more readily and prevents the overemphasis of a line that is actually very short but appears long because of the arrangement of the flow diagram.

There are many methods of accomplishing good arrangement. Pumps can be placed below other equipment as shown in Fig. 5-5 or all equipment may be arranged above a common base line. Process lines can be shown on one drawing and utility lines on another. Or, both utility and process lines can be shown on the same drawing. The latter is done so that the full picture can be seen on one drawing. Often, however, this results in overcrowding especially if utility connections become excessive, such as on large compressor installations.

The flow diagram, to be read and understood easily, must not appear crowded. Often a number of separate drawings are needed to represent a process. Lines passing from equipment on one sheet to that on another are maintained at a constant elevation so that the continuation of the line on the next sheet can be found with ease.

## CONFIDENTIAL INFORMATION

Because of their completeness, process flow diagrams, engineering flow diagrams, and at times block diagrams, contain confidential information. They should never be distributed indiscriminately or allowed to be reproduced by unauthorized persons. If unscrupulous individuals are permitted to copy a flow diagram, they are able to obtain unfair advantage by avoiding the time, effort, and expense required to originate a complete flow diagram. Such practice violates all known codes of ethics.

# CHAPTER 6

# PLOT PLANS

After the engineering flow diagrams are completed and before detailed piping, structural, and electrical design can begin, the layout of process units in a plant and the equipment within these process units must be planned. This planning is done by a group of experts including the project engineer, piping and structural designers, and the process engineer. The results of their work are plot plans: the master plot plan and the unit plot plans. The master plot plan (Fig. 6-1) shows the location of each process unit, roadway, and building. The unit plot plan (Fig. 6-2) shows the location in plan view of each piece of equipment within a single process unit.

The preparation of a plot plan is the most important function in the design of a process plant. One key to good operation, economical construction, and efficient maintenance is well-planned, functional arrangement of equipment and buildings. A key to employee satisfaction and good public relations is a neat and esthetically pleasing plan.

Each plant differs in many ways and no two plant sites are exactly alike; hence, there is no one ideal plot plan. It is possible, however, to describe some of the more important rules which insure a satisfactory arrangement.

## MASTER PLOT PLAN

Figure 6-1 is a sketch of a typical master plot plan. Dimensions have been omitted for clarity. Generally the plant is divided into blocks

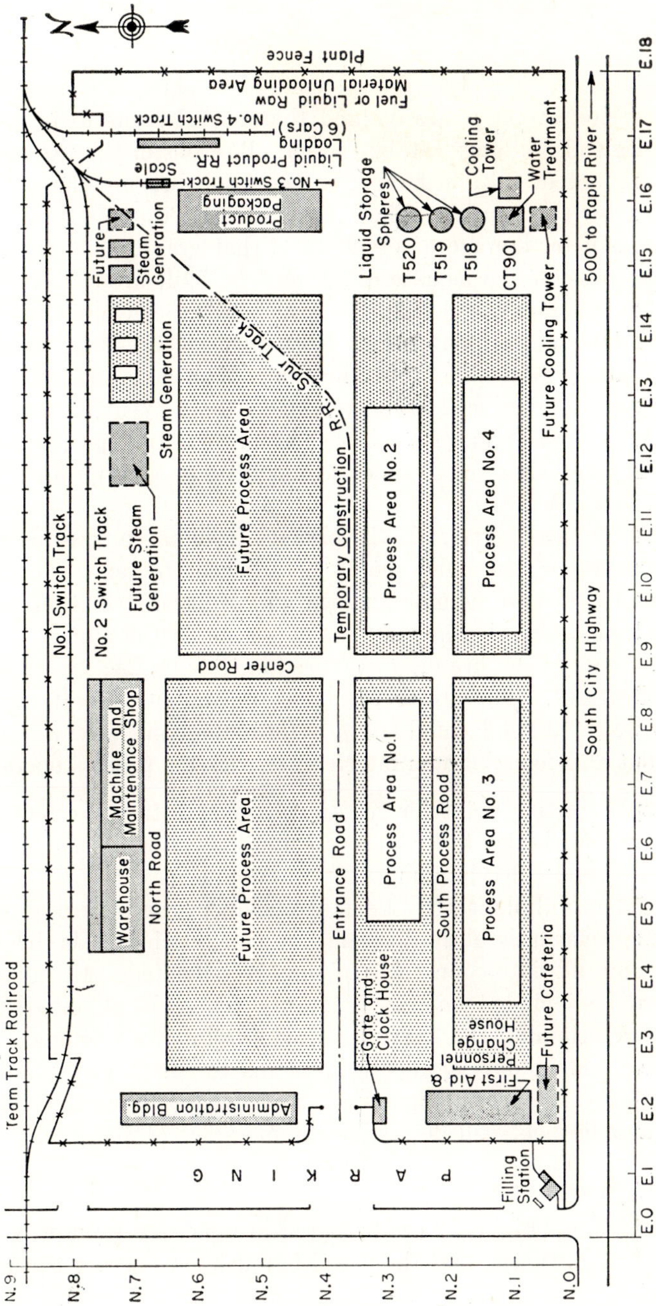


Fig. 6-1. Master plot plan.

which are most conveniently separated by roadways. Numerous factors influence the location of roadways and the arrangement of blocks. Some of these are:

PLANT TERRAIN. The drainage system can be planned from a contour map of the site. Next roadways and the general areas to be used for process units and buildings can be chosen. Portions of the site requiring the least amount of fill and having the best soil-bearing characteristics are selected for these areas. The highest area on the site may be used advantageously for storage of products so that gravity loading of tank cars will be possible. Adequate dikes must be built around these tanks to prevent rapid spread of fire if tanks overflow. (See Chapter 24.)

EXISTING HIGHWAY AND RAILROAD FACILITIES. Offices and warehouses must be readily accessible from the main highway, and warehouses, storage yards, and product loading areas must be convenient to the nearest railroad. If the plant is to have harbor facilities, product storage should be in the vicinity of the docks, though far enough to minimize the danger of dock fire.

TYPES OF PROCESS UNITS. If the product from one process unit is fed directly to another, the units should be adjacent to reduce piping and pumping costs. Operation, maintenance, and utility distribution are often simplified by locating similar units in one section of the plant.

SAFETY. Safety rules are fully discussed in Chapter 24.

PLANT SERVICES. The power plant, shops, warehouse, cafeteria, and change house should be located not only for maximum efficiency and convenience but also for minimum interference with the process operations.

One simple starting point for general planning is to make rectangular cardboard cutouts scaled to each process unit, power plant, machine shop, tankage area, and office building. These cutouts may be moved over a drawing of the total plant site. Each arrangement can be further studied by preparing schematic block diagrams on which each area is shown as a rectangle and all interconnecting piping is indicated. Utility piping should also be included on this diagram. These block diagrams are helpful in determining the space required for interunit piping and are thus a great aid in developing the final master plant plot plan. The economics of construction, operation, and future expansion for each proposed arrangement will determine the most acceptable plan.

## UNIT PLOT PLANS

Unit plot plans (Fig. 6-2) are prepared in much the same way as the master plot plan, except that the detail is much greater due to the large number of elements making up the process unit. Scale models begun at this stage and developed as design progresses are most helpful. Briefly, the steps in preparing unit plot plans are:

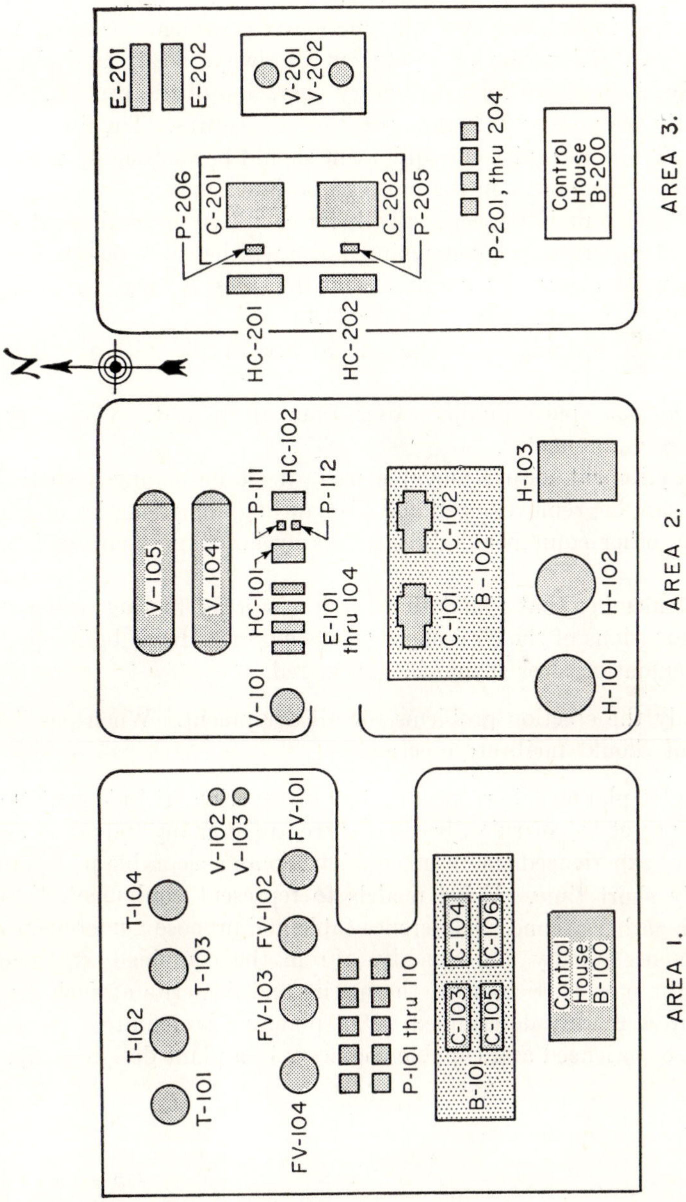


Fig. 6-2. Unit plot plan.

1. List all major items of equipment such as vessels, heat exchangers, pumps, and compressors. Indicate the size and weight of each.

2. Decide which equipment should be elevated. Usually the elevation of equipment is dictated by pump suction requirements or other process requirements. Elevation of equipment is always costly and should be done only if necessary for satisfactory operation of the process.

3. Study the process flow and operating procedures. The flow sequence and function of each piece of equipment should be thoroughly understood so that its arrangement in the plot can be functional.

4. Determine maintenance methods for each item of equipment so that equipment requiring frequent maintenance will be easily accessible.

5. Study all operating hazards so that the safest arrangement of equipment can be devised. (See Chapter 24.)

6. Plan the spacing of equipment in accordance with the following rules:

*a.* Do not space pumps closer than three feet. Closer spacing hampers maintenance.

*b.* Equipment with removable parts must be arranged so that the parts can be removed without dismantling long lengths of pipe or moving other equipment. Free access by hoisting apparatus is necessary.

*c.* Remember that foundations such as spread footings often exceed the dimensions of the equipment they support. Hence both equipment and foundation sizes should be considered.

7. Study the erection problems for all equipment. When possible, its placement should facilitate erection.

Good plot planning is more a matter of experienced judgment than of following a set list of rules, because there are so many values to be met. A group of experienced engineers can develop an acceptable plot plan in a relatively short time. Using models to represent equipment, these experts can gather around a conference table and propose numerous possible arrangements vividly and quickly. From the combined experience of each of the men participating, the merits and demerits of each arrangement can be readily determined. Plot planning is definitely the job for the most experienced and capable engineers in a plant design group.

CHAPTER 7

# SCHEDULING THE PROJECT

Careful planning and scheduling are vital to any project as complex as the design and construction of a process plant. After the completion of preliminary process design work, flow diagrams, and plot plans, a project schedule can be developed. This project schedule will then become a guide for planning and recording progress throughout the job.

## PURPOSE OF THE PROJECT SCHEDULE

The schedule should indicate the start and completion dates for engineering, drafting, and procurement, as well as construction operations for each major item of equipment. If properly prepared and followed, completion of the project on time will be assured.

In addition to assuring reasonably prompt completion of a construction job and efficient coordination of engineering, schedules provide the only method for contracting firms or design groups of operating companies to anticipate manpower requirements. Contracting organizations, in particular, require careful planning in order to maintain reasonably continuous employment for their personnel. The most economical utilization of personnel can be realized when a minimum of three and preferably four projects are being handled consecutively. A single project usually results in slack or idle periods for specialized personnel. The same problem arises with parallel project operations (two or more projects being engineered simultaneously) since the fundamental steps occur simultaneously.

The ideal work schedule for any type of engineering organization which contracts services in process plant design would be obtained by spacing consecutive contracts to permit a 30- to 45-day interval between them. Ideal conditions rarely exist in practice since the contractor is usually confronted with a promise to "start work immediately" regardless of his internal work program. The contractor's engineering schedules, however, aid him in planning the best conditions possible.

Once the preliminary schedule is made, organization of personnel and planning for the job are greatly expedited. The job schedule, however, must be a vital instrument. As work proceeds, more accurate information becomes available and hence more realistic and more detailed scheduling is possible.

## THE SCHEDULE SHEET

A schedule prepared in graphical form as shown for a crude distillation unit in Fig. 7-1 has been found most convenient. The symbols are used to indicate start and completion dates. Opposite each item two horizontal graphs are sketched. The top graph shows the planned schedule. The bottom graph indicates the actual progress of work up to the time of issue (April 8). The second graph for each item shows the progress up to that time. If required, the time scale can be enlarged to indicate the progress by weeks instead of months.

Schedules such as these can be prepared on printed forms and reproduced for distribution to executive personnel of the contractor and customer. This graphical presentation permits key personnel to determine rapidly the progress of the job as well as plan future work in accordance with the proposed schedule.

## HOW SCHEDULES ARE MADE

The key equipment and material items that are known to control job progress are listed as in Fig. 7-1. The flow diagram and plot plan are most convenient sources of information for preparing this list. The plot plan provides a logical basis for division of the project into sections. The example in Fig. 7-1 is for one such section or unit.

### Equipment Delivery Times Control Completion

The next step in preparation of a schedule is determining the earliest completion date. This date is usually controlled by the item of equipment having the longest delivery time.

Once the probable completion date is selected, engineering, drafting,

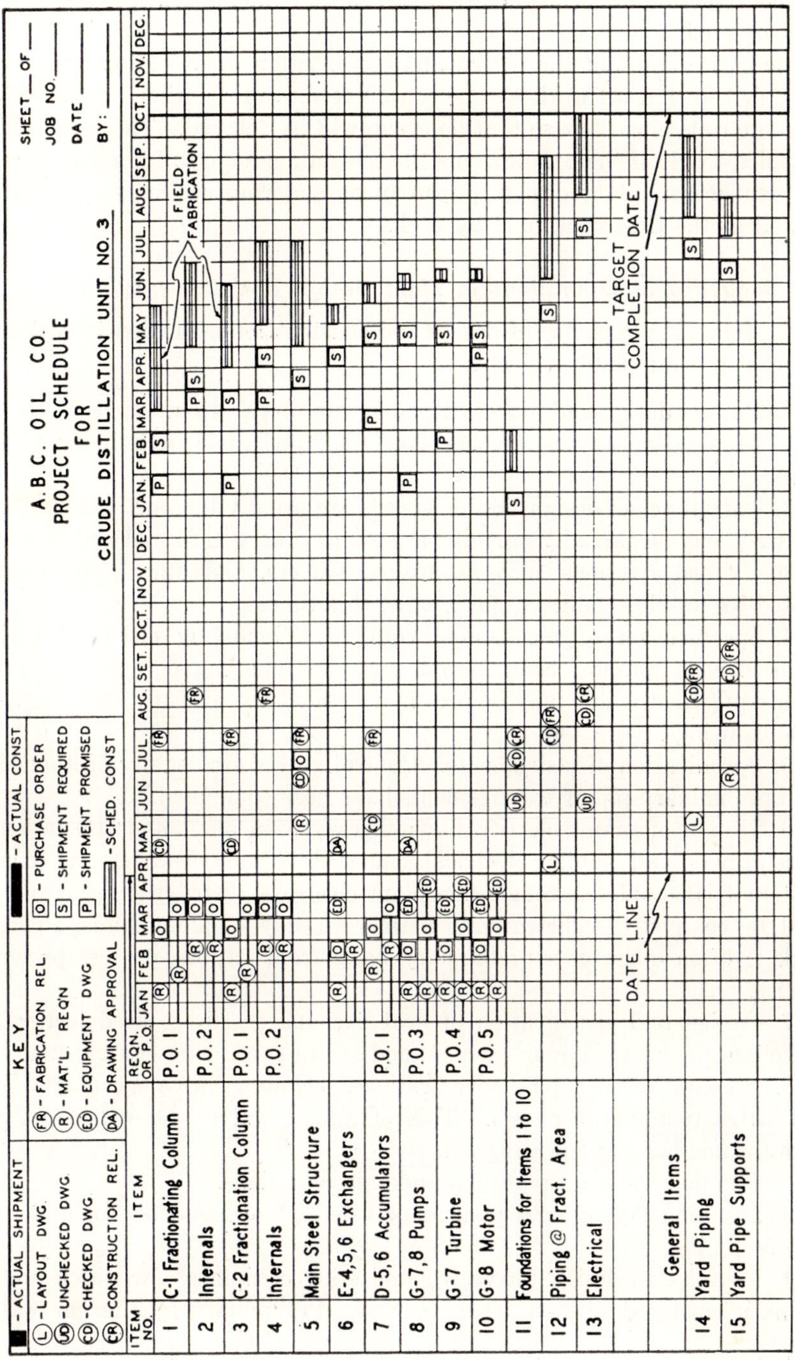


Fig. 7-1. Project schedule: crude distillation unit. [Reprinted by permission, M. Mattozi, *Oil & Gas Journal, 51*, No. 48, 100 (1953).]

and construction are scheduled in detail so that the master schedule may be completed.

The entire effort of the design and construction organization must be scheduled so that all work will be coordinated. Information, equipment, and materials must flow so that no group will be delayed. This goal will be realized only if the people doing the scheduling have full knowledge of the sequence of plant design, procurement and construction operations, as well as the time consumed by each.

### *Sequence of Operations in Plant Design and Construction*

The following sequence of design and construction operation is typical for most process plant projects.

1. Process design and process flow diagram preparation.

2. Preparation of engineering flow diagrams and preliminary plot plans.

3. Partial vessel design. Major vessels are usually long-delivery items and they may be partially procured as soon as process information is complete. Vessel internals such as mixers, stirrers, and other equipment related to vessels should be specified at this time.

4. Specification of mechanical equipment including electrical and steam drivers, pumps, compressors, material handling equipment, and other equipment which may be procured without the preparation of complete drawings.

5. Specification of instruments. Control valves are usually postponed until preliminary piping arrangements have been completed.

6. Design of major electrical switchgear, main plant and substation transformers, and smaller electrical gear such as motor starters, as soon as item 4 is complete. Single-line plant electrical distribution diagrams must be completed sufficiently to permit a study of plant loads. This work is often done on a preliminary basis from various process power studies.

7. Design and specification of heat exchangers, direct fired heaters, miscellaneous heater appurtenances, and steam generating equipment.

8. Preparation of piping specifications, including valve specifications and valve item number lists. This permits a valve take-off directly from the engineering flow diagram.

9. Finalizing general plot plan and area plot plans.

10. Planning piping arrangements and listing* of materials for underground piping.

11. Foundation design for vessels and other equipment for which preliminary vendor date have been received. Preliminary foundation ma-

* The word "listing," as used in this outline, includes procurement.

terial listing. Listing of all major equipment anchor-bolts. Where piling is to be used it is probable that piling arrangements may be complete at this time and pile driving could start. Preliminary construction operations are sometimes started at this point.

12. Structural steel design or design of other structures required for process operations. Architectural studies of plant buildings.

13. Electrical conduit arrangements, underground electrical material listing.

14. Foundation design for independent electrical equipment under item 6.

15. Receipt of certified dimension drawings on items 3 through 7.

16. Completion of piping arrangements sufficiently to determine and fix vessel nozzle orientations.

17. Release of vessels for fabrication. Vessel fabricators will have been selected by means of 3 above.

18. Release of foundation drawings to construction forces for erection.

19. Release of structural steel drawings for procurement and fabrication of structural steel: release of all materials for construction of elevated reinforced concrete structures by the construction forces.

20. Construction start: clear site, preliminary grading and foundation excavations.

21. Completion of preliminary piping material listing including valves and miscellaneous piping appurtenances for procurement.

22. Completion of miscellaneous major structural steel for pipe support and listing of all miscellaneous items for field construction use in pipe support manufacture.

23. Start of instrument installation drawings: listing of all materials required for instrument installation.

24. Completion of piping design and release of piping for fabrication.

25. Completion of architectural design and material listing for buildings. This phase may be completed much earlier for buildings not used for process operations.

26. Completion of electrical drawings and material listing.

27. Completion of procurement of all major equipment. The procurement of miscellaneous materials will continue.

28. Delivery of certain items under 4 through 8 to jobsite begins.

29. Completion of underground installations.

30. Completion of foundations.

31. Delivery of major items, structural steel, and start of above ground erection.

32. Installation of piping.

33. Installation of electricals.

34. Installation of instrumentation.

35. Testing.
36. Insulation installation.
37. Clean-up.
38. Operation start-up procedures.
39. Acceptance of plant by customer.

A brief study of this list will indicate the dependence of each operation upon one or more of the previous operations. A complex interdependence exists in which design groups depend on information from other design groups, fabricators depend on information from the designers (delivery can be delayed if this information is late); and finally, the construction crews depend on information from designers and material from fabricators.

## Example of Schedule Preparation

As in example of schedule preparation consider the items in Fig. 7-1. The piping layouts in most process plants require the major portion of drafting manhours. Piping designers and draftsmen, therefore, must be supplied as early as possible with vendor's drawings of equipment and vessel sketches.

Vessel sketches based on design calculations and applicable codes are prepared by the vessel design group. These sketches are then released to the procurement department accompanied by a requisition or other document describing pertinent design requirements and guarantee desired. The release of the requisition to procurement for the purpose of obtaining bids is designated by the symbol R on Fig. 7-1. (Referring to the bottom portion of item 1 it is seen that the time of requisitioning for column C-1 was behind schedule.) Since the fractionating columns (items 1 and 3) are more important in piping layout, the time for requisitioning these is scheduled two weeks before the less critical accumulators (item 7).

Vendor's drawings of equipment must be received by the piping design group at about the same time as vessel sketches. Early solicitation of bids (indicated by R) and placement of an order (O) will assure on-time receipt (ED) of vendor's drawings. These dates depend, therefore, on the rapid finalization of process design specifications and efficient procurement operations.

Vendor's drawings and vessel sketches, along with flow diagrams and plot plans, are transmitted to the piping design group for use in preparing piping layouts for major lines. The completion date (L) of this layout is estimated from the number of drawings required and the number of layout men assigned to the job.

The piping layout contains information sufficient for the preliminary design of main steel structures and for orientation on vessel nozzles which in turn permit requisitioning of fabricated steel, R (item 5), design of foundations, UC (item 11), and design of the electric distribution system.

Piping design continues during this period and reaches a stage of completion such that checking of drawings can begin. Vessel drawings are checked (CD) and sent to the fabricator so he can prepare shop drawings. After approval of these shop drawings the fabricator is "released" to fabricate the vessel (FR).

Equipment drawings are also checked and returned to the vendor with or without comments. Then, checking of structural steel, foundation, piping, and electrical drawings follows in the order named. In scheduling this work it must be remembered that vendor's and vessel drawings are needed for structural steel checking. For foundation drawings, structural steel, vendor's and vessel drawings are needed, and for piping and electrical drawings all previously checked drawings are required.

This example serves to illustrate the principles of schedule preparation. Detailed suggestions are given below for estimating delivery times and manhour requirements needed in the preparation of a workable schedule.

## INFORMATION FOR SCHEDULE PREPARATION

### Manufacturing and Delivery Times

The delivery date of equipment should influence all work throughout the design organization, including process design. Major equipment, the basic design of which will not be greatly affected by later computation, should be designed in detail as early as possible since usually such items will have the longest delivery. If extraordinarily large machines or heavy-walled vessels are required, these items should be specified or sketched as early as possible and released in a preliminary form so that procurement operations may proceed.

In order to determine probable maximum delivery times the equipment to be procured must be studied and a list of all major items prepared. The following list indicates probable delivery times for various types of equipment. Current quotations are the most reliable source of delivery times, but the following list indicates probable values for various types of equipment and can be used for estimates.

Typical Delivery or Manufacturing Time for Various Types of Equipment (1955)

| Vessels | |
|---|---|
| *a*. Forged or heavy wall chambers, greater than 3″ thick | 12–14 months |
| *b*. Large heavy plate vessels, greater than 1″ thick | 10–12 months |
| *c*. Drums or other lightwall vessels | 6–8 months |
| *d*. Small vessels | 4–6 months |

| | |
|---|---|
| *e*. Internals, trays, or special equipment | 4–6 months |
| MACHINES | |
| *a*. Large capacity pumps, turbines, compressors, any machine over ten tons in weight | 10–18 months |
| *b*. Medium capacity pumps | 8–10 months |
| *c*. Low pressure, low temperature pumps, compressors (cast-iron and smaller cast steel units) | 6–8 months |
| HEAT EXCHANGE APPARATUS | |
| *a*. Large condensers, high pressure or high temperature apparatus | 8–12 months |
| *b*. Ordinary shell and tube heat exchangers | 6–8 months |
| *c*. Double pipe exchangers (extended surface) | 4–6 months |
| HEATERS | |
| *a*. Direct fired heaters (erected) | 6–8 months |
| *b*. Package units of various types | 4–6 months |
| PIPING AND ACCESSORIES | |
| *a*. Fabricated pipe (from time approved drawings are received by fabricator) | 6–8 months |
| *b*. Random pipe, carbon steel (dependent on size and quantity) | 1–2 months |
| *c*. Random pipe, alloy or stainless (depending on size) | 1–6 months |
| *d*. Valves, alloy, high pressure, or special | 6–10 months |
| *e*. Control valves, alloy and other | 1–6 months |
| *f*. Valves, cast and forged steel (depending on size) | 1–3 months |
| *g*. Valves, cast-iron and small brass or bronze | 1–3 months |
| *h*. Expansion joints and special support appurtenances | 1–6 months |
| *j*. Fittings, alloy and carbon steel (depending on material) | 1–6 months |
| STRUCTURAL STEEL. (From time approved drawings are received by fabricator.) | 1–4 months |
| ELECTRICAL EQUIPMENT | |
| *a*. Electrical materials, motors, cable, conduit (depending upon size and quantity) | 1–8 months |
| *b*. Electrical switchgear (depending upon size) | 2–12 months |
| INSTRUMENTS | |
| *a*. Prefabricated complete panelboards | 5–8 months |
| *b*. Single instruments | 2–4 months |

## Manhour Requirements for Engineering Drafting and Construction

Once the delivery dates on all important items of equipment have been determined, the manhour requirements for the engineering drafting, procurement, and construction phases must be calculated so that a realistic schedule can be developed and the proper number of personnel assigned to the job.

A discussion of manpower requirements is best accomplished by use of a typical illustration. Assume that an owner decides to construct a plant, the total cost of which has been estimated at 10 million dollars. The completion date desired is 18 months after the signing of the contract. Material delivery expectations make this completion date possible. Typical labor costs and manhour requirements might be:

| | Cost | Manhours | Start Work | Months Worked |
|---|---|---|---|---|
| (1) Specialty engineering | $ 10,000.00 | 2,500 | 1st month | 5 |
| (2) Project engineering | 20,000.00 | 3,100 | 1st month | 18 |
| (3) Process engineering | 30,000.00 | 7,500 | 1st month | 4 |
| (4) Design drafting, Material listing, Specification preparation, etc. | 662,000.00 | 134,900 | 2nd month | 14 |
| Total engineering | 722,000.00 | 148,000 | | |
| (5) Procurement | 34,000.00 | 9,000 | 1st month | 10 |
| (6) Construction labor | 2,744,000.00 | 914,000 | 5th month | 14 |
| Total labor | $3,500,000.00 | 1,071,000 | | |

The various operations in this tabulation would be scheduled somewhat as follows:

Specialty Engineering. This function is usually spread over the first 5 to 8 months of the contract time. Personnel may be used within the first week after contract is signed. They perform specialized engineering in preparation of data for procurement of such items as general mechanical equipment, pumps, compressors, instruments, special vessels, heat exchangers, condensers, various types of heaters, and electrical switchgear.

With the exception of vessels, all other equipment in this category may be procured by specification or without the preparation (by purchaser) of design drawings. Thus work in this group may proceed as soon as the process design department has established design conditions for equipment. Selection of certain instrumentation, such as control valves, is delayed for several months since process piping size and arrangement will often influence valve sizes.

Project Engineering. The project engineer and his assistants will be continuously employed throughout the contract.

Process Engineering. Sufficient personnel will be assigned to complete process design as early as possible. Process engineering must precede nearly all other types of design. The design-drafting department usually will be given information on various parts of the project which has been accumulated by the project engineer, so that their work may begin before all process engineering has been completed.

Manhour requirements for process engineering depend upon the type of process, the availability of fundamental data, the availability of process engineering personnel, and the policies of the firm performing the work. Owner-operating firms develop and accumulate process data by experimentation and experience with operating units. When the decision is made to install a plant or process, the data are given to an engineering-contracting firm for resolution into the various steps of unit-operations and finally into forms which may be used for detailed design. For such cases and for the process design of known processes, time required for process engineering by the contractor is often very small. If the contractor is asked to perform research and development work and complete process design, the time required may be considerable.

It would be unwise to attempt a list of suggested time requirements for process design. Each job is usually unique and the chief process engineer should be consulted when the schedule is being prepared. It is well to remember that time requirements for projects which require considerable research and development are the most difficult to estimate. Difficult problems often do not conform readily to schedules. A conservative estimate is preferable for such problems.

Design-Drafting. Requirements are described below in greater detail. This work encompasses all functions after the process and specialty engineering phases. The design-drafting for the plant illustrated starts the second month after the contract is signed, and will reach a peak in the fifth to the eighth month.

Procurement. Procurement must proceed with each design stage or vendor prints will be received late causing a delay in detailed design. The lack of vendor prints can as effectively stop work as lack of equipment. Procurement thus must place all orders with dispatch, in order to receive vendor prints as early as possible and secure early delivery of equipment. Procurement usually starts in the first month and may or may not continue through construction. Certain procurement at the jobsite must continue as long as the job is in progress.

Construction Labor. In the illustration, construction starts on the fifth month after the contract signing. The delay allows time for a considerable amount of material and equipment to arrive soon after groundbreaking.

It is extremely important that construction work does not begin too soon. If begun early, idle periods will be encountered in which it will be necessary to await delivery of equipment and materials. Such conditions naturally increase cost.

The time required for construction of a given project usually can be estimated by experienced construction firms. Mattozi[1] has given a construction schedule chart (Fig. 7-2) based on an analysis of a number of projects that will permit estimation of the construction time when the total manhours of labor is known. This manhour requirement is usually

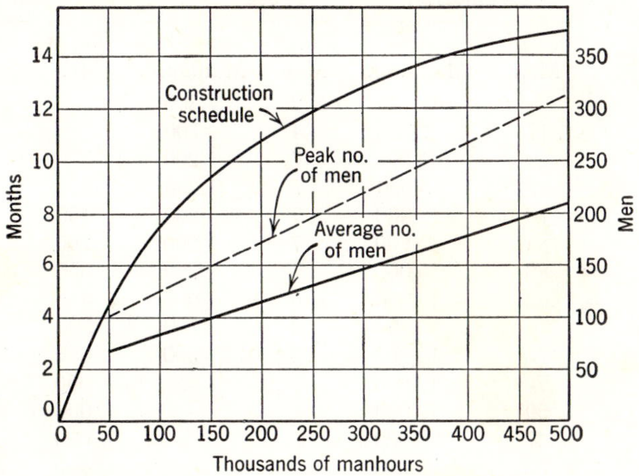


Fig. 7-2. Construction schedule. [Reprinted by permission, M. Mattozi, *Oil & Gas Journal, 51,* No. 48, 100 (1953).]

given in the original job estimate or proposal, either as an actual manhour figure or an estimated labor cost. Manhours can be calculated from the labor cost by dividing by the average hourly wage rate. If such information is not available Mattozi[1] suggests as a guess 18,000 manhours for each $100,000 worth of materials and equipment.*

## Planning Personnel Distribution

Table 7-1 gives the personnel distribution for the example project by month.

The increase in engineering and drafting labor during the peak part of the job is about 500% above that at the start. Operating on the basis of 40 hours per week, the engineering payroll may amount to some $14,000.00 per week for the 6th and 7th month. Firms contracting engineering must, therefore, be prepared to accept considerable expense if a large staff is to be maintained.

Schedules for the various engineering phases are rarely precise. Probably the most efficient or practical operation for firms engaged in engineering services is to understaff and work overtime during peak work loads. Another alternative is staff versatility with movement of personnel

* Based on 1946 material and labor costs

into the various operations as required. Many smaller firms operate very effectively on the latter principle.

TABLE 7-1. Typical Personnel Distribution

| Time | Engineering | | Procurement | | Construction | |
|---|---|---|---|---|---|---|
| Month | No. of Men | Manhours | No. of Men | Manhours | No. of Men | Manhours |
| 1 | | | 2 | 300 | | |
| 2 | 17 | 2,900 | 4 | 700 | | |
| 3 | 29 | 5,000 | 7 | 1,200 | | |
| 4 | 47 | 8,000 | 7 | 1,200 | | |
| 5 | 70 | 12,000 | 7 | 1,200 | 150 | 26,000 |
| 6 | 88 | 15,000 | 7 | 1,200 | 200 | 34,000 |
| 7 | 88 | 15,000 | 7 | 1,200 | 350 | 60,000 |
| 8 | 82 | 14,000 | 6 | 1,000 | 470 | 81,000 |
| 9 | 76 | 13,000 | 3 | 500 | 590 | 100,000 |
| 10 | 70 | 12,000 | 3 | 500 | 650 | 110,000 |
| 11 | 70 | 12,000 | | | 650 | 110,000 |
| 12 | 59 | 10,000 | | | 600 | 102,000 |
| 13 | 47 | 8,000 | | | 600 | 102,000 |
| 14 | 35 | 6,000 | | | 400 | 68,000 |
| 15 | 12 | 2,000 | | | 300 | 51,000 |
| 16 | | | | | 200 | 34,000 |
| 17 | | | | | 150 | 26,000 |
| 18 | | | | | 59 | 10,000 |
| | | 134,900 | | 9,000 | | 914,000 |
| Total manhours shown | | | | | | 1,057,900 |
| Other engineering services | | | | | | 13,100 |
| | | | | | | 1,071,000 |

### *Calculating Drafting Manhour Requirements*

A number of methods have been devised by various engineering-contracting firms for determining design-drafting manhour budgets. Most of these methods are based on experience in the particular firm and are not ordinarily useful when applied to other firms that have different methods. The distribution of work by percentage shown below however is applicable to nearly any type of unit processing fluids. In general mechanical equipment, vessels, and piping drafting consume from 55 to 65 per cent of the total manhours for any type process unit. It can usually be assumed that, for those processes in which piping requirements are reduced, corresponding increases will be reflected in mechanical equipment or vessels or both. It must be understood that the following percentages apply to single units. If a project consisting of several units is to be estimated each unit or element must be considered separately.

| Design Category | Percentage of Total Engineering Manhours | Possible Variation ± |
|---|---|---|
| *A*. Foundations | 7 | 2.0 |
| *B*. Structures and structural steel | 15 | 3.0 |
| *C*. Mechanical equipment* | 3 | 1.0 |
| *D*. Vessels | 7 | 2.0 |
| *E*. Piping | 55 | 11.0 |
| *F*. Electrical | 6 | 2.0 |
| *G*. Instrumentation | 7 | 2.0 |

* Includes pumps, compressors, heat exchangers, and any type mechanical equipment not shown.

A preliminary or overall estimate of design-drafting time may sometimes be made by a percentage method working backward from the particular process unit cost. For instance, materials for a certain unit have been estimated to cost $4,000,000.00. For this particular type unit as based on similar past contracts two manhours of combined engineering, design drafting, and material listing operations will be required for each $100 of material cost. Thus, an estimated budget of 80,000 manhours could be set.

By using the above percentages the tentative distribution of work might be as follows:

| Category | Manhours |
|---|---|
| *A* | 5,600 |
| *B* | 12,000 |
| *C* | 2,400 |
| *D* | 5,600 |
| *E* | 44,000 |
| *F* | 5,600 |

This appears to be a simple and straightforward analysis, but such estimates are not reliable unless the initial manhour estimate is accurate. In order to actually prepare a manhour budget a list of equipment must be prepared. From it a detailed list of drawings may be prepared. By applying values of required manhours per drawing according to category the total manhour budget and detailed budget, by category, can be determined.

As an illustration of the method used in estimating the number of drawings required, assume that the following items of equipment are a partial list of the items required for a process plant. The list is separated into two parts, *D*, Vessels and *C*, Mechanical Equipment.

| | Item Number | |
|---|---|---|
| *D*. Vessels | | |
| | T-101 | 10′ diam × 60′ long vertical vessel |
| | T-102 | 8′ diam × 45′ long vertical vessel |
| | T-103 | 9′ diam × 55′ long vertical vessel |

*C*. Mechanical equipment

| | |
|---|---|
| C-101 | Engine driven compressor 650 HP |
| C-102 | Engine driven compressor 430 HP |
| C-103 | Engine driven compressor 520 HP |

From these items a tentative drawing list can be made immediately.

Drawings for Vessels

| | | |
|---|---|---|
| *D*. Vessels | 3 | Vessel drawings (all details are usually on a single drawing for each vessel) |
| *A*. Foundations | 1 | For above, T-101, 102, 103 |
| *B*. Structural | 3 | Ladders & platforms for T-101, 102, 103 |
| *F*. Electrical | 3 | Electrical & instrument conduit & lighting |
| *G*. Instrument | 1 | Schematic drawings of instrument installation |
| Insulation | 1 | Insulation details for T-101, 102, 103. Insulation is not shown on the percentage list since it is generally negligible with respect to total design-drafting. |
| Subtotal | 12 | Drawings for *D*. |

For machines such as C-101, 102, 103 a shelter of some kind would be mandatory.

Drawings for Mechanical Equipment

| | | |
|---|---|---|
| *A*. Foundations | 3 | Compressor building, plan, section, details |
| Foundations | 4 | Compressors C-101, 102, 103, foundations |
| *B*. Compressor building | 1 | Arrangement plan |
| Compressor building | 1 | Structural steel |
| Compressor building | 1 | Roof plan & elevations |
| Compressor building | 1 | Architectural details |
| *D*. Miscellaneous vessels for compressor auxiliaries | 6 | Compressor bottles, starting air bottles, water bottles, water system, fuel, lubrication system |
| *E*. Piping | 6 | Process piping |
| Piping | 2 | Cooling water & lubrication systems |
| Piping | 1 | Starting air system |
| *F*. Electrical | 3 | Plans, sections, details power & lighting |
| *G*. Instrumentations | 1 | Installation details |
| Subtotal | 30 | Drawings for *C* |
| Total for *D* and *C* | 42 | Drawings |

From the original list of 6 items, T-101,102,103 and C-101,102,103, 42 drawings have been listed. By applying a manhour value per drawing, total manhours can be estimated as follows:

| Type Drawing | Manhours/Drawing |
|---|---|
| *A* | 200 |
| *B* | 150–200 |
| *C* | 150–250 |
| *D* | 150–200 |
| *E* | 300–400 |
| *F* | 175 |
| *G* | 250 |

| Category | No. Drawings | Manhours/Drawing | Total Manhours |
|---|---|---|---|
| *A*. Foundations | 8 | 200 | 1,600 |
| *B*. Structural | 7 | 200 | 1,400 |
| *C*. Mechanical | – | | |
| *D*. Vessels | 9 | 200 | 1,800 |
| *E*. Piping | 9 | 400 | 3,600 |
| *F*. Electrical | 6 | 175 | 1,050 |
| *G*. Instrumentation | 2 | 250 | 500 |
| (Insulation) | 1 | 200 | 200 |
| Total | 42 | | 10,150 |

This example is very incomplete since no piping was included in the first list for the three vessels. Other drafting which would be included in a complete list is process and utility piping for the vessels, utility piping for heating and plumbing in the compressor building and mechanical details for vessel appurtenances and for hoisting apparatus in the building.

The reliability of any data such as the percentages of work distribution or manhours per drawing depends greatly upon operational methods and extent of statistical data. Such data is subject to many variables even within an organization and cannot be applied with confidence unless continuously tested. Statistical data on labor should be accumulated regularly and regularly revised.

In scheduling drafting manhours one must realize that the total manhours required for each operation cannot always be continuous. It might be assumed for instance, that the 1800 manhours estimated for vessels in the previous illustration could be completed in one month by 10 men. Such thinking cannot be applied since certain vessels will require the maximum manhours while others will probably be under the minimum. Furthermore vessel design usually proceeds to near a 60% point when it must be stopped to await finalization of piping. A further 10–20 per cent may then be used completing orientation of nozzles or supports to suit piping. After this, another period of waiting must ensue while the vessel fabricator completes his shop drawings. Vessel designers then must check the fabricator's drawings for strength and accuracy before original drawings may be completely released since the fabricator's design practices sometimes affect the original design. Several months may elapse, therefore, between initial design effort and completion, most of which is consumed in awaiting data either from the originating organization or the fabricator. For this reason several jobs in various stages of completion will serve to maintain a relatively constant work load.

It is necessary in scheduling work for an organization to attempt the maximum use of all personnel. One aid in accomplishing this is the preparation of a bar graph or monthly time chart showing the various projects in an organization, their starting times, and the anticipated time for each phase of work. Then a drawing sheet count can be utilized to determine the possible maximums or minimums of manpower. New jobs should be scheduled to start during anticipated minimum work loads.

Regardless of the number of men available only one person can be employed on one drawing at any particular time. It is, therefore, not good practice to use large drawing sheets. Better use of manpower can be realized by reduction in drawing size. This will increase the number of drawings and thus increase the possible use of manpower. There is, of course, a practical limit in size reduction, but the use of larger scales serves to increase the number of drawings required.

## SCHEDULING DEPARTMENTS

Contractors as well as operating companies that do a considerable amount of their own engineering design and drafting both find it desirable to establish a permanent scheduling department. This department must be staffed with men having adequate background in all phases of the design and construction of a process plant. The scheduling of a multimillion dollar project is hardly the job of a low-paid clerk. It requires a man of broad experience who is thoroughly familiar with four stages of process plant construction: namely, process design, mechanical design and drafting, procurement, and erection.

Contracting firms have found the use of a scheduling department advantageous since such firms are ordinarily engaged in the execution of more than one job. It is preferable under such situations to have the scheduling done by a group of men who have an overall viewpoint and are not interested in only one particular job. If a project engineer schedules his own job, there is a tendency to favor that job. It is essential, however, that the project engineer assist the scheduling group in the preparation of the project schedule.

## REFERENCE

1. Mattozi, M., *Oil and Gas Journal, 51,* No. 46, 304; No. 47, 180; No. 48, 100; No. 49, 102 (1953).

# CHAPTER 8

# ENGINEERING DESIGN AND DRAFTING

After the project engineer has received process specifications and general mechanical specifications,* preliminary plot plans, and project schedules, the work of the specialty engineers and design draftsmen may begin. It is the job of these men to convert the calculations and decisions of process and project engineers into drawings and specifications which are indispensable in the procurement, fabrication, and erection of process plant facilities.

Engineering design and drafting efforts are usually divided among groups of specialists in the various phases of the design. The work of these specialists may be divided into two major categories: equipment specialty engineering and design drafting.

## EQUIPMENT SPECIALTY ENGINEERING

Equipment specialty engineers in cooperation with process engineers develop the detailed mechanical specifications for process equipment. In

*A portion of the general mechanical specifications are often supplied by the company for whom the plant is being built (see Chapter 11). These are incorporated with the usual specifications of the design organization and become, upon approval by the customer, the general guide for all mechanical design including piping, structural, electrical, and equipment design.

addition they aid the procurement division in selecting the equipment which most accurately fulfills the process specifications. The following specialty groups are maintained:

Pumps and Compressors. Engineers who specialize in pump and compressor selection are included in this group. They maintain close contact with pump and compressor manufacturers and not only select the best equipment for a given job but also advise process and project engineers on application and installation problems.

Heat Exchangers. This group is staffed by experts on heat exchanger design who can check quotations and specifications on all heat exchange equipment by making design calculations for comparison. Some contracting firms manufacture their own heat exchangers. In such cases this group does the design calculations.

Vessel Design. Vessels must be built to conform to the requirements of the process, and any plant design organization must have a group of expert vessel designers. These men design all pressure vessels, tanks, stacks, and ducts in accordance with customer specification and applicable codes. Vessel designers also prepare or supervise the preparation of vessel sketches and layouts of vessel internals.

Instruments. This fast moving field requires engineers who devote full time to instrumentation. They advise the process and project engineers on instrumentation during the development of flow diagrams and also specify and select all instruments for the project.

Furnaces. Contracting firms that build furnaces maintain a large staff of furnace designers who make detailed process calculations necessary for proper sizing of furnace and heater installations. They also direct the design drafting on furnaces. Organizations which do not manufacture furnaces have a small group of furnace experts who prepare furnace specifications and select the best furnace for a given job.

Piping Materials. Piping materials experts aid in the preparation of piping and insulation specifications for each job. They list all piping materials and accessories shown on piping drawings and select the materials in accordance with the specifications and applicable codes.

Miscellaneous Equipment. Each process plant project is unique in some way and special equipment not handled by other groups must also be selected or designed. Equipment such as dryers, filters, materials handling apparatus, autoclaves, centrifuges, evaporators, and mixers, may be required in many projects. Process engineers and the project engineer can often collaborate in the development of specifications and in the selection of such equipment. Some companies, however, have found it advantageous to maintain a group of experienced chemical and mechanical engineers who specialize in the wide variety of chemical plant equipment not common to all installations. In addition to their more routine work these men frequently develop the mechanical design of equipment that must be built to order because it is not regularly manufactured.

## DESIGN DRAFTING

The development of a complete plant design after equipment selection has been completed is so intimately associated with drafting that the term design drafting is often used to describe the four major categories: piping design, structural design, electrical design, and instrument installation design. The design work in these categories is largely accomplished by drafting techniques.

### Piping Design

Piping design is important in process plant engineering. In the engineering of nearly any process plant, piping design and drafting will usually consume from 40 to 60 per cent of the total engineering (design-drafting) manhour budget.

Piping itself is intimately related to process engineering. Designers who lay out the piping for a plant must learn and understand the various process requirements better than personnel in other phases of the work. For this reason organizations providing engineering services for the process industries depend on their piping design personnel to lead the other phases of the design work.

Piping designers do the following:

*a.* Layout, arrange, and design all piping in accordance with the specifications and the applicable codes.

*b.* Orient pressure vessel nozzles.

*c.* Check drawings other than piping for clearances with structural steel, foundations, and other types of equipment.

*d.* Study all piping arrangements for stress.

*e.* Prepare stress calculations.

*f.* List and specify all mechanical (manufactured) expansion joints.

*g.* Show all thrust and other forces on piping drawings and tentatively locate all anchors and supports.

*h.* Complete all formal calculations in a reproducible form.

*i.* Design, select, and list all detailed pipe supports, hangers, etc.

### Structural Design

Personnel in this group design all foundations, all structural steel, and/or reinforced concrete structures, and all industrial-type buildings in conjunction with architectural designers included in this group. Since foundations are largely located below ground, personnel in this group must work closely with piping personnel designing the underground installation. It is common practice with certain organizations to show all underground work such as foundations, piping, and electrical conduit on the same drawings. This procedure is extremely advantageous both in initial design and during erection. Designers in this group must, of course, conform to all applicable national codes and the particular specification for the work. The work consists of:

*a.* Preparation of design sketches and formal calculations.

*b.* Preparation of all structural steel detail drawings.

*c.* Preparation of building drawings and design of heating, ventilation, and plumbing.*

*d.* Listing and specification of all materials.

### Electrical Design

Personnel in this group are responsible for the complete electrical design required for the project according to the particular specifications and the various national codes governing electrical installations. Specifically the work consists of:

*a.* Preparation of all electrical conduit layouts and final drawings.

*b.* Preparation of all other electrical drawings for substations, switchgear arrangements, etc. (*Note:* Where reinforced concrete work is required for foundations, manholes, and supports of electrical equipment, electrical designers usually prepare outline drawings only. Final concrete design is done by the structural designers. Usually the electrical designers complete whatever structural steel drawings are required for switchracks, substations, etc. If the structure is involved or extensive, this work may be completed or checked by structural designers. Manhole designs for underground electrical systems must be designed to provide adequate space and suitable arrangement of electrical equipment by electrical designers. Structural designers set wall thicknesses, specify, and locate reinforcing steel.)

*c.* Preparation of all electrical equipment specifications. For major plant switchgear and plant transformer substations, specifications are usually prepared in conjunction with the project engineer.

*d.* Preparation of all instrument electrical conduit drawings in conjunction with instrument (system) design personnel.

*e.* Complete listing of all electrical materials, cable, conduit, etc.

*f.* The horsepower, rotation, and other mechanical requirements of motors are often specified by personnel in the mechanical group. Complete motor specifications, however, are prepared in conjunction with or by electrical design or engineering personnel.

* Some organizations maintain a separate group of heating and ventilating air-conditioning engineers, but often this work is subcontracted.

### Instrument Engineering

Instrument engineering for the process plant is more fully covered in Chapter 20. The selection of instruments and preparation of specifications for instruments may be handled by instrument specialists before the drafting operations and need not be directly associated with the engineering required for installation. The so-called instrument drafting personnel have the following duties:

*a.* Make general arrangement drawings for instrument installation showing the location of all instruments by symbol and the various runs of instrument piping in general.

*b.* List all instruments showing the proper piping drawing number or other drawing number on which the primary instrument is located, the item number, and type of instrument (indicated by symbol), type of capillary, or other connecting piping, air or process fluid, or electrical connection from primary instrument to controller and controlled element.

*c.* Prepare detailed drawings of the installation of all instruments locally mounted, except control valves which are shown by personnel in piping design. The hook-up of control valves with special appurtenances is usually detailed on the instrument installation drawings. (*Note:* These drawings are all schematic except for connection details to equipment.)

*d.* Prepare control panel detail drawings for fabrication and erection of the control panel and prepare back-of-board drawings for the control room installation. Prepare initial board-front layout when graphic panels are used.

*e.* List all miscellaneous piping or other materials used except electrical wiring materials and electrical conduit. This may be done in conjunction with or by the electrical group. Personnel from the electrical group work in conjunction with personnel from this group to avoid duplicating the material take-off.

*f.* Check adequacy of vendor's drawings. Instrument vendor's drawings suffice for instrument dimension detail. Drawings prepared by the instrument group show instruments drawn diagrammatically or in outline. Locally mounted instruments, such as liquid level devices, which connect physically to other equipment must be shown dimensionally for installation.

*g.* Check piping drawings for proper location of orifices and manual and automatic control devices, thermo-wells, pressure points, etc. Check all drawings for proper location of other primary devices.

### Types of Process Plant Drawings

Two major types of drawings are required for process plant projects: drawings prepared by the designers and bv the manufacturers. Nearlv all drawings that are prepared by the process plant designer are used in procurement; either by direct release to vendors for price or for material listing (take-off) and subsequent procurement.

1. Drawings prepared by the process plant designer.

*a*. Drawings prepared for the installation of equipment, either fabricated at the site of construction or manufactured or fabricated elsewhere; also used for certain material listing.

*b*. Drawings prepared for the manufacture of equipment by others with initial or general design by purchaser.

*c*. Drawings prepared for the construction of facilities; also used for material listing.

2. Manufacturer's drawings and data.*

*a*. Drawings prepared by the manufacturer for his own use in manufacture or fabrication; sometimes called shop drawings;

*b*. Erection drawings prepared by the manufacturer for use by the purchaser in erecting the equipment purchased;

*c*. Dimension drawings, data sheets, catalog tear sheets prepared by the manufacturer for use by the purchaser in design.

### Installation Drawings

The peculiarity of the process plant requires the preparation of drawings to be used primarily in locating equipment to a greater extent than drawings showing the details of manufacture, construction or erection. It would be impossible, for instance, to construct a catalytic-cracking unit without some graphical illustration of the various pieces of equipment in position. For any such installation the locations of equipment must be carefully planned and located by means of drawings to determine space requirements. Thus, all parts of a major plant must be drawn in considerable detail in position and to scale. Connection dimensions must be shown on the drawings to permit erection.

The importance of location or installation drawings becomes more apparent when it is realized that structural steel, pumps, compressors, vessels, and other equipment must properly fit anchor bolts set in the concrete foundations. The dimensions of the equipment and the locations must be known before design drawings of the foundations and supports can be completed. The construction forces must have these design drawings before construction starts.

### Vendor-Prints

Proper handling of vendor-prints (manufacturer's data and drawings) is vital to the success of a project. The lack of such data or misuse by

* These drawings and data are commonly referred to as *vendor-prints*.

careless distribution methods often results in serious errors, which increase the cost of engineering, procurement, and erection. While it would be undesirable to attempt to describe the variety of methods and procedures used by the various engineering organizations, it is believed that the details of the use of vendor-prints and some discussion of distribution methods will be helpful.

The three types of prints submitted by vendors are shop drawings, erection drawings, and dimension drawings and data sheets.

Shop drawings are not ordinarily furnished by the vendor except for fabricated piping, pressure vessels and tanks, structural steel, and other items designed in accordance with the purchaser's special requirements. Such drawings are not required by the purchaser for use in plant design, and with certain equipment they would divulge the vendor's "know-how." Shop drawings of mechanical equipment usually consist of a large number of drawings showing each piece or part of the assembled item. Their number alone would present a problem.

Vendor-prints designated as erection drawings are used by the purchaser to identify manufactured equipment and indicate erection sequence.

The third type of vendor-print, the dimension drawing and data sheet, is used by the purchaser as a source of data for plant design. For manufactured mechanical equipment, the vendor does not ordinarily furnish the purchaser with all the details used in actual manufacture but instead furnishes only the information necessary for installation. This information consists of a single outline drawing showing in plan and elevation all piping or other connection dimensions, disassembly dimensions, anchor-bolt locations, total weight and weight of parts which may be disassembled, and overall dimensions for clearance purposes. For certain reciprocating machines, direction, frequency, and degree of operating forces may be included. Flow diagrams for lubrication and cooling systems and wiring diagrams where required should be furnished.

Individual data sheets are furnished by the vendor for auxiliary equipment supplied with large machines. The purchaser should carefully list all such auxiliary equipment by subitem numbers on data which is prepared by and distributed in the purchaser's organization so that these data may be identified by the engineering, procurement, and erection groups.

Manufacturers of large machines do not ordinarily stock or warehouse these machines. Each machine is manufactured on the basis of an order. If the manufacturer has previously manufactured identical or similar equipment, preliminary data may be supplied with the proposal. Proposal data, however, is usually only approximate and cannot be used in design of the process plant. Often a month or more may elapse between order placement and receipt of preliminary data for major equipment.

Vendor-prints are required for all types of equipment except that which is commonly manufactured in quantity and in standard fixed dimensions. It would be unnecessary, for instance, to request vendor-prints on ordinary forged steel flanges since these are manufactured to dimensions and specifications fixed by various codes. The same would be true for ordinary valves or any similar commonly manufactured equipment not specially designed for the single application.

### *Filing and Distribution of Vendor-Prints*

The handling of vendor-prints in a design organization should be carefully planned and an exact procedure should be followed. Lack of proper identification is the most common source of error. Such errors cause wasted effort and may delay the release of the manufacturer to proceed with fabrication.

The most important requirement in vendor-print handling is identification with the proper equipment item number. Specific instructions should be issued to vendors, preferably starting with the inquiry, requesting the inclusion of the equipment item number on their drawings. With large organizations it is possible to have design and procurement proceeding parallel on similar work utilizing almost identical equipment. Casual identification in such cases may be misleading. The use of a purchase order number for identification by the vendor is insufficient since more than one item may be included on an order.

All drawings or other data received should be date stamped upon arrival, then checked by some single person or group, preferably the same group whose function is distribution of internal as well as external data. In some organizations such groups are called production departments and have the responsibility for receipt and recording of design data; the filing of all record or permanent data; the identification of data and the proper distribution of data both internally and externally. It is also common practice to have all reproduction services such as blue-printing under the same department. This makes the flow of all design data the responsibility of one department which is imperative for the large organization. Clerical personnel when trained to the routine can usually follow the steps.

The production department should have in its possession a card showing the item number, vendor, and other descriptive data for each piece of equipment requiring a vendor-print. The card is originated by the personnel originally specifying the equipment. The descriptive data are added to the card as it progresses through various groups concerned with the particular equipment item. By the enforcement of simple rules such as continuous display of cards by each temporary holder, the movement of prints through all departments can be followed.

Vendor-prints should be checked for proper identification. If the vendor has failed to include identification, the data and attachments should be referred to the originating department for identification. Usually the data can be identified by inspection of the purchase order or the original specification.

Vendor-prints may be submitted for preliminary use, for purchaser's approval, or as certified prints. After identification an inspection of the record card will indicate if the prints are for preliminary use or for approval. Final certified prints must include the vendor's certification and thus can be easily noted. After determination of status, the prints should be stamped, *Preliminary, For Approval,* or *Certified.* All vendor data should be so stamped regardless of the degree of importance. Once the method is instituted design-drafting personnel will always look for the stamp and misuse of data will be avoided. This procedure assures that the correct item number and status will always appear on the vendor-prints. If missent or lost they can be easily identified.

Distribution, names of recipients or departments, and dates should all be recorded on the record card. Since all movement will be through the production department the status and location of all data can be determined at any time.

Vendor-prints submitted for approval require special handling. Since manufacturers are often instructed to withhold fabrication until approval is received such drawings or data must be handled with the utmost dispatch. The open card system mentioned above is very advantageous since it forces action. Drawings and data submitted for approval can be assigned a colored card to indicate urgency. Approval drawings or data may be checked by designers or specialist groups. The final approval stamp should only be affixed by the project engineer. After approval all copies should be returned to the Production Department for recording, internal distribution, and return to the Procurement Department and to the vendor.

The receipt of final certified vendor-prints releases the purchaser's design completely. Dates for release of construction are often fixed as a set interval after receipt of certified vendor data. These data must, therefore, be recorded and handled in a manner similar to that already discussed.

Another function of the Production Department should be the physical removal of all superseded vendor data. This may be a difficult procedure at times since design personnel often have what they consider good reasons to retain superseded data. If this practice is permitted each designer will shortly be maintaining his own separate file, but without control. Data which is superseded should always be destroyed. All files of prints from first to last issue should be maintained but only in the Production Department and should not be released except by some authority.

## Drafting Practices

Each of the specialty groups and design drafting groups produces drawings of some type for use in further design and in construction. In addition process engineers and project engineers require a certain amount of drafting assistance in the preparation of flow diagrams, sketches, and plot plans.

Standard equipment purchased from manufacturers requires no drafting. The specialty engineer need only obtain certified vendor-prints from the manufacturer for transmittal to other design groups. Such equipment includes pumps, compressors, motors, instruments, and furnaces and heat exchangers (if they are being purchased direct from a manufacturer).

The design of piping, vessels, structural steel, foundations, electrical systems, and instrumentation requires a number of drawings to correctly portray the design calculations and decisions. Some typical methods used for these drawings and process plant drawings in general are given here.

### *Piping Design*

Process piping is usually drawn single-line except for the large sizes (12 in. and above) which are drawn double-line, showing the width of the pipe. Details and notes on drawings and instructions in the pipe specifications indicate where pipe is to be bent or fabricated from fittings. Fittings are rarely shown in detail on drawings, symbols being used throughout. The diameter of flanges is always drawn to scale so that clearances may be readily determined. The conventional symbols are used for indicating valves. Face-to-face dimensions of valves may or may not be shown on piping drawings depending upon practices followed.

The proper scale for piping drawings varies. The use of a small scale such as $\frac{1}{4}$ in. = 1 ft usually results in a seriously congested drawing, particularly when the piping is for involved processes. When all piping is extraordinarily large, 6 in. and above, $\frac{1}{4}$ scale may be acceptable. Details for smaller piping should be increased in scale, however, and shown elsewhere. Many firms insist on $\frac{3}{8}$ in. scale being the minimum for all piping drawings.

Piping may be shop fabricated or field fabricated. All piping $2\frac{1}{2}$ in. and below which is ordinarily screwed is field fabricated. All piping 3 in. and above is usually welded and flanged and, therefore, shop fabricated.

Each piece of pipe to be shop fabricated is given a consecutive serial number on the erection drawing prepared by the purchaser. The piece number usually includes the drawing number on which the piece is shown. All such numbers may be listed in a bill-of-material or other reproducible form for issue to the fabricator (vendor). Receipt of the list of piece numbers releases the fabricator to proceed with fabrication.

Details and dimensions shown on piping drawings are considerably reduced when piping is shop fabricated since the fabricator must prepare shop drawings called detail sheets for his own use. These sheets are not usually submitted for approval or checking by the purchaser except for alloy or heat treated pipe. A complete check of all the detail sheets would constitute a recheck of all piping drawings. Since the fabricator (vendor) must prepare his details to fit the designer's (purchaser's) erection drawings the burden of accuracy is with the fabricator and he is subject to cash penalties for errors in fabrication requiring field changes.

Since the process plant designer must prepare piping drawings to show location and arrangement, the fabricator does not furnish additional erection drawings, other than the detail sketches. These shop sketches prepared by the fabricator may be submitted to the purchaser as a check on pipe fabrication invoices. Fabricated piping is often purchased on a tonnage basis, in which case each load of pipe delivered must be weighed on public scales with certified gross and tare weights. When fabricated on the basis of cost per piece, however, schedules of material and labor costs are the basis for invoicing together with the shop sketches.

### Piping Stress Analysis

No detailed drafting is required for this part of piping design except simple sketches. The location of anchors and supports and all forces are usually marked on prints of the piping drawings for use of personnel designing pipe supports. The forces are not ordinarily added to the drawings for reproduction. Anchor points for pipe are shown by means of a symbol placed at the point of anchor.

A schedule of various types of mechanical expansion joints is sometimes prepared under the direction of personnel in this group. The schedule lists the various types of joints to be procured, giving dimensions (not vendor data) and the required thrust and stress characteristics. Manufacturer's shop drawings or dimension sheets are usually required since only the manufacturer can supply certain necessary data. Upon completion of procurement and receipt of drawings, final data may be added to the expansion joint schedule for use in erection.

### Pipe Supports

The various types of pipe supports such as shoes and hangers (see Chapter 18) are usually indicated by symbols or numbers on the piping drawings. These numbers refer to standard detail sheets. If extraordinary forces are involved the design of both the attachment and the support may be checked by structural designers. Usually all necessary materials for the various attachments are listed on the standard detail sheets.

### Piping Materials

A certain amount of sketching is required by personnel listing piping and fittings. The work consists primarily of the preparation of lists, of piping materials which must include all valves and other piping accessories, such as bolting and gasket materials. Piping material listing or take-off is limited to field-fabricated pipe when part of the piping is shop fabricated since the fabricator lists his own material. When all pipe is field fabricated, all materials must be listed, assuming the process plant designer is also erecting the plant.

### *Vessel Design*

Pressure vessels, tanks, drums, fractionating columns, and other separating equipment are usually designed and manufactured in accordance with codes as described in Chapter 12. The code design of such equipment specifies materials, temperature stress, and manufacturing or construction practices.

The purchaser must prepare an outline drawing showing the diameter and length of shell, the location and type and size of all connections or openings, thickness and type of materials for shell and heads, all internal and external appurtenances required for attachments, and operating conditions for design. Since vessel design involves total shell stresses including those caused by wind forces on vertical towers, the diameter and number of anchor bolts must also be specified. Design-drafting requirements for vessel procurement are satisfied upon the completion of the vessel outline drawing.

Vessel internals such as fractionating column trays, after process requirements are specified, are often structurally designed by other firms specializing in tray manufacture. This is particularly true when stainless or alloy tray materials are used. Vessel designers, however, may be required to prepare nominal layouts of trays according to process requirements. Other internals such as mixing equipment are often mechanically designed for support and location by the vessel designers. Vessel design personnel usually prepare schematic drawings showing vessel insulation details.

The vessel manufacturer is responsible for construction of the vessel in accordance with the proper code requirement when the vessel specification requires code construction. He must, in effect, guarantee that the equipment meets requirements for the allowable working pressures and temperatures. He must, therefore, check the purchaser's design.

Since pressure vessel fabricators prepare shop drawings for manufacturing use, these drawings are also used for vendor-print purposes. Purchasers of pressure vessels generally require the vendor to submit prints of shop drawings for approval prior to release for fabrication.* A considerable period of time usually elapses between initial vessel inquiry and release for fabrication since all piping layout and connections must be fixed before vessel nozzles can be located or oriented. For large vessels this period of time is usually required in any case by the vessel fabricator for procurement of plate, heads and appurtenances.

### *Structural Design*

The location of process equipment may be based on process, safety, or maintenance requirements. Structures, supports, shelters, and foundations are, therefore, of a secondary nature and their design must depend upon layout and arrangement of the various pieces of process equipment.

Design of structural systems is usually based upon standardized requirements of the manual of practice published by the American Institute of Steel Construction (see Chapter 23). The process plant designer or purchaser must prepare outlines or single-line arrangement drawings showing all structural members, their sizes and connection points. Connection points are usually what are known as working points in industrial practice.

Information must be included such as top-of-steel, minimum clearances, handrail dimensions, grating or floor plate boundaries, ladders, stairs and platform or other support connections.

The vendor or structural steel fabricator uses purchaser's drawings for weight estimates in preparation of a price proposal, and for preparation of his shop drawings. Each piece of steel must be detailed for fabrication in the shop. This practice is similar to methods followed by the piping fabricator. Structural steel is often sold on a tonnage basis. The total or final price is usually based on calculated weights of material shipped. When orders are placed on a lump-sum basis, all purchasers' drawings must be complete at time of inquiry.

In addition to structural details the steel fabricator also prepares (single-line) erection drawings, following standardized piece-marking procedures used by the industry. These drawings are used by the purchaser in erection. The vendor's shop drawings are not ordinarily submitted to the purchaser for either approval or checking unless the vendor

* After drawings are approved the manufacturer is informed to proceed with the fabrication of the item in question. This notification is referred to as a release for fabrication.

requests an approval from the purchaser for some change in structural member size. Changes may be required in order to provide clearance for ladders or piping. The details are, however, supplied for the purchaser's use or record.

It is well to point out that all general practices followed in ordinary building construction are not acceptable in process plant design. To illustrate it usually makes little difference if structural member substitution by a fabricator results in a member having strength equal to or greater than the original. But replacements or changes made by the vendor for process plant structures should have prior approval from the purchaser.

In addition to the general outline drawings prepared by the purchaser, notes on drawings and specifications should be included to advise the fabrication or erection method to be used (riveted, bolted, or welded), and the galvanizing or painting specifications. When galvanizing is required, purchaser's drawings are affected because all miscellaneous connections usually made in the field must be done in the shop before galvanizing. Field welding or other changes are not permissable since the galvanizing would be disturbed. If changes must be made, the member must usually be removed and regalvanized.

### *Electrical Design*

Schematic single-line wiring or circuit diagrams must be prepared for the process plant showing all electrical equipment, load distribution, transformers, and switchgear. Electrical designers use these diagrams to lay-out and show in detail all conduit runs and accessory equipment. All conduit, circuits, and equipment are assigned numbers or symbols.

In addition to the detail drawings showing conduit runs and connections, schedules must be prepared which list by number all circuits and conduits, the drawing on which the conduit is shown, conduit length and size, and cable length and size. Other tabulations are also prepared showing the proper sequence and setting for power control apparatus.

Drawings prepared for the electrical installation are extremely schematic and without dimensions except on details showing terminal points of conduit runs, the arrangement of conduit in banks, entrances of conduit to buildings, or exact arrangements required for clearance. For these drawings area plot plans are needed as well as vendor's certified dimension prints for all process equipment requiring electrical connections.

Usually electrical drawings prepared by the process plant designer are used only for erection of the plant and for material listing. Most electrical equipment is always procured by means of written descriptions.

### *Instrument Engineering*

Design-drafting operations for instrumentation installation are very similar to those for the electrical installation in that drawings are usually schematic and are used primarily for material listing. Installation details, again schematic, are usually prepared for those instruments directly connected to other equipment or for instruments located in a control center. The latter drawings are dimensioned but not necessarily drawn to scale. Tabulations similar to those prepared for the electrical installation are prepared for all instrument control circuits both electrical and pneumatic. Vendor data for instrumentation and area plot plans are required for instrument design drafting.

## Drawing Scales

A frequent time waster in many organizations is the arbitrary assignment of fixed scales for certain types of drawings regardless of the variation in design or scope. No doubt some of the fixed scale thinking is derived from certain architectural practices in which one tracing may be used as a so-called overlay to check for interferences. This procedure is rarely practical for process plant drawings where drafting to a precise scale (measured) or in great detail is undesirable. A usual scale for piping drawings is ⅜ in. = 1 ft. The next lowest readable scale is ¼ in. = 1 ft. For congested drawings ⅜ in. may be too small. The best scale is one which provides a clear picture or which fulfills the intent of drafting by providing a graphic illustration of data.

The actual scope of work covered by a drawing has little to do with economy except for some saving in printing costs. The highly congested drawing completely covered in each square inch of space, as is the usual result when ¼ in. scale is used for process piping, is in reality a tremendous time waster. It is inefficient in the initial drafting, in checking, in material listing, and in erection.

Process plant drafting does not usually require absolute precision such as would be found in machine design. In nearly all design work the practice of scaling drawings to determine dimensions is strictly limited to estimating. Most drawings for the process plant must be made to some certain scale in order to show relative sizes and locations of equipment, but extreme precision is not required nor justified unless the design is purely a development of some mechanical device. The majority of design drawings for the process plant are prepared for erection purposes rather than for manufacture.

In drafting practice the centerlines of equipment only are located as precisely as possible to scale, so that discrepancies are not cumulative. All dimensions which are necessary for erection, construction, or clearance are shown on the drawing. Scaling, except for some estimating purpose, is unnecessary. It will probably be obvious in any case that with scales such as 3/8 in. = 1 ft, accuracy closer than one or two inches would be impractical.

## Modern Drafting Techniques

Although many advances have been made in all branches of science, relatively few improvements have been made in the most basic of all engineering functions, the preparation of drawings. No methods have yet been devised to completely eliminate the painstaking manual procedure of drafting. For the process plant, the cost of labor consumed in drafting operations alone often reaches 5% of the total erected plant cost. Thus it is extremely important to utilize time-saving procedures in this phase of engineering, particularly for firms contracting engineering services to industry.

Since the discovery of blue-printing the only major change made in drafting procedure has been the substitution of pencil techniques for inked drawings. Several decades ago all drawings made for reproduction were first prepared in pencil, then traced in ink. Improvements in drafting procedures, materials and reproduction processes permit production of prints from pencil drawings equal to prints from inked drawings. Few process plant engineering firms any longer use inked drawings except at times for final flow diagrams or for certain standard drawings.

Linen drawing sheets prove much more satisfactory under the severe services found in process plant design-drafting and reproduction than the various tracing papers. The much better working surface of the linen sheets will save sufficient time to warrant the slight additional cost.

A number of methods common to the printing industry are being used to advantage in drafting operations by the more progressive firms. Many time-saving possibilities are offered by the offset printing process. Details and instructions that must appear on all drawings can be reproduced on the linen drawing sheet by this process instead of being hand drawn.

Various frequently used details are sometimes printed on small sheets and distributed along with larger drawings as standards. Although drafting labor is saved, references to the sheets must be made and the printing and handling of an additional drawing consumes time. Such details may be reduced in size by a simple photographic process and reproduced directly on the drawing sheet by offset printing. Although the size may be reduced, the reproductions produce sharp legible prints. The same process may be used for reproducing catalog pages, manufacturer's dimension drawings, or any descriptive data required for erection of the plant.

In general any detail which can be printed directly on an original drawing will save drafting manhours. Most of these various printing processes and reproduction methods are extensively used by the major manufacturing firms. Any detail manually added to a drawing must be checked. Hence printed data should always be less costly.

A machine similar to the typewriter which permits typing directly on drawings is extensively used for adding standard notes to drawings. The laborious drudgery of hand lettering lengthy schedules and lists may thus be eliminated. Any capable typist can be trained to operate the machine so that a further cost reduction in labor is possible. Such machines usually repay their initial cost in a few months when properly used.

Undoubtedly many of the firms engaged in contracting engineering services to industry could profit by investigating drafting and reproduction methods used by some manufacturing firms. With design-drafting costs for the process plant approaching $0.07 per minute per person it would appear that any method which offers labor saving possibilities should be welcomed.

# PART II

# BUSINESS AND LEGAL PROCEDURES

THE project engineer not only works with engineering problems but also deals with procurement operations, office procedures, and contract interpretations.

CHAPTER 9

# PROCUREMENT OPERATIONS

To purchase the needed equipment and materials at reasonable cost and to have them delivered at the proper time is one of the most important phases of process plant construction. The procurement department of an engineering organization does this work, but not without the help and advice of the project engineer. Even though many purchasing agents in the process field have engineering backgrounds it is impossible to expect them to be well informed on all the phases of technology. It is, therefore, necessary for them to seek the advice of the project engineer whenever engineering judgment is needed. In like manner the project engineer must depend on the purchasing department for up-to-date product information and prices, contacts with vendors, and above all for the procuring of materials on schedule. It is important, therefore, to be conversant with the organization and operation of a purchasing department for process plant construction procurement.

Procurement for a process plant project is different from the purchasing operations required for a manufacturing plant, where the buying process covers only a relatively few standard commodities which are bought in quantity at regular intervals. Personnel engaged in purchasing for a process plant project must have a wider experience, and must be aware of possible sources for many more kinds of equipment.

## ORGANIZATION AND OPERATION OF A PROCUREMENT DEPARTMENT

The NAPA (National Association of Purchasing Agents)[5] has recommended a variety of organizations for procurement departments. One typical of procurement for process plant construction may consist of a purchasing agent (purchasing manager), assistant purchasing agent (one or more), buyers, clerks, and typists. The purchasing agent or manager with the help of his assistant formulates purchasing policy and directs the many functions of the department.

Two methods of buying are in common use: contract and commodity buying.

### Contract versus Commodity Buying

For the process plant project, procurement is usually done by contract or commodity buying methods, or by a combination of the two. In contract buying one or more buyers are assigned to a specific project (contract). In commodity buying each buyer handles certain materials or equipment regardless of the project for which they are to be used. Both methods have merit.

The commodity buying system is more often used in large organizations handling a number of projects. Commodity buyers tend to become specialists in the particular commodity handled, and can remain informed on changes in that field. Whether or not such specialization is necessarily desirable or can be used to its best advantage in procurement for the construction project is a controversial question. When commodity buying is used, one or more project coordinators must be designated so that procurement will follow a definite schedule or program.

While a contract buying system requires more versatile or widely experienced buyers it has some advantages in that buyers are more interested in the particular contract and thus may be more capable of concerted effort or teamwork.* Contract buying is usually much more suited to the smaller organization.

Commodity buying is most efficient when the policies of the organization permit overbuying. Large quantities of a given material can then be purchased at a discount. Surplus problems will result from this method, however, unless the organization is prepared to store the excess materials. The proper care of surplus materials, freight, handling, and storage, mean additional costs, not always chargeable to the particular contract. Generally the purchaser of the plant will stipulate that all materials or equipment are to be "new." If a contracting firm is the initial buyer the customer may raise an objection to the use of stored excess material or equipment. This is particularly true of such items as valves or other equipment which may rust and thus appear used.

### Procurement of Equipment Requiring Engineering Participation

Few basic materials may be purchased without the preparation of drawings either by the purchaser's engineering department or the vendor.

* This assumes that undesirable competition is avoided between various buyers working on the different contracts.

One of the primary reasons for preparing drawings is to aid in procurement of materials or equipment by providing concise and readily understood representations of design data.

The commercial or buying phase and the engineering phase should never be separated completely. It is usually necessary to continue engineering participation up to the time of order placement of specialized operating equipment, such as pumps, compressors, electrical switchgear, and instruments

The specifications and drawings for process equipment which constitute the vital procurement data must be prepared by engineering groups. In this work engineers usually find the assistance of sales engineers valuable in the preliminary stages of specification development. In fact it is sometimes more practical depending upon the scope of the project for specialized engineering groups* to develop the design and final selection of equipment, such as pumps and instruments, directly with the vendor. The procurement organization then only handles the final commercial and clerical operations since pricing will have been decided earlier.

For the mutual benefit of both vendors and purchasers it is advisable to restrict the number of employees authorized to deal with vendors. If this is not done, the situation becomes hopelessly confused. It is further necessary that engineers who contact vendors be aware of the business terminology and legal aspects of bidding so that correct quotations are assured.

## Procurement of Off-The-Shelf Materials

Commonly manufactured, or off-the-shelf items, are ordinarily handled by the procurement department, since the placement of orders for this equipment is usually a routine matter of price and delivery after specifications and data are complete. The number of acceptable manufacturers for the common items, such as piping materials, often presents a problem. If many known and reliable manufacturers exist, it wastes time to continue sending inquiries to all the acceptable sources for the same type of equipment. One analysis of bids from all acceptable manufacturers should be sufficient basis for placement of successive orders with the same vendor until the price structure changes significantly. Blanket orders or agreements to purchase all one type of material from a given manufac-

* Perhaps one of the most common and usually expensive errors in engineering judgment is the tendency to design each item without regard to the availability of materials, equipment, or costs. The only remedy is experience. Project engineers should either have a wide knowledge of manufacturer's practices, or have made available to them the combined experience of a number of specialists. Many engineering organizations recognize this necessity and train engineering personnel along specialized lines. The project engineer then, can depend upon the experience of specialists for precise specifications, which meet process requirements but yet fall into an economic pattern. (See Chapter 8.)

turer, however, should not be made for long periods even on a single project. The procurement operations on a major process plant will usually continue over at least one year, and prices may fluctuate during that time. The procurement department must know price trends and reserve the right to take advantage of these changes for the sake of economy.

### Expediting and Inspection

Expediting and inspecting are often directed by the procurement department. The primary function of project expediting, however, is expediting movement of materials to meet the construction schedule, and need not be directed by those engaged in the buying operation.

## PROCUREMENT PROCEDURES

The procurement of materials for a process plant project can be divided into several phases.

1. Inquiry—A formal request for a quotation of price.
2. Quotation—The formal reply to the request for price.
3. Bid comparison—Objective study of all quotations received for a particular item.
4. Purchase order—The formal order sent to the successful bidder.
5. Inspection—The examination of materials during fabrication and upon completion to insure highest quality.
6. Expediting—The acceleration of progress on manufactured equipment by the purchaser.

### The Inquiry

Actual detailed operations of the inquiry phase are: (*a*) accumulation of data in the form of specifications or drawings, (*b*) selection of vendors to whom inquiry is to be directed, and (*c*) typing and mailing. Item (*a*) is a responsibility of engineering, (*b*) evolves from past experience or engineering development, and (*c*) is a clerical operation.

Many companies have standard inquiry forms. The form invariably has *this is not an Order* printed plainly on the sheet to avoid confusion. Drawings and specifications where applicable are attached to the inquiry. The method of quoting is outlined and requests for delivery date promise, terms of payment, and a price breakdown are made. The date on which the quotation is required is given, and written quotations are requested.

The number of inquiries solicited for a given item depends on company policies. If procurement is being done by a contracting firm for a customer, the customer's desires will be followed. In fact many process firms furnish contractors with a list of acceptable vendors and specify the minimum number of inquiries to be solicited. Three inquiries is the usual minimum since three quotations will insure selection of a vendor on the basis of price, delivery, and quality of product.

### The Quotation

The seller (vendor) submits his quotation on a special form upon which a number of general conditions and definitions are printed. These are printed in small type, but should not be disregarded since they are a legal part of the document.

The specific information pertinent to the equipment being quoted is typed on the quotation form and should include the following:

*Quantity and Description of Material Being Quoted.* The description may be brief especially if it is referred to a specification which gives a complete description. Additional descriptive literature and drawings are also attached if these aid in better describing the item.

*Price.* The price should be given as both a unit price (price per item) and a total price (price for total number of items and also price for the total order).

*Estimated Weights.* This information is necessary in planning best shipping methods and in comparing bids.

*Type of Quotation.* (*a*) Firm price—Price quoted is binding on the seller if accepted within a given time limit (usually 30 days); (*b*) Escalation clause—Many quotations during the days of initial material shortages and indeterminable costs of World War II contained escalation clauses which permitted the upward adjustment of prices due to increases in labor, material, and transportation between the date of placing of the order and shipment.

*Trade and Quantity Discounts.* Certain quantity discounts and discounts applicable to buyers who will resell (such as contractors) are regularly given. The amount of the discount should be given on the quotation. Any discount that is treated by the vendor with the greatest of secrecy should be looked upon with suspicion.

The use of trade discounts on materials listed in catalogs is particularly advantageous for both vendor and buyer. A catalog may be printed once a year or less with complete descriptions and so-called list prices (which are prices above the actual anticipated selling price) given for each item. The actual price can then be varied to allow for changing costs by revising the discount sheet and not the entire catalog.

Discounts are often quoted as a series of percentage discounts such as 20, 10 and 5. To calculate the actual price the list price is multiplied by (100 − Discount)/100 in this case, by 0.80, 0.90, 0.95. A series of discounts is used because one discount may not apply to every situation, but various combinations of discounts can be devised which will be applicable.

*Terms.* The terms of payment including the amount of cash discount along with the F.O.B. (free-on-board) terms should be listed. A typical statement involving these items might be "2%, 10 days, 30 days net, F.O.B. Philadelphia, Pa." This means that if the bill is paid within 10 days a 2% discount on the net price will be allowed. The payment in any event is due 30 days after receipt of the bill, in which case the full net price must be paid. Failure to take advantage of the cash discount is very poor business practice and can be shown most forcefully by calculating the equivalent interest rate per annum if the buyer elects to wait until the end of the thirty-day period. In the case of 2%, 10 days, the interest equivalent is 36 per cent.[9]

**TABLE 9-1. Abbreviated List of Trade Terms Used in Quotations**

1. *C.I.F.* An export trade term. The seller's quoted price includes the cost of goods, marine insurance, and all transportation charges. Prices quoted are usually in the currency of the destination country.

2. *C. & F.* Same as C.I.F., except that insurance is not included.

3. *Ex-Dock.* Price quoted includes cost of goods and additional costs required to transport and then unload goods at a given port of importation, including payment of duty.

4. *Ex-Factory, Ex-Mill, Ex-Warehouse.* Price quoted is for materials at the point of origin (manufacturer's plant).

5. *F.O.B.* Free on board. The terms F.O.B. must be qualified to avoid misunderstanding.

5(*a*). *F.O.B.* (named inland carrier). Price quoted includes cost of goods and cost of loading goods on the carrier. (Truck, railroad car, etc.) Seller is responsible for goods until loaded on carrier.

5(*b*). *F.O.B.* (named inland carrier). Freight prepaid to named point. Quoted price includes cost of goods, loading, and freight to named point. The seller, however, does not assume responsibility for the goods after leaving the point of departure.

5(*c*). *F.O.B. Vessel* (named port of shipment). Quoted price includes cost of goods and delivery in proper condition on board a vessel and a named port of shipment. Seller is responsible for goods until they have been loaded on board the vessel.

5(*d*). *F.O.B. Factory.* Not a very definite term. May mean that cartage to railroad depot must be paid by purchaser. Term should be avoided.

6. *F.A.S. Vessel* (named port of shipment). Free along side of vessel at port of shipment. Quoted price includes cost of goods and the transportation and unloading charges necessary to deliver goods within reach of the ships loading tackle.

The F.O.B. terms in the illustration mean that the goods will be delivered on board the cars at Philadelphia, Pennsylvania. The purchaser must pay the transportation charges from this point to his location. In addition, the legal title to the goods passes to purchaser at the F.O.B. point, and all losses which occur in transit must be recovered by the purchaser from the carrier. In some instances quotations indicate the F.O.B. point and then state that freight will be allowed. In this case the purchaser gains title to the goods at the F.O.B. point but the seller pays the freight charges. These and other traffic terms such as F.A.S. (free-along-side ship) which are used in the purchasing and shipping of materials must be thoroughly understood by those responsible for purchasing. Complete definitions together with legal implications are available.[3,6,8] A brief summary is given in Table 9-1.

*Delivery Date.* All important to many construction projects is the expected delivery date. The approximate delivery, usually expressed as days, weeks, or months from date of order, should appear on all quotations.

In addition to the above essentials a quotation should be dated, identified by some number, and signed by an authorized representative of the company. Many quotations have a space for signing of the quotation by the buyer which makes it in effect a legal agreement to buy. In general, however, most organizations prefer to issue a formal purchase order (see below).

### Bid Comparisons

After the bids or quotations have been received the vendor can best be selected by preparing a tabulation and analysis of the various significant factors in each quotation. Factors which must be considered are:

1. Specification.
2. Price and escalation.
3. Delivery.
4. Terms of payment.
5. Conditions (policy).
6. Guarantee.
7. Freight.
8. Labor clauses (manufacturing labor).
9. Shipping method.

The project engineer should list and evaluate all design features. When a detailed analysis is made and the features of each vendor's offering are compared in an orderly tabulation, differences which would have remained obscure become quite clear.

Even the price requires close scrutiny. What may appear to be a low price may in reality be high when compared on an equivalent basis. Such factors as shipping charges, fixed prices or escalation and terms (see below) must be considered and each quotation must be adjusted to the same basis for comparison.

Often an order may be awarded to a company because in addition to having a good product, its technical service, promptness in supplying drawings, and its general attitude of helpfulness, though perhaps intangible, far outweigh the slight additional cost.

Many companies have standard forms designed to aid in the orderly tabulation of bids. All such tabulations which require engineering judgment should be submitted to the project engineer for approval. The purchasing department should be particularly careful to indicate all terms and transportation factors and should point out the difference in these items. Copies of the quotations should be supplied with the tabulation to aid in the engineer's study. The contractor's project engineer will often submit these tabulations to the customer's project engineer for his approval.

## The Purchase Order

After selection of the vendor, a purchase order based on the successful quotation is issued. Purchase order forms are used for this purpose.* The information on these forms may be divided into three categories—heading, main portion, and standard printed notes. The location of each of these on the purchase order has been specified by the American Association of Purchasing Agents.[5]

HEADING. Gives company name and address and spaces for vendor's name and address, shipping address, and instructions, F.O.B. point, terms, date of order, purchase-order serial number, and date material is wanted.

MAIN PORTION. The main portion of the form, usually preceded by a reference to the vendor's quotation, lists the quantity, description, and price for each item. The total price of the order is also given. If the order can be referred to specifications and drawings, the description can be brief and the specification and drawings transmitted as a part of the order. When available the approximate weight of each item is listed since this information often proves useful to recipients of purchase order copies, such as the engineering department.

PRINTED NOTES. A number of standard notes, terms and conditions of the order are also printed on the purchase order. These include instructions for mailing the bill of lading (carrier's receipt), packing list (list of contents, dimensions, and weight of each shipping package or crate), and invoice (list of items shipped or on order showing prices and other charges).† Instructions for identifying all shipments, letters, and documents with the purchaser's order number and equipment item number are also given. It is very important that the vendor follow these requests; otherwise considerable delay and confusion is caused in identifying the materials.

* Forms for changes to the original order (change order) are also in wide use. The wording *Change Order* is printed prominently on the face to call attention to the changes and differentiate it from the original order.

† Orderly transaction of business depends on the prompt receipt of accurate copies of all these items. The construction department materials men use the packing lists to check incoming material. The accounting department must check the invoices against the order and certify them to the comptroller's office for payment.

The purchase order form should also include all the purchaser's standard notes on policy printed (not typed) directly on the original document; otherwise, loose addendums or attachments will be required. Each loose piece of paper represents an added cost in handling and can easily be lost in transmittal. The following notes are standard with most procurement organizations.

*Guarantee.* Some type of guarantee clause should be included for all equipment and material purchased. Extensive and elaborate, guarantee clauses usually require additional correspondence, with subsequent waivers or amendments. The best guarantee is always the integrity of the vendor. The following is typical. "All equipment or materials furnished on this order shall be guaranteed, for a period of not more than twelve (12) months from date of delivery (or date of operation or use) against operational failure, due to manufacturer's design; or mechanical failure, due to manufacturer's workmanship, when the equipment or material is being operated or used in accordance with the original specifications. All parts or materials found to be defective within the specified time interval shall be replaced without cost to the purchaser, F.O.B. manufacturer's plant."

A primary vendor (one who receives order) does not guarantee manufactured parts of equipment furnished to him by another manufacturer, beyond that other manufacturer's guarantee. The primary vendor will usually acknowledge the guarantee clause by stating that the overall guarantee is limited to the guarantees of the subvendors. Manufacturers, however, accept a certain moral responsibility for any equipment sold since their reputation is at stake. But no manufacturer will ordinarily accept any responsibility for loss of profit in production caused by failure of his equipment, and practically all guarantees are limited to replacement of the item which fails.

*Inspection.* A paragraph should be included stating, in effect, that "all equipment and material furnished shall be subject to shop and field inspection by the purchaser or his delegated agents, and the vendor shall notify the purchaser at least——days in advance of its readiness for shop inspection. If the purchaser does not wish to inspect certain equipment the vendor will be so notified, but shall understand that any such inspection release does not in any way relieve him of the responsibilities set forth in the guarantee clause." Another clause frequently used is that "purchaser's inspectors, or representatives shall have access at all times to manufacturer's plants or shops for the purpose of inspecting equipment during the process of manufacture."

*Quality.* A statement is included on the quality of materials. "All materials furnished or used in the manufacture of purchased items shall be new and of first quality and in exact accordance with the applicable specifications."

*Labor Clauses.* A typical labor clause might be: "The vendor agrees that all labor used in the manufacture of the specified items furnished shall be employed and paid under the applicable regulations of the Fair Labor Standards Act of 1938, and its subsequent or future amendments. The vendor further agrees that all invoices submitted to the purchaser shall bear a written certification of his compliance with the above regulations."

A number of laws exist regarding discriminatory practices. The particular area in which the work is to be done and the type of labor used all should be investigated previous to order placement with unknown firms. Serious difficulties may be caused by the use of an open-shop firm if closed-shop workers are erecting the equipment. Shop fabricated pipe may be a source of many labor entanglements, particularly if the pipe is not fabricated in the area where it is to be erected. The piping specifications should include a requirement for a stamp certification on each piece of pipe giving the manufacturer's association with the applicable labor union.

*Subcontractor Clauses.* The purchase order is frequently used as an instrument in placing subcontracts. The standard notes in the purchase order need not repeat the phrases in the subcontract. It is sufficient to state: "When equipment, materials, or services are being furnished under a subcontract, the executed subcontract form shall apply and become a part of this order." The subcontract form itself is legally an addendum or attachment to the order.

*Shipping Instructions.* The point of origin and the destination of material covered by the purchase order must be listed for use by both vendor and purchaser. The method of transportation is selected early in the negotiations if possible and should be indicated on the purchase order. When time is limited the purchaser may request truck shipment, railway express, or air express. If shipping methods cannot be decided at time of order placement, the statement, "Shipping instructions will be issued later," should be included on the order. The vendor must be informed of shipping methods for large equipment so that assembly of the equipment for shipment can be planned.

*Patent Protection.* Since the patent law holds the user of a patented item just as liable for suit by the patentee as the seller who has knowingly or unknowingly infringed, the patent protection clause is inserted in purchase orders to protect the purchaser. It requires that the vendor compensate the purchaser and/or ultimate user of the material for all costs which may develop due to suits filed by the party claiming infringement.[2]

Many other special terms and conditions peculiar to the particular transaction may be included as part of the purchase order. All such notes should be clearly phrased and above all should be kept to the minimum. It would seem that clauses protecting the buyer should be limited to those protecting against fraud and contingencies beyond the control of the buyer. Certainly man will never devise clauses so cleverly worded that they will protect against practices which are legal but nevertheless unethical and damaging. The best protection is a reputable manufacturer.

### *"Acceptance"* Order

Legally, the purchase order becomes a contract only after acceptance by the seller. It is, therefore, common practice to send an "acceptance" copy of the purchase order to the seller so that he may sign and return it. The signee for the vendor must be an authorized agent of the manufacturer.

## INSPECTION

An important phase of procurement for process plants is the inspection of equipment during and upon completion of manufacture. The advantages gained through careful development of specifications and drawings by the purchasing company can be lost by careless workmanship and errors made in the manufacturer's shops. Particularly on major items of equipment, therefore, it is worthwhile for the purchaser to send his inspector to the vendor's plant. The privilege to inspect equipment at the plant is usually agreed upon prior to the placement of the order and is included in the purchase order. No reputable manufacturer will deny inspection privileges and many welcome a competent inspector's suggestions.

It is the inspector's job to witness tests on mechanical equipment such as pumps and compressors, to observe and approve fabrication methods of vessels, pipe and structural steel, and to encourage the best workmanship on equipment being manufactured for his employer. He should be a person of considerable practical know-how with an ability to spot mistakes and also an ability to get along well with people. A large portion of a competent inspector's work consists in preventing mistakes before they occur.

The inspector's reports are made directly to the purchasing department. The project engineer should be notified about rejected equipment, but adjustments on rejected items are handled by the purchasing department. If inspection is made at the site of construction, arrangements are made for replacement of rejected material and for returning it to the vendor. It is sometimes possible, in the case of equipment such as pumps and filters for the vendor to correct faults in the field. For this work a technical representative of the vendor is sent to the jobsite at the vendor's expense.

Because of the high cost of inspection, only major equipment items can be inspected prior to shipment. Stock items such as valves and instruments must be checked upon delivery at the construction site.

## EXPEDITING

The primary function of expediting is to assure delivery of purchased materials at the jobsite in accordance with the project schedule. Material such as valves and fittings are available from supplier's warehouses and can be delivered at any time. The progress of specially designed equipment, however, such as pressure vessels or large compressors, must be followed continuously by employees of the purchaser called expediters. Manufacturers estimate probable delivery dates for equipment to be fabricated in their shops, but these estimates depend on prompt delivery of materials from their suppliers, efficient scheduling of shop work, and early receipt of drawings from the purchaser. The purchaser's expediter can do much to assure that estimated delivery dates will be met by working with both the manufacturer and his own organization.

His work can be explained more clearly by describing the progress of a large steel pressure vessel from the inquiry stage to delivery.

An inquiry is sent to three or more fabricators and it includes specifications and a preliminary sketch of the vessel. The vessel diameter, plate thickness, type of heads, and number and size of nozzles are given. From this information the vendors are able to develop their quotations which give costs and estimated delivery time.

During the inquiry period (two to three weeks) engineering work proceeds in the purchaser's offices and by the time the order is placed the metal specifications and vessel size and thickness have been confirmed. If changes have been made, a revised sketch is attached to the purchase order. The fabricator is informed by a note on the order that he is "released to purchase all materials on the basis of the preliminary sketches." The expediter receives a copy of the purchase order at the same time it is sent to the vendor and it then becomes the responsibility of the expediter to see that delivery is made on time.

After the fabricator has been told to proceed with the purchase of materials, the purchaser's expediter visits the fabricator's shop and obtains copies of all material orders. He then checks by telephone or personal visit with the steel mill to be certain that a definite rolling schedule for the steel plate has been arranged. He also confirms the delivery dates for all other equipment ordered by the manufacturer.

When the materials are delivered to the fabricator, rolling of plate to form the shell can begin; but fabricators are reluctant to do further work until final design drawings are received from the purchaser. It is difficult and costly to start work on a large vessel and then be forced to discontinue for a short period until some design change is made. The next important job of the expediter, therefore, is within his own organization.

He must be certain that the vessel design group completes the final detailed drawings on schedule so the fabricator will not be delayed.

After the manufacturer receives final drawings and fabrication begins, the progress of the work can be reported by the purchaser's inspector who must observe the shop operations daily. The expediter need only visit the shop occasionally, but he continues to follow the progress of the vessel until it is delivered. When the inspector's report indicates that work is being delayed, the expediter will visit the shop, determine the cause of the delay, and help the fabricator overcome it. If the fabricator has a number of orders from other companies, the expediter must make more frequent visits to be certain that his job is not unduly delayed because of work on other orders.

In summary an expediter's duties may be briefly outlined as follows:

1. Accumulate data which may affect delivery.
2. Anticipate delays and bottlenecks and resolve these directly with the vendor.
3. Assist vendor in obtaining priorities.
4. Assist vendor in resolving his procurement problems.
5. Change delivery schedules when necessary.
6. Correspondence with the subvendors and suppliers of the primary vendor.
7. Be aware of changes of engineering schedules in his own organization, and to relay this information to the vendor or fabricator when necessary.
8. Expedite his own organization to complete vital drawings on schedule.

Although delivery on time is essential, early delivery may be a disadvantage. Materials obtainable from stock should not be delivered until needed or extensive warehousing facilities will be necessary. Specially built equipment such as a vessel must be lifted twice if delivered before foundations are ready. The expediter, therefore, must aid in avoiding such early deliveries.

An expediter must have a working knowledge of fabricating procedures, procurement methods, sources of supply, and shipping practices. In addition he must be able to work effectively with people in his own and the fabricator's organizations. Since his primary duty is to encourage people to finish a job on time, his actions must be diplomatic in order to avoid resentment.

## GENERAL PURCHASER-VENDOR PRACTICES

The following are common practices in the relation between purchasers and vendors:

1. In general all transactions between purchaser and vendor should be kept on a formal basis, i.e., in writing.

2. Instructions to vendors shall always be sent in the form of a changed purchase order or addendum to original order. This should be a document similar to the purchase order, carrying the signature of the purchasing agent and identified with the original purchase order by some similar numerical designation.

3. It is good policy for only procurement personnel to discuss prices, materials, or equipment with vendors during the inquiry stage. Since it is unethical to disclose any vendor's proposal to another vendor, the fewer persons having this information, the less chance there is for accidental disclosure.

4. Price re-negotiation may or may not be permitted after a formal proposal is received. This is usually a matter of policy with the purchaser. If one vendor discovers an error in the purchaser's inquiry or for some reason is allowed to revise his quotation, all other vendors participating should be allowed the same privilege.

5. Sufficient time should be allowed for receipt of proposals before deciding the successful bidder.

6. A good policy is to require that all changes, additions, or deletions be handled by the procurement department only. Under certain conditions this policy may require additional time, but it avoids confusion. It should be the complete responsibility of the procurement department to negotiate the commercial part of an order (price, terms, and delivery), regardless of the technical requirements.

If equipment is being manufactured in accordance with the purchaser's design, the fabricator may ask a number of technical questions which can only be answered by the engineering department. These questions should always be addressed to the procurement department and then relayed to the engineering department. A formal reply is then forwarded by engineering to procurement for transmittal to the vendor. In this manner the procurement department can retain full responsibility for all changes in price caused by design alterations.

7. Order cancellations of fabricated equipment or materials of special design may be required because of changes in design or scope of work which are the fault of the purchaser. Cancellations may also be made because of failure on the part of the vendor to manufacture the equipment properly. Cancellation clauses are written into the purchase order, by stating simply, "This order may be cancelled within 30 days after placement without cost to the purchaser." This statement is used when the intent of the purchaser is to limit the vendor's action until the purchaser is ready to proceed with the work. When work has been started by the vendor and it is decided to cancel the equipment because of

changes in scope of the project, the vendor must be formally notified to "immediately stop work and incur no further costs to the account of the purchaser." The vendor is then asked to submit cancellation charges accompanied by proof of his costs up to the time of cancellation. The purchaser may accept delivery of materials purchased by the vendor for use in completing the order, or the materials may be retained in the vendor's warehouse and their value deducted from the cancellation charges.

Even if equipment is cancelled because of the vendor's failure to perform satisfactorily, cancellation charges are incurred by the purchaser. It is the purchaser's responsibility to have some knowledge of the vendor's capabilities.

Cancellation should never be undertaken without complete knowledge of the costs involved. These include the vendor's labor costs, material costs, loss of profit, and the purchaser's loss of time and engineering manhours.

## PROJECT ENGINEERING AND PROCUREMENT COOPERATION

The many duties and responsibilities of the procurement department can be met only through cooperation between procurement and project engineering groups. Mutual understanding of the duties and problems of each and an awareness of their interdependence are needed.

Project engineers should remember that changes in design produce delays in fabrication. A change that may actually be rather simple can become quite complex when large organizations are involved. Drawings, orders, and suborders must be altered before the actual change reaches the manufacturer's shop. Only alterations which are absolutely necessary should be made during the fabrication stage.

## REFERENCES

1. Harriman, Norman F., *Principles of Scientific Purchasing,* McGraw-Hill Book Co., New York, 1928.
2. Heinritz, Stuart F., *Purchasing,* Prentice-Hall, Inc., New York, 1947.
3. Johnson, E. R., G. C. Huebner, and G. L. Wilson, *Transportation Economic Principles and Practices,* D. Appleton-Century Co., New York, 1940.
4. Lewis, Howard T., *Procurement Principles and Cases,* Richard D. Irwin, Inc., Chicago, 1949.
5. *National Association of Purchasing Agents Handbook of Purchasing Policies and Procedures,* Volumes I and II, National Association of Purchasing Agents, New York, 1939 and 1942.
6. *Revised American Foreign Trade Definitions,* National Foreign Trade Council, 111 Broadway, New York, 1941.
7. Rindsfoss, C. E., *Purchasing,* McGraw-Hill Book Co., New York, 1915.
8. Stufflebeam, G. T., *The Traffic Dictionary,* Simmons-Broadman Publishing Co., New York, 1950.

# CHAPTER 10

# OFFICE PROCEDURE

Engineering is only one of the functions of the project engineer. In addition to technical duties, he must manage his office. The staff of this office usually consists of assistant project engineers, a secretary, and a file clerk. It is, of course, only a small part of a large engineering office which includes other project engineers or project managers and their staffs as well as departments such as accounting, purchasing, drafting, and engineering specialties.

There is a constant interchange of ideas and instructions between the project engineer's office and the other departments, as well as direct contacts and correspondence with people outside the company. Therefore, he also must develop facility in holding conferences, correspondence, and filing. A discussion of each of these functions will be of value since poor execution of any of these will produce mistakes and faulty design no matter how competent the engineering.

## CONFERENCES

The project engineer must exchange ideas with many people during a working day not only by means of formal policy-making conferences, but also by informal conferences or get-togethers with men from other departments, customers' representatives and sales engineers.

As is always the case in dealing with people the intangible, and for the most part unteachable, qualities such as humility, open-mindedness, and courtesy are important.

Humility is one of the most essential personal traits in any field of science. Any true scientist is aware of the many assumptions and gaps of knowledge in his field and never presumes knowing everything. Being humble, however, does not mean being obsequious. Tentative positions can be held without reticence.

Open-mindedness is important in dealings with both individuals and groups. If two ideas of equal technical merit are proposed, one by you and another by a person who must execute the idea, it will be advantageous to use the other person's idea. It is logical to assume that he will do a better job using his own method.

In large conferences, too, open-mindedness pays dividends. In these situations it becomes the ability to listen to others. Often in conferences an atmosphere is created which seems to compel each participant to perform. By resisting this temptation and remaining open-minded, one can contribute immeasurably as the conference proceeds by offering carefully considered conclusions based on an analysis of the problem and the ideas presented by the other men.

Most people have had the merits of courtesy drilled into them since childhood. These lessons are often forgotten when authorized visitors arrive during busy moments. Sales engineers, in particular, are given improper treatment at times when their arrival is inopportune. Competent sales engineers have considerable knowledge about their particular fields, and aid the engineer in keeping informed on new developments. Time spent with them is not wasted even though no purchase of their products is anticipated at the time.

### Formal Conferences

The formal conference has become an important tool in modern industry. For the process plant project it permits a sharing of ideas between key personnel and specialists which would take weeks if handled by correspondence.

To be successful, however, a formal conference must have a capable leader and a definite purpose which should be known several days before the meeting. The purpose must be a specific one if much is to be accomplished. Conferences on general subjects require too many people, accomplish nothing, and often create misunderstandings.

When need arises for the discussion of a specific phase of work, only those people who can contribute most to the discussion should be requested to attend. The conference leader should then outline the course of discussion and see that the original plan is followed but refrain from strong statements of his own ideas.

Complete notes of the conference should be made. Few memories are infallible, and it is desirable to have records of agreements and decisions so that responsibilities are clearly outlined and work assignments completely understood. The conference room should be well lighted, comfortable, quiet, and furnished in good taste. A blackboard is always useful in group discussions and should be part of the permanent equipment of the room.

The conference technique is often misused in modern business. Too many unnecessary conferences are convened. Some organizations have so many that there is little time for work. If one man can make a decision, it is senseless to take the time of other personnel. Conferences should be called only when a decision must be made requiring the knowledge represented by many backgrounds.

## TECHNICAL WRITING

The engineering graduate is often criticized for his lack of ability in expressing his ideas. Generally the undergraduate engineer takes one or two courses in report writing and these provide his only guidance in written expression.

These courses are invaluable, but good self-expression in any form can never be developed solely through courses that teach the techniques of the art. Good self-expression, rather, is a by-product of thoughts and experience beyond the limits of the individual's specific field.

It is unfortunate that the rigorous curriculum of a formal engineering education leaves little time for reading and study in other fields. Many practicing engineers realize this lack and use some of their free time for reading nonscientific literature.

Writing done by the project engineer must be suited for widely varying audiences. Such writing includes:

TECHNICAL REPORTS. (*a*) Contract proposals; (*b*) descriptions of plants visited and their operations; (*c*) detailed conference notes; (*d*) technical instructions such as operating instructions.

TECHNICAL ARTICLES (written for publication in trade or professional journals).

CORRESPONDENCE. (*a*) Interoffice; (*b*) design instructions; (*c*) various types of letters outside the company.

### Technical Reports

The preparation of a technical report as such has been described in many forms[3,4,5,6] and is usually a part of the engineer's training. Each firm has its own method of composition. Generally the first page is a letter to a superior often called a letter of transmittal. This is followed by a title page, an index, a summary, and then the main body of the report. The engineer-contractor's proposal is one type of report.

The form of report used by a given firm must be followed by its employees. There is no one perfect form. Usually the one used by a firm has evolved as the company developed, and it particularly suits that organization. Management and other personnel become accustomed to the form and much time is saved because of this familiarity.

Within the confines of the form the engineer is free to exercise his ingenuity and writing talent.

The most important consideration in report writing is the audience. In writing a report for the president of a large chemical company, for instance, an engineer should realize that most of this man's time is spent in management. He no longer practices his technical skills and its intricacies are no longer part of his thinking. If it is necessary to discuss theory in the report, background should be supplied and, if possible, it should be described in terms understandable to the layman.

### Technical Articles

Periodical technical journals are distribution centers for new theories and ideas. Technical men must read articles or abstracts of articles in these journals in order to be familiar with the latest developments, and, when possible, should contribute new ideas and experimental data to the general knowledge by submitting papers for publication.

The quality of such contribution reflects on both the author and his employer, and great care should be exercised in their preparation. Assuming the information presented in the article is reliable, the next most important aspect is the mode of presentation.

Technical writing should be free from long, ponderous sentences and should be as clearly written as possible. Since the subject matter is often difficult to comprehend, the least a writer can do is make the individual sentences easy to understand.

### Correspondence

Correspondence makes up the major portion of a project engineer's writing. If correspondence is excessive, the project engineer is little more than a "paper shuffler."

Nearly all large business firms are confronted with a tremendous volume of paper. Although in these modern times communications have been greatly simplified, the written word is still the only legally acceptable method of indicating responsibility. Recording reasons for actions by one party as instructed by another in many organizations is practiced largely because of a lack of individual trust, initiative, and usually self confidence. The phrase "get it in writing!" is a common statement in industry. From this idea much of the needless correspondence evolves.

External correspondence is necessary but can be reduced by careful planning. Internal or inter-office correspondence is used for issuing instructions or defining responsibility. When verbal instructions will suffice, however, it is wasteful to insist on written memoranda. Of course, judgment must be used in deciding when verbal instructions are adequate. Certainly complex instructions involving numerical data should be written, but a simple request or order seldom need be written.

Another major source of useless correspondence in business can be traced to excessive restriction of authority. Without authority or without clear-cut instructions the individual must seek to "cover himself" at every step by written proof of his actions.

All business organizations can profit by an objective analysis of their correspondence procedures. Much inter-office correspondence can be eliminated altogether and that which is necessary can be limited to an original and one copy for the job file. Forms which can be filled in by hand or by a few simple typewriter manipulations save both time and expense. These are discussed below.

### *Interoffice Correspondence*

The interoffice letter, note, or memorandum is usually identified by a printed heading on some type of form. Interoffice or Interdivision correspondence in engineering service organizations is often also made readily identifiable by use of colored stationery. Some typical forms are shown here.

**Example of Interoffice Letter**

| | | | | |
|---|---|---|---|---|
| (1) | Date: | January 2, 1955 | (6) | XYZ Chemical Co.<br>Interoffice |
| (2) | To: | R. D. Plow (2) | | |
| (3) | From: | S. S. Shear | (7) | Approp. No. 2246 |
| (4) | Copies: | E. D., M. N., R. J. | (8) | File 4, 15, 6 |
| (5) | Subject: | Dwg. 2246-02-E1, Revisions | | |

(9) Drawing 2246-02-E1 is to be revised according to the attached marked print. When revision is complete and checked, kindly return two (2) prints to the writer prior to general distribution.

(10) S. S. Shear Proj. Engr.

(11) Attachment 1
(12) SSS/ef

The numbers are used here for purposes of discussion, but do not appear on the actual letter. One peculiarity of the interoffice letter is the omission of the salutation, such as *Dear Sir,* or *Gentlemen* and the closing, such as *yours truly*. There is, however, identification material in items 5, 6, 7, and 8.

The example is a transmittal letter common to larger firms. It provides a record for several files and also in this case shows the status of a drawing. All numbered items except 9, 10, 11, 12 are printed on the letter. Item 6 identifies the firm and is necessary since copies of such letters are sometimes attached to formal external correspondence. Initials of recipients of copies, (Item 4), are often used on interoffice communications. Files (Item 8) refer to the particular record files for the department or project and aid the file clerks in filing the letter. Second sheets (when the letter is longer than one page) omit all printed headings except 6. The date and project number is added by typing. Copy sheets are similar to second sheets showing item 6 only.

### Memo Forms

Memorandum forms are sometimes used for interoffice communication. They are generally lettersize ($8\frac{1}{2}$ x $5\frac{1}{2}$). The firm name appears on these forms, and items 1 and 2, 6 and 7 are usually included on the heading. Such memoranda may be handwritten notes.

### General Instruction Letters

General instructions to personnel are usually issued on some type of reproducible form so that distribution to several people can be made. The following form is sometimes used.

**Example of a General Instruction**

XYZ Chemical Co.
Eng. Dept.
Letter No. 2832

Date: January 2, 1955

Subject: Use of Variable-type Machines

To: Chief Draftsman, All Project Engineers, Design Section Leaders, Engineering Division Department Heads, Plant Maintenance Dep't.

Four (4) variable-type machines have been installed in room 308 (third floor north-end of Engineering Building) and have been in use for several weeks during operator training. We are enclosing copies of a sheet showing the various sizes of type form, numerals, and symbols which are available. You may call Ext. 325 for extra copies.

The machines have been provided to reduce hand lettering on drawings and, by means of the smaller type, to reduce size of printed forms. Supervisory personnel are expected to become familiar with the application of these machines so that they can be used to the fullest extent.

General instructions are being prepared and will be issued shortly.

______________________________
A. D. Hochstad, Chief Engineer

This general letter includes a letter number which simplifies filing and identification when revisions are necessary. Such letters are often issued on special forms so that they can be identified easily. The use of different-colored stationery is probably the simplest method of identification for the large firm.

### Transmittal Letters (Interoffice)

A large amount of data must be transferred through an organization. Transmittal letters or forms save typing time. The following form only requires filling in or checking the applicable phrases.

**Example of Transmittal Letter**

Date:

To:

From: Project No.____________________

We are attaching ________copies of:________________________________

Drawing No.____________________

B/M No.____________________

Vendor Print____________________ Vendor____________________

Approved____________________ Kindly revise as noted ____________

Approved as noted________________ Return to Vendor ____________

Release for fabrication ____________ Release for construction____________

For your use in design ____________ For your use and record ____________

For preliminary use only ____________

When revised, please return________________prints to writer.

### Design Instruction Letters

In addition to the general instruction letter it is desirable to provide a special form for engineering and design changes. Engineering-contracting firms use different names for this correspondence such as "Design Instruction," "Engineering Memo," or sometimes "Design Letter." The most important requirement for such memoranda is a visible difference from other types of correspondence so that they will receive special handling. The following is a typical design letter.

### Example of a Design Instruction

(1) ABC ENGINEERING CORP., PRINCIPIA, N. J.

(2) Customer XYZ CHEM. CO.
(3) Date 11/21/42 (4) Contract 6–25316–2 (5) Design Inst. No. E–22
(6) DES. SECT. E (7) Sheet 1 of 1
(8) Dwg. (9) Vendor X (10) Mat'l B/M
(11) Spec. X

It has been agreed in a conference on 11/21/57 that all electric motors shall be as manufactured by Gyre Electric Co. The applicable catalog numbers are:

| H.P. | Volt. | Cat. No. |
|---|---|---|
| Up To 5 hp | 220/110 | E–A–1222 |
| 6 to 30 hp | 440 | E–B–2136 |
| 31 to 50 hp | 440 | E–C–3145 |
| 51 and above | 2300 | D–4460 |

A. L. Gerund, Proj. Eng.

(12)

| G | F | B | S | P | V | H | M | E | I | Proc. | EF | CF | C | Total |
|---|---|---|---|---|---|---|---|---|---|---|---|---|---|---|
| | | | | | | | | 1 | | 1 | 1 | 1 | 2 | 6 |

Items 1, 2, 3 are self-explanatory. Item 4 shows the contract number as assigned by the contractor, "ABC Eng. Corp." This type of numerical designation is finding wide use in industry where automatic accounting systems are used. In this instance the first number refers to the type of general work such as design and erection of a chemical plant. The first digit of the second number refers to the type of contract, the next four digits are the consecutive contract number. The last number indicates the part of the project.

Item 5 gives the subject and number of the instruction. The subject in this case is *E* (electrical design); 6 shows the basic design section or group to which the instructions are directed primarily; 7 is a sheet numbering method designed to prevent misplacing of sheets.

Items 8, 9, 10, and 11 provide a means of designating the type of information in general. In this case, 9, vendor, and 11, specification, are checked because the information concerns a vendor and is a specification. If a drawing or bill of material were involved, items 8 or 10 would be checked.

Item 12 provides a simplified method for indicating the distribution and numbers of copies of the Design Instruction. The letters refer to design sections or departments as follows:

| | | | |
|---|---|---|---|
| G | General | B | Buildings |
| F | Foundations | P | Piping |
| S | Structural, steel or concrete | H | Heat apparatus |
| V | Vessels | E | Electrical |
| M | Mechanical equipment | Proc. | Procurement |
| I | Instrumentation | CF | Contract file |
| EF | Engineering file | | |
| C | Customer | | |

In this case copies are checked for E, Electrical; Proc., Procurement; EF, Engineering File; CF, Contract File; and C, Customer.

### *Commercial Correspondence*

The types of letters written to outside firms may be (*a*) correspondence between an owner and contractor; (*b*) between purchaser and vendor; (*c*) various engineering or procurement instructions to subcontractors, vendors, and consulting firms.

#### Owner-Contractor Correspondence

Operating firms usually carefully restrict external correspondence to executive officers or key personnel. Correspondence between an owner and contractor originating with the owner usually will all be signed by an executive or in the name of the executive by the originator (see example).

Contracting firms to the contrary cannot restrict the writing of correspondence since many functions must occur simultaneously and any such restriction would be impractical. The project manager, however, should see copies of all correspondence as it passes into the general contract file.

All correspondence between the owner and contractor will ordinarily be directed to the contractor's project engineer regardless of subject. Some examples follow:

**Example of a Letter from Owner to Contractor**

XYZ Chemical Co.
Dinwiddie, Florida

E. G. Martin, Chief Engineer

Letter Number 168
February 2, 1956

ABC Engineering Corp.
1223 8th St.
Principia, N. J.

Attn. Mr. A. L. Gerund

Subject: App'n No. 8-2324, ABC Contr. 6-25316-2
Des. Inst. E-22.

Gentlemen:

In order to effect a possible saving in driver costs it has been decided that the specification for motors above 5 H.P. will include the words "or equal" after the Gyro Electric Co. catalog number.

For other acceptable motor manufacturers please refer to our Approved Vendor List transmitted in our letter No. 55, November 16, 1955. It is suggested that only two (2) bids be requested in addition to proposals from Gyro.

Yours truly,
XYZ Chemical Co.
E. G. Martin, Chief Engineer

By______________________________
S. S. Shear, Project Engineer

SSS/ef
Copies E.D., M.N., R.J.
App'n 8-2324

**Example of a Letter from Contractor to Owner**

ABC Engineering Corp.
1223 Eighth St.
Principia, N. J.

(1) Contract 6-25316-2

(2) Letter No. 85

(3) Files 5, 7, 15.
February 6, 1956

(4) Mr. E. G. Martin, Chief Engineer
XYZ Chemical Co.
Dinwiddie, Fla.

(5) Attn. Mr. S. S. Shear, Proj. Eng.

(6) Subject: Letter 168 and Design Instruction E-22. Telephone Conversation 2/2/1956.

Gentlemen:

This will confirm our telephone conversation in which we advised you that your letter No. 168 (February 2, 1956) would not affect the original specification of one manufacturer, Gyro Electric Co., for motor drivers.

Following our usual practice the procurement department obtained three bids for motors based on preliminary data. This was unnecessary since your original specifications had indicated only a single manufacturer. Proposals received were analyzed, as is our practice for all equipment. Gyro was found to have an acceptable delivery, and the prices were lower than other bidders on all items. The analysis in this case was not transmitted to you because the low bid was received from the manufacturer that you had originally specified.

A letter of intent was issued to the vendor January 2, 1956 in order to place the equipment on an early schedule. Our confirming purchase order, based on final specifications, has now been issued to the Gyro Electric Company. If our priority is not changed, delivery can be expected in early June.

Yours truly,
ABC Engineering Corp.

______________________________

A. L. Gerund, Project Engineer

ALG/rs

Correspondence between an owner and contractor is usually numbered as shown in item 2. The letter is addressed to the chief engineer and to the attention of the project engineer, items 4 and 5. Item 6, the subject, is described in detail. When the contract number is not placed elsewhere, it should be included in the subject.

## Letters to Vendors, Letters of Intent

Letters to vendors concern changes in design and materials, extra charges, delivery dates, and statements of intent.

Ordinarily vendors are notified of order placement by receipt of a purchase order. Preparation and distribution of a purchase order is time consuming, and when it may delay delivery, a letter of intent, as in the accompanying example, is written in order to place equipment on order before the issuance of a formal purchase order.

### Example of a Letter of Intent

ABC Engineering Corp.
1223 Eighth St.
Principia, N. J.

(1) Contract 6-25316-2

(2) Files 7, 15.
January 2, 1943

Gyro Electric Company
2286 Enoch Ave.
Portland 6, Pa.

Subject: Proposal 3427, October 23, 1942

Gentlemen:

Please accept this communication as a letter of our intent to purchase equipment essentially as described in your proposal 3427, provided:

a. Prices quoted are firm.
b. Delivery promise of six (6) months or better as based on our priority rating will be maintained.
c. You agree to proceed with engineering, design and manufacture immediately by placing this equipment in your schedule.

Please acknowledge receipt of this letter immediately. Our formal purchase order will be issued together with complete specifications at an early date.

Yours very truly,
ABC ENGINEERING CORPORATION

______________________________

N. A. Mayberry, Jr.,
Purchasing Agent

NAM/pe

**Example of a Letter to Vendor on Design Data**

ABC Engineering Corp.
1223 Eighth St.
Principia, N. J.

Contract 6-25316-2
Files 7, 15.
February 16, 1956

Robertson Tubing Co.
897 Field St.
Portland 6, Pa.

Subject: Our P.O. 5316-87,
your letter of February 12, 1956

Gentlemen:

The stainless steel material substitutions requested in your letter of February 12 have been submitted to our engineering department and approval has been granted. Type 309 is acceptable where Type 316 has been specified.

Our original specification will be revised and

transmitted; however, please proceed immediately on the basis of this letter.

Yours very truly,
ABC ENGINEERING CORP.

______________________
N. A. Mayberry, Jr.,
Purchasing Agent

*RL/NAM/me

External Transmittal Letters

A transmittal letter form is sometimes used for transmittal to outside firms.

**Example of an External Transmittal Letter**

ABC Engineering Corp.
1223 Eighth St.
Principia, N. J.

Date:
To:

Contract No.______________
Letter No.______________
File______________

Gentlemen:

We are enclosing____copies of preliminary final revised Drawing No. B/M No.

For approval________
For your use and record________

Kindly return____copies with your approval or comments as soon as possible.

Yours truly,
ABC Engineering Corp.

______________________

Kindly return one copy of this transmittal for our record upon receipt of enclosures.

### *Telegrams, Cables, Teletype*

The telegram (domestic), cable (foreign), and teletype are used for important written communication requiring immediate action. Tele-

*This letter was prepared by someone else but was issued over the purchasing agent's signature.

grams are transmitted over Western Union lines and teletype messages, over telephone company lines.

Both Western Union (telegram) and the local telephone system (teletype) will install teleprinters or machines over which messages may be sent or received in any office.

### Telegrams

No charge is made for installation or use of the Western Union teleprinter, but a firm must be well known or must have some background which would indicate its use. Unless there is an existing shortage of machines in the particular area, there is not much difficulty in having a teleprinter installed.

The keyboard of the machine is almost identical with a standard typewriter. The operator signals the local office, and when acknowledged, the message is typed on a continuous (roll) sheet. The paper is 8½ inches wide, similar to standard stationery. The message may be reproduced in several copies. The local office receives the message and relays it to a local office in the city of destination. If the recipient has a teleprinter, it is reproduced on his machine. If the recipient has no machine, the message is usually telephoned and then a copy is mailed. When the recipient is in a metropolitan area the message may be delivered by messenger.

The telegram message charge is based on the number of words and type of message. The straight wire which is most expensive is transmitted immediately regardless of time of day. Other types are the day letter which is delivered in sequence during working hours and the night letter which is transmitted at night and delivered the following morning.

#### Example of a Business Telegram

```
ABC ENGINEERING CORP.  WUX
PRINCIPIA, N. J.

  (1)        (2)
2324-832  RELET EIGHT THREE SEND MORE INFORMATION ON
PRICE AND DELIVERY.  PREFER ORIGINAL SPECIFICATION
TWO ZERO SIX DASH ONE.  LETTER FOLLOWS.
                             (3)  XYZ CHEMICAL MARTIN
```

The abbreviation WUX after the company name indicates that the company has a teleprinter. Item 1 is a numbering system sometimes used for cost accounting purposes. The number 2324 is the appropriation number as is used by the accounting department in charging costs against the appropriation. The second number, 832, is a consecutive telegram number and has nothing to do with charges but may be used for filing and reference.

Item 2, "Relet," is a word condensation meaning "reference to letter." "Retel" is used to refer to a telegram, "Rephone" is sometimes used for reference to phone messages. Such condensations are counted as one word. The local telegraph office will supply cost data and acceptable abbreviations and condensations. Numbers after the telegram are spelled out since transmitted numbers may be changed or mixed in sending and spelling numbers eliminates the possibility of such error.

Cables are similar to telegrams, but more attention should be paid to phraseology of cables because of the higher cost per word. Regardless of cost, however, the message of a telegram or cable should be clear. Saving the cost of a word and sacrificing clarity may mean sending another wire or cable.

### Teletype

The teletype teleprinter is similar to the Western Union teleprinter except the procedure is different. The operator signals the local teletype office giving call letters and number of distant party desired. The operator then must wait until a connection is made with the distant office before a message can be transmitted. Teletype messages can be transmitted only to another machine. These machines operate much the same as telephonic communications. No relaying or transfer of messages occurs. Keys struck on sender's machine type the message on sender's and recipient's machines simultaneously. Typed conversation may be carried on since transmission is instantaneous.

Charges for messages are on a time basis similar to long distance toll calls. A nominal monthly charge is made for teletype machines regardless of service.

Teletype form may be varied to suit the sender. The number of words affects price only as it affects the sending time. A message can only be transmitted after connection is made with recipient's receiver, thus the address or salutation can be simplified. If a firm has several branch offices the teletype is extremely useful in transmitting identical messages to all connected receivers simultaneously.

### Leased Wire Services

Either teletype or telegram systems offer continuous or instantaneous service by means of leased wires between points. If a leased wire is obtained, the machine instantaneously operates connected machines. The cost is high, since this amounts to continuously holding a line. The leased wire may prove economical, however, when the number of messages per day to a given office is very large.

Long Distance Telephone versus Teletype and Telegram

The typed communication provided by teletype and telegram is in many instances more satisfactory for business use than telephonic communication. No record of a telephone conversation is obtained unless recorded by a third party or a machine.

There are times when a long distance call, though more expensive, is preferable. If a personal or conversational touch is needed to settle some problem the telephone is superior. Hearing a person's voice often gives better insight into the particular situation. Unfortunately some employees abuse this privilege and carry on reunions and personal conversations with distant employees before getting to the business of the call.

### *Style of Written Correspondence*

Correspondence functions as a substitute for conversation and should be as natural as the dignity of the subject permits. Pompous and hackneyed phrases should be avoided, and letters should be short and to the point.

Most people compose letters that are relaxed and easily understood when they dictate such letters to a secretary, or better, to a dictating machine. After an engineer practices the art of dictating, his dictated letter sounds much less stereotyped and more conversational than a letter he composes on paper.

The use of a dictating machine has many advantages. Thoughts can be recorded as they occur and the secretary need not be interrupted. The transcription of dictated material, however, requires an above average secretary; one who can spell, punctuate, and paragraph accurately.

### *Secretarial Help*

A competent secretary is a necessity for the smooth operation of a project engineer's office. An intelligent person preferably with some background in science or related fields as well as a good knowledge of English usage is required.

The quality of correspondence is largely governed by her ability to transcribe dictated material with neatness and accuracy. This must be done with minimum disturbance of other staff members. A secretary with initiative will prove invaluable in catching errors. In addition to all these qualities the secretary should be pleasant and neat in appearance since she must also serve as receptionist for the project engineer and his staff.

## FILING SYSTEMS

A filing system for any type of business operation can become extremely cumbersome unless it is kept in a consistent and logical manner. The fundamental purpose in keeping files is to provide a record; the importance of records varies, of course, with the type of business.

### General File

Any engineering organization should maintain some type of general file where policy and other general matters applicable to all operations may be recorded. The following is a typical filing system used for a process plant project general file. The numerical designations are used to indicate file categories on all communications so that file clerks need not be required to decide on the proper file.

- 1.00 Engineering, Departmental Policies.
  - 1.01 Process Engineering.
  - 1.02 Mechanical Engineering.
  - 1.03 General.
- 2.00 Design Data.
  - 2.01 Mechanical Equipment.
  - 2.02 Electrical.
  - 2.03 Instrumentation.
  - 2.04 Vessels.
  - 2.05 Piping.
  - 2.06 Structural Steel.
  - 2.07 Foundations.
  - 2.08 Architectural.
- 3.00 Specifications.
  - 3.01 Mechanical Equipment.
  - 3.02 Electrical.
  - 3.03 And others.

These files can be extended as much as desired. Additional sections might be, 4.00 Material Listing; 5.00 Construction Operations; 6.00 Procurement; 7.00 Contract Policies. The advantage of the decimal system is apparent. As an illustration, 2.04 could be subdivided:

- 2.04 Vessels.
  - 2.0401 Forged Chambers.
  - 2.0402 Plate Vessels.
  - 2.0403 Thin Plate Tanks.
  - 2.0404 Bins.
  - 2.0405 Elevated Storage Tanks, Water; etc.

or, 2.01 Mechanical Equipment, extended:

- 2.0101 Pumps, Reciprocating.
- 2.0102 Pumps, Centrifugal.
- 2.0103 Pumps, Rotary.

2.0104 Pumps, Positive Displacement.
2.0105 Turbines, etc.

The only information which should be filed in the General File is that which may be useful in establishing general policies. As an illustration assume that a certain type of pipe or plate material specification is being discontinued because of manufacturer's policy. It should be of immediate interest to all persons concerned that the particular specification has become obsolete. A general letter would be circulated to the interested department heads; a copy of this information would be filed in 2.04. The actual filing of data in some single location such as a general file enables anyone interested in the history of the various policies to find wanted information without disturbing a particular department. Hence, the general file has some features of a library.

### Project File

Any firm engaged in contracting services should establish some single type of filing system which should be used for all projects. Such systems have the decided advantage of being familiar to all personnel, and the training of clerical personnel in filing becomes a minor task. If each individual establishes his own private system of filing, there will result as many systems as there are individuals, each of which will be completely foreign to all other persons except those directly involved.

The following project filing system is simple and easily learned.

1.0 Contract.
  1.01 Estimate.
  1.02 Contract Extras.
2.0 Design Instructions (May be in loose leaf binders).
  2.01, 2.02, 2.03 (Divide the same as for general file).
3.0 Correspondence to and from Customer (Except 1.0 and 6.0).
4.0 Correspondence to and from Construction Department.
5.0 Progress and Commitment Reports.
  5.1 Engineering, 5.2 Procurement, 5.3 Construction.
6.0 Accounting Correspondence.
7.0 Procurement.
  7.1 Shipping, Traffic, Expediting, 7.201 (Divide same as General File 2.00).
8.0 Conference Notes.
9.0 Daily File (copies of all outgoing correspondence filed according to date).

The last file, 9.0 is important since it provides a positive means of locating a letter when only the approximate date of writing is known.

## HUMAN RELATIONS AND PERSONNEL MORALE

The office functions that have been discussed and most of the technical phases of the project engineer's duties all involve contact with other people. High personnel morale, and skillful human relations make the many personal contacts in an office pleasant and profitable.

Few individuals can ignore the fact that nearly every function in life is based on relations with others. An engineer may be extraordinarily capable in his field and yet be completely unsuccessful or limited in his usefulness because of his inability to work with others. There are doubtless as many approaches to the problem of human relations as there are personalities, and there is no simple solution for all situations since no two persons nor groups of people react in the same pattern. It is certain, however, that few people are callous to a consistently pleasant personality or to thoughtfulness on the part of another individual.

Organization morale is affected by many factors. It may be produced by salary inequities, poor working conditions, incompetent executive personnel, or a lack of incentive for advancement. Salary problems are always difficult for the engineer who sometimes feels that bargaining is unethical.

Although working conditions are often not as important to the engineer as to other types of personnel, any environment which is pleasing to the individual should improve his general attitude and thus his production.

The usual complaint of the young engineer is the feeling of being "buried" in a large organization. The group leader who consistently takes over all details or fails to distribute responsibility is only creating a bad morale problem which often results in high labor turn-over. Many times the interest of younger personnel can be held by added responsibility. Bottle-necks in engineering organizations can often be traced directly to a single individual, who fails to understand the advantage both to himself and the organization, of maximum work distribution.

The engineer's admirable faculty of becoming absorbed in a particular problem often results in misunderstandings. Thus engineers sometimes gain the unfortunate reputation of being oblivious to everything but their work. They become accustomed to reducing everything to some factual or logical basis. Human relations, however, are seldom factual or logical. Unless the engineer makes an effort toward understanding the requirements of compatibility he may find himself isolated by his personality regardless of his technical ability.

## REFERENCES

1. Crouch, W. G., and R. L. Zetter, *A Guide to Technical Writing,* 2nd Ed., Ronald Press Co., New York, 1954.

2. Howell, A. C., *Handbook of English in Engineering Usage,* 2nd Ed., John Wiley & Sons, New York, 1940.

3. Kerekes, F., and R. Winfrey, *Report Preparation,* Iowa State College Press, 2nd Ed., Ames, Iowa, 1951.

4. Kobe, K. A., *Chemical Engineering Reports,* 3rd Ed., Hemphills Book Store, Austin, Texas, 1950.

5. Mills, G., and J. Walter, *Technical Writing,* Rinehart Bros., New York, 1954.

6. Weil, B. H., *The Technical Report,* Reinhold, New York, 1954.

CHAPTER 11

# CONTRACTS AND CONTRACTORS

In simple terms a contract is an agreement between two or more parties by which they legally bind themselves to do certain things. The form of contract used varies with different firms and with the type or scope of service required. Contracts can be made for engineering services only or for complete engineering and construction including the purchase and erection of materials. A contract may be a simple instrument such as a purchase order or it may be a lengthy and extensive document. In all cases the documents become effective when formally signed by officers of each firm.

The purpose of a contract is to protect both parties and to assure that certain duties will be performed in a prescribed way. It is a common error on the part of the layman to assume that construction contracts are primarily for the owner's protection against the contractor. Contracts may be written to favor either party. An agreement of any kind, however, is binding on all parties to the agreement.

## ENGINEERING AND CONSTRUCTION FIRMS (CONTRACTORS)

The construction industry is a most important and necessary part of the national economy, and renders a unique service to industry as a whole by maintaining staffs of highly developed specialists in every field of engineering.

The majority of process firms have found it advantageous to engage the services of an engineering and construction firm when new facilities are required. An experienced contracting organization can always do a better and more expeditious job of engineering and construction than engineering and construction forces of the operating company which must usually be hastily expanded for a new plant project.

Erroneously held by some people is the belief that contracting firms are often unscrupulous. Major engineering-construction firms, however, operate much the same as any other large business organization. Continued failures, negligence, bad customer relations, or poor performance in general would shortly ruin any contractor's business since his reputation is his most important asset.

Most criticisms of contractors are directed toward their service charges. No commercial organization can operate without profit. Consideration of contractor's fees with respect to the production value of an erected plant will usually show that instead of being exorbitant they are in reality more likely to be modest. A large percentage of process plants are designed to pay-out their total cost in from three to five years, including normal replacement maintenance. However, few plants are constructed which do not continue to operate with minor modifications for ten to twenty years. It would seem more equitable to consider the contractor's fee over the life of the plant rather than as a lump sum earned in 18 or 24 months. When spread over the many years of normal operation the fee seems small indeed.

## SELECTING THE CONTRACTOR

A number of statutes are in effect in each state governing the methods of letting contracts when public or governmental funds are involved. These methods are not usually followed in private industry, but the practice of receiving bids or proposals in the form of estimates for work or services is followed by practically all firms. The selection of a contractor should be based on an analysis of proposed prices obtained by issuing inquiries and on the contractor's reputation or know-how in certain types of work.

The preparation of data to be used in obtaining proposals from contractors is usually a costly undertaking for the owner since the information will become the basis for the contractor's proposal, and therefore must be as complete as possible to avoid misunderstandings. The best way to assure completeness is for experienced design engineers to prepare the data. Enough information must be given the contractor to design the type of plant desired in the manner preferred by the owner.

Many process firms have extensive mechanical and process specifications which have been compiled to provide a certain uniformity in purchasing and engineering design of commonly used equipment. In such cases copies of these should be given to the contractor preparing a proposal so that they may be incorporated into his preliminary design and estimate. In addition complete descriptions of the owner's and contractor's duties, the proposed plant site, transportation facilities, utilities, available raw materials, and processes to be used should be provided.

The extent of process information required will vary with the project. In some cases the contractor will act as a licensor and provide a complete process design. Often, however, the process firm has developed its own process. In that case the process firm must either supply the contractor with a complete process design or furnish enough process data so the contractor can make a preliminary design.

Contractors must prepare a preliminary design in order to make an estimate of the cost for use in their proposals. This proposal should contain the following information:

*a.* General description of the project.
*b.* Scope of the contractor's work.
*c.* Description of each portion of the plant with accompanying flow diagrams and plot plans.
*d.* Estimated utilities.
*e.* Auxiliary equipment to be constructed (warehouses, shops, etc.).
*f.* List and brief description of major equipment (pumps, vessels, heat exchangers).
*g.* Mechanical specifications to be followed.
*h.* Statement of product quality and production rate.
*i.* Proposed cost and fees.

These are some of the major items. All information necessary to describe the proposed design and the contractor's intent should be included. Clear, uncomplicated illustrations, charts, and schematic flow diagrams are invaluable. Executives who must examine proposals are busy and greatly appreciate such aids to rapid comprehension.

Contracts should never be awarded on the basis of price alone. Price proposals from the larger process plant constructors are usually within a few percentage points of each other for a given installation. The owner should use extreme care in accepting a low proposal from an unknown firm. Although the contractor may be sincere in his offer, inexperience can cause errors which later may result in an untenable situation.

Inexperienced owners sometimes make the error of assuming that any contractor is qualified to construct any plant if he has had construction experience. Contractors are also sometimes guilty of the same thinking. Before selecting an unknown contractor, the owner should be certain of the contractor's ability. He should obtain from the contractor names and qualifications of personnel who will be assigned to the contract, and investigate by visual inspection similar installations built by the contractor. Discussions with other owners are also advantageous. For the major project the owner should make every effort to determine the contractor's limitations, his financial background, the condition of his equipment, and his general business and labor relations. Financial status should also be considered in the selection of contractors for major projects since the contractor must usually make extensive expenditures before reimbursement is forthcoming.

Although contracts may be relatively simple or extremely lengthy, it is doubtful if the length or restrictive terms of a contract have much effect on the general type of work a contractor performs. The best insurance for the owner is the intent and integrity of the contractor, as proven by past performance. It can probably be shown that most of the problems between owners and contractors are the result of ineptitude on the part of either the owner's personnel or the contractor's personnel which lead to unresolved misunderstandings and finally become major issues. The contracting organization, just as any other group, is subject to the common frailties and human errors. Reliable firms, however, will always make every effort within reason to satisfy their customers since their reputation is at stake. Furthermore, unless the contractor is completely incapable, basic design and construction will always be within the limits of the various national codes of practice.

## THE BASIS OF A CONTRACT

Contracts may be entered into for the development, engineering, procurement, and construction of a process plant or for only one of these phases. Contracts may be between an owner and a contractor or between a prime contractor acting as an agent for an owner and many subcontractors. The bases for contractual agreements are practically unlimited. The following descriptions of various services are those generally used for process plant design and erection. All are assumed to be between an operator or *owner* and contracting firm or *contractor*.

1. Contractor as a consultant; contractor performs consulting or supervisory services only with engineering procurement and construction by others.
2. Owner to perform all engineering and procurement, contractor to perform erection services only; this may be varied to have erection performed by several contractors.
3. Owner to perform all process design furnishing data to contractor; contractor to perform all detailed engineering, procurement, and erection.
4. Contractor to perform primary research and development, engineering, procurement, erection, and initial operation: this is sometimes called a "turn-key" contract.

## TYPES OF REIMBURSEMENT (Contract Classification)

The classification of contracts in accordance with the mode of reimbursement is the most familiar method and the most important from a legal and business point of view. Contracts are, in fact, usually named according to the reimbursement policy. A brief description of the more important of these contract forms is given below. Reference to some detailed works in this field[2,3,4,5] will prove valuable to those requiring more information, but the need for legal advice on all such matters should be obvious.

### Cost-Plus Contract

The term cost-plus simply means that the contractor is to be reimbursed for all costs applicable to the contract plus a percentage of these costs for his overhead and profit. Usually the percentage varies with each type of contract and service. The reimbursed costs for engineering services may be as much as 215–220%, or more, of the cost of engineering labor. Other costs such as travel, communications, reproductions, clerical labor, are generally in the order of 105 to 115% of the actual cost. The higher percentage on engineering labor is based on the sale of contractor's know-how and the cost of the contractor's key personnel.

Construction costs include skilled and common labor wages at 110 to 125%, construction supervision charges on a similar basis as for engineering, and heavy construction tool charges based on fixed rental rates for machines. Actual cost plus a small percentage is usually charged for small tools and other expendable equipment.

If the contractor procures all plant materials and equipment he charges at a percentage between 105 and 115% of the actual purchase price to cover purchasing costs. If the owner pays for all personnel costs in procurement, the percentage based on the cost of materials or equipment may be much lower.

The usual criticism of the cost-plus contract is that since the contractor's profits increase directly with his costs, the owner has no control over the costs and the contractor accordingly increases his costs as much as possible. This idea is erroneous. Under the cost-plus contract the owner's auditors and engineers have continuous access to all the contractor's records. In addition the owner exercises authority over all the contractor's personnel and the activities related to the contract. In fact the contractor acts almost as a division of the customer's company.

For these reasons unless the owner's personnel are completely inexperienced there is little possibility for extraordinary cost excesses. The owner usually also retains full control of all commitments or disbursements, thus the owner actually is more responsible for costs than the contractor. If the owner is experienced in engineering and construction operations, there is little possibility for excessive costs because of the cost-plus feature.

Many of the larger process firms prefer the cost-plus contract to other types since it gives them much more freedom and control over the contractor. There is considerable advantage to the owner in the cost-plus contract when process design is incomplete or many changes are to be made in the original design. In reality, the cost-plus contract, if properly executed and if placed with a reliable contractor, can be as effective and as economical as the lump-sum, or guaranteed maximum or other types of fixed price contracts.

Experienced contractors, however, generally prefer a fixed-price arrangement, since this form of contract, after specifications are accepted, permits the contractor more freedom. In the case of a cost-plus operation a much stricter accounting of costs is necessary and the contractor's personnel are in some respects only in the employ of the owner. The cost-plus contract, however, is generally the only satisfactory method of operation when the work cannot be specifically anticipated or many construction problems exist which are difficult or impossible to estimate at the start of the job.

### Cost-Plus with Fixed-Fee Contracts

The cost-plus contract with a fixed fee is similar to the regular cost-plus form except that the contractor can only earn a specified fee. All other payments are reimbursement for direct costs only. There are many possible variations of the two cost-plus type contracts.

### Lump-Sum or Fixed-Price Contracts

The lump-sum contract allows nothing over or under a fixed single sum. The contractor agrees to perform certain services which are specified in detail by the owner in an inquiry or by the contractor in a proposal. If any changes are made by the owner after the contract is in effect, an adjustment may or may not be made, depending upon contract terms. If costs are increased, however, the contractor usually can demand an extra payment. Such contracts often cause bad relations between the parties because of conflicting interpretations of original specifications or drawings.

The fixed-price contract for the process plant requires the preparation of a detailed estimate. Preliminary data may be furnished by the owner or contractor. Owners sometimes request contractors to accumulate their own data for the proposal, or to employ other firms to prepare data for inquiry purposes.

The contractor's proposal or the owner's inquiry for the lump-sum or fixed-price job must be in much greater detail than for other types of contracts which have provisions for contingencies.

Like other forms, the lump-sum contract is subject to criticism. Since the contractor is limited to a fixed price, all possible savings increase his profit. Thus the contractor may be accused of effecting savings at the expense of quality. Generally, however, specifications for the various parts of the process plant are very complete, and there is little possibility for extensive change from the original proposal.

### Guaranteed Maximum Contracts

The guaranteed maximum cost contract is similar to the lump-sum contract except that the guaranteed price may apply to material (and equipment) only, or labor only. Escalation on either material or labor may be permitted or the contract may be written to exclude all escalation.*

Guaranteed maximum contracts are often written to allow the contractor a bonus depending on the amount of charges under the maximum. The owner participates to a considerable extent with the guaranteed-maximum contract since he must be continuously aware of all costs.

The contract types that have been discussed represent the most common types used for process plant projects. Many others which incorporate features from several of the common types are used.

## THE CONTRACT FORM

The most conventional type of contract consists of two parts. The first part is called the agreement. It includes an outline of the general scope of work, usually by references to a proposal or specifications and drawings, the contract price or terms of payment, and the necessary formal signatures witnessed and notarized. The signatures may sometimes be on the last page of the document. The second part consists of general conditions. Such items as completion date, cancellation, subcontracts, insurance, and many other terms are discussed. A third part may also be added to include exhibits, certifications, and special descriptions of certain items included in the second part.

Rather than attempt a detailed discussion of contract features, more can be learned by actually studying typical contracts. Two contracts typical of those used for cost-plus process plant projects are given below.

Contrary to popular belief, engineering contracts are not difficult to understand. They must necessarily be wordy so that no misunderstanding is possible. Many years of legal experience and court decisions have

* Escalation is a clause in a contract that permits a change in unit price for material, equipment, or labor after a specified fixed period of time. A contract signed in January, for instance, may allow increases in costs after that date up to a certain percentage of the original quoted price.

produced certain standard phrases which have proved immune to misinterpretation and, hence, are strong protection to both parties.

## Typical Cost-Plus Contract for Engineering Services Only

### I. AGREEMENT

This Agreement entered into this ________ day of ______________________, by and between XYZ Company, licensed to operate in ________________, hereinafter designated CONTRACTOR and PDQ Chemical Corporation, a Delaware Corporation, with its principal offices in Chicago, Illinois, hereinafter designated CUSTOMER.

WITNESSETH: That it is agreed by and between the parties hereto as follows:

*Article 1. Scope of Work*

The CONTRACTOR shall perform all the necessary engineering and design as required in the preparation of all drawings; the listing of all materials; the preparation of all specifications; in accordance with the CONTRACTOR'S proposal, attached hereto as Exhibit I.

*Article 2. Time of Completion*

The CONTRACTOR agrees to proceed with the work immediately and to diligently execute and perform the work with the utmost dispatch. Unless prevented by circumstances beyond his control as described below, Article 13*b* of Part II, General Conditions, the CONTRACTOR proposes to complete the work within twelve (12) months after start.

*Article 3. CONTRACTOR'S Compensation*

The CUSTOMER will reimburse the CONTRACTOR for all direct costs as shown below under Direct Costs, Articles 1, 2, 3, and 4, of Part II, General Conditions.

### II. GENERAL CONDITIONS

*Article 1. Direct Costs, Labor, and Laboratory Services*

*a.* Salaries and wages of all personnel engaged in the work, including, but not limited to, engineers, designers, checkers, draftsmen, tracers, material listers, vari-type operators, requisition typists, reproduction machine operators, stenographers and others of the CONTRACTOR'S personnel who engage in work on this specific contract.

*b.* Salaries and wages of CONTRACTOR'S Laboratory Technicians and direct costs, as approved by CUSTOMER for use of CONTRACTOR'S laboratory equipment and/or laboratory services.

*Article 2. Direct Costs, Materials*

*a.* Materials used in reproduction operations.

*b.* Rental costs for special equipment.

*Article 3. Direct Costs, Miscellaneous*

*a.* Communication costs including telephone, teletype, telegraph, cable, postage.

*b.* Traveling and living expenses of CONTRACTOR'S personnel when travel is requested and authorized by CUSTOMER.

*c.* Miscellaneous insurance costs incurred by CONTRACTOR solely as a result of this contract.

*d.* Consulting service costs, when authorized by CUSTOMER; as costs for labor, material or equipment rental.

*Article 4. Direct Costs, Plus Percentage*

The costs billed the CUSTOMER by CONTRACTOR for salaries and wages shall be increased by fixed percentages according to the schedule of costs under Exhibit II, except for overtime hours as discussed under Article 10*e*, Part II. All other costs shall be increased by a percentage according to the Schedule of Costs under Exhibit III.

*Article 5. CONTRACTOR'S Billing Procedure*

*a.* All invoices submitted to the CUSTOMER by the CONTRACTOR shall be substantiated by reference to specific payrolls for labor costs; and by reference to vendor invoices for materials.

*b.* Duplicates of insurance certificates shall be submitted for CONTRACTOR'S reimbursement of insurance premiums.

*c.* Expense accounts for employees shall be substantiated by hotel statements, receipts, travel ticket stubs, or other transportation receipts as proof of expense.

*d.* On or before the 5th day of each month, CONTRACTOR will furnish CUSTOMER an estimate (in quadruplicate) of total expenses on labor and material (listed separately) for the next two following months. One of these months shall be for the month in which the estimate is tendered. The estimate shall include manhours in the categories, engineering, drafting, and clerical with estimated total dollar cost itemized with manhours.

*e.* On or before the 5th day of each month, CONTRACTOR shall invoice CUSTOMER for costs incurred in the previous month. Invoices shall be submitted in quadruplicate and separately for each category of cost: labor, material, communications, travel expense, and insurance. All such costs shall be computed according to the Schedule of Costs under Exhibits II and III.

*Article 6. CUSTOMER Payments*

*a.* When found to be in accordance with the terms of this agreement CUSTOMER shall reimburse the CONTRACTOR for all billed costs, on or before the 5th day of the month following the month in which invoices are submitted by the CONTRACTOR.

*b.* The CUSTOMER, if he elects to do so, may withhold from payments to CONTRACTOR 5% of billed labor costs up to completion of contracted work and full acceptance by CUSTOMER of CONTRACTOR'S work. If the CUSTOMER elects to pay CONTRACTOR in full as invoiced, within the interval for payment specified above, CUSTOMER shall be allowed a reduction in billed labor costs of one-half (½) of one per cent (1%) on labor charges. All monies withheld by the CUSTOMER shall be due and payable thirty (30) days after receipt of final regular billing from CONTRACTOR and acceptance of work by CUSTOMER.

*Article 7. CONTRACTOR'S Labor and Cost Accounting Procedure*

CONTRACTOR shall maintain daily records of time expended by CONTRACTOR'S hourly rated personnel and weekly records of time expended by monthly salaried personnel. CONTRACTOR shall maintain separately records of all costs applicable to the contract including payrolls, material invoices, expense accounts, communication records, and any other records which may be required for substantiation of costs. All such records shall be kept in accordance with methods acceptable to CUSTOMER'S auditors, and be made available to the CUSTOMER for auditing as required. CONTRACTOR shall not be reimbursed for costs incurred in the preparation or maintenance of such records.

*Article 8. CONTRACTOR'S Estimates*

Estimates prepared by CONTRACTOR for CUSTOMER are understood to be CONTRACTOR'S best analysis of work to be done or completed during the next two (2) consecutive months as based on his normal schedule of the work. The CONTRACTOR does not represent that such estimates are to be considered exact with respect to amounts billed to the CUSTOMER.

*Article 9. CONTRACTOR'S Personnel*

*a.* The CONTRACTOR shall assign a project engineer or project engineers to the work acceptable to and as agreed upon with the CUSTOMER. Personnel so assigned shall not be removed from the work nor assigned other duties by the CONTRACTOR without approval from the CUSTOMER.

*b.* The CONTRACTOR shall provide the CUSTOMER with a list of all key personnel to be used on CUSTOMER'S work. The list shall include the qualifications of each person named.

*c.* The CUSTOMER may at any time upon written request to the CONTRACTOR remove any assigned CONTRACTOR'S personnel from CUSTOMER'S work.

*d.* The CUSTOMER reserves the right to control CONTRACTOR'S assignment of personnel, below the level of key design personnel with respect to the number of persons used in CUSTOMER'S work at any time. The CONTRACTOR shall reduce his personnel as requested within one working day after receipt of notice from CUSTOMER for personnel employed in CONTRACTOR'S home office. Salaries or wages of CONTRACTOR'S personnel on assignment for CUSTOMER away from CONTRACTOR'S home office shall be to CUSTOMER'S account up to time of return to CONTRACTOR'S home office.

*Article 10. Travel Expenses, Holidays, Vacation Periods, CONTRACTOR'S Personnel*

The CUSTOMER will reimburse the CONTRACTOR for out-of-pocket expense of personnel living away from home, traveling, or otherwise undergoing expense at the request of the CUSTOMER, up to the amounts described in Schedule of Costs, Exhibit III. The travel time for such CONTRACTOR'S personnel shall be from time of departure from CONTRACTOR'S home office city to time of return to point of departure.

*Article 11. Holidays, Vacation Periods—CONTRACTOR'S Personnel*

*a.* The CONTRACTOR will absorb all costs in connection with paid holidays, or vacation periods normally allowed CONTRACTOR'S personnel employed on CUSTOMER'S work when such periods fall within the duration of this contract.

*b.* The CUSTOMER may request of the CONTRACTOR that certain personnel be worked over week-ends or on holidays. When mutually agreed with the CONTRACTOR the CUSTOMER shall reimburse the CONTRACTOR for such costs according to the personnel policies of the CONTRACTOR, and such payments shall be handled according to overtime or premium payments.

*c.* The CUSTOMER may request the CONTRACTOR to postpone vacation periods for certain personnel; however, the CUSTOMER will not be responsible for vacation period payments to CONTRACTOR'S personnel in the event an employee's vacation is postponed, unless and except mutual agreement is reached for specific employees between CONTRACTOR and CUSTOMER.

*Article 12. Employment of New Personnel, Severance, and Termination Payments, CONTRACTOR'S Personnel*

*a.* All CONTRACTOR'S personnel shall be considered to be in the employ of the CONTRACTOR and the CUSTOMER accepts no responsibility for payments to such personnel according to CONTRACTOR'S policies if terminated by the CONTRACTOR.

*b.* Where CONTRACTOR employs personnel according to CONTRACTOR'S estimate of required work, CUSTOMER assumes no responsibility for duration of such work nor in any way accepts any obligation to CONTRACTOR or personnel as to length of work or employment.

*c.* CONTRACTOR shall have complete freedom as to employment of personnel according to his past policies, except CUSTOMER will not accept increases in personnel labor rates over average rates shown for various categories of personnel in schedule of costs, Exhibit II.

*d.* Changes in rates of CONTRACTOR'S key personnel, if mutually agreed upon between CUSTOMER and CONTRACTOR, may be accepted by CUSTOMER; however, the CUSTOMER reserves the right to reject any and all re-classifications, rate changes, or any changes in rate structure, which would tend to increase costs.

*Article 13. Completion of Work, Overtime*

*a.* CONTRACTOR shall make every effort to complete work as early as possible. CUSTOMER shall not hold CONTRACTOR responsible for work hold-up due to lack of data to be supplied by or through efforts of CUSTOMER. CONTRACTOR accepts all responsibility for accumulation of data or the expediting of data from manufacturers for equipment listed in proposal under Exhibit I, as developed and/or designed by CONTRACTOR.

*b.* In the event actions beyond the CONTRACTOR'S control, strikes, fires, or any other occurrences not the result of CONTRACTOR'S actions affect CONTRACTOR'S work to seriously delay completion of the work, this agreement may be cancelled by mutual consent of both parties.

*c.* CONTRACTOR may work certain personnel overtime hours if it is mutually agreed upon by both parties that such work is necessary. CONTRACTOR will not be reimbursed for excess manhours nor overtime pay to personnel where prior consent is not obtained from CUSTOMER; however, CONTRACTOR may, if he so desires, work personnel overtime hours up to a maximum of 10% of normal working hours, with such excess hours and premium costs charged to his own account.

*d.* CONTRACTOR agrees to work personnel overtime hours as requested by CUSTOMER up to a maximum of 20% in excess of normal hours per week per employee. The CONTRACTOR shall be reimbursed by CUSTOMER for all resulting costs for overtime work.

*e.* Plus-percentages shown in Schedule of Costs, Exhibit II, for salaries and wages shall not be applied to premium payments to CONTRACTOR'S personnel for overtime hours.

*Article 14. Cancellation of Contract*

*a.* CUSTOMER may cancel this agreement at any time upon giving the CONTRACTOR written notice ten (10) working days in advance of such action.

*b.* In the event of cancellation, the CUSTOMER will allow the CONTRACTOR a reasonable period of time to prepare records or any of the data involved for transfer to CUSTOMER. The period of time required for turning over such records or data shall in no case exceed thirty (30) days from date of notice of cancellation.

*c.* The CUSTOMER will reimburse the CONTRACTOR for salaries and wages of mutually agreed upon personnel who may be engaged in any of the work after notice of cancellation through the above thirty-(30) day termination period.

*d.* The CONTRACTOR shall also be reimbursed for other reasonable costs such as may be expected for materials or labor in connection with the transfer of data to CUSTOMER.

*e.* Plus-percentages shown in Schedule of Costs, Exhibit II, for salaries and wages shall not be applied to labor costs for manhours expended after CONTRACTOR gives or receives notice of cancellation.

*f.* CONTRACTOR may cancel contract by (1) giving CUSTOMER written notice of such action thirty (30) days in advance and (2) in the event only of failure on the part of CUSTOMER to fulfill his part of this agreement. Such failure may be considered to be in effect if the CUSTOMER refuses to reimburse the CONTRACTOR promptly, or fails to reimburse the CONTRACTOR after receipt of CONTRACTOR'S invoices within the agreed period of time.

*g.* In the event of cancellation by the CONTRACTOR, all forms shall apply as written under this Article II as applicable for cancellation by the CUSTOMER, except that the cancellation notice shall be given to the CUSTOMER by the CONTRACTOR thirty (30) days in advance of such action.

*Article 15. CONTRACTOR'S Drawings, Calculation Sheets, Data*

*a.* All drawings, sketches, tracings, reproductions of any drawing or tracing used in the work are the property of the CUSTOMER and shall be turned over to the CUSTOMER at any time upon receipt of written demand by the CUSTOMER.

*b.* All calculations, copies of calculations, rough drafts, or other data used in the work prepared by the CUSTOMER for use of the CONTRACTOR or prepared by the CONTRACTOR are the property of the CUSTOMER and shall be turned over to the CUSTOMER at any time upon receipt of written demand by the CUSTOMER.

*c.* All material lists, bills of material, or other relative data are the property of the CUSTOMER and shall be turned over to the CUSTOMER upon receipt of written demand by the CUSTOMER.

*d.* Cancellation of the contract by the CUSTOMER or the CONTRACTOR shall constitute a request for the return of all data covered in the above paragraphs *a*, *b*, and *c* of this Article 15, of Part II, of this agreement and the CONTRACTOR shall proceed immediately accordingly except as follows under *e*.

*e.* In the event the CUSTOMER fails to reimburse the CONTRACTOR within sixty (60) days after presentation and receipt of CONTRACTOR'S invoices CONTRACTOR may withhold all data, drawings, and other pertinent parts of the work described in this Article 15 up to such time as CUSTOMER meets his obligations or some mutual agreement is reached by both parties.

*Article 16. Insurance*

The CONTRACTOR shall provide insurance coverage in accordance with instructions from the CUSTOMER: (*a*) for all records or data, or any of the CUSTOMER'S property located on the CONTRACTOR'S premises or in his possession; (*b*) for any automobile used in CUSTOMER'S work; (*c*) for any employee of the CONTRACTOR traveling at the request of the CUSTOMER.

The CUSTOMER agrees to reimburse the CONTRACTOR for all such insurance premiums.

*Article 17. Patents, Royalties*

*a.* The CONTRACTOR agrees to save the CUSTOMER harmless from all or any costs arising from patent infringements in any design utilized by the CONTRACTOR except for process design initiated by the CUSTOMER or design in accordance with CUSTOMER'S explicit instructions.

*b.* The CONTRACTOR will immediately bring to the CUSTOMER'S attention designs requested by CUSTOMER which in CONTRACTOR'S opinion would constitute infringement.

*c.* The CUSTOMER will negotiate for and be responsible for the payment of all royalties, fees, assessments, or other costs arising out of use of licensed processes or methods or materials the use of which require licensing.

*Article 18. CONTRACTOR'S Qualification, Engineering Design*

*a.* The CONTRACTOR represents that he is a qualified designer of the equipment and processes for which he accepts responsibility in Exhibit I, Proposal.

*b.* The CONTRACTOR accepts full responsibility for all basic mechanical, structural, and electrical design methods used in engineering work performed by him.

*c.* The CONTRACTOR accepts full responsibility for process designs included in Exhibit I.

*Article 19. Guarantee, CONTRACTOR'S Liability*

*a.* The CONTRACTOR guarantees all equipment basically designed by him against failure from faulty design for a period not to exceed one (1) year from date of first use in operation, when such equipment is erected and used in the service for which it was designed and intended mechanically or process-wise. In the event equipment is not put into operation the CONTRACTOR'S time guarantee is limited to eighteen (18) months from date of completion of work or from receipt of equipment at point of erection, whichever may first occur.

*b.* The CONTRACTOR'S guarantee is limited to equipment designed by himself. Equipment designed or manufactured by others is limited by the guarantee of the designer and/or manufacturer, except that the CONTRACTOR will prepare specifications for such equipment to obtain the best possible guarantee in every respect for the CUSTOMER'S best interest.

*c.* The CONTRACTOR'S liability in the event of mechanical failure of equipment designed by himself shall be limited to replacement or repair of any such equipment without cost to CUSTOMER.

*d.* The CONTRACTOR'S liability as to process production shall be limited to the costs of changes or modifications in the drawings or specifications as required to meet the process requirements in accordance with the CONTRACTOR'S proposal attached hereto as Exhibit I.

*e.* The CONTRACTOR will not be held liable for CUSTOMER'S loss-of-profit or for cost of raw material, operating labor, and other production costs in the event of process failure.

*f.* The maximum liability with respect to costs absorbed by CONTRACTOR under the above paragraph *d* of this Article 19 shall not exceed fifty (50) per cent of the total of sums paid the CONTRACTOR as fees and/or profit over and above direct costs according to the schedules of costs under Exhibits II and III.

*Article 20. Assignment of Contract*

Neither party shall assign this Agreement in whole or in part without the consent of the other.

The above paragraphs include most of the terms ordinarily found in contracts for engineering services. The policies of the organizations involved and scope of the contract affect the terms and conditions.

One of the bases for any agreement must be some statement of intent on the part of both parities. If either party can show that the intent of the other was fraudulent an agreement can usually be dissolved. Another requirement of any contract or agreement must be an understanding or "meeting of the minds," as it is sometimes called, which is essentially understanding between the parties of their obligations.

Generally most contracts include statements, such as the following, to indicate the general understanding and intent.

*Article 21. Intent*

The CONTRACTOR and the CUSTOMER mutually agree that these documents form the basis of a working agreement whereby the CONTRACTOR is to perform certain services diligently and as economically as possible and the CUSTOMER is to reimburse the CONTRACTOR for all normal and reasonable costs plus a percentage which constitutes the CONTRACTOR'S overhead and profit. It is further agreed that neither CONTRACTOR nor CUSTOMER shall profit from errors in nor omissions from these documents resulting in interpretations different from the basic intent.

*Article 22. General Understanding*

CONTRACTOR and CUSTOMER, by execution of these documents, represent each that they have examined the contents in detail and understand fully the scope of the obligations herein and are in complete agreement with all the terms and conditions.

## Exhibits

Reference was made in the contract to several exhibits. The following will serve to illustrate or outline the more important requirements.

### *Exhibit I—Proposal*

A proposal for an engineering contract such as this is usually divided into two parts consisting of (*a*) Process and (*b*) Mechanical specifications.

Process specifications describe the process in detail including raw material and finished product specifications, heat and material balances, and necessary flow diagrams to describe the process graphically. If parts of the process are designed by others, the proposal should clearly state the responsibilities.

The mechanical specifications list and describe all equipment by means of sketches and written descriptions. General specifications for purely mechanical equipment, such as pumps or compressors, are included in sufficient detail to enable manufacturers to estimate prices. If possible manufacturer's names and catalog numbers are included so that identification is fixed although the specificiations may not be complete in all details. Preliminary plot plans showing the scope of work, sketches of pressure vessels, and specification sheets for heat-exchange equipment, as well as specifications for various parts of the work such as the structural, piping, electrical, and instrumentation phases are included.

Most of the specifications included in a proposal are not complete since a considerable part of engineering which will follow upon signing of the contract consists of the development of specifications. It is not uncommon for a CUSTOMER to prepare complete detailed specifications or have them prepared by an engineering firm for the CUSTOMER'S use in obtaining bids from other firms.

The degree of completeness of specifications usually depends upon the terms of the contract or types of bids or proposals to be obtained. For cost-plus contracts, such as the one described above, mechanical specifications may be brief while fixed-price contracts require detailed specifications. If complete specifications are supplied by the customer, the phraseology in the beginning of the contract is somewhat different. In the above contract, for instance, Article 1 of Part I would be changed, where the words "CONTRACTOR'S proposal" appear, to read "Specifications."

### *Exhibit II—Schedule of Costs, Labor*

| Personnel | Salary or Wage Range | Billed Plus Percentage |
|---|---|---|
| 1. Project manager | 600–700/month | Paid wages plus 150% |
| 2. Project engineer | 500–600/month | Paid wages plus 150% |
| 3. Specialty engineer | 500–600/month | Paid wages plus 150% |
| 4. Design engineer | 500–600/month | Paid wages plus 125% |
| 5. Section leader | 3.15–3.45/hour | Paid wages plus 125% |
| 6. Designer & layout | 3.00–3.25/hour | Paid wages plus 125% |
| 7. Senior draftsman | 2.85–3.15/hour | Paid wages plus 110% |
| 8. Draftsman | 2.50–2.85/hour | Paid wages plus 110% |
| 9. Tracer, B. P. machine operator | 1.50–1.75/hour | Paid wages plus 50% |
| 10. Specification writers | 2.85–3.15/hour | Paid wages plus 110% |
| 11. Material lister | 2.65–2.85/hour | Paid wages plus 100% |
| 12. Variable-typewriter operator | 1.25–1.50/hour | Paid wages plus 100% |
| 13. Typist-specifications or bills of material | 1.15–1.25/hour | Paid wages plus 50% |
| 14. Stenographic-clerical | 1.10–1.50/hour | Paid wages plus 50% |
| 15. Procurement personnel, buyers, clerical | 1.35–1.95/hour | Paid wages plus 50% |
| 16. Laboratory technicians | 2.85–3.15/hour | Paid wages plus 125% |

For all other personnel in engineering categories receiving monthly salaries reimbursement shall not be less than paid salary plus 110%. For all other personnel in design, drafting, engineering, or clerical operations on weekly or hourly basis reimbursement shall not be less than paid salary plus 75%.

CONTRACTOR'S personnel specifically excluded from above categories, or personnel whose salaries or other costs shall be absorbed in CONTRACTOR'S fee or profit are: executive officers, department heads, estimators, office managers, accountants, billing clerks, secretaries to executives, telephone operators, receptionists, elevator operators, messengers, and janitors. With the exception that where CUSTOMER specifically requests the services of such personnel, mutual agreement shall be reached on the percentage of salary or plus percentage which shall be reimbursed to CONTRACTOR.

### Overtime Payments

Salaried employees, or employees not paid on an hourly basis may receive overtime payment for hours worked up to a maximum of 30% of normal monthly salary. Such overtime is calculated on the basis of hours over 8 per day or 40 per week. The rate per hour for monthly salaried personnel is computed by multiplying the fixed monthly salary by a factor, 0.006. This is essentially equal to the nominal hourly rate the CONTRACTOR shall be reimbursed for such overtime pay on the basis of rates noted above with respect to the calculation of plus percentages. Hourly rated, weekly paid employees are paid for all hours worked over 8 per day or 40 per week, at a rate of 1½ times their fixed hourly rate. CONTRACTOR shall be reimbursed for all premium wages paid such employees; however, the above plus percentages shall be applied only to hours worked at the fixed hourly rates. As an illustration, assume a person in category 8 earning $2.50/hour worked 44 hours. CONTRACTOR'S billing to CUSTOMER would be 46 × 2.50 or $115.00 plus (44 × 2.50 × 1.10) or $236.00.

### *Exhibit III—Schedule of Costs Material or Services*

1. *Communications.* Actual costs of telegrams, cables, teletype, local (where chargeable) and long-distance telephone calls, plus 10%. Equipment use or rental costs shall not be included in costs to CUSTOMER. CONTRACTOR shall maintain accurate records of all such costs in accordance with CUSTOMER'S requirements.

2. *Postage, Parcel Post, Air Mail, Air Parcel Post.* Actual costs of all such charges plus 10%. Records shall be kept as above for 1.

3. *Telephone Service.* Where CUSTOMER requests telephone service or installation of telephones or other communication equipment, CONTRACTOR shall be reimbursed for all such costs plus 10%.

4. *Office Space and Equipment.* CONTRACTOR agrees to furnish office space together with all necessary office furniture without charge, for two (2) of the CUSTOMER'S engineers. Secretarial or stenographic assistance labor shall be reimbursable in accordance with item 14 in Exhibit II. Such personnel assigned to CUSTOMER shall be understood to be in the employ of the CONTRACTOR and shall observe all common rules and practices as required by CONTRACTOR.

5. *Supplies.* All necessary supplies, except as included below under 6, required for the work shall be furnished by the CONTRACTOR and no costs for such shall be billed to the CUSTOMER.

6. *Reproduction Services.* CONTRACTOR shall be reimbursed for all costs incurred in preparing or printing, blueprints, blue-line prints, photocopies, ditto, mimeograph, either when performed by himself or by an outside firm for CUSTOMER'S requirements. Such costs shall be in accordance with standard local

commercial practices plus 5%. Labor costs for blueprint or other machine operators shall be reimbursed in accordance with item 9 of Exhibit II, when such personnel are worked overtime at CUSTOMER'S request. When CUSTOMER elects, he may accept charges for all materials used in reproduction and reimburse CONTRACTOR for labor used in these operations as noted above in item 9 of Exhibit II. CONTRACTOR shall be reimbursed for costs of all materials used plus 5%.

7. *Consultation Service.* CONTRACTOR shall be reimbursed for any services required of outside firms as agreed upon with CUSTOMER. Such costs shall be billed to CUSTOMER plus 5%.

8. *Laboratory Service.* Where CONTRACTOR provides laboratory services, costs for such services shall be agreed upon prior to start of work. Such costs shall include all charges and no plus percentages will be added to the agreed upon cost. Where CUSTOMER requests special laboratory service, billed rates for personnel shall be computed according to item 16 of Exhibit II.

9. *Travel Costs.* CUSTOMER will reimburse CONTRACTOR for out-of-pocket travel costs of CONTRACTOR'S personnel travelling at CUSTOMER'S request up to a maximum of $12.00 per day for employees who remain in a single location for 30 days or more continuously. For personnel making overnight or short trips the rates shall be reasonably greater; however, in no case shall living expenses billed to CUSTOMER exceed $15.00 per day.

Allowed expense items shall include payments for first class railroad fares plus pullman, air travel, bus, or taxi or any public carrier fares, laundry, and other normal living expenses. Receipts for expense items, hotel bills, ticket stubs, and any other substantiation of payment shall be obtained as far as possible and held on file by CONTRACTOR for CUSTOMER'S inspection when requested. Fixed rates specified above shall apply to travel within the continental limits of the U. S. For foreign travel per diem rates shall be adjusted to meet local conditions.

Telephone or other communication costs shall be reimbursable; however the employee shall use collect calls if possible so that records may be kept of these costs.

Automobile travel costs when insurance requirements are met shall be reimbursable at $0.10 per mile. All costs incurred by CONTRACTOR for travelling expense shall be reimbursable according to the above plus 5%.

## Typical Engineering and Construction Contract

The complete engineering and construction contract is usually similar to the form of contract for engineering services except that it includes additional clauses under Part II.

The following contract is similar to those prepared for cost-plus with fixed-fee contracts. The first page of the document consists of a title:

Agreement between
ABC Chemical Company and XYZ Engineering Company
Including Engineering, Procurement, and
Construction Services
At The Colorado Plant

The second page, or pages, consists of an index similar to that following. Indexes are most helpful for the longer contracts.

## INDEX

General Agreement.
Definitions.
Article I. Scope of Work.
Article II. Contractor's Costs and Terms of Payment.
Article III. Direct Costs.
Article IV. Consumable Construction Supplies, Field Purchasing.
Article V. Construction Materials and Tools, Field Purchasing.
Article VI. Construction Labor.
Article VII. Transportation.
Article VIII. Completion of Work.
Article IX. Cancellation of Contract.
Article X. Protection of Materials and Property.
Article XI. Insurance.
Article XII. Guarantees.
Article XIII. Assignment of Contract.
Article XIV. Surplus Materials.

Exhibit I. Proposal.
Exhibit II. Estimate.
Exhibit III. Engineering and Other Labor.
Exhibit IV. Construction Material and Labor.

## GENERAL AGREEMENT

This Agreement made and entered into this_______day of_____________ by and between ABC Chemical Co., (address), hereinafter referred to as ABC and XYZ Engineering and Construction Co., (address), hereinafter referred to as CONTRACTOR, witnesseth that:

Whereas ABC desires CONTRACTOR to perform certain services more completely described in the hereto attached, listed, or referenced drawings and specifications under the Scope of Work Article I;

Now therefore, for and in consideration of the payments hereinafter specified to be made to the CONTRACTOR by ABC, and in consideration of the terms herein included the parties hereto do agree as follows:

## DEFINITIONS

1. *Contract Documents.* All papers, drawings, documents attached hereto or included herewith by reference constitute the contract documents and, if separate hereto, are to be considered as part or parts of this agreement the same as if attached hereto.

2. *Contractor.* Party to agreement, XYZ Engineering Company, its executive officers, project engineer, construction superintendent.

3. *ABC.* Party to agreement, ABC Chemical Co., its executive officers, project engineer, inspectors, field construction engineers.

4. *Place of Work, Construction.* The expressions "jobsite," or "field" are understood to mean the place of erection of facilities. Albert County, State of Colorado, ABC Chemical Co., Colorado Plant. (Mail address should be included.)

5. *Place of Work, Engineering.* Contractor's home or branch offices. (Mail address or addresses should be included.)

6. *Work.* The term work shall be understood to include labor or material or both.

7. *Local Laws.* The laws of the place or building shall govern.

8. *Jobsite.* The term jobsite shall be understood to mean place of erection or construction as designated above in 4.

9. *Drawings.* Drawings or prints shall be understood to mean reproductions from drawings prepared by either party but identified by written matter as a part of the work (see 6 above). Tracings shall be understood to mean original drawings from which reproduced copies or prints have been obtained. Drawings may be said to be identified when their identifying numbers are included in the contract documents or addenda thereto.

10. *Supplies, Construction Materials, and Materials.* (*a*) Supplies are understood to mean those materials of a consumable nature exclusive of tools, such as emery cloth, sandpaper, drinking cups, clerical supplies, fuels, lubricants, and any other such items which are necessary for construction but do not necessarily enter into the permanent plant or remain as a part of the plant after construction is complete; (*b*) construction materials are the heavier materials used in construction such as scaffolding, form lumber, form ties (but not reinforcing rod); construction buildings together with all their parts including electrical lighting, plumbing, heating or ventilating equipment; temporary electrical service connections for electrically driven construction equipment, miscellaneous piping, and miscellaneous bulk materials of all kinds used for construction but removed after construction is completed; (*c*) materials are generally understood to mean permanent materials such as the component of concrete, or stone, sand, and cement; brick, mortar, miscellaneous fill materials, welding rod, rivets, miscellaneous bolts.

11. *Construction Tools.* (*a*) Small tools, consumable: screwdrivers, pliers, pipe and other wrenches less than eighteen in. (18″) in length; hack saws, hack saw blades, chisels, hammers, welding sets, welding (oxyacetylene) hoses, etc.; (*b*) hand tools: hand tools are generally understood to mean the larger tools such as axes, shovels, spades, wheel barrows, cement buggies, dollies, sledge hammers, etc.; (*c*) driven tools: electric or pneumatic, drills, grinders, hammers, rivet-guns, caulking tools, impact wrenches, etc.; (*d*) machine tools, small: portable electric or pneumatic hoists, pipe and conduit bending and threading machines, portable air compressors, welding machines, etc.; (*e*) construction tools, large: steam or electric driven hoisting engines, guy, and other types of derricks, including cables, sheaves, hooks, slings, selfdriven cranes, grading machines, tractors, trucks including all their appurtenances. Heavy machine tools installed (fixed base) for use during construction such as lathes, drill presses, metal saws, etc.

12. *Security.* The word security shall be understood to mean ABC's plant security or protection organization, or the organization provided by CONTRACTOR to police the construction areas.

13. *Premises.* Premises shall be understood to mean ABC's property either the immediate construction area or adjacent, or other areas owned by ABC or used by ABC.

14. *Trade Names, Terminology.* Where trade names are used in descriptions or specifications, they shall be as applicable as formal descriptions. It is understood that the contractor is experienced in the type of work to be performed and expressions common to the trade shall apply and be acceptable.

15. *Equipment.* Where the word equipment is included herein it is understood to mean process or permanent plant equipment as installed or erected or purchased for use of ABC.

*Article I. Scope of Work*

The CONTRACTOR shall furnish all necessary designs, instructions, drawings, specifications, labor, materials, and equipment as are required to procure, fabricate, erect, and install, ready for use or operation, all equipment to provide the facilities and processes herein described in accordance with CONTRACTOR'S proposal attached hereto as Exhibit I.

*Article II. CONTRACTOR'S Costs and Terms of Payment*

1. ABC shall reimburse CONTRACTOR for all costs indicated as direct costs as described below plus the percentages indicated for the various parts of the work.

2. CONTRACTOR'S fee shall be computed as a summation of the plus percentages added to CONTRACTOR'S direct costs and shall not exceed eight hundred, and fifty thousand dollars ($850,000.00), except as noted below.

3. Estimated Selling Price: the CONTRACTOR estimates the costs of labor and material and equipment to complete the facilities described in Exhibit I, plus the above fee to be not more than eight million, six hundred and twenty-five thousand dollars ($8,625,000.00) in accordance with the detailed estimate included herein as Exhibit II. CONTRACTOR does not guarantee estimate but agrees that the fee in item 2 above shall be a maximum when the estimated total cost is equal to or exceeds the amount above.

4. Fee Increase or Change. The CONTRACTOR'S fee in item 2 above shall be increased or changed only. (*a*) When and if the above estimated total selling price exceeds the actual costs of the work, the CONTRACTOR'S fee shall be increased by an amount equal to twenty per cent (20%) of the difference between CONTRACTOR'S estimate and actual total cost. This increase shall be understood to be a bonus payment; (*b*) when and if the scope of work herein shall be increased by ABC by additions or changes increasing CONTRACTOR'S work or equipment or materials in mutually agreed upon changes, the CONTRACTOR'S fee shall be increased by an amount equal to ten per cent of the actual costs of such increases or changes. All such extras shall be kept separate from total costs of original estimate and shall not affect the bonus payment above in section *a*; (*c*) when and if the scope of work herein shall be reduced by ABC as changes or modifications, the actual cost reduction shall be computed as a percentage of the above estimated selling price item 3 and CONTRACTOR'S fee shall be reduced by a like percentage. The maximum reduction in CONTRACTOR'S fee due to reduction in scope of work shall not exceed one hundred and eighty-five thousand dollars ($185,000.00), except in the event of cancellation under Article IX.

5. Accounting and Billing Procedure.

(*Note:* See Articles 5 and 6 of the Example Engineering Services Only Contract. The billing procedure for any contract where owner pays direct costs plus a percentage consists in presentation of the contractor's invoice plus substantiating date in the form of attached papers or reference to the contractor's records. Generally for labor charges copies of actual payrolls must be included. Usually papers such as vendors' invoices are held by contractor since obtaining sufficient copies for owner as well as contractor is sometimes difficult. In order to prove an expenditure for equipment the contractor ordinarily has three instruments: (1) contractor's purchase order, (2) a field receiving report, and (3) a copy of the vendor's invoice. These are sufficient in most cases since if the contractor acknowledges that he has the vendor's invoice he is committed to pay it. It is possible, of course, to go one step further and obtain a cancelled check, but this is not ordinarily done since the contractor will usually wish to retain his cancelled checks as his own positive record.

Billing on an engineering and construction contract is similar to that for an engineering contract, except emphasis is placed on all labor rather than engineering labor. The amount withheld by owner would probably be 10% rather than 5% as shown for the engineering contract. Usually some fixed amount is billed as the engineering fee with each regular billing. The time of billing may be monthly, bi-monthly, or continuous as invoices are accumulated by contractor. The amount withheld may be as much as 25%, but 10% is the usual amount.)

*Article III. Direct Costs*

Direct Costs; direct costs shall be limited to the following: (*a*) engineering costs; all labor costs for all personnel engaged as engineers or used in engineering services as described in Exhibit III, plus fixed percentage of 125%; (*b*) all costs of process plant or other equipment purchased by CONTRACTOR for installation at the jobsite plus a fixed percentage of 6%; (*c*) construction material and labor costs; all labor costs for construction personnel and all costs of construction supplies or materials as described in detail in Exhibit IV, plus a fixed percentage of 8%.

*Article IV. Consumable Construction Supplies, Field Purchasing*

(*Note:* When the contract is essentially cost-plus or when the CONTRACTOR is reimbursed for all construction costs, the precise differentiation between consumable construction supplies is relatively unimportant. A clause similar to the following is sometimes included:

It is understood that all materials or supplies (any materials) purchased by the contractor for the work become the property of ABC.

This clause indicates that all or any materials or equipment purchased (and delivered to the jobsite), including small tools, become the property of ABC and therefore may not be removed from the jobsite by the CONTRACTOR. The word purchased also tends to separate those items which are the property of the CONTRACTOR from ABC's property.

Consumable supplies are defined to some extent under Definitions. The following clause would apply here.)

All minor materials and supplies of a consumable nature required for the work shall be purchased at the jobsite by the field-buyer. In general all materials and equipment shall be purchased by the CONTRACTOR'S procurement organization in the CONTRACTOR's home office, however certain materials may be more expeditiously purchased locally. All such field-purchased materials shall be recorded by means of field requisitions and purchase orders and properly specified. Wherever possible such materials shall be charged against a specific part of the work or costs allocated to several parts. The receipt of all such materials on the jobsite shall be substantiated by completed receiving reports signed by an authorized employee of the CONTRACTOR. The CONTRACTOR shall maintain proper indexes and cross-indexes to facilitate the work of ABC's auditors. Purchases in excess of two hundred dollars ($200.00) shall require prior approval from (*a*) the CONTRACTOR'S home-office purchasing agent, and (*b*) from the ABC's representative located at the jobsite. It is agreed that prior blanket approvals may be obtained for bulk materials such as sand, gravel, stone, ready-mixed concrete, fuels, lubricants, and other similar materials, provided CONTRACTOR has obtained competitive prices from suppliers.

*Article V. Construction Materials and Tools, Field Purchasing*

1. Materials used in construction such as foundation and form materials or equipment shall be specified and listed so far as is practicable in CONTRACTOR'S home office. When such materials are purchased in large bulk quantities, the major or blanket orders shall be placed from the CONTRACTOR'S home office after competitive bidding.

(*Note:* This does not necessarily conflict with the statement under Article IV since major orders for total requirements of cement or stone may be placed on a cubic yard basis or by railroad car lots. Notification of readiness for delivery of such material may best be issued from the field. When essentially all concrete is delivered to the jobsite ready-mixed, blanket orders for price basing may be placed by the CONTRACTOR'S home office or a field office but notification to deliver must be issued only by the field office.)

2. Tools owned by CONTRACTOR used in the work will be billed to the PURCHASER on a rental basis. Rates for such tools shall not exceed rates for equal equipment available in the area.

3. Rental rates for special equipment shall be set prior to use by mutual agreement of the parties hereto.

4. ABC may at any time question the use of equipment and request its removal from ABC's premises, or to a storage point.

5. ABC agrees to pay all freight and other charges incidental to bringing equipment to jobsite from points within the continental U.S.

6. All materials and equipment except CONTRACTOR'S tools shall be understood to be the property of ABC when received at the jobsite but it shall be the responsibility of the CONTRACTOR to protect and maintain all such materials or equipment up to time of formal acceptance by ABC. The CONTRACTOR shall erect temporary storage sheds or warehouses, or provide shelter as may be required to protect all such material or equipment.

7. The CONTRACTOR shall employ sufficient watchmen, whose qualifications shall be acceptable to ABC to provide plant protection during the construction period.

8. Fire protection practices shall be followed by CONTRACTOR and regular practice fire drills shall be held. Storage points for flammable materials or liquids shall be isolated and arranged as agreed upon between the parties hereto. The transportation, handling, and storage of all flammable or hazardous materials shall be in strict accordance with ABC's instructions titled General Plant Safety Practices.

*Article VI. Construction Labor*

1. The CONTRACTOR agrees to comply with all local and federal rules,

regulations, orders, directives, or statutes applicable to wage and employment practices, and to act in accordance with the policies of ABC on matters which affect area practices and which might tend to set precedents. Construction labor wage rates in Exhibit IV form the basis for CONTRACTOR'S estimate in Exhibit II. The CONTRACTOR will not enter into agreements with various labor organizations employed in the work which tend to increase such rates without previous notification to ABC.

2. The CONTRACTOR will maintain suitable temporary sanitary and shelter facilities for use of its employees. CONTRACTOR'S employees shall not trespass on nor use sanitary or other of ABC's plant facilities.

3. The CONTRACTOR will employ a competent registered nurse and provide suitable examination and first-aid facilities continuously at the jobsite during construction. Provisions will also be made for prompt medical aid and hospitalization in the nearest facility for such services.

4. CONTRACTOR'S employees when on ABC's premises shall comply with all general rules and regulations as required by ABC. CONTRACTOR'S employees shall not leave CONTRACTOR'S work area nor visit other parts of ABC's premises without permission, or necessary authority, unless in the line of duty and/or when accompanied by an employee of ABC.

5. Disregard for rules and practices as required by ABC by an employee of the CONTRACTOR shall be sufficient justification for dismissal.

6. Employees discharged for cause by CONTRACTOR or by ABC's request shall not be re-employed by CONTRACTOR for work under this agreement.

7. CONTRACTOR'S employees shall be processed for identification and be provided with identification badges acceptable to ABC. The processing of CONTRACTOR'S personnel to meet security regulations shall be carried out in strict accordance with instructions from ABC. The CONTRACTOR will maintain at the jobsite a separate file on each employee regularly employed in CONTRACTOR'S work or intermittently visiting the jobsite. All employees visiting the jobsite shall follow instructions with regard to security and identification the same as if regularly employed at the jobsite. Any of ABC's security officers may at any time request proper identification and/or the employment history of any personnel.

8. Certain of CONTRACTOR'S personnel may be permitted to bring personal automobiles within ABC's plant premises. Infractions of rules or regulations with regard to use of automobiles will automatically revoke any such permit. In general all construction personnel shall use the ABC plant gate designated for CONTRACTOR'S entrance. When other entrances are used CONTRACTOR'S personnel must obtain a special pass from security.

9. Vendor's representatives or other persons visiting CONTRACTOR'S offices for business purposes shall comply with ABC's plant regulations with regard to visitors and must be identified by delegated CONTRACTOR'S personnel. The CONTRACTOR shall be responsible to ABC for such persons. Unless special permission is obtained such persons are not to be allowed access to any part of the work except CONTRACTOR'S offices.

*ARTICLE VII. Transportation*

CONTRACTOR shall route all materials over carriers designated by ABC by means of traffic maps provided for the CONTRACTOR.

*Article VIII. Completion of Work*

1. CONTRACTOR agrees to diligently proceed with all parts of the work for ABC's best interest.

2. Unless prevented by causes beyond his control CONTRACTOR proposes to complete all the facilities herein included not more than eighteen months after execution of this agreement.

3. CONTRACTOR agrees to schedule his work for completion in accordance with the precedence agreed upon with ABC and described in Exhibit I.

4. Whenever any part of the work shall be completed and ready for operation ABC may take over that facility and operate it; however, the taking over or operation of any part of the completed work shall not necessarily constitute acceptance on the part of ABC, and CONTRACTOR shall be obligated to complete any unfinished parts of the facility and shall be responsible for errors or omissions discovered after operation by ABC the same as if the facility had not been put into operation.

5. When all parts of the work are completed to the satisfaction of ABC, then ABC shall provide CONTRACTOR with an authorized acceptance at which time all funds or payments due and/or withheld shall be due and payable to CONTRACTOR.

*Article IX. Cancellation of Contract*

1. This agreement may be cancelled by ABC at any time upon giving CONTRACTOR written notice ten (10) days in advance of such action. The CONTRACTOR shall be allowed a reasonable time for transfer of all parts of the work in his possession to ABC. The period of time allowed after notice of cancellation shall be known as the CANCELLATION period. In no case shall the CANCELLATION period exceed thirty (30) days from date of receipt of notice of cancellation.

2. If cancellation is effected after CONTRACTOR has made commitments for more than fifty per cent (50%) of the estimated value of materials according to CONTRACTOR'S estimate as shown in Exhibit II, then CONTRACTOR'S fee, eight hundred and fifty thousand dollars ($850,000.00) shall be due and payable within thirty (30) days after notice of cancellation is received.

3. If cancellation is effected before CONTRACTOR has made commitments up to fifty per cent (50%) of the estimated value of materials according to CONTRACTOR'S estimate as shown in Exhibit II, then CONTRACTOR'S fee, eight hundred and fifty thousand dollars ($850,000.00) shall be reduced to the actual plus percentage on labor and material earned up to the time of cancellation.

4. CONTRACTOR shall continue to furnish all necessary personnel and make every effort to expediently transfer data and other forms, drawings, specifications, purchase orders as prepared for and used in the work to ABC. CONTRACTOR shall be reimbursed for all such costs in accordance with the schedules of costs and plus-percentages applicable thereto up to time of notice by ABC to CONTRACTOR of completion of cancellation and transfer of data and materials to ABC.

5. In the event of cancellation ABC shall have the right to immediately enter upon the premises at any time after notice of cancellation and take over any part or parts of the work it so chooses, together with all materials, tools either owned by itself or by CONTRACTOR, and complete the work. The CONTRACTOR shall be paid reasonable rentals as agreed upon with ABC for use of his tools or if he so elects he may remove such tools from ABC's premises provided (*a*) the removal of such tools will not affect completion of work; (*b*) such removal is agreeable to ABC; (*c*) such removal is charged to the CONTRACTOR'S cost. CONTRACTOR'S tools used by ABC for completion of the work will be dismantled, loaded and/or otherwise prepared for

shipment by ABC. All freight handling or other costs after loading on carrier or leaving ABC's premises shall be to CONTRACTOR'S account whether handled, moved, loaded, or shipped by CONTRACTOR or ABC.

6. All materials or equipment on order at time of cancellation will be handled up to delivery and invoicing by either CONTRACTOR or ABC as decided at time of cancellation of contract. CONTRACTOR will be reimbursed for all invoices due or paid by himself after cancellation in accordance with the terms herein applying to earned plus percentage. When requested to do so by ABC, CONTRACTOR shall transfer to ABC all open orders or orders on which materials have not been delivered. When practical vendor shall be notified of transfer by CONTRACTOR and shall change his invoicing procedure accordingly. In the event such invoices are submitted to CONTRACTOR for payment, mutual agreement will be reached between the parties hereto regarding the most expedient method of handling. Whenever CONTRACTOR is required to process and pay such invoices he shall be reimbursed for such costs plus the percentage fixed herein.

7. In the event that cancellation of this agreement is due to ABC's decision to discontinue the work or for other causes beyond ABC's control with respect to completion of the work all agreements herein shall apply the same as the preceding paragraphs of this Article IX with the following exceptions:

7.1. The CONTRACTOR shall act immediately to cancel all purchase orders for materials or equipment undelivered, advising each vendor by telegram or teletype of the intention to cancel such order and further instructing each vendor to stop work as of the time of receipt of such instructions. Vendor shall advise of cancellation charges and if such charges will exist he shall advise CONTRACTOR in detail of the reasons for such costs. The CONTRACTOR will then advise ABC immediately of such charges and request a statement from ABC as to acceptance.

7.2. ABC will reimburse CONTRACTOR for all such costs whether vendor cancellation costs or other costs which result from cancellation.

7.3. In general ABC will take delivery on any items on which cancellation will be 100% of purchase price but CONTRACTOR shall advise ABC of such charges before vendor is notified to continue.

*Article X. Protection of Materials and Property*

1. In addition to the conditions of Article V above, CONTRACTOR shall at all times act for the best interests of ABC and shall protect the property, equipment, and tools of ABC as well as other contractors or subcontractors employed in CONTRACTOR'S work area or areas to the best of his ability.

2. The CONTRACTOR shall be considered the prime contractor on ABC's premises and he shall be responsible for the work and actions of all other contracting or subcontracting firms employed by himself. The CONTRACTOR shall notify ABC immediately of any harmful practices by other contractors not employed by himself. In the absence of the authorized representative of ABC the CONTRACTOR shall act at his own discretion to prevent or avoid any actions by others which may result in damage to property or endanger work or personnel.

*Article XI. Insurance*

The CONTRACTOR shall provide all or any insurance as may be required by ABC and he shall be reimbursed for all premiums paid for such insurance.

*Article XII. Guarantees*

1. The CONTRACTOR guarantees all work, materials, and equipment designed or fabricated by himself, against failure from faulty design or workmanship for a period not to exceed one (1) year from first date of installation or operation but not more than eighteen (18) months from date of delivery to jobsite, saving the ABC from all costs of replacement, except freight costs, when the equipment is being or has been used in the method and manner for which it was designed. The CONTRACTOR cannot guarantee mechanical equipment or other equipment purchased as specialty items manufactured by others beyond the guarantee of the manufacturer of the equipment; however, the CONTRACTOR agrees to make every effort to obtain the best guarantees possible for all such equipment.

2. The CONTRACTOR does not guarantee any equipment against failure or damage due to corrosive effects from fluids or other materials unless the equipment is specifically designed to withstand the conditions caused by the particular fluid as a part of the process.

3. The CONTRACTOR agrees to make any mechanical modifications, changes, or corrections in the process which are necessary to provide the production and quality of products as outlined in the CONTRACTOR'S proposal, Exhibit I, without cost to ABC, provided the raw materials or fluids which are to be processed have the characteristics which were the basis for design, and provided the operation of the processes is maintained for sufficient periods of time required to prove production.

4. The CONTRACTOR will not be held responsible for loss of profit or for operational costs or for costs of fluids, materials, or utilities used in acceptance tests.

*Article XIII. Assignment of Contract*

This contract shall not be assigned by either party to this agreement except and unless mutual agreement is reached between the parties hereto.

*Article XIV. Surplus Materials*

1. The CONTRACTOR agrees to make every effort to avoid the accumulation of unreasonable surplus stocks of materials; however, when a surplus is found to exist the CONTRACTOR will immediately notify ABC of the existence of such surplus, giving details of the original order, cost, and specification. ABC will then advise CONTRACTOR in writing whether the material is to be transferred to ABC's warehousing facility or whether it is to be sold. If equipment or material is to be sold, CONTRACTOR will first attempt to return it to the original vendor. If the order cannot be returned then it may be disposed of for the best price CONTRACTOR may obtain. When equipment or material can be returned to vendor, the CONTRACTOR may use his own judgment in accepting vendor's re-stocking charges without further notification to ABC.

2. All surplus materials remaining or accumulated at time of acceptance of completed plant by ABC will be accepted by ABC and moved or stored as directed.

3. All equipment found to be surplus will be protected by painting and stored in accordance with instructions from ABC. Equipment wilfully allowed to deteriorate, rust, or otherwise become damaged through negligence of CONTRACTOR'S employees shall become the property of CONTRACTOR, and all costs of such material or equipment including CONTRACTOR'S earned fee on such items shall be deducted from payments to CONTRACTOR.

### Exhibits for Engineering and Construction Contracts

Several Exhibits have been referred to in the above contract. The discussion at the end of the engineering contract has described a typical proposal in some detail. For such a contract as the above, the proposal would be in greater detail, describing in addition to the process and process equipment all facilities, types of architecture, and any points which may be peculiar to the particular installation.

Exhibit II, the estimate, would in this case be quite detailed. The following form is sometimes used. The first part of such an estimate or summary and the estimate for unit *A* is included below.

### Exhibit II—Estimate, dollars

1. Combined Estimate

| Facilities | Engineering and Clerical | Material | Labor | Total |
|---|---|---|---|---|
| Unit *A* | 166,000 | 3,026,000 | 804,000 | 3,996,000 |
| Unit *B* | 93,000 | 1,700,000 | 510,000 | 2,303,000 |
| Unit *C* | 25,000 | 470,000 | 120,000 | 615,000 |
| Power services | 6,000 | 112,000 | 33,000 | 151,000 |
| General utilities | 12,000 | 207,000 | 48,000 | 267,000 |
| Plant buildings | 18,000 | 340,000 | 85,000 | 443,000 |
| Subtotals | 320,000 | 5,855,000 | 1,600,000 | 7,775,000 |
| Fee | | | | 850,000 |
| Total combined estimates and fee | | | | 8,625,000 |

2. Estimate—Unit *A*

| | Materials | Labor | Total |
|---|---|---|---|
| Foundations | 95,000 | 115,000 | 210,000 |
| Structural steel and building | 184,000 | 126,000 | 310,000 |
| Process equipment | 1,890,000 | 230,000 | 2,120,000 |
| Piping | 619,000 | 226,000 | 845,000 |
| Electricals | 96,000 | 59,000 | 155,000 |
| Instrumentation | 142,000 | 48,000 | 190,000 |
| | 3,026,000 | 804,000 | 3,830,000 |
| Engineering and clerical | | | 166,000 |
| Total, Material, Labor, Unit *A* | | | 3,996,000 |

In actual practice estimates are much more detailed. The above estimate for Unit *A* is condensed. The item "process equipment" includes all mechanical equipment, pressure vessels, tanks, heat exchange, and condensing equipment and insulation. Estimated costs of each of these would be shown separately. General items such as supervision and clerical operations for construction are also separated in the usual estimate exhibit.

The various exhibits attached to a cost-plus contract usually include much detailed information on construction craft rates, consumable supplies, and contractor's tool rental rates.

### Lump-Sum Contract Forms

The lump-sum or fixed-price contract is much less involved, and a typical form for such contracts has not been included. Since a fixed price must be established, the original specifications, descriptions of the work, and types of equipment are described in detail in the contractor's proposal or owner's inquiry. The contract then need only carry the standard general clauses.

## CONTRACTS AND ENGINEERS

The contract phase of project engineering requires a competent legal department. Men who are trained in both engineering and law are ideal members of this group.

All project engineers, however, should have an understanding of contracts since it is their duty to execute many of the terms of a contract. In developing this understanding, the engineer should avoid an over-critical attitude toward the legal phraseology. It should be remembered that the long sentences, big words, and hackneyed legal expressions are part of the conservatism of legal practice which have proved their durability through decades of legal decision.

## ETHICS AND THE CONTRACT

There is never any intention among reputable contractors or owners to deceive through fine print or confusing terminology. The contract should be thought of more as a written agreement between gentlemen. It is written because memories are poor, and written records do much to preserve friendly business dealings. Legally the contract contains provisions for cancellation and penalties for failure in performance, but these actions are costly to both parties in both time and money.

Above all, the integrity and technical competence of the personnel of both organizations are the only criteria for successful project plant projects. The contract protects against willful negligence and illegal acts. It cannot protect either party from unethical practices or incompetence.

## REFERENCES

1. Eshbach, O. W., *Handbook of Engineering Fundamentals,* 2nd Ed., John Wiley & Sons, Inc., New York, 1952.

2. McCullough, L. B., and J. R. McCullough, *The Engineer at Law,* Vol. I, Iowa State College Press, Ames, Iowa, 1946.

3. Sadler, W. C., *Legal Aspects of Engineering,* John Wiley & Sons, Inc., New York, 1940.

4. Sadler, W. C., *The Specifications and Law on Engineering Works,* John Wiley & Sons, Inc., New York, 1948.

5. Simpson, L. P., and E. R. Dillavon, *Law for Engineers and Architects,* 3rd Ed., West Publishing Co., St. Paul, Minn., 1946.

# PART III

# DETAILS OF ENGINEERING DESIGN AND EQUIPMENT SELECTION

IN order to participate intelligently in selection of equipment and design of process facilities, foundations and structures the project engineer must understand to some extent the details of engineering design and equipment selection.

# CHAPTER 12

# VESSELS

The design and fabrication of the many sizes, shapes and types of vessels required in process plants is a complex task requiring the co-operation of process and project engineers, vessel designer, and fabricator. Vessel design requires the skill of an experienced engineer who specializes in the field. A working knowledge of the various codes, fabrication properties of materials, and procurement problems requires full-time attention to this one field. The greatest need is for project and process engineers to familiarize themselves sufficently with basic design and fabrication procedures so that an intelligent exchange of ideas with the vessel designer can be effected.

The purpose of this chapter shall be to develop background information which will aid the project engineer in directing the design and fabrication of process vessels. It is not possible to develop all the details of vessel design and no attempt will be made to repeat information that is now presented in various codes and standards which are available to anyone.*

## VESSEL FABRICATION

### The Shell

Pressure vessels and tanks for process plants are usually of cylindrical shape. Bending rolls are used to form the cylinder.

* The codes referred to in this section, unless otherwise stated, are the API-ASME and the ASME codes for "Unfired Pressure Vessels."[1,2]

A bending roll is shown in Fig. 12-1. The top roll imposes the bending forces on the plate and the distance between the top and bottom rolls controls the diameter of the cylinder. The rolling may be done cold on thin plates of smaller size. Prior to rolling, the ends of the plates are shaped to the desired radius in a press using a radius-forming die. By this procedure, perfectly round vessels can be made. After rolling, the edges of the plates are planed to prepare them for welding.

Fig. 12-1. Bending rolls: 244 inches long, 96-inch OD shell being rolled from ⅜-inch plate. (Courtesy of Lukens Steel Company.)

## Welding

Welding has become the most common means of joining metals. Bursting tests invariably prove the weld to be stronger than the shell itself. Such welds, however, must be made by competent welders. Elaborate tests to qualify welders are given in the various vessel codes.[1,2]

A number of very excellent welding processes are in use. Shielded arc welding is perhaps the most common method used in vessel fabrication. An arc is produced between a metal electrode and the base metal. The heat created causes melting of the electrode and portions of the base metal near the arc. The weld is formed by the fusion of the metal from the electrode and the base metal. The welding rods (electrodes) are coated with a flux which, when heated, produces an inert gas blanket around the arc and deposits a protective coating of slag over the weld to prevent oxidation. Arc welding can be manual or automatic. For large longitudinal and circumferential seams automatic welding machines prove particularly advantageous. For circumferential seams the cylinder is moved and the torch remains stationary. The opposite procedure is followed for longitudinal seams.

Acetylene gas welding is common, especially for welding thin plates and small attachments. In addition, acetylene welding is a necessity in any vessel fabrication shop for many utility purposes. A few of its uses are: flame hardening of small parts, cutting, and tack-welding sections prior to automatic arc welding. Gas welding is also used in brazing which is the joining of metals with a nonferrous filler without melting the base metal. Other welding procedures such as atomic-hydrogen, flash resistance, thermit, and submerged-melt welding are used for special types of work. For a detailed discussion of welding techniques, the reader is referred to standard works on the subject.

Welding of vessel seams generally requires the application of several layers of weld material. It is necessary that great care be exercised in making such "multipass" welds. The surface of each weld pass must be thoroughly cleaned and examined before proceeding with the next pass to insure good penetration and freedom from cracks, slag inclusion, and other weakening defects. A faulty weld can be disguised by a satisfactory final weld pass.

The welder's torch has also become an invaluable tool for cutting metal to any desired shape. Mechanically guided cutting torches are used extensively and jobs that would be tedious when handled by ordinary mechanical methods can be done with ease when a cutting torch is used.

## Shell Plates

Plates for the fabrication of vessels are usually ordered by thickness. Plates made by a sheared-plate mill are preferred because they are of better quality and are available in a greater number of sizes. They can be obtained in a wide variety of thicknesses, the most common being from $\frac{3}{16}$ in. to 1 in. in increments of $\frac{1}{16}$ in., 1 in. to $1\frac{1}{2}$ in. in increments of $\frac{1}{8}$ in., and $1\frac{1}{2}$ in. to 4 in. in increments of $\frac{1}{4}$ in. Lengths up to 800 in. are produced in the thinner plates. The maximum width obtainable is around 195 in., although widths over 100 in. carry an extra width charge. If both the circumference and length of a shell are greater than 195 in., it becomes necessary to employ two or more plates to fabricate the vessel (Fig. 12-2). Since extra charges on width start at 100 in., it may prove economical to use more than one plate in certain instances for sizes requiring plates over 100 in. wide. A comparison between plate width extras and the additional costs of fabrication with multiple plates must be made.[14]

### Vessel Heads

Ideally, a spherical-shaped pressure vessel is the most desirable since it withstands higher pressures for a given metal thickness. Spherical vessels are used in the storage of volatile liquids and gases, but they are not suitable for the construction of the usual process vessels. The cylindrical-shaped vessel is the next best design and is used most extensively. The heads for such vessels may be flat, ellipsoidal (elliptical dished), dished, hemispherical, or conical. Examples of such designs are given

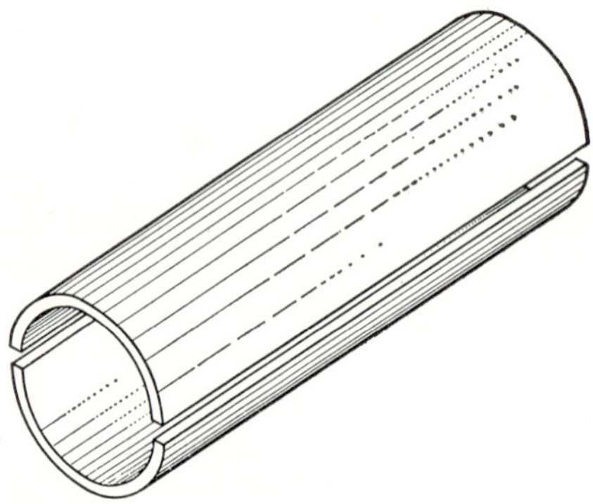

Fig. 12-2. Shell fabricated from more than one plate.

in Fig. 12-3. Flat heads can be employed economically only on small vessels (3 ft or less) or on vessels operating at low pressures. Flat heads may be welded to the vessel or installed as blind flanges on smaller vessels.

Heads other than flat heads are manufactured by pressing or spinning large flat plates to the desired shape. Sizes from twelve inches to twenty-one ft in diameter can be manufactured by spinning. Pressing is more generally confined to smaller heads and is only economical when there is a demand for a large number.

Often the process requirements dictate the style head to be used. A conical head, for instance, would be used for the bottom head of a settling drum. If no certain style is dictated, however, the choice between hemispherical, standard dished, and ellipsoidal heads is a question of economics. As the thickness required to withstand a given pressure increases in the order hemispherical, ellipsoidal, and dished, the cost of forming decreases in the same order. Although it is difficult to generalize, it may be said that in the range of low to moderate pressures the standard dished head is used. At higher pressures (above about 200 psig) it proves to be more economical to employ the ellipsoidal (elliptical dished) head with 2 : 1, ratio of major to minor axis. A wide assortment of standard heads is manufactured, and it is always desirable to use these whenever

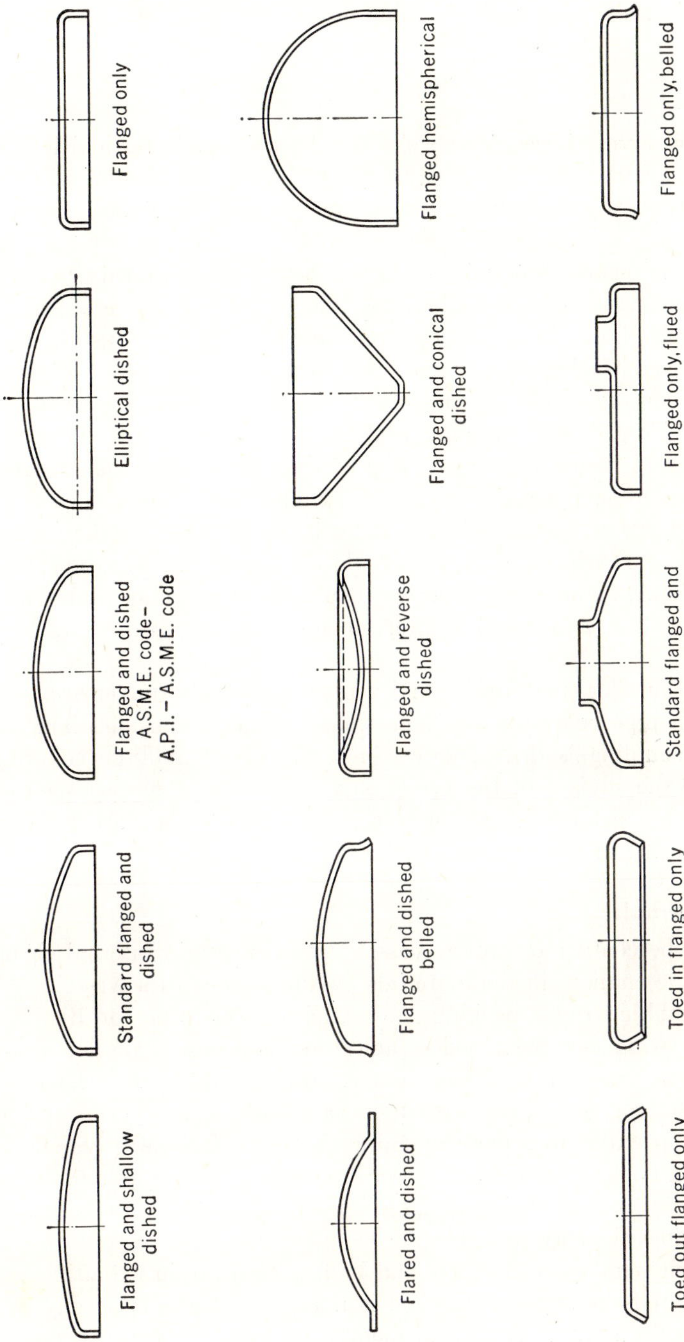


Fig. 12-3. Types of heads for cylindrical pressure vessels. (Courtesy of Lukens Steel Company.)

possible. The use of a nonstandard head requires the manufacture of additional dies needed in the forming process, and thus entails additional expense.

## Nozzles

All process vessels require nozzles for the fluids to enter and leave the vessel, for drains, handholes, and manholes. These nozzles may be fabricated of pipe, pipe couplings, forged steel nozzles, cast steel fabricated plate, or other suitable material in accordance with the code requirement. Although the codes[1,2] permit the use of both fusion welded or threaded attachments, the use of threaded attachments is generally avoided.

Cutting of holes into the shell of the vessel for the installation of nozzles weakens the vessel; and, therefore, reinforcement around the nozzle is required. Detailed rules for such reinforcement are given in the codes.[1,2] Forged steel welding necks of lengths sufficient to protrude beyond vessel insulation are mass-produced by several manufacturers. The lengths of such nozzles have been standardized, and it is customary in designing vessels to adhere to standard lengths, whether the nozzle be purchased or fabricated from pipe. So-called welding-type nozzles are also available. These are made with integral reinforcing necks and do not require the additional reinforcement necessary for welding neck nozzles.

Connections for small drains and vents one inch and under are usually made with pipe couplings. A hole smaller than the inside diameter of the pipe coupling is drilled or cut into the vessel and the coupling is welded to the surface of the vessel only. The use of nipples instead of couplings is considered poor practice, since the threads are constantly exposed during installation and can easily be damaged.

## Vessel Internals

A large percentage of process vessels requires certain internal elements to effect the changes in the materials passing through the vessel. These include bubble trays, packing, and packing supports for distillation towers, heating bundles, baffles, agitators, and other similar devices. Usually, the vessel internals are not manufactured by the vessel fabricator, since most such items are of special design and require a technical know-how peculiar to a particular organization. Bubble trays and caps represent a typical example of vessel internals which are purchased from a manufacturer specializing in these items. The vessel fabricator, however, provides the tray supports.

Lightweight alloy steel plates and bubble caps are in extensive use at this time, and the support problem is much less than in the day of the heavy cast iron tray. A typical modern-type bubble tray arrangement showing the trusses for support and bubble cap assemblies is given in Fig. 12-4. Grid and perforated plates are becoming popular, and their support is even less of a problem because of the lightweight of such trays.

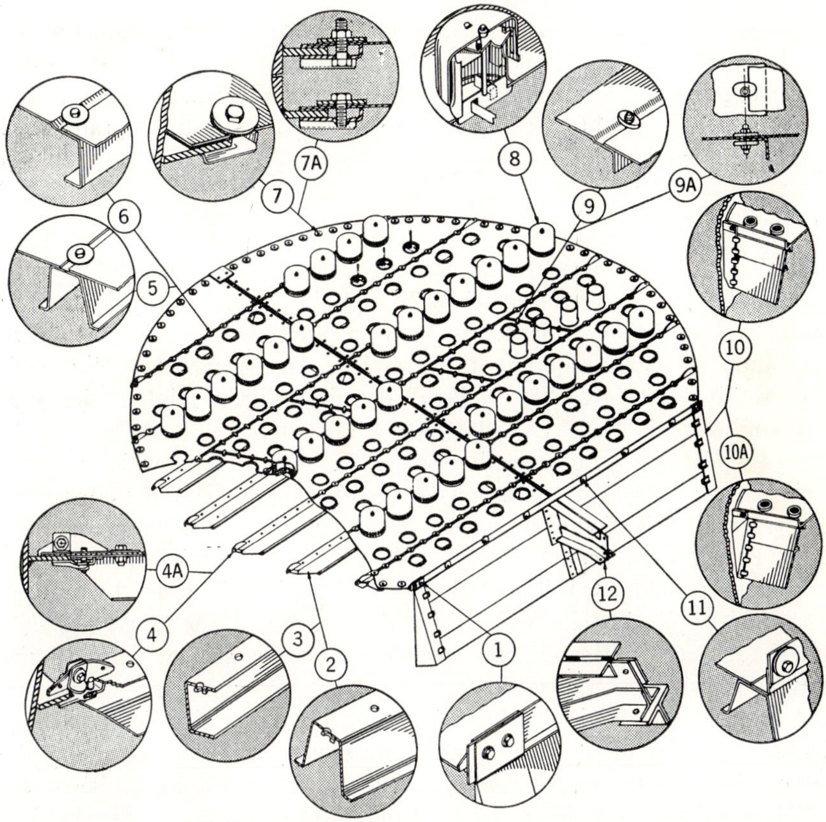


Fig. 12-4. Bubble tray. (Courtesy of Fritz W. Glitsch & Sons, Inc.)

Key:

1. Adjustable weir seal plate.
2. Trapezoidal minor truss.
3. Channel-shaped minor truss.
4 & 4A. Top-type truss hangers.
5. Thermal expansion joint between tray floors and supporting trusses.
6. Integral minor support in truss.
7 & 7A. Peripheral tray clamps.
8. Removable cap and riser assembly.
9 & 9A. Hold-down mechanism for internal tray manways.
10 & 10A. Downcomer clamping bars.
11. Adjustable weirs.
12. Fabricated structural major beam.

It is desirable for all vessel internals to be designed so that they may be removed from the manways in the vessel. Large pieces of equipment must be arranged so that they may be taken apart and removed in sections.

### External Appurtenances

In addition to the nozzles, manholes, and handholes that must be connected to the average process vessel, certain other exterior attachments are usually required. These include angles for the support of insulation, lugs and various brackets for the support of platforms and ladders which are to be attached to the vessel, and skirts for the support of the tower at the required elevation. These items are supplied by the vessel fabricator and welded to the vessel prior to shipment. Structural grade steel is normally employed.

### Stress Relieving

Residual stresses are produced around the welded seams of vessels. Such stresses weaken the vessel at these points, particularly in thicker vessels (above 1¼ in. thick). In these instances and in all cases where maximum safety is required, the vessel is stress-relieved to eliminate such residual stresses. The codes[1,2] permit a higher joint efficiency if stress relieving is used.*

Stress relieving is an annealing or heat treating process which relieves residual stresses. The operation varies, depending on the material of construction. For instance, carbon steels are heated to approximately 1100° F and then cooled slowly. Most austenitic stainless steels are heated to around 2000° F and then cooled rapidly in order to prevent carbide precipitation characteristic of these metals in the range of 800 to 1600° F. Precipitation of the carbides in stainless steel destroys the corrosion resistance.

A well-equipped vessel fabrication shop has large electrically heated furnaces capable of receiving almost any size vessel that can be shipped in one piece. Large vessels that must be welded in the field can be stress-relieved around the welds using gas fired portable furnaces. Several low temperature stress-relieving processes are also used. These employ special oxyacetylene torches which apply heat to the area at the welds, followed by a water spray.

Because of the wide differences in physical properties of metals, it is imperative that the proper stress-relieving procedure be outlined by a competent metallurgist.

### Radiographing

When the greatest confidence in the soundness of a construction of a vessel is desired, radiographing (X-raying) of all welds is done in order to detect defects. Because of the greater knowledge of weld quality, the codes[1,2] permit a lower safety factor in the form of a higher joint effi-

* See Table 12-1.

ciency. It is particularly advisable to radiograph the large longitudinal and circumferential seams of all important vessels.

### Shop and Field Fabrication

Because of the superior equipment and working conditions in a fabricator's shop, field fabrication should be kept to a minimum. Large diameter vessels (20 to 30 ft), however, must be field fabricated. Rolled sections (arcs) of the shell 7 to 8 ft wide are shipped to the jobsite. When planning any vessel over 10 ft diameter by 35 ft long, railroad agents should be consulted. Sizes above this which can be shipped depend largely on the railroad facilities serving the area. Lengths up to 60 to 80 ft have been handled using special cars, but such shipments must be carefully planned early in the job.

### Vessel Inspection

The vessel codes require inspection by some public inspector, such as state, municipal, or insurance company inspectors, before the vessel can be certified as complying with the codes. Prior to this inspection, the purchaser should have inspected the vessel during all phases of its fabrication. This includes inspection of the plates and heads as they arrive at the fabricator's shop, study of mill test reports,* and the following of all operations including the final hydrostatic test of the completed vessel.†

## THEORY OF PRESSURE VESSEL DESIGN

Rules for the design and fabrication of unfired pressure vessels have been carefully outlined by two major codes which are used by the process industries. These are the ASME and the API-ASME codes for unfired pressure vessels.[1,2] They are based on the judgment of experts. All process vessels should be constructed in accordance with the rules of one of these codes.

Before proceeding with a discussion of design equations a brief theoretical background is desirable. The formulas given in the codes for determining the minimum shell and head thicknesses are very simple and easy to use. Many occasional users, therefore, forget the limitation and the basis of these equations. Such indiscriminate use, which has been encouraged by the large numbers of so-called rapid vessel design calculation charts, lead to poor design.

* Mill test reports give the physical and chemical properties of the particular steel that was used by the steel mill in manufacturing the plates being used for the vessel.

† Vessels are hydrostatically tested to one-and-a-half times the maximum working pressure. At this test pressure all welded joints are subjected to impact by hammer, the weight of which is equal approximately to the thickness of the shell in tenths of an inch. A number of variations of these tests are possible depending on the vessel size, thickness, material, and other similar factors.

### Thin-walled Vessels

Thin-walled vessels as defined by the ASME Code have thicknesses less than one-half of the inside radius.* Such vessels constitute the majority of those encountered in process plants although high pressure operation, such as that necessary for ammonia synthesis reactors (10,000 to 12,000 psi), require thicker walls and the usual analysis for thin walls is not applicable. For thin-walled vessels equal stresses along the entire

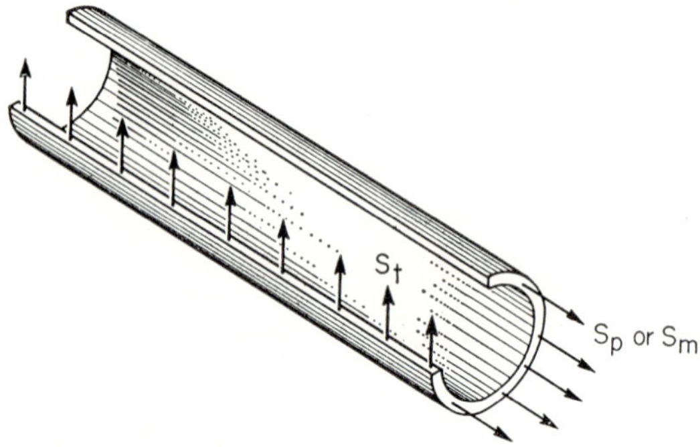


Fig. 12-5. Stresses acting on a cylinder under internal pressure.

thickness of the wall can be assumed. The pressure acting within such a cylindrical vessel produces longitudinal and circumferential or tangential stresses as shown in Fig. 12-5. These stresses can be calculated by assuming equilibrium between the opposing forces. The derivation for the simple cylinder is familiar to many and is given in any standard work on strength of materials. Den Hartog[5] has presented a most enlightening generalized derivation from which equations for any shell produced by revolution of an element can readily be deduced. His general equation is as follows:

$$\frac{S_m}{R_m} + \frac{S_t}{R_t} = \frac{P}{t}$$

where $S_m$ = meridional stress, the stress in a plane with the center line of rotation for the membrane and acting perpendicular to the center line

$S_t$ = tangential stress, acts in a tangent to the circular cross section perpendicular to the center line

$R_t$ = normal distance between the center line and a point on the shell (for a cylinder, the radius)

$R_m$ = radius of curvature in meridional plane, infinity for a cylinder

$P$ = internal pressure, psig

$t$ = thickness of shell

* The API-ASME Code specifies one-tenth of inside diameter.

In the case of a sphere, $R_m$ and $R_t$ are equal to $R$ the radius of the sphere, and because of this symmetry both $S_m$ and $S_t$ are the same.

Hence,
$$\frac{S}{R} + \frac{S}{R} = \frac{P}{t}$$

or
$$S = \frac{PR}{2t} = \frac{PD_m}{4t}$$

where $D_m$ = mean diameter

In like manner, using the same generalized equation, Den Hartog has demonstrated Biezeno's classical construction of a pressure vessel head of such curvature that the membrane stresses are the same over both the head and the cylinder and the thicknesses are constant. It is interesting to note that the popular 2 : 1 elliptical head very nearly approaches this ideal shape.

For a cylinder $R_m$ is infinite and $R_t$ is equal to the radius of the cylinder. Hence

$$\frac{S_m}{\propto} + \frac{S_t}{R} = \frac{P}{t}$$

$$S_t = \frac{PR}{t} = \frac{PD_m}{2t}$$

The meridional stress, which in the case of a cylinder is called the longitudinal stress ($S_p$), is obtained by considering the shell in the direction of the center line. From Fig. 12–5 it is seen that the total stress at the end of the cylinder is $\pi D_m t S_p$. Opposing this is the pressure acting in the longitudinal direction, $P\pi D_m{}^2/4$.

Hence,
$$\pi D_m t S_p = \frac{P\pi D_m{}^2}{4}$$

or
$$S_p = \frac{PD_m}{4t}$$

Thus the tangential stress is the larger and need be the only one considered in calculating the wall thickness required for a thin-walled cylinder to withstand a given internal pressure. The tangential stress in a cylinder is often referred to as the hoop stress.

## Codes

Some very excellent rules and regulations governing the design and fabrication of pressure vessels and tanks have been published in the form of codes. The following is a representative, but not complete, list of codes and organizations which have formulated rules for vessel and tank construction.

API-ASME Code; Unfired Pressure Vessels for Petroleum Liquids and Gases.
ASME Boiler Code Section VIII; Rules for Construction of Unfired Pressure Vessels.
American Petroleum Institute; Specifications for Oil Storage Tanks, Section No. 12-C.
Underwriters Laboratories, Inc.
National Board of Fire Underwriters.
American Factory Mutual Fire Insurance Companies.
American Bureau of Shipping Rules for Classification and Construction of Steel Vessels.

Of the above named organizations and codes, the code published by the American Society of Mechanical Engineers (ASME) and that published jointly by the ASME and The American Petroleum Institute (API)* are the most widely used rules for vessel design and construction. Knowledge of the historical background of these codes is of value in their interpretation. Samans[11] has compared the recent editions of the two codes most effectively using an historical approach.

The American Society of Mechanical Engineers, between the years 1920–1925, formulated the Boiler Construction Code, Section VIII of which appeared in 1925. This section was the first edition of *ASME Code for Unfired Pressure Vessels*. Although it represented the first complete compilation of such regulations, it was based mainly on practices existing in the steam power and boiler industries. At about this same time the petroleum industry began its remarkable march of progress, and the need for vessels that could withstand high pressures and temperatures became more and more common in that industry. Because of the narrow range of temperatures covered by the ASME code at that time, engineers in the petroleum refining industry considered the ASME code of 1925 inadequate for their use. Their thoughts were presented before the Boiler Code Committee in 1925, and it was agreed that the petroleum industry should publish its own practices and methods of pressure vessel design for the benefit of both the petroleum industry and all other process industries. A joint committee of the API and the ASME was formulated. The API-ASME code which they developed was first issued in 1934 and was followed by revised editions in 1936, 1938, 1943, and 1951. Since the issuance of the first API-ASME code, the ASME code has been extensively revised. The latest 1952 edition of Section 8 contains complete rules for the use of cast iron, nonferrous materials, and high alloy steels as well as rules for carbon and low alloy steels covered by the API-ASME code and previous editions of the ASME code. The API-ASME code, in fact, now refers to the rules for high alloy steels and clad steels by refer-

*As book went to press it was learned that the API-ASME Code will be discontinued. It is believed, however, that many organizations will continue to use it for several years to come.

ence to the ASME code and thus avoids repetition of these regulations.

It is not surprising that certain differences have developed between the two codes, many of which are minor and result from differences in wording that have produced variations in interpretation. Perhaps the most important variation is the definition of qualified inspectors; the ASME code seems to have improved the system of inspection.

The ASME code now contains a more complete list of permissible materials and a more recent method for calculating thickness of vessels subject to external pressure. These differences are due to variations in philosophy that occurred between the groups responsible for formulating the rules.

Situations continually arise in the design and fabrication of pressure vessels which make interpretation of the code difficult. The Boiler Code Committee, therefore, accepts inquiries about code interpretation and meets monthly for discussion. Decisions made in these meetings are sent to the inquirer as well as published subsequently in *Mechanical Engineering*, the official publication of the ASME. In like manner, questions can be addressed to the American Petroleum Institute concerning the interpretation of the API-ASME code.

### Working Formulas

The theoretical formulas presented above for thin-walled pressure vessels have proved reliable by actual test, and are used in both the API-ASME and ASME codes. The API-ASME simply adds a corrosion allowance and an efficiency factor for the longitudinal joints and presents the formula as follows:

$$t = \frac{PD_m}{2SE} + C \tag{1}$$

where $P$ = maximum working pressure or design pressure, psig
$D_m$ = mean diameter in in.
$S$ = allowable working stress, psi (values of $S$ are given in both codes[1,2])
$E$ = efficiency of longitudinal joint (values of efficiency vary from 50 to 95% depending on type of joint and whether stress relieving and/or radiographing are used. See Table 12-1 for typical values)
$C$ = corrosion allowance in in.
$t$ = thickness in in.

The use of a mean diameter instead of an inside diameter is an attempt to more nearly estimate the mean stress value across the thickness of the shell. A more convenient equation in terms of an inside diameter can be

**TABLE 12-1.† Maximum Allowable Efficiencies for Arc and Gas Welded Joints**

| Type of joint | Limitations | Basic joint efficiency[1] per cent | Radio-graphed[2] | Thermally stress-relieved[3] | Maximum joint efficiency per cent |
|---|---|---|---|---|---|
| Double-welded butt joint | None | 80 | . . . | . . . | 80 |
| Single-welded butt joint with backing strip[4] | Longitudinal joints not over 1¼ in. thick. No thickness limitation on circumferential joints. | | . . . | * | 85 |
| | | | * | . . . | 90 |
| | | | * | * | 95 |
| Single-welded butt joint without backing strip | Circumferential joints only, not over ⅝ in. thick. | 70 | . . . | . . . | 70 |
| | | | . . . | * | 75 |
| Double full-fillet lap | Longitudinal joints not over ⅜ in. thick. Circumferential joints not over ⅝ in. thick. | 65 | . . . | . . . | 65 |
| | | | . . . | * | 70 |
| Single full-fillet lap joints with plug welds | Circumferential joints only not over ⅝ in. thick; and for attachment of heads not over 24 in. outside diameter to shells not over ⅝ in. thick. | 60 | . . . | . . . | 60 |
| | | | . . . | * | 65 |
| Single full-fillet lap joints without plug welds | For the attachment of heads convex to pressure, to shells not over ⅝ in. required thickness, only with use of fillet weld on inside of shell; and for attachment of heads having pressure on either side, with fillet weld on outside of head flange only, to shells not over 24 in. inside diameter and not over ¼ in. required thickness. | 50 | . . . | . . . | 50 |
| | | | . . . | * | 55 |

[1] Spot examinations of all vessels required by Par. UW-52, except where specifically exempted therein.

[2] See Par. UW-11.

[3] See Par. UW-10.

[4] May be considered the equivalent of a double-welded butt joint within the limits given. (See Par. UW-35(b) and Par. UA-6.)

* Indicates that these operations have been performed.

† Reproduced by permission, *Rules for Construction of Unfired Pressure Vessels,* 1952 ed., Am. Soc. Mech. Engrs. (UW numbers refer to code paragraphs.)

developed from eq. 1 by calling

$$D_m = \frac{D_1 + (D_1 + 2t)}{2} \qquad \text{(where } D_1 = \text{inside diameter)}$$

Hence

$$t = \frac{PD_1}{2SE - P} + C \tag{2}$$

The API-ASME code lists eq. 2 and a similar one based on the outside diameter. The ASME approaches the problem of obtaining a more accurate mean stress value by adding 0.6 of the thickness to the radius giving the equation,

$$t = \frac{PR}{SE - 0.6P} \tag{3}$$

where $R$ = inside radius in in.

Comparison with equation 2 will indicate only a slight difference if $R$ is changed to $D_1$ by multiplying numerator and denominator by 2.

Similar comparisons between the basic theoretical equations and the working equations presented by the codes can be given for each of the vessel shapes. It is helpful for the project engineer to become familiar with the background of each frequently used formula as given by the codes so that the limitations may be clearly understood.

In particular it should be recognized that shell thickness formulas are based on the thickness required for the stresses created by internal pressure. Loadings such as dead loads (due to weight of vessel and contents), wind loads, temperature stresses, and other loads which may occur must be considered separately. Often the wind load on a high tower will govern the thickness of the lower courses of the tower.

Frequently used vessel design equations are summarized for convenience in Table 12-2. Users of this table are cautioned against blind use of formulas. Until the basis and limitation of these equations become familiar to the user, reference to the original source should be made.

The formulas for thin-walled vessels were based on the assumption that the circumferential stress is uniformly distributed across the wall. This assumption is not true, for thick-walled vessels. Various equations are used for designing such thick vessels. Perry[10] gives a good review of some of these. Accurate calculation is still not possible, however, since all the formulas developed thus far are based on some erroneous assumption. Experimental studies on thick cylinders are being conducted by several large chemical companies engaged in high pressure operations. Verification of calculation procedures or development of new equations should result from this work. At present it is necessary to use existing data and employ sizable safety factors to allow for the limited knowledge of thick-wall vessel design.

**TABLE 12-2. Summary of Formulas for Design of Thin-Walled Pressure Vessels (Internal Pressure)**

| Purpose | ASME | API-ASME |
|---|---|---|
| 1. Thickness of shells (Internal pressure) | | |
| (a) Cylindrical | $t = \frac{PR}{SE - 0.6P} + C$ | $t = \frac{PD_1}{2SE - P} + C$ |
| (b) Spherical | $t = \frac{PR}{2SE - 0.2P} + C$ | $t = \frac{PD_1}{4SE - P} + C$ |
| 2. Heads | | |
| (a) Ellipsoidal (2:1) | $t = \frac{PD_1}{2SE - 0.2P} + C$ | $t = \frac{PD_m}{2SE} + C$ |
| (b) Dished (standard ASME with Knuckle radius 6% of crown) | $t = \frac{0.885PR_c}{SE - 0.1P} + C$ | $t = \frac{0.885PR_c}{SE} + C$ |
| (c) Hemispherical | $t = \frac{PD_1}{4SE - 0.4P} + C$ | $t = \frac{PD_m}{4SE} + C$ |
| (d) Conical | $t = \frac{PD_{c_i}}{2\cos\alpha(SE - 0.6P)} + C$ | $t = \frac{PD_c}{2SE\cos\alpha} + C$ |
| 3. Thickness of cylindrical shells based on combined loads | | |
| (a) Combined dead weight, internal pressure, and wind loads | Windward side $t = \frac{2P_{wh}{}^2}{\pi D'S} - \frac{W}{\pi D_m S} + \frac{PD_m}{4S}$ Leeward side $t = \frac{2P_{wh}{}^2}{\pi D'S} + \frac{W}{\pi D_m S} - \frac{PD_m}{4S}$ | |
| 4. Buckling | $t = \frac{2P_{wh}{}^2}{\pi D'S_B} + \frac{W}{\pi D_m S_B}$ | |

**Nomenclature**

$C$ = corrosion allowance, in.
$D$ = outside diameter of shell, in.
$D'$ = outside diameter of shell including insulation, in.
$D_1$ = inside diameter of shell, in.
$D_c$ = mean diameter of a cone at the point of consideration, in.
$D_{c_i}$ = inside diameter of a cone at the point of consideration, in.
$D_m$ = mean diameter of shell, in.
$E$ = joint efficiency, fraction (see codes)
$h$ = distance from top of vessel to point under consideration, ft

$I$ = moment of inertia, $(\text{in.})^4$
$M$ = bending moment due to wind, lb-ft
$P$ = working pressure, psig
$P_w$ = wind pressure, psf
$R$ = inside radius of shell, in.
$R_c$ = radius to which the head is dished measured to centerline at thickness of crown plate, in.
$S$ = allowable working stress (see tables in code), psi
$S_B$ = allowable stress for buckling, $2 \times 10^6(t/D)$, psi
$S_0$ = stress due to dead weight, psi
$S_p$ = longitudinal stress, psi
$S_w$ = stress due to wind, psi
$t$ = thickness of shell or head, in.
$W$ = weight of vessel, lb
$\alpha$ = one-half of the included angle of a cone at the centerline of the conical head

The calculation of thicknesses for shells subjected to external pressure (vacuum vessels) has been presented in the form of charts in both codes. The detailed design and application of stiffening rings is also outlined.

### Shell Stresses Due to Dead Weight and Wind Loads

The codes for pressure vessels state that wind loads and loads due to dead weight of the vessel and its contents should be considered in analyzing the stresses which occur in a given vessel, but no specific rules governing this are stated in the code.

On the windward side of a vessel the stresses due to wind and internal pressure (in this case, longitudinal stress) act to produce a tension on the vessel wall, while the dead weight stress produces compression and therefore is opposite in sign. On the leeward side, the stress due to wind and weight both produce compression, while the longitudinal stress due to internal pressure is again acting in tension and is opposite in sign. These two conditions can be expressed mathematically as follows:*

Windward: $$S = S_w - S_0 + S_p \tag{4}$$

Leeward: $$S = S_w + S_0 - S_p \tag{5}$$

where $S$ = allowable stress
$S_w$ = stress due to wind load
$S_0$ = stress due to dead weight
$S_p$ = longitudinal stress, $PD_m/4t$

For estimating purposes the allowable stress in the above formulas may be taken as the joint efficiency multiplied by the allowable working stress listed in the codes for the material in question.

In addition to the conditions considered thus far, the wind and dead weight loads can also produce buckling on the leeward side of a cylindrical vessel. To be on the safe side, it is advisable to assume that the internal

* The above analysis is based on the most commonly used and simplest theory, the maximum stress theory. Other theories in use include the maximum strain, maximum shear, and modified strain-energy. Brummerstedt[3] reviews these theories.

pressure is zero for this calculation, since internal pressure increases the stability of a vessel. The allowable stress for buckling is equal to the stress due to wind load plus that due to dead weight load ($S_B = S_w + S_0$). L. H. Donnell[6] has done considerable experimental work on the buckling of shells and has developed the following empirical equation for evaluating the allowable stress.

$$S_B = E\left[\frac{0.6(t/R) - 10^{-7}(R/t)}{1 + 0.004\,(E/S_y)}\right] \tag{6}$$

where $t$ = shell thickness
$R$ = shell radius
$E$ = modulus of elasticity
$S_y$ = yield point

Jorgensen[7] has simplified this formula for the usual carbon steel shell to $S_B = 2 \times 10^6(t/D)$ which was derived using a factor of safety of 4 on the stress calculated from Donnell's formula.

The allowable stress formulas can be further expanded as follows:

Wind load $= S_w = \dfrac{MD'12}{2I}$ (based on the well-known bending stress formula, bending moment/section modulus)

where $M$ = bending moment
$= (1/2)P_w \times (D'/12)h^2$ lb of force-ft
$P_w$ = unit wind pressure,* psf
$D'$ = outside diameter of shell in in. Includes insulation and/or other projections
$h$ = distance from top of vessel to point being considered, ft
$I$ = moment of inertia
$=$ approximately $\pi(D')^3t/8$, (in.)$^4$
$t$ = thickness of shell in in.

Hence

$$S_w = \frac{MD'(12)(8)}{(2)\pi(D')^3t} = \frac{48M}{\pi(D')^2t}$$

$$= \frac{48\,P_wD'h^2}{24\pi(D')^2t} = \frac{2P_wh^2}{\pi D't} \tag{7}$$

Dead load:

$$S_0 = \frac{W}{\pi D_m t}$$

where $W$ = weight of vessel plus internals and liquid
$D_m$ = mean diameter of shell in in.

* The unit wind pressure is related to the wind velocity by an empirical formula developed by the U.S. Weather Bureau. 100 mph wind is approximately equivalent to 30 psf; a 125 mph wind, to 40 psf.

Hence substituting in eqs. 4 and 5

$$\text{Windward allowable stress} = S = \frac{2P_wh^2}{\pi D't} - \frac{W}{\pi D_mt} + \frac{PD_m}{4t}$$

or

$$t = \frac{2P_wh^2}{\pi D'S} - \frac{W}{\pi D_mS} + \frac{PD_m}{4S} \tag{8}$$

$$\text{Leeward allowable stress} = S = \frac{2P_wh^2}{\pi D't} + \frac{W}{\pi D_mt} - \frac{PD_m}{4t}$$

or

$$t = \frac{2P_wh^2}{\pi D'S} + \frac{W}{\pi D_mS} - \frac{PD_m}{4S} \tag{9}$$

$$\text{Buckling stress} = S_B = \frac{2P_wh^2}{\pi D't} + \frac{W}{\pi D_mt}$$

or

$$t = \frac{2P_wh^2}{\pi D'S_B} + \frac{W}{\pi D_mS_B} \tag{10}$$

The above three formulas and that for the circumferential stress should be used to determine the controlling thickness. Since the operating weight of a vessel depends on the thickness finally chosen, the weight must first be assumed. Several short cuts to this estimate have been suggested. Jorgensen[7] has developed a formula: $W = 15.6Dht + 0.163hD^2$ for insulated towers having cast iron trays. Nelson[9] suggests calculating the weight based on the thickness $t$ and height $h$, using a metal density of 0.283 lb/cu in., and then increasing this value by 18 per cent to allow for liquid, insulation, trays, etc. Employing this procedure eq. 8 for the windward side becomes

$$t = \frac{2P_wh^2/\pi D' + PD_m/4}{S + 4h} \tag{11}$$

eq. 9 for the leeward side becomes

$$t = \frac{2P_wh^2/\pi D' - PD_m/4}{S - 4h} \tag{12}$$

and eq. 10 for buckling becomes

$$t = \frac{2P_wh^2}{\pi D'(S_B - 4h)} \tag{13}$$

Eq. 11 through 13 provide a method for estimating thickness for typical towers. Eq. 8 through 10, however, should be used when more accurate information is required and the thickness can then be determined by successive approximation of the weight.

In extremely tall towers the thickness based on the internal pressure alone is often not adequate to withstand the combined stresses produced by wind and weight. Since these stresses increase from top to the bottom of the tower, it is usually economical to increase the thickness of the lower

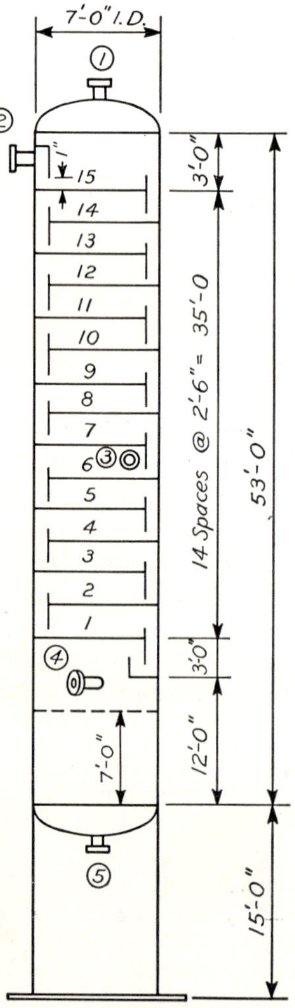


| | |
|---|---|
| Item No. | T-101 |
| Service | Plant B Feed Stabilizer |
| Inside Diameter | 7'-0" |
| Length (Tangent to Tangent) | 53'-0" |
| Operating Temperature °F | 550 |
| Operating Pressure psig | 100 |
| Design Temperature °F | 600 |
| Design Pressure | 120 |
| Corrosion Allowance | 1/8" |
| Weir Height | 3" all trays |
| Downcomer Clearance | 1" all trays |

NOZZLES

| No. | Size (Inches) | Service |
|---|---|---|
| 1 | 16 | Overhead Vapor |
| 2 | 3 | Reflux |
| 3 | 12 | Feed Inlet * |
| 4 | 16 | Reboiler (Make Tangential) |
| 5 | 8 | Bottoms |

* Design internal distributor to provide even distribution across length of tray.

Note: Provide manways, gage glass, and level control connections.

Fig. 12-6. Process vessel sketch.

section of the tower. Fabrication procedures and plate costs must be considered in selecting the thicknesses. Theoretically, the thickness of a tall shell should increase steadily from top to bottom. Such a design is, of course, impractical. In some cases a single-shell thickness equal to the maximum required at the bottom may prove the most economical. In others several thicknesses may be indicated. Each case must be considered separately. Generalizations are to be avoided.

The above principles and equations are briefly illustrated in the following example. The calculations represent only the preliminary thickness calculations in pressure vessel design. Detailed stress calculations when necessary must be made by a competent vessel designer skilled in engineering mechanics.

EXAMPLE: A process sketch of a fractionating tower is given in Fig. 12-6. Prepare the mechanical design assuming a maximum wind velocity of 125 miles/hour (equivalent to 40 psf wind pressure). Design in accordance with the API-ASME Code. Use ASTM A-70 Firebox Grade Steel and double-welded butt joints. No stress relieving or radiographing is required.

THICKNESS OF SHELL REQUIRED FOR INTERNAL PRESSURE

$$t = \frac{PD_1}{2SE - P} + C$$

From API-ASME Code: $E = 0.80$
$S = 13{,}750$

$$t = \frac{(120)(7)(12)}{(2)(13{,}750)(0.80) - 120} + 0.125 = 0.46 + 0.125 = 0.585 \text{ in.}$$

Nearest Practical Thickness $= \frac{5}{8}$ in.

CONSIDERING STRESSES AT BOTTOM TANGENT LINE

Estimated operating weight (not including skirt), $D_m = 7.05$ ft

| | | Pounds |
|---|---|---|
| Cylinder: $(1/12)(5/8)(\pi)(7.05)(53)(489)$ | = | 30,000 |
| Heads (two at 1622) | = | 3,244 |
| Trays (alloy steel trays 26 psf) $\frac{(\pi)(7)^2(26)}{4}$ | = | 15,000 |
| Liquid on trays (base on water and 3 in. depth) $(15)[(\pi)(7)^2/4](3/12)(62.4)$ | = | 9,000 |
| Liquid in bottom of tower (7 ft) $\frac{(7)(\pi)(7)^2(62.4)}{4} + (44.8)^*(624)$ | = | 19,650 |
| Insulation (3 in. plastic insulation) | = | 8,600 |
| | | 85,494 lb |
| Weight of water when full during hydrostatic test | = | 133,000 |
| Weight of vessel full of water | = | 189,844† |

THICKNESS BASED ON WINDWARD STRESSES AT BOTTOM TANGENT

$$S = (13{,}750)(0.8) = 11{,}000 = \frac{2P_w h^2}{\pi D' t} - \frac{W}{\pi D t} + \frac{PD}{4t}$$

* Volume of 7′–0″ ID 2 : 1 elliptical head.
† 85,494 − 9,000 − 19,650 + 133,000 = 189,844.

$$t = \frac{(2)(40)(53)^2}{(\pi)(7.5)(12)(11{,}000)} - \frac{85{,}494}{(\pi)(7.05)(12)(11{,}000)} + \frac{(120)(7.05)(12)}{(4)(11{,}000)}$$

$$= 0.072 - 0.0292 + 0.232 = 0.2748 \text{ in.}$$

THICKNESS BASED ON LEEWARD STRESSES

$$S = \frac{2P_w h^2}{\pi D' t} + \frac{W}{\pi D_m t} - \frac{PD_m}{4S}$$

Inspection indicates this not to be controlling.

THICKNESS BASED ON BUCKLING. Use weight full of water as worst condition.

$$S_B = 2 \times 16^6 \left(\frac{0.500}{(7)(12)}\right) = 11{,}900 \text{ or } (11{,}900)(0.8)$$

$$= 9500 \text{ approximately (based on corroded thickness, } 0.625 - 0.125 = 0.500)$$

$$t = \frac{2P_w h^2}{\pi D' S_B} + \frac{W}{\pi D_m S_B}$$

$$t = \frac{(2)(40)(53)^2}{\pi(7.5)(12)(9500)} + \frac{189{,}844}{(\pi)(7.05)(12)(9500)} = 0.14 \text{ in.}$$

All of the above values are less than the uncorroded thickness based on the circumferential pressure; hence the latter is controlling.

## Design Conditions

The allowable working stress to be used in the various formulas is given in tabular form in each of the codes. It is the minimum tensile strength of the particular metal divided by a safety factor. This factor is 5 for the ASME Code up to 650° F, and 4 for the API-ASME Code. The allowable stress corresponding to the design temperature should be selected from the applicable tabulation. The maximum working pressure or design pressure and the design temperature must be selected by the vessel designer in cooperation with the project or process engineer. As specified by the codes, the design pressure should certainly not be less than the maximum relief valve settings. It is customary to use a design pressure of approximately 20% above the maximum operating pressure anticipated. Unusual operating conditions may require selection of higher design pressures. In a similar manner, the design temperature should be chosen at a reasonable value above the normal operating temperature, such as 50°. Higher margins of safety may be used in processes where greater increases in temperature are possible. It should be remembered that every possible contingency cannot be considered in the design, since the cost would then become prohibitive. But the vessel should be designed to withstand all reasonable possible contingencies, and suitable pressure-relieving devices should be installed to protect the vessel and the personnel in case of unusual situations.

## DEVELOPMENT OF A VESSEL DESIGN

### Setting Basic Dimensions

The size and shape of most process vessels are determined by the process engineer. By training and experience he is the logical person to make such decisions. The diameter and height of a distillation tower are set within limits, by the distillation design calculation. The volume of accumulator drums, reactors, and settling tanks are similarly dictated by the process calculation. Anyone who has made such calculations, however, realizes that there is considerable leeway possible in the final choice. For example, an accumulator drum must have a certain volume, but an infinite number of length to diameter ratios can be employed to give that volume. Other factors being equal, it is necessary to choose the most economical design. Brummerstedt[4] has shown that a length to diameter ratio of 3 : 1 or 4 : 1 is most economical for low pressure tanks in the range from 50 to 75 psig, and length to diameter ratio of from 4 : 1 to 6 : 1 is preferable for pressures above 100 psig. Other factors, such as plant layout, appearance, and foundation conditions must also be considered. Above all, the process engineer must specify vessels so that standard materials can be employed throughout. The use of standard heads, for instance, will avoid extra charges for special patterns.

The process engineer will do well to consult with the vessel designer during the early stages of vessel planning so that the most economical design can be realized. Engineers concerned with the detailed mechanical design and fabrication of a vessel are not in a position to change basic dimensions, since it is assumed that these dimensions are required for the process. It is, therefore, imperative that the basic dimensions be set intelligently by the process engineer so that they will fulfill both process requirements and the requirements of economical design.

### Materials of Construction

Selection of the materials of construction for the vessel must be a cooperative effort between the vessel designer, project engineer, and process engineer. The metals employed must resist corrosive or erosive conditions and must also possess sufficient strength at the operating temperature and pressure to permit a practical design. Wise selection of materials will insure low maintenance costs and low first costs.

Any attempt to list recommended materials would be incomplete and misleading. Selection of materials for corrosion resistance in particular is an empirical science which depends on the exact conditions of use including even the presence of trace impurities in the environment surrounding the metal. Engineers should use so-called "handy" material selection charts with caution. Experimental results, previous plant experience under similar conditions, or the opinion of a corrosion expert are the only safe guides.

### Transmitting Process Specifications

The most efficient means for transmitting information about vessels and tanks is by sketch or drawing. The process engineer can make a simple sketch of the vessel which indicates the operating condition, the number and size of nozzles and types of internals required (Fig. 12-6). He also should recommend the material of construction (particularly in the cases involving corrosive fluids) and the thickness of insulation. Any requirements which are unique or unusual should be discussed with the vessel designer. This includes the proper corrosion allowance to be used in determining the thickness of the vessel.

### Detailed Mechanical Design

After receiving the process sketch, the vessel designer can proceed with the detailed mechanical design of the vessel. The design is invariably based on one of the codes (API-ASME or ASME); the choice of code is usually dictated by the industry for which the vessel is being designed. If there is no preference on the part of the ultimate user, the designer is free to use his judgment. Many states and cities have codes or special requirements that must be fulfilled, although a large number of these are based directly on the API-ASME or ASME codes. A summary of these requirements for the United States and Canada is contained in *Synopsis of Boiler Codes, Rules and Regulations by States, Provinces, and Cities,* National Bureau of Casualty and Surety Underwriters.[15] There is, of course, no rule forbidding construction superior to that outlined by the codes, and in many cases the ultimate user or the designer elects to follow certain rules in order to realize a better design. In such cases, specifications are written by the designer, which include appropriate references to the code as well as the additional requirements which are desired.

After completion of the initial mechanical design, which includes calculation of the required thickness of the shell and heads, a preliminary sketch of the vessel is made which shows the major details including the size and number of nozzles. Exact location or orientation of the nozzles is not attempted. This initial sketch can be made in such a form that it will eventually, after subsequent revisions, become the final and complete vessel drawing, or it can be made on a smaller-size sheet as a temporary working drawing. In either event, it finds its greatest use as a means of transmitting preliminary information to vessel fabricators for the purpose of obtaining quotations. Since steel plate and other materials required for fabrication of vessels often have long delivery times, it is necessary to place the order for the vessel as soon as possible, and it is not feasible to wait until a final design has been made. Once the order has been placed, the vessel design can be developed to an extent such that the fabricator can purchase the major material items, such as shell plate and heads.

The plant design proceeds simultaneously with the development of the vessel design and as piping drawings are completed, the orientation of nozzles and manholes and exact locations of vessel internals can be finalized. The vessel drawing is then completed after being checked by the piping designers for orientation and possible interferences, the structural steel designers for location of lugs, the process engineer for process requirements, and the project engineer for economical design and process requirements.

Vessel drawings pass through the hands of a number of different engineers for checking and comments. It is important that the engineer who is asked to check such a drawing check only those features for which his training and experience qualify him. It is a waste of time for a process or project engineer to concern himself with a detailed check of the mechanical design of the vessel.

An example of a finished vessel drawing is given in Fig. 12-7. It will be noted that important information such as design conditions, operating conditions, code, and material specifications appear on the drawing. The weight of the vessel when both empty and full of water should be included since this helps the structural and foundation groups plan their work.

It has been found by most companies that nozzle details and support details, such as saddles and tower base plates, are most conveniently standardized. It is unnecessary, therefore, to repeat these details on each drawing; instead, a standard drawing is included as an attachment to the individual vessel drawings.

The fabricator prepares shop drawings from the vessel drawing received from the purchaser's engineering department. These shop drawings are working drawings for use by the welders, boilermakers, and machinists. Standard welding symbols, a portion of which is shown in Fig. 12-8, are used to guide the welders in their operations.

It proves worthwhile for the purchaser's engineers to check the fabricator's drawings against the original vessel drawings since it is possible for the fabricator to misinterpret certain features of the original drawings. This checking should be done without delay so that the fabricator can continue his work. Last-minute changes in the vessel design or specifications should be avoided unless absolutely necessary. Even the smallest change can cause considerable delay, confusion, and expense

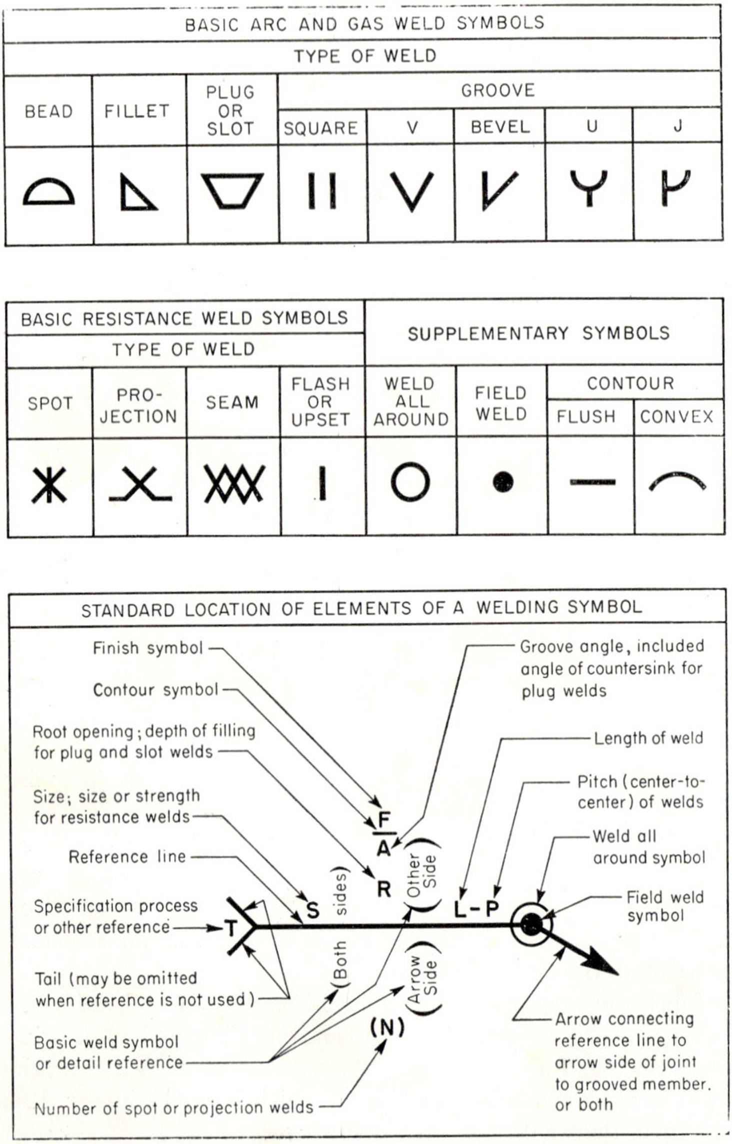


Fig. 12-8. Standard welding symbols. (Redrawn by permission, *Welding Handbook*, American Welding Society.)

## REFERENCES

1. *API-ASME Code for Unfired Pressure Vessels,* Am. Soc. Mech. Engrs., New York (1951).

2. *ASME Code for Unfired Pressure Vessels,* Am. Soc. Mech. Engrs., New York (1952) and (1956).

3. Brummerstedt, E. F., *Nat'l. Petroleum News, Tech. Sec., 35,* No. 44, R-539 (Nov. 3, 1943).

4. Brummerstedt, E. F., *Nat'l. Petroleum News, Tech. Sec., 36,* No. 18, R-282 (May 3, 1944).

5. Den Hartog, J. P., *Advanced Strength of Materials,* McGraw-Hill Book Co., New York, 1952.

6. Donnell, L. H., *Trans. Am. Soc. Mech. Engrs., 56,* 795 (1934).

7. Jorgensen, S. M., *Petroleum Refiner, 24,* 381 (1945).

8. Marin, Joseph, *Strength of Materials,* Macmillan Co., New York, 1948.

9. Nelson, J. G., *Petroleum Refiner, 27,* 565 (1948).

10. Perry, John H., *Chemical Engineer's Handbook,* 3rd Ed., McGraw-Hill Book Co., 1950.

11. Samans, Walter, *Alco Products Review, 2,* No. 2, 17 (1953) American Locomotive Co., New York.

12. Siemon, Karl, *Manual for the Design of Ferrous and Non-Ferrous Pressure Vessel and Tanks,* 4th Ed., Edwards Brothers, Inc., Ann Arbor, Michigan, 1952.

13. Spring, H. M., Jr., *Pressure Vessels for Industry,* McGraw-Hill Book Co., New York, 1947.

14. *Steel Plates and Their Fabrication,* Lukens-Steel Co., Coatesville, Pa. (1947).

15. *Synopsis of Boiler Laws, Rules, and Regulations by States, Provinces, and Cities (U.S.A. and Canada),* Nat'l. Bureau of Casualty and Surety Underwriters, Boiler and Machinery Division, 60 John St., New York.

16. Timoshenko, S., *Strength of Materials,* Part II, Van Nostrand Co., Inc., New York, 1941.

17. Timoshenko, S., *Theory of Elastic Stability,* 439–497, McGraw-Hill Book Co., New York, 1936.

# CHAPTER 13

# HEAT EXCHANGERS

Chemical processes invariably require the addition or removal of heat. Careful design and application of heat exchangers, therefore, has become an important consideration in plant design.

The complete design of heat exchangers is seldom the job of the process engineer or the project engineer. Such equipment is generally purchased from the heat exchanger manufacturer and is designed and guaranteed by the manufacturer. The project engineer, however, must understand the methods of designing and fabricating heat exchangers. An understanding of these techniques permits close cooperation between the project engineer and the manufacturer's representatives, a cooperation that can save both time and expense.

## HEAT EXCHANGER TYPES

Many different types of heat exchangers are manufactured. A few of the most widely used heat exchanger types will be discussed below. Special designs can always be devised, which may be advantageous for a particular application. It is, however, preferable to use standard designs or so-called "stock" items wherever possible. Exchanger manufacturers are able to produce certain exchanger types and sizes on an assembly-line production basis. Any deviation of consequence from these stock designs will require special operations and, therefore, increases in cost.

## Double-Pipe Heat Exchanger

As the name implies, the double-pipe heat exchanger consists of two concentric pipes. One fluid flows in the internal pipe and the other in the annulus between the inside and outside pipes. Such exchangers are most conveniently arranged in the form of hairpins, similar to the extended surface exchanger of Fig. 13-1. The usual practical length of these hairpins is twenty ft. Such exchangers are very easily made in practically any shop and continue to prove useful where very small surface requirements exist (approximately 100 sq ft).

## Double-Pipe Extended Surface Exchangers

One of the advantages of the double pipe exchanger is its low cost. A desire to benefit from this low cost and at the same time to have the advantages of larger surfaces prompted the development of the extended surface tube. Typical examples of such tubes are shown in Fig. 13-2. The longitudinal fin is used most frequently in double-pipe exchanger design (Fig. 13-1). Transverse fins are employed principally for cross-flow arrangements in either extended surface tube-and-shell exchangers or in air-cooled exchangers which are growing more popular in areas where water scarcity is a problem. Use of fin tubes is particularly desirable for gases, viscous liquids, or streams of small flow rate. Such fluids produce high resistances to heat flow, which are partially overcome by the larger effective area introduced by the fins. For surface requirements below 1,000 sq ft the use of extended surface double-pipe exchangers often effects a considerable saving. Particular consideration should be given to such units for surfaces below 500 sq ft. The double-pipe sections can be arranged in rows and connected in series (Fig. 13-3).

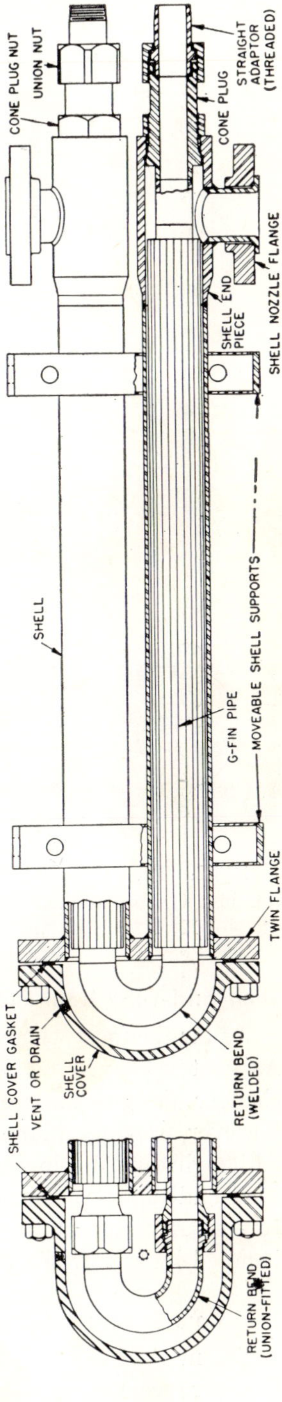


Fig. 13-1. Extended surface double-pipe exchanger; G-fin. (Courtesy of Griscom-Russell Company.)

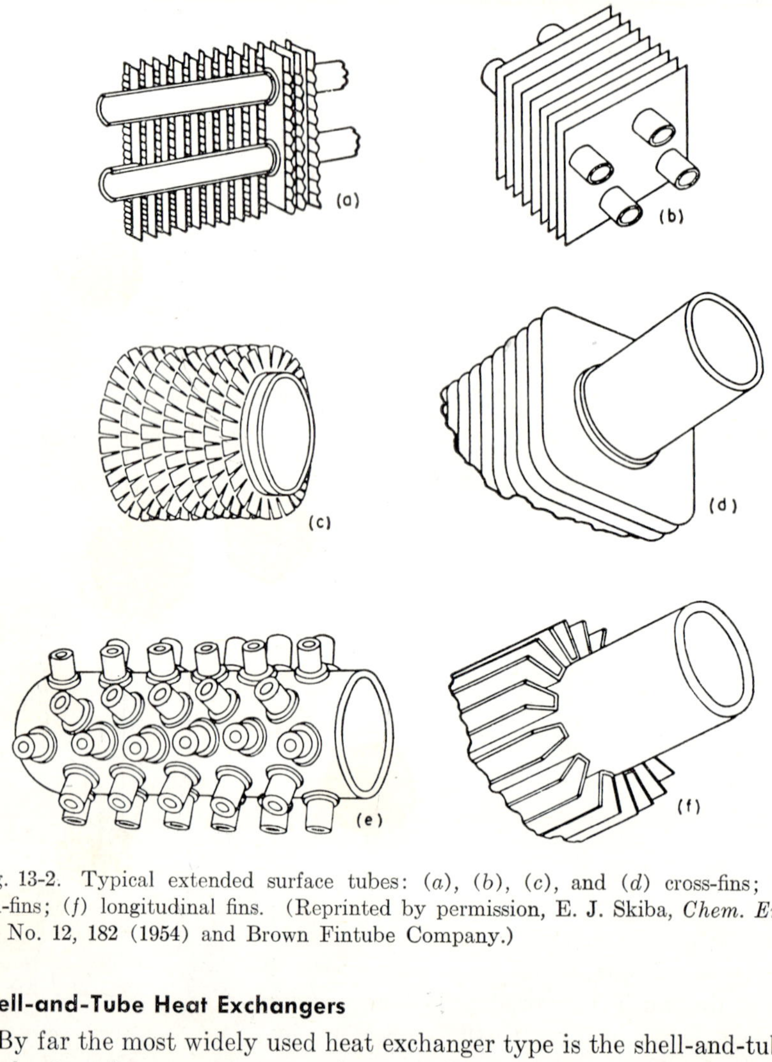


Fig. 13-2. Typical extended surface tubes: (*a*), (*b*), (*c*), and (*d*) cross-fins; (*e*) pin-fins; (*f*) longitudinal fins. (Reprinted by permission, E. J. Skiba, *Chem. Eng.*, *61*, No. 12, 182 (1954) and Brown Fintube Company.)

## Shell-and-Tube Heat Exchangers

By far the most widely used heat exchanger type is the shell-and-tube. As heat exchange surface requirements become large, double-pipe exchanger designs require excessive space. The shell-and-tube design, however, provides a large surface area per unit volume of space occupied. Although differing in certain design details, there are several standard shell-and-tube exchanger types fabricated by all exchanger manufacturers. Approximately ninety per cent of all heat exchange duties can be handled by one or the other of the basic designs.

### *Counterflow 1-1 Exchanger*

The most efficient exchange of heat can be accomplished by a counterflow arrangement in which one fluid flows in an opposite direction from the other fluid. In shell-and-tube equipment this is most nearly approached by the so-called "1-1 exchanger" which consists of one shell pass and one tube pass (Fig. 13-4). If a 1-1 exchanger is supplied with a float-

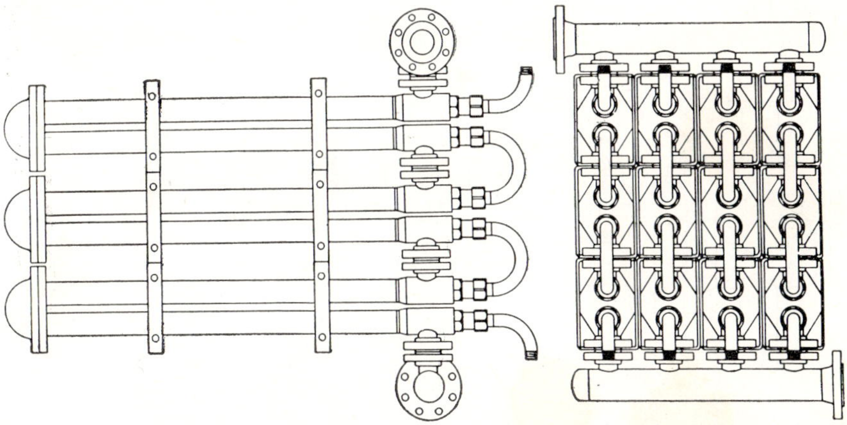

**Arrangement of four parallel banks of three sections connected in series. Tube connectors are used to connect the elements of each bank in series, with 90° threaded adaptors being supplied for the tube inlet and outlet connections. Inlet and outlet shell manifolds connect the four banks in parallel.**

Fig. 13-3. Typical arrangement of double-pipe exchanger sections: G-fin. (Courtesy of Griscom-Russell Company.)

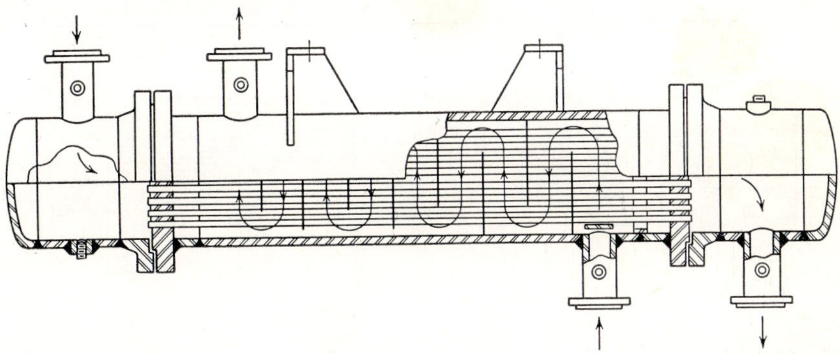

Fig. 13-4. One-one exchanger: single-pass shell, single-pass tubes, fixed tube sheet. (Courtesy of Engineers and Fabricators, Inc.)

ing head on the tube bundle (Fig. 13-9), the outlet nozzle on the floating-head side must pass through a packing gland, since it must be free to move with the floating head. The 1-1 exchanger is employed when tube-side friction losses must be kept to a minimum and when temperature requirements are such that true countercurrent flow must be used.

### 1-2 and 2-4 Exchangers

The 1-2 exchanger, as illustrated in Fig. 13-5, consists of one shell pass and two or more tube passes (even number of tube passes). This design is perhaps the most common of all shell-and-tube exchangers and

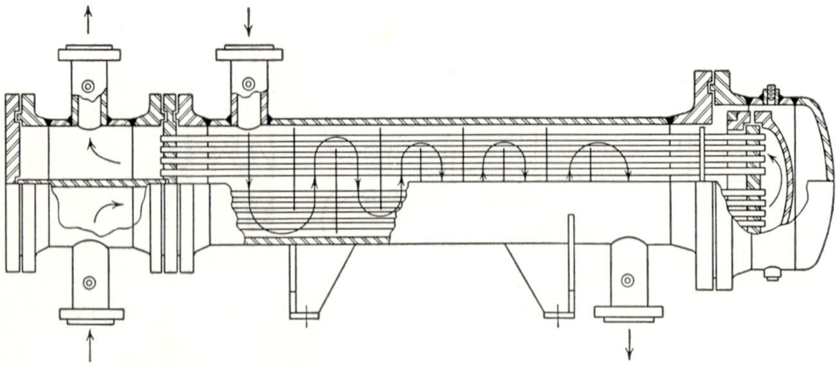

Fig. 13-5. One-two exchanger: single-pass shell, two-pass tubes, removable tube bundle. (Courtesy of Engineers and Fabricators, Inc.)

it is the simplest to maintain. In the 2-4 exchanger, as illustrated in Fig. 13-6, two shell passes in a single shell are provided by installation of a longitudinal baffle. This exchanger may have two or more even-numbered tube passes.

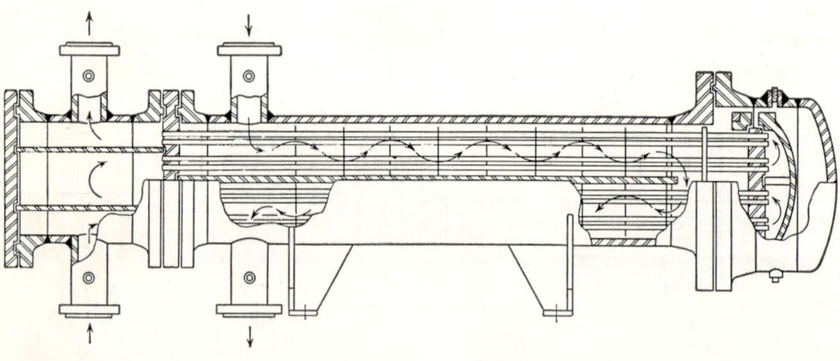

Fig. 13-6. Two-four exchanger: two-pass shell, four-pass tubes, removable tube bundle, cross baffles arranged for side-to-side flow. (Courtesy of Engineers and Fabricators, Inc.)

Shells with more than two shell passes are impractical but the advantages of additional shell passes can be obtained by connecting exchangers in series. In this manner two 1-2 exchangers connected in series would be equivalent to one 2-4 exchanger. The two-shell pass exchanger is more difficult to maintain, since the longitudinal baffle requires a more complex design. The two-pass shell is best used for clean fluids.

### *Reboilers*

The primary source of heat supply to most distillation columns is the reboiler. A reboiler is a heat exchanger designed to supply heat and permit easy control of distillation tower conditions. One of the earliest

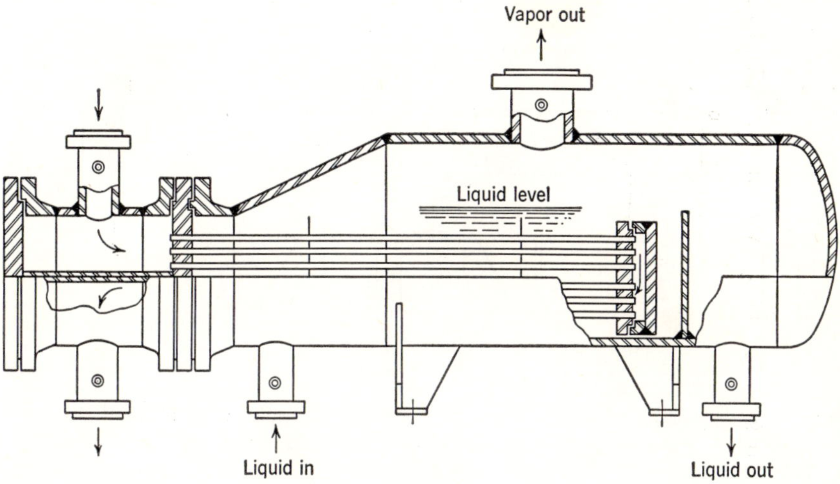


Fig. 13-7. Kettle-type reboiler: removable two-pass tube bundle. (Courtesy of Engineers and Fabricators, Inc.)

methods for supplying heat to the bottom of the distillation column was to insert an exchanger bundle in the bottom of the tower. This is still done in some cases but is not very popular because maintenance is more of a problem. Large bundles in towers are difficult to design and cumbersome to service.

Kettle-type reboilers (Fig. 13-7) are constructed similar to small evaporators. A weir is located at the end of the bundle so that the liquid completely covers it. A large vapor space is provided. All the heat furnished by the kettle-reboiler is returned to the tower by the overhead vapors from the reboiler. The kettle-reboiler, therefore, acts as a separation stage similar to a theoretical plate.

Thermo-syphon reboilers, as illustrated in Fig. 13-8, are the most commonly used in the process industry. The thermo-syphon design is similar to a standard shell-and-tube heat exchanger except that the inlet and outlets are usually located in the middle of the shell so that flow can be divided equally between the two halves of the exchanger. This divided flow arrangement insures low pressure drops on the shell side.

The liquid from the bottom of the distillation tower passes to the thermo-syphon reboiler and is partially vaporized. This less dense liquid-vapor mixture then passes back to the tower. The rate of circulation is governed by the pressure drop through the system and the driving force produced by the difference in liquid density on one side and

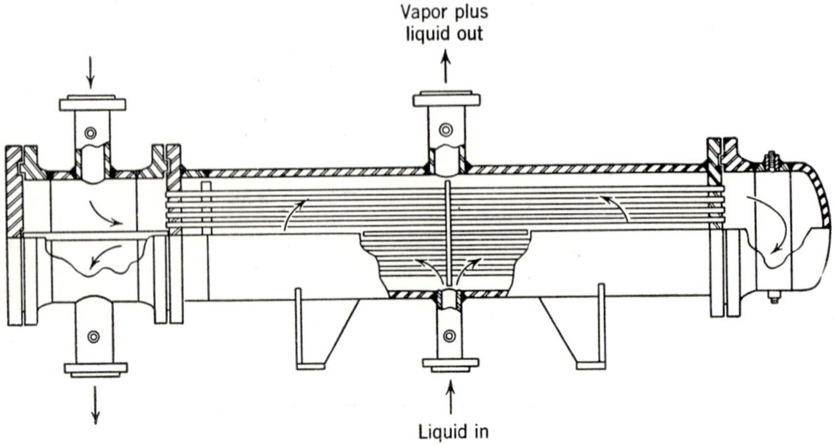


Fig. 13-8. Thermosyphon reboiler: fixed tube sheet, split flow in shell, two-pass tube bundle. (Courtesy of Engineers and Fabricators, Inc.)

liquid-vapor density on the other side of the system. Careful consideration must be given to the design of piping for thermo-syphon reboilers so that circulation will not be limited by excessive pressure drop in the piping system. A safety factor of at least two should be applied to all pressure drop calculations connected with reboiler piping design.

### Condensers

Condensers are usually standard 1-2 exchangers with vapor belt or annular distributors on the vapor inlet nozzle. This design permits better distribution of the vapor and eliminates excessive entrance losses. Other modifications are made such as the arrangement of inlet and outlet nozzles for split flow in order to keep pressure drop to a minimum.

### *Vertical Exchangers*

Although the majority of heat exchangers are placed in the horizontal position, certain situations arise in which a vertical exchanger is preferable. A vertical condenser is often preferable to the horizontal condenser when subcooling is required. Subcooling can be accomplished in a horizontal condenser by providing a loop seal so that a portion of the tubes is covered by liquid, but the amount of surface covered is not proportional to the height of liquid. Many operators believe that subcooling in vertical condensers is better controlled than in the horizontal condensers. The vertical unit, however, is more difficult to service and requires a clear space above it equal to its length so that the tube bundle can be removed.

Horizontal reboilers are the most widely used, but vertical reboilers are preferred for certain installations. In vertical reboilers the process stream is usually passed inside the tube. This is an advantage for fouling fluids since they are more economically handled inside the tubes.

### *Shell and Tube Heat Exchanger Nomenclature and Parts*

In Fig. 13-9 the important parts of a heat exchanger are shown and the usual terminology employed in describing these parts is given. In addition to the floating head design illustrated, several other types of heads can be provided. If expansion is not a problem, fixed-tube sheets and fixed-head exchangers can be used. These exchangers are cheaper than the floating head exchangers but provide no means for alleviation of differential expansion between the shell and tubes. Two types of floating heads are in use. If the head is the split ring type, the floating head cover must be removed before the tube bundle. The "pull-through" floating head is designed so that the head diameter is less than the diameter of the shell. The tube bundle can then be removed without removing the floating head cover. The pull-through arrangement, however, reduces the number of tubes that can be placed in a shell of a given diameter.

In order to avoid the use of a floating head and internal gasketing, it is possible to employ U-tubes to make up a tube bundle. The U-bends expand freely with temperature change. Since cleaning of such bundles is a difficult task, their use is restricted to only clean fluids.

The segmental baffles shown in Fig. 13-9 are the most widely used means for providing proper distribution of the shell-side fluids. Orifice-type baffles form annular spaces around each tube through which the shell fluid passes and can be used for clean, nonfouling fluids. The limited usefulness of the orifice-type baffle, as well as prejudices against it by many former users, has practically made it obsolete.

### Coil-in-Box Cooler

When adequate water supply is available, the submerged pipe coil, or coil-in-box cooler, is one of the cheapest cooling surfaces. Material to be cooled passes through a series of straight pipes connected by return bends. The entire coil is placed in a large trough or box constructed of concrete, wood, or steel. Cool water usually enters the bottom of the box and passes over a weir from the opposite end. It is simple to build pipe coils, and the boxes are usually large enough to accommodate addi-

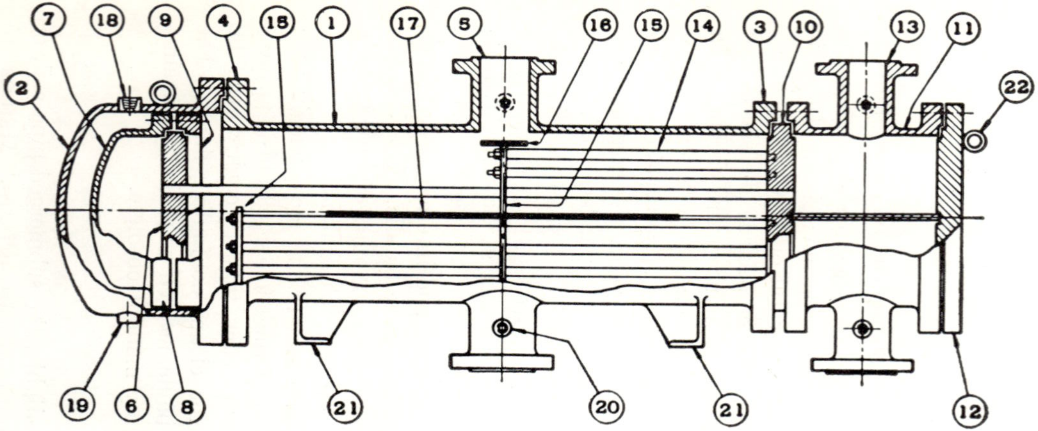


NOMENCLATURE

1. Shell
2. Shell Cover
3. Shell Channel End Flange
4. Shell Cover End Flange
5. Shell Nozzle
6. Floating Tubesheet
7. Floating Head
8. Floating Head Flange
9. Floating Head Backing Device
10. Stationary Tubesheet
11. Channel
12. Channel Cover
13. Channel Nozzle
14. Tie Rods and Spacers
15. Transverse Baffles or Support Plates
16. Impingement Baffle
17. Longitudinal Baffle
18. Vent Connection
19. Drain Connection
20. Test Connection
21. Support Saddles
22. Lifting Ring

Fig. 13-9. Heat exchanger parts and nomenclature. [Reprinted by permission, *Standards of the Tubular Exchanger Manufacturers Association*, 3d ed. (1952).]

tional lengths of coils as other cooling loads become necessary. This flexibility, together with the low cost and ease of replacing corroded or eroded tubes, makes the coil-in-box cooler popular for cooling corrosive or erosive streams. They are particularly useable when the outlet temperature is not critical, as in cooling materials to be pumped to storage. Temperature control of the coil-in-box cooler is much more difficult than for the usual shell-and-tube exchanger.

### Pipe Coils

One of the easiest, although not always the most efficient, means of supplying heat to or removing heat from a tank or a piece of pilot plant equipment is with a helical pipe coil. Such coils are easily fabricated, and can be placed in a small tank made from large diameter pipe and used as a sample cooler or as a pilot plant cooler. Pipe coils are also used for heating contents of tanks and reactors.

### Trombone or Trickle Coolers

The cooling of extremely corrosive materials is often a difficult and expensive job. The trombone, or trickle, cooler has been used with marked success. It consists of straight lengths of pipe connected in series by return bends. Water is allowed to trickle over the pipes and is collected in a trough beneath the unit. Some of the water evaporates as it moves down the pipes. The pipes can be fabricated of practically any material that will withstand the fluid flowing inside. Ceramic tubes have been particularly popular for corrosive materials such as wet HCl.

### Air-Cooled Exchangers

Many portions of the world suffer from serious water shortages. The development of the air-cooled exchanger has permitted process plant construction even in areas where water scarcity is a definite problem. These units usually consist of a large number of tubes with transverse fins. The fluid to be cooled passes through the tubes and air is forced past the tubes by either an induced or forced draft fan. Because of the high resistance of gas films these air coolers require large surface areas, and therefore a large space, per Btu of heat transferred.

## FABRICATION OF SHELL-AND-TUBE HEAT EXCHANGERS

### Standards

The Tubular Exchanger Manufacturers' Association has published detailed standards for the design and construction of shell-and-tube heat exchangers.[23] The mechanical standards have been divided into three parts representing three different classes of heat exchangers, Classes *R*, *C*, and *A*. Class *R* is designed to meet the requirements of petroleum refinery and high pressure petrochemical operations. Class *C* is designed to meet general process conditions involving moderate pressures and temperatures and relatively noncorrosive fluids. Class *A* is designed for the use of alloy materials of construction for handling particularly corrosive or high temperature fluids. Class *R* specifies larger corrosion allowances than Class *C*, and in general, has more rigid construction requirements.

Overall tube lengths are specified by these standards as eight, twelve, or sixteen feet. Three-quarter and one-inch diameter tubes are the most common, although the use of one and-a-quarter and one and-a-half, as well as smaller sizes is permitted where special conditions warrant.

Shell sizes have also been standardized. Sizes up to twenty-three inch, inside nominal diameter, are made of pipe. Above twenty-three inch, they are fabricated from rolled plate.

The minimum pitch of tubes is set at one and-a-quarter times the outside diameter of the tube. Minimum baffle spacing is one-fifth of the shell inside-diameter, or two inches, whichever is greater. Vent and drain connections are three-quarter inch. These and many other standards, including dimensional tolerances, are carefully specified by the Exchanger Manufacturers' Association. One of the greatest benefits of such standardization is in the simplicity of stocking spare parts.

## Fabrication Procedure

Various parts of a shell-and-tube heat exchanger are illustrated in Fig. 13-9. Reference to this figure will aid in understanding the following discussion. The shell portion of the heat exchanger, being a pressure vessel, is fabricated in the same manner as that discussed in Chapter 12. These shells are fabricated in accordance with the API-ASME or ASME Code for Pressure Vessels. Automatic welding is used almost exclusively on the longitudinal seams and also on most of the circumferential seams. The large flanges needed for the channel and shell covers require careful facing operations.

Assembly of the tube bundle is an interesting operation. Tube sheets are cut from steel or brass plate. The cutting operation is usually a mechanical operation for brass, but welding-torch cutters are used for steel tube sheets. The tube holes are then drilled in the tube sheet after being laid out in their proper pitch and orientation. Then grooves are milled into the tube holes. Holes are also drilled in the cross baffle to accommodate the tubes. Baffles, tube sheet, tubes, tie rods, and spacers are all brought to an assembly jig where the tube sheets are placed securely at the correct spacing for a given exchanger. Tie rods are then screwed into the tube sheets. Baffles are slipped onto the support rods and firmly located in their proper place. Tubes are then passed through the tube sheet and baffle, and the tubes are fixed in place by an expanding operation. They are set in a preliminary fashion by forcing a plug against the tubes. The plug prevents the tube from turning when the roller expander is inserted.

A cross-section of an expanding operation is illustrated in Fig. 13-10. The tubes are expanded into grooves in order to produce a stronger joint. Expanding of tubes requires experienced operators, since overexpansion will unnecessarily thin the metal and produce rapid tube failure.

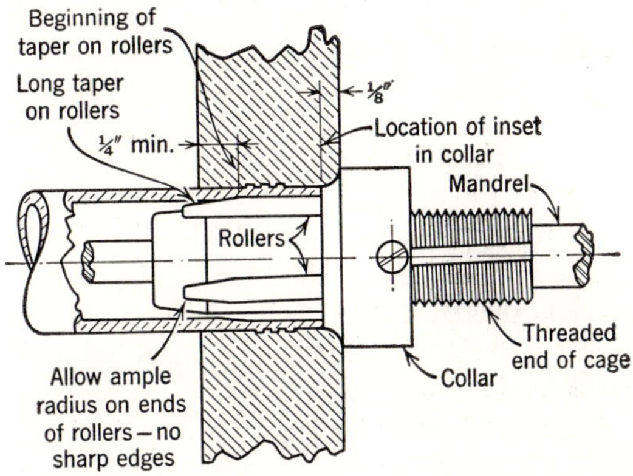


Fig. 13-10. Tube expanding operation. (Courtesy of Bridgeport Brass Company.)

Tubes may be a number of different metals: stainless steel, various copper alloys, including bronze, brass, or admiralty metal. Aluminum, Monel, nickel, and other materials are widely used in severe services. It is not uncommon to find conditions where the fluids both inside and outside the tubes are extremely corrosive, and in addition, require a different metal on each side. Tubes which combine two metals can be used to meet this problem. The metals are permanently bonded together, one facing the inside and the other the outside. Combinations include steel and admiralty, aluminum with copper or brass, and alloy steels with copper alloys.

Using the same pitch (center to center tube dimension) it is possible to place more tubes in a given shell employing a triangular pitch (locating tubes at the apices of equilateral triangles) than by employing a square pitch (locating tubes at the corners of a square). The use of the square pitch, however, is recommended whenever fouling fluids are to be encountered outside of the tube. This arrangement affords larger cleaning lanes and, therefore, simpler maintenance. For cases in which fouling fluids must be handled inside the tubes, sizes smaller than ¾" OD and No. 14 BWG* are usually avoided.

* BWG refers to Birmingham Wire Gage and is a measure of tube thickness. Handbooks list the thickness corresponding to the various BWG numbers.

After completion of the bundle, it is brought to a test rack where a hydraulic test is applied. Bundles are then lowered vertically into the exchanger shells and the final hydraulic test of the exchanger is made. After the outside of the shell is painted with a rust-preventative paint and all flanges are covered to prevent damage, the unit is ready for shipment.

## HEAT EXCHANGER DESIGN

The basic equation for heat exchanger design is perhaps familiar to most engineers who have had an introductory course to the subject of heat transfer.

$$A_O = \frac{q}{U_d \Delta T} \quad \text{or} \quad \frac{qR_d}{\Delta T} \tag{1}$$

where $A_O$ = required effective surface of the heat exchanger based on the outside surface area of the tubes, in sq ft

$q$ = heat transferred in Btu/hr

$\Delta T$ = mean temperature difference between hot and cold fluid (this is the driving force)° F

$U_d$ = design overall coefficient of heat transfer in Btu/(hr)(° F)(sq ft) based on outside surface area

$R_d$ = design overall resistance $1/U_d$, based on outside surface area

The amount of heat transferred is directly proportional to the surface area and the driving force $\Delta T$, and is inversely proportional to the overall resistance to the flow of heat $R_d$. The total resistance $R_d$ is composed of a number of resistances in series, and, as electrical resistances, they are additive. These resistances are:

1. Resistance of fluid film on inside of tube.
2. Resistance of scale on inside of tube, $r_i$.
3. Resistance of tube wall.
4. Resistance of scale on outside of tube, $r_O$.
5. Resistance of fluid film on outside of tube.

Hence
$$\frac{R_d}{A_O} = \frac{1}{h_i A_i} + \frac{1}{r_i A_i} + \frac{L_w}{A_m k_w} + \frac{r_O}{A_O} + \frac{1}{h_O A_O}$$

or
$$R_d = \frac{A_O}{h_i A_i} + r_i \frac{A_O}{A_i} + \frac{A_O L_w}{A_m k_w} + r_O + \frac{1}{h_O}$$

where $h_i$ = inside coefficient, Btu/(hr)(sq ft)(° F)

$h_O$ = outside film coefficient, Btu/(hr)(sq ft)(° F)

$r_i$ = inside scale resistance or fouling factor, (° F)(sq ft)(hr)/Btu

$r_O$ = outside scale resistance or fouling factor, (° F)(sq ft)(hr)/Btu

$L_w$ = thickness of tube, ft

$k_w$ = conductivity of tube, Btu/(hr)(° F)(ft)
$A_m$ = average of $A_i$ and $A_O$, sq ft
$A_O$ = outside surface area of tube, sq ft
$A_i$ = inside surface area of tube, sq ft

Since $R_d = 1/U_d$ by definition, then

$$U_d = \frac{1}{\dfrac{A_O}{h_i A_i} + r_i \dfrac{A_O}{A_i} + \dfrac{A_O L_w}{A_m k_w} + r_O + \dfrac{1}{h_O}} \tag{2}$$

The overall coefficient $U_d$ is the more familiar term in commerce and will be used here.

The major problem in heat exchanger design is the evaluation of the individual resistances that appear in eq. 2. Estimation of the inside and outside film, especially the outside coefficient, have been the most perplexing problems.

### Inside Film Coefficient

Nusselt[16] showed by dimensional analysis that the inside film coefficient was related to the properties of the fluid and the diameter of the tube as follows:

$$\frac{h_i D}{k} = \alpha \left(\frac{DG}{\mu}\right)^a \left(\frac{C_p \mu}{k}\right)^c \left(\frac{L}{D}\right)^i \tag{3}$$

where $D$ = inside diameter, ft
$G$ = mass velocity, lb/hr-sq ft
$\mu$ = viscosity of fluid, lb/ft-hr
$k$ = thermal conductivity, of fluid, Btu/(hr)(° F)(sq ft/ft)
$C_p$ = heat capacity of fluid, Btu/(lb)(° F)
$L$ = length of tube, ft

$\dfrac{DG}{\mu}$ = Reynolds' number

$\dfrac{C_p \mu}{k}$ = Prandtl number

It is important to realize that each of the terms are dimensionless and consistent units must be employed within each term.

$\alpha$, $a$, $c$, and $i$ are constants which must be determined experimentally, and, following Nusselt's derivation in 1910, considerable experimental work was done to evaluate consistent constants that would apply to all fluids. Most of the early work was done with air and water, and the correlations were based on properties developed at the temperature of the film. This temperature would appear to be correct theoretically, but it was tedious to employ in design since it had to be estimated by trial and error. In 1929 Morris and Whitman[14] attempted a correlation based on a number of fluids and used the average bulk temperature instead of film temperature. They obtained a relation that was consistent for fluids being cooled and another relation for fluids being heated. Dittus and Boelter[5] followed with their well-known equation based on bulk temperature and flow in the turbulent region (above Reynolds number of 10,000) where the ratio $L/D$ had no effect.

Heating: $$\frac{h_i D}{k} = 0.024 \left(\frac{DG}{\mu}\right)^{0.8} \left(\frac{C_p \mu}{k}\right)^{0.4}$$

Cooling: $$\frac{h_i D}{k} = 0.026 \left(\frac{DG}{\mu}\right)^{0.8} \left(\frac{C_p \mu}{k}\right)^{0.3}$$

The desire to find one equation which would express both heating and cooling encouraged further investigation. Sieder and Tate[22] reasoned that the relation between the viscosity in the bulk of the fluid to that at the wall produces a change in the velocity distribution of the individual fluid streams. One type distribution is obtained in heating and another in cooling. In cooling the viscosity at the wall is greater than in the bulk of the fluid, and the opposite is true for heating. They decided to introduce another dimensionless term, the ratio of the viscosity in the bulk of the fluid to that at the wall, $\mu/\mu_w$. The exponent was determined experimentally, and the following equations applicable to both heating and cooling were developed from the data:

Streamline flow (Re = 2100 or less)

$$\frac{h_i D}{k} = 1.86 \left(\frac{DG}{\mu}\right)^{1/3} \left(\frac{C_p \mu}{k}\right)^{1/3} \left(\frac{L}{D}\right)^{-1/3} \left(\frac{\mu}{\mu_w}\right)^{0.14} \tag{6}$$

Turbulent flow (Re = 10,000 or above)

$$\frac{h_i D}{k} = 0.027 \left(\frac{DG}{\mu}\right)^{0.8} \left(\frac{C_p \mu}{k}\right)^{1/3} \left(\frac{\mu}{\mu_w}\right)^{0.14} \tag{7}$$

The transition region which lies in between Reynolds' numbers of 2100 and 10,000 was not easily represented by an equation. Instead it was found simpler to plot eq. 6 and 7 as follows:

$$\frac{DG}{\mu} \text{ versus } \left(\frac{h_i D}{k}\right) \left(\frac{C_p \mu}{k}\right)^{-1/3} \left(\frac{\mu}{\mu_w}\right)^{-0.14}$$

This produces a single straight line on logarithmic coordinates for the turbulent region and a series of lines for various values of $L/D$ in the streamline region. The transition region was then determined by extrapolating the two sets of curves and checking with experimental data.

The Sieder-Tate curves and equations have become the standard and are widely used in heat exchanger design for all fluids. Other equations have been developed which are more accurate for specific fluids. The Sieder-Tate curves, however, have proven the most widely applicable and are included in the Standards of the Tubular Exchanger Manufacturers Association.[23]

The Sieder-Tate equations can be rearranged by multiplying each side by $c_p\mu/DGc_p$ and the form

$$\frac{h}{C_pG} = A\left(\frac{C_p\mu}{k}\right)^{-2/3}\left(\frac{DG}{\mu}\right)^{-0.2}\left(\frac{\mu}{\mu_w}\right)^{0.14}$$

results. This form is particularly useful for gases since $c_p\mu/k$ tends to be constant over normal temperature ranges for a gas, and the problem of finding values of gaseous conductivities, which is a difficult one, is eliminated.

## Outside Film Coefficient

### *Double Pipe Exchanger*

The film coefficient outside the tubes presents a more perplexing problem. In the simple case of a plain double-pipe exchanger the familiar engineering procedure of employing an equivalent diameter $D_e$ in place of the term $D$ is followed using the equation for the inside film coefficient. This equivalent diameter is defined as four times the hydraulic radius which may be four times the cross sectional area divided by the wetted perimeter or the heated perimeter. The former is the same equivalent diameter as is used in pressure drop calculation and is $D_2 - D_1$. ($D_2$ is the inside diameter of the outer pipe and $D_1$ the outside diameter of the inner pipe.) Based on the heated perimeter, $D_e$ becomes $(D_2{}^2 - D_1{}^2)/D_1$. Attempts at correlation of data using both types of equivalent diameter produced equal accuracy. Until more data become available, however, the use of the same equivalent diameter required for pressure drop calculations $(D_2 - D_1)$ seems most convenient.

### *Extended Surface*

The outside coefficient of an extended surface exchanger is complicated by temperature variations along the fin. Some weighted value of the heat transfer coefficient for the combination of fin and plain-tube surface must be obtained. It is difficult to establish a satisfactory base area on the outside of the tube because of the many variations of fin arrangements that are possible. Hence, all extended surface calculations are based on the inside surface.

$$\frac{1}{U_i} = \frac{1}{h_i} + \frac{1}{h_{f_i}} \tag{8}$$

where $h_{f_i}$ is the fin-side coefficient based on the inside tube surface area. $h_{f_i}$ is related to the average outside coefficient for both fin and bare surface, $h_f$, by the expression

$$q = h_{f_i} A_i \, \Delta T = h_f (A_{f_e} + A_O) \, \Delta T$$

or,

$$h_{f_i} A_i = h_f (\Omega A_f + A_O) \tag{9}$$

where $A_{f_e}$ = an effective fin area equivalent to the total fin area $A_f$ multiplied by the fin efficiency $\Omega$

$A_O$ = outside tube surface area exclusive of area beneath the fins

$q$ = heat transferred

The fin area is not as effective as the outside tube surface and the fin efficiency serves to correct this area to an equivalent amount of outside surface. Murry[15] and Gardner[9] have derived an expression for the fin efficiency

$$\Omega = \frac{\tan h \, mb}{mb}$$

$$\text{where } m = \left(\frac{h_f P}{k a_x}\right)^{1/2} \tag{10}$$

$k$ = conductivity of fins and tube, Btu/(hr)(° F)(sq ft/ft)

$P$ = perimeter of fins, ft

$a_x$ = cross sectional area of fins, sq ft

$b$ = height of fin, ft

The fin efficiency, therefore, provides a means of accounting for the variation of material of construction, size, and arrangement of fins.

DeLorenzo and Anderson[4] correlated data for a number of extended surface arrangements using the Sieder-Tate form of the equation plotting:

$$\left(\frac{h_f}{C_p G}\right)\left(\frac{c_p \mu}{k}\right)^{2/3}\left(\frac{\mu}{\mu_w}\right)^{-0.14} \quad \text{versus} \quad \frac{D_e G}{\mu}$$

*De* is the equivalent diameter defined as

$$4\left[\frac{\frac{\pi}{4}(D_2^2 - D_1^2) - \text{cross sectional area of fins}}{\text{total wetted perimeter of fin side}}\right]$$

where $D_1$ = outside diameter of inside tube

$D_2$ = inside diameter of outside tube

This correlation enables calculation of $h_f$ which can be converted to $h_{f_i}$ using Eq. 9. Knowing $h_i$, the inside film coefficient, the clean inside coefficient $U_i$ can be calculated. The design value of $U_i$ must be determined by adding the appropriate scale or fouling factors. The fouling factor on the fin side has the same complex relationship to the resistance based on the inside area as does $h_f$. Hence the factor must be added to $1/h_f$ and the resulting design $h_f$ used to determine a design $h_{f_i}$ from Eq. 9. Then

$$\frac{1}{U_{d_i}} = \frac{1}{h_{f_i}} + \frac{1}{h_i} + r_i$$

### Shell and Tube Exchangers Outside Coefficients

The outside or shell side coefficient of a shell and tube heat exchanger has proved the most difficult coefficient to correlate. It was early agreed that for forced convection outside of tubes an equation of the same form as that used for flow inside the tubes would apply.

$$\frac{hD_O}{k} = a\left(\frac{D_O G}{\mu}\right)^n \left(\frac{C_p\mu}{k}\right)^m \tag{11}$$

where $D_O$ = outside diameter of the tube, ft

But because of the variations in baffle and tube arrangement it has been a real problem to correlate data for all shell and tube exchangers with one equation. Colburn[2] proposed the equation

$$\frac{h_O D_O}{k} = a\left(\frac{D_O G_{max}}{\mu}\right)^{0.6} \left(\frac{C_p\mu}{k}\right)^{1/3} \tag{12}$$

where $a$ = 0.33 for staggered tube arrangement and 0.26 for in-line tube arrangement

Essentially these same data are presented in the Standards of the Tubular Exchange Manufacturers Association.[23] $G_{max}$ is there called $G_{eff}$ (effective mass velocity) which is defined as the pounds per hour flowing divided by an effective cross flow area. This effective cross flow area represents the factor most difficult to define and no attempt is actually made to do so in the Standards. It must somehow take into account design clearances, baffle spacing, leakages, and by-passing. Such factors vary with the manufacturer. In our present state of knowledge each manufacturer has developed some means of obtaining an effective area that permits good correlation of design data with actual operating performance. Curiously enough it has not been until recently that a systematized approach to the problem of shell side coefficients for commercial exchangers* has been made. A research program at the University of Delaware under the joint sponsorship of the ASME, API, TEMA, and many exchanger manufacturers and operating companies is in progress. It involves the study of idealized models in order to isolate the variables, and will be followed by a study of full-scale commercial

* Short[20] has made a study on a laboratory exchanger which has been of great aid to exchanger designers.

exchangers. It is hoped that useful and more reliable correlations will be produced from this work. For the reader who does not have access to the various confidential company charts or procedures for empirically estimating the outside coefficient, the equation presented by Donohue[7] is suggested as the most reliable. Donohue proposed the equation:

$$\frac{hD_O}{k} = 0.22\left(\frac{DG_e}{\mu}\right)^{0.6}\left(\frac{C_p\mu}{k}\right)^{0.33}\left(\frac{\mu}{\mu_w}\right)^{0.14} \tag{13}$$

where $G_e$ is defined as $W/S_e$ and $W$ = lb/hr of fluid flowing

$$S_e = \sqrt{(\text{crossflow area})(\text{baffle hole area})} = \sqrt{S_B S_c}$$

The crossflow area $S_c$ is arbitrarily based on the inside diameter of the shell.

$$S_c = [(ID) - (D_O)(N)][B] \tag{14}$$

where $B$ = baffle spacing, ft
$N$ = number of tubes at or near inside diameter
$ID$ = inside diam of shell, ft

The baffle hole area is:

$$S_B = \text{Area of segment} - \text{area occupied by tubes}$$

Donohue's correlations based on this effective area exhibit reasonable accuracy when applied to commercial exchangers. There are many other correlations for arriving at an effective area involving much more tedious calculations. The order of accuracy of such correlations, however, does not appear to justify the additional effort required.

### Condensing Coefficients

The condensation of vapors on the shell side of horizontal exchangers is an important operation. Unfortunately the theory, although voluminous, is very inadequate. It can be said with impunity that the design of a process condenser, particularly one involving mixtures of vapors is an art. The novice can best use typical overall coefficients and avoid a large amount of calculations, the interpretation of which is difficult and questionable. Using a highly idealized mechanism Nusselt[17] considered the condensation of pure vapors on both horizontal and vertical surfaces. The following expressions were derived based on viscous flow.

Horizontal tubes

$$h = 0.725(k_f^3\lambda\rho_f^2 g/D_O\mu_f\,\Delta t)^{1/4} \tag{15}$$

or

$$h\left[\frac{\mu_f^2}{k_f^3\rho_f^2 g_c}\right]^{1/3} = 1.51\left[\frac{4G'}{\mu_f}\right]^{-1/3} \tag{16}$$

Vertical tubes

$$h = 0.943\ (k_f^3\lambda\rho_f^2 g_c/\mu_f L\,\Delta t)^{1/4} \tag{17}$$

$$h\left(\frac{\mu_f^2}{k_f^3\rho_f^2 g_c}\right)^{1/3} = 1.47\left(\frac{4G'}{\mu_f}\right)^{-1/3} \tag{18}$$

where $\lambda$ = latent heat of condensation, Btu/lb

$g_c$ = 32.17

$\Delta t$ = temperature difference between saturated vapor and condensing surface, ° F

$D_O$ = OD of tube, ft

$L$ = tube length, ft

$W$ = lb/hr

$G'$ = Mass rate of condensate, (lb/hr)(sq ft)

$$\text{Vertical tube: } G' = \frac{W}{\pi D_O}$$

$$\text{Horizontal tube: } G' = \frac{W}{L}$$

$\rho$ = density of condensate, lb/cu ft

Subscript $f$ refers to properties at the film temperature.

Because of the similarity of eq. 17 and 18 they are often written as one equation using an average constant 1.50 and an appropriate value of G′. Horizontal tubes in vertical rows present an additional problem since splashing increases turbulence. In addition thicker films develop on each successive tube which makes the lower tubes less effective. If the tubes are staggered, results would be expected to differ from that obtained for the in-line arrangement.

McAdams,[13] Donohue,[6] Kern,[12] and others have proposed methods for calculating the $G'$ for banks of tubes to be used in the Nusselt equation. Kern suggests the following:

$$\text{Vertical tubes:} \quad G' = \frac{W}{N_t D_O}$$

$$\text{Horizontal tubes:} \quad G' = \frac{W}{N_t^{2/3} L}$$

where $N_t$ = total number of tubes

$W$ = total condensate flow rate, lb/hr

For simple vapors approximate condensation coefficients for the outside of horizontal tubes and for the inside and outside of vertical tubes can be predicted from the equations of Nusselt. For turbulent flow of the condensate and high vapor velocities there result larger coefficients than predicted from the Nusselt equation. Methods for calculating coefficients under these conditions have been discussed.[13]

Often vapors enter a condenser superheated. Part of the condenser acts as a gas cooler and that portion in gas cooling service operates with a much lower transfer coefficient. Such situations are best handled by calculating each section separately. A weighted overall transfer coefficient and a weighted $\Delta t$ are reported.

$$(U_{\text{wt}}) = \frac{(U_x)(A_x) + (U_G)(A_G)}{A_x + A_G}$$

$x$ refers to the condensing section
$G$ refers to the gas cooling section

$$\Delta t_{\text{wt}} = \frac{q}{\dfrac{q_x}{\Delta t_x} + \dfrac{q_G}{\Delta t_G}}$$

where $q$ = total heat transferred
$q_x$ = heat transferred in condensing section
$q_G$ = heat transferred in gas cooling section
$\Delta t$ = mean temperature difference

The condensation of a single vapor in the presence of noncondensables or the condensation of mixed vapors is a difficult process to analyze. In each case vapor must diffuse through a gaseous film before reaching the condensing surface and the phenomenon is one of heat transfer combined with mass transfer. Rigorous methods for considering both diffusion and heat transfer in the determination of a condensation coefficient have been described.[3] These methods are admittedly tedious and involve trial-and-error solutions. Many "short-cut" methods have been proposed, therefore, involving certain simplifying assumptions which were known to be incorrect.

The methods in current use can be divided into two classes. One considers heat transfer alone, assuming that the heat flows through a series of resistances, including a vapor film, a condensate film, tube wall, dirt film, and a cooling liquid film. The other, the more rigorous, considers both heat and mass transfer. It is neither possible nor worthwhile to review the various methods which have been proposed. Curiously enough, the accuracy of any of the available methods is poor. The best procedure is to adopt the method most convenient for one's own use and employ it continuously. Calculations based on this method should be compared at every opportunity with operating data so that the designer can develop a sense for the relationship between the actual coefficient and a calculated coefficient. The members of the Tubular Exchanger Manufacturers' Association are well aware that existing condenser calculations are inadequate and attempts are being made to remedy the situation.

In the condensation of multicomponent mixtures the heavy materials condense first at the higher temperatures and are subcooled as they progress toward the end of the exchanger. The exchanger designer should be supplied with a condensation curve for use in designing such exchangers (Fig. 13-11). The temperature of the condensing mixture is plotted versus the heat removed from the mixture and is obtained from flash vapor

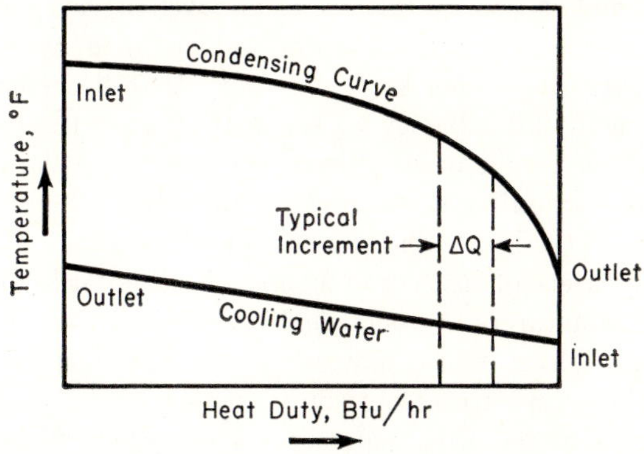


Fig. 13-11. Condensation curve.

equilibrium calculations. This curve is a great aid in designing a condenser, since it indicates the desuperheating and subcooling regions and also is used in obtaining a weighted temperature difference, as described below.

## Boiling Coefficients

The calculation of boiling coefficients in the design of reboilers and similar equipment is, at this stage of development, practically impossible. Most designs are based on some allowable heat flux ($q/A$ in Btu/(hr)(sq ft), since $q/A = U\ \Delta t$, $U$ can be determined from the allowable heat flux and an acceptable value of $\Delta t$ which is based on practice. For organic liquids in forced circulation Kern[12] suggests a maximum of 20,000 Btu/(hr) (sq ft) and 12,000 for natural circulation. As in condensers, often a portion of the reboiler is used in the transfer of sensible heat, and in these cases the sensible heat region and the boiling region are considered separately.

## Pipe Wall Resistance

The pipe wall resistance given in eq. 2 is readily calculated, since the conductivities of most metals used in heat exchangers are well known. It is most often very small in comparison to the other resistances and can be neglected.*

* Not true for liquid metal heat transfer where pipe wall resistance may be controlling.

### Fouling Factors

During normal use the inside and outside tube walls of an exchanger become coated with a deposit of dirt, rust, or algae, depending upon the fluids flowing through the exchanger. These deposits materially reduce the capacity of an exchanger and must eventually be removed. Most plants have found it economical to clean exchanger tubes during the normal shutdown or turn-around period, which in many plants may be only once a year. It is, therefore, necessary that the exchanger be designed to operate at full capacity with a coating of dirt and scale equivalent to that which would occur during use between cleanout times. A fouling resistance or fouling factor must be included in the overall transfer coefficient for both the inside and the outside of the tube. Based on experiences of many users of heat exchangers, the Standards of the Tubular Exchanger Manufacturers' Association[23] presents a list of recommended fouling factors. These are for a wide variety of fluids but not all conceivable situations, and are at best only guesses.

The fouling resistance often represents a large percentage of the total resistance in most exchangers. It is therefore understandable that young engineers become discouraged when faced with the problem of tedious calculations for inside and outside tube coefficients when they realize that the large resistances due to fouling are only estimates.

### Mean Temperature Difference

If it is assumed that the overall heat transfer coefficient is constant over the entire exchanger, that a steady state exists, that heat capacity is constant, that no changes in phase occur (except under isothermal conditions), and that flow through the exchanger is either parallel or countercurrent, it can be shown that the mean temperature difference in Eq. 1 corresponds to a logarithmic mean of the difference in temperature between the hot and cold fluid at one end of the exchanger and the hot and cold fluid at the other. This is expressed mathematically as

$$\Delta t_m = \frac{\Delta t_2 - \Delta t_1}{\ln \dfrac{\Delta t_2}{\Delta t_1}}$$

where ln = natural logarithm
$\Delta t_m$ = mean temperature difference
$\Delta t_1$ = temperature difference at one end of exchanger
$\Delta t_2$ = temperature difference at other end

Many charts and nomographs have been prepared to expedite this calculation. Often, however, the use of certain charts requires more effort than the small work involved in simply making the indicated calculation on a slide rule.

Shell and tube exchangers which have more than one tube pass no longer operate in true countercurrent flow. The log mean temperature difference, therefore, is not applicable. Correction factors have been developed which can be applied to the log-mean temperature difference to yield a true temperature difference for use with various arrangements of shell and tube heat exchangers, and other exchangers such as cross flow and film-type or trickle coolers. These factors are plotted on convenient curves which are reproduced in the Standards of the Tubular Exchanger Manufacturers' Association. ($\Delta t_m = F_T LMTD$ where $F_T$ is the correction factor obtained from the chart.)

In the condensation of mixed vapors the temperature progressively drops throughout the condensation. Since it was assumed in the derivation of the log-mean temperature difference that any phase change was isothermal, the use of the log-mean temperature difference is not justified. Instead a weighted temperature difference must be used which is calculated from the condensation curve for the mixed vapors. Referring to Fig. 13-11, the cooling water is assumed to flow countercurrent to the condensing mixtures. Small increments of temperature are chosen along the temperature scale and a log-mean temperature difference is calculated over each increment. This procedure is justifiable since with small temperature increments the variation in vapor temperature is slight. The ratio of the heat extracted in the increment chosen divided by the log mean is calculated. The sum of these incremental values is related to the weighted temperature difference:

$$\Sigma \frac{q}{\Delta t} = \Sigma UA = \frac{q_T}{\Delta t_w}$$

where $\Delta t_w$ = weighted $\Delta t$
$q_T$ = total heat load

### Caloric Temperature or Average Fluid Temperature

In the expression for the log-mean temperature it is assumed that the overall transfer coefficient remains constant throughout the exchanger. This is seldom true, and Colburn attempted to correct for this assumption by employing average temperatures which can be used in evaluating the fluid properties in the film coefficient equation. These average temperatures are often called caloric temperatures and are defined as:

Hot fluid: $$T = T_1 + F_c(T_2 - T_1)$$

Cold fluid: $$t = t_1 + F_c(t_2 - t_1)$$

where $F_c$ = factor unique for a given system, subscripts 1 and 2 refer to opposite ends of the exchanger

The factor, $F_c$, which is unique for a given system, can be obtained from the generalized chart for hydrocarbons in TEMA.[23] For nonviscous liquids (below one centipoise) at moderate temperature ranges, the value of $F_c$ is 0.5, and the caloric temperature then is simply the average of the inlet and outlet temperatures. Such is the case for most inorganic solutions.

Sieder[21] has suggested the following rules: (1) For water-cooled units, use an $F_c$ of 0.3 for the shell side hot fluid; (2) for steam-heated units, use $F_c = 0.55$ for the shell side fluid being heated; (3) for oil-to-oil exchangers use 0.45 for both hot and cold fluids. The average temperatures thus calculated are used to evaluate the properties appearing in the heat transfer film coefficient equation.

### Pipe Wall Temperature

In the Sieder-Tate equations it will be noted that a value of the pipe wall temperature is required in order to evaluate $\mu_w$. Sieder[21] points out that this can frequently be assumed to be equal to the outlet water temperature in oil-water coolers. It may also be estimated by simple calculation based on film coefficients which were calculated assuming $\mu/\mu_w = 1.0$. The following equations are used:

Cold fluid in tubes: $t_w = t + R(T - t)$

Hot fluid in tubes: $t_w = T - R(T - t)$

where $t_w$ = pipe wall temperature
$t$ = average cold fluid temperature
$T$ = average hot fluid temperature

$$R = \frac{h_O}{h_i(A_i/A_O) + h_O} \quad \text{for standard tubes}$$

$$R = \frac{h_i h_{fi} A_i}{h_f(A_f + A_O)(h_i + h_{fi})} \quad \text{for finned tubes}$$

### Pressure Drop

The problem of determining the pressure drop in a shell-and-tube heat exchanger is complicated by the progressive change in temperature of the fluids. Sieder and Tate,[22] however, discovered that for flow inside tubes the usual isothermal friction factor could be used, if it is divided by a correction, $\phi = (\mu/\mu_w)^{0.14}$ for Reynolds' numbers above 2100, and $(\mu/\mu_w)^{0.25}$, for Reynolds' numbers below 2100. In addition to the actual friction loss in the tubes, there are losses in the channel and floating heads due to abrupt changes in direction, and losses at the entrance and exit.

Kern[12] suggests that the entrance and exit losses be estimated equal to the loss of one velocity head and that the channel and floating head losses be assumed to be equal to four velocity heads per pass.

An accurate shell-side pressure drop is just as difficult to calculate as an accurate shell-side coefficient. A number of equations have been proposed. The equation presented by Chilton and Genereaux[1] is perhaps the most familiar.

$$\Delta P = \frac{4f\rho u^2_{\max}N}{(2g_c)(144)} = 0.000432(f\rho u^2_{\max}N)$$

where $\Delta P$ = pressure drop, psi per pass
$u_{\max}$ = velocity based on flow through space between tubes
$\rho$ = fluid density
$g_c$ = 32.17
$N$ = number of horizontal rows
$f$ = friction factor

This equation gives the pressure drop across the tube bundle. Values of the friction factor were determined by Grimison[10] and these data as later analyzed by Colburn are presented in TEMA Standards[23] correlated with the Reynolds number. Just as for flow inside the tubes, the friction factor obtained from the chart is divided by the correction $(\mu/\mu_w)^{0.14}$.

In addition to the cross flow pressure drop a pressure drop also occurs as the fluid passes through the baffle opening. This opening can be considered as an orifice with a discharge coefficient of 0.7, as shown by Donohue.[7] Pressure drop through the baffle opening then becomes:

$$\Delta P = \frac{\rho u^2}{144 g_c}$$

where $u$ = velocity through baffle opening, ft/sec
$\rho$ = density, lb/cu ft
$g_c$ = 32.17
$\Delta P$ = pressure drop, psi per baffle opening

The total shell-side pressure drop is the sum of pressure drops of cross-flow, baffle opening, and the entrance and exit. Research scientists at the University of Delaware are studying and correlating shell-side pressure drop data. These data should prove valuable, since the accuracy of existing methods is questionable.

## DESIGN PROCEDURE

Designers of heat exchangers must be given certain information before they can begin the design. In order to prevent delay, the project engineer should see that each manufacturer who is bidding on an exchanger is

## Project Engineering of Process Plants

### Exchanger Specification Sheet

|  | | | |
|---|---|---|---|
| 1 | | | Job no. |
| 2 | Customer | | Reference no. |
| 3 | Address | | Inquiry no. |
| 4 | Plant location | | Date |
| 5 | Service of unit | | Item no. |
| 6 | Size | Type | Connected in |
| 7 | Surface per unit | Shells per unit | Surface per shell |

**Performance of one unit**

| Line | | Shell side | Tube side |
|---|---|---|---|
| 10 | Fluid circulated | | |
| 11 | Total fluid entering | | |
| 12 | Vapor | | |
| 13 | Liquid | | |
| 14 | Steam | | |
| 15 | Non-condensables | | |
| 16 | Fluid vaporized or condensed | | |
| 17 | Steam condensed | | |
| 18 | Gravity—liquid | | |
| 19 | Viscosity—liquid | | |
| 20 | Molecular weight—vapors | | |
| 21 | Specific heat—liquids | Btu/lb | Btu/lb |
| 22 | Latent heat—vapors | Btu/lb | Btu/lb |
| 23 | Temperature in | °F | °F |
| 24 | Temperature out | °F | °F |
| 25 | Operating pressure | lb/sq in. | lb/sq in. |
| 26 | Number of passes | | |
| 27 | Velocity | ft/sec | ft/sec |
| 28 | Pressure drop | lb/sq in. | lb/sq in. |
| 29 | | | |
| 30 | | | |

|  | | |
|---|---|---|
| 31 | Heat exchanged—Btu/hr | M.T.D. (Corrected) |
| 32 | Transfer rate—service | Clean |

| Line | Construction | | |
|---|---|---|---|
| 33 | | | |
| 34 | Design pressure | lb/sq in. | lb/sq in. |
| 35 | Test pressure | lb/sq in. | lb/sq in. |
| 36 | Design temperature | °F | °F |

|  | | | | | | |
|---|---|---|---|---|---|---|
| 37 | Tubes | No. | OD | BWG. | Length | Pitch |
| 38 | Shell | | ID | OD | Thickness | |
| 39 | Shell cover | | | | Floating head cover | |
| 40 | Channel | | | | Channel cover | |
| 41 | Tube sheets—stationary | | | | Floating | |
| 42 | Baffles—cross | | Type | | Thickness | |
| 43 | Baffle—long | | Type | | Thickness | |
| 44 | Tube supports | | | | Thickness | |
| 45 | Gaskets | | | | | |
| 46 | Connections—shell—in | | Out | | Series | |
| 47 | Channel—in | | Out | | Series | |
| 48 | Corrosion allowance—shell side | | | | Tube side | |
| 49 | Code requirements | | | | Tema class | |
| 50 | Weights—each shell | | Bundle | | Full of water | |
| 51 | Note: indicate after each part whether stress relieved (S.R.) and whether radiographed (X-R) | | | | | |
| 52 | Remarks: | | | | | |

Fig. 13-12. Exchanger specification sheet. [Reprinted by permission, *Standards of the Tubular Exchanger Manufacturers Association,* 3rd Ed. (1952)].

provided with all the necessary information to complete a satisfactory design. The Tubular Exchanger Manufacturers' Association has prepared a Standard Exchanger Specification Sheet[23] which serves as a handy reminder of the information required. Most exchanger manufacturers use this form to transmit information on the heat exchanger they are proposing. The same sheet is also valuable as an inquiry form (Fig. 13-12) since the upper portion of the sheet describes the performance of the unit and has space for information that must be supplied by the ultimate user. The following essential data* are required.

*Fluid Circulating.* Common materials such as water need only be identified by name. Complete descriptions of less common materials are necessary, however. If the fluid is a complex mixture, its composition should be given and if it is composed of several phases (vapor and liquid), these should be defined and the amount of each established, including the amount of vaporization or condensation expected in the cooling or heating process. Generalized correlations are available for estimating most physical properties of hydrocarbons when the specific gravity and boiling point are known. It is preferable, however, to provide the designer with as much physical and chemical data as is possible.

*Specific Gravity or Density for Liquids.* This information is required for the calculation of mass flowing and the estimation of other properties from generalized charts.

*Molecular Weight.* Molecular weight is important in defining the properties of a gas.

*Viscosity.* As seen in the discussion on theoretical aspects of heat transfer, viscosity data are perhaps the most important physical data required. Viscosity is significant in both heat transfer and pressure drop equations. At least two values of viscosity at different temperatures within the operating range should be given. Additional values can normally be obtained by interpolating on a logarithmic plot of temperature versus viscosity, assuming a straight-line relationship with temperature. For hydrocarbons more accurate estimation is possible using ASTM kinematic viscosity-temperature graph paper (Chart D, ASTM D-341).

*Latent Heats and Specific Heats.* Although the duty of the exchanger (heat exchanged in Btus per hour) is included in the tabulation, latent heats and specific heats should also be given. The manufacturers' designer can then check the duty calculations. Specific heats are also used in the calculation of transfer coefficients. Where possible, several values at temperatures in the operating range should be given.

*Corrosion.* The designer should be given any information on the corrosive characteristics of the fluid and any unique hazards associated with the transporting of the fluid.

* A more detailed discussion of this problem has been given by Oten.[18] Some of his suggestions have been incorporated in the above discussion.

The procedure for the design of a heat exchanger has been outlined in several books on heat exchange. Kern[12] in particular, has given some excellent illustrative problems which show the philosophy and factors involved.

Since the calculation of film coefficients depends upon the type and design of exchanger, it is necessary to assume an exchanger type and size in order to determine the overall coefficient by calculation. The general procedure is as follows:

1. Specify process conditions. In addition to the quantities flowing, this includes specifying the inlet and outlet temperatures of all streams.

Charts for the determination of optimum cooling water outlet temperature have been prepared. In general, however, the economic factors are not well defined, and use of such charts is questionable. Certain conditions, however, make the selection of the outlet water temperature simple. For many waters it is necessary to keep the flow rate above four ft per second to prevent the growth of algae and it is further necessary, as suggested by Kern, to avoid heating the water to temperatures much higher than 120° in order to prevent excessive corrosion. Where cooling tower water is employed the operation of the cooling tower should be considered in conjunction with the selection of the outlet water temperature. When heat is being exchanged between two process streams, the optimum temperatures must be determined by an economic study, balancing exchanger costs and utility costs. This study must include the other exchangers, coolers, and heaters required for the two streams being considered, since alteration in the outlet and inlet temperatures of the heat exchanger will affect the size of the other units.

2. Route the fluids. Certain general rules for the routing of fluids have been developed through the years. High pressure fluids and corrosive fluids should be routed through the tubes, since it is cheaper to design a tube for high pressure than an entire shell and only the tube need be resistant to corrosion. High-fouling fluids, including water, should be routed through the tubes. Because of the flexibility possible in baffle arrangements, extremely large volumes of fluids and very small volumes of fluids, as well as viscous fluids, are best passed through the shell. When low pressure drops are required the shell side should be used.

3. Assume a design overall coefficient and calculate a trial area.

4. Select the length, arrangement, pitch, and size of tubes. Standardization where possible is desirable. This permits ease of maintenance, requires the stocking of fewer parts, and makes for more attractive plant appearance. It should be remembered that short tubes are much more expensive per square ft than longer tubes (twelve or sixteen ft).

5. Determine the minimum number of tube passes based on the total number of tubes corresponding to the trial area. The number of tube

passes depends on the allowable pressure drop and cost of construction. A reduction in the number of tube passes below the minimum will result in a lower transfer rate and thus higher surface areas. At this point a check of the tube side pressure drop should be made to be certain that the number of tube passes chosen is satisfactory.

6. Select the number of shell passes which will give a correction factor ($F_T$) for the log-mean temperature difference of 0.75 to 0.8 or greater. Below these values of $F_T$ the curves for the correction factor become quite steep and a small change in temperature conditions causes a large change in correction factor and consequently the ultimate mean temperature difference. In this region, therefore, it is reasoned that inherent errors in the assumptions made in the derivation of the correction factor become significant and thus the region below 0.75 to 0.8 should be avoided. If the $F_T$ correction is too low for a one-shell pass, often a two-shell pass arrangement will provide a larger $F_T$ which will be acceptable. It is to be remembered that two one-shell pass exchangers connected in series are equivalent to a two-shell pass exchanger.

7. Select a standard exchanger to meet as closely as possible the requirements of items 3, 4, 5, and 6. Determine the tube side transfer coefficient.

8. Proceed to the shell side and assume a trial baffle spacing using a small spacing for a small amount of fluid and a large spacing for a large amount. These spacings according to TEMA may be varied from one-fifth the shell diameter to one shell diameter in order to meet the heat transfer and pressure drop requirements on the shell side.

9. Determine the shell side coefficient.

10. Calculate the overall clean coefficient and then a design coefficient based on proper fouling factors.

11. Determine required area. Compare this to the surface area of the unit selected. The ideal choice is of course the minimum area that will produce the required heat transfer. If the exchanger selected does not meet the requirements, it is necessary to assume a higher surface area unit and repeat the procedure.

These steps require trial and error in many portions of the calculation. For the novice or one who rates exchangers only occasionally, the job proves tedious. Frequent calculation of exchangers, and particularly calculation of exchangers in similar services, enables an engineer to develop know-how or perhaps a clairvoyance which eliminates to a large degree the laborious trial-and-error procedure.

## Short-Cut Methods of Heat Exchanger Design

Many so-called short-cut methods for the design of heat exchangers have appeared from time to time in the literature. A number of such articles present curves or nomographs for determining film coefficients. Hutchinson[11] has compiled charts in graphical form for the determination of film coefficients for a wide assortment of compounds. In addition, many articles have presented a series of charts designed to reduce the labor in heat exchanger calculations. Unfortunately, however, the occasional user of such charts finds it just as difficult and time-consuming to design an exchanger using these charts as to follow the method previously outlined.

Sieder[21] has presented a number of heat transfer tables in booklet form which enable rapid estimation of film coefficients. The data are presented in tabular form in order to eliminate mistakes that are so often made when using graphs or nomographs. The sources of all data are plainly given so that one may evaluate the information intelligently. These tables are particularly recommended for engineers who only occasionally need to check heat exchanger ratings, such as the project engineer.

The only truly short-cut method of heat exchanger design is the one in which the overall coefficient is assumed based on previous experience and from which an area is calculated. Such a method is valid for rapid estimation and comparison of heat exchanger quotations.

The selection of an overall coefficient can often be made with a fair degree of accuracy, particularly when good operating information is available on exchangers in similar services. It would behoove the project engineer to keep a record of typical overall transfer coefficients for various exchanger services.

Fair and Rase[8] have presented a table of typical resistances for a representative number of fluids. These values given in Table 13-1 are

TABLE 13-1. Design Resistances for Shell-and-Tube Heat Exchangers*

| Fluid | Liquid (No phase change) | Boiling liquid | Condensing vapor |
|---|---|---|---|
| Aromatic liquids | | | |
| Benzene, toluene, ethylbenzene, styrene | 0.007 | 0.011 | 0.007 |
| Dowtherm | 0.007 | .... | .... |
| Inorganic solutions | | | |
| $CaCl_2$ Brine (25%) | 0.004 | .... | .... |
| Heavy acids | 0.013 | .... | .... |
| NaCl Brine (20%) | 0.0035 | .... | .... |
| Misc. dilute solutions | 0.005 | .... | .... |
| Light hydrocarbon liquids | | | |
| $C_3$, $C_4$, $C_5$ | 0.004 | 0.007 | 0.004 |
| Chlorinated hydrocarbons | 0.004 | 0.009 | 0.007 |
| Miscellaneous organic liquids | | | |
| Acetone | 0.007 | .... | .... |

TABLE 13-1 (Continued)

| Fluid | Liquid (No phase change) | Boiling liquid | Condensing vapor |
|---|---|---|---|
| Amine solutions | | | |
| Saturated diethanolamine and monoethanolamine ($CO_2$ and $H_2S$) | 0.007 | .... | .... |
| Lean amine solutions | 0.005 | .... | .... |
| Oils | | | |
| Crude oil | 0.015 | .... | .... |
| Diesel oil | 0.011 | .... | .... |
| Fuel oil (bunker C) | 0.018 | .... | .... |
| Gas oil | | | |
| Light | 0.0125 | .... | 0.015 |
| Heavy (typical of cat. cracker feed) | 0.014 | .... | 0.018 |
| Gasoline (400° EP) | 0.008 | 0.010 | 0.008 |
| Heating oil (domestic 30° API) | 0.010 | .... | .... |
| Hydroformate | 0.006 | .... | .... |
| Kerosine | 0.009 | .... | 0.013 |
| Lube oil stock | 0.018 | .... | .... |
| Naphthas | | | |
| Absorption | 0.008 | 0.010 | 0.006 |
| Light virgin | 0.007 | 0.010 | 0.007 |
| Light catalytic | 0.006 | 0.010 | 0.007 |
| Heavy | 0.008 | 0.011 | 0.0085 |
| Polymer ($C_8$'s) | 0.008 | 0.010 | 0.008 |
| Reduced crude | 0.018 | .... | .... |
| Slurry oil (fluid cat. cracker) | 0.015 | .... | .... |
| Steam | | | 0.001† |
| Water | | | |
| Boiler water | 0.003 | .... | .... |
| Cooling tower (untreated) | 0.007‡ | .... | .... |
| Condensate (flashed) | 0.002 | .... | .... |
| River and well | 0.007‡ | .... | .... |
| Sea water (clean and below 125°F.) | 0.004 | .... | .... |

* Reprinted by permission, *Petroleum Refiner, 33,* No. 7, 121 (1954).

† Condensing resistances for steam may be much higher when noncondensibles are present as is often the case with low pressure exhaust steam.

‡ A value of 0.007 is suggested when doubt as to the quality of water exists.

| Gases in turbulent flow | Resistance |
|---|---|
| Air, CO, $CO_2$, and $N_2$ | 0.045 |
| Hydrocarbons (light through naphthas) | 0.035 |

based on operating data and current design practice. To obtain an estimated overall design coefficient for a shell and tube exchanger, proceed as follows:

1. Read values of resistances for the two fluids. Note that no distinction is made between shell and tube sides.

2. Calculate the overall design coefficient by taking the reciprocal of the sum of these two resistances.

Other exchanger types are estimated by multiplying the overall design coefficient for shell and tube exchanger by 0.25 for extended surface exchangers and 0.30 for coil-in-box exchangers. The method is recommended for most standard exchangers. Unusual designs, very large or very small exchangers, and partial condensers cannot be estimated by this method.

## EVALUATING EXCHANGER QUOTATIONS

In evaluating heat exchanger quotations the more obvious factors such as price, delivery time, terms, and manufacturer's reputation should be considered first. In addition the important mechanical and design factors should be carefully scrutinized. A tabulation should be made for each quotation comparing the materials of construction, various design features, shell and tube thicknesses, surface areas, and other special features. It is possible that designs for the same service by different manufacturers will show large variations in surface area with comparable variations in price. In such cases it is necessary for the project engineer to estimate a surface area for the service in question for comparison.

## REFERENCES

1. Chilton, T. H., and R. P. Genereaux, *Trans. Am. Inst. Chem. Engrs., 29,* 161 (1933).

2. Colburn, A. P., *Trans. Am. Inst. Chem. Engrs., 29,* 174 (1933).

3. Colburn, A. P., and T. P. Drew, *Trans. Am. Inst. Chem. Engrs., 33,* 197 (1937).

4. DeLorenzo, B., and E. D. Anderson, *Trans. Am. Soc. Mech. Engrs., 67,* 697 (1945).

5. Dittus, F. W., and L. M. K. Boelter, *Univ. of Calif. Pub. in Eng., 2,* 443 (1930).

6. Donohue, D. A., *Ind. Eng. Chem., 39,* 62 (1947).

7. Donohue, D. A., *Ind. Eng. Chem., 41,* 62 (1949).

8. Fair, J. R., and Howard F. Rase, *Petroleum Refiner, 33,* No. 7, 121 (1954).

9. Gardner, K. A., *Trans. Am. Soc. Mech. Engrs., 67,* 621 (1945).

10. Grimison, E. D., *Trans. Am. Soc. Mech. Engrs., 59,* 583 (1937).

11. Hutchinson, F. W., *Industrial Heat Transfer,* The Industrial Press, New York, 1952.

12. Kern, D. Q., *Process Heat Transfer,* McGraw-Hill Book Co., New York, 1950.

13. McAdams, W. H., *Heat Transmission,* 3rd Ed., McGraw-Hill Book Co., New York, 1954.

14. Morris, F. H., and W. G. Whitman, *Ind. Eng. Chem., 20,* 234 (1928).

15. Murray, W. M., *J. Appl. Mechanics, 5,* A78 (1938).

16. Nusselt, W., *Mitt. Forsch. Arb. Ing., 89,* 1 (1910).

17. Nusselt, W., *Z. Ver. deut. Ing., 60,* 541, 569 (1916).

18. Oten, P. S., *Chem. Eng. Progr., 44,* 411 (1948).

19. Sakiadis, B. C., and J. Coates, Louisiana State University Eng. Exp. Station Bulletin No. 34, Baton Rouge, La., 1952.

20. Short, B. E., University of Texas Publication No. 4324, 1–55 (1943).

21. Sieder, E. N., *Heat Transfer Tables,* American Locomotive Co., New York, 1952.

22. Sieder, E. N., and G. E. Tate, *Ind. Eng. Chem., 28,* 1429 (1936).

23. *Standards of Tubular Exchanger Manufacturers Association,* 3rd Ed., Tubular Exchanger Manufacturers Association, Inc., New York, 1952.

# CHAPTER 14

# PROCESS PUMPS

The modern process plant would be inoperative were it not for the pumps which maintain the steady flow of fluids through the plant and make continuous processes possible. Improper and ill-considered selection of pumps could create serious bottlenecks and cause unsatisfactory operation of the plant. Close supervision, therefore, by the project engineer of selection, purchase, and engineering of pump installations is imperative.

Since pumps are produced by the manufacturers in standard sizes, it is only necessary to select the size and type that most nearly fits the service in question. Close collaboration with manufacturers' representatives is helpful and strongly urged. To obtain the maximum benefit from such collaboration and to properly judge the manufacturer's recommendations, it is essential that certain basic fundamentals of pump operation be understood.

The pumps most common to the process industries (centrifugal, reciprocating, and rotary) will be discussed in this chapter giving particular emphasis to the theory and information required for intelligent pump selection. The lack of comment on many special types of pumps designed for specific services does not imply that these are not to be considered. Indeed, certain services absolutely require a special design and manufacturers' representatives should be asked for recommendations in these cases.

## RECIPROCATING PUMPS

Although the use of a reciprocating action to produce pumping had its origin some 2000 years ago, it was not until 1840 that Henry R. Worthington invented the direct-acting, reciprocating, steam-driven pump. These steam-driven pumps became extremely popular and in the years to come motor-driven units also came into use.

### Application

In the early twentieth century, the centrifugal pump began to encroach upon the field of the reciprocating pump. The uniform discharge of the centrifugal pump gave it a distinct advantage, especially in the process industries where steady flow rates are a necessity for satisfactory operation. The applications of centrifugal pumps were limited only by design restrictions on capacity and pressure. Improvements in design, however, have gradually eliminated these restrictions and there are now few applications belonging exclusively to the reciprocating pump. Modern reciprocating pumps are best suited to high head and low capacity. Fig. 14-1 presents a qualitative picture of the range of usefulness of various pump types. It is often found advantageous to employ steam-driven reciprocating pumps for temporary installations or for intermittent and varying conditions such as tank area transfer pumps. In such installations the ruggedness, flexibility, and economical characteristics of these pumps may be used to an advantage, and their unsteady discharge is no disadvantage. The reciprocating pump is often preferred for the pumping of highly volatile liquids (especially where gassing conditions are possible), slurries, and viscous liquids.

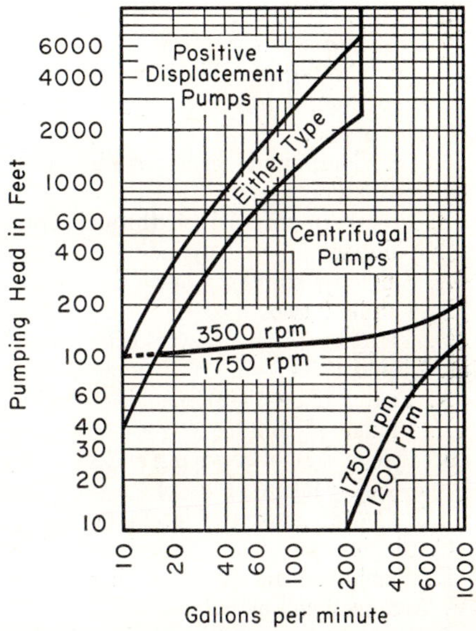


Fig. 14-1. Pump selection chart. [Adapted from R. L. Jacks, *Chem. Eng. Progr.* *49*, 234 (1953), by permission.]

### Description

The following tabulation gives an abbreviated list of terms used to classify reciprocating pumps. The majority of all such pumps used in the process industries are steam-driven, direct-acting, duplex or simplex, double-acting or single-acting. The double-acting is more common except for high-pressure service for which the single-acting pump (usually the plunger type) is more frequently employed.

### *Reciprocating Pump Classification*

(Based on terminology of the Hydraulic Institute)

I. Drive
  A. Steam
    1. Direct-acting; steam piston directly connected to liquid piston through piston rod.
    2. Crank and flywheel; pump with crankshaft on which the flywheel is mounted. Used for large installations, and with compound cross arrangements.
    3. Types of steam ends
      *a*. Simple; one high pressure steam cylinder for each liquid cylinder.
      *b*. Compound—tandem; a high- and low-pressure cylinder on same center line.
      *c*. Compound—cross; a high and low cylinder arranged side by side with cranks 90° apart.
  B. Power; motor driven through crankshaft
II. Number of liquid cylinders
  A. Single or simplex; one
  B. Duplex; two
  C. Triplex; three (power-driven only)
III. Liquid ends
  A. Piston (Fig. 14-2), packing carried on piston
  B. Plunger; packing ring is stationary; preferred for high-pressure operation because of ease of servicing and visibility of packing gland
    1. End-packed (Fig. 14-3)
    2. Center-packed (Fig. 14-4)
  C. Single-acting; one pumping stroke per revolution
  D. Double-acting; two pumping strokes per revolution
IV. Arrangement of cylinders
  A. Vertical; usually for designs with three or more liquid cylinders
  B. Horizontal

Reciprocating pumps have become standardized over the years. The various types of pumps and their parts are illustrated in Figs. 14-2, 14-3, and 14-4. Brief comment on certain of the these items is in order.

### *Materials of Construction*

The following are the most frequently used material designations for steam pumps as given by the *Standards of the Hydraulic Institute*[9]*.

*Bronze Fitted* (Symbol *BF*). *a*. A bronze fitted pump consists of bronze piston rods (except end packed plunger pattern) iron liquid pistons or plungers, bronze or rubber liquid valves, bronze liquid valve seats, guards,

* Reprinted by permission from the *Standards of the Hydraulic Institute,* copyright 1955 by the Hydraulic Institute, 122 East 42nd St., New York 17, N. Y.

and springs, iron or steel liquid cylinders; *b*. Bronze fitted piston pattern pumps include bronze lined liquid cylinders. Bronze fitted plunger pattern pumps include bronze bushed plunger glands and throats.

*Fully Bronze Fitted* (Symbol FBF). A fully bronze fitted piston pump consists of bronze piston rods (except end packed plunger pattern), bronze liquid pistons or plungers, bronze or rubber liquid valves, bronze liquid valve seats, guards, and springs, iron or steel liquid cylinders. Fully bronze fitted piston pattern pumps include bronze lined liquid cylinders. Fully bronze fitted plunger pattern pumps include bronze bushed plunger glands and throats.

*Acid Resisting* (Symbol *AR*). All parts of the pump in direct contact with the liquid pumped are made of corrosion resisting materials of suitable properties for the specific application.

*All Bronze* (Symbol *AB*). All parts of the pump coming in direct contact with the liquid pumped are made of bronze.

*All Iron* (Symbol *AI*). All parts of the pump coming in direct contact with the liquid pumped are made of ferrous metal.

*Standard Fitted.* Individual manufacturers may use the term *standard* or *regular fitted* to designate the material used in pumps of their regular production.

### Discussion of Reciprocating Pump Types

#### Duplex Vs. Simplex

Because of the smoother flow characteristics of the duplex pump, it is preferred to the simplex. The steam slide valve action on a duplex pump is such that short stroking (failure of pump to make full stroke) can occur. This condition in effect increases the clearance volume and is a disadvantage on pumps operating as wet vacuum pumps or in highly volatile liquid pumping service. Under such conditions the lack of full displacement permits the accumulation of excessive volumes of vapor in the cylinder. The simplex design does not have this defect and is, therefore, indicated for such services.

#### Piston Vs. Plunger

Pistons are differentiated from plungers by their shape and the mode of packing. Referring to Figs. 14-2 and 14-3 it can be seen that a piston is a flat disk while a plunger is more like a long barrel. In addition, piston-type pumps carry packing on the piston itself to reduce leakage from the discharge to the suction side of the piston. The plungers in a plunger pump operate through stationary packing in the heads of the cylinder and are classified as end-packed or center-packed according to the packing location. The end-packed plunger pump requires no packing box on the piston rod while the center-packed pump, like the piston pump, does (see Figs. 14-3 and 14-4).

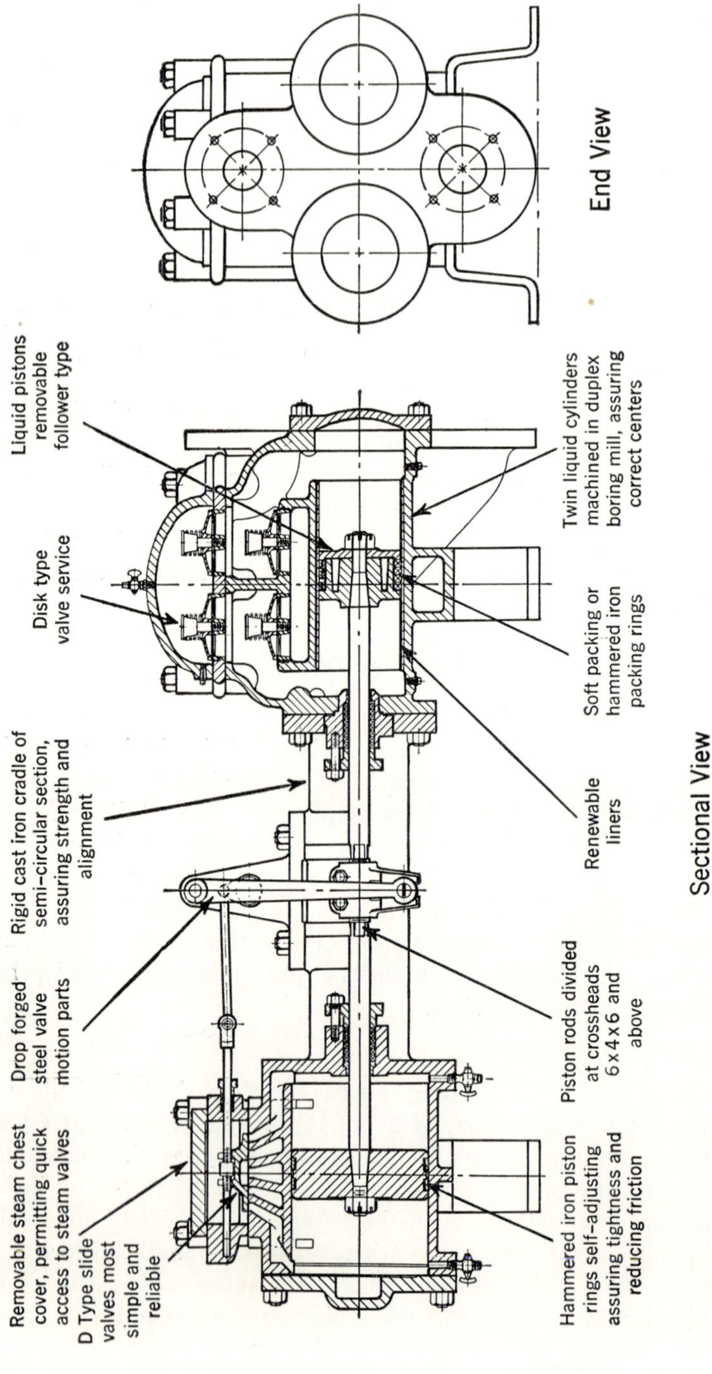


Fig. 14-2. Horizontal duplex piston pump. (Courtesy of Worthington Corporation.)

Because all packing leakage on plunger pumps is clearly visible to the operator and because of the ease of repacking or adjusting the packing, the plunger design, although more expensive, is preferred for severe

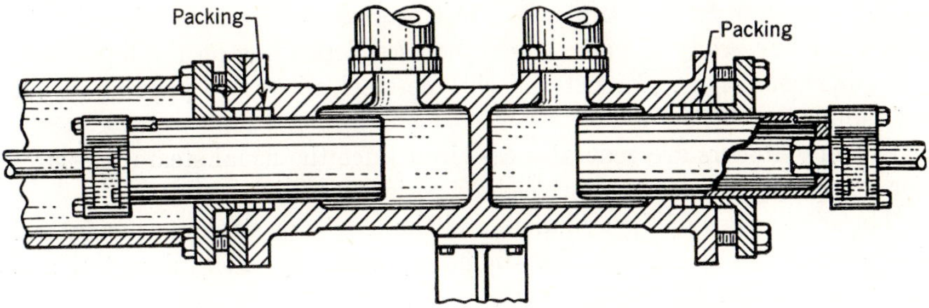


Fig. 14-3. Portion of end-packed plunger pump.

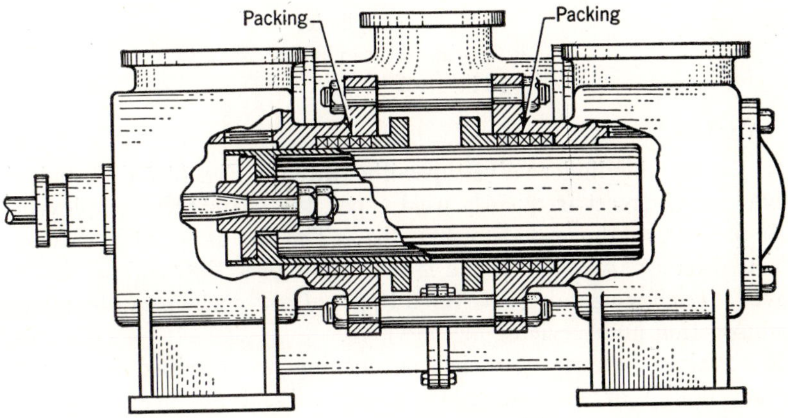


Fig. 14-4. Portion of center-packed plunger pump.

services such as high pressure operation. Many users prefer the plunger pump for differential pressures above 400 psi, especially in high temperature service. For pressures above 1000 psi the plunger pump is almost universally specified when the reciprocating pump is indicated.

### Reciprocating Pump Parts

(See Fig. 14-2.)

#### Cylinders

Liquid cylinders are cast iron for most general service pumps, but at high pressures and temperatures cast or forged steel is required, the latter being used for extremely high pressures.

The cylinder should be designed for the full stalling pressure which is the pressure at which the forces on the liquid piston and steam piston are equal.

Wherever excessive wear or corrosion is expected a cylinder liner is either bolted or pressed in place. Most process pumps are equipped with liners, and a variety of material can be used depending on the liquid and pumping conditions. Nickel and chrome-steel alloys are popular in hot oil pumps.

Steam cylinders are generally cast iron since the usual steam pressures for pump operation seldom exceed 250 psig.

The dimensions of the cylinders and the length of stroke are expressed in this order: steam cylinder diameter, water cylinder diameter, and length of stroke.

EXAMPLE:

10 × 4½ × 10
10 in. steam cylinder
4½ in. water cylinder
10 in. stroke

### VALVES

Of the liquid cylinder valve types illustrated in Fig. 14-5 disk and wing valves are the most frequently used types, the latter being employed for high pressures (above 300 to 400 psi). The discharge disk valves are mounted on a valve plate which is removable once the head has been removed. Each wing valve is covered by a *pot cover* which can be removed for inspection and servicing of the valve.

Ball valves because of their free opening are specified for pumping viscous liquids or slurries. Many other valve types are available for particular pumping problems.

The steam cylinder valve arrangement is standard. Slide valves, direct connected or operated through auxiliary pistons, or balanced piston valves are used. The latter is preferred for use with superheated steam.

### PISTON RODS

Rods are usually alloy steel of high tensile strength and hardness, and are preferably made in two parts to insure ease of dismantling.

### PACKING

Proper choice and installation of packing is a must for satisfactory pump operation. The steam pistons are usually sealed with hammered-iron snap rings. The liquid pistons are provided with bakelite rings for temperatures below about 250 to 350° F. and hammered-iron rings for higher temperatures.

Other piston packings such as combination duck and rubber, bronze, and monel, are used for special conditions. The Standards of the Hydraulic Institute[9] give detailed recommendations.

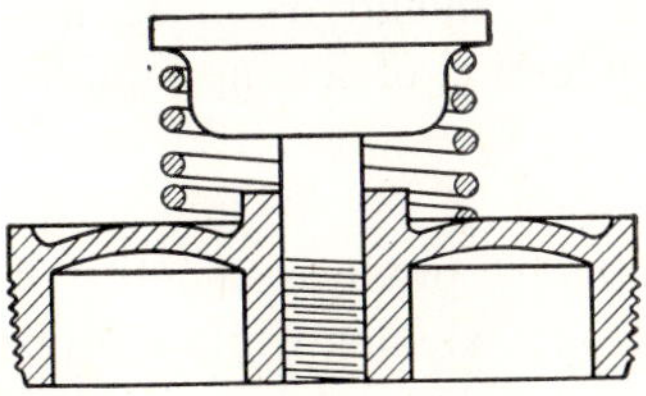

*Disk valve.* Flat circular valve of rubber or metal, guided by stem fastened to valve seat.

*Wing valve.* Circular valve with guides cast or forged integral, with or without facing. Facing may be rubber, leather, fiber, or other material.

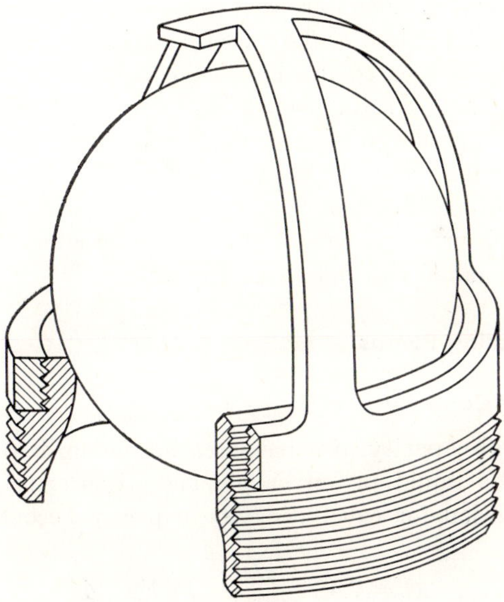

*Ball valve.* Spherical valve of rubber or metal.

Fig. 14-5. Reciprocating pump, valve types. (Reprinted from *Standards of the Hydraulic Institute,* copyright 1955 by Hydraulic Institute, 122 E. 42nd St., New York 17, N. Y.)

Leakage along the piston rod is prevented by the stuffing box packing which is usually graphite impregnated asbestos on the steam end. On the liquid end the stuffing box packing varies with the type of fluid and pumping conditions. Asbestos, rubber, semi-metallic, and metallic packings are used. For severe conditions such as those encountered in the pumping of hot oils metallic packing is indicated.

Stuffing boxes on pumps operating at high temperatures (above 400° F.) should be water cooled. Pumps operating under a high vacuum on the suction side or those pumping volatile liquids should be fitted with a lantern ring in the stuffing box for the introduction of sealing liquid.

#### Air Chambers

Air chambers installed on reciprocating pumps produce smoother flow. The air compressed during the pumping stroke expands while the piston is reversing and thus forces liquid from the chamber to the discharge line. The volume of these chambers should be six to eight times the displacement of simplex pumps and three to four times that for duplex pumps.

#### Flanges

Flanges on pump cylinders are cast as an integral part of the cylinder and may be faced as raised or flat-faced flanges. Care must be exercised in specifying the proper companion flange to the pump flange. If the pump flange is flat faced, the companion flange must be specified as flat faced. The strain produced in tightening a raised face against a flat-faced flange on the pump may crack the casing. The Hydraulic Institute recommends the specification of tongue and groove type joints when temperature and pressure conditions require greater gasket pressures than those obtainable with flat-faced flanges.

## Sizing Reciprocating Pumps

### *Theoretical Capacity*

The theoretical capacity of a reciprocating pump depends on the displacement of the liquid piston or plunger. Expressed as gallons per minute the equation for calculating the capacity becomes

$$G = 7.48 \frac{\pi D^2}{(4)(144)} PCF - Z$$

or

$$G = 0.0408\, D^2PCF - Z$$

where $G$ = gpm
$D$ = diam of plunger or piston, in.
$P$ = piston speed, ft/min
$C$ = number of cylinders
$F$ = factor, 0.5 for single acting, 1.0 for double acting
$Z$ = correction for volume occupied by the piston rod during a pumping stroke

The correction $Z$ depends on the design of the pump. In most cases the piston rod extends through the inboard end of the pump and is attached to one end of the piston. For this design:

Single-acting pumps

$$Z = 0$$

Double-acting pumps

$$Z = 0.0204 d_r^2 PC$$

where $d_r$ = diam of rod, in.

Certain designs such as end-packed plunger pumps require no correction.

### Actual Capacity

The theoretical capacity is never met in practice because of leakage through piston packing, stuffing boxes, or valves. The amount of air or other gas carried into the pump also reduces the theoretical flow rate. As is so often the case in engineering, the relation between the actual and theoretical capacity is expressed as an efficiency called the volumetric efficiency ($E_v$):

$$E_v = \frac{G_a}{G} \times 100$$

where $G_a$ = actual capacity
$G$ = theoretical capacity

### Slip

Another term often used is *slip.*

$$\text{Slip} = 1 - \frac{E_v}{100}$$

Hence

$$\text{Slip} = \frac{G - G_a}{G} = \frac{\text{Leakage}}{\text{Theoretical Displacement}}$$

If the leakage can be assumed constant with speed, the slip will increase with decreasing speeds. At high speeds, however, the leakage increases.

The slip is usually less than 10% (0.10). A value of 3% (97% volumetric efficiency) is often used for calculating piston diameters for pumps operating at normal speeds. When in doubt, however, 10% is a safe design figure.

### Recommended Piston Speeds

Based on years of operating experience, the Hydraulic Institute[9] has recommended in its standards design speeds which were chosen at speeds that would produce the minimum wear on the pump.

Typical values corresponding to the length of stroke commonly used for simplex and duplex are reported in Table 14-1. Although a pump

**TABLE 14-1. Recommended Piston Speeds***

Simple and Duplex Direct-Acting Steam Pumps

| Length of stroke | Speed, ft/min |
|---|---|
| 3 | 37 |
| 3½ | 42 |
| 4 | 46½ |
| 5 | 53 |
| 6 | 59 |
| 7 | 64 |
| 8 | 68 |
| 10 | 75 |
| 12 | 81½ |
| 14 | 87 |
| 15 | 89½ |
| 16 | 92 |
| 18 | 96½ |
| 20 | 100 |
| 24 | 106½ |
| 28 | 111½ |
| 36 | 120 |

Viscosity or Temperature Correction

| Viscosity, SSU | or Temperature, °F | Multiply basic speed by |
|---|---|---|
| 250 | 70 | 1.00 |
| 1000 | 83 | 0.89 |
| 2000 | 105 | 0.80 |
| 3000 | 132 | 0.74 |
| 4000 | 165 | 0.69 |
| 5000 | 210 | 0.65 |
| 6000 | 260 | 0.62 |
| 7000 | 330 | 0.58 |
| 8000 | 415 | 0.55 |
| 9000 | 520 | 0.52 |
| 10,000 | 650 | 0.50 |

Note: Base correction on temperature or viscosity, whichever is controlling. For intermediate values, interpolate linearly.

* Adapted by permission of the *Standards of the Hydraulic Institute,* Copyright 1955 by the Hydraulic Institute, 122 East 42nd Street, New York 17, N. Y.

with a short stroke can make more revolutions per minute, one with a long stroke can run at a higher piston speed. The values given are average values based on a wide variety of pumps and conditions. Individual manufacturers' recommendations should be used when obtainable. For pumps operating at pressures above 300 psig it is often desirable to base the selection on a speed 0.8 of normal to reduce danger of shock.

### Water Horse Power (Liquid Horse Power)

Water or liquid horsepower is the output horsepower of the pump and is calculated in the same manner for all types of pumps. Basically the calculation originates from the general energy equation which is most conveniently applied by an energy balance between the suction and discharge of the pump. Kinetic and static effects are usually negligible and the work done by the pump is:

$$W = \frac{(P_D - P_s)144}{\rho} (G_a) \left(\frac{1}{7.48}\right) (\rho) \left(\frac{1}{33{,}000}\right)$$

$$= \frac{(P_D - P_s)(G_a)}{1715}$$

where $P_D$ = discharge pressure, psia
$P_s$ = suction pressure, psia
$W$ = work, hp
$G_a$ = actual gpm pumped
$\rho$ = density, lb/cu ft

The friction effects caused by entrance and exit losses and losses within the pump are all included in the evaluation of the mechanical efficiency.

### Mechanical Efficiency

The term mechanical efficiency $E_m$ is defined as the ratio of work done in the liquid end to that done by the driving end of reciprocating pumps and generally represents the overall efficiency including the hydraulic losses and mechanical losses. The volumetric losses are evaluated in the calculation of $G_a$.

Steam driven pumps

$$E_m = \frac{\text{WHP}}{\text{IHP}}$$

Power driven pumps

$$E_m = \frac{\text{WHP}}{\text{BHP}}$$

where WHP = water hp
IHP = indicator hp
BHP = brake hp

Typical values of mechanical efficiencies $E_m$ for reciprocating pumps as reported by Jacks[5] are given in Table 14-2.

**TABLE 14-2. Mechanical Efficiencies of Steam-Driven Reciprocating Pumps***

| | Mechanical Efficiency | | |
|---|---|---|---|
| | To 300 psi Differential | | Above 300 psi Differential |
| Stroke | Piston | Plunger | Plunger |
| 3 | 55 | 50 | 41 |
| 4 | 60 | 55 | 44 |
| 6 | 65 | 60 | 51 |
| 8 | 68 | 65 | 54 |
| 10 | 72 | 68 | 57 |
| 12 | 74 | 70 | 59 |
| 16 | 76 | 73 | 63 |
| 18 | 78 | 75 | 65 |
| 20 | 79 | 77 | 66 |
| 24 | 80 | 78 | 67 |

* Reprinted by permission. R. L. Jacks, *Chem. Eng. Progr.*, *49,* 234 (1953).

### *Relation Between Pressure in Steam Cylinder and Water Cylinder*

The force exerted by the steam cylinder is

$$F_s = (\text{MEP})\left(\frac{\pi D_s^2}{4}\right)$$

where $D_s$ = steam cylinder diam, in.

MEP = mean effective pressure determined from the indicator card diagram in psi

The most common process pump has only one steam cylinder per water cylinder. In such pumps the steam exerts full pressure during the entire stroke and the indicator diagram is thus a rectangle. The mean effective pressure, therefore, is simply the height of this rectangle, which is the steam line pressure minus the back pressure (discharge or low pressure steam line pressure). The value of back pressure when not definitely known can be assumed to be 16 psia.

Values of MEP for various steam engine drives can be approximated as follows:[8]

| Type | Formula |
|---|---|
| Simple cylinder | $\text{MEP} = P - b$ |
| Compound | $\text{MEP} = 2P - \dfrac{P}{R} - bR$ |
| Triple-expansion | $\text{MEP} = 3P - 2\dfrac{P}{R_1^{1/2}} - bR_1$ |

where $P$ = steam line pressure at pump, psia
$b$ = back pressure, psia

$$R = \frac{\text{intermediate-pressure cylinder area}}{\text{high-pressure cylinder area}}$$

$$R_1 = \frac{\text{low-pressure cylinder area}}{\text{high-pressure cylinder area}}$$

Recommended values of $b$ to be used when back pressure is not known:

| | |
|---|---|
| Simple | 16 psia |
| Compound | 6 psia |
| Triple-expansion | 5 psia |

In like manner the force exerted by the liquid cylinder, since it is always a single cylinder, would be

$$F_w = (P_D - P_s)\left(\frac{\pi D_w^2}{4}\right)$$

where $P_D$ = discharge pressure, psia
$P_s$ = suction pressure, psia
$D_w$ = liquid cylinder diam. in.

Because of the mechanical friction effects the water cylinder force is less than the steam cylinder force. The ratio of these two forces is the mechanical efficiency.

$$(F_s)\left(\frac{E_m}{100}\right) = F_w$$

or

$$(\text{MEP})\left(\frac{\pi D_s^2}{4}\right)\left(\frac{E_m}{100}\right) = (P_D - P_s)\left(\frac{\pi D_w^2}{4}\right)$$

Hence

$$P_D - P_s = \left(\frac{D_s}{D_w}\right)^2 (\text{MEP})\left(\frac{E_m}{100}\right)$$

This equation may be used to calculate the steam cylinder diameter when the pressures and water cylinder diameter are known. The maximum discharge pressure $P_D$ that a given pump produces can also be calculated for any known steam pressure. Kent[6] recommends that 90% of the usual values of mechanical efficiencies be used in the above calculation.

### Controlled Volume Pumps

A very important group of pumps in the process industries is the controlled volume pump (Fig. 14-6) which is used extensively in feeding accurately controlled streams into a process and in proportioning the flows of several streams to reactors and treating or blending tanks. These pumps act not only as pumps but also as positive volumetric metering devices. They are plunger-type reciprocating pumps with ball-type valves on both suction and discharge. Units are built for pumping against discharge pressures up to 25,000 psig. Both simplex and duplex designs

Fig. 14-6. Controlled volume pump. (Courtesy of Milton Roy Company.)

can be obtained. A special duplex design has been offered which produces a constant flow rate without pulsation. This is accomplished by causing one plunger to decelerate at the same rate that the other plunger accelerates.

The units are equipped with variable speed drives and micrometer adjustments for accurate variation of the piston speed and thus of the flow rate. The maximum safe speed of these pumps exceeds that recommended for the usual reciprocating pump, although in pumping liquefied gases lower speeds are recommended.

Controlled volume pumps can be obtained either with electric or air-motor drives. The air motor is often used in hazardous locations in preference to the more expensive explosion-proof electric motor.

These pumps are widely used as chemical feeders, and are constructed of various metals for the different services. Materials used include cast irons and steels as well as corrosion-resistant alloys and plastics.

### Other Reciprocating Pumps

Special designs of reciprocating pumps are often required in the process industries. The use of such corrosion resistant materials as stoneware, porcelain, and rubber require unique design features. One of the more familiar special designs is the diaphragm pump (see Fig. 14-7.) The plunger of this pump is attached to a flexible rubber diaphragm, which is the only material in contact with the fluid being pumped. The up and down movement of the diaphragm produces the pumping action. These pumps are well suited for pumping heavy slurries such as the sludge from a Dorr thickener.

## CENTRIFUGAL PUMPS

The development of the centrifugal pump has been phenomenal, and it is now used for almost every type of service. Indeed, most manufacturers produce centrifugal pumps ranging from capacities of 5 gpm and 2 to 5 psi differentials to large multistage pumps capable of pumping 2800 to 3000 gpm at 3000 psig discharge pressures. Many designs are built for successfully handling fluids up to 850°F, highly volatile fluids, and slurries.

Special pumps have been designed for even more impressive operating conditions. Pumps at the Colorado River aqueduct can pump 90,000 gpm at 444 ft of head and those at the Grand Coulee Irrigation Project, 607,000 gpm at 310 ft of head. Special installations of centrifugal pumps in series have been designed for pressures above 5000 psi.

### Classification of Centrifugal Pumps

The confusion that exists in many fields because of misleading nomenclature has not escaped the field of centrifugal pumps. Originally a centrifugal pump was a pump that produced its pumping action by means of centrifugal force. As the field grew, however, new impeller designs produced only part or none of the pumping action by centrifugal force. All of these pumps, however, are classed as centrifugal pumps since there is a basic relation between them.

In its broad sense then a centrifugal pump consists of an impeller which produces a pressure head by rotating in a casing. The various classes of these pumps are best defined according to the impeller design which varies from radial to axial flow types. The classifications given below conform to those suggested by the Hydraulic Institute (see Fig. 14-8).

1. *Radial type.* This impeller directs the flow of fluid by centrifugal force radially to the periphery of the impeller. The velocity head thus created is then largely converted to pressure head in the discharge portion of the casing. The vanes of these impellers are usually backward curved and have a single curvature. The radial impeller is by far the most common type used in process plants.

Specific speed: 500 to 3000, single suction<br>
up to 6000, double suction

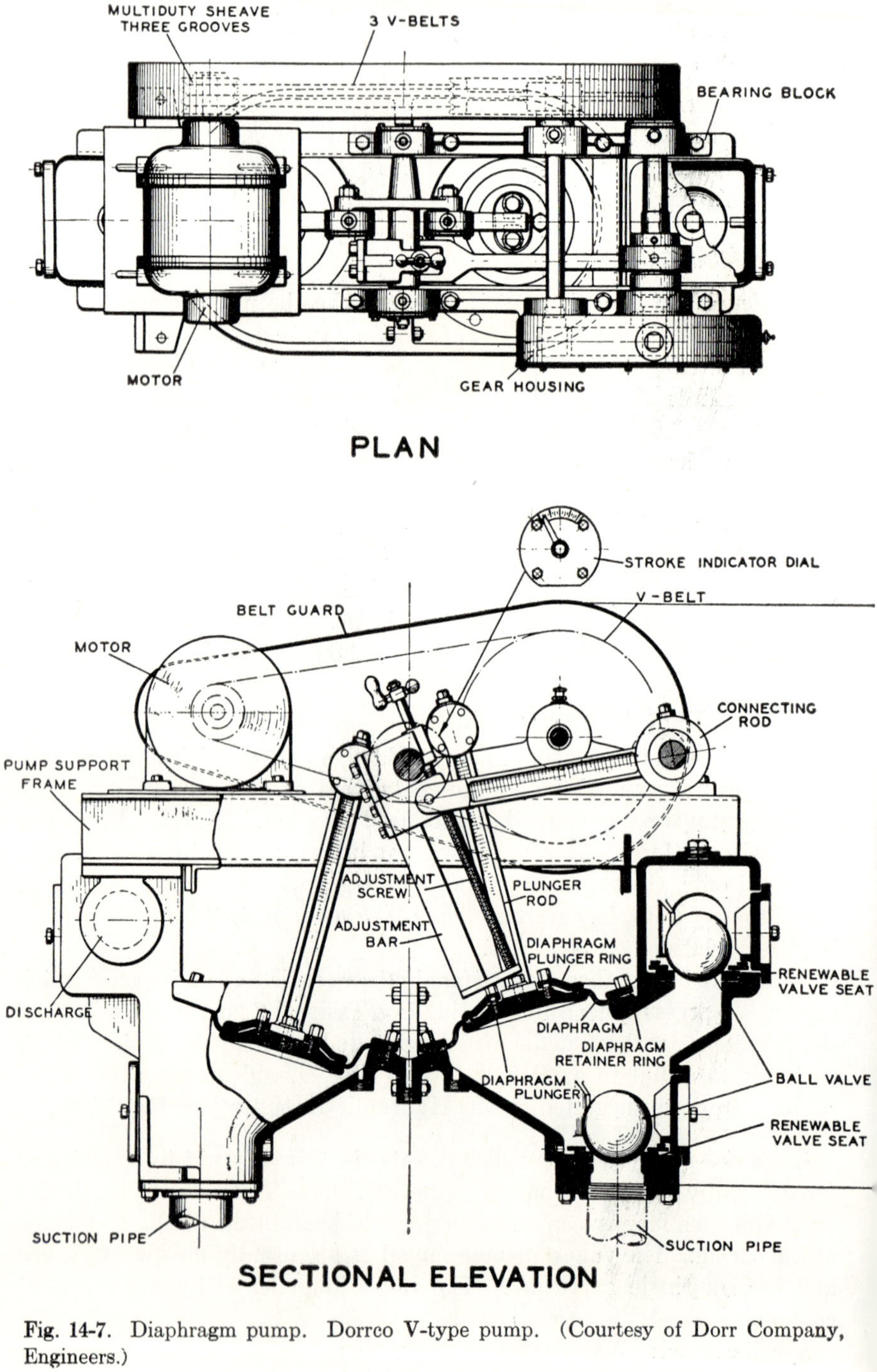


**Fig. 14-7.** Diaphragm pump. Dorrco V-type pump. (Courtesy of Dorr Company, Engineers.)

2. *Mixed Flow.* Head is developed with thin impeller partly by centrifugal force and partly by the push of the vanes. This is accomplished by making the vanes doubly curved or screw-shaped so that the discharge is a combination of axial and radial flows.

Specific speed: 4500 to 9000

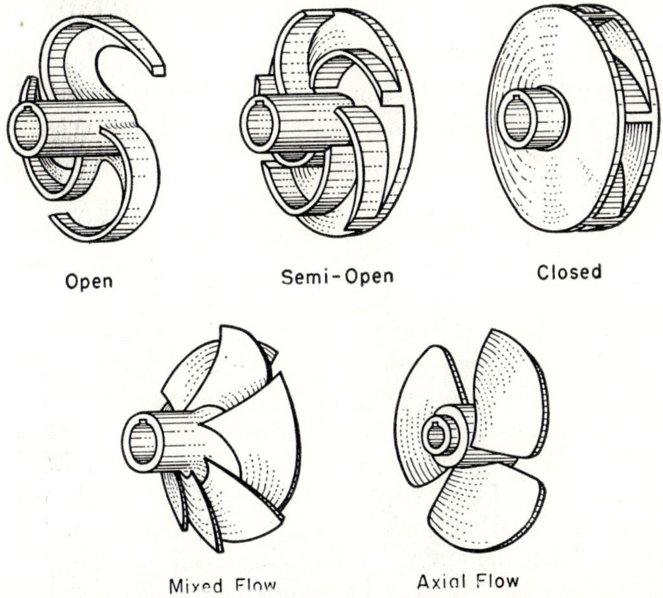


Fig. 14-8. Impeller types. (The three impellers in the top row are radial.)

An intermediate type between the mixed-flow and the radial is the Francis-type impeller. It has an axial inlet and a radial outlet.

Specific speed: 1500 to 4500

3. *Axial Flow or Propeller Type.* Most of the head produced by this impeller is due to the pushing or lifting action of the vanes. The fluid enters and leaves the impeller in an axial or nearly axial direction.

Specific speed: 9000

In proceeding from the radial type to the axial flow impellers the characteristics change from high head and moderate flow rates to extremely high flow rates and low heads.

### Description of Pump Types and Parts

In the accompanying illustrations several of the more common centrifugal pumps are illustrated and parts are identified. The discussion will be confined primarily to the radial impeller design since it is by far the most common in process pumps.

The unique requirements of the process industries which often require the pumping of fluids at high pressure and temperature and under corrosive conditions have caused the development of complete lines of pumps specifically designed for process plants, called process pumps. These can be divided into three major types: general service, hot oil, and chemical pumps.

### General Service Pump

These are moderate-priced pumps designed for services which do not require the special alloys and mechanical design features necessitated by high temperature or corrosive conditions.

### Hot Oil Pumps

Some manufacturers refer to their line of hot oil pumps as process pumps although it would seem desirable to reserve this term for the description of all pump types employed in pumping process streams.

Hot oil pumps are specifically designed for pumping hydrocarbons and various other chemicals or solutions under conditions of high temperature (up to 800° F.). These conditions require special construction to prevent undue strain caused by uneven temperature distribution in the pump body and supports. In addition stuffing box design and materials of construction must be given special consideration. Generally, these pumps are designed so that the entire rotating element( bearing bracket, shaft, and the impeller assembly) can be removed without disconnecting any piping or moving the pump case or driver. A spacer type coupling (Fig. 14-9) is used, and the pump cover can be unbolted. The rotating element and bearing assembly can then be removed as a unit. This makes it possible to take the rotating part of the pump to the shops for repairs.

### Chemical Pumps

Pumps in this category are constructed of corrosion-resistant materials such as alloy steels, rubber-lined steel and even plastics, glass, or stoneware. Because of the unusual properties and high price of such materials, the mechanical design of these pumps is unique, but simplicity of design is stressed in order to reduce the use of expensive alloys.

The casting of certain alloy steels is often a difficult problem. In order to avoid strains and shrinkage which cause porosity in castings, tapped holes and bosses are eliminated by clamping the two portions of the casing. This arrangement has the further advantage of eliminating the passage of bolts through gaskets where corrosion can occur unnoticed.

A typical chemical pump with a casing constructed of plastic is shown in Fig. 14-10. Metal mounting plates are used to clamp the two portions of the plastic casing together. These plates protect the plastic and as in pumps constructed of alloys, the bolts do not pass through the gasket.

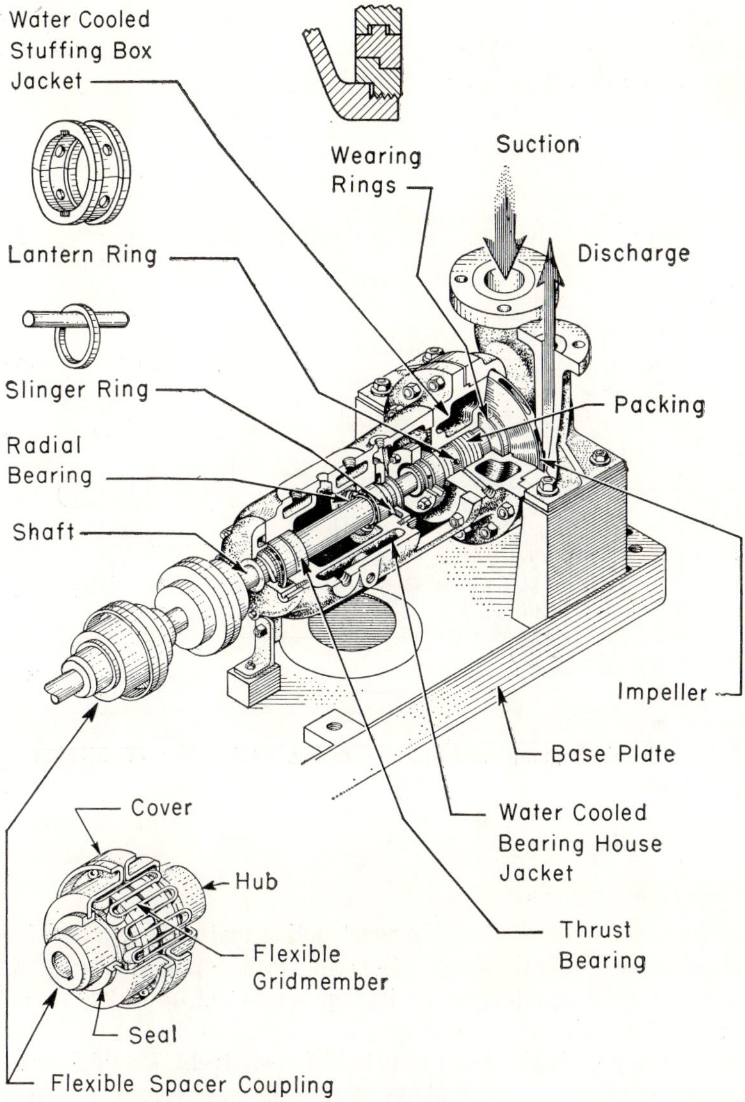


Fig. 14-9. Typical process pump (hot-oil pump).

All chemical pumps should be designed so that leakage and corrosive action can be easily detected, and so that dismantling and servicing of the pump is simple. There should be suitable means for removing stuffing box leakages without damaging the nonalloy portions of the pump.

### Rotation of Pump

In specifying motors and planning pump installations it is necessary that the designation of rotation be standardized. It is customary to describe the rotation of the pump as clockwise or counterclockwise from a position at the driver end, facing the pump.

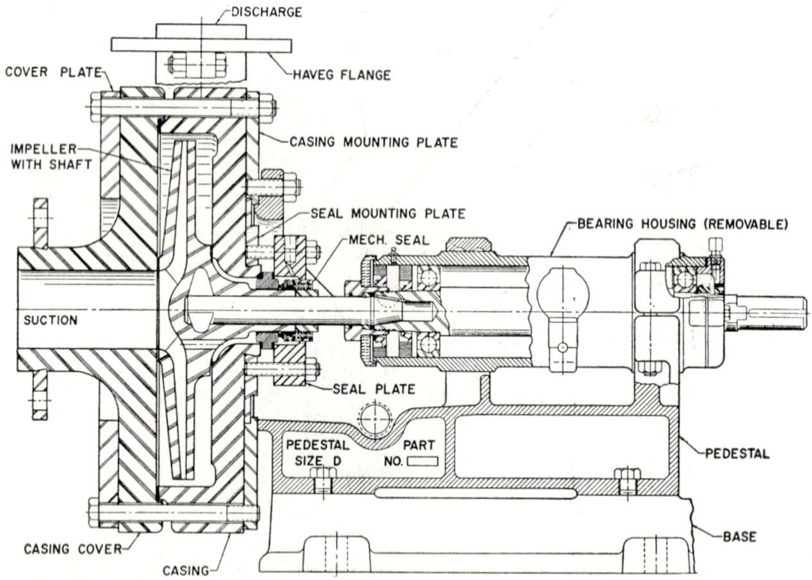


Fig. 14-10. Chemical pump with plastic casing. (Courtesy of Mission Pump Company.)

### Pump Casings

Casing Materials

The type service determines the construction materials required for the casing of a centrifugal pump. General service pump casings which operate with cold liquids are cast iron or semi-steel up to 300 to 450°F. and bronze for all-bronze pumps.

Hot oil pump casings are constructed of cast steel. For high pressure multistage pumps operating above 1500 psi, however, a forged steel "barrel" must be used. The internal assembly is fitted into this barrel and each stage has a separate inner casing. The outer casing or barrel is under full discharge pressure and the inner casing is under a compressive force due to the differential pressure between stages. The "barrel" is designed as a pressure vessel. The casings for chemical pumps vary, depending on the service.

### Volute, Circular, and Diffusion Casings

Although several manufacturers make single-stage pumps with circular casings and claim an advantage in pumping abrasive liquids, the majority of all single-stage pumps are built with volute casings designed so that the average velocity through all sections will be the same.

Many multistage pumps have a diffusion casing. Diffusers (see Multistage Impellers) located between stages produce a gradual change from velocity head to pressure head.

### Vertically or Horizontally Split

Pump casings are either vertically (Fig. 14-9) or horizontally split (Fig. 14-11). Hot oil pumps are usually constructed with vertically split casings. This permits the support of the pump at the shaft centerline so that it can expand or contract without danger of misalignment.

Casings split on the horizontal centerline are most common in pumps having impellers mounted between the bearings (double-suction or low pressure multistage). The upper half is then removable for ready access in case of repairs without disturbing the piping since the suction and discharge nozzles are located in the bottom portion.

### Suction and Discharge Nozzle Arrangement

Although general service and chemical pumps ordinarily are center suction and top discharge, it is possible to order alternate discharge nozzle locations at 45° and 90° with the horizontal in any of four quadrants.

In addition to center suction and top discharge designs hot oil pumps are also built with top suction and top discharge. The latter is more often the standard arrangement. Both the suction and discharge nozzles are well supported by the pedestal, and the pump casing is not subjected to stresses created by pipe expansion. On center suction designs the suction line must be supported as it enters the pump in order to prevent undue stress on the casing and possible cracking. Top-suction, top-discharge pumps also eliminate the necessity of special vents and thus simplify piping.

### Casing Mounting

The casing of the pump may be cast integral with a rigid foot support or box type frame having bolt holes for mounting. Many pumps, particularly in services above 500° F, are supported on pedestals (saddle mounting) at or near the centerline of the pump. As mentioned previously, the vertical split hot oil pump (Fig. 14-9) has a support of this type at the horizontal centerline allowing free expansion and contraction without disturbing the alignment.

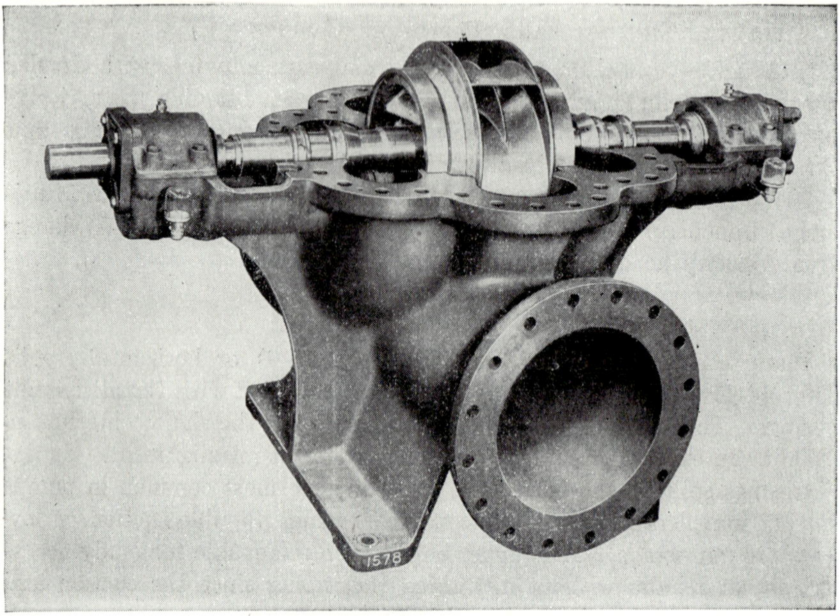


Fig. 14-11. Double-suction centrifugal pump. (Courtesy of Pacific Pump Company.)

Many manufacturers offer water-cooled pedestals for pumps operating above 500° F. Stepanoff[10] cautions that water cooling on turbine driven units might do more harm than good in certain cases since the turbine mounting is never water cooled and often reaches the same temperature as the pedestal. In this case misalignment would result if the pump pedestals were water cooled.

### Single-Suction and Double-Suction Pumps

Although double-suction pumps can be made vertically split so that the casing can be supported at the shaft centerline, such an arrangement is less convenient for servicing. Hence practically all double-suction pump casings are horizontally split. It then follows that hot oil pumps are generally single suction.

Essentially a double-suction impeller (Fig. 14-11) appears to be two single-suction impellers touching back to back. The liquid enters both sides of the impeller. Since the same suction and discharge pressures act on both sides of a double-suction impeller, the theoretical axial thrust is zero. Actually, a small thrust bearing is used on these pumps. The single-suction pump has suction and discharge pressure acting on the front and only discharge pressure on its back side. An axial force outward is thereby created which requires a larger thrust bearing. Some single-suction pumps are built with a balancing hole through the impeller for admitting suction pressure to the back side of the impeller, thus achieving a more balanced design. In any event thrust bearings are designed for the full thrust expected from a single-suction design. Because of the added cost for thrust bearings, the double-suction design has advantages on large pumps operating at moderate temperatures.

### *Radial Impeller Types*

Impellers are cast in one piece and are made of cast iron, bronze, alloy such as 11–13% chrome steel, or other materials suitable for the liquid to be pumped. The impeller is keyed to the shaft and held secure.

Several radial impeller types are shown in Fig. 14-8. They are closed, semi-closed or semi-open, and open. The closed impeller is usually installed on hot oil pumps and is very common on all types of pumps handling clear liquids. The impellers of this design have side walls called shrouds which totally enclose the vanes except at the periphery and the suction. Closed impellers are installed to provide a running contact between the impeller and the casing, thus minimizing leakage from discharge to suction.

Semi-open impellers have only one wall or shroud, the back one. Open impellers have no shrouds, although ribs or partial shrouds are provided for mechanical strength, especially on the larger sizes.

Liquids containing solids in suspension or viscous liquids are best handled by open or semi-open impellers. These impellers can be cleaned easily and operate with a minimum of clogging. The open impeller requires the simplest casting and the least metal; and, therefore, is the lowest in cost, a particular advantage when special alloy construction is required as in chemical pumps.

Because of the leakage encountered in the semi-open and open impellers, which increases with wear, the efficiency of these pumps after prolonged service is not as great as for the closed impeller. The open and semi-closed impeller designs, however, give a slightly improved initial efficiency due to reduction in disk friction.

Although the semi-closed and open impellers are cheaper to fabricate, it does not necessarily follow that a semi-closed impeller pump, for instance, is cheaper than a closed impeller pump for the same service. Indeed, hot oil pumps with semi-closed impellers are more expensive than their closed impeller counterpart. This is due to the mechanical difficulties in making the clearance between the semi-closed impeller and the casing a minimum in order to reduce leakage and maintain efficiencies.

### *Wearing Rings*

Most pumps with closed impellers (hot oil, etc.) are provided with wearing rings as illustrated in Fig. 14-9. Since these rings are removable and renewable, a worn pump in which the running clearance between the impeller and the casing has become excessive can be restored to satisfactory operation at a small cost by installing new rings.

The majority of all hot oil pumps have both casing and impeller rings. Some are furnished with casing rings only, but the impeller hubs are oversized in such pumps so that impeller rings can be installed when the hub has become worn. General service pumps with closed impellers usually come equipped with casing rings only although some designs have no wearing rings.

Absence of wearing rings is a disadvantage on large pumps where the cost of replacement of an impeller compared to that of a wearing ring is great. Because of the mechanical problems involved with alloy construction, wearing rings are often not used in closed-impeller chemical pumps. Care is exercised, however, in the design and manufacture to provide horizontal wearing surfaces on each side of the impeller.

Since clearances between wearing surfaces are very small (0.01 in.) it is necessary that the two matching surfaces be made of different materials or at least of the same materials with different surface hardnesses.

Closed impeller pumps designed for handling abrasive fluids are equipped with a flushing chamber. Injection of clean liquid into this chamber prevents the entrance of abrasive material into the clearance between the rings.

### Single Stage and Multistage Impellers

Up to about 600 ft of head (600 ft-lb of force/lb of mass)* differential single-stage pumps are used almost exclusively. Above this head, however, multistage impellers must be used. Multistage pumps have two or more stages connected in series and can be pumps of the volute or diffuser-type (Figs. 14-12 and 13). In the former, each stage discharges into the next through a volute. In the diffuser pump the liquid discharges into a diffuser after leaving each stage. The diffuser converts velocity head into pressure head and then discharges the fluid to the suction side of the next stage.

The diffuser-type pump is characterized by higher efficiencies although in recent years volute designs have been introduced that have similarly high efficiencies. Diffusers permit operation over a wider range of operating conditions at a sustained high efficiency.

Kristal and Annett[7] report that at reduced capacities volute multistage pumps can develop large radial thrusts on the impeller which have been known to cause shaft failures, but radial loads on diffuser-types are

* The term *head* is often expressed in ft. Actually it is a work term on a pound of mass basis. The correct units are foot-pounds of force (ft-#) per pound of mass (lb).

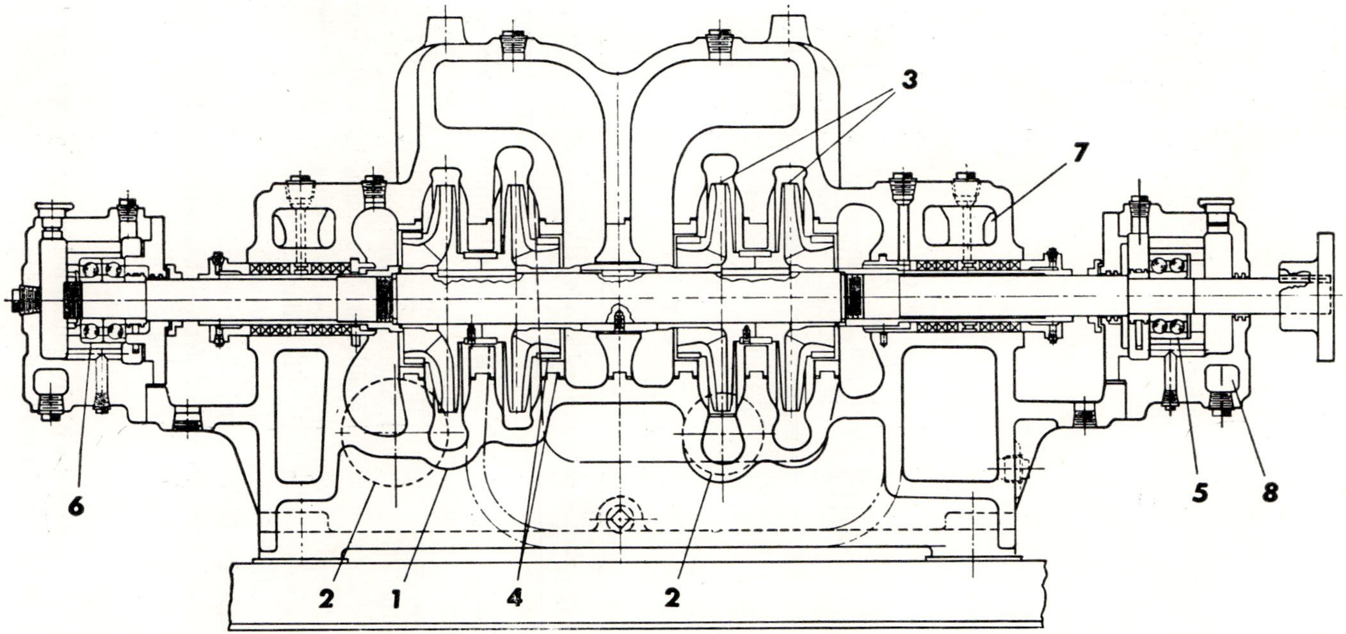


1. Axially split case, volute, staggered for radial balance.
2. Side suction and discharge.
3. Opposed impellers for balanced axial thrust.
4. Removable and renewable case and impeller wearing rings.
5. Double-row, deep-grooved ball radial bearing.
6. Angular contact ball thrust bearing.
7. Water cooled packing boxes.
8. Water cooled bearing housings.

Fig. 14-12. Volute-type multistage centrifugal pump. (Courtesy of Pacific Pump Company.)

balanced. This objection has been largely overcome, however, on some volute pump designs by the use of a double volute with outlets 180° apart. Opposed forces thus created insure radial balance at any capacity. On other designs radial unbalance is partly overcome by staggering volutes radially.

Volute multistage pumps are popular especially in the two and three stage designs. In high pressure-high temperature hot-oil pumps the diffuser type predominates.

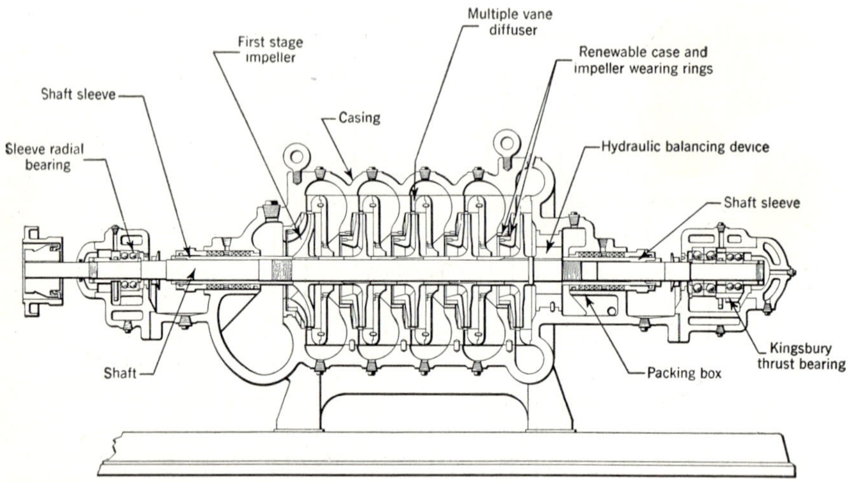


Fig. 14-13. Diffuser-type multistage centrifugal pump. (Courtesy of Pacific Pump Company.)

The problem of balancing axial thrust is important in multistage pumps. Volute type designs have an advantage because hydraulic balance can be approximated by arranging part of the impellers facing in one direction and part in the other (Fig. 14-12). Diffuser-type pumps cannot be so arranged because of the mechanical problems. The stages in the diffuser design must, therefore, be arranged all in the same direction and axial thrust overcome by means of a hydraulic balancing device employing a balancing chamber which is connected to the suction line (Fig. 14-13). The area of the balancing disks is such that the hydraulic axial thrust of the pump is balanced by the pressure differential between the back face of the last stage impeller and the balancing chamber.

### Bearings

All pumps, whether equipped with hydraulic balancing devices or not, are constructed with both a thrust bearing and a radial bearing. The thrust bearing is locked while the radial bearing is free to move axially with the shaft. General service and many chemical pumps employ grease-lubricated ball bearings for both radial and thrust bearing. Ball bearings are also used in most hot oil pumps but they are oil lubricated.

The bearing housing on hot oil pumps provides an oil reservoir in which the correct oil level is maintained by a constant level oiler. Slingers or rings attached to the shaft continuously discharge the oil in troughs leading to the bearing. Bearing housings on such pumps are water cooled.

Large high-pressure multistage pumps often employ sleeve-type radial bearings and Kingsbury thrust bearings since these bearings are superior for high speed, heavy duty service. Lubrication of Kingsbury bearings in these large pumps is accomplished by an oil pump, gear driven from the pump shaft. The oil is circulated through a small cooler for temperature control.

### *Shaft and Shaft Sleeves*

To withstand the forces acting on it, the shaft of a centrifugal pump must be made of high tensile steel or alloy (SAE 1035, SAE 4140, SAE 2340, and various alloys such as 11–13% chrome). In all but very small pumps the shaft is protected against wear and corrosion by a sleeve over the portion of the shaft which extends through the stuffing box. The choice of sleeve material depends on the nature of fluid being pumped.

### *Stuffing Box*

As can be seen in Fig. 14-9 the shaft of a pump passes through a stuffing box as it enters the casing. The proper design of this stuffing box is an important factor in the successful operation of a pump, since it prevents leakage both in and out of the pump.

The stuffing box can either be packed as shown in Fig. 14-9 or equipped with a mechanical seal (Fig. 14-14). Packed boxes may be packed solid or provided with a seal cage or lantern ring located in the middle of the course of packing (Fig. 14-9).

When a pump is operating at low suction pressures air leakage into the pump through the stuffing box is prevented by using a lantern ring—seal cage—(Fig. 14-9). Sealing liquid is injected into the lantern ring and is distributed throughout the packing producing a cooling and lubricating action. A small portion of the liquid leaks to the outside through the stuffing box gland.

The liquid being pumped can be used as a sealing liquid by piping discharge liquid to the lantern ring. If the liquid being pumped, however, is corrosive or erosive another sealing liquid must be used. In petroleum refineries, for instance, a special sealing oil is used.

For high temperature operations a circulating-type stuffing box can be employed so that the external sealing fluid can be cooled and recirculated. The circulating-type stuffing box is designed so that cool sealing liquid enters one side of the stuffing box, circulates through it, and then passes to a cooler from which it is recirculated.

Sealing liquid (seal oil) systems are often installed in plants having large numbers of pumps that require sealing fluid. The oil can be piped to the pumps in a common header and returned to a cooler through another header. The flow of sealing oil into the pump is controlled by a pressure regulator which maintains a constant differential between the suction and seal cage pressure. In addition to the sealing oil, hot oil pumps are often equipped with a water jacket around the stuffing box.

Whenever the lubricating liquid leaking through a stuffing box is at a high enough temperature to cause flashing of flammable, noxious, or disagreeable vapors, a smothering gland may be used instead of the usual packing gland. The smothering gland is arranged so that cold water enters at the point where sealing liquid touches the gland. The water cools the sealing liquid thereby preventing vaporization.

Most general service and chemical pumps are designed to receive four or five rings of conventional packing; hot oil pumps receive from seven to ten. Proper packing is a necessity. If the shaft cannot be turned by hand, the stuffing box has been improperly packed.

General service pumps are usually packed with asbestos-impregnated graphite. Packing for chemical pumps must be selected for the specific liquid to be handled. In high temperature service such as that encountered with hot oil pumps metallic packing is used. Metallic packing is intended for use with hardened steel shaft sleeves. Only a soft grade of packing should be used with shafts which have bronze sleeves.

Packing problems have become increasingly difficult as the process industries resort to higher temperatures and pressures and to the pumping of volatile materials. A number of mechanical seals have been developed by various manufacturers for these severe services (Fig. 14-14). These are used in lieu of packed stuffing boxes and most of the hot oil and chemical pumps are designed to receive either packing or the mechanical seal. The mechanical seal is more expensive, but often service conditions warrant its use.

Mechanical seals consist of two highly polished surfaces in sliding contact. The contact is maintained by springs which are adjusted to exert a force sufficient to prevent leakage of the fluid being pumped. The surfaces are of dissimilar materials. One surface is stationary and is connected to the frame while the other rotates and is fastened to the shaft. The surfaces are lubricated by the fluid being pumped or by an auxiliary fluid if that being pumped is nonlubricating. The mechanical seal effectively stops all stuffing box leakage even under high stuffing box pressures. It is, therefore, particularly indicated for use in services where leakage cannot be tolerated such as the pumping of light hydrocarbons which are near the boiling points or the pumping of corrosive and toxic materials.

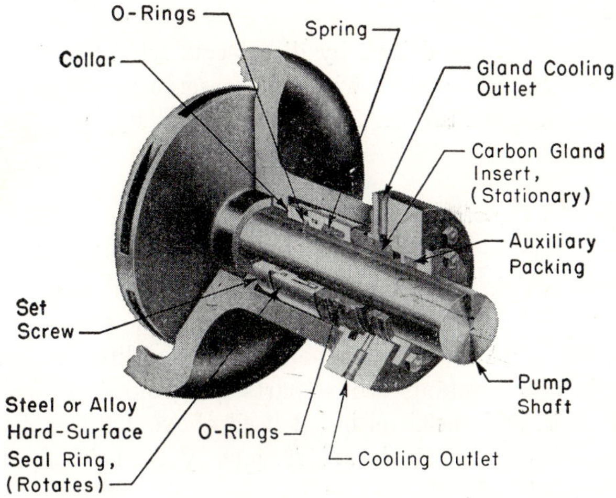


Fig. 14-14. Mechanical seal. (Courtesy of Durametallic Corporation.)

For highly corrosive or high temperature liquids a double mechanical seal is employed. It is designed so that a cool lubricating liquid can be injected across the seal faces preventing contact by the pumped liquid.

### Couplings

Flexible couplings (Fig. 14-9) of the steel-spring-grid, gear, spacer, and pin and rubber bushing type are used. These couplings can compensate for slight misalignments but are not intended as universal joints, and great care in alignment of the shafts is necessary. The flexible coupling is usually furnished by the pump manufacturer, but he must be told the dimensions of driver shaft so the coupling can be properly bored.

### Base Plates (Bed Plates)

Pump manufacturers usually supply a base plate on which the pump is mounted and which is drilled to receive the motor or turbine specified for the service. The purchaser must supply the pump manufacturer with certified dimensional drawings furnished by the driver manufacturer (for economic reasons it is usual practice to purchase pumps and drivers separately). The equipment is shipped separately, for mounting in the field. The base plates for small pumps are usually built of steel channel. Base plates for larger pumps are cast iron and have a rimmed trough sloping toward one end for draining leakage to waste. Grout holes are also provided. On extremely large installations it may be more economical to mount the pump and driver on separate steel pads or sole plates on the integral foundation, rather than provide a common base plate.

#### Drivers

Centrifugal pumps are driven by direct-connected motors or turbines. When two pumps are installed with one to serve as a spare, a motor is often used for one and a turbine for the other thus supplying the maximum flexibility.

#### Materials of Construction

A distinct advantage of the centrifugal pump is the wide variety of materials from which its various parts can be constructed. Casings can be made of materials ranging from glass and stoneware to the toughest alloy steels. General statements, therefore, are of little value. It might be said that in process plants bronze parts are avoided in services operating above 250° F. and cast iron is not used above 350 to 400° F. Alloys such as 11–13% chrome steels are frequently used for operating temperatures above 400° F.

### Centrifugal Pump Calculations

#### Horsepower and Efficiency

The water horsepower is defined and calculated for centrifugal and reciprocating pumps in like manner. The brake horsepower, defined as the energy input to the pump shaft, is related to the water horsepower by an efficiency. This efficiency for centrifugal pumps is an overall value and includes mechanical, hydraulic, and volumetric efficiencies.

$$E = \left(\frac{\text{WHP}}{\text{BHP}}\right)(100)$$

#### Performance Curves

The graph in Fig. 14-15 illustrates a typical head-capacity or performance curve for a centrifugal pump at a constant speed and is called a characteristic curve. The corresponding efficiencies, brake horsepowers, and the minimum values of net positive suction head required at each flow rate are also given. These curves are furnished by the pump supplier for each size and type of pump that he manufacturers. The head-capacity curve represents the work per unit mass of fluid pumped called head (ft-lb of force/lb of mass) at each possible flow rate. The head, which is often called feet of head, can be converted to the differential pressure under which the pump operates by multiplying by the density.

The curves are usually based on the pumping of clear, cool water. The characteristics for pumping other liquids are not necessarily the same as for water.

The characteristic curve is not to be confused with the so-called test curve. Test curves are obtained from an actual test of a particular pump

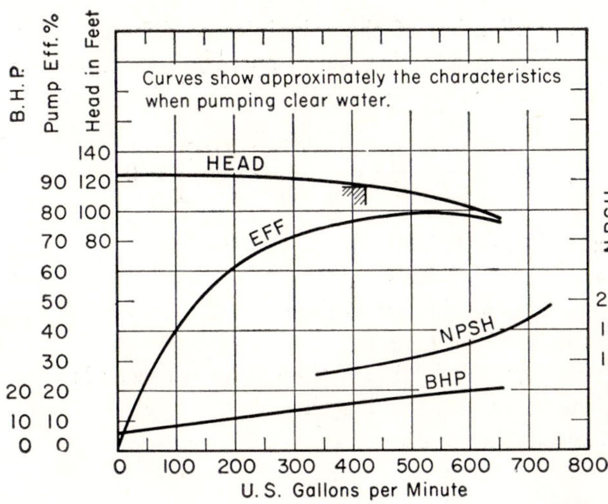


Fig. 14-15. Performance curve for a centrifugal pump. (Adapted from Allis Chalmers Company data, by permission.)

and may be purchased with that pump for an additional charge. Characteristic curves represent the average performance of the type pump purchased and small variations between individual pumps produced on the same pattern are possible. It is generally not necessary or economically sound to request test curves except for large expensive pumps in which case the test cost will be only a small fraction of the total cost.

The mark

which appears on characteristic curves designates the head and capacity for which the pump was sold. The mark

$\wedge$

indicates the guaranteed efficiency. In general pumps are selected so that the operating point corresponds to or is close to the best efficiency point (the maximum efficiency obtainable, usually designated as BEP).

As the capacity of a centrifugal pump is increased, a point is reached at which fluid velocity in the casing is so great that local pressures in the casing decline to the vapor pressure or below. This point is called the cut-point or break-point because vapor bubbles are formed and a rapid decline in head, efficiency, and capacity results. For this reason performance curves should never be extrapolated.

### *Affinity Laws*

Through the application of the principles of dimensional analysis on the physical properties affecting pump operation the following relationship is obtained.*

$$f\left[\frac{Q}{(g_cH)^{1/2}D^2}, \frac{nQ^{1/2}}{(g_cH)^{3/4}}, \frac{\rho QD}{\mu D^2}\right] = 0 \quad (1)$$

where $Q$ = capacity, cu ft/unit time
$H$ = pump head, ft-#/lb
$n$ = speed, revolutions/unit time
$D$ = impeller diam, length
$g_c$ = lb-ft/(time)$^2$/#, 32.17 if time units are seconds
$\rho$ = density, mass/cubic length
$\mu$ = viscosity, absolute
# = pound of force
lb = pound of mass

Call the dimensionless expressions $\pi_1$, $\pi_2$, and $\pi_3$ in the order in which they appear. For true dynamic similarity, $\pi_1$, $\pi_2$, and $\pi_3$, must be constant for similar pumps or for the same pump at different speeds. $\pi_3$, which will be recognized as a Reynolds' number since $Q/D^2$ is proportional to the velocity, cannot be maintained constant with changes in pump speed or size if the same liquid is pumped. It has been shown, however, that for practical applications the Reynolds' number can be neglected in considering the conditions for dynamic similarity.

Thus the conditions for dynamic similarity are constant values of $\pi_1$ and $\pi_2$. The form of eq. 1 can be altered by appropriate mathematical operation to give the following:

$$f\left[\frac{Q}{(g_cH)^{1/2}D^2}, \frac{nD}{(g_cH)^{1/2}}\right] = 0$$

Calling $nD/(g_cH)^{1/2} = \pi'_2$ it can be stated that for conditions of similarity of flow $\pi_1$, $\pi_2$, and $\pi'_2$ are constant.

It has been shown[10] that for similar pumps or for the same pump at different speeds the velocity diagrams at the impeller are geometrically similar when the pumps are operating at the same efficiency.

* This development is based on the more complete work of Stepanoff.[10]

## Application of Affinity Laws

The following conclusions can now be made.

For the same pumps or geometrically similar pumps $Q/(g_cH)^{1/2}D^2$, $nD/(g_cH)^{1/2}$, and $nQ^{1/2}/(g_cH)^{3/4}$ have unique values corresponding to each efficiency and these values remain constant regardless of the impeller diameter or pump speed.

The constancy of the various dimensionless groups expressed above is an ideal case and not completely true. It is, however, a useful approximation and several relationships can be derived that are of considerable value to the chemical engineer in determining operational characteristics of a pump when the pump diameter or speed is changed.

*For the Same Pump Operating at Different Speeds:*

$$D = \text{constant}$$

Hence at any given efficiency:

*Capacity*

$$\frac{Q}{(g_cH)^{1/2}D^2} \quad \text{and} \quad \frac{nD}{(g_cH)^{1/2}} \quad \text{are constant}$$

or

$$\frac{Q}{(g_cH)^{1/2}D^2} \div \frac{nD}{(g_cH)^{1/2}} = \text{constant}$$

Hence

$$\frac{Q}{nD^3} = \text{constant} = K \tag{2}$$

But $D$ is constant

$\therefore\ Q = K'n$ where $K'$ and $K$ are constants

or

$$\frac{Q_2}{Q_1} = \frac{n_2}{n_1} \tag{3}$$

*Head*

$$\frac{nD}{(g_cH)^{1/2}} = \text{constant} = J$$

or

$$\frac{n^2D^2}{H} = J' \quad \text{where } J \text{ and } J' \text{ are constants} \tag{4}$$

Hence at constant diameter

$$\frac{H_2}{H_1} = \left(\frac{n_2}{n_1}\right)^2$$

*BHP*

$$\text{BHP} \propto QH \propto n^3 \qquad \text{at constant } D$$

$$\frac{\text{BHP}_2}{\text{BHP}_1} = \frac{n_2^3}{n_1^3} \tag{5}$$

The application of these laws is best illustrated by the following example. Remember that these laws apply only for efficiencies of the same value.

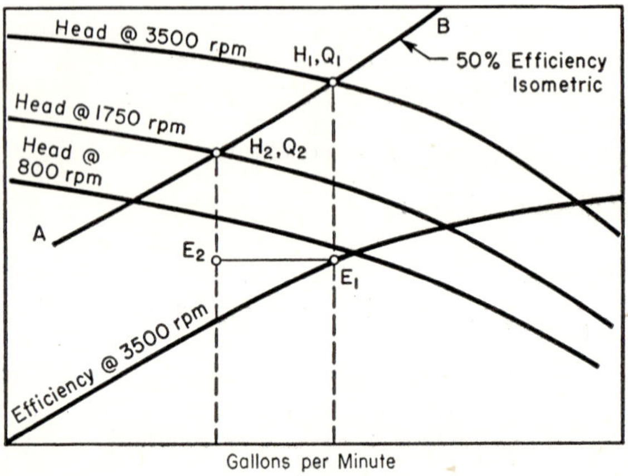


Fig. 14-16. Estimation of performance curve for new conditions.

EXAMPLE: Given the performance curve at 3500 rpm illustrated in Fig. 14–16. It is desired to calculate a new performance curve at 1750 rpm.

SOLUTION:

1. Select any efficiency and read off corresponding values of $Q$ and $H$.
2. Solve for $H_2$ and $Q_2$

$$H_2 = H_1 \left(\frac{1750}{3500}\right)^2$$

$$Q_2 = Q_1 \frac{1750}{3500}$$

3. Plot these as one point on the Head capacity curve.
4. Plot the value of the selected efficiency opposite $Q_2$ as a point on the efficiency curve for 1750 rpm.
5. Enough additional points are selected to enable sketching in the desired curve.
6. BHP curve is calculated in a similar manner.

A whole series of curves can thus be calculated for various speeds. The line $AB$ in Fig. 14-16 indicates a line of constant efficiency (50%). There are numerous such lines corresponding to each value of efficiency. It is only on these lines that the affinity laws may be applied.

*For geometrically similar pumps with different impeller diameters but same speeds:**

At the same efficiency points:

*Capacity*

From Eq. 2 above,

$$Q = KnD^3$$

At constant speed

$$Q = K''D^3 \qquad \text{where } K \text{ and } K'' \text{ are constants}$$

or

$$\frac{Q_2}{Q_1} = \left(\frac{D_2}{D_1}\right)^3 \tag{6}$$

*Head*

At constant speed from eq. 4 above

$$\frac{H_2}{H_1} = \left(\frac{D_2}{D_1}\right)^2 \tag{7}$$

*BHP*

In like manner

$$\frac{\text{BHP}_2}{\text{BHP}_1} = \left(\frac{D_2}{D_1}\right)^5 \tag{8}$$

*Application:* A series of curves can be prepared for different diameters when one curve is known in a manner similar to that described above.

CHANGING IMPELLER DIAMETERS IN THE SAME CASING. Even more important to the chemical engineer than the variation of head and capacity in a series of geometrically similar pumps are the changes produced by reducing the impeller diameter of a given pump. A smaller impeller may be installed or the existing impeller turned-down on a lathe. The new impeller and casing combination is no longer geometrically similar to the original arrangement, but the effect of the diameter change can be approximated when it is assumed that the reduction in diameter reduces the discharge peripheral velocity without affecting the efficiency. The peripheral velocity is directly proportional to the impeller diameter and hence $Q \propto u \propto D$. Head is proportional to $u^2/2g_c$ (where $u$ is peripheral velocity).

These approximations yield the following relations which apply at equal efficiencies just as the previous equations based on the affinity laws.

$$\frac{Q_2}{Q_1} = \frac{D_2}{D_1} \tag{9}$$

* Each style or type of pump is manufactured in many different sizes. If the relation between the casing and impeller dimensions in the different sizes of one type is the same, the pumps are said to be geometrically similar.

$$\frac{H_2}{H_1} = \left(\frac{D_2}{D_1}\right)^2 \tag{10}$$

$$\frac{\text{BHP}_2}{\text{BHP}_1} = \left(\frac{D_2}{D_1}\right)^3 \tag{11}$$

These relations are not as accurate as the affinity laws and should be used with caution since the efficiency is reduced with large reductions in impeller diameter. In practice it is preferable to undercut and test the impeller before cutting further (*i.e.*, if it is calculated that the new diameter should be 75% of the original, cut for 80% and test it first). To compensate for some of the inaccuracies inherent in these calculations, corrections to the calculated diameter ratio are presented by Stepanoff[10] for both radial and Francis (high specific speed) impellers. Reductions of about 90% of those calculated for radial impellers and 80% for Francis type are suggested.

EXAMPLE: A pump with a 16 in. radial impeller operates at 1750 rpm and produces 2600 gpm at 200 ft at the best efficiency point. What reduction in the impeller diameter will be required to reduce the flow rate to 2000 gpm at the BEP? What will be the corresponding head?

$$\frac{D_2}{D_1} = \frac{2000}{2600} = 0.77$$

Reduce (100)(0.23)(0.9) = 20.7%

or $$D_2 = (16)(1 - 0.207) = 12.7 \text{ in.}$$

*New Head*:

$$H_2 = H_1\left(\frac{D_2}{D_1}\right)^2 = (200)\left(\frac{12.7}{16}\right)^2$$

$$= 178 \frac{\text{ft-\#}}{\text{lb}}$$

SPECIFIC SPEED

One of the dimensionless expressions in the Affinity Laws is

$$\frac{nQ^{1/2}}{(g_cH)^{3/4}} = \text{constant at a given efficiency}$$

It is often used by pump men in the form $nQ^{1/2}/(H)^{3/4} = n_s$ and is called the specific speed. As such it has lost its dimensionless properties because $g_c$ has been omitted. $Q$ and $H$ are arbitrarily expressed, therefore, in gpm and ft, respectively.

Unless otherwise stated the specific speed of a pump is given for the maximum efficiency point (best efficiency point, BEP) which has the same numerical value for similar pumps.

Do not attempt to attach physical meaning to the specific speed. It is merely a convenient number and has come to be used as a type characteristic for geometrically similar impellers. The specific speed varies from zero to infinity for any given impeller and has a constant value for different speeds and sizes of impellers operated at the same efficiency. When used as a type number it is given for the best efficiency point only.

The first and most important step in the mechanical design of a pump is the selection of a specific speed and the impeller type that will most efficiently handle a pumping problem. Impeller design constants used for the calculation of impeller diameter, inlet and outlet impeller areas and other factors in pump design have been correlated with specific speed. In addition specific speed has been related to the suction conditions and the net positive suction head (see below).

Since specific speed is a type characteristic it can be used in selecting a pump in a given series when data on only one of the pumps in the series are available.

EXAMPLE: A pump of a specific speed of 554 has an 8 in. impeller and operates at 3500 rpm pumping 63 gpm at 187 ft of head at the best efficiency point. A new pump geometrically similar is to be built for an operation at 520 gpm and 300 ft of head at its best efficiency point. What will be its speed and impeller diameter?

$$n_s = \frac{nQ^{1/2}}{H^{3/4}}$$

$$554 = \frac{n(520)^{1/2}}{(300)^{3/4}}$$

or

$$n = 1750 \text{ rpm}$$

The affinity laws apply at the same efficiency, hence

$$\frac{Q_1}{n_1(D_1)^3} = \frac{Q_2}{n_2(D_2)^3}$$

$$\frac{63}{(3500)(8)^3} = \frac{520}{(1750)(D_2)^3}$$

or

$$D_2 = 20.3 \text{ in.}$$

### *Corrections to Performance Curve for Viscous Liquids*

The Reynolds' number could be neglected in comparing characteristics of centrifugal pumps for nonviscous liquids such as water. In the case of viscous liquids, however, the Reynolds' number plays a significant role. The viscosity actually affects the location of the head-capacity curve as well as the efficiency curve because of increased frictional losses.

The usual performance curve is based on water as the liquid pumped. Exhaustive experimental tests on many types of pumps have yielded correction factors which can be applied to the head ($H_w$) and efficiency ($E_w$)

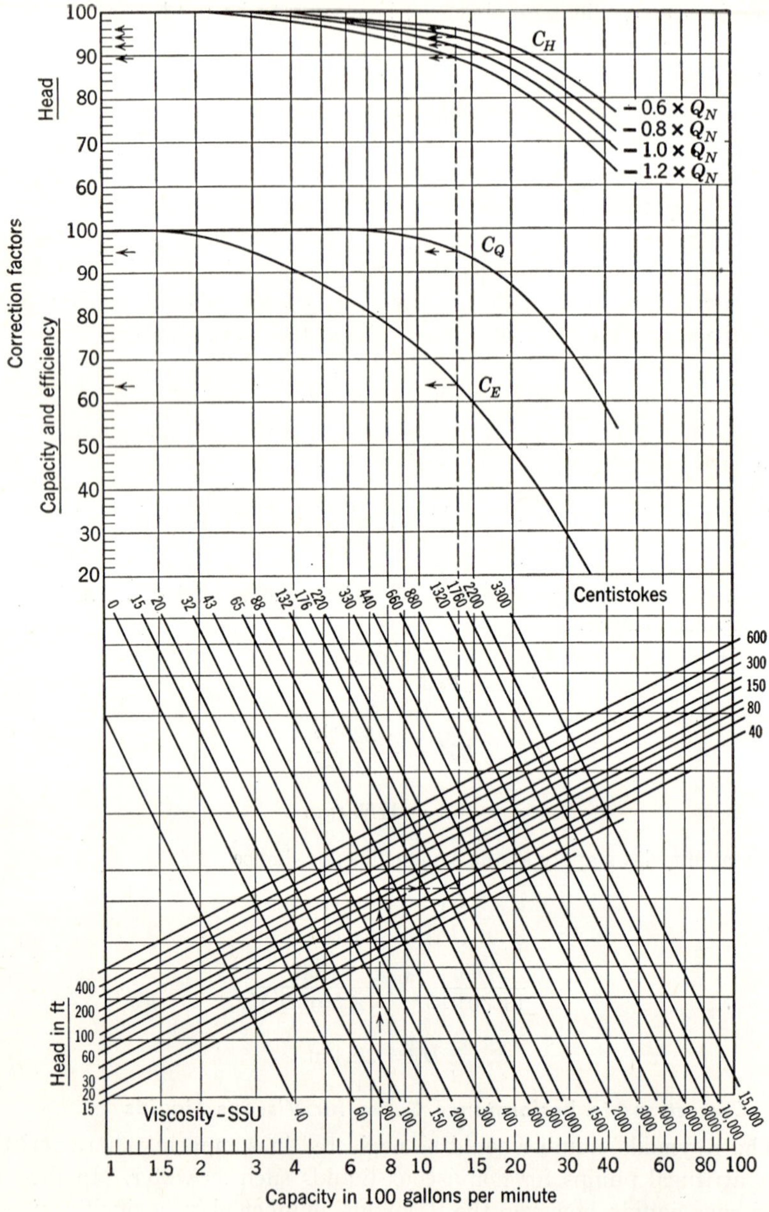


Fig. 14-17. Viscosity correction curves for centrifugal pump performance. (Reprinted from *Standards of Hydraulic Institute,* copyright 1955, Hydraulic Institute, 122 E. 42nd St., New York 17, N. Y.)

*USE OF FIG. 14-17:*

Given the complete performance characteristics of a pump when handling water, locate the water capacity at maximum efficiency ($1.0 \times Q_{nw}$). Calculate $0.6 \times Q_{nw}$, for water at the BEP to obtain values for the viscous liquid. These factors are correlated with the Reynolds' number and are given in the Standards of the Hydraulic Institute[9] or in *Centrifugal and Axial Flow Pumps,* by Stepanoff.[10]

The following relations apply:

$$H_v = C_H H_w$$

$$E_v = C_E E_w$$

where $H_w$ and $H_v$ = head for water and the viscous liquid, respectively
$E_w$ and $E_v$ = efficiency for water and the viscous liquid, respectively
$C_H$ and $C_E$ = corrections for the head and efficiency

Since the specific speed is constant at the best efficiency point:

$$\frac{nQ_w^{1/2}}{H_w^{3/4}} = \frac{nQ_v^{1/2}}{H_v^{3/4}}$$

or

$$\frac{Q_v}{Q_w} = \left(\frac{H_v}{H_w}\right)^{3/2}$$

Thus $Q_v$ (capacity when pumping the viscous liquid) corresponding to $H_v$ can be calculated. A second point can be established at the shut-off point (0 gpm) since viscosity has no effect at 0 flow and hence the efficiency (zero at zero flow) and head remain the same at shut-off. These two sets of points permit the sketching of corrected curves for the viscous liquid. The Hydraulic Institute[9] has developed a chart (Fig. 14-17) from which more than two points can be estimated so that a more accurate curve can be drawn.

### NPSH (Net Positive Suction Head)

Perhaps no other subject in pump design and selection has been discussed more profusely and with less reason than the calculation of NPSH. The calculation of NPSH is simple and requires no detailed explanation.

NPSH is simply the total suction pressure at the suction nozzle of the pump minus the vapor pressure of the liquid at the pumping temperature

---

$0.8 \times Q_{nw}$, and $1.2 \times Q_{nw}$. Read the values of $C_E$ (efficiency correction), $C_Q$ (capacity correction), and $C_H$ (head correction) for all four capacities by entering the chart at the water capacity and reading upward to the corresponding head $H_W$, then horizontally (either left or right) to the desired viscosity curves of the liquid in question, and then proceed upward to the various correction curves. Multiply each head, capacity, and efficiency for water by the corresponding correction to obtain the corrected values.

Plot corrected head and corrected efficiency against corrected capacity. Draw smooth curves through these points. The head at shut-off can be taken as approximately the same as that for water.

expressed in foot-pounds of force per pound of mass (usually referred to as NPSH in feet of liquid pumped).

$$\text{NPSH} = h_s - h_{vp}$$

or

$$\text{NPSH} = h'_s - h'_{vp}$$

where $h_s$ = total suction head in ft-#/lb absolute (includes atmospheric pressure)

$h_{vp}$ = absolute vapor pressure in ft-#/lb

$h'_s$ and $h'_{vp}$ = ditto except gage pressures in ft-#/lb

Vapor pressure is most generally reported as an absolute pressure. So-called Reid vapor pressures are gage values.

NPSH is important in the selection and installation of pumps which handle fluids near or at the boiling point. If sufficient head above the vapor pressure is not provided, operation of the pump will be impossible or unsatisfactory. As the pressure at the suction of the pump nears the vapor pressure of the liquid, vapor bubbles form in the pump casing and are subsequently collapsed at the discharge portion of the casing. This action, called cavitation,* can cause severe pitting of the impeller and casing. In addition as vaporization increases pumping ceases or becomes erratic. A centrifugal pump can only produce the head in ft-#/lb indicated by the performance curve. This corresponds to a very low pressure for low density material such as vapors and thus the pump cannot produce the discharge pressure required by the system.

The NPSH most suitable for successful operation as reported by the Hydraulic Institute[9] is correlated with specific speed for hot water and correction factors are given for hydrocarbons. Manufacturers recommend the NPSH required for their pumps. These values usually appear on the performance curve furnished with the pump. (See Fig. 14-15.)

### *Specific Gravity and Head*

The so-called head of a centrifugal pump is actually a unit of work being correctly expressed as ft-lb of force per lb of mass (ft-#/lb), although it is usually just referred to as ft of head. If this fact is clearly understood, the effect of specific gravity or density of the pumped fluid on pump operation is easily followed. The actual values of head on the performance curve are not affected by density since head is expressed as the work per unit mass of fluid that the pump will accomplish for a given volume flow. To calculate the work done per unit time, however, the volume of flow must be converted to mass by multiplying by the density. In addition, the differential pressure in psi also depends on

* When cavitation occurs in a pump, noisy operation and undue vibration of the pumping unit are observed.

the density.

$$\text{psi} = \left(\frac{\text{ft-\#}}{\text{lb}}\right)\left(\frac{\text{lb}}{\text{cu ft}}\right)\left(\frac{1}{144}\right)$$

### Operating Points of Centrifugal Pumps

For any given piping arrangement a system head curve can be superimposed on the head-capacity curve for the pump. The system head curve is composed of the potential (static) head, pressure head, and friction head. Only the latter varies with flow. Such a plot is shown in Fig. 14-18. The point at which the two curves cross (Point *A*) is termed the operating point for the system. This is the only flow rate which the pump will deliver unless the characteristics of the system are changed. The usual way to accomplish this change is by throttling the discharge valve. Throttling adds an additional quantity of head on the system and moves the curve to the left thus reducing the capacity of the pump to some point *B* which depends on the amount of throttling.

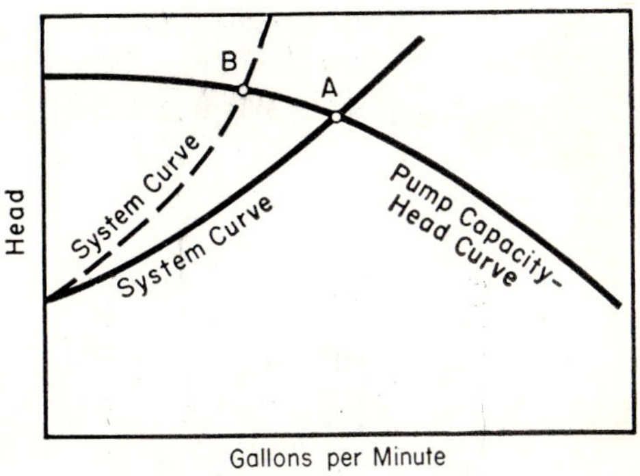

Fig. 14-18. Operating point of a centrifugal pump.

### Series and Parallel Pump Arrangements

When it is impossible to meet the head requirements with a single pump two or more pumps in series may be used. The flow rate through each pump is the same at any given time for pumps connected in series. The heads of the head-capacity curves for the two pumps are additive. The delivery for a given system can be determined by the intersection of the system curve with the new combined performance curve.

To increase capacity beyond that available with single pumps, parallel arrangement can be used. In parallel arrangement the pumps produce equal heads, and the new head-capacity curve is determined by adding the capacities. Parallel arrangement is often used in pumping stations where demand varies widely. All but one of the pumps can then be cut out at low demand periods. The satisfactory operation of pumps in parallel requires that the pumps have stable characteristic curves so that wide swings or surges in pumping rates can be avoided. A stable curve is one having a head which constantly increases toward zero capacity.

## ROTARY PUMPS

Rotary pumps are positive displacement pumps which, through rotating action, trap liquid by mechanical means at the suction side and release it at the discharge side. Among the numerous types of rotary pumps manufactured are gear, screw, lobe, vane, and cam pumps. A typical gear pump is shown in Fig. 14-19*a*. The unmeshing of the gears produces a vacuum which draws the liquid into the spaces between the teeth. The liquid is conducted to the other side of the pump where the meshing of the gears forces the liquid into the discharge line. Pumps such as these produce pressures up to 1000 psi and flow rates to 300 gpm.

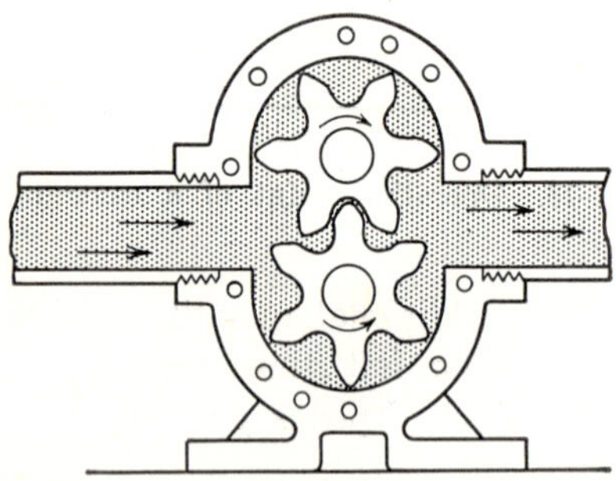


(*a*)

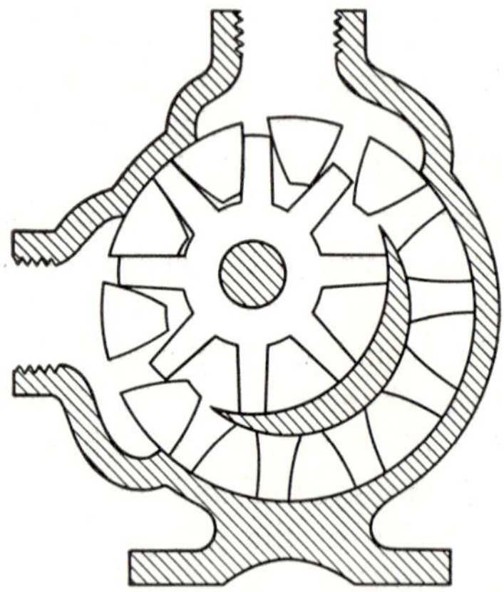

(*b*)

Fig. 14-19. Gear Pumps. (*a*) External gear pump. (Courtesy of Geo. D. Roper Company); (*b*) internal gear pump. (Courtesy of Viking Pump Company.)

Fig. 14-19*b* illustrates another type of gear pump, called an internal gear pump. The crescent shape on the inside of this pump acts as a seal between the suction and discharge ports. These pumps are manufactured in sizes producing flow rates as high as 1000 gpm and pressures of 200 psi.

Theoretically most rotary pumps should deliver the same volume regardless of discharge pressure. A certain amount of running clearance, however, is necessary in all designs and this causes a small amount of leakage (slip). The slip is inversely proportional to the pressure differential of the pump and is also affected by the viscosity of the liquid being handled. Usually, however, the capacity varies only a small amount with discharge pressure.

Numerous rotary pumps other than the gear type are manufactured.

Each has special advantages for particular applications. Because of the close clearances required for certain rotary pumps, abrasive fluids cannot be successfully handled. This does not apply, however, to all rotary pumps. Certain special designs of helical pumps can handle solutions containing large solid particles.

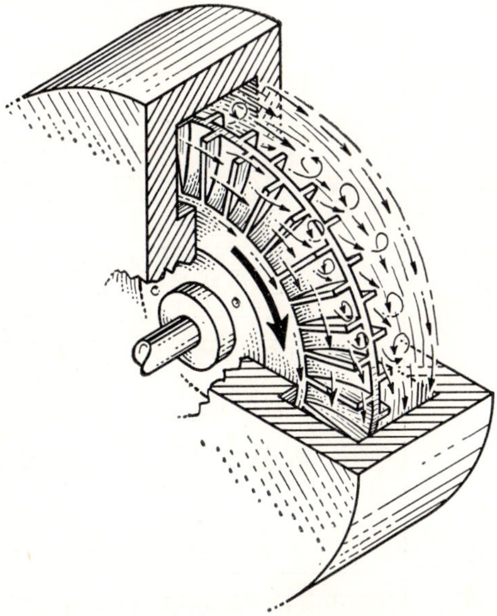

Fig. 14-20. Peripheral pump impeller showing flow pattern.

Rotary pumps are most often advantageously applied in process plants for pumping viscous liquid which would be difficult for centrifugal pumps to handle. Viscosities as high as 400,000 centistokes can easily be pumped. Rotary pumps are manufactured which are capable of discharge pressures up to 5000 psi. Engineers engaged in the selection of pumps should not fail to investigate the many special designs of rotary pumps, particularly those for difficult pumping jobs in the low capacity range (below 500 gpm).

## PERIPHERAL OR "TURBINE" PUMPS

Peripheral or "turbine" pumps (Fig. 14-20) develop high heads by re-circulating the liquid to the vanes. In this manner even small capacity pumps can be designed for high heads, and it is in this service that they find their greatest use.

## SPECIFYING AND SELECTING PUMPS

The successful selection of the best pump for a given service requires close cooperation with the manufacturer's representative. The widely used standard specification sheet aids in the orderly presentation of process information (Fig. 14-21). Because the specification sheets are written in abbreviated form, however, engineers are prone to provide sketchy or inadequate information. Full details of the pumping problem must be given.

In employing safety factors on the required capacity and head, it is preferable to give both the operating and the calculated design values. It often happens that the arbitrary safety factor used produces a design value which necessitates the selection of a more expensive pump. In

**PROCESS SPECIFICATION**
**FOR PUMPING EQUIPMENT**

| | | | |
|---|---|---|---|
| PROCESS DESIGN | PJC | PLANT | CPLA-2 |
| CHECKED BY | RLJ | SPEC. NO. | BR-000 |
| DATE | 3/20/52 | ITEM NO. | P-105 |

| | | |
|---|---|---|
| **GENERAL** | | |
| SERVICE | | Furnace Feed |
| | | |
| NO. OF PUMPS REQUIRED | IN REGULAR USE | 1 |
| | AS SPARES | 1 |
| | | |
| | | |
| **PROCESS REQUIREMENTS PER PUMP** | | |
| FLUID HANDLED | | Gas Oil |
| QUALITY OF FLUID | CORROSIVE OR NON–CORROSIVE | Slightly Corrosive |
| | CORROSIVE COMPOUNDS | Sulfur |
| | SOLIDS ( IF ANY ) | None |
| QUANTITY HANDLED G/H AT 60°F AND 760 MM. | | 18,350 |
| SPECIFIC GRAVITY AT 60°F AND 760 MM | | 0.882 |
| PUMPING TEMPERATURE | °F | 150 |
| VISCOSITY–CENTIPOISES AT PUMPING TEMP. | | 8.5 |
| | | |
| | | |
| **SUCTION CONDITIONS (AT PUMP)** | | |
| PRESSURE | P. S. I. ( ABS. OR GA. ) | 0 psig |
| VAP. PRES. AT PUMP TEMP. P. S. I. ( ABS. OR GA ) | | Negligible |
| SP. GR. AT PUMP TEMP. AND SUCTION PRESS. | | 0.840 |
| NET POSITIVE HEAD ABOVE VAP. PRESS. FT. | | 40.4 |
| | | |
| | | |
| **DISCHARGE CONDITIONS (AT PUMP)** | | |
| PRESSURE | P. S. I. ( ABS. OR GA. ) | 220 psig |
| SP. GR. AT PUMP TEMP. AND SUCTION PRESS. | | 0.840 |
| CAPACITY AT DISCH. COND. | GPM | 294 |
| | | |
| | | |
| **DESIGN CONDITIONS** | | |
| DIFFERENTIAL PRESSURE | P. S. I. | 220 psig |
| | FT. | 605 |
| LIQUID HORSE POWER | | 37.7 |
| MAXIMUM TEMPERATURE | °F | 250 |
| MAX. SUCTION PRESS. P. S. I. ( ABS. OR GA. ) | | 10 psig |
| MAX. ALLOWABLE PUMP ΔP,P.S.I. | | 290.2 |
| | | |
| **REMARKS** | | |
| ( SPECIAL REQUIREMENTS OF | | |
| PUMPS OR DRIVERS, ETC. ) | | |
| TYPE OF PUMP RECOMMENDED | | Centrifugal |
| | | |
| TYPE OF DRIVER RECOMMENDED | | |
| REGULAR | | Motor |
| SPARE | | Turbine |
| | | |
| | | |
| | | |
| | | |

Fig. 14-21. Process specification form for pumping equipment. [Reprinted by permission, R. L. Jacks, *Chem. Eng. Progr. 49,* 234 (1953).]

such cases a reduction of the arbitrary safety factor will yield lower design conditions for which a less expensive pump can be selected. Having both design and operating values, the pump engineer can spot such situations and inform the project engineer at once.

Extra consideration must be given to selecting a pump which is to serve as a spare for two pumps operating in different services.[11] The work per unit mass (head) required for each service as well as the differential pressure and capacity must be compared. This is particularly necessary for centrifugal pumps since the head developed by a centrifugal pump remains the same for a given capacity regardless of the specific gravity. The spare must of course be chosen for the maximum condition.

## INFORMATION TO BE SUPPLIED BY MANUFACTURER

The manufacturer should be requested to supply performance curves, installation and operating instructions, and certified dimensioned drawings for all pumps purchased so that accurate piping layout and foundation drawings can be made.

## PUMP INSTALLATIONS

### Location

Pumps should be located as close as possible to the source of fluid to reduce the length and complexity of the suction line. Free space for servicing both around the pump and above the pump should be provided. In the case of reciprocating pumps this space should be adequate to permit the removal of pistons and rods. In addition the area around the pumps should be free of undue safety hazards.

If the pumps are in a pump house, adequate air changes should be provided. Special exhaust ducts are necessary at the stuffing boxes of pumps handling volatile or poisonous materials. Where possible, therefore, pump houses should be avoided. In general they create additional expense and hazards and have little real advantage.

When practical, process pumps should be installed in a line opposite hazardous process units. A roadway or free area can then be located between the unit and the pumps so that they can be serviced without undue interference in plant operation.

This in-line arrangement makes electrical conduit installation and steam piping much simpler although the process piping to and from the pumps can become complicated. If the process piping would be too complex for a certain pump, it may be preferable to place that one pump within the process unit area. In general, pumps taking suction from the area and discharging to another area are most amenable to the remote in-line arrangement. In operations where fire is not a hazard pumps should be grouped close to the units being served.

## Foundations

All pumps regardless of type require substantial foundations which are capable of absorbing the vibrations created by the pump. Foundations are generally concrete poured on solid ground which requires in most cases that they extend beneath the surface of the top soil (see Chapter 22). They extend above grade or floor level a minimum of one foot. Foundations for pumps to be installed in-line are often made a constant height above grade. If the pumps vary greatly in size, however, the foundation heights above grade may be varied so that the pump centerlines are at the same level. When so arranged, servicing of the pumps will be easier, since they will always be at one height above ground.

Pump foundations should be poured early to allow adequate curing time before the pump is installed. Approximately one inch of space between the baseplate flange and the top of the foundation should be allowed for grout. The foundation bolts of the size specified by the pump vendor and of sufficient length (it is better to have them too long since they can always be cut) are held firmly in the foundation. They are encased by a pipe sleeve of three to four times the diameter of the bolt. This permits movement of the bolt to conform to the baseplate holes.

After alignment of the pumping unit, a thin mortar consisting of one part cement and two parts sand and water sufficient to make a fluid plastic mass, is injected under the base of the pump. This operation is called grouting and insures firm attachment of the pump base to the foundation. A wooden form is built around the baseplate to hold the fluid grout. Holes provided in the baseplate of the pump permit the escape of air and enable tamping of the grout mixture.

## Alignment

Detailed instructions for the alignment of all pump types are furnished by the manufacturer. Proper alignment is essential to satisfactory pump operation. If the motor or turbine operates at a temperature much different from the pump, final alignment should be checked under operating conditions or proper allowance made during the initial alignment.

## Suction and Discharge Piping

Adequate provisions should be made for expansion in the design of all hot oil pump piping so that none of this expansion force is exerted on the pump casing. Clever arrangement of the piping so that it can expand freely will often obviate the need of costly expansion joints. But suction and discharge lines must be anchored or supported so that the pump casing will not be required to withstand expansion thrusts.

Suction lines should be as short as possible and designed to keep friction losses to a minimum. This is accomplished by using an adequate line size (usually one or two sizes* larger than the suction nozzle on the pump), long radius elbows, and a minimum number of bends.

Pockets where gas or air can accumulate should be avoided in suction lines of centrifugal pumps. Eccentric reducers with the sloping side installed downward should be used at the suction nozzle so that no air pocket can form.

Suction lines should be sloped, where possible, one-fourth inch to the foot toward the pump when the pump is below the source, and in the opposite direction when pump is above the source. Elbows entering double suction pumps should be installed in the vertical. When installed horizontal or at an angle, the result is unequal distribution of flow to the two sides of the impeller.

Unless pumps are manifolded in parallel, a gate valve is usually not used in the suction line. A strainer, however, approximately three to four times the cross-sectional area of the suction line is desirable and should be used at least during plant start-up when loose materials may be in the lines. When gate valves are installed in the suction line, it is recommended that they be installed with the stem horizontal to eliminate chances for vapor collection in the top of the valve. This practice, though often recommended, is not followed generally.

To assure minimum turbulence at the pump suction, a straight run of pipe approximately three pipe diameters in length should be used between the suction nozzle and the last elbow.

Common suction headers for a series of pumps operating in parallel should be designed for equal velocities at all points (this approximates equal pressure drops for any path and assures good distribution).

Discharge lines can be made smaller than the discharge nozzle and sometimes are. The available head and economic considerations decide the size to use (see Chapter 18). Velocities in these lines range from 3 to 15 ft/sec.

A check valve is placed between the discharge nozzle and the gate valve to prevent liquid backup in case of pump failure.

A relief valve should be installed on the discharge of positive displacement pumps between the pump and the gate valve. It should be set to open at 10% in excess of the maximum working pressure of the pump. A relief valve is not necessary for steam-driven reciprocating pumps which have a stalling pressure less than the maximum allowable pressure on the liquid cylinder.

Pumps having flat-face flanges require the use of flat-face companion flanges regardless of piping specification.

* This, obviously, is a rule-of-thumb. The size should be governed by the friction calculations and NPSH required.

### Auxiliary Piping

Auxiliary piping (cooling water and gland oil piping) particularly to centrifugal pumps is a small but extremely important item. Manufacturers' suggestions on this phase of the piping should be heeded. All valves should be accessible and the arrangement of the piping should be such that servicing and inspection of the pump will not be impaired.

Provision for piping of stuffing box leakage and other drainage away from the pump must be provided. Such leakage is often collected in a common sump for all pumps from which it is then pumped to waste by an automatically controlled sump pump.

## REFERENCES

1. Addison, Herbert, *Centrifugal and Other Rotodynamic Pumps,* Chapman and Hall, Ltd., London, 1948.

2. Carter, R., I. J. Karassik, and E. F. Wright, *Pump Questions and Answers,* McGraw-Hill Book Co., New York, 1949.

3. Church, A. H., *Centrifugal Pumps and Blowers,* John Wiley & Sons, Inc., New York, 1944.

4. Finch, V. C., *Pump Handbook,* National Press, Milbrae, Calif., 1948.

5. Jacks, R. L., *Chem. Eng. Progr.,* 49, 234 (1953).

6. Kent, R. I., *Mechanical Engineer's Handbook II,* 11th Ed., John Wiley & Sons, Inc., New York, 1936.

7. Kristal, F. A., and F. A. Annett, *Pumps,* 2nd Ed., McGraw-Hill Book Co., New York, 1953.

8. Nickel, F. F., *Direct Acting Steam Pumps,* McGraw-Hill Book Co., New York, 1915.

9. *Standards of the Hydraulic Institute,* 10th Ed., Hydraulic Institute, New York, 1955.

10. Stepanoff, A. J., *Centrifugal and Axial Flow Pumps,* John Wiley & Sons, Inc., New York, 1948.

11. Taylor, Irving, *Chem. Eng. Progr., 44,* 336 (1948).

# CHAPTER 15

# COMPRESSORS AND VACUUM PUMPS

Transporting gases and vapors is as important as the movement of liquids in many chemical plants since some processes require that feed gases be compressed to elevated pressures. The compressor station is a plant within a plant composed of complex and expensive machines which require not only skilled operation but also careful engineering in selection and installation. Compressor manufacturers usually provide the services of their engineers to the customer or prospective customer to aid him in selecting the proper units. Here again, however, the project engineer must know certain fundamental facts if for no other reason than to enable him to ask pertinent questions and carry on intelligent conversations with the manufacturer.

A listing of compressor types is similar to that for pumps. Because of the compressibility of gases, however, the design of a compressor is far more complicated. Furthermore, the use of reciprocating pumps has declined to relatively small proportions, while the reciprocating compressor still does the major portion of all gas compression, although since 1945 centrifugal compressor usage has increased remarkably.

## RECIPROCATING COMPRESSORS

Briefly, as defined by the Compressed Air and Gas Institute[3] a reciprocating compressor is one in which the compressing element is a piston following a reciprocating motion in a cylinder. Because of the many types of reciprocating compressors being built today it is difficult to prepare a rational list of general types. The Compressed Air and Gas Institute[3] suggests the following listing:*

1. Single-Acting. Compression effected in one end of cylinder only.
2. Double-Acting. Compression occurs in both ends of cylinder.
3. Single-Stage. Compression from initial to final stage occurs on each working stroke of the piston.
4. Multistage. Gas compressed from initial to intermediate pressure in first cylinder and is completed in one or more subsequent cylinder.
5. Vertical. Compressing element in vertical position.
6. Horizontal. Compressing element in horizontal position (most common in process industries).
7. Angle. Multicylinder compressor with cylinder axes at an angle with each other (particularly common to small and portable air compressors). This terminology is based entirely on the compressor cylinders. The use of angle as a type now more generally refers to large direct connected machines in which the compressor cylinders are horizontal and the power cylinders are at an angle with the compressor cylinders (see below).
8. Straight line. (Single frame)—Horizontal or vertical units with a single frame having one or more cylinders in line.
9. Duplex. Two parallel cylinders connected to a common crank shaft.
10. Air cooled. Cooled by natural atmospheric air circulation (small compressors only).
11. Water cooled. Water circulated through jackets surrounding the cylinders (most process compressors).

The large high-speed multicylinder compressors which were introduced in the early 1940's have become common to process plants but their complexity makes the above terminology inadequate in many respects. It would seem preferable to describe the modern process compressors on the basis of the type of drive.

## Reciprocating Compressor Types

### *Direct-Connected Steam-Driven*

The majority of early compressor installations were steam-driven units. Use of these units is now largely confined to small (10-300 hp) low-pressure installations, single stage or two-stage tandem units which produce maximum pressures of about 150 psi and 500 psi, respectively. These compressors operate at 250 to 500 rpm and are most popular as air compressors. The steam end is similar to that on steam-driven reciprocating pumps.

Larger (200 to 1500 hp) steam-driven compressors are generally of the horizontal, duplex type in single-stage or two-stage designs (Fig. 15-1). These can be built for high-pressure operation up to 5000 psi. Choice of such a unit will depend largely on the steam balance in an existing or

* Reprinted by permission *Compressed Air Handbook,* 2nd Ed., Compressed Air and Gas Institute, McGraw-Hill Book Co., N.Y. (1954).

anticipated plant. A disadvantage of large steam-driven compressors is the relatively large floor space consumed by such units, which probably accounts for their decline along with the advent of the gas engine driven compressor and the availability of cheap natural gas (see below).

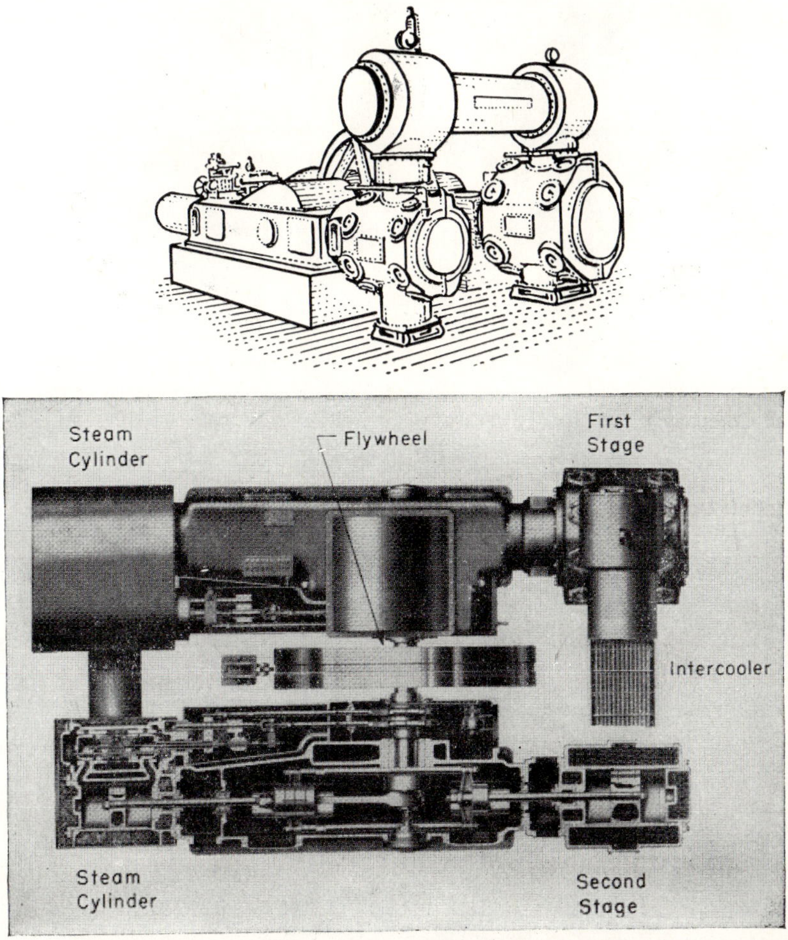


Fig. 15-1. Plan and pictorial view of two-stage, steam-driven reciprocating compressor. (Photograph by courtesy of Ingersoll-Rand Company. Pictorial adapted by permission.)

### *Motor-Driven*

There is a wide variety of motor-driven reciprocating compressor designs ranging from small single cylinder models (5 to 150 hp) to large opposed multicylinder compressors (150 to 4500 hp). The smaller units are arranged for flat belt or V-belt drive while the larger units are designed for direct-connected synchronous motors (Figs. 15-2 and 15-3).

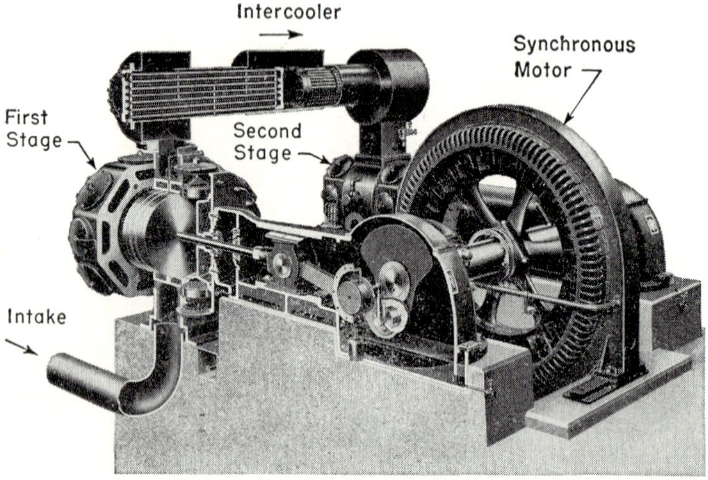


Fig. 15-2. Two-stage, synchronous motor-driven compressor. (Courtesy of Ingersoll-Rand Company.)

Fig. 15-3. Typical multicylinder motor-driven compressor installation. (Courtesy of Cooper-Bessemer Corporation.)

Units such as that illustrated in Fig. 15-3, composed of four to six compressor cylinders require the most careful design. Inertia forces can be great in such installations, and these must be carefully balanced in the design. The opposed arrangement of cylinders in Fig. 15-3 minimizes these forces. Complete balance, however, is never achieved and due consideration of the magnitude of unbalance must be given in the design of compressor foundations.

Synchronous motors can be conveniently designed to operate at the low speeds required for reciprocating machines. If a synchronous motor is to be employed in a hazardous atmosphere, however, it is necessary to separate the motor from the compressor by enclosing the motor in a brick or concrete room. A motor shaft extension which is supported by an outboard bearing is used. The room is ventilated so that the pressure inside is higher than in the compressor room, and any air and gas leakage will go out instead of into the motor room. In many hazardous atmospheres, however, it has been found satisfactory to encase the motor collector ring in a sheet metal housing and maintain a positive pressure in this housing using an inert gas such as nitrogen. Such an arrangement is not particularly convenient and for lower horsepower installations (below 450 hp) it is possible to employ induction motors economically which can be made explosion-proof. Induction motors, however, require reduction gears because of their higher speeds (compressor maximum rpm is generally around 500).

#### *Turbine-Driven*

Steam turbines instead of electric motors can be used to drive the compressors described above. Steam turbines are used when power costs are high and steam costs are low.

#### *Engine-Driven*

Large compressors designed for motor drive can also be driven by diesel, gas-diesel,* or gas engines. These engines are connected through a flywheel to the compressor unit. It is desirable, however, to employ the compact "angle-type" compressors described below when diesel or gas fuel is indicated. For certain high-pressure, high-capacity applications beyond the range of the "angle-type" (above 2500 hp) it is necessary to use the engine drive connected to a "motor-driven" type unit.

#### *Angle-Type*

An angle-type compressor has horizontally mounted compressor cylinders which are driven by power cylinders mounted at an angle of approximately 90° to the horizontal.

* The gas-diesel engine can operate on either gas or diesel fuel and is indicated where an interruptable source of gas supply exists.

The early gas engines were simply modified steam engine compressors. They were improved during the 1920's and many large installations of horizontal, 4-cycle gas engine-driven, double-acting compressors were installed especially for gas transmission service. These units are still produced by some manufacturers, but have declined in popularity because of the advent of angle-type compressors.

Perhaps one of the most spectacular developments in compressor design took place in the late 1930's and early 1940's when the 2-cycle gas-driven, angle-type compressor was introduced. It had three immediately apparent advantages over its lower speed horizontal predecessors; namely, higher thermal efficiency, lower (approximately one-half) space requirement, and lower cost per horsepower installed.

There are two designs of angle-type compressors. One is the so-called "right angle" (Fig. 15-4) with the power cylinders arranged side by side; it has the advantage of easy accessibility to all power cylinders from one platform. A second type, the V-angle with the power cylinders in a V-arrangement (Fig. 15-5) has the advantage of more cylinders per unit length of engine. Both types give excellent service and have proved very popular. The "right-angle" type has 5, 6, 8 or 10 power cylinders and the "V-angle" type has 6, 8, or 10 cylinders. Both deliver from 250-300 hp per cylinder. The operating speed ranges around 300 rpm and as many as six to seven stages of compression can be incorporated on one compressor.

Both two- and four-cycle engines are manufactured. Four-cycle engines operate extremely well at wide variations in load below the rated capacity. Two-cycle machines operate satisfactorily high above rated capacity. In two-cycle operation so-called scavenging air is pre-compressed in an auxiliary scavenging air compressor which is connected to the crankshaft. This air enters the cylinder toward the end of the expansion stroke and aids in expelling or "scavenging" the remaining gases. The four-cycle engine accomplishes the intake and compression of air in each cylinder and thus requires no scavenging air compressor.

There are certain inherent advantages to each type which are discussed fully in mechanical engineering literature. The two-cycle engine became popular since it required smaller space and less weight per horsepower and operated very economically at the low speeds required for reciprocating machines. The four-cycle engine requires no scavenging air compressor and recent designs afford savings in space. The four-cycle operation inherently has high efficiency, and the two-cycle has more uniform crank action since every down stroke is a power stroke. The four-cycle can be controlled from 100% to 50% of operating speed while the range for a two-cycle is 100% to 60%. In the final analysis installed cost comparisons along with comparison of engineering features will dictate the choice of any style unit.

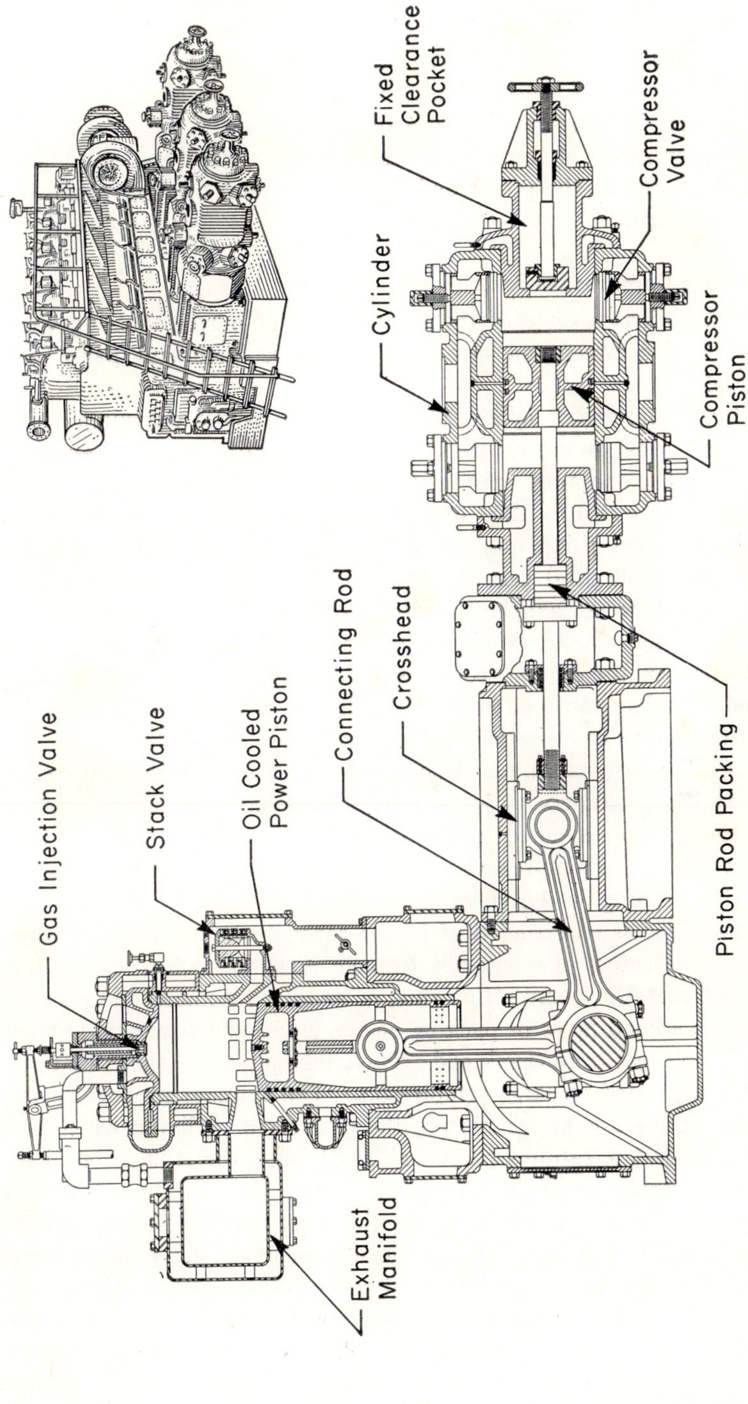


Fig. 15-4. Transverse section and pictorial view of a right-angle, gas-driven compressor. (Section, courtesy of Clark Brothers Company. Pictorial, adapted by permission.)

Diesel-driven, angle-type compressors are used where gas is unavailable. Gas-diesel units are employed if the gas supply may be interruptable (seasonal availability). The gas-driven and certain gas-diesel models are spark ignited with low voltage ignitors. In hazardous areas a low fire hazard ignition system should be specified.

The power cylinders of angle-type compressors are water cooled. Lubrication is force fed to the power end by a gear pump directly connected to the crankshaft.

### Reciprocating Compressor Parts

Fig. 15-5 indicates some of the major parts of a reciprocating compressor. Brief comment on several of the more important parts may prove helpful.

*Cylinder.* Compressor cylinders are built to handle pressures up to 25,000 psig. Most manufacturers have many cylinder styles and are able to select the one most appropriate for any particular service. Low pressure cylinders (up to 1200 to 1500 psig) are generally cast iron or so-called semi-steel (high quality cast iron). In the medium pressure range (1500 to 2500 psig) cast steel construction is used, and in the high pressure range (above 2500 psig) forged steel. The cylinders are designed for water cooling. Water is circulated around the cylinder barrel and valves to remove the heat of compression. For low compression ratios, air cooled cylinders can be used. The cylinders of most compressors are lined with cast iron or other removable liners.

*Pistons.* The pistons are cast iron or fabricated steel depending on the size and type of cylinders. Two or more snap-type rings of cast iron or other material recommended by the manufacturer prevent leakage past the piston during the compression stroke. Pistons are flat-faced, rounded, or conical depending on the valve arrangement.

The piston rod is constructed of forged carbon steel (alloy steel when corrosion is expected) and extends through a water-cooled stuffing box packed with metallic, composition ring, or another recommended packing material. A second layer of packing on the crank end serves as a wiper to minimize oil loss from crank end. The packing next to the cylinder is force-feed lubricated on large compressors. The cylinder support or distance piece forms a chamber next to the compressor cylinder which can be vented so that hazardous gases leaking through the stuffing box may be conducted safely out of the area.

*Compressor Valves.* In 1915 the introduction of pressure actuated suction and discharge valves sparked development of the modern compressor. There are numerous different types of these valves. All operate on essentially the same principle. The pressure difference across the valve causes flexible metal strips or thin plates to move against or away

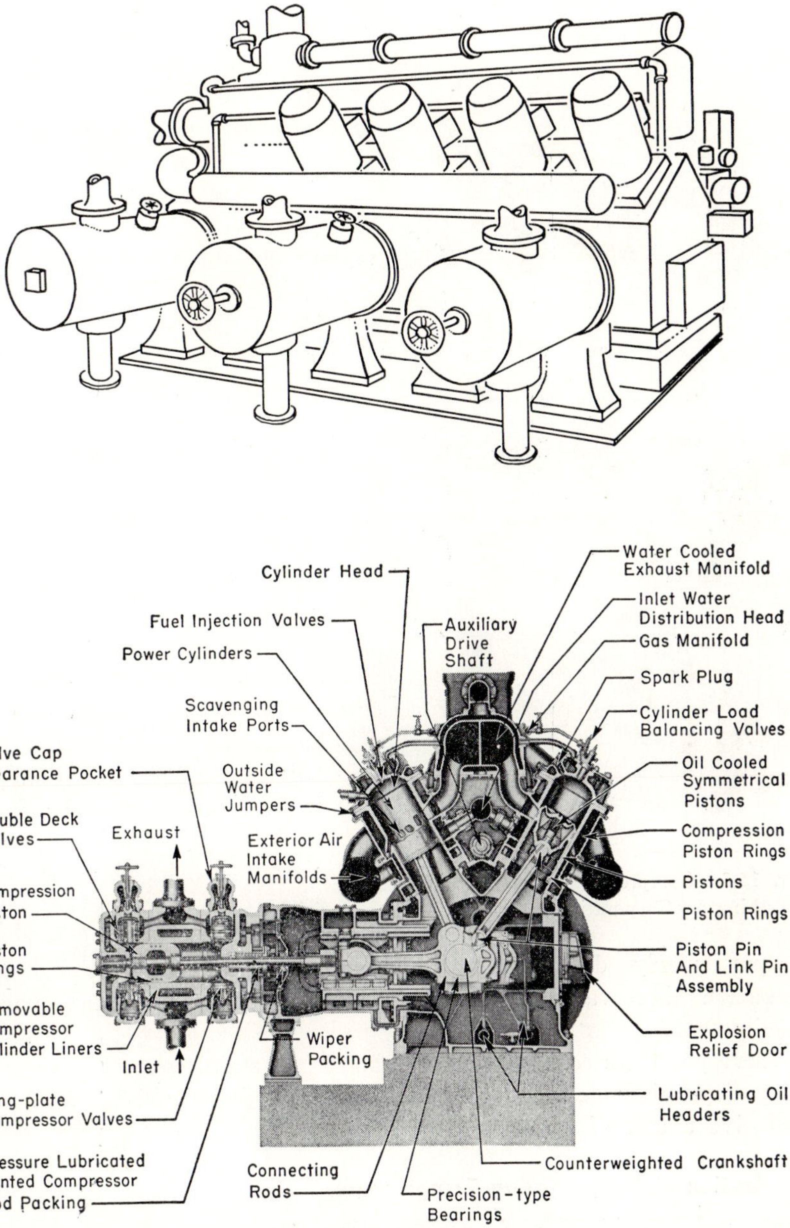


Fig. 15-5. V-angle gas-driven compressor. (Section, courtesy of Cooper-Bessemer Company. Pictorial, adapted by permission.)

from a series of concentric or parallel rectangular openings. The differential pressure required to accomplish the movement is increased on certain designs by the application of a counteracting force in the form of spring tension (helical or leaf spring) or weights. This counteracting force cushions the valve action.

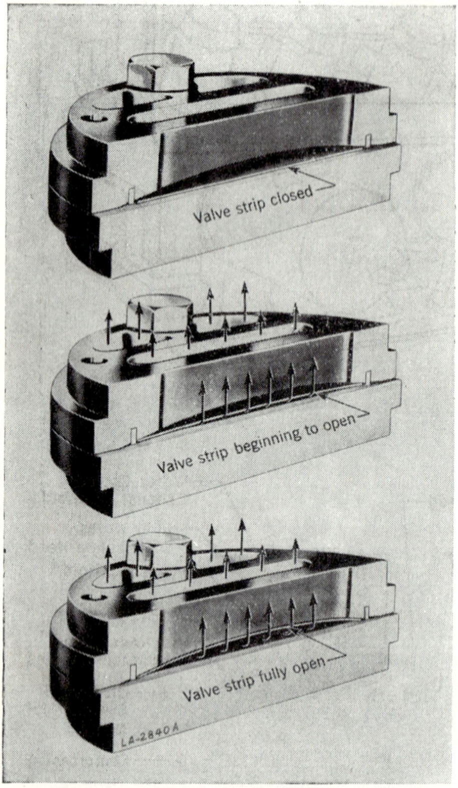


Fig. 15-6. Compressor valve. (Courtesy of Worthington Corporation.)

The valve shown in Fig. 15-6 consists of thin steel strips which cover a number of parallel ports. The strips are flexible and open or close depending on the pressure differential across the valve. For example, the strips on the suction valve move to an open position on the suction stroke when the pressure in the suction line exceeds that in the cylinder.

The goal in valve design is gentle opening and closing and sufficient valve area to enable low gas velocity through the valve. It is sometimes desirable to employ a double-deck valve assembly which is simply two valves, one mounted above the other and arranged to operate in parallel. This design increases the effective valve area and reduces valve losses. It is particularly desirable for high-capacity, low-compression ratio operation where valve losses may be a large portion of the horsepower requirements.

Valve parts are constructed of carbon steel, stainless steel or chrome alloy depending on the type valve, the operating conditions and the corrosive nature of gas.

## Lubrication of Reciprocating Compressors

The proper lubrication of a reciprocating compressor is a complex problem and careful attention must be given to the compressor manfacturer's recommendations as well as those of oil company lubricating specialists.

A typical gas or diesel-driven, angle-type compressor may have three separate lubrication systems. A force-feed type lubricator supplies special cylinder oil under controlled feed rates to the compressor cylinders and piston rod packing. Its function is to create an effective piston seal and to protect the compressor valve parts. Too much oil, however, will lead to carbon formation on the valves causing valve leakage. Another lubricator supplies the power cylinders and scavenging blower. In addition, there is a continuous recirculating lube oil system which consists of a rotary pump driven off the compressor crankshaft, an oil filter, and a cooler. This system supplies lubrication for the crankshaft, bearings, and connecting rods and also serves as a coolant for the power piston. Oil is pumped from beneath the crankshaft, through the filter and then the cooler to all parts of the system. A pressure regulator maintains the proper lubricant pressure.

All reciprocating compressors require lubrication of a similar nature. On some small units certain lubrication streams can be combined. The Compressed Air and Gas Institute[12] gives some valuable suggestions on lubricants, but manufacturers' recommendations are more specific and, hence, more useful.

All reciprocating compressor cylinders require oil lubrication,* and suitable separators must be provided at the discharge of the compressors to remove unwanted oil from gas streams.

### Reciprocating Compressor Regulation

The gas delivered by a reciprocating compressor can be regulated by one or a combination of several methods. The method used depends largely on the type drive. Regulation in most processes is conveniently based on the maintenance of a constant compressor discharge or suction pressure. Whether to use the suction or discharge pressure largely depends on the process for which the compressors are being selected. Often in process plants it is desirable to maintain a given pressure in a system into which the compressor discharges. In this instance discharge pressure is used as the sensing element. When suction pressure is above atmospheric pressure it is sometimes preferable to control the suction pressure.

*Start and Stop Control.* Commonly used for plant air compressors; a pressure switch starts the motor when the air pressure falls below that required and stops the motor when it reaches the desired pressure.

*Driver-Speed Control.* Direct-acting, steam-driven compressors can best be controlled through variation in the speed by means of governors of the throttling type or automatic steam valve cut-off type. Gas, gas-diesel, and diesel-driven compressors are controlled from 100 per cent

* It is possible to obtain at additional cost special nonlubricated cylinders having pistons fitted with carbon rings, but this is justified only when compressed gases must not come in contact with oil.

to 50 or 60 per cent of capacity by means of a governor (can be automatically actuated) which operates by controlling the amount of the fuel admitted to the power cylinders. In most processes this range of control is entirely adequate, although suction valve unloaders and clearance pockets as described below are an aid to smoother operation if suction or discharge pressures vary. Where this variation is rapid automatic control of the unloaders and clearance pockets is indicated.

*Clearance Pockets.* Clearance pockets are usually provided in compressor cylinders as shown in Figs. 15-4 and 15-5. The clearance valves can be arranged for automatic operation or manual operation. When opened, the increased clearance added to the cylinder reduces the capacity of the machine. It has been demonstrated by York[14] that although the volumetric efficiency is reduced by the opening of a clearance pocket, compression efficiency is essentially unaffected (the gas in the clearance pocket is compressed but returns this energy in the reexpansion step). Even on compressors not designed for clearance control, there are clearance plugs which permit manual change in compressor capacity if operating conditions change permanently.

*Suction Valve or "Free Air" Unloaders.* The suction valve unloader as shown in Fig. 15-7 (No. 2 and No. 3) holds the suction valves open with steel fingers fastened to the pneumatically operated unloader pistons. This action prevents the piston from compressing gas during the particular stroke in which the unloading occurred. Compressor capacity is thereby reduced.

*Step-Control.* So-called five-step or, for small compressors, three-step capacity control is employed on constant speed motor-driven compressors. The capacity is controlled in five (100, 75, 50, 25, 0%) steps or three (100, 50, 0%) as the case might be, through the use of clearance pockets, suction valve unloaders or a combination of both. Fig. 15-7 illustrates a typical five-step control employing two unloaders and one clearance pocket on a double-acting cylinder. A governor is actuated by the compressor discharge pressure controller. The governor selectively energizes or de-energizes one or more of the solenoids which control air to the unloaders (No. 2 and No. 3) or clearance pocket valves (No. 1). Complete unloading is necessary for starting synchronous motor-driven units under load.

### Selection of Reciprocating Compressor Drive

The selection of the type reciprocating compressor for a given plant is primarily dependent on the type of drive most suitable for the plant fuel and power conditions. Fuel, steam, and electrical costs in the area or plant must be evaluated before an intelligent decision can be made. Each situation is unique in itself. If power costs are extremely low, the

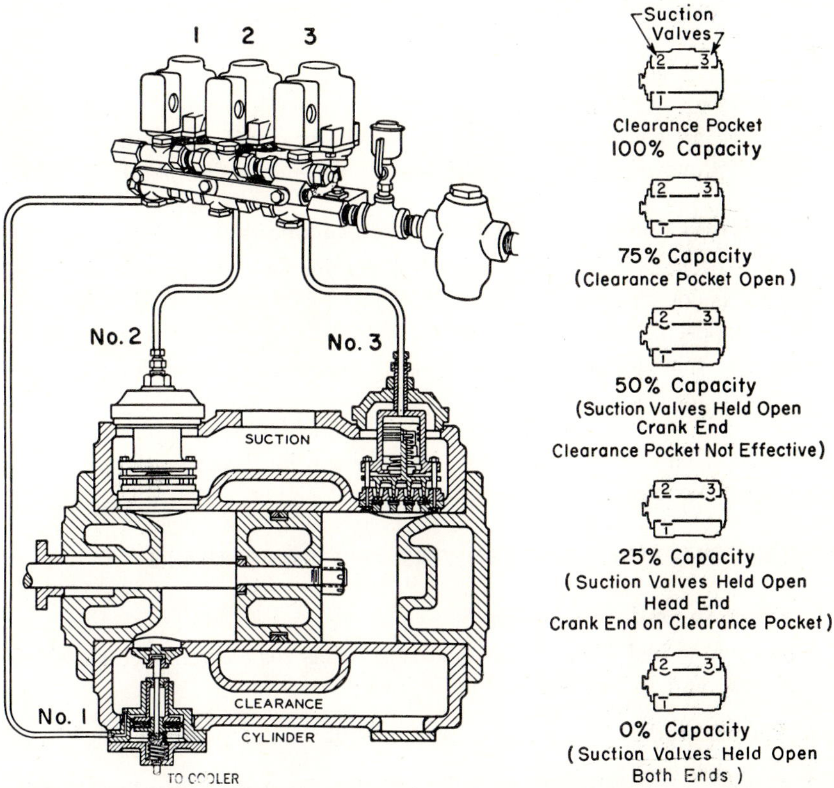


Fig. 15-7. Five-step clearance control. (Courtesy of Worthington Corporation.)

motor drive should be used. Compressors of all sizes can be obtained for motor drive and the advantage of low power cost should not be overlooked just because a number of other plants constructed in different parts of the country have used other types of drives.

Gas-driven compressors have become particularly advantageous for refineries and petro-chemical plants since supplies of gas are produced as "waste" after the heavier portions are removed for use in the process. Chemical plants operating in areas having large uninterruptable gas supplies also find the gas-engine drive advantageous. In areas where gas is sold at a low price only during the warm months the gas-diesel engine drive should be considered.

When low cost steam* is available or when low pressure process steam must be obtained from high pressure steam, turbine-driven compressors

* Unfortunately plant steam costs are often unrealistic book values and do not reflect actual conditions.

are sometimes indicated. The cost of reduction gears and steam piping, however, must not be overlooked in such a decision. On small and medium-sized compressor installations direct steam-drive often proves advantageous. It is important to make a thorough economic study combined with an engineering evaluation of the ease of operation and suitability of the various types of drivers for the particular plant situation.

### Reciprocating Compressors in Hazardous Locations

In the process industries equipment for hazardous areas is designed to be explosion-proof as discussed in Chapter 24 Such design is practically impossible when compressor installations employ gas or diesel drives and synchronous motors. In such cases explosion-proof construction would be prohibitive. It is possible, however, to build these units with low fire hazard construction, using shielded spark plugs for gas engines and an enclosed collector ring with an inert gas or air purge for synchronous motors.

## CENTRIFUGAL AND AXIAL BLOWERS AND COMPRESSORS

### Classification and Terminology

The Compressed Air and Gas Institute[13] has for convenience classified centrifugal and axial machines as follows:

1. *Blowers* (turbo-blowers). Discharge pressure not exceeding 40 psig (based on air).
2. *Compressors* (turbo-compressors). Discharge pressure exceeding 40 psig.
3. *Fans.* Church[2] has defined these low speed centrifugal machines as operating in the very low pressure range below 1 psig and with large volumes.

Much of the terminology of the centrifugal machines is similar to that already presented for centrifugal pumps, and, therefore, only brief comment need be made concerning most of the terms listed below:

1. *Single-Stage.* One impeller or one row of vanes
2. *Multistage.* Two or more impellers or rows of vanes
3. *Centrifugal.* Have radial impellers and the flow through the machine is in a radial direction. These impellers may be open, semi-closed, and closed. They may have straight or backward curved blades. In multistage units diffusers are located between stages for efficient conversion of velocity head to pressure head (Fig. 15-8).
4. *Axial.* Flow of gas is in an axial direction. This motion is created by vanes attached to a rotating drum (Fig. 15-9). Stationary vanes separate each row of rotating vanes and serve in the same manner as diffusers on centrifugal machines.
5. *Stage Group.* Defined by the Air and Gas Institute as any group of individual impellers and diffusers operating as a unit on a single shaft, in a single casing and without cooling between individual impellers.
6. *Multistage Group.* Two or more stage groups driven by a single driver with intercoolers between groups.

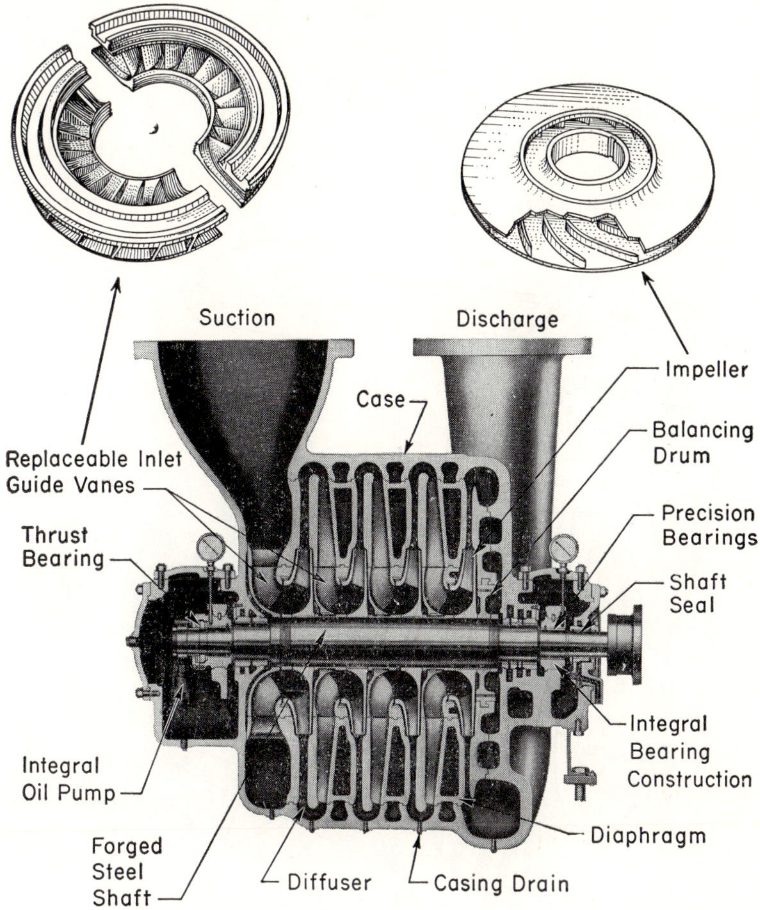


Fig. 15-8. Multistage centrifugal compressor. (Courtesy of Clark Brothers Company. Pictorial inserts adapted by permission.)

7. *Characteristic Curves.* Plots of the discharge pressure versus the inlet capacity.
8. *Performance Curves.* Show the shaft horsepower versus the inlet capacity.
9. *Equivalent Air Pressure.* A term which has been used in comparing centrifugal or axial compressors and unfortunately appears well entrenched. It is the pressure in psig that a compressor will develop operating at the rated capacity when handling air at 14.4 psi absolute and 60° F.

As with centrifugal pumps, centrifugal and axial compressors or blowers accomplish the pumping or compressing action by the rotation of vanes or impellers. The velocity imparted to the gas by the impeller is converted to pressure in a volute or a diffuser. At first glance a multistage centrifugal compressor (Fig. 15-8) might be mistaken for a multistage centrifugal pump. The impellers of the compressor, however, generally decrease in width from the low to high pressure side because of the decrease in fluid volume as the gas becomes compressed.

Some experts believe that the centrifugal compressor is beginning a period of popularity similar to that of centrifugal pumps in the past

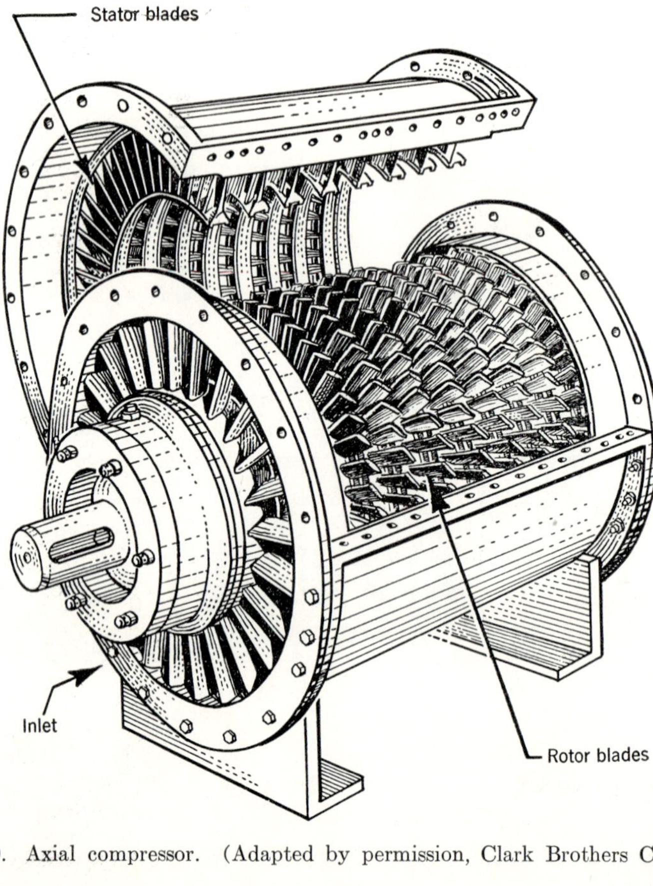


Fig. 15-9. Axial compressor. (Adapted by permission, Clark Brothers Company.)

decade. The centrifugal machine has definite advantages over the reciprocating machine for extremely high volumes and in the lower pressure range (see below). At lower volumes the efficiency of the centrifugal compressor drops off so rapidly that it becomes uneconomical. The range of applicability, however, is continually being increased as new design techniques are developed.

### Impeller Types

Before discussing the various types of centrifugal fans, blowers, and compressors a few comments on impellers which will be common to all these may prove helpful. As with centrifugal pumps the vane shape of the impeller influences the head capacity characteristics. Many shapes and types are built. The three important centrifugal impeller types are shown in Fig. 15-10: (*a*) Forward curved; (*b*) backward curved; (*c*) radial.*

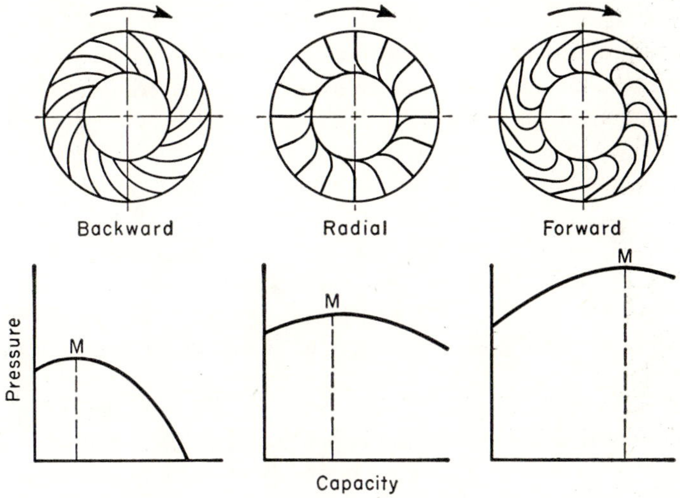


Fig. 15-10. Impeller types and typical characteristics.

Typical† characteristic curves for these impeller shapes are shown. It can be seen that a maximum head is reached at point $M$. As with pumps, operation at or near this point produces unstable operation or surging since the machine will "hunt" between two capacities for one discharge pressure. The maximum or critical point for the forward-curved blade illustrated is so far removed from the origin that the efficient operating range is in the unstable region. For a given diameter, however, the forward curved impeller can handle a greater volume of gas than the other types. It has, therefore, become popular in some fan applications when operation is at high capacity and low pressure. Such operation is in the range where the unstable region can be most conveniently avoided. For high pressures and high speed the backward curved and radial designs have proved the most efficient and reliable and are, therefore, used almost exclusively on centrifugal blowers and compressors.

* The term *radial* as here employed implies radial arrangement of the impeller blades about the axis. These may be paddle-wheel steamer type or may be curved at inlet end. Since all centrifugal machines impart a radial motion to the gas as opposed to the axial machine, the term radial is also used to describe centrifugal blowers and compressors.

† These curves are only representative of the general type. Pronounced variation can be obtained by alteration of the impeller blade curvature.

Because of the high speeds employed for most blower and compressor applications careful static and dynamic balancing of the impeller is imperative. Factory tests at about 25% over the rated speed should be made before delivery so that balance can be checked.

### Single-Stage Blowers

Single-stage blowers are regularly furnished in two types, both of which have overhung impellers similar to those of centrifugal pumps. The impeller may be mounted on an extended shaft of the driver or the impeller may be mounted on a separate shaft supported by two sleeve bearings and attached to the motor shaft by a coupling, as in most centrifugal pumps. These units operate most economically at flow rates above 8000 standard cfm and at discharge pressures from 2 to 9 psi. The casings are constructed of cast iron but can be made of special alloys for corrosive service or for high inlet pressures.* The casing is shaped in the form of a volute with a tangential discharge and a single or double suction inlet. By rotating the casing the discharge nozzle can be located in any desired position.†

For a blower developing higher heads a diffuser (see Chapter 14) passage is located in between the impeller and volute. Impellers may be open-type with radial blades or closed-type with radial or backward curved blades. Those with radial blades are cast in one piece and are generally designed for high tip speeds and thus a high compression ratio.

Single-stage blowers are used extensively in the process industries. They are employed as gas boosters in processes as varied as hydroforming and water gas production. They are most frequently used as sources of large volumes of low pressure air for flotation, pneumatic conveying, ventilation, and for foundry cupolas and other ovens and furnaces.

### Multistage Blowers and Compressors

To obtain higher differential pressures multistage units have been designed (Fig. 15-8). Fully enclosed impellers are used exclusively in these units. They are keyed to the shaft. Large impellers have machined alloy disks and cover plates with vanes, generally backward curved, formed from stainless steel plate which are riveted or welded in place. The fully enclosed impeller provides better support for the vanes and is better suited for multistage design.

Each impeller is surrounded by a vaneless type diffuser (Fig. 15-8) for converting velocity head to pressure head. At the inlet to each impeller guide vanes direct the gas to the impeller eye. The first stage guide vanes can be made adjustable so that the characteristics of the compressor can be modified by changing the angle of the vanes.

* Single stage casings have been designed for operation at 900 psi.

† Number of positions is limited only by the number of bolt holes in the mount

The stationary separation walls between stages are called diaphragms (Fig. 15-8). They are constructed with a passageway for conducting the gas from the diffuser of the previous stage to the inlet guide vanes of the next. These guide vanes may be cast integral with or attached to the diaphragm. If high compression ratios are necessary in each stage, temperature rise will be great. Diaphragms may be designed for internal cooling of the gases between stages by flow of cooling water through a baffled passage in the diaphragm. Normally, however, American centrifugal compressors are uncooled.

The casings are cast iron or steel depending on the operating conditions. They are horizontally split permitting easy access for repairs. All stationary elements such as the diaphragms are also horizonally split.

Multistage centrifugal blowers and compressors are widely used in many process plants for the movement and compression to moderate pressures of large volumes of air or gas. Perhaps the most familiar application of multistage blowers is supplying air to blast furnaces and Bessemer converters at 30 psig. The blast-furnace blowers are usually rated at 100,000 standard cfm while the Bessemer blowers pump 30,000 to 40,000. Similar units are employed for copper convertors. Many applications of centrifugal blowers in other process industries have evolved, such as (1) compressing flue gas from furnaces or lime kilns to 10 to 15 psi for brine treating; (2) supplying regeneration air for fluid catalytic cracking units (20,000 to 600,000 cfm at 15 to 25 psig); (3) boasting gases in any process involving recirculation; and (4) aerating sewage. Most multistage blower applications require from two to five stages.

As the differential pressures become greater the term centrifugal compressor instead of blower is employed. The growing popularity of centrifugal compressors in the past few years has been astounding. They are designed for shaft extension at both ends, permitting the coupling together of up to three compressors so that one driver can be employed with resulting savings in initial costs and space. Such an arrangement is shown in Fig. 15-11. This unit was designed for oxygen compression with three compressors in series compressing oxygen from 14.7 psia to 350 psig with intercooling between compressor bodies. Units for handling from 2000 cfm to 120,000 cfm are manufactured.

Economy at high flow rates has made the centrifugal compressor a standard choice for large refrigeration systems such as those required for dewaxing plants. This same characteristic has brought about the use of centrifugal compressors as an initial stage in the compression of large volumes of gas such as feed gas to ammonia convertors or natural gasoline plants. In such instances the centrifugal compressor compresses the gas from its initial low pressure to a pressure of from 40 to 150 psig. The lower volume of gas at the higher pressure allows reduction in the size of the reciprocating compressor which receives the gas from the centrifugal and compresses it to the high pressures required for these processes. Other typical applications include the production of 100 psi plant air, compression of gas for a vapor recovery unit, compression of butane in alkylation units, and application as pipeline booster compressors.

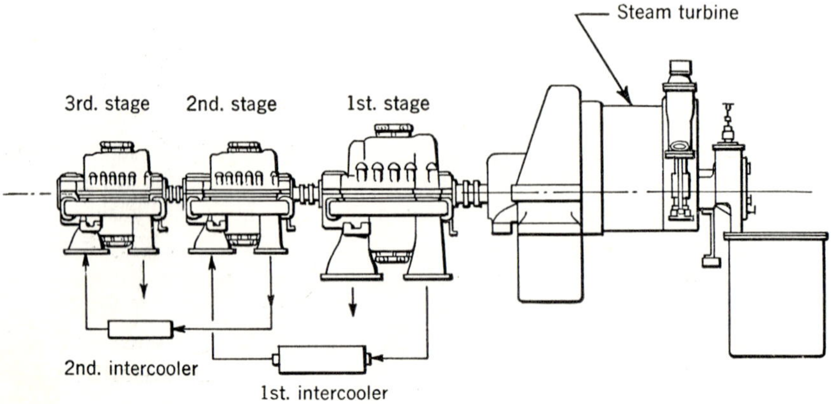


Fig. 15-11. In-line arrangement of centrifugal compressors. (Adapted by permission, Clark Brothers Company.)

## Miscellaneous Features of Centrifugal Machines

*Shaft Seals.* In low pressure air blowers where small leakages are not a disadvantage, a labyrinth seal is employed around the shaft where the shaft enters the casing. These seals are thin strips of stainless steel or other suitable material which are designed to have a minute radial clearance between the shaft and the strip. The net leakage through several such strips in series is small. Labyrinths are also installed around the impeller inlet to prevent leakage between stages.

When handling gases or air at high pressure, stuffing boxes prevent leakage between shaft and casing. Several types of stuffing boxes are manufactured. A patented water seal is often used. A collar or sleeve with a small impeller machined on the sleeve rotates with the shaft, and in a chamber filled with water. The centrifugal force generated builds up water pressure and creates an effective seal. This seal does not function when the blower is not operating. Manual tightening of the sleeve during this period prevents leakage. Another popular seal for gas compressors is an oil seal which operates without metal-to-metal contact. The sealing is accomplished by an oil film under pressure. The sealing oil is supplied by an oil pump either driven off the shaft or by a separate motor. This seal operates satisfactorily at high pressures. Seals constructed of carbon rings held against the shaft by springs are also employed. They operate satisfactorily for pressures up to 50 psig.

Mechanical seals similar to those described for pumps are used where no leakage can be tolerated. The principle disadvantage of the mechanical seal is the large friction loss produced by direct contact of sealing surfaces.

*Bearings.* A balancing drum (Fig. 15-8) is installed at the discharge end of the shaft to partially balance out a major portion of the axial thrust. A Kingsbury type thrust bearing (Fig. 15-8), however, is also provided on the opposite end to handle any unbalanced axial loads which might develop. The main bearings that carry the rotor are usually journal bearings, which are babbitt-lined. Ball bearings are employed on small low pressure blowers.

*Lubrication.* Low pressure blowers equipped with ball bearings require occasional grease lubrication. Compressors and blowers with journal bearings, however, require pressure lubrication. Some high speed compressors have an elaborate means for supplying lubrication to insure satisfactory operation. A gear pump submerged in the suction bearing reservoir and driven off the shaft pumps oil through a cooler and filter to the bearing housings and returns it to the reservoir.

### Axial Compressors

For volumes of air or gas above 150,000 standard cfm the axial compressor should be considered. The axial compressor as illustrated in Fig. 15-9 consists of vanes attached to a drum which rotates. These vanes are shaped to produce an axial movement of the gas. This movement is converted to pressure head by stationary vanes attached to the casing. The vanes are longer at the inlet than at the outlet.

These machines are highly efficient when handling large volumes of gas. In comparison with the centrifugal (radial) compressor the axial unit has a very steep pressure volume curve and is, therefore, better for constant volume applications. These machines are widely used in gas turbine installations and in aircraft. Further developments and applications in the process industries will be watched with great interest.

### Drives for Centrifugal and Axial Blowers & Compressors

Motors and steam turbines are the usual drive for centrifugal and axial blowers and compressors. The high operating speeds required (4000 to approximately 10,000 rpm) make high-speed, direct-connected steam turbines most advantageous. Electric motors in many cases must operate through a speed increasing gear to supply the required rpm for most centrifugal machines. High capacity axial machines can be designed to operate at speeds comparable to motor operation.

Gas-turbines are being used as centrifugal compressor drives. Further development may increase the use of centrifugal compressors since it will enable competition with gas-driven reciprocating machines in areas where cheap gas is available. Several gas turbine-driven centrifugals have been installed.*

### Regulation of Centrifugal and Axial Compressors and Blowers

If the compressor is turbine-driven, varying turbine speed is the most satisfactory and economical method of control. For constant speed drives it is preferable to control by throttling the suction. For a constant discharge pressure and a given mass flow rate the horsepower consumed is less for throttling the suction than for throttling of the discharge. This has been very clearly discussed by Kassarik.[10]

A second method of regulation for constant speed machines is the use of adjustable first stage inlet guide vanes. These may be operated through gears and linkage so that they automatically change the pressure-volume characteristics of the compressor by changing the inlet angle. The inlet guide vane method is more efficient in power utilization than suction throttling.

Because of the relatively flat characteristic curve of centrifugal machines it is preferable to use actual flow rate as the sensing device for actuating the controls on a centrifugal. This flow is generally measured by a venturi meter.

Centrifugal machines are ideally suited in many applications as first stage units feeding reciprocating compressors. In such instances it is necessary to control the discharge pressure of the centrifugal to insure smooth operation of the reciprocating machines. This may be done by means of a pressure controller in the discharge line of the centrifugal machine. It is important to be certain that one control does not work against another in designing the complex controls for such installations.

### Fans

Fans can be built as centrifugal or axial flow units. The axial designs are called propeller-type units. The same principles already discussed apply. They are used most frequently in process plants for ventilating, for conveying and removing dusts, for supplying draft to boiler furnace draft fans, for circulating hot air or gases, and for similar applications.

### Rotary Compressors and Blowers

Gases are compressed and moved in rotary compressors and blowers by the positive action of the rotating elements. There are several popular designs of these positive displacement machines. The sliding-vane type

* One large installation employs 10 gas-turbine driven centrifugals for reinjection of natural gas into a gas field at 1935 psig. A series-parallel arrangement of compressors is used.

(similar to the sliding-vane pump in Fig. 14-19) has an eccentrically mounted rotor to which sliding vanes are fitted. Two-impeller blowers have two mating lobed impellers which rotate in opposite directions (Fig. 15-12). The impellers do not touch and, therefore, no internal lubrication is required. This is an advantage in the compression of air.

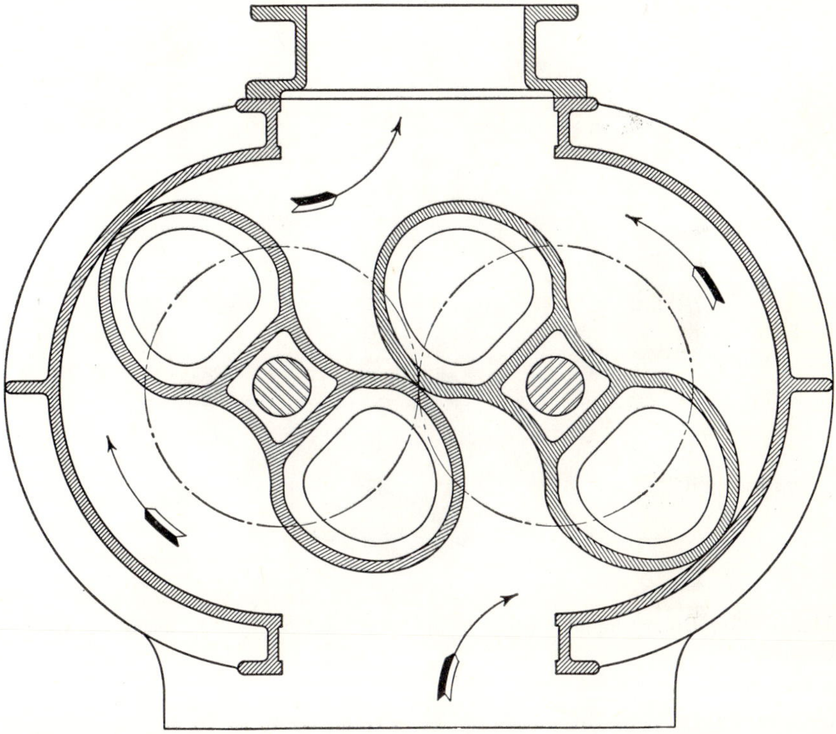

Fig. 15-12. Two-impeller blower. (Courtesy of Roots-Connersville Blower Division of Dresser Industries.)

The two-impeller blower has been used for many years and has proved a rugged piece of machinery. It operates at low speeds around 450 rpm and at differentials up to 11 to 15 psi. This blower is used extensively for supplying low-pressure air to flotation cells, cupolas, and pneumatic conveying systems. For volumes below 10,000 standard cfm it appears favored over the centrifugal blower because of lower first cost. To control capacity on constant speed units, the discharge is by-passed to the suction or the gas is vented.

A third popular type of rotary positive displacement compressor is the rotary-liquid-piston type as shown in Fig. 15-13. The multiblade rotor of this machine revolves in an elliptical case, partially filled with water or other liquid. The rotor revolves at sufficient speed to throw the liquid against the casing walls. The elliptical shape of the casing forces the liquid to fill the buckets of the rotor on the sides and recede on the top and bottom. This motion draws in gas at the inlet ports, and compresses

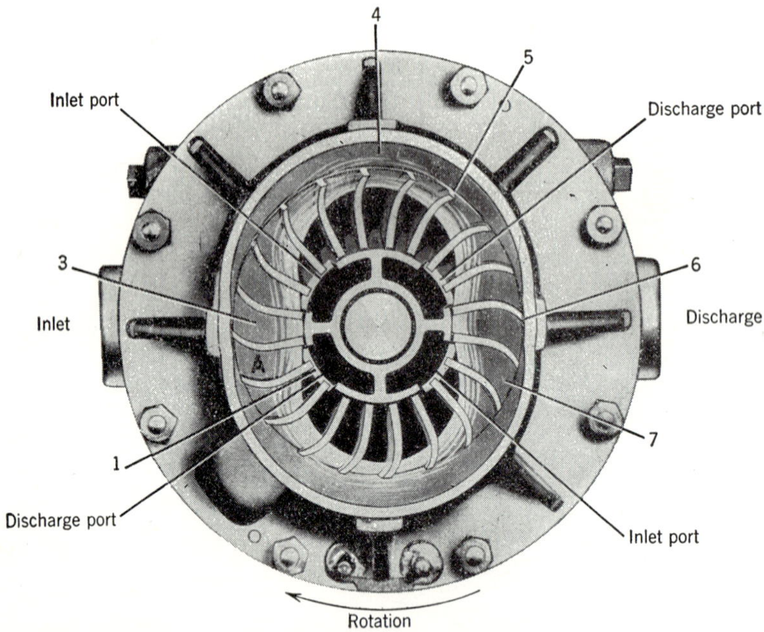


Fig. 15-13. Rotary-liquid piston compressor. (Courtesy of Nash Engineering Company.)

and discharges it at the outlet ports. Pressures up to 75 psi and volumes up to 50 cfm are atttained. Unusual gas pumping problems can often best be handled by these units since the liquid can be varied to suit almost any situation. Dry chlorine, for instance, can be compressed without moisture contamination using concentrated sulfuric acid as the liquid pumping agent.

## CHOOSING COMPRESSOR TYPE FOR A GIVEN JOB

It is difficult and unwise to generalize about the use of various types of equipment for a given application. Careful economic studies should be made for each case.

For high differential pressures and low to moderately high volumes the reciprocating compressor is the best choice. High volumes and moderate differentials can be handled most economically in centrifugal machines. For pressures too low for economical reciprocating compressor

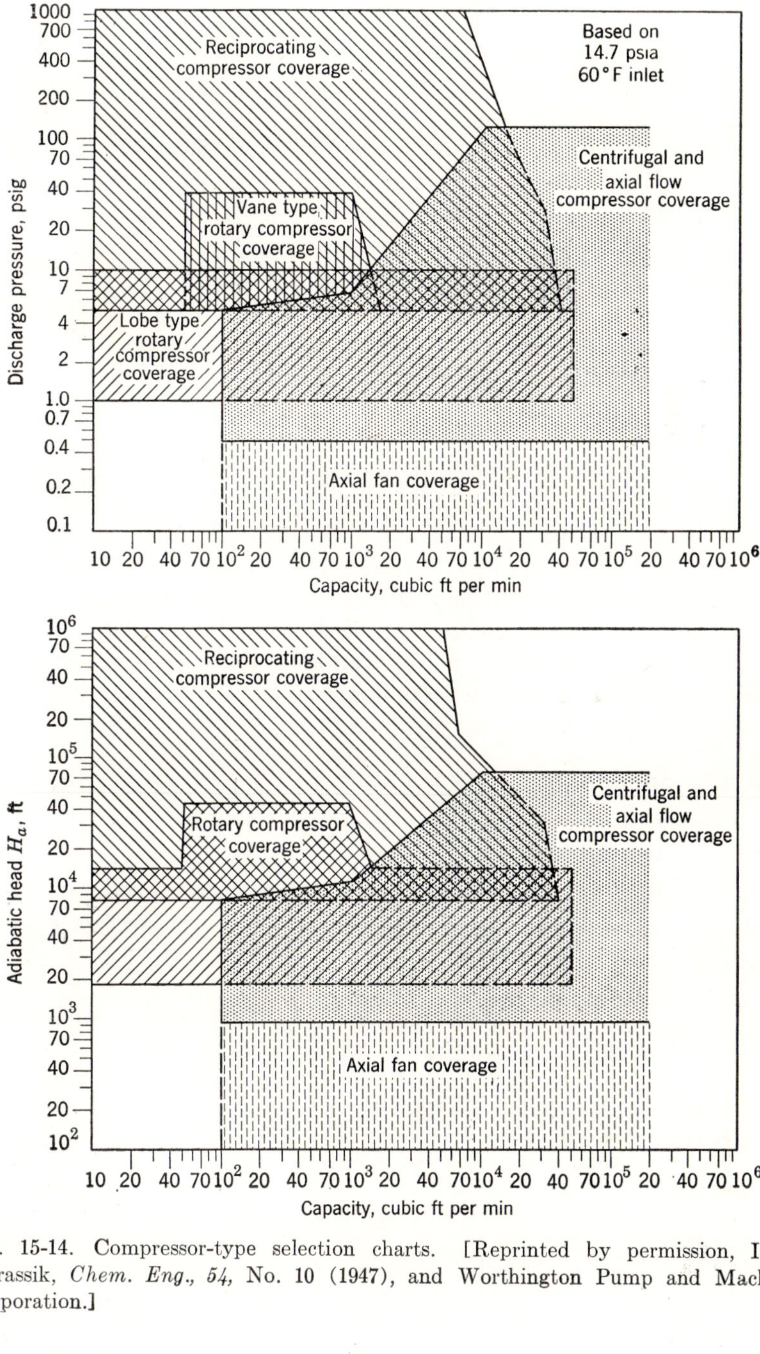


Fig. 15-14. Compressor-type selection charts. [Reprinted by permission, I. J. Karassik, *Chem. Eng.*, *54*, No. 10 (1947), and Worthington Pump and Machine Corporation.]

operation and volumes too small for centrifugals the rotary compressor such as the lobe-type appears advantageous. All these types, of course, overlap. Karassik[10] has presented several charts (Fig. 15-14) illustrating the fields of application. These are for illustrative purposes, of course, and should not be used blindly in choosing compressors or blowers.

Maintenance costs should not be overlooked in the selection of compressors. It is well known, for instance, that centrifugal machines require less maintenance than reciprocating machines. The unbalanced forces characteristic of reciprocating machines cause greater strain and wear on moving parts. The repair of a centrifugal machine, however, is more complex than that of a reciprocating machine. It has been found advantageous to maintain spare centrifugals in the plant warehouse for compressors in important services.

## VACUUM PUMPS

Most of the equipment described thus far can be built for use as vacuum pumps. The principles of operation are the same except that the inlet pressure is below atmospheric pressure which necessitates certain design changes. Volumes are large at these low pressures and compressor cylinders, for instance, must be correspondingly increased in size. Steam jet ejectors with direct contact condensers and in conjunction with barometric condensers are extensively employed for production of vacuum especially for vacuum distillation units. Applications of various types of vacuum-producing equipment are illustrated in Fig. 15-15.

## COMPRESSOR ACCESSORIES

Low capacity single-stage reciprocating compressors and centrifugal blowers require few accessory items. Air cooled cylinders, grease, or simple oil lubrication suffice. Large compressors so typical of modern chemical and petroleum plants demand a large number of accessory items. Several of the more important items are:

I. Reciprocating compressors

A. Auxiliary lubricating and cooling oil system. Provides forced lubrication and piston cooling oil in addition to lubricators operated from compressor crankshaft.

1. Lube oil gear pump (motor driven).
2. Lube oil cooler (shell and tube).
3. Oil strainers.
4. Oil filter.

B. Intercoolers and aftercoolers. Shell and tube or extended surface (fin) exchangers for cooling compressed gases between stages and after final discharge (aftercooler). In high pressure installations less expensive construction can be realized by passing the compressed gas through the tubes.

C. Cylinder cooling water. Each compressor cylinder should be provided with a separate valve for controlling flow. Closed cooling water systems often

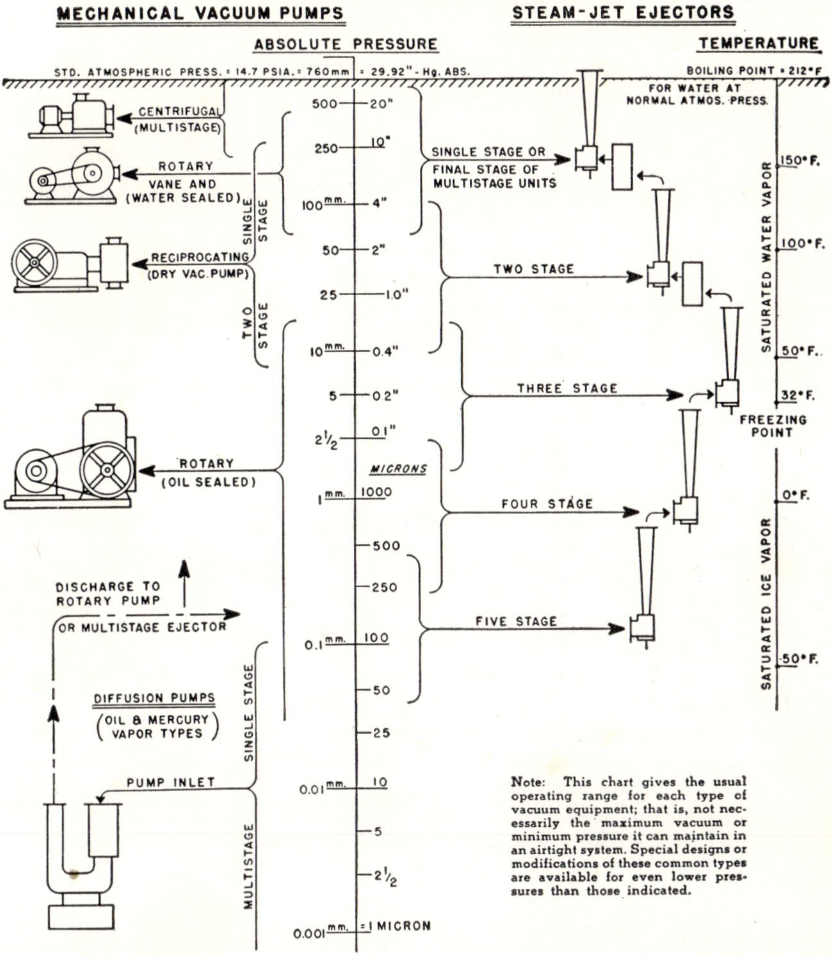


Fig. 15-15. Vacuum-equipment selection chart. (Courtesy of Ingersoll-Rand Company.)

prove feasible. Carefully treated water is employed and large volumes of water are used with a temperature rise of only ten degrees or less.

In a closed system if the water has been properly treated, higher temperature water can be used, 130–150° F, without excessive corrosion or scale formation. This high temperature permits the successful application of air cooled exchangers for cooling the water in areas where water is scarce.

D. Knockout drums. Condensate in compressor cylinders can have disastrous effects. If condensate is expected to form, provision should be made for its complete removal after each cooling stage.

E. Suction and discharge surge drums or bottles. Because of the pulsating action of reciprocating machines surge drums on both the suction and discharge side of each cylinder are necessary. These are designed to have a volume sufficient to limit pressure fluctuation to 5% of the pressure at the drum, and have been found to be most satisfactory when built with a length to diameter ratio of 2 : 1. Compressor manufacturers are prepared to supply the recommended sizes for their machines.

F. Instrumentation

1. Pressure gages.
2. Temperature indicators.
3. Pressure recorders and controllers to actuate governor or clearance pockets and suction valve unloaders.
4. Suction pressure regulating valve.
5. Relief valves.

II. Centrifugal conmpressors

A. Auxiliary lubricating system.

1. Pump. Many installations are equipped with a submerged gear pump in the suction bearing reservoir. An auxiliary pump can be supplied and is preferred by many operators.
2. Oil cooler.
3. Filter.

B. Seal oil pump.

C. Intercooler and interstage separator. Used between two compressors operating in series.

D. Instrumentation.

1. Pressure gages and temperature indicators.
2. Flow controllers (venturi) on suction to control speed of drive or suction pressure.
3. Suction pressure controller.

III. Drives

A. Gas engine or gas-diesel for reciprocating compressors.

1. Starting air compressors (250 psig).
2. Starting air surge tanks.
3. Diesel oil tanks, pumps, and filter.
4. Engine oil gear pump, cooler, filter, strainer.
5. Fuel gas regulating valve.
6. Fuel gas surge tank.
7. Scavenging air compressor.
8. Air filter for engine air. Can be preceded by individual cooling towers which cool water for the engine while heating the combustion air.
9. Cooling water pump.
10. Automatic controls.

B. Turbines for centrifugal compressors.

1. Askania-type controller for speed control.
2. Surface condenser and associated steam jet ejector and condensate pumps for condensing turbines.
3. Pressure gages, thermometers, and steam flow recorders.

C. Motors. Compressor motors on large units are usually expensive synchronous motors and require various protective devices and other appurtenances not common to small motors. Some of these are:

1. Supply of direct current from a motor generator set or selinium rectifiers to provide a field for the motor.
2. Differential protection system to guard against short circuit.
3. Instantaneous relay to protect against short circuit.
4. Induction pattern relay to guard against locked rotor overload.

5. Thermal relay to protect against operating overloads.
6. Motor temperature indicator and alarm.

Some of these motor accessories are rather expensive but are entirely justified on the 1000 hp and larger motors encountered in compressor practice.

A knowledge of the characteristics of the power supply and continuity of operation required in the process can aid the electrical engineer greatly in the design of a system for maximum reliability of operation and maximum motor life. The project engineer should supply this information as well as information on the effect of short power interruptions on the process.

## COMPRESSOR INSTALLATION

Installation is a simple matter for small low speed machines but with the larger units many complex problems arise. Because of the unbalanced forces characteristic of large reciprocating machines great care in foundation design and piping layout is essential. The high speed operation of centrifugal machines makes accurate alignment imperative in order to avoid damaging vibrations.

Compressor manufacturers furnish recommended foundation drawings for their units based on firm soil. Poor bearing soil demands the attention of an expert foundation and soil specialist to insure the provision of adequate foundations. Mass ratios of from 2 : 1 to 4 : 1 (concrete to compressor and driver weight) are customarily used for reciprocating units. High speed centrifugals require comparable ratios. Compressors foundations are best designed separate from building foundations. Separation and insulation of the foundation from adjacent flooring and building helps eliminate vibration in the building.* When several reciprocating compressors are located in one building it is desirable to provide a common mat under all the foundations.

The building housing the compressors should be well ventilated to insure elimination of hazardous gas mixtures. Servicing space should be provided as recommended by compressor manufacturers. Trolley beams are needed for hoists and must be at sufficient height and of sufficient capacity to lift compressor parts.

Adequate pipe sizes should be used to eliminate undue pressure drop especially in the suction headers to reciprocating compressors. In compressor practice a velocity of 30 to 35 ft/sec has been found most satisfactory. It is unwise to undersize compressor piping.

To eliminate excessive friction loss changes in piping direction must be avoided as much as possible, and where they are necessary long radius elbows should be used. The approach to the inlet of a centrifugal compressor should have at least two diameters of straight pipe. This is an absolute necessity for centrifugal compressors. Sharp bends should also

* In the case of centrifugals it also prevents outside vibrations from reaching the compressor.

be avoided on reciprocating machines at the inlet. The discharge line of a centrifugal should be a very long bend, about 10 degrees.

On reciprocating units overhead piping is not recommended since in order to prevent excessive vibration anchoring problems become very difficult. The anchoring and supporting of discharge piping is, in any event, a problem which deserves careful attention. Vibration in improperly anchored systems can cause fatigue and bursting of the line due to fatigue. Piping for centrifugal compressors must be carefully supported and anchored to avoid strains on the compressor casing.

All compressor piping should be thoroughly cleaned of scale and foreign matter. During initial operation, particularly, strainers should be installed in the suction lines.

## COMPRESSION THEORY

To the engineer who must only occasionally be concerned with the calculation of compressor performance, the formulations and nomenclature familiar to the compressor industry often appear baffling. If, however, he understands several basic principles of thermodynamics, he realizes that all the many compressor design equations are based on these principles.

### Thermodynamic Energy Functions and Work of Compression

Although reciprocating compressors combine in their operation a flow and nonflow process, the overall result is a flow process. Hence, compressor equations for all compressor types can be based on the equation for flow work obtained from the general energy balance. It is convenient to base all calculations on the ideal reversible process. In this manner rigorous calculations can be made with relative ease and then corrections to the actual case applied in the form of efficiencies.

### The Overall Energy Balance

The overall energy balance in differential form assuming potential, surface and electrical effects to be zero or negligible can be expressed as follows:

$$d(PV) + d\left(\frac{mu^2}{2g_c}\right) + dU = dq - dW \tag{1}$$

where $P$ = pressure, #/sq ft
$V$ = volume, cu ft
$m$ = mass, lb
$u$ = velocity, ft/sec
$g_c$ = factor 32.17 for converting from absolute to engineering units
$U$ = internal energy, ft-#

$q$ = heat added from surroundings,* ft-#
$W$ = work done on the surroundings, (work of expansion), ft-#
# = pounds of force

Now the exact differential for the energy function $U$ (internal energy) is $dU = T\,dS - P\,dV$, and for a reversible process $dq = T\,dS$ (where $S$ is the entropy and $T$ the absolute temperature). Hence, with the restriction of a reversible process (ideal frictionless operation) Eq. 1 becomes:

$$V\,dP + P\,dV + \frac{m(u_2^2 - u_1^2)}{2g_c} + T\,dS - P\,dV = T\,dS - dW$$

or

$$-dW = V\,dP + \frac{m(u_2^2 - u_1^2)}{2g_c}$$

or

$$-W = \int_1^2 V\,dP + \frac{m(u_2^2 - u_1^2)}{2g_c} \quad (2)$$

Subscripts 1 and 2 refer to the initial and final state.

### Theoretical Work of Compression

For the usual case of compression where kinetic energy effects are negligible eq. 2 becomes:

$$-W = \int_1^2 V\,dP \quad (3)$$

For fans and other differential pressure blowers $V$ may be considered constant and work is equal to $V(P_2 - P_1)$. Eq. 2 and 3 are true for reversible flow processes only. Irreversibilities such as friction effects increase the total work that must be done. Some authorities[9] refer to the $\int_1^2 V\,dP$ term as the *useful shaft work.*

It is possible to develop eq. 3 for reversible flow work by considering the area in an ideal $PV$-diagram. This method is perhaps more concise, but unfortunately the limitations of the expression are obscured when the general energy balance is not used directly in the development.

Eq. 3 can be related directly to thermodynamic energy functions when the exact differentials of certain of these functions are considered under the restrictions of constant temperature (isothermal) and constant entropy (isentropic). These exact differentials, the development of which can be found in any standard chemical engineering thermodynamic textbook, are:

$$dG = -S\,dT + V\,dP$$

$$dH = T\,dS + V\,dP$$

* This convention is consistent with most works on thermodynamics.

where $V$ = volume

$P$ = pressure

$T$ = absolute temperature

$S$ = entropy

$G$ = free energy

$H$ = enthalpy

### Reversible Isothermal Compression

Applying $$dG = -S\,dT + V\,dP$$

$$-S\,dT = 0 \text{ at constant } T$$

$$\therefore V\,dP = dG$$

or $$-W = \int_1^2 V\,dP = \Delta G = \Delta H - T\,\Delta S \tag{4}$$

### Reversible Isentropic Compression*

Applying $$dH = T\,dS + V\,dP$$

$$T\,dS = 0 \text{ at constant } S$$

$$\therefore V\,dP = dH$$

or $$-W = \int_1^2 V\,dP = \Delta H \tag{5}$$

Eq. 4 and 5 indicate two distinct methods for determining the reversible work of compression. First, if the relation between $P$ and $V$ is known, for the isothermal or isentropic process, the integration of the $\int_1^2 V\,dP$ term will yield the work. Secondly, if values of enthalpy and entropy at the initial and final conditions are available, the work can be calculated as $\Delta H - T\,\Delta S$ for isothermal work and $\Delta H$ for isentropic. Indeed, the ease by which isentropic reversible work can be calculated from a Mollier diagram for a given gas has been a major factor in the adoption of isentropic work as the standard or base for comparison. Knowing the initial temperature and pressure and the outlet pressure, one only need follow a constant entropy line to the outlet pressure to obtain the final enthalpy and temperature. The final minus the initial enthalpy is the Btu equivalent of the work and is converted to ft-lb by multiplying by 778. (1 Btu = 778 ft-lb). Hougen and Watson[9] have presented generalized

* A reversible isentropic process is an adiabatic one since $dq = T\,dS$ for a reversible process. It is often common, therefore, for isentropic compression to be called adiabatic. Such a practice is not recommended since all adiabatic processes are not necessarily isentropic or reversible.

charts that permit the calculation of enthalpies and entropies when Mollier diagrams or tables of enthalpy are not available.

If the gas obeys the ideal gas law, solutions to the integration of $\int_1^2 V\,dP$ can be simplified as shown below. The use of Mollier diagrams when possible, however, is strongly urged since no assumption of ideality is necessary in their use and the results are for the actual gas.

### *Equations Based on Ideal Gas Law*

The Ideal Gas Law is expressed as:

$$PV = nRT \tag{6}$$

where $P$ = pressure #/sq ft
$n$ = number of lb-moles
$V$ = volume cu ft
$R$* = Gas Law constant in consistent units 1544 (#/sq ft)(cu ft)/(lb-mole) (° R)

Equation 6 may be written:

$$P\frac{V}{n} = RT = Pv \tag{7}$$

where $v$ = molal volume, cu ft/lb-mole.

Reversible Isothermal Compression (Ideal Gas)

Employing eq. 7, $v = RT/P$ the work per mole $(-W_m)$ is:

$$-W_m = \int_1^2 v\,dP = RT \ln \frac{P_2}{P_1} = 1544T \ln \frac{P_2}{P_1} \tag{8}$$

Reversible Isentropic Compression (Ideal Gas)

Similarly for the isentropic case:

$$-W_m = \int_1^2 v\,dp = \int_1^2 RT\frac{dP}{P} = (\Delta H')(778) = 778C_p\,dT \tag{9}$$

where $\Delta H'$ = enthalpy change for an ideal gas, Btu/lb-mole.
$C_p$ = molal heat capacity at constant pressure, Btu/(lb-mole)(° F)

or

$$\left(\frac{R}{778}\right)\left(\frac{dP}{P}\right) = \frac{C_p\,dT}{T}$$

$$\frac{R}{778} = 1.987 \text{ Btu/(lb-mole)(° } R)$$

*It seems preferable to remember one gas constant and make conversions as required. In compression calculations less chance for error occurs if all units are kept in feet, pounds, and degrees Rankine

Call this $R'$ the gas law constant in heat units.

$$\therefore R' \, d \ln P = C_p \, d \ln T$$

or

$$R' \ln \frac{P_2}{P_1} = C_{p_m} \ln \frac{T_2}{T_1}$$

where $C_{p_m}$ = arithmetic mean of inlet and outlet heat capacity.

Hence $$T_2 = T_1 (P_2/P_1)^{R'/C_{p_m}}$$

Now for an ideal gas $C_{v_m} + R' = C_{p_m}$ and calling $C_{p_m}/C_{v_m} = k$ (Ratio of heat capacity at constant pressure $C_{p_m}$ to that at constant volume $C_{v_m}$) the following is obtained:

$$T_2 = T_1 \left(\frac{P_2}{P_1}\right)^{C_{p_m} - C_{v_m}/C_{p_m}} = T_1 \left(\frac{P_2}{P_1}\right)^{(k-1)/k} \tag{10}$$

Eq. 10 provides a rapid means for calculating the theoretical final compression temperature for an ideal gas when the initial temperature, compression ratio, and the heat capacity ratio are known:

From eq. 9 above:

$$-W_m = \int_1^2 C_p \, dT = C_{p_m}(T_2 - T_1)$$

Substituting eq. 10,

$$-W_m = C_{p_m} T_1 \left[\left(\frac{P_2}{P_1}\right)^{(k-1)/k} - 1\right] \quad \text{in Btu/lb-mole}$$

It is convenient to convert $C_{p_m}$ to an expression in terms of $k$.

$$C_{p_m} = \frac{C_{p_m} R'}{R'} \quad \text{and since} \quad R' = C_{p_m} - C_{v_m}$$

Then $$C_{p_m} = \frac{C_{p_m} R'}{C_{p_m} - C_{v_m}} = \frac{\left(\dfrac{C_{p_m}}{C_{v_m}}\right)(R')}{\left(\dfrac{C_{p_m}}{C_{v_m}} - 1\right)} = \frac{k}{k-1} R'$$

Hence $$-W_m = \frac{k}{k-1} R' T_1 \left[\left(\frac{P_2}{P_1}\right)^{(k-1)/k} - 1\right] \quad \text{Btu/lb-mole} \tag{11}$$

or $$-W_m = \frac{k}{k-1} R T_1 \left[\left(\frac{P_2}{P_1}\right)^{(k-1)/k} - 1\right] \quad \text{ft-\#/lb-mole} \tag{11a}$$

or since $P_1 v_1 = R T_1$

$$-W_m = \frac{k}{k-1} P_1 v_1 \left[\left(\frac{P_2}{P_1}\right)^{(k-1)/k} - 1\right] \quad \text{ft-\#/lb-mole} \tag{12}$$

Based on the assumptions of these derivations the values of $k$ must be based on an average of $k$ at the inlet and outlet. These equations are true for an ideal gas and the $k$ can be evaluated from the heat capacities at 0 psia and the average temperature between the inlet and discharge. The discharge temperature is, of course, not known at first but a rough guess usually suffices. For most compressor problems variation in specific heat is small with temperature and values at room temperature are satisfactory.

Eqs. 8, 11, and 12 all have units of work per mole (ft-lb per lb-mole). There are many variations of these equations, but all are basically the same although the multiplicity of compressor formulae in the literature often cause confusion. It is advisable, however, to understand the development of one form of the equation such as those above and then conversion can be made to any other units that may be convenient. For example to obtain the total work per unit time the equations need only be multiplied by the lb-moles/minute of feed to the compressor. The work per unit weight* (ft-lb of force per lb of mass) can be obtained by dividing eq. 12 by the molecular weight of the gas. Horsepower required for compression can be calculated by dividing the value of ft-lb per minute obtained from one of the equations above by 33,000 ft-lb per minute per horsepower.

Since the equations are developed on a molal basis the molal heat capacity should be employed.

$k$ can be calculated as $C_{p_m}/(C_{p_m} - R')$.

### Polytropic Compression Coefficient

It is often argued that since actual compression is neither isentropic nor isothermal, because of certain internal irreversible effects, it is preferable to calculate a work of compression based on the expression $PV^n = \text{constant}$, where $n$ is defined as the polytropic constant. The value of $n$ for a given gas can be determined from compressor performance tests using indicator card diagrams which are in effect plots of pressure versus volume.

$$\text{Log } P + n \log V = \log \text{constant}$$

Hence the slope of the indicator diagram line will give a value of $n$. These values are necessarily approximate, however.

The expression for work in the polytropic case is

$$-W_m = \frac{n}{n-1} P_1 v_1 \left[ \left( \frac{P_2}{P_1} \right)^{(n-1)/n} - 1 \right] \tag{13}$$

* The work per unit weight (ft-#/lb) as with pumps is unfortunately often referred to as *Head* particularly in centrifugal compressor parlance. The term ft of head should be remembered to be actually ft-lb of force/lb.

The horsepower determined from this equation is an estimation of the indicated horsepower.

Because it is so easy to calculate isentropic horsepower from enthalpy diagrams, the use of the isentropic base for all calculations has much merit. The calculation of compressor work is at best an approximation and the value obtained must be corrected to brake horsepower by the application of an appropriate efficiency factor. It would seem preferable, therefore, to base all calculations on the simpler isentropic case and use efficiencies based on the theoretical isentropic basis. For reciprocating machines the compression actually approaches isentropic operation because of the water-jacketed cylinder, but for uncooled centrifugal compressors that is not the case although the isentropic basis is still recommended (see below).

## Compressor Efficiencies

It is desirable at this point to define the various efficiencies used in the compressor industry.

### *Volumetric Efficiency*

As with reciprocating pumps the actual volume of gas taken into the cylinder of a reciprocating compressor is less than the displacement of the cylinder. The ratio of actual volume measured at inlet conditions to the displacement is called the volumetric efficiency.

The relation between volumetric efficiency and the clearance ($C$) expressed as a fraction of the total displacement ($D$) per stroke is easily derived. The clearance of the cylinder is filled with compressed gas which re-expands on the return stroke. It is this re-expanded gas volume that reduces the amount of gas that can be received by the suction stroke. For ideal gases the clearance volume at suction ($V_{1_c}$) is related to the clearance volume at discharge ($V_{2_c}$) by:

$$P_2 V_{2_c}^{\,k} = P_1 V_{1_c}^{\,k}$$

or

$$V_{1_c} = V_{2_c}\left(\frac{P_2}{P_1}\right)^{1/k} = CD\left(\frac{P_2}{P_1}\right)^{1/k}$$

$$\text{Fractional volumetric efficiency} = E_v = \frac{\text{Inlet volume}}{\text{Displacement}}$$

Hence

$$E_v = \frac{D + CD - CD\left(\dfrac{P_2}{P_1}\right)^{1/k}}{D}$$

where $D + CD$ = volume compressed measured at inlet conditions

or
$$E_v = 1 - C\left[\left(\frac{P_2}{P_1}\right)^{1/k} - 1\right]^* \tag{14}$$

### Compressor Operating Efficiency

Compressor efficiency can be defined in a number of ways depending on the basis chosen. An efficiency value without reference to the basis, therefore, is meaningless. The use of compressor efficiency, in any event, is an attempt to account for the internal irreversibilities and mechanical losses which occur in actual compression. These losses differ for reciprocating and centrifugal compression. For reciprocating compressors, these losses are fluid friction losses including losses through valves and inlets and outlets, mechanical friction in moving parts such as the piston and piston rod. Losses for centrifugal compressors include fluid friction, leakage along shaft and between stages, disk friction and mechanical losses in the bearings and packing. Some of the more frequently used efficiencies are listed below.

Isothermal efficiency:

$$\frac{\text{Reversible isothermal work} \times 100}{\text{Actual shaft input work}} = E_I$$

The actual shaft work is equivalent to the brake horsepower.
Isentropic efficiency:

$$\frac{\text{Reversible isentropic work}}{\text{Actual shaft input work}} \times 100 = E_s$$

Mechanical efficiency:

$$\frac{\text{Indicator horsepower}}{\text{Actual shaft input work}} = \frac{\text{Indicator horsepower}}{\text{Brake horsepower}} \times 100 = E_m$$

Polytropic efficiency:

$$\frac{\text{Polytropic work}}{\text{Actual shaft input work}} \times 100 = E_p$$

When the polytropic exponent is determined from an actual indicator diagram the polytropic efficiency based on the $n$ value thus determined is approximately equal to the mechanical efficiency.

#### Centrifugal-Compressor Efficiency

Some manufacturers refer to polytropic efficiency for centrifugal machines as hydraulic or temperature rise efficiency. Uncooled centrif-

* An empirical correction for valve leakage is applied by most compressor manufacturers. 0.01 $(P_2/P_1)$ is subtracted from the value of $E_v$ obtained from eq. 14

ugals more nearly approach polytropic compression than isentropic. It is, therefore, argued, and with merit that the isentropic efficiency of a given machine will vary markedly with the gas. This is true because all the losses (except mechanical losses which are only a small per cent of the total) heat the gas being compressed and increase the deviation from isentropic conditions. These various deviations require different isentropic efficiencies for each gas considered in one machine.

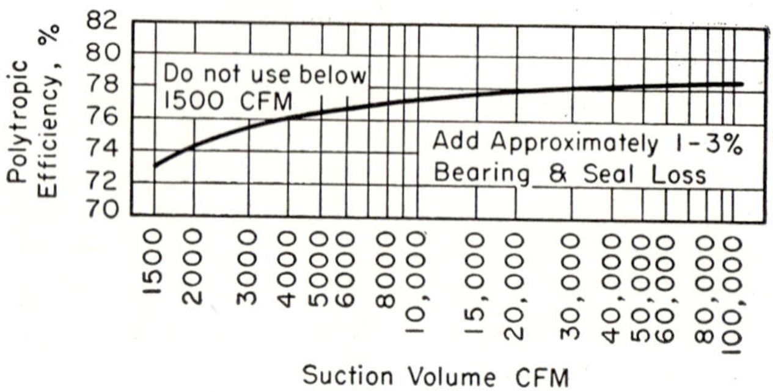


TABLE OF CORRECTION FACTORS DUE TO COMPRESSION RATIO

| Compression Ratio | 1500 | 2000 | 3000 | Inlet Volume in CFM 4000 | 5000 | 7500 | 15,000 | 30,000 | 60,000 & over |
|---|---|---|---|---|---|---|---|---|---|
| 1.35 | 1.0 | 1.0 | 1.0 | 1.0 | 1.0 | 1.0 | 1.0 | 1.0 | 1.0 |
| 1.75 | .983 | .990 | .990 | .994 | .996 | .996 | .996 | .996 | 1.0 |
| 2.25 | .976 | .977 | .986 | .985 | .991 | .995 | .995 | .995 | 1.0 |
| 2.80 | — | .969 | .980 | .980 | .989 | .992 | .993 | .994 | 1.0 |
| 5.00 | — | — | .965 | .970 | .972 | .985 | .991 | .996 | .997 |
| 10.00 | — | — | — | .944* | .955* | .960* | .985* | .989* | .995 |
| 15.00 | — | — | — | — | .935* | .957* | .971* | .986* | .993 |

Notes: Asterisk indicates figures applying only to high molecular weight hydrocarbons.
Factors apply on one compressor body with six or less impellers.

Fig. 15-16. Suction volume versus polytropic efficiency. (Adapted by permission, Clark Brothers Company.)

In Fig. 15-16 typical polytropic efficiencies for centrifugal compressors are plotted versus the inlet volume. This curve represents an estimate for any gas. The polytropic efficiencies, of course, also vary with the gas being considered although not as much as the isentropic. It is, therefore, possible to select a polytropic efficiency for a given machine which can be used for any gas from a curve such as Fig. 15-16. The advantages of working with the isentropic basis, however, are too great to be overlooked. The polytropic efficiency may be converted to an isentropic efficiency for the gas being considered by means of Fig. 15-17, and this makes it possible to use Mollier charts or enthalpy tables.

## Reciprocating Compressor Efficiency

Since reciprocating compression so nearly approaches adiabatic compression there should be no hesitation in basing all calculations on the isentropic base, and employing isentropic efficiencies to correct theoretical work to actual work. Table 15-1 presents some typical isentropic reciprocating compressor efficiencies which may be considered to be equivalent to the mechanical efficiencies as well.

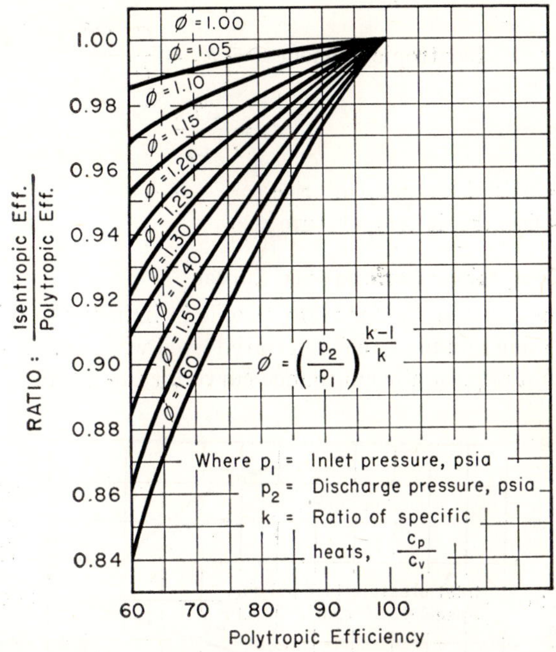


Fig. 15-17. Relationship between isentropic and polytropic efficiency for centrifugal compressors. (Adapted by permission, Clark Brothers Company.)

**TABLE 15-1. Reciprocating Compressor Efficiencies**

$$\text{bhp} = \frac{\text{Isentropic hp}}{\text{Efficiency}}$$

| Compression ratio | Engine-driven* |
|---|---|
| 1.1 | 50–60 |
| 1.2 | 60–70 |
| 1.3 | 65–80 |
| 1.5 | 70–85 |
| 2.0 | 75–88 |
| 2.5 | 80–89 |
| 3.0 | 82–90 |
| 4.0 | 85–90 |

*Multiply the above values by 0.95 for motor-driven compressors.

### Temperature Rise with Compression

The actual final gas temperature from reciprocating compressors can be estimated from eq. 10, or better, a Mollier diagram, since the machine approaches isentropic compression closely. For uncooled centrifugal compressors, however, the isentropic temperature rise is not comparable to the actual since most of the losses in a centrifugal serve to heat the gases being compressed. An approximation of the actual rise can be obtained from the following relation:

$$\text{Isentropic efficiency} = E_s = \frac{C_p\,\Delta T_S}{C_p\,\Delta T_a} \tag{15}$$

or

$$\Delta t_a = \frac{\Delta T_S}{E_s}$$

where $\Delta T_a$ = actual temperature rise
$\Delta T_S$ = isentropic temperature rise

### Centrifugal Compressor Performance Curves

It has become customary for many people in industry to think in terms of air compression and most machines are tested with air. The purchaser, however, should request a performance curve for the gas to be compressed

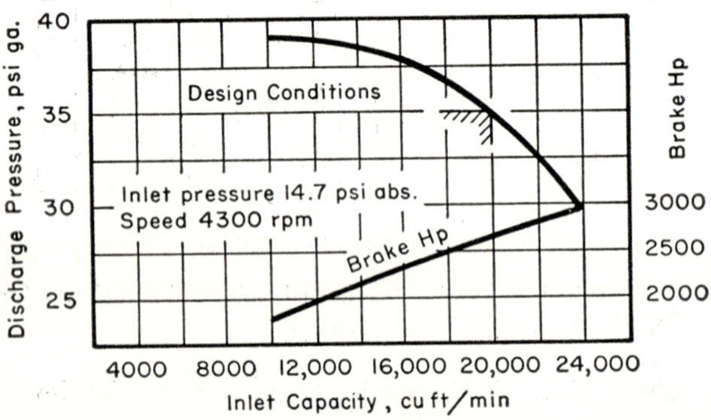


Fig. 15-18. Typical centrifugal compressor performance curve, volume versus discharge pressure. (Adapted by permission, I. J. Karassik and Worthington Corporation.)

based on the design inlet conditions. Volume is generally plotted on such curves in cfm at inlet conditions versus discharge pressure in psia. In addition brake horsepower is also shown. These performance curves (Fig. 15-18) are given for several speeds if the speed of the driver is variable.

Another more useful curve of isentropic head and efficiency can be plotted from Fig. 15-18 that will be independent of inlet pressure variation. This is done by choosing various points on Fig. 15-18 and calculating the corresponding isentropic head and then the isentropic efficiency (Fig. 15-19).

Fig. 15-19 can be used to develop volume versus discharge pressure and bhp curves for any inlet pressure. If the inlet pressure remains constant and the inlet temperature changes, the isentropic head remains constant if the compression ratio changes to fulfill this constancy (eq. 11). Discharge pressure curves for inlet temperature changes, therefore, can be calculated from the new compression ratio.

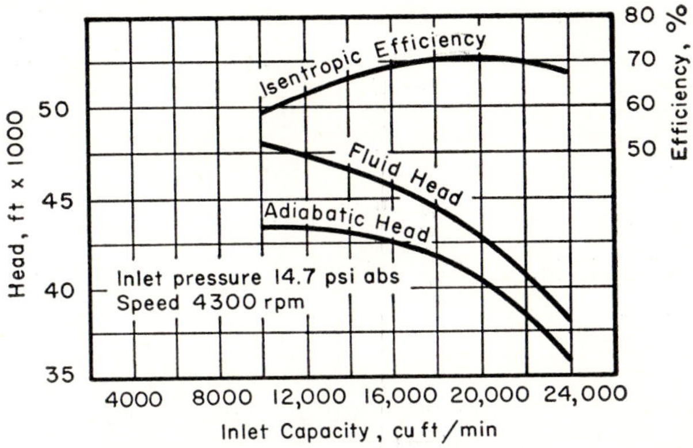


Fig. 15-19. Typical centrifugal compressor performance curve, volume versus isentropic head. (Adapted by permission, I. J. Karassik and Worthington Corporation.)

The same relations developed for centrifugal pumps expressing the effect of speed changes on performance curves can be applied, although with less accuracy, to centrifugal compressors. These are applicable at the same efficiency points:

$$\frac{Q_1}{Q_2} = \frac{N_1}{N_2} \qquad \frac{H_{s_1}}{H_{s_2}} = \left(\frac{N_1}{N_2}\right)^2 \tag{16}$$

where $Q$ = inlet volume
$N$ = rpm
$H_s$ = isentropic head, ft-#/lb

### Centrifugal Compressor Impeller Size and Speed

The work produced by one stage of a centrifugal compressor is a function of the kinetic energy and can be expressed in ft-# per lb as follows:

$$W_I = K\frac{u^2}{g_c}$$

where $W_I$ = ft-#/lb
$u$ = peripheral velocity, ft/sec
$g_c$ = 32.2 conversion factor (poundals per lb of force)
$K$ = overall pressure coefficient varies between 0.5 and 0.65 depending on impeller and stage design.

One large manufacturer designs the major portion of its centrifugal machines for a normal peripheral velocity of 770 ft/sec. Hence for an average value of $K = 0.55$, 10,000 ft-lb of force/lb per stage is a good average value for estimating the required number of stages. The number of stages is thus set by the overall compression ratio and the work per stage. The peripheral velocity is related to the impeller diameter and the rpm as follows:

$$60\,u = \frac{\pi DN}{12}$$

or

$$u = \frac{\pi DN}{720}$$

where $D$ = diam in in.
$N$ = rpm

Hence

$$N = \frac{u720}{\pi D} = \frac{720\sqrt{W_I g_c}}{\pi D\sqrt{K}} = \frac{1300}{D}\sqrt{\frac{W_I}{K}} \tag{17}$$

A standard impeller size can thus be chosen and the required rpm calculated.

### Multistage Compression

Multistage compression as discussed in this section shall refer to compression with complete intercooling between stages. This is accomplished by passing the compressed gases through water-cooled intercoolers between stages of a reciprocating compressor or between two centrifugal compressors. In this manner the gases can be cooled to near inlet temperature after each stage of compression and the compression process thereby approaches isothermal operation for which the work of compression is the least. (That isothermal operation theoretically requires the least work for a given compression ratio can be seen by comparing eq. 8 and 11.)

The total work of compression in a multistage operation is simply the sum of work in the various stages. Assuming $k$ and the suction temperature to each stage is constant and setting the differential of total work with respect to intermediate pressure equal to zero the optimum compression ratio per stage is found to be the $S$ root of the overall compression ratio, $(P \text{ outlet}/P \text{ inlet})^{1/S}$, for each stage where $S$ is the number of stages. This is the well known rule of equal work (equal compression ratio) for each stage. Strictly this derivation holds true for ideal gases only. On the basis of equal compression ratio per stage the total work becomes for $S$ stages:

$$-W_m = S\left(\frac{k}{k-1}\right)(RT_1)\left[\left(\frac{P_2}{P_1}\right)^{(k-1)/Sk} - 1\right] \tag{18}$$

It is preferable to calculate the hp for each stage separately since approximately 5 psi should be added to the discharge of each stage in order to allow for friction losses in interconnecting piping and intercoolers. For mechanical reasons, it is desirable to have approximately the same work in each stage. With actual gases this does not necessarily conform to equal compression ratio per stage.[6,14]

Dodge[4] reports that maximum pressure ratio in any stage of multistage machines is between 3 and 5. Each case should be considered on its own merits, however, and the economical compression ratio determined in cooperation with the compressor manufacturer. Typical compression ratios used are:

| | |
|---|---|
| Large pipeline compressors | 1.2 to 2.0 |
| Process compressors | 1.5 to 4.0 |
| Small pilot plant high pressure | 6:1 |

The final decision is based on many factors. It is, of course, desirable to use as few stages as possible. The limitation is the gas outlet temperature. The discharge temperature of the gas from a cylinder should be kept below 300° F. Compressor manufacturer's must estimate the actual discharge temperature since it depends on the cylinder cooling water arrangement, although a rough idea for estimating can be obtained from the isentropic calculation (eq. 10).

## Calculations for Actual Gases

Most of the equations presented thus far are for ideal gases. At high pressures some gases deviate markedly from ideal behavior. The perfect gas law has been empirically corrected to apply for actual gases by a multiplier called the compressibility factor.*

$$Pv = zRT$$

The compressibility factors of all gases have been found to correlate (Fig. 15-20) within engineering accuracy with reduced temperature $T/Tc$ and reduced pressure $P/Pc$, where $T$ and $P$ are the absolute temperature and pressure and $Tc$ and $Pc$ the absolute critical temperature and pressure of the gas. For mixtures it is satisfactory to calculate the reduced temperature and pressure employing pseudo-critical temperatures and pressures. These are determined from the molal average of the critical temperatures and pressures of the components.

* It is often referred to as the supercompressibility factor. Such superlatives, however, do not alter its value or meaning.

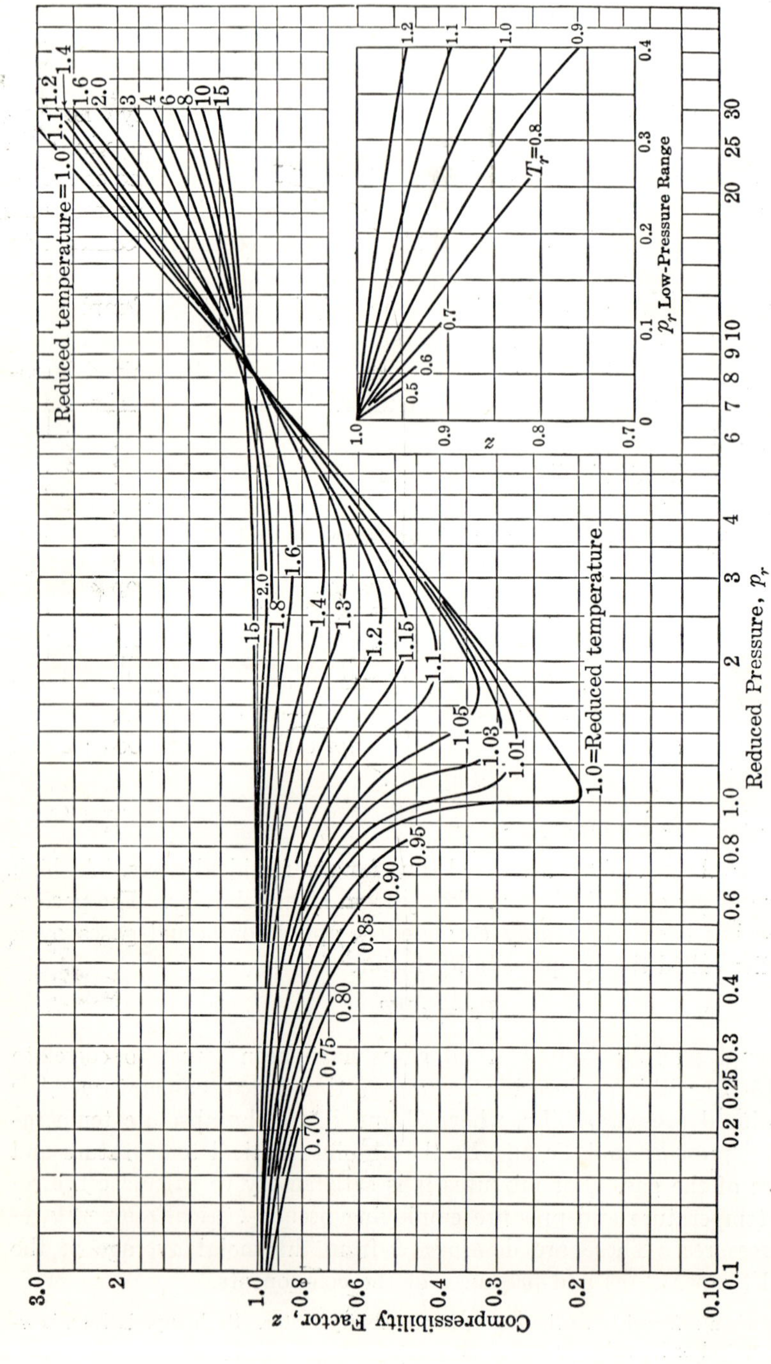


Fig. 15-20. Compressibility factor chart. [Reprinted by permission, O. A. Hougen and K. M. Watson, *Chemical Process Principles*, Vol. II, p. 489, John Wiley & Sons, Inc. (1947).]

Compressibility can be thought of as a correction factor by which the ideal gas volume must be multiplied to produce an actual volume.

EXAMPLE: Calculate the volume occupied by 1600 lb of methane at 60° F and 2000 psig. Critical temperature of methane is −116.5° F and critical pressure is 673 psig.

SOLUTION:

$$T_R = \frac{520}{343.5} = 1.52$$

$$P_R = \frac{2015}{673} = 3.00$$

From Fig. 15–20, $z = 0.78$

$$\text{Basis } \frac{1600}{16} = 100 \text{ lb-moles of methane}$$

$$V = \frac{(100)(379)(14.7)(0.78)}{(2015)} = 216 \text{ cu ft}$$

The volumetric efficiency of reciprocating machines is affected by the deviations from ideality, and failure to consider these deviations may lead to mis-sized compressor cylinders. Eq. 14, which expresses the relation between volumetric efficiency, clearance and compression ratio, can be derived for actual gases by using compressibility corrections. The expanded volume, at suction conditions is:

$$V_{1_c} = V_{2_c} \frac{z_1}{z_2} \frac{T_1}{T_2} \frac{P_2}{P_1}$$

Usually $T_2$ is not known but can be approximated from the ideal gas relationship

$$T_2 = T_1 \left(\frac{P_2}{P_1}\right)^{1-(1/k)}.$$

Hence

$$V_{1_c} = V_{2_c} \frac{z_1}{z_2} \left(\frac{P_2}{P_1}\right)^{1/k}$$

Hence

$$E_v = 1 - C\left[\frac{z_1}{z_2}\left(\frac{P_2}{P_1}\right)^{1/k} - 1\right] \tag{19}$$

where 1 refers to suction conditions and 2 to discharge conditions.

The project engineer must supply compressor manufacturers with accurate values of compressibilities for the gases to be compressed so that cylinders may be sized correctly.

If an enthalpy diagram is available for the gas being compressed, the theoretical isentropic work and temperature can be calculated accurately and rapidly. The difference in enthalpy in Btu per pound mole between initial and final conditions at constant entropy is equivalent to the isentropic work for an actual gas (see illustrative examples). If an enthalpy diagram is not available, generalized enthalpy charts can be used such as those of Hougen and Watson.[9] The ideal gas enthalpy can be corrected to actual gas enthalpy by using these generalized charts, but the calculation requires trial and error and is time consuming.

In most cases the theoretical isentropic horsepower need not be known with great accuracy and approximate methods are justifiable. Numerous methods have been proposed.[3,5,13a,14] Ridgway[13a] studied the effect of compressibilities of gases on the indicator card diagram and proposed correcting the ideal gas horsepower as calculated by equations 11, 11*a*, or 12 to actual gas horsepower by multiplying by the arithmetic average of the suction and discharge compressibilities. This same procedure can also be deduced by consideration of eq. 9. Compression work for the actual gas can be estimated as follows:

$$-W_m = \int_1^2 v\,dP$$

$$\approx \frac{z_1 + z_2}{2} \int_1^2 RT \frac{dP}{P}$$

For an ideal gas $\int_1^2 \frac{RT\,dP}{P} = (\Delta H')(778)$

$$\therefore W_m \approx \frac{z_2 + z_1}{2} \Delta H'(778)$$

or

$$-W_m = \frac{z_1 + z_2}{2} \frac{k}{k-1} P_1 v_1 \left[ \left(\frac{P_2}{P_1}\right)^{(k-1)/k} - 1 \right] \qquad (20a)$$

and

$$-W_m = \frac{z_1 + z_2}{2} \frac{k}{k-1} RT^1 \left[ \left(\frac{P_2}{P_1}\right)^{(k-1)/k} - 1 \right] \qquad (20b)$$

Since Ridgway's approximation is based on correcting the ideal gas horsepower, $k$ should be evaluated at 0 psi, and $v$ should be an ideal gas specific volume $(RT_1/P_1)$.

The method has been popular among compressor engineers because of the ease by which values for ideal gas horsepower can be corrected to horsepower for an actual gas.

## Illustrative Examples of Calculation Techniques

There is a distinct advantage in using a molal basis for all compressor calculations. The basic quantity is then independent of pressure and temperature. All the formulas presented previously are on the mole basis, and the examples to follow illustrate their use.

EXAMPLE 1: A centrifugal compressor is to be selected for compressing 462 lb per minute of pure nitrogen at 80° F and 14.7 psia to 40 psia.

Determine the following: (*a*) Brake horsepower; (*b*) Number of stages; (*c*) Impeller diameter; (*d*) Speed in rpm.

*Properties of Nitrogen*

Molal heat capacity: 6.97 Btu/(lb-mole)(° R)
Molecular weight: 28

$$k = \frac{6.97}{(6.97 - 1.987)} = 1.40 \qquad \text{and} \qquad \frac{k-1}{k} = \frac{1.4 - 1}{1.4} = 0.286$$

*a*. Basis: $\frac{462}{28} = 16.5$ lb-moles per minute.

$$\text{Work} = \frac{k}{k-1} RT_1 \left[ \frac{(P_2)}{(P_1)}^{(k-1)/k} - 1 \right]$$

$$= \frac{1.40}{1.40 - 1} (1544)(540)[(2.72)^{0.286} - 1]$$

$$= 966{,}000 \text{ ft-\#/lb-mole}$$

or

$$= \frac{(966{,}000)(16.5)}{33{,}000} = 483 \text{ hp isentropic}$$

$$\text{Inlet volume} = (16.5)(379)\left(\tfrac{540}{520}\right) = 6500 \text{ cfm}$$

From Figs. 15-16 and 15-17

$$\text{Isentropic efficiency} = (0.767)(0.955) = 0.733$$

$$\text{bhp} = \frac{483}{0.733} = 660 \text{ hp}$$

Because of the low pressure operation there is no need to correct for deviations from ideality. An assumption of ideality is completely within range of accuracy.

*b*. Number of stages: use 10,000 ft-#/lb per stage.

$$\frac{966{,}000}{(28)(10{,}000)} = 3.45 \qquad \text{(Use 4 stages)}$$

$$\text{Work per stage } \frac{966{,}000}{(28)(4)} = 8{,}630 \text{ ft-\#/(lb)(stage)}$$

*c*. The selection of the impeller diameter depends considerably on the required capacity of the machine. Manufacturer's data indicate an 18 in. impeller for the inlet capacity of 6500 cfm.

From eq. 17:

*d*. $$N = \frac{1300}{18}\sqrt{\frac{8630}{0.55}} = 9050 \text{ rpm.}$$

EXAMPLE 2: 1800 standard cfm (60° F and 14.7 psia) of natural gas composed of 80.2 mole per cent $CH_4$, 10.3% $C_2H_6$, 6.5% $C_3H_8$, 2.0% $i$-$C_4H_{10}$, and 1.0% $n$-$C_4H_{10}$ is to be compressed in a multistage reciprocating compressor. The gas enters the compressor at 85 psig and 85° F and is to be compressed to 2400 psia.

Determine the following, first assuming an ideal gas and then an actual gas. Assume complete intercooling between stages and 10% clearance for all cylinders: (*a*) Number of stages. (*b*) Horsepower per cylinder. (*c*) Total brake horsepower. (*d*) Displacement required for each cylinder.

Basis: $\frac{1800}{379} = 4.75$ lb-moles/min.

**Properties of gaseous mixture**

| Component | Mole, % | Molecular wt | Fractional mol wt | Critical pressure, psia | Fractional Critical pressure | Critical temperature, °R | Fractional critical temperature |
|---|---|---|---|---|---|---|---|
| $CH_4$ | 80.2 | 16 | 12.8 | 673 | 539.0 | 343 | 275.0 |
| $C_2H_6$ | 10.3 | 30 | 3.1 | 717 | 73.9 | 550 | 56.7 |
| $C_3H_8$ | 6.5 | 44 | 2.9 | 642 | 41.7 | 666 | 43.3 |
| $i$-$C_4H_{10}$ | 2.0 | 58 | 1.2 | 529 | 10.6 | 735 | 14.7 |
| $n$-$C_4H_{10}$ | 1.0 | 58 | 0.6 | 544 | 5.4 | 766 | 7.7 |
| | 100.0 | Average = 20.6 | | Average $P_c$ = | 670.6 psia | Average $T_c$ = 397.4° R | |

$$\text{Specific gravity} = \frac{20.6}{29} = 0.71$$

| Compound | Molal* $C_p$ at 150° F | Fractional $C_p$ = ($C_p$)(mole-fraction) |
|---|---|---|
| $CH_4$ | 8.97 | 7.19 |
| $C_2H_6$ | 13.78 | 1.42 |
| $C_3H_8$ | 19.58 | 1.27 |
| $i$-$C_4H_{10}$ | 25.82 | 0.52 |
| $n$-$C_4H_{10}$ | 26.16 | 0.26 |
| | Average $C_p$ | 10.66 |

* 150° F assumed as an average temperature.

$$k = \frac{10.66}{10.66 - 1.987} = 1.23$$

$$\frac{k-1}{k} = \frac{1.23 - 1}{1.23} = 0.187$$

Often values of $k$ are tabulated. These can be used for most calculations since the variation with temperature is usually not great.

*Assuming ideal gas*

*a.* Assume three stages.

$$\text{Comp. ratio} = \left(\frac{P_2}{P_1}\right)^{1/S} = \left(\frac{2400}{100}\right)^{1/3} = 2.9$$

Gas discharge temperature from cylinder (eq. 10)

$$T_2 = 545(2.9)^{0.187} = 665° \text{ R or } 205° \text{ F.}$$

With two stages the discharge temperature becomes 275° F. which is high.

Use three stages.

| *Suction* | *1st stage discharge* | *2nd stage* | *3rd stage* |
|---|---|---|---|
| 100 psia | 290 psia | 840 psia | 2400 approx. psia |

*b.* Horsepower per stage. (Correction for the increased pressure needed to overcome intercooler $\Delta P$ will be neglected.)

$$\text{Work} = \frac{k}{k-1} RT_1 \left[ \left( \frac{P_2}{P_1} \right)^{(k-1)/k} - 1 \right]$$

First stage:

$$\frac{1.23}{1.23 - 1.0}(1544)(545)[(2.9)^{0.187} - 1] = 990{,}000 \text{ ft-\#/lb-mole per stage}$$

Efficiency from Table 15-1 = 82%

$$\text{bhp/cylinder} = \frac{(990{,}000)(4.75)}{(33{,}000)(0.82)} = 174 \text{ hp}$$

*c.* Total bhp = (174)(3) = 522 hp

*d.* $E_v = 1 - C\left[\left(\frac{P_2}{P_1}\right)^{1/k} - 1\right] = 1 - 0.10\,[(2.9)^{1/1.23} - 1] = 0.86$

or 86% for each cylinder.

Displacement of first stage:

$$\left(4.75\,\frac{\text{lb-moles}}{\text{min}}\right)\left[\frac{(379)(14.7)}{(100)}\,\frac{\text{std cu ft}}{\text{lb-mole}}\right]\left[\frac{(545)}{(520)(0.86)}\,\frac{\text{cu ft}}{\text{std cu ft}}\right] = 322 \text{ cfm}$$

Displacement of second stage:

$$(4.75)(379)\left(\frac{14.7}{290}\right)\left(\frac{545}{520}\right)\left(\frac{1}{0.86}\right) = 111 \text{ cfm}$$

Displacement of third stage:

$$(4.75)(379)\left(\frac{14.7}{840}\right)\left(\frac{545}{520}\right)\left(\frac{1}{0.86}\right) = 38.4 \text{ cfm}$$

From the displacements calculated above, standard cylinder sizes and compressor speed can be selected. Exact clearances for the machine selected are given by the manufacturer. It is particularly important on gas-engine driven compressors to select cylinder sizes so that the engine will maintain a relatively constant load. In this way fuel economy will be high. Mast[12] has presented an interesting description of the cylinder selection process.

*Actual gas*

*a.* The number of stages should not be altered by deviations from ideal gas.

*b* and *c.* As a first assumption assume equal compression ratios per stage. Enthalpy charts for various natural gases are available. One such chart for a natural gas 0.7 specific gravity is reproduced in Figure 15-21. It can be used for the gas in question since the properties are very similar. Referring to Figure 15-21 the compression in any stage is followed on a constant entropy line from inlet to discharge pressure. Intercooling is traced on a constant pressure line.

| | Btu/lb mole | bhp | Discharge temp ° F |
|---|---|---|---|
| First stage | 1750–450 = 1300 | 177* | 215 |
| Second stage | 1520–350 = 1170 | 160 | 219 |
| Third stage | 1110–(−60) = 1170 | 160 | 225 |

In order to obtain equal work per stage a trial and error solution involving the selection of a lower first stage discharge can be readily accomplished on the graph. For this illustration the values obtained will be assumed satisfactory.

*d.* Displacement (illustrated for last stage only).

$$E_v = 1 - C\left[\frac{z_2}{z_1}\left(\frac{P_2}{P_1}\right)^{1/k} - 1\right]$$

$$1 - 0.10\left[\frac{0.89}{0.85}(2.9)^{1/1.23} - 1\right] = 1 - 0.15 = 0.85$$

$$\text{Displacement} = \frac{(4.75)(3.79)(14.7)(545)(0.85)}{(840)(520)(0.85)} = 32.6 \text{ cfm}$$

In like manner: First stage – 318 cfm
Second stage – 104 cfm

The deviations from the ideal gas case can be very large in some instances. At high pressures, particularly, calculations should be made in which these deviations are considered.

*Alternate solutions* (assuming no Mollier diagram available) *for a, b, and c*

1. Approximate method

(Ideal horsepower)(Average compressibility factor)

Calculate third stage:
Estimated Discharge Temperature 205° F

Suction $T_R = \frac{T_1}{T_c} = \frac{545}{397} = 1.37$ $P_R = \frac{P_1}{P_c} = \frac{840}{671} = 1.25$

Discharge $T_R = \frac{T_2}{T_c} = \frac{665}{397} = 1.68$ $P_R = \frac{P_2}{P_c} = \frac{2400}{671} = 3.58$

Compressibility factors (Fig. 15-20)
Suction $z = 0.85$
Discharge $z = 0.89$
Average $z = 0.87$

$$\text{bhp} = (174)(0.87) = 152 \text{ each cylinder}\dagger$$

*c.* Total bhp = (152)(3) = 456 hp

2. Another solution involving generalized enthalpy entropy charts which can be employed when Mollier diagrams are not available is well illustrated by Hougen and Watson.[9]

* $\frac{(1300)(778)(4.75)}{(33{,}000)(0.82)} = 177$ hp.

† The difference between the results obtained using the enthalpy chart and these are due in part to inaccuracies in reading the charts.

### Refrigeration

The principles of compression already discussed apply equally as well in the refrigeration cycle. The isentropic work of compression is easy to calculate since the American Society of Refrigeration Engineers and the Bureau of Standards have compiled Mollier diagrams for most industrial refrigerants. The entire refrigeration cycle of compression, condensation, and expansion can be readily followed on such diagrams. In fact, the only difficult phase of refrigeration calculations is mastering the terminology of the industry. An illustration in point is the unit of capacity. A ton of refrigeration per 24 hours is an arbitrary definition which is equivalent to the removal of 200 Btu of heat per minute from the system.

## REFERENCES

1. Baumeister, T., *Fans,* McGraw-Hill Book Co., New York, 1935.

2. Church, A. H., *Centrifugal Pumps and Blowers,* John Wiley & Sons, Inc., New York, 1944.

3. *Compressed Air Handbook,* Compressed Air and Gas Institute, New York, 1947.

4. Dodge, B. F., *Chemical Engineering Thermodynamics,* McGraw-Hill Book Co., New York, 1944.

5. Edmister, W. C., *Ind. Eng. Chem., 32,* 373 (1940).

6. Elrod, H. G., Jr., *Ind. Eng. Chem., 37,* 789 (1945).

7. *Fan Engineering,* publication of the Buffalo Forge Co., Buffalo, N.Y., 1940.

8. Gill, T. T., *Air and Gas Compression,* John Wiley & Sons, Inc., New York, 1941.

9. Hougen, O. A., and K. M. Watson, *Chemical Process Principles,* Part II, John Wiley & Sons, Inc., New York, 1947.

10. Karassik, I. J., *Chem. Eng., 54,* No. 10, 110; No. 11, 132; No. 12, 127 (1947); *55,* No. 1, 118; No. 2, 134 (1948).

11. Kearton, W. J., *Turbo-Blowers and Compressors,* Pitman and Sons, Ltd., London, 1926.

12. Mast, B. T., *Petroleum Engr., 22,* No. 6, D-14 (1950).

13. O'Neil, F. W., *Compressed Air Data,* Compressed Air Magazine, New York, 1939.

13*a*. Ridgway, R. S., *The Effect of Supercompressibility of Natural Gas Upon Compressor Performance,* presented at California Natural Gas Association, Los Angeles, March 1, 1945.

14. York, R., Jr., *Ind. Eng. Chem., 34,* 535 (1942).

15. Young, V. W., *Basic Engineering Thermodynamics,* McGraw-Hill Book Co., New York, 1952.

# CHAPTER 16

# MOTORS AND TURBINES

Motors and steam turbines are the most common drivers for machinery in a process plant, but the choice between them is not a simple one. Power and steam costs and availability must be carefully considered for each particular situation. Although electric motors are inherently more efficient, steam turbines are more reliable. The operation of an electric motor depends on equipment such as switch gear and transformers which, though usually reliable and good for many years of service, can fail without warning. The steam turbine, on the other hand, requires only a constant supply of steam and such supplies in a well-designed plant are seldom interrupted without previous warning.

In some areas power costs may be so high that steam turbines should be used wherever possible. Certain services, such as lower-speed pumps, however, are not very efficiently handled by turbines. Since the turbine is primarily a high-speed machine (3,000 to 5,000 rpm), expensive gear reducers are required when used to drive low-speed apparatus. In such cases low-speed motors can be employed without the use of gear reducers.

Even when the power costs indicate that electric motors should be used in a plant, turbines may be advantageously used for some services. Because of their greater reliability, they are very often specified as drives for spare pumps and other similar machinery. Large noncondensing turbines can often be used for the dual purpose of driving a large pump or compressor and for supplying lower pressure process steam for heating purposes in the plant. High-pressure steam is supplied to the inlet of the turbine and the exhaust steam is used for heating purposes. Unlike exhaust steam from reciprocating machinery, turbine exhaust steam is free from oil and other contaminants.

Although the detailed design of electric motors and turbines is entrusted to specialists, the project engineer benefits from understanding some of the fundamental principles of operation and from knowing the types of motors and turbines that are manufactured.

## ELEMENTARY PRINCIPLES OF MOTOR OPERATION

When the poles of a magnet are made to face each other, as shown in Fig. 16-1, the lines of magnetic force in the air gap proceed in the direction of north to south and in a straight line. An electric conductor which is carrying a current also produces a magnetic field around the wire. The direction of this field is recalled by most students when they visualize grasping the conductor in the right hand with the thumb pointing in the direction of current flow. The other four fingers, then, indicate the direction of the magnetic field around the conductor. For illustrative purposes, the cross-section of a conductor having a dot in the middle indicates that the current is flowing toward the reader, while that having a cross indicates, as in Fig. 16-2, that it is flowing away from the reader.

If the conductor shown in Fig. 16-2 is placed in the gap between the two poles of the magnet of Fig. 16-1, the field shown in Fig. 16-3 results. The force lines above the conductor are implemented by those from the flux of the magnet while those lines below the conductor oppose the forces from the magnet. This situation produces a resultant force in the illustration of Fig. 16-3 in the downward direction. If the conductor is held stationary and the magnet is free to move, this resultant force will cause the magnet to move. Conversely, the magnet can be held stationary and the conductor will then move. Although oversimplified, these same basic principles explain the operation of most motors.

## MOTOR STANDARDS

Perhaps no other item of process equipment has been more carefully standardized than the electric motor. The National Electric Manufacturers' Association publishes, and regularly revises, complete standards for motors and generators. These standards are often referred to as NEMA standards.[9] They include standard ratings, dimensions, tests, and classification of all motor types. In addition, the National Electric Code (a standard of the National Board of Fire Underwriters, the National Fire Protection Association, and the American Standards Association) outlines installation methods for motors and gives complete instructions for their use in hazardous and semi-hazardous locations. This code has been conveniently interpreted and summarized in hand-

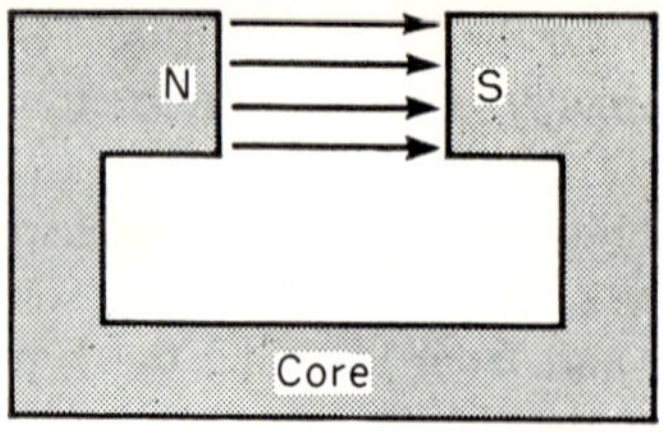


Fig. 16-1. Magnetic field around a permanent magnet.

Fig. 16-2. Direction of magnetic field in an electrical conductor.

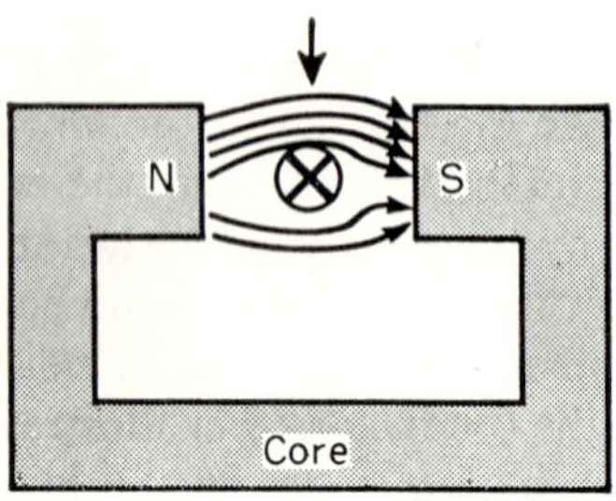


Fig. 16-3. Distortion of magnetic field by introduction of conductor.

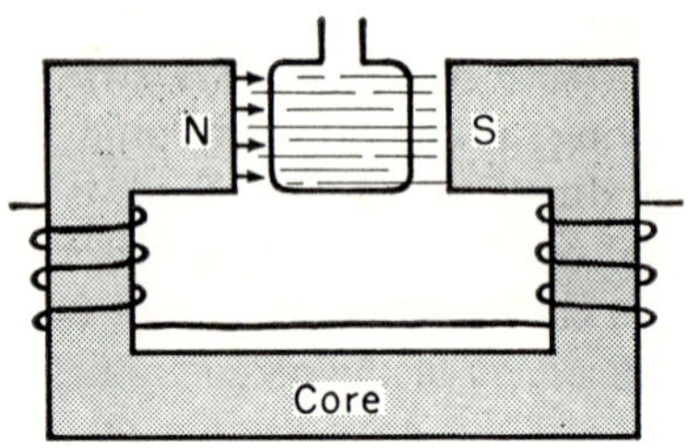


Fig. 16-4. Simplified direct-current motor.

book form.[1] The Underwriters' Laboratory, Inc., has issued standards for the construction of various parts and fittings for electric motors and appurtenances.

## MOTOR TYPES AND APPLICATIONS

### Direct Current Motors

A stronger magnetic field can be produced by replacing the magnet shown in Fig. 16-3 with an iron core around which is wrapped wire in the form of a coil. As current is passed through this wire, a strong magnetic field is produced in the air gap.

If the single conductor, shown in Fig. 16-3, is replaced by a loop, as illustrated in Fig. 16-4, the principles of a direct-current motor can be illustrated. The loop wire is called the armature (or at least serves to illustrate the function of the armature) and the helically-wrapped iron core is called the field coil. If a direct current is permitted to flow in both the field windings and the armature, magnetic fields will be set up in both. The south pole of the armature will be attracted to the north pole of the field coil causing the armature to rotate. At the instant the north and south poles are opposite each other a device called a "commutator" reverses the flow of current in the armature, causing it again to move. The process is repeated continuously. In a standard motor the armature is made up of a number of loops and may consist of two or more poles.

This discussion is admittedly an oversimplified description of the direct current motor. The reader is referred to more detailed accounts contained in standard works in electrical engineering.[4,11,12]

There are three basic direct-current motor types:

### Shunt-wound Motor

A shunt-wound, direct-current motor is one in which the field coils and armature are connected in parallel. With this arrangement it is possible to permit only a small portion of the current to pass through the field circuit. Shunt-wound motors operate at practically constant speed regardless of the load.

### Series-wound Motor

A series-wound motor has the field and armature circuits connected in series. As the load decreases, the armature current decreases and the motor increases speed. If such a motor is arranged for belt drive and the belt breaks suddenly producing no load, the motor will speed up to dangerously high rpm, causing complete destruction of the motor. Series motors have extremely high starting torques and are adaptable to such services as crane motors and drives for mobile equipment.

### The Compound-wound Motor

The compound-wound motor has two field windings. One is connected in parallel with the armature, and the other in series. As might be suspected, these motors combine features of both shunt- and series-wound motors. They have relatively high starting torques and will speed up as the load becomes lighter but not to excessively high speeds. They are particularly useful on apparatus producing varying loads and where constant speed is not required.

Direct-current motors cannot be started under full-current loads. It is necessary to employ a starter with a large resistance so that the current can be increased gradually until the motor reaches its normal speed.

Despite the larger initial cost of the direct-current motor, compared with an alternating-current motor, there are cases in which the greater ease of speed control makes the direct-current motor a desirable choice. The speed of a d-c motor is easily controlled, either by changing the voltage at the armature or by changing the strength of the magnetic field. Reducing the strength of the magnetic field increases the speed of the motor while reducing the voltage at the armature decreases the speed of the motor. Both of these operations are accomplished by introducing variable resistance in series with the field coils or the armature windings, respectively.

## Alternating-Current Motors

Alternating-current motors may be divided into two general types: induction and synchronous motors.

### *Induction Motors*

The induction motor is by far the most widely used motor in the process industries. Current flowing in the stator or nonrotating part of the induction motor produces a magnetic field which causes, by means described

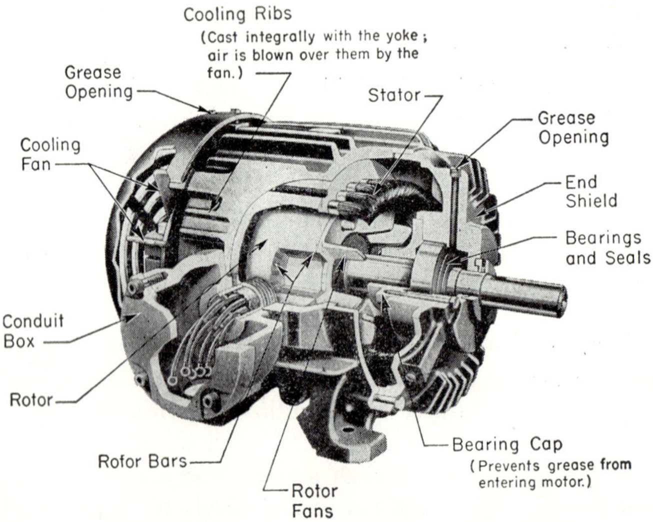


Fig. 16-5. Explosion-proof, squirrel-cage motor. (Courtesy of Allis-Chalmers Company.)

below, the rotating portion of the motor to move. The stator, or armature, is called the primary, since it is directly connected to the power supply and the rotor is termed the secondary.

The most popular induction motor has a rotor which resembles a revolving squirrel cage and has thus been termed the "squirrel-cage" motor. Such a rotor consists of a cylinder into which are embedded numerous copper bars (Fig. 16-5). These bars are all joined together by means of metal rings. Application of a polyphase current to the stator or armature of an induction motor produces a revolving magnetic field, due to the alternating current and the phase relationship of polyphase current. The revolving magnetic field cuts across the copper conductors on the spirrel cage and sets up voltages and magnetic fields in the rotor, the poles of which are attracted by the opposite poles of the revolving field.

The rotor of an induction motor must rotate at a lower speed than the revolving magnetic field. If it should rotate at the same speed, the revolving field would not cut across the field of the rotor and no voltage could then be induced in the rotor. The difference in the speed of the rotor and the speed of the revolving field is termed "slip" and is generally expressed as per cent slip.

The speed of the revolving field can be calculated when it is realized that the electromotive force passes through a complete cycle for each pair of poles per revolution. Hence, the speed of the revolving field is—

$$\text{rpm} = \frac{60f}{P/2} = \frac{120f}{P}$$

where $f$ = frequency in cycles per second
$P$ = number of poles

This speed is termed the synchronous speed of the motor. The actual rotor speed of an induction motor is this synchronous speed corrected for slip.

$$\text{Actual speed} = (\text{synchronous speed})\left(\frac{100 - \%\ \text{slip}}{100}\right)$$

The squirrel-cage induction motor is unsurpassed in sturdiness and economy through a wide range of horsepowers. By altering the design of the squirrel-cage rotor, principally by changing the location of the rotor bars with respect to the periphery, it is possible to obtain motors with varying characteristics, from low to high torque, and from low to high slip. In general an induction motor with normal torque and normal starting current is used. In starting large equipment, particularly equipment under load, such as compressors, it is often necessary to specify high torque, low starting current motors. In this design a higher resistive material is used in the rotor bars. It is important for the project engineer to realize that there are a number of induction motor types manufactured and the electrical engineer should be permitted to specify the type which has the best electrical characteristics for the job.

Wound-rotor induction motors are sometimes used instead of the squirrel-cage rotor. A wound-rotor motor is composed of insulated coils arranged to form poles on the rotor and the same in number as on the stator. This motor can be started with low current by adding external resistance to the rotor circuit. In this same manner the speed of the motor can be controlled down to about half its normal value. The wound-rotor motor, however, is more expensive than the squirrel-cage type; and is, therefore, only used on systems which would require unacceptably high starting currents.

The stator of a single-phase induction motor can only have one winding and thus the magnetic field does not move, although the polarity changes. If the rotor in such a motor is turned by hand and made to cross the magnetic field, rotation will then proceed automatically as it continues to cut across the magnetic field of the stator. Hand-starting is, of course, not desirable and automatic means have been designed. The most common means of starting the single-phase induction motor is employed in the split-phase motor which is equipped with an auxiliary winding located to differ in its magnetic position from the main winding. The main or primary winding and the auxiliary winding produce a torque on the rotor upon application of current, causing the rotor to move. As the rotor attains normal operating speed the auxiliary circuit is opened. Several different designs of split-phase motors are manufactured.[4,11,12]

It is interesting to note that in polyphase induction motors the rotation of the motor can be changed by interchanging the connections to any two of the terminals. A single-phase motor, on the other hand, will rotate in the direction in which it is started by the auxiliary starting device. To change the rotation of a single-phase motor requires a change in the starting device.

Single-phase motors are most frequently used in small horsepower and particularly fractional horsepower applications. Although polyphase fractional horsepower motors can be built more cheaply and are of simpler construction, it is often not convenient to bring polyphase current to locations requiring small motors. A small pump or other small machine may be needed in a remote portion of a plant. In such a case single-phase current used for lighting in that area is usually available and the motor may be connected to existing lines without the added expense of installing three-phase power lines.

### Synchronous Motors

As the name implies, the rotor of a synchronous motor rotates at the same speed as the stator winding field. The magnetic field in the rotor, therefore, must be induced by means other than crossing lines of flux created by the primary or stator. The rotor field current is generated by a small direct-current generator called an exciter, which is either separately driven or directly connected to the synchronous motor.

A synchronous motor is started as an induction motor. The excitation then begins, and the poles of the rotor move in step with the revolving armature field at which time synchronous speed is reached.

The speed of a synchronous motor depends upon the frequency of the power supply and thus provides dependable constant-speed service. This constant speed, however, is not its greatest advantage. The primary advantages are:

1. They can be constructed to operate at very low speeds (below 500 rpm) and high efficiencies and are, therefore, useful for operation of slow-moving machinery such as reciprocating compressors.

2. At low speed ratings and at high horsepower ratings the cost of synchronous motors is often less than induction motors.

3. A synchronous motor can be designed to operate with a leading power factor which is advantageous when the inductive load in a plant is large. The leading power factor of the synchronous motor tends to cancel the lagging power factor of the induction motor.

#### *Starting Alternating Current Motors*

Small single-phase motors are invariably started by applying full voltage across the line. All squirrel-cage induction motors can be started at full-line voltage. Standard or cross-the-line manual or magnetically operated starting equipment is manufactured for this service. Low speed synchronous motors are often started under full-line voltage, but higher speed synchronous motors are preferably started on reduced voltage to lower the current drawn from the power circuit. Resistance and auto-transformer reduced-voltage starters are used for such service and are also used for starting squirrel-cage induction motors when it is desirable to reduce the starting current in order to avoid affecting other loads on the power circuit.

## MOTOR DESIGN AND SELECTION

### Classification

The National Electrical Manufacturers Association[9] has classified motors* in accordance with mechanical protection and methods of cooling as follows:

### Classification of Motors According to Mechanical Protection and Methods of Cooling

Abbreviated list applicable to process plants. Adapted from standards of the National Electrical Manufacturers Association.[9]

I. Open machine

An open machine is one having ventilating openings which permit passage of external cooling air over and around the windings.

A. Drip-proof machine

A drip-proof machine is an open machine in which the ventilating openings are

* For brevity several less important categories have been omitted from the NEMA list.

so constructed that drops of liquid or solid particles falling on the machine at any angle not greater than 15 degrees from the vertical cannot enter the machine either directly or by striking and running along a horizontal or inwardly inclined surface.

B. Splash-proof machine

A splash-proof machine is an open machine in which the ventilating openings are so constructed that drops of liquid or solid particles falling on the machine or coming towards it in a straight line at any angle not greater than 100 degrees from the vertical cannot enter the machine either directly or by striking and running along a surface.

II. Totally enclosed machine

A totally enclosed machine is one so enclosed as to prevent exchange or air between the inside and the outside of the case but not sufficiently enclosed to be termed airtight.

A. Totally enclosed nonventilated machine

A totally enclosed nonventilated machine is a totally enclosed machine which is not equipped for cooling by means external to the enclosing parts.

B. Totally enclosed fan-cooled machine

A totally enclosed fan-cooled machine is a totally enclosed machine equipped for exterior cooling by means of a fan or fans integral with the machine but external to the enclosing parts.

C. Explosion-proof machine

An explosion-proof machine is a totally enclosed machine whose enclosure is designed and constructed to withstand an explosion of a specified gas or vapor which may occur within it and to prevent the ignition of the specified gas or vapor surrounding the machine by sparks, flashes, or explosions of the specified gas or vapor which may occur within the machine casing.

D. Dust-explosion-proof machine

A dust-explosion-proof machine is a totally enclosed machine whose enclosure is designed and constructed so as not to cause the ignition or explosion of an ambient atmosphere of the specific dust, and also not to cause the ignition of the dust on or around the machine.

E. Water-proof machine

A water-proof machine is a totally enclosed machine so constructed that it will exclude water applied in the form of a stream from a hose, except that leakage may occur around the shaft provided it is prevented from entering the oil reservoir and provision is made for automatically draining the machine. The means for automatic draining may be a check valve or a tapped hole at the lowest part of the frame which will serve for application of a drain pipe.

### Totally Enclosed Motors in the Process Plant

Kropf[7] has presented an interesting table (Table 16-1) showing the relative costs of enclosures for motors and controls. The differential between splash-proof and drip-proof motors is relatively small and of little consequence in small motors. The same can be said in comparing the totally enclosed and the explosion-proof motor.

In process plants having a preponderance of hazardous areas it has often been found advantageous to stock only explosion-proof motors. This procedure makes all motors exchangeable throughout the plant, and avoids unnecessary and disastrous errors. Certain plants, however,

**TABLE 16-1. Relative Costs of Motor and Control Enclosures***

| Enclosure | Approximate relative cost |
|---|---|
| For motors | |
| Open—drip-proof | 100% |
| Splash-proof | 109 |
| Tot. encl. fan-cooled | 145 |
| TEFC expl. proof | 165 |
| For controls | |
| Open (panel mtg.) | 100% |
| General purpose | 117 |
| Dust tight | 161 |
| Weather-proof | 203 |
| Expl.-proof (dusts) | 240 |
| Expl.-proof (vapors) | 316 |

* Reprinted by permission, Kropf, Victor J., *Chem. Eng.*, *58*, No. 7, 123 (1951).

having units located indoors with no hazardous atmosphere problem can use the drip-proof motors and save considerable money.

Kropf emphasizes the desirability of locating starter equipment remote from hazardous areas or in specially ventilated rooms to avoid the use of explosion-proof starters and controls. The reason for this can be readily seen in Table 16-1.

### *Hazardous Locations*

The National Electric Code[1] has defined various hazardous locations. Motors and control equipment are manufactured for the various classes of hazardous locations listed below.

#### Hazardous Locations

Abbreviated list applicable to process plants. [Adapted by permission, *National Fire Codes,* Vol. V (National Electrical Code), National Fire Protection Assoc., Boston (1953).]

Class I. Highly inflammable gases or vapors.

Division 1. Hazardous atmospheres exist continuously or can occur during course of normal operation.

Division 2. Flammable liquids handled and processed but normally confined.

Division 3. Locations where ignitable deposits of paint and other finishes can accumulate.

Class II. Combustible dust.

Division 1. Dust suspended in air continuously or during course of normal operation.

Division 2. Suspension of dust not likely but where deposits of dust can occur and be ignited by sparks.

Class III. Combustible fibres or flyings.

Division 1. Locations in which ignitable fibres are handled or used.

Division 2. Locations in which easily ignitible fibres are stored.

Equipment for these three classes of atmospheres can be tested and approved for the following atmospheric types:

Group *A*. Atmospheres containing acetylene.

Group *B*. Hydrogen or manufactured gas.

Group *C*. Ethyl ether vapor.

Group *D*. Gasoline, petroleum, naphtha, alcohols, lacquer-solvent, vapors, acetone, natural gas.

Group *E*. Metal dust.

Group *F*. Carbon black, coal or coke dust.

Group *G*. Grain dust.

It is possible from this list to decide the group into which almost any normally occurring hazardous material will fall.

## Mechanical Features

The highly competitive field of motor construction has produced continuous improvement in motor design. A typical modern motor is illustrated in Fig. 16-5. In addition to the electrical components already discussed, the shaft which carries the rotor is supported by bearings mounted in the end shields. These bearings can be oil-lubricated sleeve-type or grease-lubricated ball bearing. Ball bearings are particularly popular because of the small amount of maintenance required and the inherent cleanliness associated with such bearings. Electric motors are direct-connected to moving machinery, such as pumps and centrifugal compressors by means of a flexible coupling. Belt, gear, and chain drives are also available. In addition, there are gear motors which have integral gears built in the motor housing. These gear motors can be designed for variable speed.

## Motor Ratings

Horsepower, the most important rating of a motor, represents the motor's power output. Standard horsepowers have been specified by NEMA as 1/20, 1/12, 1/8, 1/6, 1/4, 1/3, 1/2, 3/4, 1, 1½, 2, 3, 5, 7½, 10, 15, 20, 25, 30, 40, 60, 75, 100, 125, 150, and 200. (In the fractional sizes 1/4, 1/2, and 3/4 are the most common.) In addition, the motor speed in rpm at full load and various percentages of full load, the voltage, full-load amperage, number of phases, and frequency of alternating current should also be stated.

The thermal characteristics of a motor are also important and appear on the nameplate of all motors as an allowable temperature rise above the ambient temperature for continuous operation. The ambient temperature for most general purpose motors should not exceed 40° C, and the usual allowable temperature rise is 40° C. Totally enclosed motors are rated for a temperature rise of 55° C and with special insulation these ratings go up to 75° C. New types of special purpose motors are insulated with silicone base materials and can be operated with higher temperature rises. Operation of motors for prolonged periods above the normal rated horsepower will produce temperature rises in the motor above the guaranteed allowable; and, thereby, shorten the life or completely destroy the usefulness of the motor.

Most motor manufacturers will be producing motors to conform to the revised NEMA Standards of 1952. Change to the new design should be complete late in 1955. These new designs will be smaller in size for a given horsepower rating.

**TABLE 16-2. Suggested Motor Voltage Ratings***

| | Integral Horsepower Motors | | Fractional Horsepower Motors | | |
|---|---|---|---|---|---|
| Nominal system voltage | Poly-phase | D. C. or 1 phase | Poly-phase | D. C. or 1 phase | Motor Horsepower |
| 120 | 110 | 115 | 110 | 115 | Up to 5 |
| 120/208 Y | 110/220 | 115 | 110/220 | 115 | Up to 7½ |
| 240 | 220 | 230 | 220 | 230 | ½ to 200 |
| 480 | 440 | ... | 440 | ... | ½ to 1,000 |
| 600 | 550 | ... | 550 | ... | ½ to 1,000 |
| 2,400 | 2,300 | ... | ... | ... | 50 and larger |
| 4,160 | 4,000 | ... | ... | ... | 150 and larger |
| 4,800 | 4,600 | ... | ... | ... | 250 and larger |
| 6,900 | 6,600 | ... | ... | ... | 500 and larger |
| 13,800 | 13,200 | ... | ... | ... | 1,000 and larger |

* Reprinted by permission, Kropf, Victor J., *Chem. Eng., 58,* No. 7, 123 (1951).

Note: These are permissible horsepower ratings for indicated voltages. Selection of motor voltage depends on motor costs and the particular distribution system. A typical selection for a plant having 120, 480, 2400, and 4180 feeders and 13,800 primary voltage might be:

| Motor Horsepower | Motor Voltage |
|---|---|
| Fractional | 110 single-phase |
| 1–150 | 440 |
| 150–1000 | 2300 or 4000 |
| 1000 and larger | 13,200 |

## Selecting Motors

Although the selection of a motor for a given service should be handled by a competent electrical engineer, it is necessary that the project engineer supply adequate information so that the electrical expert can reach intelligent conclusions. The operating characteristics of the machine to be driven must be clearly outlined. The horsepower, the speed of rotation (rpm) and speed variations, if required, all must be specified. In addition, the possibility of overload, the variability of loads, the magnitude

| Typical Applications | AC Motors: Normal Torque Squirrel Cage | High Torque Squirrel Cage | High Slip Squirrel Cage | Wound Rotor | Synchronous | Low Torque Single Phase | Medium Torque Single Phase | High Torque Single Phase | DC: Shunt Wound | Compound Wound | Series Wound |
|---|---|---|---|---|---|---|---|---|---|---|---|
| Agitators —½ to 15 hp. | ■ | | | ■ | | | ■ | | ■ | | |
| Attrition & ball mills—20 to 900 hp. | | ■ | | ■ | ■ | | | | ■ | ■ | |
| Banbury mixers—200 to 900 hp. | ■ | | | | ■ | | | | | | |
| Beaters—up to 200 hp. | | ■ | | ■ | ■ | | | | | | |
| Blowers—up to 500 hp. | ■ | | | ■ | ■ | | ■ | | ■ | | |
| Bucket elevators—5 to 25 hp. | | | ■ | ■ | | | | | | ■ | ■ |
| Chippers—up to 1,500 hp. | | | ■ | ■ | ■ | | | | | | |
| Compressors—up to 600 hp. | | ■ | | ■ | ■ | | | ■ | | ■ | |
| Conveyors—3 to 100 hp. | | ■ | ■ | ■ | | | | | ■ | ■ | |
| Cranes & hoists—3 to 150 hp. | | | ■ | ■ | | | | | ■ | ■ | ■ |
| Crushers—5 to 300 hp. | | ■ | | ■ | ■ | | | | | ■ | |
| Extractors—3 to 100 hp. | | | ■ | ■ | | | | | | ■ | |
| Fans—up to 150 hp. | ■ | | | | ■ | ■ | | | ■ | | |
| Grinders (pulp)—1,000 to 4,000 hp. | | | | | ■ | | | | | | |
| Grinders and granulators—¼ to 30 hp. | ■ | | | | | | ■ | | ■ | ■ | |
| Hammer mills—20 to 200 hp. | | ■ | ■ | ■ | ■ | | | | | ■ | |
| Jordans—up to 400 hp. | ■ | | | | ■ | | | | | | |
| Kilns—20 to 100 hp. | | | | ■ | | | | | ■ | | |
| Mixers—2 to 200 hp. | ■ | | | ■ | ■ | | ■ | | ■ | | |
| Pulverizers—10 to 250 hp. | | | | ■ | ■ | | | | | ■ | |
| Pumps (centrifugal)—up to 1,000 hp. | ■ | | | | ■ | | ■ | | ■ | | |
| Pumps (reciprocating)—up to 200 hp. | | ■ | | ■ | ■ | | | ■ | ■ | ■ | |
| Shredders—5 to 300 hp. | ■ | | | ■ | ■ | | | | ■ | | |
| Stokers—5 to 50 hp. | ■ | ■ | | ■ | | | ■ | | | | |

**Starter Types**

| | |
|---|---|
| Squirrel cage | Magnetic, full or reduced voltage. |
| Wound rotor | Manual, semi-magnetic or full magnetic with secondary resistance. |
| Synchronous | Magnetic full or reduced voltage. |
| Single phase | Manual or magnetic full voltage. |
| Direct current | Manual or magnetic with field and armature resistance. |

Fig. 16-6. Typical process plant motor applications. [Reprinted by permission, Victor J. Kropf, *Chem. Eng., 58,* No. 7, 123 (1951).]

of starting torques, and if intermittent service is anticipated, the frequency of starting should be given. The atmospheric conditions anticipated including the possibility of hazardous or corrosive gases should be defined.

The electrical engineer must also be provided with a complete description of the power supply, power costs, and a list of other power users in the plant so that an intelligent choice of control equipment and motors can be made for maximum economy and maximum power factor. The choice of the proper voltage to be applied to a given motor is an important one and a suggestion of desirable motor voltage ratings has been presented by Kropf[7] and is reproduced in Table 16-2. These suggestions are based on combined experience of a number of plants and can be termed values in general use.

A discussion of the proper motor for various applications in process plants is a lengthy and involved one, and generalizations are dangerous,

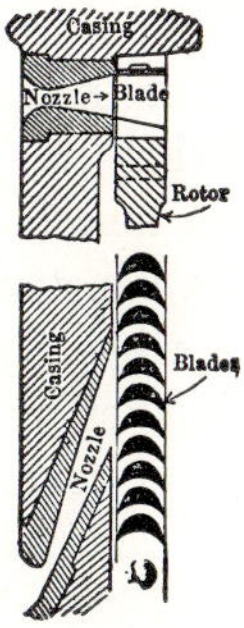


16-7

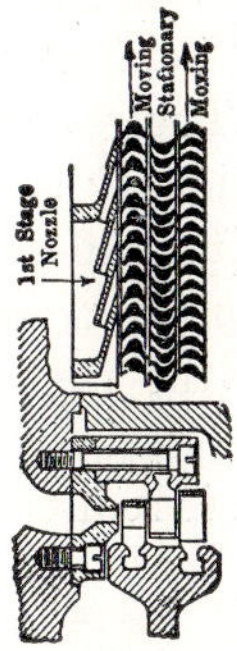


16-8

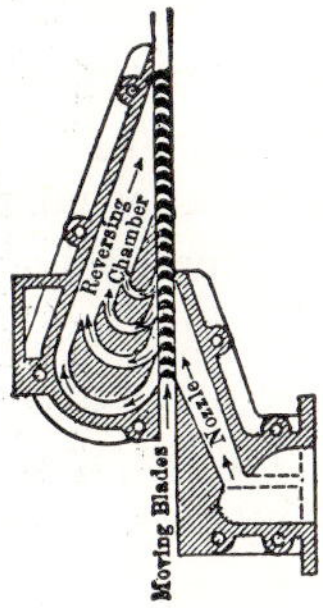


16-9

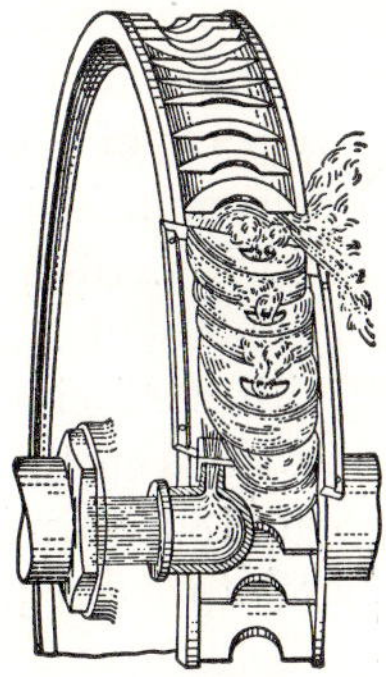

16-10

Fig. 16-7. Single-velocity stage impulse turbine. [Reprinted by permission, R. T. Kent, *Mechanical Engineers' Handbook,* Vol. II, 12th Ed., John Wiley & Sons (1950).]

Fig. 16-8. Two-wheel turbine showing stationary guide buckets. (Same reference as 16-7.)

Fig. 16-9. Re-entry nozzle. (Same reference as 16-7.)

Fig. 16-10. Helical flow stage. (Same reference as 16-7.)

if not impossible. Kropf,[7] however, has presented an interesting chart which is helpful in understanding some of the more common applications in process plants for the various motor types (Fig. 16-6).

## STEAM TURBINES

### Classification

Turbines may be broadly classified into two subdivisions based on the mode of transmission of heat energy into mechanical energy. The impulse turbine produces movement of a rotor by the force of steam impinging on the bucket or blade. The reaction turbine is caused to revolve by the reaction of steam issuing from jets attached to a revolving disk. There are many examples of each major classification. The following brief listing represents some of the more important types.

#### *Impulse Turbines*

(a) SINGLE-VELOCITY STAGE IMPULSE TURBINE. This is perhaps one of the most common types of turbines. It is shown diagrammatically in Fig. 16-7. Full expansion of the steam takes place in the nozzle, and the revolving buckets receive the steam without further pressure drop and transform the kinetic energy into work.

(*b*) VELOCITY-COMPOUNDED TURBINE. In order to use more efficiently the available thermal energy, velocity compounded turbines have been developed which redirect the steam issuing from the first row of buckets into a second and even third row of moving buckets. In this manner most of the velocity of the steam is dissipated. The re-directing of the steam can be accomplished in several ways. Stationary guide buckets may be used as in Fig. 16-8, or a re-entry nozzle (Fig. 16-9) or the helical-flow stage (Fig. 16-10). The latter design employs a specially constructed wheel with semi-circular buckets cut into the rim of the wheel. The steam expands in the nozzle and enters the semi-circular bucket, makes a complete 180° turn in the bucket, and then is redirected by the reversing chamber into subsequent buckets.

### *Reaction Turbines:*

Actually a practical true reaction turbine is not manufactured. In fact, the insistence on a rigid definition of impulse and reaction turbines is not advised since the last stages in multistage impulse turbines produce small amounts of reaction. What is often referred to as the reaction turbine should more properly be called the impulse-reaction turbine.

The "reaction" turbine is constructed of rows of moving blades attached to a wheel or rotor and an equal number of fixed blades attached to the casing. The fixed blades act as nozzles. In passing through the fixed blades the steam undergoes a pressure drop; and, therefore, an increase in kinetic energy or velocity. In entering the rows of moving blades an impulse is produced just as in the impulse type turbine. These moving blades, however, are designed so that a pressure drop will also occur as the steam passes through the blade passages, producing a reaction. The turbine is thus driven by a combination of reactive and impulsive forces.

Most turbines are of the axial-flow type, that is, the direction of steam flow is approximately parallel to the axis of the turbine. One notable exception is the Ljungstrom turbine, a reaction-type with radial flow. It is noted for high capacity and efficiency.

### *Single and Multistage Turbines**

Single-stage turbines are built in sizes up to 1500 hp in speeds ranging from 600 to around 7000 rpm. They are invariably designed to operate as noncondensing units and have become particularly popular for many

* The turbine stage as defined by Kent[6] consists of "that part of the machine in which a drop of pressure occurs with the generation of kinetic energy together with such succeeding blades and passages where no further drop in pressure occurs."

process plant applications. One of the advantages of a turbine is that it produces clean lower pressure steam which can be employed as process steam (heating, etc.). The well-designed plant, therefore, can employ a number of single-stage turbines as pump drives and at the same time provide the necessary quantities of low pressure process steam for use throughout the plant.

The more expensive multistage turbine is generally needed on large loads, for which case the steam economy is an important factor. These high-efficiency turbines are then often operated as condensing units for driving generators and large centrifugal compressors. In power stations many ingenious means have been devised for obtaining maximum steam economy for each unique situation. A straight condensing cycle can be employed, or a feed heating cycle in which steam is withdrawn from the turbines at various points for heating feed water.

#### *Extraction Turbines*

Modern boiler plants produce extremely high-pressure steam, and it is necessary to reduce the pressure of the steam from the boilers before piping it to parts of the plant for process use. Extraction turbines prove extremely useful for this purpose. Steam is withdrawn automatically from these turbines at intermediate stages for use as process steam. The remaining steam passes from the exhaust of the turbine to a condensing system. By this means it is possible to generate large quantities of power and at the same time provide the required amount of process steam at several pressure levels.

### Turbine Application

Noncondensing single-stage turbines make up the largest group of turbines for process use. They are generally designed to operate at steam pressures up to approximately 600 psi and at exhaust pressures from 0 to 100 psi. Special noncondensing, so-called "superposed" turbines are available for operation at steam pressures up to 2000 lb and exhaust pressures up to 600 lb. These turbines have proved particularly advantageous for power stations in which high-pressure steam is produced. The lower-pressure steam can then be used in the standard multi stage condensing units. Condensing turbines are designed for steam pressures up to 900, and in some cases 2000 psi with exhaust conditions at 28 in. of mercury vacuum. These units are most frequently used for large-scale industrial power generation and for driving large compressors and pumps.

### Mechanical Features

A typical multistage turbine is shown in Fig. 16-11. Casings are horizontally split with the steam inlet and outlet in the lower half for easy servicing. Low pressure casings are constructed of cast iron and the higher pressure units are built of steel. The rotating element is often called the wheel. The wheels are keyed into the shaft and must be carefully balanced to avoid excessive vibration. The blades on the wheels are usually made of the high-grade alloy steel, certain of the stainless steels being the most popular. Ultimate choice of blade material, of course, depends upon the temperature of the steam and blade design.

The shaft in most standard units is forged carbon steel which is often metallized with an alloy such as a stainless steel around the packing gland to minimize wear at this point. Diaphragms, which make up the stationary blading of multistage turbines, are generally constructed of a stainless steel. The diaphragms are horizontally split in the same manner as the casings and for the same reason.

The nozzle material varies, depending upon the conditions of operation. Cast brass and bronze, monel, stainless steel, all have been used.

Self-oiling bearings are provided on small turbines while forced lubrication designs are necessary for the larger units. These bearings are horizontally split and babbitt-lined. They may be removed from the unit without disturbing the casing or rotor. Although the end thrust on an impulse turbine is small, a thrust bearing is provided to maintain the rotor in its correct position. The thrust bearing is usually a ball-bearing type, although Kingsbury thrust bearings are also used.

Packing glands are necessary where the shaft leaves the casing, and these are most generally carbon ring glands. The rings are constructed in several segments so that they can easily be removed. In multistage turbines interstage glands are also required. These may be carbon-ring type or labyrinth type glands. Since condensing turbines operate at subatmospheric pressure it is necessary to seal the glands to prevent air leakage into the turbine casing. This is most conveniently accomplished by the use of steam. It is piped to the glands and a small amount continuously bleeds into the gland housing.

## Turbine Control

The speed of a turbine can be controlled by throttling the steam to the turbine or by controlling steam flow to the nozzles in a multistage turbine. For nozzle control the first-stage nozzles are divided into several groups each of which is served by a separate nozzle valve. The nozzle valves are designed to operate either open or shut (at full-steam pressure) and thus the throttling action is kept to a minimum. Nozzle valves are opened one by one as the load increases, and at full load all the valves are open. At fractional loads only some of the valves are open.

The advantage in nozzle control is that throttling losses are eliminated. Nozzle control shows economies on the large multistage machines in which prolonged partial loading is anticipated. Most small and medium sized turbines typical of process plant applications are designed for steam throttling. Often, however, hand-controlled valves are used to cut-out excess nozzling so that maximum fractional load economy will be realized.

The speed sensing device, or governor, may be the common centrifugal type which consists of two rotating weights direct-connected to the turbine shaft or connected through gears. The centrifugal force on the rotating weights is resisted by tension in connecting springs. The position of the weights is indicative of the speed; and the movement, therefore, can be transmitted to the turbine-regulating device. In place of the centrifugal governor, a so-called "hydraulic" speed governor can be employed. It consists of an oil pump driven off the turbine shaft. Changes in turbine speed produce changes in oil pressure which can be transmitted to the turbine controller.

The transmission of the governor signal on small turbines is by direct means through levers connected to the steam-throttling valve. On larger turbines the signal is transmitted through a small pilot valve. The pilot valve is supplied with oil from an oil pump which in turn operates a piston to open or close the steam-throttling valve or the various nozzle valves if automatic nozzle control is employed.

In process plants it often becomes desirable to control outlet pressure or differential pressure of pumps and compressors rather than use the speed of rotation as the primary control. In such cases a diaphragm-operated pressure regulator is used. It is set for the required pressure or differential pressure and automatically operates the steam throttling valve so that the turbine will rotate at the speed necessary to produce the desired pump or compressor discharge pressure. A standard centrifugal-type governor is usually included, however, so that it can control the turbine should the pressure regulator demand excessive speeds.

Emergency governors are installed on all turbines. Such governors consist of a weight eccentrically mounted, which is designed to swing out by centrifugal force at speeds of 10% above normal. In swinging out it trips a device which closes the main steam valve and stops the turbine.

### *Turbine Calculations*

In preliminary estimates and economic studies and for the comparison of bids it is often necessary to estimate turbine performance. Calculations are simple. Turbines operate in the reverse fashion to compressors and the same principles apply. A Rankine cycle is assumed for the ideal case and an overall efficiency is applied to correct for actual conditions. The steam at its inlet conditions is expanded isentropically (constant entropy) to the final condition. The difference in enthalpy represents the theoretical work done per lb of steam. Since Mollier diagrams for steam are readily available, this calculation is most conveniently and rapidly done on a Mollier diagram.

$$\text{Work} = H_2 - H_1 \text{ Btu/lb of steam} \tag{1}$$

where $H_2$ = inlet steam enthalpy
$H_1$ = outlet steam enthalpy

$$\text{or} \qquad \frac{H_2 - H_1}{2544}, \qquad \frac{\text{hp-hr}}{\text{lb of steam}} \tag{2}$$

The reciprocal of eq. 2 is called the theoretical steam rate or often the water rate. The units are lb of steam per hp-hr. The actual steam rate

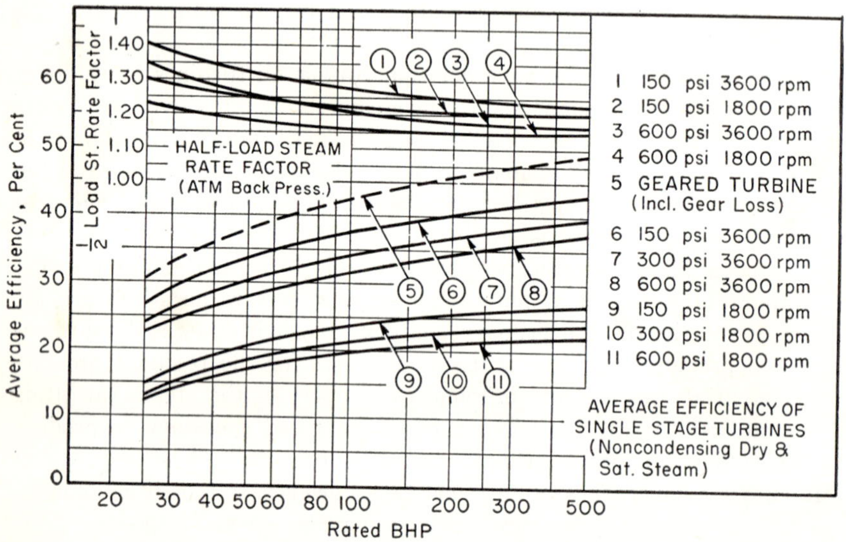


Fig. 16-12. Average efficiency of single-stage turbines. [Adapted from *DeLaval Engineering Handbook* (1947), DeLaval Steam Turbine Company, by permission.]

can be estimated by the application of an appropriate overall turbine efficiency.

$$\text{Actual steam rate} = \frac{\text{Theoretical steam rate}}{\text{Turbine efficiency}} \tag{3}$$

Values for estimating purposes are given in Figs. 16-12 and 16-13. These curves represent typical efficiencies at full load and present a correction factor for obtaining the efficiency at one-half load.

It has been observed that turbines controlled by throttling exhibit a linear relationship between steam consumption in lb per hr (not unit

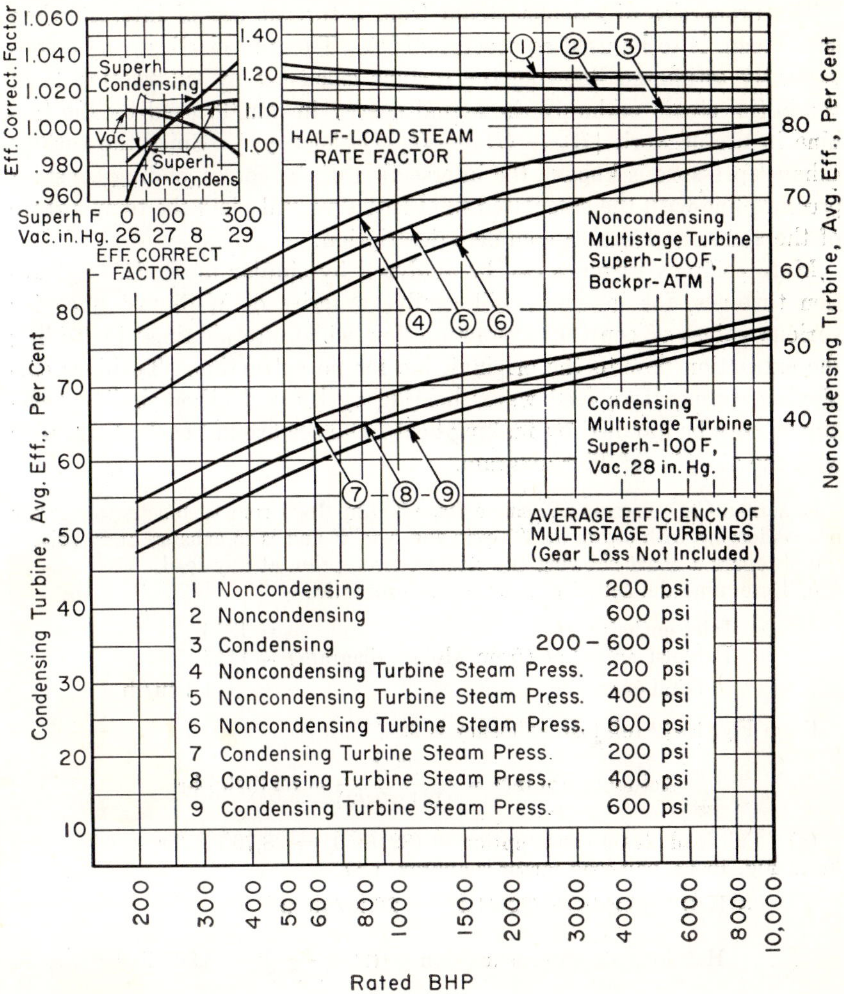


Fig. 16-13. Average efficiency of multistage turbines. [Adapted from *DeLaval Engineering Handbook* (1947), DeLaval Steam Turbine Company, by permission.]

steam rate in lb/hp-hr) and the load in convenient units. This straight line, which is called the Williams line, affords a simple means for estimating steam consumption for any loading of a given turbine when two values are known.

Nozzle-governing turbines do not produce the straight line relationship, but rather a series of Williams' lines of varying slopes interconnected by steps corresponding to each nozzle valve opening. A straight line relationship, however, will produce conservative values of steam flow for estimating purposes.

The properties of exhaust steam from a turbine are of interest when such steam is to be used as process steam. These properties can be conveniently estimated knowing the turbine efficiency by subtracting from the initial steam enthalpy the actual energy in Btu/lb used by the turbine in useful work $[H_2 - (H_2 - H_1)(\text{Efficiency})]$. Since the turbine exhaust pressure is known, the intersection of the corresponding constant pressure line with the calculated final enthalpy will yield the temperature of the steam and thus a complete description.

More complex turbines can be analyzed by similar methods. Extraction turbines, for instance, will exhibit a series of Williams' lines for various values of constant extraction flow which are obtained by adding the extraction flow to the original line for no extraction. Turbines employing regenerative feed water heating cycles and other similar techniques may be handled by making heat balances around each steam user with the aid of a Mollier diagram.

EXAMPLE CALCULATION: A single stage 350 hp 1800 rpm turbine is to operate noncondensing with steam at 385 psig and 520° F and is to exhaust at 60 psig.

*a.* Prepare a chart showing the steam consumption at any load.

*b.* Determine the actual exhaust steam properties—

| | |
|---|---|
| Inlet enthalpy $H_2$ | = 1256 |
| Exhaust enthalpy (from Mollier diagram) | = 1115 |
| Total energy | 141 Btu/lb |

From Fig. 16-12 full-load efficiency is 22%

$$\text{Actual steam rate} = \frac{2544}{(141)(0.22)} = 82 \text{ lb/hp-hr}$$

(*a*) Full load steam consumption = (82)(350) = 28,700 lb/hr
From Fig. 16-12, half-load factor is approx. 1.17

$$\therefore \text{Half load rate} = (82)(1.17) = 96 \text{ lb/hp-hr}$$

$$\text{Half load steam consumption} = (96)\left(\frac{350}{2}\right) = 16{,}800 \text{ lb/hr}$$

A performance curve or Williams' line is plotted for this unit from these two points. See Fig. 16-14.

(*b*) Estimated final enthalpy $= H_2 - (H_2 - H_1)(\text{Efficiency})$

$$1256 - (141)(0.22) = 1225 \text{ Btu/lb}$$

Intersection of 1225 enthalpy line on Mollier diagram with exhaust pressure of 60 psig corresponds to 390° F or approximately 80° F of superheat.

## GAS TURBINES

The gas turbine produces useful work by expanding hot gases from the combustion of fuel and compressed air (see Fig. 16-15). Part of the energy extracted by the turbine is used to compress the air for the combustion portion of the cycle.

The first common use of gas turbines in the process industries was as drives for air compressors supplying regeneration air to Houdry fixed-bed catalytic cracking units. Hot products of combustion from the catalyst regeneration cycle were expanded in the gas turbine. Any surplus power from the turbine was utilized by generating electric power with a small generator connected to the turbine shaft. The use of fixed-bed

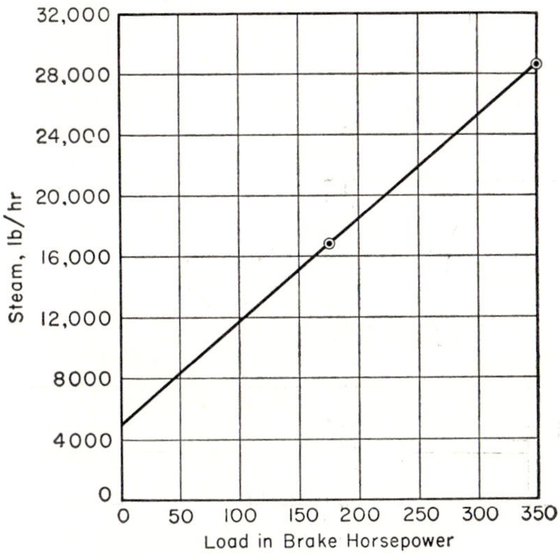


Fig. 16-14. Williams line plot.

units declined, however, and gas turbines gradually disappeared from process plants. As described by Miller[10] the Houdry process was unique in its adaptability to the gas turbine since the regeneration temperature and thus the gas temperature to the turbine had to be maintained below 1100° F by auxiliary cooling to prevent deactivation of the catalyst. Materials of construction for turbine temperature were not difficult to find. For the ordinary gas turbine cycle, however, combustion temperatures are high and the gases must be cooled somewhat before entering the turbine. As high temperature-resistant alloys became more common, not as much cooling was required and the design of gas turbines became more simple. Once again, therefore, gas turbines are being used in many process plants for certain applications in which they effect economies compared to motor or steam turbine drives.

A simple gas-turbine cycle is illustrated in Fig. 16-15. It consists of a multistage turbine expander, associated combustion chamber for the fuel gas, a centrifugal air compressor, and an electric generator. Compressed air is fed to the combustion system and the hot combustion gases produced are cooled by dilution with air before entering the turbine. The expanding gases just as in a steam turbine produce rotation of the turbine which in turn drives the compressor and generator. Separate speed reducers may be used on each shaft of the turbine so that the compressor and generator may each be operated at maximum efficiency.

The steam turbine has proved economical as a combination supplier of electric power and process steam and as a drive for centrifugal compressors. Process steam is removed at one portion of the cycle. In a

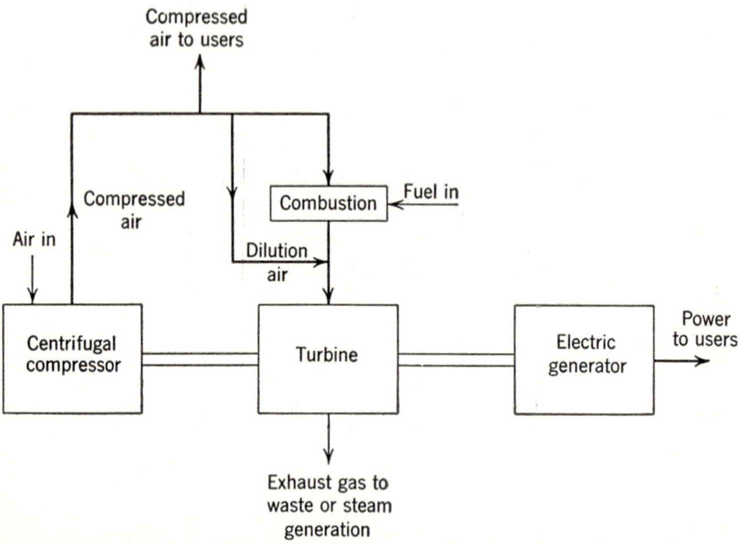


Fig. 16-15. Simple gas turbine cycle.

similar manner the gas turbine has operated economically as a combination supplier of electric power and process air and as a centrifugal compressor drive. Whenever cheap gaseous fuel is available such as natural gas, producer gas, or waste gases, such as products of combustion from a process, the gas turbine should be considered. Some typical applications are:

1. Compressing of natural gas
   *a*. Pipeline boaster compressors.
   *b*. Reinjection of natural gas into oil bearing formations.
2. Supplying large amounts of process air
   *a*. Ammonia synthesis; air for production of synthesis gas.
   *b*. Nitric acid manufacture; air for oxidation of ammonia.
   *c*. Tonnage oxygen and nitrogen production.
   *d*. Catalyst regeneration; air for burning coke from catalyst (attractive when regeneration is conducted under pressure).

3. Utilizing waste gases for production of power and compressed air

*a.* Manufacturing nitric acid; waste gas from the acid absorption column at 80 psig is heated to 900° F by exchange with the convertor reaction products and then expanded in a gas expander turbine. Air is compressed for the process and electric power is generated from the energy thus recovered.

*b.* Regenerating catalysts; products of combustion from the regeneration expand in a turbine. Compressed air and electric power are generated.

Thermodynamic calculations for gas turbines systems follow the same principles outlined for steam turbines and compressors. Theoretical power consumed by the air compressor can be based on isentropic compression. The gas turbine's theoretical power production is equal to the difference in enthalpy between inlet and outlet gases. Miller[10] has given the following typical efficiencies:

| | % |
|---|---|
| Compressor isentropic efficiency | 85 |
| Combustion efficiency | 98 |
| Turbine efficiency | 85 to 90 |

He further suggests as approximations 2 to 5% pressure drops in the mixing and combustion steps and a 1 to 3% of turbine enthalpy drop lost in operation of turbine auxiliaries.

Many ingenious cycles have been developed to increase thermal efficiencies. These include generation of steam from the turbine discharge gases and introduction of this steam into the gases entering the combustion system. Nevertheless, energy calculations become no more difficult since simple energy balances around each portion of the equipment are always possible. Operating tests, however, must be made to provide data on thermal losses and efficiencies.

## REFERENCES

1. Abbott, A. L., *National Electric Code Handbook,* 7th Ed., McGraw-Hill Book Co., New York, 1952.

2. Church, A. H., and H. Gartman, *DeLaval Engineering Handbook,* DeLaval Steam Turbine Co., Trenton, N.J., 1947.

3. Drinkle, L. R., F. Hodik, and E. T. Grost, *Alternating Current Motors,* American Technical Society, Chicago, 1948.

4. Gray, A., and G. A. Wallace, *Principles and Practice of Electrical Engineering,* 6th Ed., McGraw-Hill Book Co., New York, 1947.

5. Kearton, W. J., *Steam Turbine Theory and Practice,* 5th Ed., Pitman Publishing Co., New York, 1948.

6. Kent, W., *Kent's Mechanical Engineers' Handbook,* Vol. II, John Wiley & Sons, Inc., New York, 1950.

7. Kropf, V. J., *Chem. Eng.,* No. 7, 123 (1951).

8. Miller, B., *Chem. Eng., 62,* No. 1, 175; No. 2, 187 (1955).

9. *NEMA Standards for Motors and Generators,* National Electrical Manufacturers Association, New York, 1949.

10. Neman, L. E., A. Keller, J. M. Lyons, and L. B. Wales, *Modern Turbines,* John Wiley & Sons, Inc., New York, 1944.

11. Shoults, D. R., C. J. Rife, and T. C. Johnson, *Electric Motors in Industry,* John Wiley & Sons, Inc., New York, 1942.

12. Willis, C. H., and H. M. Chandler, Jr., *Introductory Electrical Engineering,* D. Van Nostrand Co., Inc., New York, 1952.

CHAPTER 17

# OTHER PROCESS EQUIPMENT

In the preceding chapters of this section, equipment common to all process plants has been discussed. In addition there is other important equipment which the project engineer must help select for certain processes.

To make intelligent selections the project engineer must have some familiarity with the theory or empirical method used in design and the kind of information required by the manufacturer before recommendations can be made. Various types of process equipment are discussed briefly in this chapter and references for more intensive study are given. Information required by the vendor is listed for each type of equipment considered in this chapter.

## AUTOCLAVES

The sizing of chemical reactors is the specific job of the process engineer. Chemical kinetic data together with bench scale and pilot plant data are used to determine the size, style, and shape of a given chemical reactor. The reactors are then built to order.

Certain processes, especially those requiring batch operation, such as some polymerization reactions, are best handled in autoclaves. The demand for autoclaves with attached agitators and heat transfer surface has influenced several manufacturers to develop a complete line of such equipment and to offer a wide range of standard sizes many of which are available from stock. In the interest of economy, therefore, it becomes necessary to adapt one's needs to these standard size autoclaves. The volumetric capacity of the autoclave must be set by the process engineer since the autoclave manufacturer is not familiar with the reaction, and it is often preferable to avoid revealing reaction date outside the company.

The autoclave manufacturer can and should be asked to develop the heat transfer surface, agitator, and other mechanical features. The high-pressure packing problems, thick-wall pressure vessel design, and fabrication methods with various alloys are all best handled by the manufacturer since his experience with these problems is superior. Autoclaves may be heated by hot water, steam, direct firing, hot oil, Dowtherm, and electric heating. Each of these systems, especially the direct firing, Dowtherm, and electric heating, requires considerable "know-how" which autoclave manufacturers have acquired through many years of experience.

It is necessary to provide the manufacturer with information needed for the selection and design of an agitator system, the heat transfer system, and the vessel itself. Manufacturers are aware of the secrecy necessary with some processes, and it will be satisfactory under these conditions to describe the reaction in terms of some similar familiar fluid. Physical properties, however, should be reasonably accurate.

As techniques of applied kinetics are improved, it may be possible to conduct many reactions continuously in "pipe-line" reactors* which have been traditionally carried out in autoclaves as batch reactions.

### Information Required by Vendor

1. Physical and chemical properties

*a.* Density of reaction mix at operating conditions.

*b.* Viscosity of reaction mix at operating conditions.

*c.* Toxicity and flammability of reaction mix.

*d.* Any other facts or description of material (or similar material) that it is possible to give.

*e.* If autoclave is to be used for several different reactions, give range of physical properties.

2. Operating conditions

*a.* Maximum pressure inside autoclave.

*b.* Maximum temperature (give time-temperature cycle or rough limits). Some idea of rate of temperature rise is necessary for careful pressure vessel design.

*c.* Type heating preferred.

*d.* Jacket working pressure (pressure and temperature of heating medium).

*e.* Is cooling required? Describe cooling medium.

*f.* Describe extent of agitation required. Suggest agitator style preferred.

*g.* Describe automatic temperature and pressure control system.

3. Materials of construction: Suggest materials based on experience with fluid to be handled

4. Miscellaneous

*a.* List size, number, and preferred locating of connections required.

*b.* Power characteristics.

* Orifices are installed in the pipe to produce the necessary mixing.

*c*. Is explosion-proof equipment required?
*d*. Auxiliary equipment required.
*e*. Location of equipment (describe).
*f*. Special attachments required, if any.

## REFERENCES

1. Hougen, O. A., and K. M. Watson, *Chemical Process Principles,* Volume III, John Wiley & Sons, New York, 1947. (Applied Kinetic Theory.)

2. Eldridge and Piret, *Chem. Eng. Prog. 46,* 290 (1950). (Stirred tank reactors in series (calculations).)

## CENTRIFUGES

There are two primary types of centrifuges. One may be called a centrifuge and the other a centrifugal filter. The centrifuge effects separation by causing a liquid or solid to move through a continuous liquid phase under the influence of centrifugal force. In the centrifugal filter, the solid phase is supported by a filtering medium (e.g., perforated screen) and centrifugal force causes the liquid to pass through the solid. The first is similar to gravitational settling and Stokes' Law can be applied while the second is similar to filtration and the resistance of the filter cake plays an important role.

For the settling type machine (centrifuge) it can be shown that $Q = 2v_g\Sigma$, where $Q$ is the rate of liquid flow, $v_g$ the velocity of particle sedimentation and $\Sigma$ is a constant for a given style and size machine. Ambler[1] has indicated that the performance of two centrifuges of different size operating on the same system will be similar when $v_{g_1} = v_{g_2}$. Hence $Q_1/\Sigma_1 = Q_2/\Sigma_2$, and this relationship provides a means for comparing small and large scale units. Correlations, however, have not been completely satisfactory, although steady progress is being made in this field.

The greatest success has been realized in the prediction of bowl centrifuge performance from laboratory data. Some encouraging results have been obtained with other styles, however, particularly when the small scale unit is operated at a speed such that it produces roughly the same centrifugal force as will be produced by the large unit.

Centrifugal filter theory is even less advanced than that for centrifuges and tests on large scale units are mandatory unless the material to be filtered is exactly the same as that for some successfully operating unit.

Even in cases for which small scale test results are not reliable for extrapolation, they are of value in indicating the nature of the problem and aid in planning subsequent test on larger equipment.

On the basis of the foregoing discussion it would seem that it is essential that the centrifugal manufacturer be supplied with a sample of the material to be processed. A sample size of about 5 gallons for preliminary small scale work is satisfactory. Later, larger samples may be required or it may prove desirable to conduct large scale tests at the plant site.

### Information Required by Vendor

1. Physical data
   *a.* Complete analysis of feed to centrifugal.
      1. Weight per cent solids and liquids.
      2. Chemical composition of each constituent.
      3. Possible variations in composition and relative amounts.
   *b.* Density and viscosity of liquid components at operating temperature.
   *c.* Density and viscosity of feed at operating temperature.
   *d.* Operating temperature (maximum possible often desirable because high temperature lowers viscosity).
   *e.* Density of solids.
   *f.* Particle size and size distribution of solids (typical screen analysis desirable if crystals are to be filtered).
   *g.* Fragility and shape of particles.
   *h.* Gravity settling rates, if available.
   *i.* Abrasiveness of solids.
2. Operating conditions
   *a.* Purpose of operation: clarification, separation, crystal dehydration and washing, etc.
   *b.* Disposition of components and quality of each phase (moisture content of crystals, purity of liquids, etc.).
   *c.* Capacity (quantity per hour and per day).
   *d.* If crystals are being separated and are to be washed, suggest wash liquid, need for separating from mother liquor, and purity of crystals desired.
   *e.* Description of the proposed operation of the centrifuge in relation to the entire process is helpful.
3. Materials of construction
   Centrifuges are constructed of certain standard materials as listed in manufacturer's catalogs. Possible reaction of the feed or components of the feed with any of these materials should be noted and alternate materials suggested.
4. Miscellaneous
   *a.* Floor space available.
   *b.* Power characteristics and type motor required (explosion-proof, etc.).
   *c.* Need for vapor-tight centrifuge construction if materials handled are toxic or flammable.

## REFERENCES

1. Ambler, Charles M., *Chem. Eng. Progr., 48,* 150 (1952). (Evaluation of centrifugal performance and description of one manufacturer's equipment.)

2. Ambler, Charles M., New Jersey Section Meeting, Am. Inst. Chem. Engrs., May 5, 1953. (Further evaluation techniques.)

3. Perry, J. H., *Chemical Engineer's Handbook,* 3rd Ed., 992–1013, McGraw-Hill Book Co., New York, 1950. (Equipment description.)

4. Smith, J. C., *Ind. Eng. Chem. 39,* 474 (1947). (Centrifuge selection techniques and description of equipment.)

## COOLING TOWERS

The performance of cooling towers is based on a simultaneous transfer of heat by diffusion and convection. Merkel expressed this fact mathematically in 1925 using the principles of mass and heat transfer. Many variations of his original expression have been presented. A typical equation for a counterflow tower may be written as follows:

$$Z = \frac{C_pL}{K_xa} \int \frac{dt}{H^* - H}$$

where $Z$ = height of tower in ft
$C_p$ = heat capacity of the water
$L$ = water rate lb/(hr)(sq ft of tower cross section)
$a$ = water surface per unit tower volume, sq ft/cu ft
$t$ = temperature of water, ° F
$H^*$ = enthalpy of saturated air at water temperature, Btu/lb of dry air
$H$ = enthalpy of main mass of air, Btu/lb of dry air
$K_x$ = an overall transfer coefficient having the units lb dry air/(hr) (sq ft of water surface)

Since it is ordinarily impossible to determine the value of $a$ separately, the combination $K_xa$ is determined experimentally by performance measurements on operating cooling towers. The term $K_xa$ contains all the empirical factors of design. Manufacturers have obtained reliable values of $K_xa$ from performance tests on the cooling towers they manufacture.

The integration indicated in the above equation can be done simply by either graphical or numerical means when a saturation curve of humid enthalpy versus temperature is available. Below this curve a straight line or operating line is plotted based on the inlet and outlet enthalpies $H_1$ and $H_2$ as shown in Fig. 17.1. Values of $H^* - H$ are then readily obtainable from this plot for the conditions of the problem and the integration can be accomplished.

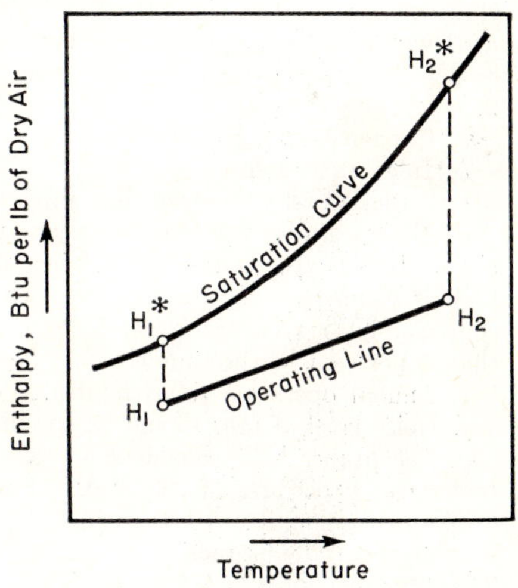

Fig. 17-1. Graphical integration of $dt/H^* - H$.

This technique permits determination of $K_xa$ for a given tower from performance data, estimation of tower performance under varying conditions, and calculation of tower heights for which $K_x a$ is assumed. The $H^* - H$ term is the driving force and the more humid the air, the closer the operating line moves to the saturation curve and thus the larger the tower must be for a given duty.

The use of this simple but valuable theoretical equation in design depends upon accurate prediction of design values of $K_x a$. Only cooling tower manufacturers and perhaps certain operating companies with large amounts of data on one manufacturer's tower are equipped to make such predictions. It, therefore, becomes necessary to supply the manufacturer with one's requirements and permit him to design a unit to satisfy the conditions of operation. Because of the magnitude of cooling tower installations, their construction should be regarded as a subcontract and detailed specifications should be written covering the grade of materials and workmanship together with operating conditions.

## Information Required by Vendor

1. Type of cooling tower

Induced draft because of the lesser tendency for recirculation of discharged air is becoming the most common. Forced draft, however, is more suitable for highly corrosive waters since the fan and moving parts do not contact the saturated vapors as in the case of the induced draft design. Other types are used for special applications. (See reference below.)

2. Operating conditions

*a.* Capacity, gpm of water.

*b.* Inlet water temperature.

*c.* Outlet water temperature desired.

*d.* Design ambient wet bulb temperature. The design wet bulb temperature should be selected on the basis of economics and should not necessarily be the highest value recorded for the area. One common basis is the so-called 5 per cent wet bulb. This is a wet bulb temperature which is not exceeded more than 5 per cent of the time elapsing between June and September.

*e.* Annual operation time (hours).

*f.* Noise level of fans should be specified.

*g.* Acceptance tests should be specified in accordance with the latest cooling tower test procedure of the ASME Power Test Code for atmospheric water cooling equipment.

3. Materials of construction

The type of lumber, filling, hardware, piping connections, basin, fans, and motors. These details can be developed in cooperation with the manufacturer but should be included in the specifications. The term redwood, for instance, is not completely descriptive. Redwood is graded in accordance with the specifications of the California Redwood Association and the proper grade* must be specified.

4. Miscellaneous

*a.* Space limitations.

* So-called "heart" redwood is superior for cooling tower construction.

*b.* Data on prevailing winds.

*c.* Sketch of space available for cooling tower indicating surrounding equipment sizes.

*d.* Location of nearby cooling towers.

*e.* Elevation above sea level.

## REFERENCES

1. Baker, D. R., and L. T. Hart, *Petroleum Refiner, 31,* No. 10, 97–103 (1953). (Theory.)

2. Degler, H. E., *Petroleum Engr., 22,* No. 3, p. C-29; No. 4, p. C-15; and No. 5, p. C-25 (1950). (Descriptive.)

3. Kern, D. W., *Process Heat Transfer,* pp. 563–609, McGraw-Hill Book Co., New York, 1950. (Theory, calculations, and equipment descriptions.)

4. Perry, J. H., *Chemical Engineer's Handbook,* 3rd Ed., 789–797, McGraw-Hill Book Co., New York, 1950. (Charts for estimating cooling tower size and fan power.)

## CRUSHERS AND GRINDERS

Classification of equipment has always been considered a primary step in the systematic study and understanding of any field. Often, however, classification becomes more of a problem than understanding. This is true in the case of crushing and grinding equipment. For many years grinding equipment has been classified in accordance with the type of product. But hammer mills, for instance, can be applied successfully to production of all types of product such as coarse, fine, and superfine. The project engineer, instead, needs a description of each type of grinder currently manufactured including information on its field of application and its advantages and disadvantages. Such information is far more helpful than some ponderous attempt to classify equipment.

Empiricism reigns supreme in the field of crushing and grinding. Theoretical equations for calculating power requirements as proposed by Rittinger and Kick have proved far less accurate than the more recent semi-empirical correlations of Bond and Wang.[2] Although effort is being made to develop basic grinding theory, the ultimate selection of a crusher or grinder for the grinding of a given material depends upon test work or upon plant operating data on the same or similar material.

Since no single company manufactures all types of crushing and grinding equipment, a manufacturer should not be expected to give an unbiased recommendation of the type equipment to employ for a certain problem. It is necessary, therefore, that the chemical engineer make a preliminary selection of at least several types of grinding machines before making detailed inquiry to the manufacturers. The equipment manufacturers should then be permitted to conduct tests on the material to be crushed so that the most favorable operating conditions can be determined for the particular type grinder. Smith[6] has prepared a most helpful report to aid the engineer in such preliminary selections.

The particle size of feed and product, capacity, and the size distribution and other physical properties of the product must be known before attempting preliminary selection. This information, together with more detailed data as outlined below, should be furnished the manufacturer to aid in the final selection of a unit. Samples of materials to be ground should be sent to the manufacturer. A five- to twenty-five-pound sample is required for most pilot plant units. If a larger scale test is to be made, at least 100 pounds is advisable.

### Information Required by Vendor

1. Physical data
   *a.* Description of material to be handled.
      1. Chemical properties.
      2. Similarity to other well-known materials.
      3. Size of pieces.
      4. Hygroscopic, corrosive, inflammable, toxic, etc.
   *b.* Bulk density of feed (indicate density determination method).
   *c.* Bulk density of finished product (indicate density determination method).
   *d.* Sieve analysis or particle size of desired product (give lower and upper limits).
   *e.* Per cent moisture content (if other liquids present in the material, name and give percentage).
   *f.* Solvents for material being ground. Suggest solvent most desirable for cleaning equipment.
   *g.* Abrasive properties.
2. Operating conditions
   *a.* Feed rate (minimum and maximum).
   *b.* Operating temperature (maximum and minimum; list temperatures at which material melts, decomposes, or undergoes some other unwanted change).
   *c.* Can material be ground wet?
   *d.* Is drying permissible before grinding?
   *e.* Is heating or cooling of material allowable?
   *f.* Type of classification desired after grinding.
3. Materials of construction

Suggested materials for corrosion resistance should be given. In supplying this information it should be remembered that these materials must also be resistant to abrasion and erosion.

4. Miscellaneous
   *a.* Power supply.
   *b.* Space limitations.
   *c.* Description of proposed location.
   *d.* Safety precautions now in effect if material is hazardous.

## REFERENCES

1. Bond, F. C., *Trans. AIME, 193,* 484 (1952). (Theory.)
2. Bond, F. C., and Jen-Tung Wang, *Trans. Am. Inst. Mining Met. Engrs., 187,* 871 (1950). (Theory and correlations for power estimation.)
3. Erickson, Henry W., *Chem. Eng. Progr., 49,* 63 (1953). (Description and selection.)

4. Foote, J. Howard, *Chem. Eng. Progr., 49,* 68 (1953). (Unusual grinding techniques described.)

5. Piret, Edgar J., *Chem. Eng. Progr.,* 49, 56 (1953). (Review of theoretical developments.)

6. Smith, Julian C., *Chem. Eng., 59,* No. 8, 151–166 (1952). (Description and suggestions to aid in preliminary grinder selection.)

## CLASSIFIERS

The most important application of classification equipment is in closed circuit grinding where the object is to separate finished product from the material which requires further grinding. Such applications are best developed in conjunction with the selection of a grinder. The grinder manufacturers are prepared to suggest equipment that is best adapted to their units. Texts in the field of ore and mineral dressing give detailed information on classifiers.

## CRYSTALLIZERS

A supersaturated solution must be formed in order to produce crystals from a given solution. This supersaturation can be produced by cooling, evaporation of the solvent, salting out (addition of a foreign substance which causes the material of interest to crystallize), or a combination of these methods. Knowledge of the solubility temperature relationships aid greatly in the selection of the general type of crystallizing equipment. A relatively flat solubility curve indicates evaporation while a rapidly rising (with temperature) curve indicates a straight cooling type of crystallizer.

Complete solubility data for most well-known compounds have been published, and experimental work is being done in order to better understand the mechanism of crystal formation. Such information has aided greatly in improving crystallizer design. Crystallizer operation, however, does not consist merely of the formation of crystal nuclei. Simultaneous formation and growth of crystals is common to industrial crystallizers, and these processes are further complicated by diffusional resistances. These several rate processes have also been studied but suitably accurate calculation of the rate is not yet possible.

This lack of reliable rate data makes it necessary, especially because of the many different crystallizer designs, to design and select crystallizers largely on the basis of laboratory tests and performance data. Such tests are particularly important when trace impurities are suspected since crystal formation and growth may be strongly affected by small amounts of certain impurities.

### Information Required by Vendors

1. Physical and chemical properties
   *a.* Chemical composition of feed liquor trace impurities.

*b*. Solubility temperature relations.

*c*. Heat of crystallization; specific heats of solution, mother liquor and crystals (enthalpy-concentration diagram desirable if available).

*d*. Density of feed and mother liquors.

*e*. Boiling points of feed and mother liquors.

*f*. Toxicity or flammability.

2. Operating conditions

*a*. Weight of feed liquor per unit time.

*b*. Weight of crystals to be formed.

*c*. Feed temperature and final desired liquor temperature and crystal yield.

*d*. Desired crystal size and size distribution. (This may be some truly optimum value or may depend on the wishes of the buyers of crystals, whether these wishes are logical or not.)

*e*. Operating hours per day.

*f*. Any special requirements for finished crystals should be stated.

3. Materials of construction

Suggest best possible materials to vendor.

4. Miscellaneous

*a*. Steam pressure and temperature (approximate cost).

*b*. Cooling water.

1. Source.
2. Temperature.
3. Approximate cost.

*c*. Other cooling media available.

*d*. Power supply.

## REFERENCES

1. Perry, J. H., *Chemical Engineer's Handbook,* 3rd Ed., McGraw-Hill Book Co., New York, 1950). (Theory and apparatus description.)

2. Thompson, A., *Chem. Eng., 57,* No. 10, 125 (1950). (Description and selection.)

## DRYERS

The theoretical aspects of drying have received particular attention in recent years, and the developments resulting have aided greatly in the understanding of drying fundamentals. The unique design features inherent in the many different types of dryers, however, make it mandatory that the selection and final design of a dryer be based on tests made with experimental dryers that simulate the operation of the plant-scale dryer. It is necessary, therefore, that samples of the material to be dried be sent to the manufacturer. The size of the sample required will depend upon the type dryer and the nature of the test. A large sample is usually preferred.

Drying theory has been most helpful in the correlation of test data. The transfer unit concept has been adapted to the design of several dryer types. Dryer tests are required in such cases for the determination of drying rates and optimum operating conditions. The manufacturer of a given type of dryer is the most logical party to make the dryer tests and recommend the design and size most suitable. The potential purchaser, however, must make some preliminary decisions on the type dryer desired.

Marshall and Friedman[2] have prepared an outstanding section in Perry's *Chemical Engineer's Handbook.* They give a tabular classification of commercial dryers. Their detailed discussions of dryer types, design, selection, and test procedures aid greatly in the preliminary dryer selection. An example of dryer selection is also given.

### Information Required by Vendor

1. Physical and chemical properties
   *a.* Description of material (chemical name, characteristics when wet and dry).
   *b.* Size, shape, and size distribution of particles.
   *c.* Density (bulk and true) of wet, commercially dry, and bone dry material.
   *d.* Toxicity, flammability, and corrosive nature.
   *e.* Moisture content of feed (give basis and method).
   *f.* Maximum moisture content of product (give basis and method).
   *g.* Shrinkage and decomposition tendencies of material, if known.
2. Operating conditions
   *a.* Feed rate.
   *b.* Continuous or batch operation.
   *c.* Hours operation per day.
   *d.* Treatment of material before and after drying. (Will some "before-treatment," such as forming wet material into pellets be permissible?)
   *e.* Estimate of probable drying time.
   *f.* Maximum drying temperatures permissible (when wet and dry).
   *g.* If hot air available, what is maximum temperature and humidity?
   *h.* Fresh air supply (temperature range, humidity, and cleanliness).
   *i.* Room temperature (range).
   *j.* Feeding mechanism (by manufacturer or purchaser, describe).
   *k.* Fuels available.
      1. Steam (pressure and temperature).
      2. Gas (pressure, gravity, heating value).
      3. Oil (gravity, viscosity, and heating value).
3. Materials of construction
   Specify where possible in cooperation with manufacturer.
4. Miscellaneous
   *a.* Space available (describe or provide sketch).
   *b.* Floor loading.
   *c.* Type of automatic controls desired and who furnishes.
   *d.* Power supply.
   *e.* Motors and starters (type and who furnishes).
   *f.* Who will furnish air filters and other accessories (specify).

## REFERENCES

1. Friedman, S. J., *Heating and Ventilating, 48,* No. 2, 96–110 (1951). (Dryer selection procedure.)

2. Marshall, W. R., Jr., and M. S. Friedman, in Perry's *Chemical Engineer's Handbook,* 3rd Ed., 799–875, McGraw-Hill Book Co., New York, 1950. (Dryer theory, design, testing and selection.)

## EVAPORATORS

In addition to the three so-called standard evaporator designs (natural-circulation, forced-circulation, and falling-film types) there are numerous special designs which are built to overcome certain unique problems.

The complexity of the heat transfer process in boiling and the variety of mechanical design features make the development of theoretical methods for estimating overall heat transfer coefficients for evaporation difficult if not impossible. Although methods for approximating these coefficients have been developed, previous experience and tests on pilot scale units represent the only reliable means for obtaining design coefficients. Here again the burden of design rests upon the manufacturer of the equipment.

The average chemical engineer is familiar with the heat and material balance calculations necessary for the design of a multiple effect evaporator. If he was not too fascinated by the clever algebraic manipulation possible in some of these problem solutions, he may have noticed that the overall coefficients for each effect were usually called "typical." But reliable not "typical" values of these coefficients are absolutely necessary and no amount of calculation can improve the evaporator design if inaccurate heat transfer coefficients are used.

If it is at all possible, the manufacturer should be supplied with enough sample to conduct pilot plant evaporation tests, particularly if the evaporation problem is new.

The task of the project engineer in evaporator design becomes one of evaluating the recommendations of various manufacturers on a specific evaporation problem. In any such evaluation, the basic problem is one of interpreting all quotations on a common basis. The following suggestions of items for comparison may be of help in such an analysis.

1. Utility requirements (steam, power, cooling, water).
2. Evaporator type and size of each effect and details of materials of construction.
3. Total heat transfer surface.
4. Pumps, capacity and type.
5. Accessories, size and type, materials of construction.
6. Code construction (ASME, TEMA, etc.).
7. List of all materials furnished by vendor including piping, instruments, structural steel.
8. Cost comparisons; place each offering on an equal basis by adding to or

subtracting from the total cost the estimated value of materials necessary to make the individual offering compatible with the arbitrarily chosen base design.

### Information Required by Vendor

1. Physical and chemical properties of thin and heavy liquors
   *a*. Description of material to be evaporated (chemical analysis including impurities present).
   *b*. Boiling points.
   *c*. Densities.
   *d*. Viscosities (at several temperatures).
   *e*. Per cent solids.
   *f*. Foaming characteristics.
   *g*. Flammability and toxicity.
   *h*. Nature of solids.
2. Operating conditions
   *a*. Quantity of thin liquor per unit time.
   *b*. Hours operation per day.
   *c*. Temperature of thin liquor.
   *d*. Highest temperature that liquor can be heated without causing unfavorable results.
   *e*. Quantity of solids to be separated per unit time or unit amount of feed.
3. Materials of construction

In the evaporation of many materials certain metals are not unduly corroded, but they contaminate the product sufficiently to render their use inadvisable. The manufacturer should be given such information particularly with regard to the behavior of iron or steel. Suggestions of suitable materials should be made when possible.

4. Miscellaneous
   *a*. High pressure and exhaust steam available at site.
      1. Pressure and temperature.
      2. Approximate cost.
      3. Restrictions on use if any.
      4. Ultimate use of evaporator condensate.
   *b*. Cooling water.
      1. Temperature (maximum and minimum).
      2. Source and description.
      3. Restrictions on use.
   *c*. Power supply.
   *d*. Space available (provide plot plan).
   *e*. Special equipment preferences such as types of pumps, drives, auxiliary equipment, and valves.

## REFERENCES

1. Coates, J., *Chem. Eng. Progr.*, *54*, 25 (1949). (Estimating required area, capacity and steam requirements of multiple effect evaporators.)

2. Kern, D. Q., *Process Heat Transfer*, McGraw-Hill Book Co., New York, 1950. (Calculations.)

3. Lindsey, E. E., *Chem. Eng.*, *60*, No. 4, 227 (1953). (Evaporator types, advantages and disadvantages.)

4. Storrow, J. A., *Ind. Chemist*, *24*, 311 (1948). (Calculations.)

## FILTERS

Filter manufacturers maintain well-equipped laboratories to study customer problems and to determine engineering data for design. Whenever possible samples of the material to be filtered should be sent to the manufacturer. With approximately 5 gallons of sample, bench-scale studies can be made with test filter leaves operating under vacuum. The filtration characteristics elucidated by these tests usually can be interpreted in terms of performance of the various filter types so that a size and style of filter can be recommended. Leaf tests in unusual filter problems sometimes show need for further pilot plant studies. In such cases larger samples are required for tests on actual pilot scale filters. If more convenient, filter manufacturers will make these small scale filters available to the customer at the plant site on a loan basis or for a nominal rental fee.

The majority of ordinary filtration problems, however, can be resolved by merely supplying the vendor with certain pertinent data as outlined below.

### Information Required by Vendor

1. Physical data
   *a.* Description of material to be filtered (chemical composition, of liquid and solid including *p*H).
   *b.* Per cent cake forming solids in slurry.
   *c.* Density of solids.
   *d.* Density of clear liquid at several temperatures.
   *e.* Filtration temperature.
   *f.* Viscosity of filtrate and slurry (approx.) at filtration temperature.
   *g.* Description of solids.
      1. Sandy, granular, crystalline, slimy, etc., percentage of each type.
      2. Screen analysis of dry solid.
2. Operating conditions
   *a.* Quantity to be filtered per unit time.
   *b.* Operating time per day.
   *c.* Which portion of slurry is to be recovered (filtrate, cake or both)?
   *d.* Cake characteristics if available.
      1. Is washing required?
      2. If washing is required, what is minimum purity permissible?
      3. Must wash liquid be isolated from filtrate?
      4. Give cracking tendency of cake during washing and drying.
   *e.* Describe most important goal of the filtration and list secondary goals. (Example: maximum filtrate clarity.)
3. Materials of construction
   Any experiences with metals and other materials in handling the slurry should be shared with the filter manufacturer.
4. Miscellaneous
   *a.* Normal barometric pressure.

*b*. Characteristics of electric power, wash water, and steam available at plant site.

Air filters require special consideration. Of particular importance is an analysis of particle size distribution of solids in the air or gas to be filtered. Such an analysis requires careful techniques since particle sizes are small. Air filters can be employed in removing dusts above approximately 0.3 microns while smaller sizes must be handled by electrostatic precipitators.

The purity of filtered air or gas should be specified. Tests, where possible, are strongly recommended because of the many variables which cannot be predicted.

## REFERENCES

1. Brownell, L. E., *Encyclopedia of Chemical Technology* (Kirk-Othmer editors), *6,* 506–30, Interscience Encyclopedia Inc., New York (1951). (Equipment description and theory.)

2. Dickey, G. D., and C. L. Bryden, *Theory and Practice of Filtration,* Reinhold, New York, 1946. (Equipment, application and theory.)

3. Grace, H. P., *Chem. Eng. Progr., 47,* 502–7 (1951). (Filter selection, advantages and disadvantages of various types.)

4. Miller, S. A., *Chem. Eng. Progr., 47,* 497–501 (1951). (Summary of theory from 1946–1951.)

5. Miller, S. A., *Chem. Inds., 66,* 38–48 (1950). (Filter descriptions and list of manufacturers.)

6. Perry, J. H., *Chemical Engineer's Handbook,* 3rd Ed., McGraw-Hill Book Co., New York, 1950. (Summary of theoretical and empirical equations, equipment and design.)

## HEATERS AND FURNACES

Design of a heater or cracking furnace requires thorough knowledge of heat transfer, fluid flow, and properties of metals and refractories at high pressures and temperatures. Some years ago the majority of the direct fired heaters were used to supply heat for crude distillation units, and the design was empirical. As the use of these heaters expanded and applications ranging from distillation tower reboilers to reactors became widespread, a stronger theoretical basis for design was developed, especially in the calculation of both the radiant and convection section heat absorption. Furnace design has, therefore, become a specialty. Furnace manufacturers maintain a staff that specializes in furnace design and construction. Some small furnaces can be built at the factory and shipped complete to the plant site, while the larger designs must be erected on the job.

The engineer responsible for selecting a furnace profits from knowledge of furnace design. Perhaps the most helpful paper on this subject is that by Rickerman.[3] He presents a good description of heater types and outlines in simple terms the steps involved in complete furnace design. A useful bibliography is given.

In the development of cracking furnaces or any other furnace in which a chemical reaction occurs, the process engineer should determine the size and length of tube required to produce the required reaction at a specified heat flux.

Fair and Rase[2] have demonstrated the method for such calculation based on the principles of chemical kinetics. The furnace designer completes the work by providing the mechanical aspects of the design based on the required heat flux and temperatures. This work includes selection and location of tubes and burners, selection of materials of construction and structural design.

The amount of information required by the manufacturer varies somewhat depending on the customer. Some plants prefer certain burner designs and header arrangements, for instance. Such preferences should, of course, be clearly stated during the early inquiry stage to avoid costly delays as the project proceeds.

Oils for which generalized correlations are available need not be described as fully as materials for which little published data has appeared.

## Information Required by Vendor

1. Physical and chemical properties of charge and product
   - *a.* Analysis of material charged and product (distillation analysis).
   - *b.* Density or API gravity at 60° F. (Also density at other temperatures, if available.)
   - *c.* Viscosity (at several temperatures).
   - *d.* Molecular weight of vapors at inlet and outlet.
   - *e.* Coking characteristics of charge.
2. Operating conditions
   - *a.* Charge rate.
   - *b.* Inlet temperature and pressure.
   - *c.* Outlet temperature and pressure.
   - *d.* Per cent vaporized in inlet and outlet.
   - *e.* State maximum allowable pressure drop.
   - *f.* Maximum allowable radiant heat transfer rate.
   - *g.* Furnace duty (Btu/hr absorbed).
   - *h.* Minimum velocity of flow (if important).
   - *i.* Fuel; complete description (analysis of gas desirable).
     1. Heat content (lower heating value).
     2. Pressure at furnace.
     3. Viscosity (oil).
     4. Density.
     5. Temperature.
   - *j.* Steam pressure at furnace.
   - *k.* Furnace efficiency desired (give excess air on which based).

$$\text{Efficiency} = \frac{\text{Duty}}{\text{Heat liberated by fuel}} \times 100$$

3. Miscellaneous
   - *a.* Space requirements.

*b*. Special mechanical design features desired.
*c*. Corrosion data.

## REFERENCES

1. Kern, D. Q., *Process Heat Transfer,* McGraw-Hill Book Co., New York, 1950.
2. Fair, J. R., and H. F. Rase, *Chem. Eng. Progr., 50,* 415 (1954).
3. Rickerman, J. H., *Trans. Am. Soc. Mech. Engrs., 67,* 531 (1954).

See also bibliography at end of this article.

## MATERIALS HANDLING

The scarcity and high cost of labor during World War II created a revolution in the materials handling field. Hundreds of new and ingenious devices were introduced, and the proper selection and application of materials handling equipment has become more complex. One university is now offering a degree in materials handling engineering. It is advisable, therefore, that the project engineer seek the assistance of company representatives trained in the field of materials handling. A general knowledge of the equipment, however, is necessary in order to expedite the discussion with manufacturer's representatives. Some knowledge can be gained by studying several of the references listed below.

A detailed study should be made of conveying systems to be located in warehouses and shipping departments for finished goods. The layout and flow sheet for such systems should be studied with the manufacturer's engineer so that the most efficient arrangement can be devised.

Manufacturer's literature on conveyors usually contains suggestions for the selection of standard conveyor types. Special designs must be selected by the manufacturer. It would seem desirable that the final selection of even a standard design be proposed by the manufacturer since he knows his equipment best. The printed methods for selection can easily be misinterpreted by the occasional user.

What has been said about conveyors holds equally well for other materials handling equipment. In all cases the purchaser should supply complete information about the nature of the material to be handled, its weight, the exact purpose of the handling, and the proposed layout showing free space available.

The information required by a vendor of conveyors will serve as an illustration.

### Information Required by Vendors of Conveyors

1. Physical and chemical properties
   *a*. Size and size distribution (maximum lump sizes in inches).
   *b*. Moisture content.
   *c*. Temperature.
   *d*. Density of material (bulk and real density).

*e*. Abrasiveness, chemical reactivity, stickiness, etc.
*f*. Ability of material to withstand conveyor action.
*g*. Corrosive nature.

2. Operating conditions
*a*. Temperature and humidity in conveyor area.
*b*. Maximum rate material is to be handled.
*c*. Length and location of conveyor including sketch showing proposed path of travel. Indicate adjacent equipment.

## REFERENCES

1. *Booklet No. 2 of Library of Knowhow,* Materials Handling Institute, Pittsburgh, Pa. (1953). (Methods for studying and evaluating materials handling problems.)
2. Mallick, R. W., and A. T. Goudreau, *Plant Layout Planning and Practice,* John Wiley & Sons, Inc., New York, 1950. (Includes section on materials handling.)
3. *Modern Materials Handling, 8,* No. 6, 58 (1953). Papers presented on conveyor application at Conveyor Institute, Univ. of Ill., April, 1953.
4. *Mill and Factory, 52,* No. 1, 82 (1953). (Selection of monorail systems.)
5. *Pallet Pattern Selection,* Materials Handling Labs., Inc., 795 Boylston Street, Boston 16, Mass.
6. Stocker, Harry, *Materials Handling,* Prentice-Hall, Inc., New York, 1951.
7. Strube, H. L., *Chem. Eng., 61,* No. 4, 195 (1954). (Report on conveyors and elevators, design information, application, and selection.)

## MIXING EQUIPMENT

The proper selection of the size and style mixer most suitable for a given service requires laboratory study. A complete description of the materials to be mixed must be given to the manufacturer so that synthetic mixes can be prepared. Better still an actual sample may be provided. The past five years have seen rapid development of mixing theory. Following the principles of dimensional similitude, small scale mixers can be tested under conditions dynamically similar to those required for large scale operation. Scale-up of power requirements, and impeller and tank size can be made with confidence.

The procedure for mixers involving liquids is typical. Preliminary tests are made in a small glass beaker (approximately one quart) using a scale model of a mixer. Optimum power requirements and impeller types and position are then selected. For simple mixing problems, scale-up of this information to the final plant size is possible. With more complex problems intermediate scale-up to pilot plant equipment followed by additional tests in large tanks yield more reliable data on impeller diameter and power requirements.

### Information Required by Vendor

1. Physical data
*a*. Complete description and weight per cent of each component in mix.
*b*. Viscosity and density of each liquid component.

*c.* Viscosity and density of mixture.

*d.* Solid component description.

1. Particle size and distribution (sieve analysis).
2. Density of dry solids and settled slurry.
3. Settling velocity.
4. Nature of solids: soluble, insoluble, crystalline, light in weight, etc.

*e.* Foaming tendency (describe).

2. Operating conditions

*a.* Purpose of mixing: dissolve solids, blend liquids, disperse gas in liquids, suspend solids, etc.

*b.* Type of agitation required (violent, intermediate, or mild).

*c.* Type operation.

1. Batch (give typical maximum, minimum, and normal batch sizes and time available for mixing each batch).
2. Continuous (give flow rate).

*d.* Vessel description. Supply a complete vessel drawing showing all dimensions, type heads, and operating and design conditions.

3. Materials of construction

Suggest materials most suitable for those parts of the agitator that will contact the mix. The more possibilities given, the greater freedom the manufacturer will have in selecting the most economical and mechanically sound material.

4. Miscellaneous

*a.* Electrical power characteristics.

*b.* Type motor required (splash proof, totally enclosed, explosion proof, etc.).

*c.* Space requirements if unusually limited.

Certain special information is required for solid and paste mixers such as kneaders and millers. This information includes the abrasive and shearing properties of the materials, a sieve analysis of the solid, quantity of liquid if any to be added, and the amount of heating or cooling required. A description of the purpose of the mixing, such as the dispersion of pigments, is also necessary since this governs the type mixer and blades.

## REFERENCES

1. Boutros, R. D., *Chem. Eng. Progr., 48,* 211 (1952). (Mixer installation and maintenance.)

2. Perry, J. H., *Chemical Engineer's Handbook,* 3rd Ed., 1195–1231, McGraw-Hill Book Co., New York, 1950. (Mixer types described, tables giving preferred mixer types for various mixing jobs.)

3. Rushton, J. H., *Chem. Eng. Progr., 47,* 485 (1951). (Pilot plant mixing data and scale-up procedure.)

4. Rushton, J. H., R. D. Boutros, and C. W. Selheimer, in *Encyclopedia of Chemical Technology* (Kirk and Othmer—editors), Vol. 9, 133–166, Interscience Encyclopedia Ind., New York, 1952. (Equipment description and application includes solids and pastes.)

5. Rushton, J. H. and J. Y. Oldshue, *Chem. Eng. Progr., 49,* 161 and 267 (1953). (Review and theory and applications.)

## SEDIMENTATION EQUIPMENT

A simple laboratory settling test in a graduated cylinder will often indicate that the separation of large quantities of liquid from a slurry can best be accomplished by a combination of settling and then filtration. The test work needed for the proper selection of sedimentation equipment, however, must go far beyond such laboratory settling tests. Settling problems though seemingly simple can become very complex because of the many factors involved. These include the marked effect of equipment design on settling and the many physical and chemical factors which affect settling characteristics of a given slurry.

The observations of Coe and Clevenger in 1916[2] are valid today. They observed several settling zones which, for a continuous thickener, may be described from top to bottom as a clear zone, a zone of feed consistency, an intermediate zone, and finally a zone of compression. The settling rate of the material representing each of these zones can be determined experimentally and the area required for each zone can be calculated.

The following expression is used to estimate required areas for continuous thickeners. This equation is simply a material balance.

$$AR = \frac{S(F - D)}{\rho}$$

where $A$ = area required, sq ft
$R$ = settling rate ft/hr
$\rho$ = density of liquid, lb/cu ft
$F$ = weight ratio of liquid to solid in feed
$D$ = weight ratio of liquid to solid in underflow discharge
$S$ = weight of solids treated per unit time, lb/hr

Coe and Clevenger[2] also described a method for estimating the volume and thus the height of a thickener from settling tests made in a graduated cylinder. The volume can be estimated from the retention time necessary to produce the desired underflow consistency. Similar and more reliable tests on pilot plant scale equipment have been devised, but the principles are essentially the same.

Although these techniques are simple, the settling rates and required retention time must be determined experimentally. Such experimentation involves small scale laboratory tests followed by pilot plant operation on equipment of the type selected for the particular job.

It is highly desirable to conduct these tests using an actual sample of the feed rather than a synthetic one. Small quantities of a material that will alter the properties of the suspended solids can have marked effect on the settling characteristics (e.g., wetting agents). Such materials might be overlooked in usual analytical procedures. Sedimentation equipment manufacturers maintain extensive laboratories for testing customer's samples with their equipment. In addition they have extensive files of data on operating units.

### Information Required by Vendor

1. Physical and chemical data

*a.* Complete description of feed and analysis, including *p*H.

*b.* Settling rates if available.

*c.* Toxicity, flammability, and any other hazardous property.

*d.* Quality and consistency of products desired including weight ratio of liquid to solid in underflow and maximum permissible solids in clear liquid.

*e.* Is clear liquid to be used?

*f.* Is underflow to be used?

2. Operating conditions

*a.* Feed rate.

*b.* Operating temperature.

*c.* Should heat be conserved?

*d.* Outdoor or indoor operation.

*e.* Maximum variations in feed conditions and analysis.

3. Materials of construction

Suggest materials best suited for handling corrosive substances. On large equipment particular consideration should be given to protective coatings. In some cases a small amount of corrosion may be tolerable especially if the corrosive products form a scale that prevents further corrosion.

4. Miscellaneous

*a.* Power supply.

*b.* Description of location.

*c.* Bearing value of soil.

*d.* If inside, give allowable loading of floor.

*e.* Description and contour of outdoor location.

*f.* Accessibility to railroad spur and road.

*g.* Utilities available at site prior to construction.

## REFERENCES

1. Brown, G. G. et al, *Unit Operations,* John Wiley & Sons, Inc., New York, 1950.

2. Coe, H. S., and G. H. Clevenger, *Trans. Am. Inst. Mining Met. Engrs., 55,* 356 (1916).

3. Perry, John H., *Chemical Engineer's Handbook,* 3rd Ed., McGraw-Hill Book Co., New York, 1950.

CHAPTER 18

# PIPING DESIGN

Any book on project engineering of process plants would be incomplete without a chapter on piping. At the same time it must be understood that complete information on any single phase of piping would in itself fill an entire book. The information included therefore is general. It is hoped that it will create interest, and serve as a starting point for additional reading in the listed references. An attempt has been made to cover briefly only those practices, materials, and accessories most used in piping for process plants and the several utility services generally found in all plants.

It is believed that no other single industrial product has had so much to do with the progress of the human race as does pipe. Conduits of one kind or another for transporting water were well known and used in ancient times, perhaps centuries before the beginnings of recorded history.

The reciprocating steam engine was developed to provide a means of removing seepage water from the mines of Cornwall, England. Aside from the development of the engine itself, there is no doubt that much ingenuity was required to provide some kind of pipe which would withstand the hydraulic heads. Pumps or other devices for the movement of water had been in use for many hundreds of years previous to the development of the steam pump. In 1581, Peter Morrys, a Dutchman, installed a water-driven pump for the London Bridge Water Works, to increase the London water supply. This device was built almost entirely of wood. Paddle wheels located in the water of the Thames River provided power. Rotation of the paddle wheel shaft was converted to reciprocating motion by wooden toothed gears and cranks, connected to a series of pistons, each acting independently as a pump.

Much of the piping in these early times was bored from logs or constructed of wooden staves similar to barrel staves. This pipe gave very good results. Various types of terra-cotta or other clay products were also used, chiefly for drainage as is the practice today. Cast iron pipe was in use in the seventeenth century. Some of the original cast iron pipe laid in those early years is still in use today. Lead pipe was also extensively used in ancient times. It was perhaps the first pipe made from metal.

At the turn of the twentieth century steam power was well along in its development. The earliest pressure piping practices were developing

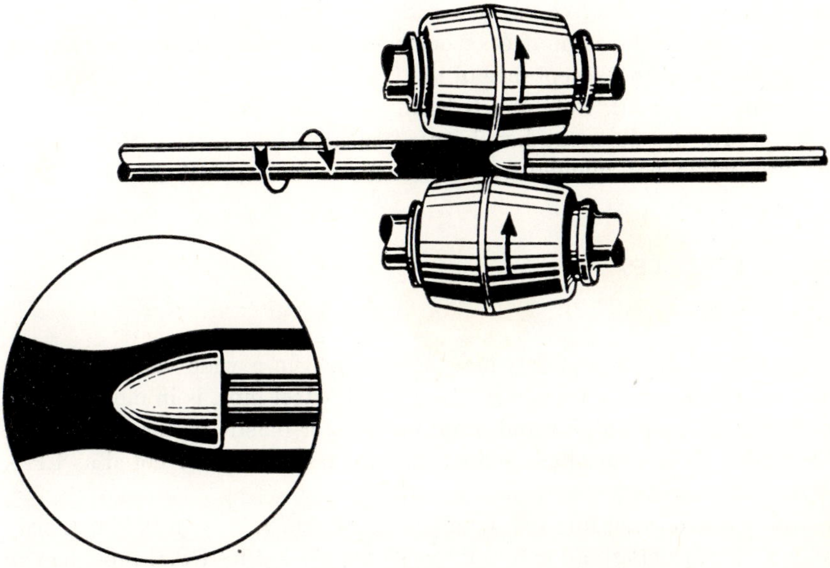

Fig. 18-1. Piercing cylindrical billet in manufacture of seamless pipe. (Seamless Steel Institute.)

to meet the needs of steam power and various hydraulic operations. Improvements in pipe manufacture continued, but each manufacturer followed the requirements of his particular customers.

In contrast to the early methods of pipe manufacture, pipe today is made by a number of different processes and of such widely varying materials as carbon steel, alloy steels, brass, glass, plastics, and rubber. The several steels and alloys steels are the most common piping materials. Their manufacture into pipe is accomplished either by forming flat plate to a cylindrical shape and welding the seam using one of the several welding processes (butt, lap, resistance, fusion, and spiral) or by piercing a heated billet which is forced through a pair of revolving rolls as shown in Fig. 18-1. This piercing operation produces a seamless pipe which is specified, because of its greater allowable stress, for a majority of process plant operations especially at high temperature and pressure.

The method is similar to that used in the manufacture of seamless tubing except that tubing tolerances are more rigid.

Starting in the twenties a number of interested consulting engineers together with various users' and manufacturers' engineers began the task of standardizing the many codes and practices then in existence. The Code for Pressure Piping[48] gives a short description of this work. The Code,[48] as it will be referred to in this chapter, can be applied to all types of piping installations, excepting certain steam generating piping and piping governed by various U.S. Government specifications. The ultimate goal of the committees working on the codes will be the inclusion of all practices and acceptance of the Code by all governmental agencies, including municipal, state and federal, as a controlling instrument for all piping. Its development has been of incalculable service to industry. Its use, while not mandatory at present, is widespread in the process industries.

## TRADE PRACTICES

### Pipe Sizes

Pipe and tubular products may be obtained in a wide variety of sizes and materials. Since carbon steel, or simply steel pipe, is in more general use than the many alloys and other metallic or plastic materials, most of the trade practices are based on steel pipe sizes, commonly noted as I.P.S. or Iron Pipe Size.

Pipe is sized according to a nominal figure, which only in the large sizes above 12 in. refers to an actual dimension. Thus 2 in. I.P.S. pipe, has an outside diameter of 2.375 in. and its inside diameter may vary from 2.157 in. to 1.503 in. according to the required wall thickness. "Pipe" refers to the 24 nominal pipe sizes from ⅛ in. to 30 in. ordinarily listed for carbon steel. (See Table 18-1.) In these sizes there are some 200 or more variations in thickness, of which perhaps 100 are rarely if ever specified for process piping, or are unobtainable except by special order.

Nominal pipe sizes listed in Table 18-1 above ⅜ in. are the sizes most generally used in process piping. The list is incomplete; 1½ in., 2½ in., 3½ in., 5 in. sizes are also available, but seldom used. The sizes below ½ in. are used in instrumentation piping, and at times for small drains. The smaller pipe sizes are not used for process piping even when flow rates permit. Because of their structural weakness continuous support or added intermediate support is required. It is also necessary to protect small pipe during the construction period. In addition, the small opening is much more subject to closure by scale or foreign matter. It has become common practice to specify heavy wall pipe (Schedule 80) for all sizes below 1½ to 2 in. in order to assure good structural stability.

TABLE 18-1. Abbreviated List of Some Standard Pipe Sizes

| Nominal size, in. | Schedule number | Outside diameter, in. | Inside diameter, in. | Weight, lb/ft. |
|---|---|---|---|---|
| 1/8 | 40 | 0.405 | 0.269 | 0.245 |
| | 80 | 0.405 | 0.215 | 0.314 |
| 1/4 | 40 | 0.540 | 0.364 | 0.425 |
| | 80 | 0.540 | 0.302 | 0.535 |
| 3/8 | 40 | 0.675 | 0.493 | 0.568 |
| | 80 | 0.675 | 0.423 | 0.739 |
| 1/2 | 40 | 0.840 | 0.622 | 0.851 |
| | 80 | 0.840 | 0.546 | 1.088 |
| | 160 | 0.840 | 0.466 | 1.304 |
| 3/4 | 40 | 1.050 | 0.824 | 1.131 |
| | 80 | 1.050 | 0.742 | 1.474 |
| | 160 | 1.050 | 0.614 | 1.937 |
| 1 | 40 | 1.315 | 1.049 | 1.679 |
| | 80 | 1.315 | 0.957 | 2.172 |
| | 160 | 1.315 | 0.815 | 2.844 |
| 1 1/2 | 40 | 1.900 | 1.610 | 2.718 |
| | 80 | 1.900 | 1.500 | 3.632 |
| | 160 | 1.900 | 1.337 | 4.866 |
| 2 | 40 | 2.375 | 2.067 | 3.653 |
| | 80 | 2.375 | 1.939 | 5.022 |
| | 160 | 2.375 | 1.689 | 7.445 |
| 3 | 40 | 3.500 | 3.068 | 7.58 |
| | 80 | 3.500 | 2.900 | 10.25 |
| | 160 | 3.500 | 2.624 | 14.33 |
| 4 | 40 | 4.500 | 4.026 | 10.79 |
| | 80 | 4.500 | 3.826 | 14.99 |
| | 160 | 4.500 | 3.438 | 22.51 |
| 6 | 40 | 6.625 | 6.065 | 18.98 |
| | 80 | 6.625 | 5.761 | 28.58 |
| | 160 | 6.625 | 5.189 | 45.30 |
| 8 | 40 | 8.625 | 7.981 | 28.56 |
| | 80 | 8.625 | 7.625 | 43.4 |
| | 160 | 8.625 | 6.813 | 74.7 |

In the stainless steels, cast iron, and nonferrous materials there are innumerable other sizes and thicknesses. Since these materials are not used to the extent of carbon steel, descriptive information will not be included; however, the references completely cover most of the ordinary materials.

Tubing is manufactured in even-numbered outside diameters with various wall thicknesses and close tolerances. It is often used for piping and is always specified according to the exact outside diameter (within tolerances). Its wall thickness is sometimes specified by a gage number. Tubing of the same dimensions and specifications as pipe may be obtained, but generally the major portion of tubing is manufactured in sizes below 2 in. diameter. Tubing is available in a much greater number of varying diameters and wall thicknesses than pipe since it has a wide use in conduit services other than for the conveyance of fluids.

### Schedule Numbers

Table 18-1 is a condensed version of a list of nominal pipe sizes. Only a portion of the range of thicknesses or weights and nominal sizes are included. All columns are self-explanatory except that titled Schedule Numbers. The schedule number is an approximation of the expression:

$$\text{Schedule number} = 1000P/S$$

where $P$ = internal pressure in psig
$S$ = allowable working stress in psi

This expression is based on the actual formula used for calculating required pipe thickness.

$$t = \frac{PD}{2S} + C$$

where $D$ = outside pipe diameter in in.
$t$ = thickness of pipe in in.
$C$ = corrosion allowance in in.

The schedule number is thus a rational expression that is approximately proportional to the ratio of the working pressure to the allowable stress and also the ratio of the corroded thickness to the diameter.

Before the use of schedule numbers the terms standard, extra strong, and double extra strong served to designate pipe thickness. Sizes through 10 in. in schedule 40 are the same as standard weight and sizes through 8 in. in schedule 80 are the same as the former extra strong. Double extra strong pipe sizes have been discontinued in some sizes and schedule 160 is used instead. The older designations are gradually disappearing from use.

## Materials of Construction

There are over 150 different materials specified by the American Society for Testing Materials (ASTM) for use in the manufacture of pipe. These include carbon steel, alloy steels, and a wide selection of nonferrous materials. The ASTM and the ASME (American Society of Mechanical Engineers)[3] have provided an invaluable service in preparing detailed specifications for the bewildering array of materials that are available.

It would seem that steel will continue to be the most important piping material. The usual severe service conditions of pressure and temperature encountered in process industries make its use mandatory in the majority of processes. Plastic piping materials, however, are making spectacular gains in low pressure and low temperature services. In the four-year period from 1950 to 1954 the sale of plastic pipe increased from 5 million to 30 million pounds a year. Polyethylene, polyvinylchloride, polystyrene, cellulose acetate butyrate, and styrene-butadiene-acrylonitrile have all been used with considerable success for water and many corrosive chemicals. Polyvinylchloride and polystyrene valves and fittings which sell for less than some alloy steel valves and fittings are offered by several manufacturers.

Plastic pipe can be installed with the minimum of effort. The light weight and forming properties of plastic make it easy to handle and its use should be considered whenever economical. It would be futile, however, to give a cost comparison of plastic pipe since the picture is changing rapidly in this new industry. The engineer should maintain an open mind and evaluate each new development on its merits. The industry is developing rapidly and as soon as badly needed standards are established for plastic pipe and fittings consumption will increase tremendously.

A number of other piping materials have proved unusually valuable for corrosive services. They include rubber-lined steel, lead-lined steel, and glass pipe. These can withstand higher pressures and temperatures than most plastic pipe and have the advantage of years of successful applications in many services.

The ASTM specifications provide thorough descriptions for each type of conventional metal piping material including the process of manufacture, chemical composition, tensile properties, bending properties, and required testing procedures. Careful reading of these specifications will aid the designer in selecting the most suitable material for a given service. The following generalizations, will serve as guideposts in the evaluation of steel piping materials described by the ASTM.* For the many

* Adapted by permission, *Piping Engineering,* 2.01, Tube Turns, Inc., Louisville, Kentucky (1947–53).

special problems in material selection, however, competent metallurgists must be consulted.

I. Process

*A*. Open hearth and electric furnace grades are superior to Bessemer because of the better quality control possible.

*B*. Deoxidized (or killed) carbon steels are of more uniform structure and are better for cold or hot forming than open (or rimmed) steels.

*C*. Silicon killed steels have higher creep stress than aluminum killed steels.

*D*. Steels oxidized with both silicon and aluminum are preferred for low temperature service.

II. Chemical composition

Affects corrosion resistance, weldability, forming properties, strength, and cost. A complete discussion would cover the whole field of metallurgy. Some generalizations of value are:

*A*. High carbon and manganese.
1. High tensile and yield points at moderate temperature.
2. Less weldable and less ductile.

*B*. Silicon.
1. Increases resistance to oxidation.
2. Improves subzero impact strength.

*C*. Phosphorous and sulfur; objectionable except in small amounts.

*D*. Molybdenum and tungsten; promotes high temperature or creep strength.

*E*. Chromium.
1. Inhibits graphitization.
2. Produces resistance to certain corrosive agents such as sulfur compounds.
3. Steel with chromium and molybdenum combine the features of high creep resistance with resistance to high temperature oxidation and corrosion.

*F*. Nickel; makes steel tough, particularly at subatmospheric temperatures.

*G*. Nickel-chromium alloys.
1. High creep strength.
2. Corrosion resistance to a number of agents.
3. Columbium or titanium act as stabilizing agents in these alloys by preventing carbide precipitation at high temperatures.
4. Molybdenum enhances corrosion resistance of these alloys.

III. Tensile properties; a comparison of the tensile strengths and yield points specified is an aid to the measure of quality.

IV. Tests; the types of tests and the requirements of the tests give evidence of the suitability of the material for a given service.

*A*. Hydrostatic; safeguard against defects.

*B*. Check analysis and tensile; assure conformance to physical and chemical specifications.

*C*. Bending; suitability for bending operations indicated.

*D*. Flattening; indicates suitability for forming and checks quality of seams on welded pipe.

Table 18-2 gives a list of some of the types of tubular products used in process piping. This list is similar to lists found in the Code except that it lacks a tabulation of allowable stresses which vary according to temperature.

**TABLE 18-2. Abbreviated List of Pipe and Tubular Materials**

| Material | ASTM specification | Ultimate tensile strength | |
|---|---|---|---|
| Seamless steel (silicon killed) | A-106 Grade A | 48,000 | |
| Seamless steel (silicon killed) | A-106 Grade B | 60,000 | |
| Seamless steel | A-53 Grade A | 48,000 | |
| Seamless steel | A-53 Grade B | 60,000 | |
| Seamless steel | A-83 Grade A | — | boiler tubing |
| Seamless steel | A-120 | — | restricted use |
| Lap-welded steel | A-53 | 45,000 | |
| Butt-welded steel | A-53 | 45,000 | |
| Electric fusion welded steel | A-155 Grade C45 | 45,000 | |
| Electric fusion welded steel | A-155 Grade C50 | 50,000 | |
| Electric fusion welded steel | A-155 Grade C55 | 55,000 | |
| Electric resistance welded steel | A-135 Grade A | 48,000 | |
| Electric resistance welded steel | A-135 Grade B | 60,000 | |

Pipe cannot be specified completely by ASTM numbers alone. It is to be hoped that at some time in the future specifications for carbon steel pipe may become more specific so that a single number will be sufficient for a complete specification. As an illustration note ASTM A-106. This is seamless pipe specification, thus, it is unnecessary to consider different classifications such as butt-welded or lap-welded when specifying ASTM A-106, but it is still necessary to specify the grade.

Usual carbon steel pipe specifications for process work are ASTM A-53 or ASTM A-106, the latter of which is only manufactured as seamless pipe. These two specifications in seamless pipe are essentially the same except ASTM A-106 is subject to more rigid certification of the components and methods of manufacture. ASTM A-106 is specified for severe service conditions for which carbon steel is satisfactory and for piping requiring special bending. It costs 4 to 7% more than ASTM A-53.

At times neither ASTM A-53 nor A-106 pipe are readily available in sizes below 2 in. In such cases ASTM A-83, which is a tubular specification, is specified when smaller seamless pipe is required since it has properties similar to both A-53 and A-106.

Many process piping fabricators stock only ASTM A-106. This practice reduces inventory and eliminates the possibility of installing lower grade piping in high pressure service.

There are a number of piping specifications that are undesirable for process fluids, but which may be acceptable for utility services, such as low-pressure air, exhaust steam (15 psig and below), water, and low-pressure condensate. Cast iron pipe and fittings are restricted to low-pressure and temperature service by the code. Therefore, cast iron piping materials are only used for certain utility services.

Alloy steel, nonferrous, and plastic pipe have become standard for certain services. In unusual or unknown services it may be necessary to develop the application of these materials by laboratory tests. It is, of course, nearly always possible to select alloy pipe which will last many years under corrosive conditions. The cost of alloy pipe, however, is far greater than the cost of several replacements by carbon steel. Nearly all processes are subject to obsolescence. It is obviously wasteful to use costly materials with unlimited life in processes which are subject to frequent modifications. In certain processes, however, it is necessary to use alloy materials to protect product quality.

The selection of materials must also be influenced by fabrication and erection costs. There are many inexpensive materials which could be used, but the methods of fabrication or the costs of installation may render their use impractical.

## PIPE JOINTS

For many years the only practical method of joining steel pipe was by connecting threaded pipe ends with couplings or other threaded devices such as fittings, valves, and unions (Fig. 18-2). Improvements in the welding of carbon steel by oxyacetylene or other flame welding processes and later the shielded-arc electric welding process reduced labor and provided a completely sealed and much stronger joint.[33] At present, threaded pipe joints are usually limited to pipe sizes 1½ in. or 2 in. and below. Larger pipe is joined by flanges at joints that may require dismantling. All other joints in continuous runs of pipe are welded. The flanged joint consists of two flanges, a gasket or device serving the same purpose, and the necessary bolts or bolt-studs.

### Threaded Joints

Joints in threaded steel pipe are made by unions and couplings. Unions serve the same purpose as flanges in the dismantling of pipe. Welded joints may be made in the smaller sizes but are usually not economical or practical.

In some services backwelding or seal welding is required after the threaded joint has been made. This is done for severe high-pressure services or for fluids difficult to contain such as certain gases which diffuse through threaded joints. Seal welding may distort the fitting or accessory if care is not taken in the procedure. Threaded joints which are to be seal-welded should be made-up without conventional pipe joint lubricants or compounds since most of these materials affect the welds adversely.

The "Railroad" type malleable-iron, or "Oil malleable" union as it is sometimes called, is extensively used for moderate temperature and pressure services. Forged steel unions, because of their greater strength, are used in all process services (below 3 in.) requiring high pressures or

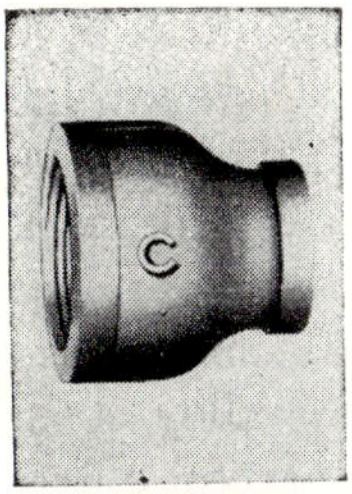

(*a*) Reducer

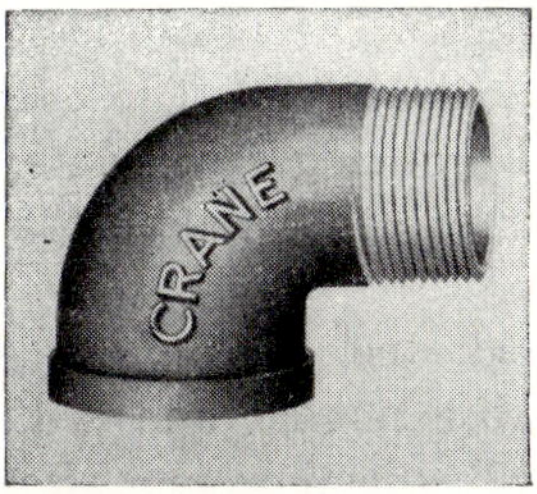

(*b*) 90° Street Elbow

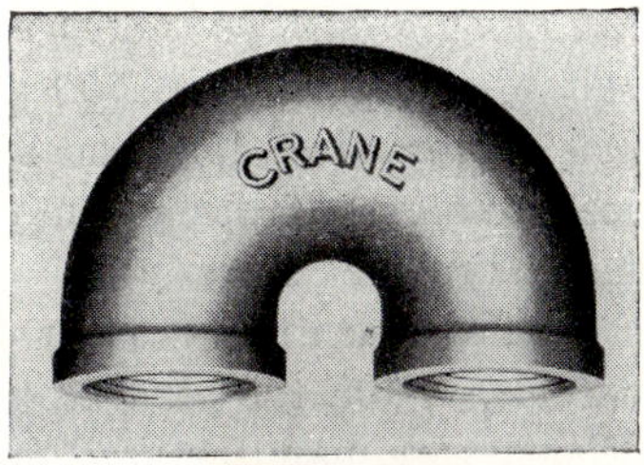

(*c*) Return Bend

(*d*) Cross

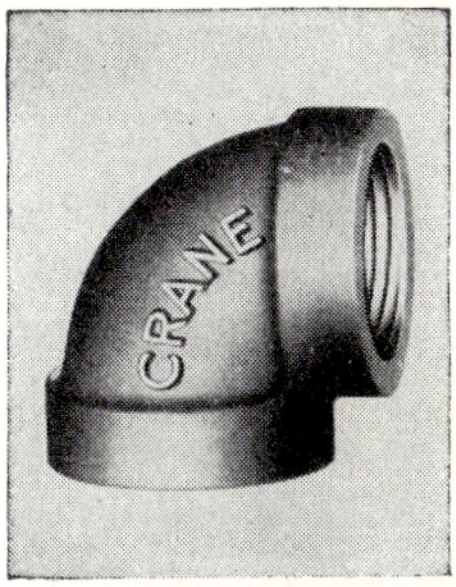

(*e*) 90° Elbow

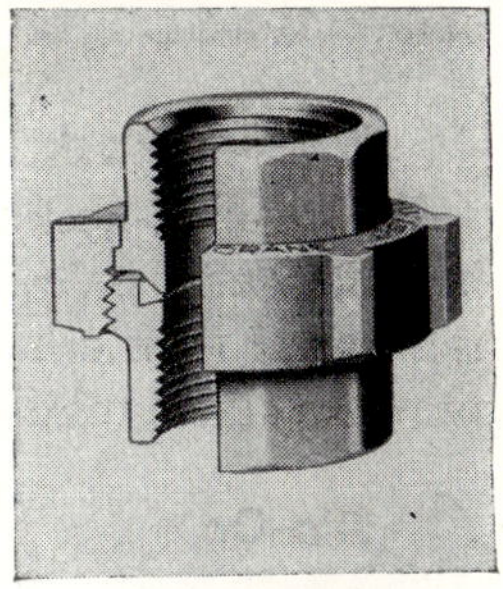

(*f*) Female Union

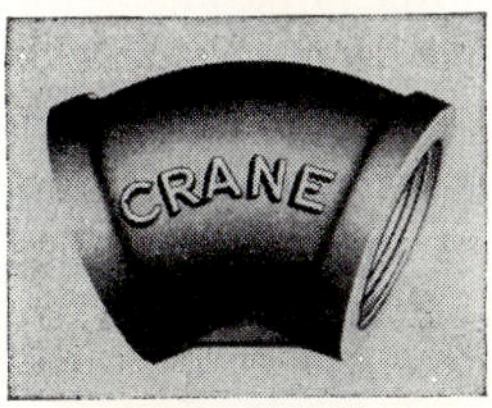

(*g*) 45° Elbow

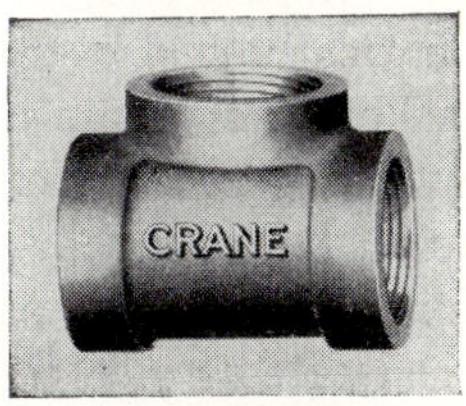

(*h*) Tee

(*i*) Coupling

Fig. 18-2. Screwed pipe fittings. (Courtesy of Crane Company.)

temperatures. These unions have accurately machined sealing services (ground-joint) and do not require gaskets (Fig. 18-2). Some ground-joint unions have a soft metal (brass or bronze) seat ring held securely on one side of the joint facing. Unions so equipped can be taken apart and reassembled many times without destroying their effectiveness. Composition or fiber inserts are manufactured but are unsatisfactory when the joint must be disassembled at frequent intervals because these inserts are usually damaged in the operation.

The pressure and temperature ratings for small fittings are given in manufacturer's catalogs.[7,13,29,31,32]

## Flanged Joints

Standard forged steel flanges have been conveniently classified as ASA* 150 lb, 300 lb, 400 lb, 600 lb, 900 lb, 1500 lb, and 2500 lb. These ratings

* American Standards Association.[4]

are nominal service pressure ratings. The 150 lb flange, for instance, is suitable for service at 150 psig and 500° F. The other ratings are for 800° F for raised face and 850° F for ring-type joints. Maximum service pressures at other temperatures are given in the Code[48] for each flange type and material.

ASTM specifications for forged steel flanges are A-181 for general service, A-105 for high temperature service, and A-182 for alloy steel. Flanges are manufactured for all the nominal pipe sizes in several different general types such as welding-neck, slip-on, lap-joint, socket-welding, and threaded flanges (see Fig. 18-3).

### Welding-Neck

The welding-neck flange is probably the most widely used type. Its installation is less costly, requiring only a single weld. The welding-neck is stronger than other flange types because its integral long-tapered hub provides added reinforcement. It is used for most severe process services.

### Slip-On

The less costly slip-on flange is also welded to the pipe. Alignment prior to welding is greatly simplified because the flange slips on to the pipe and is thereby automatically held in the proper position. Two welds are required for the attachment of a slip-on flange, making installation more costly than for the welding-neck. Costs studies by pipe fabricators and construction companies, however, have indicated that the installed costs of slip-on and welding-neck flanges are equal. Slip-on flanges are recommended only for moderate service conditions.

### Lap-Joint

The lap-joint flange is used with a stub. A single weld joins the stub and pipe end. Lap joint flanges have some advantage in fabrication since the flange proper is free to rotate. The bolt holes, therefore, do not need to be carefully aligned during fabrication.

A further use of lap-joint flanges is in alloy or nonferrous pipe services or corrosive services in which all materials in contact with the fluid must be capable of withstanding the corrosive conditions. With lap-joint flanges it is possible to utilize a stub fabricated from the higher quality material required to withstand the service and a relatively inexpensive carbon steel flange, thus effecting a considerable saving.

In the process plant it is frequently desirable to dismantle equipment for inspection or cleaning. The lap-joint flange is sometimes used since disassembly is somewhat simplified with the loose flanges. Modifications of the lap-joint flange have been used for joints in piping manufactured from plastic or other nonferrous materials, without sufficient bending

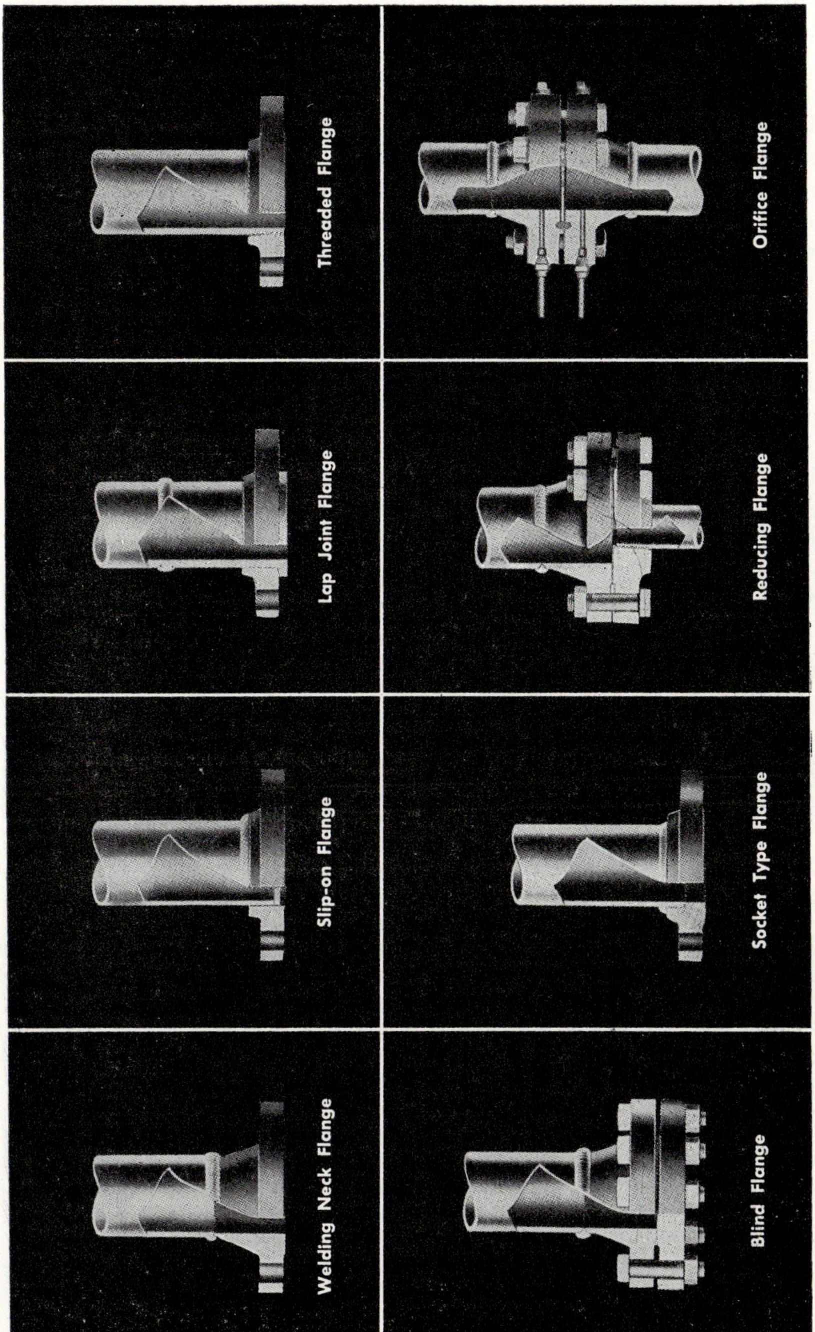


Fig. 18-3. Typical flange types. (Courtesy of Tube-Turns Company.)

strength for flanging. In these applications the flange is sometimes constructed so that it can be installed back of the flared end of the pipe.

#### Socket-Welding

The socket-welding flange is very similar to the slip-on flange. It has proved popular for small-size piping operating at high pressures. Unlike the slip-on, the pipe fits into a recessed portion (socket) of the flange. The inside of the flange and pipe thus present a smooth uninterrupted surface.

#### Threaded

Threaded flanges are used for services that cannot be welded. These include cast-iron piping, certain high pressure alloy piping that cannot be conveniently heat treated after field welding, and small piping (2 in. and below). When work must be done in a hazardous location and welding is not permitted, threaded flanges may be used. In general, however, threaded flanges are to be avoided unless absolutely essential.

#### Lens

In high-pressure work, a joint known as the "lens" joint has been successfully used. The pipe ends to be joined are machined to a concave shape, and a convex ring usually of the same material as the pipe, is placed between these ends. Flanges are screwed loosely on each section of pipe, with the pipe protruding. When the flange bolts are tightened, compression is exerted between the two pieces of pipe and the ring or lens. Threads on the pipe and the flanges are usually very coarse so that the flange is not "tight" on the pipe. This joint has some of the advantages of the lap-joint flange in that the flanges are not fixed in position and are not an integral part of the pipe.

#### Cast-Iron

Two flange ratings for cast-iron, 125 lb ASA and 250 lb ASA, have the same drilling (facing and bolt hole circle) respectively, as the 150 lb and the 300 lb ASA steel rating. Cast iron flanges are manufactured in accordance with ASTM A-126.

A raised-face steel flange is never joined against a flat-faced cast-iron flange since the shoulder caused by the raised-face portion of the steel flange can centralize or overstress the cast-iron flange. As the strength of cast-iron in bending is limited, failure could occur by overtightening the flange bolts. The Code and manufacturers of cast-iron eqiupment, therefore, recommend that the raised face on steel flanges be ground off or that flat-faced flanges be utilized wherever steel and cast-iron equipment are joined together.

The welding of cast-iron in piping services has only limited application. Though it is possible under closely controlled conditions, the costs are prohibitive. Hence, if cast-iron flanges are used with steel pipe, they are always threaded.

### Flange Facings and Gaskets

A tight mechanical joint in piping could not be effected without some plasticity of the members to be joined, since obviously if both faces of a joint were equally unimpressionable only perfect machining or grinding, could effect a sealed joint. In order to close the joint without extensive machining and polishing, gaskets are used. The gasket must be of a material which will withstand service conditions and yet be sufficiently plastic to deform under soft compression so that all interstices are closed.

All of the ASA flange classifications may be obtained with raised-face, male-and-female, tongue-and-groove, ring-joint or flat-faced facings (see Fig. 18-4). The most common facing is the raised-face. The ring-joint has a wide use in certain severe services and is the most expensive. There are several standard types of machining available for the portion of the flange in contact with the gasket. For ordinary services the serrated or smooth-face finish is satisfactory.

Gaskets may be obtained in a wide variety of materials for practically any type of service. Asbestos combined with other materials is perhaps the most common gasketing material. Ring-type joints employ metal rings that are softer than the flange material.

### Special Flanges

Many special types of flanges have been designed to meet particular requirements of a single industry. These include certain flanges common to oil fields[2] and ammonia fittings. The latter is a tongue-and-groove type, with oval, triangular, or square flanges. It was developed early in the century for ammonia refrigeration compression systems and continues to be used by some portions of that industry even though standard tongue-and-groove or ring-type flanges would serve equally well. Special flanges such as these should be avoided in process plants. They have no particular advantages and only complicate warehousing.

### Bolting Practices for Flanges

Two general types of bolts are used for flanged joints. The most common is a modification of the ordinary carbon steel machine bolt with a square or hexagon head and a hexagon nut. The other is the so-called bolt stud. It consists of a fully threaded stud with two hexagon-headed nuts.

As piping pressures and temperatures increase, carbon steel bolts can

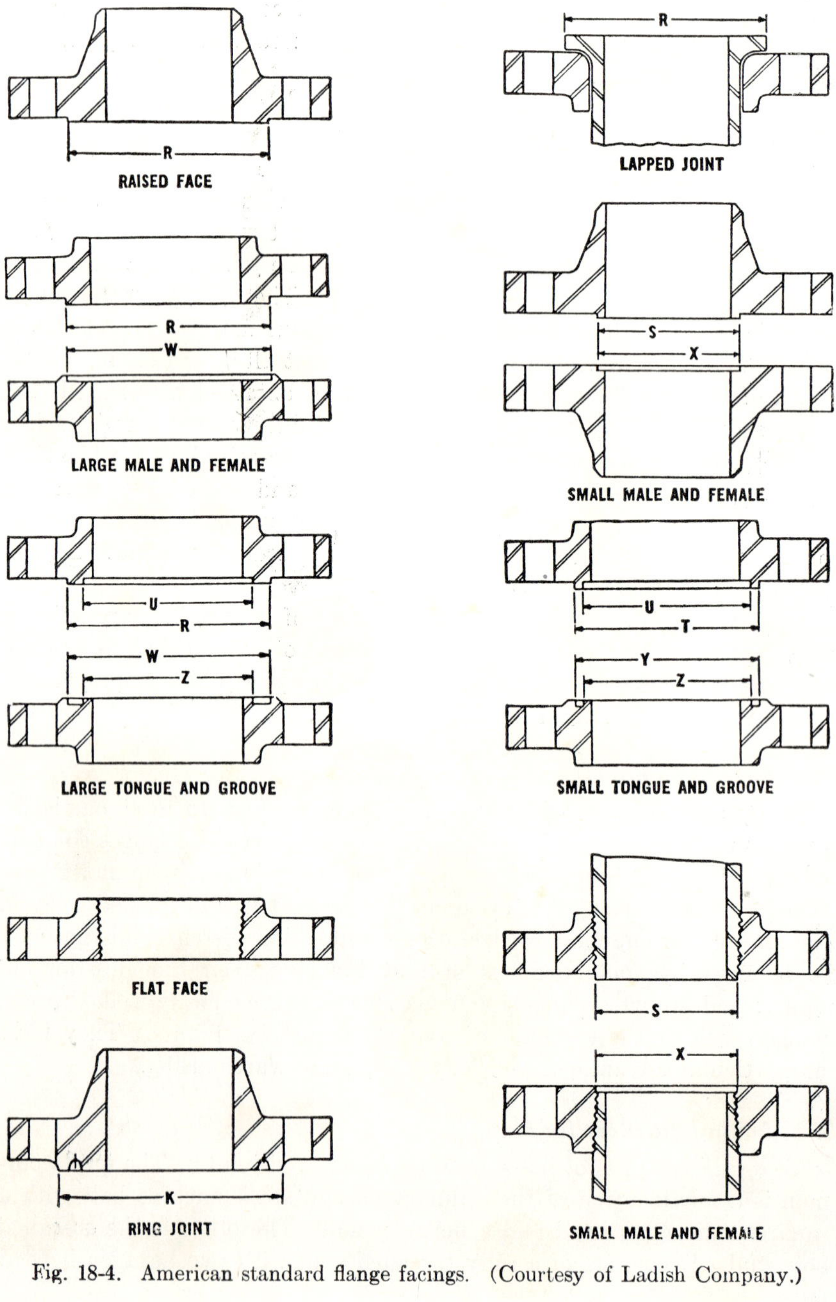


Fig. 18-4. American standard flange facings. (Courtesy of Ladish Company.)

no longer meet the stress requirements and alloy steel bolts must be used. At the present time there are four ASTM specifications for carbon steel bolts and studs, twenty for alloy steel, and twelve for nuts. Vogrin *et al*[30] indicate in their study of bolting practices that ASTM A-307 Grade B carbon steel bolting should be used for services to 450° F. Above 450° F alloy steel bolting materials A-261, A-354, and A-193 in a number of different grades are recommended for temperatures up to and above 1500°F. For low temperature services below —20°F, A-320 is recommended in seven grades to —300°F.

Vogrin recommends Class 2A fit for bolts and 2B fit for nuts in nonhigh temperature bolting. For high temperature bolting, Class 7 fit for the bolt and Class 2B fit on the nut is suggested. Table 18-3 summarizes present piping bolting practice.

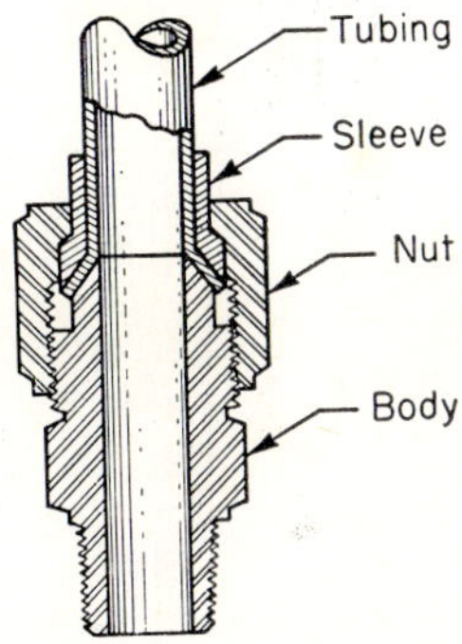

Fig. 18-5. Flared fitting.

### Miscellaneous Joints

A compression type fitting is often used for joining 1 in. and smaller steel or nonferrous tubing (Fig. 18-5). In process plants these fittings are primarily used for instrument tubing.

There are numerous other methods for joining pipe, such as the solder-joint which is used in plumbing, the leaded joint for cast-iron piping, and the integral flange sometimes used for cast and other formed materials. The leaded bell and spigot joint for cast-iron piping is still employed extensively in municipal services such as water piping, gas piping, and sewers. A bolted mechanical joint for cast-iron pipe is becoming popular since it requires less labor for installation than the leaded joint.

## PIPING FITTINGS

### Welding Fittings

In welded piping systems now used in nearly all process installations, butt-welding fittings have been developed to reduce the necessity for bending pipe. Welding fittings are manufactured under various trade names.[13,25,26] The fittings conform to standard pipe sizes and specifications and can be obtained as tees, ells, reducers and pipe-caps (Fig. 18-6). Nearly all such fittings are seamless and have standard dimensions. The center-to-face dimension on all long radius 90° weld ells, for instance, is 1½ times the nominal size. This standardization simplifies design and construction.

An ingenious process is used for the manufacture of the various bends.

TABLE 18-3. Recommended Bolting for Flanges*

| Boiting materials | | | | Flange materials and ASA flange ratings | | | | | | | |
|---|---|---|---|---|---|---|---|---|---|---|---|
| ASTM | | Temperature limit (2) ° F | Notes | Cast iron | | Carbon steel | | Carbon ½ pct. mo. all | Chromium-molybdenum all | Austenitic nickel-chromium all | Nickel steel all |
| Specification | Grade | | | 125 lb | 250 lb | 150 lb | 300-2500 lb | | | | |
| A307 | B | 450 | — | A | A | B (4) | — | — | — | — | — |
| A261 | BO | 850 | — | B (3) | B | A | B | B | B | — | B |
| A354 | BB, BC, BD | 750 | — | B (3) | B | A | B | B | B | — | A |
| A193 | B7 | 1000 | — | — | — | A | A | A | A | A(5) | A |
| A193 | B7a, B14, B16 | 1100 | (6) | — | — | A | A | A | A (7) | — | — |
| A193 | B5 | 1100 | (8) | — | — | — | — | — | B | — | — |
| A193 | B8,B8c, B8t | 1500 | — | — | — | — | — | — | — | A (7) | — |
| A193 | B8F | — | — | — | — | — | — | — | — | (9) | — |
| Services below —20° F | | | | | | | | | | | |
| A320 | L7 & L10 | —150 | — | — | — | A | A | — | — | A (10) | A |
| A320 | L | —225 | — | — | — | — | — | — | — | A (10) | — |
| A320 | L8, L8c, L8t, L8F | —300 | — | — | — | — | — | — | — | A (11) | — |

Notes

(1) Recommended bolting is designated by A and B; A indicates preferred material, with B as acceptable alternate.

(2) The temperature limit shown includes the range of temperature from —20° F.

(3) Bolting conforming to this grade may be used with 125 lb ASA cast-iron flanges, provided a full-faced gasket is used or when a ring gasket is limited to soft gum rubber or material similar in behavior.

(4) Bolting conforming to this grade is not recommended when flanges are to be used with metal or metal-jacketed gaskets.

(5) This bolting is recommended for austenitic flanges at temperatures of 450° F and below to avoid the necessity of a special pre-tightening bolt.

(6) Grades B14 and B16 provide higher strength than Grade B7a at temperatures between 850° F and 1050° F. They should therefore be used for unusually severe service conditions.

(7) Bolting in elevated-temperature service should be selected so that the allowable stress at working temperature is not less than that permitted for the flange material. For operation at temperatures above 1100° F, bolting material should be special and of material similar to the flange.

(8) This material is not commonly used. Its selection depends primarily on the necessity to provide increased corrosion or oxidation resistance.

(9) The free-machining grades of austenitic bolting materials should be limited to low-temperature services until the Codes have sufficient data on mechanical properties to publish allowable stresses.

(10) These materials are preferred for use with austenitic flanges in low-temperature service within the temperature limits indicated.

(11) Austenitic bolting should be used only in the temperature range for which the ferritic grades cannot meet impact requirements.

*Reprinted by permission, Vogrin, C. M., F. S. G. Williams, and J. S. Worth, Paper No. 52–PET–7, 7th Pet. Div. Conf., Amer. Soc., Mech. Eng. (1952).

Seamless pipe cut to a predetermined length is forced over a specially designed mandrel shape so as to produce the desired bend. Careful control insures uniform wall thickness, smooth inside wall, and no residual stresses.

All welding fittings are finished with beveled edges so that full penetration welds are assured when joining the fitting to the pipe. Fittings are available in most of the same materials utilized for pipe. Each fitting is stamped with an identifying marking, and all nonalloy steel fittings are painted to prevent corrosion in storage.

For reasons of economy many users restrict the use of tees. Instead pipe can be carefully cut and the branch connection welded into place. This method is more economical for small branch lines connecting into large pipe. Costs of ready-made tees and construction of branch connections in the field should be compared for each job and policy outlined for their use in the piping specifications.

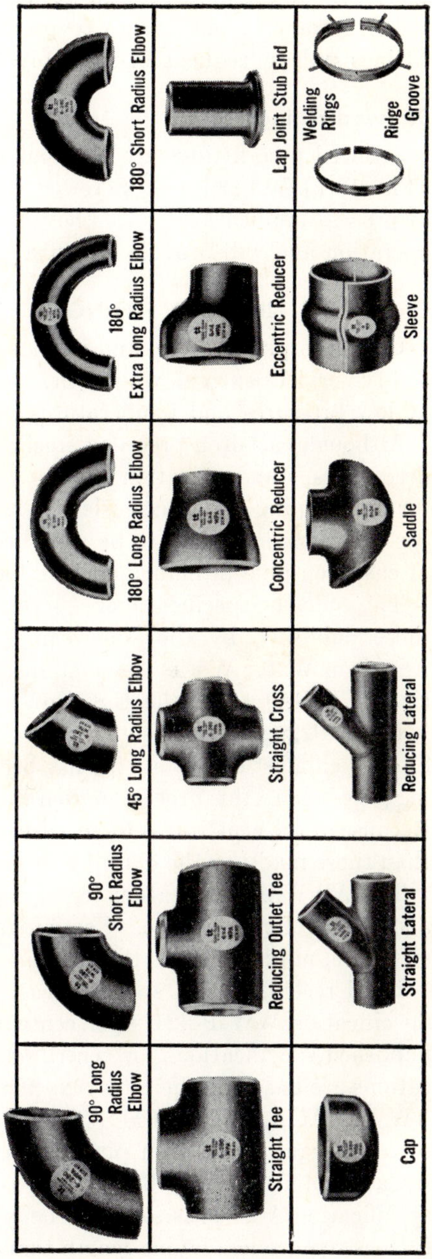


Fig. 18-6. Welding-type pipe fittings. (Courtesy of Tube-Turns Company.)

## Flanged Fittings

Cast-steel, forged-steel or alloy-steel flanged fittings are produced in nearly all sizes and have the same ASA pressure and temperature classifications as flanges.[7,31] Flanged fittings are used only when maintenance requires frequent disassembly since additional flanges, bolts, and gaskets are expensive and subject to leaks. Actually maintenance work on most piping systems is confined to the accessories, such as valves, and not on the pipe itself. Since all valves above 2 in. will normally be flanged, the joints at these valves can be used for disassembly. It is, therefore, not necessary in most cases to provide a number of additional flanged fittings unless frequent disassembly is anticipated.

### Screwed Fittings

Screwed pipe fittings are obtainable in nearly all materials and styles (Fig. 18-2).[7,13,29,31,32] Screwed forged-steel fittings have a wide use in process applications since they are satisfactory for high temperature operation and will provide an extra safety factor in case of fire.

## CAST-IRON PIPE AND FITTINGS

Cast-iron pipe, valves, and fittings are generally used for underground utility services such as water, air, fuel, and sewers, and drains operating at low pressures and temperatures.

Although cast-iron preceded steel as a piping material by several hundred years, standardization has not yet been accomplished as it has been for steel piping and accessories. Changes in the various codes and size specifications have had to be slow because of the wide use and long life of cast iron pipe in municipal installations.

The codes or specifications covering cast iron pipe and fittings include those published by the American Standards Association (ASA),[4] the American Water Works Association (AWWA),[5] the American Gas Association (AGA),[1] and the U.S. Government.[11]

Until 1922 cast-iron pipe was cast in stationary sand molds (pit cast). Since then most cast-iron pipe has been produced by the centrifugal casting process. In this process molten iron is introduced into spinning molds. Because of the centrifugal force, castings produced are denser and stronger than those made by the old pit cast methods.

The AWWA specification for cast-iron pipe covers pit-cast pipe only and is, therefore, obsolete. Centrifugally cast pipe, because of its denser structure, need not be as thick as pit-cast to withstand a given pressure.

Until recently users specified cast-iron pipe on the basis of Federal Specification WW-P-421[11] for centrifugally cast pipe. In 1953 the ASA published specifications for centrifugally cast iron pipe.[4] These specifications are based on the latest experimental evidence accumulated by the AWWA, AGA, and ASTM. They take full advantage of the strength of centrifugally cast iron pipe and should be used in all specifications.

Cast iron fittings may be specified as ASA or AWWA fittings. The ASA specification A 21.20 is for short-body fittings 3 in. through 12 in. These fittings require less laying length than the more familiar AWWA fittings. The AWWA specification for fittings, which is still in use, covers fittings from 4 in. through 60 in. The AWWA type fittings are the only available bell and spigot fittings in sizes above 12 in.

There are a number of joints for cast-iron pipe. The most widely used in process plants are bell and spigot, mechanical flange, compression, and threaded. The characteristics of these joints are briefly outlined in Table 18-4.

TABLE 18-4. Cast-Iron Joint Types

| | Description | Packing | Usual service | Remarks |
|---|---|---|---|---|
| Bell and spigot | Bell-shaped end receives straight spigot end to form joint | Jute twine and lead | Underground water piping up to 200 psi. Underground gas piping below 10 psi | Fluids must be wet so that jute remains expanded |
| Mechanical type | Straight portion of pipe fits into bell-shaped end fitted with a gland which tightens against gasket | Rubber or composition | Gas piping at moderate pressures. Also water piping | Withstands vibration and and expansion |
| Flange | ASA flanges Class 25, 125, and 250. Integral part of pipe or screwed to pipe | Standard gaskets for flat-face flanges | Above-ground water lines, and low-pressure gas | Above-ground water lines and low-pressure gas |
| Compression coupling | Plain end cast-iron pipe is held together by a compression type coupling | Rubber | Gas, oil, and water | Permits expansion and contraction |
| Threaded | Screwed couplings similar to steel pipe couplings | None | Noncritical waste and water lines | Threaded type compression joint also available |

Cast-iron pipe for process fluids such as gas or oil must have positive attachment between joint sections or it will not meet Code requirements.

A large number of joints are required for cast-iron piping since welding is not possible. Maintenance and installation costs, therefore, are higher for cast-iron systems. This disadvantage together with the service limitations imposed by the Code (not permitted above 400 psi at 300°F below ground and 150 psi at 300° F above ground) restrict the use of cast-iron pipe in the process plant.

## VALVES

The valve, which is the most important piping accessory, is available in materials and designs of almost unlimited variety.[*,7,14,17,20,29,31]

* One publication[27] helps compare various manufacturer's valves.

Plant warehousing and maintenance are greatly facilitated, however, if the types of valves stocked are kept to a minimum. It is especially economical to standardize in the smaller sizes by selecting a one- or two-valve design which will satisfy all services anticipated.

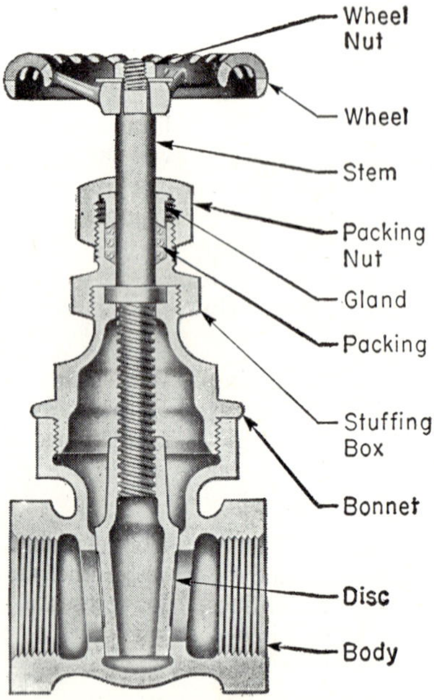


Fig. 18-7. Valve parts. (Courtesy of Crane Company.)

## Classification

Valves used in process piping may be classified as gate, globe, plug, check, and control valves. The service conditions of many of the more common types of these have been standardized as well as the face-to-face and flange dimensions of flanged valves.[41,15] In general, however, reference must be made to manufacturer's catalogs for complete specifications and descriptions of a given valve. These descriptions together with published codes and standards for valves[2,4,11,15,27,28] will aid in the selection of the most suitable valve for a given service.

## Nomenclature

All gate and globe valves have two similar basic parts, the body and the bonnet (Fig. 18-7). The body contains the orifice, the part which closes the orifice, and the necessary flanges or screwed openings for connection to the piping.

The bonnet is the upper part of the valve joined to and closing the body. It contains the gland, supports and guides, the stem, and valve wheel or closing device. There are three types of valve bonnets (Fig. 18-8). The screwed bonnet is the most inexpensive as well as the least reliable. Tight sealing of the stem requires screwing the bonnet strongly against the neck which often causes distortion of the valve neck. The union bonnet is designed to provide an easily disassembled valve. The fluid pressure acts to seal the joint more tightly. No distortion occurs even upon repeated disassembly and reassembly. The bolted-flanged bonnet has a bolted packing gland and a large stuffing box. It is the sturdiest design and is used for most larger process gate and globe valves. In such installations it usually has flanged ends. The importance of the method of attachment of the valve bonnet is indicated in paragraphs 309(*d*) and (*e*) of the Code.[48]

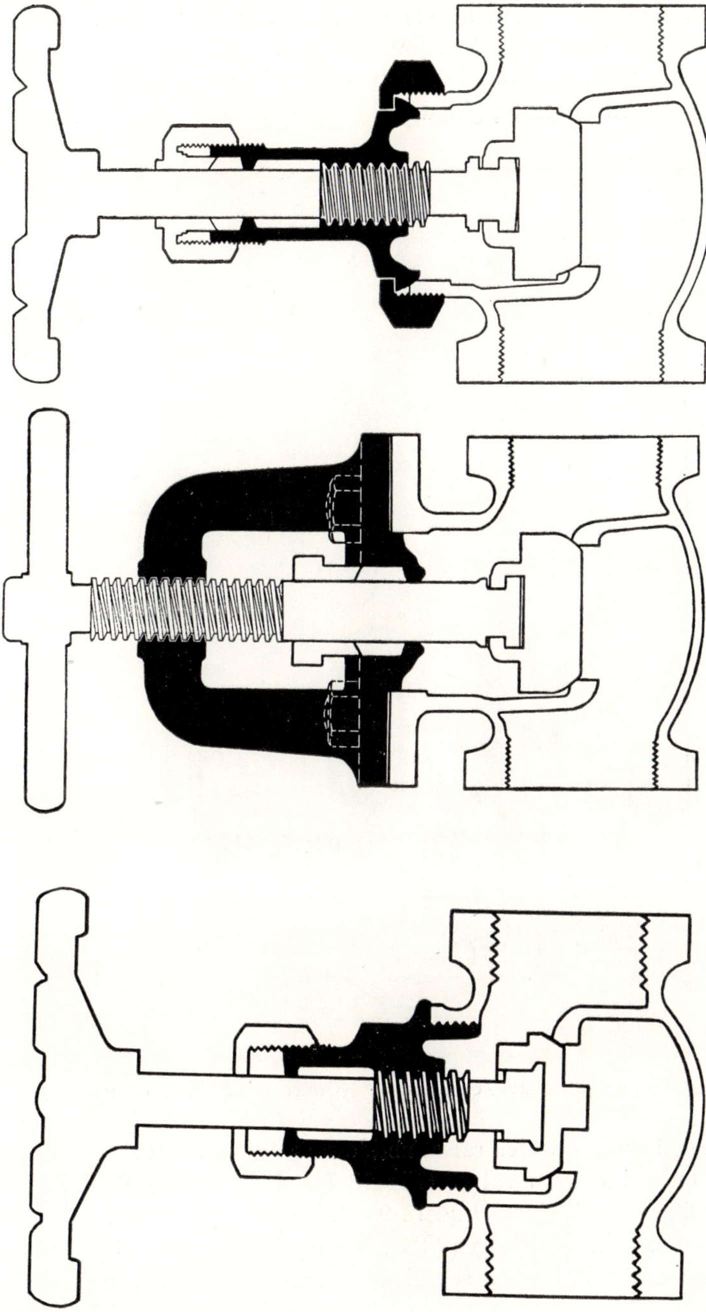


Fig. 18-8. Valve bonnet designs. (*a*) Screwed-in bonnet; (*b*) bolted flanged bonnet, outside stem and yoke; (*c*) union bonnet. (Courtesy of Wm. Powell Company.)

The valve stem raises and lowers the plug or gate which closes the orifice. The gland is the packed joint through which the stem passes. Packless valves, made in small sizes only, have a bellows type attachment which eliminates the packing gland.

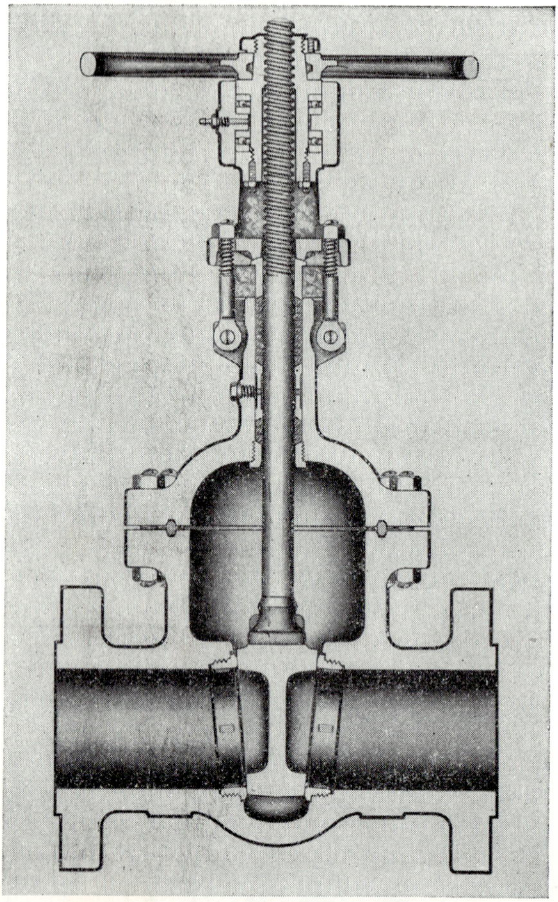

Fig. 18-9. Wedge gate valve, O.S. and Y. (Courtesy of Crane Company.)

The threaded portion which raises and lowers the valve stem is enclosed in the body on inside screw valves (Fig. 18-7). It is on the outside of the body of outside screw valves (Fig. 18-9) which are recommended by the Code for sizes $2\frac{1}{2}$ in. and larger [Par. 309(1)]. The term O.S. and Y. is a common description for these valves and means outside screw, rising stem and yoke. The outside screwed valve permits constant observation of the important moving part of the valve. In addition, the threaded sections do not come in contact with the fluid which is an advantage when handling corrosive fluids.

The valve seat is that portion of the orifice which is closed by the plug or gate. Valves may be furnished with integral seats or renewable seats. Valve trim is the term applied to the materials which form the seat, plug or gate, and the stem.

Valve ends are described as screwed, flanged, or welding. Screwed ends are the cheapest and are more rapidly installed in the small sizes. They, however, always constitute a potential source of leakage. Flanged ends provide a strong leak-proof joint, and even though more costly, their reliability and superior strength make their use almost mandatory for large process valves. Welding ends are used only for special installations requiring absolute insurance against leakage.

## Gate Valves

The gate valve (Figs. 18-7 and 18-9) is the most commonly used valve. Fluid flows directly through the gate valve axial to the piping (Fig. 18-10). The valve consists essentially of a sliding plate or gate set in grooves. The plate is moved directly across the orifice and is closed by means of a threaded stem rotated by a hand wheel. The gate may be wedge-shaped so that it will seat tightly against the sides of grooves, or it may consist of two plates expanded against the sides of the grooves. Gate valves, however, are generally used as block valves for isolating equipment or lines, or in full open or closed position services. They are rarely used for throttling, for if the valve is kept in the partially open position, the bottom of the gate becomes eroded. Throttling service is usually restricted to the more easily controlled globe valve. Since the orifice is approximately concentric and the flow is axial to the piping, gate valves have a lower pressure drop than globe valves.

## Globe Valves

The globe valve (Fig. 18-11) is different from the gate valve in that the globe-valve orifice is arranged perpendicular to or on an angle to the axis of flow. Closing is accomplished by a disk or plug attached to a stem which may be rotated to open or close the circular orifice. Fluid flowing through the valve must make approximately a 270° turn to reach the orifice (Fig. 18-10). After passing through the orifice, it turns another 270° to re-enter the original flow line. Pressure drop through the globe valve, therefore, is much higher than through the gate valve.

Globe valves are preferred for throttling services, and are almost always used if tight shut-off is required. Because of the circular construction of the seat and plug of a globe valve, it may be more precisely ground or "lapped" to provide a tighter seal than the gate valve.

Large size globe valves are ordinarily more expensive than large gate valves. Many users purchase globe valves only in the small sizes, 1½ in.

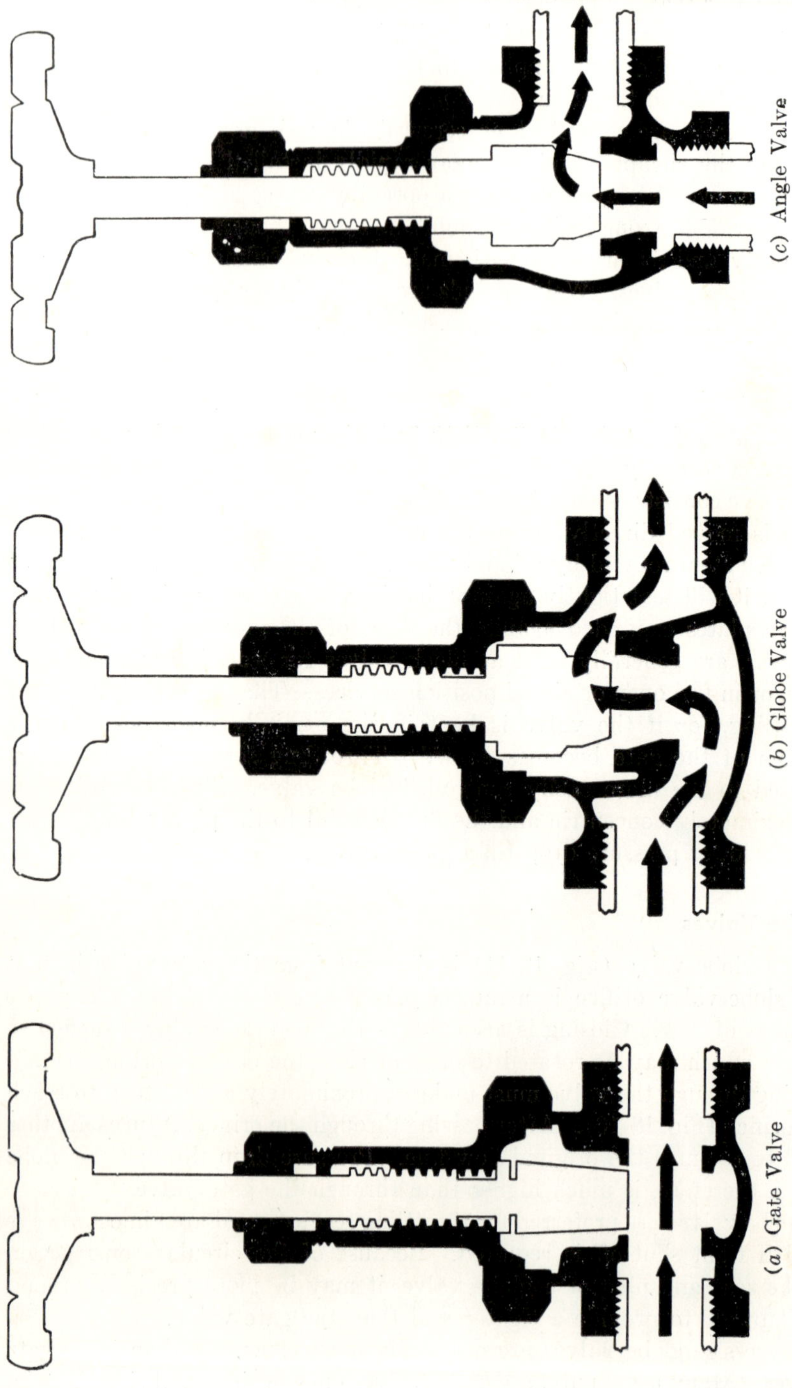


Fig. 18-10. Valves and valve flow characteristics. (Courtesy of Wm. Powell Company.)

(d) Swing Check Valve

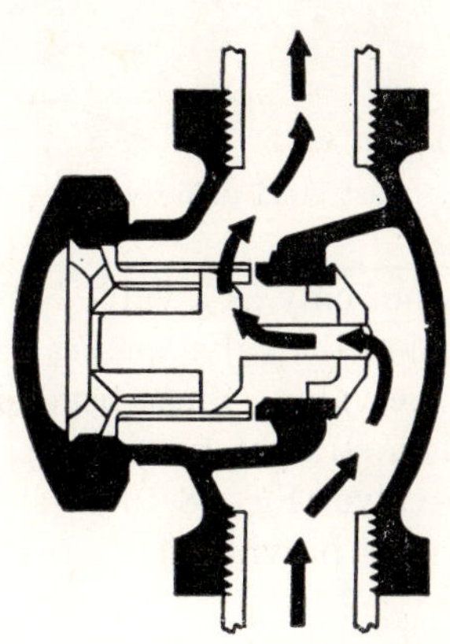

(e) Horizontal Lift Check Valve

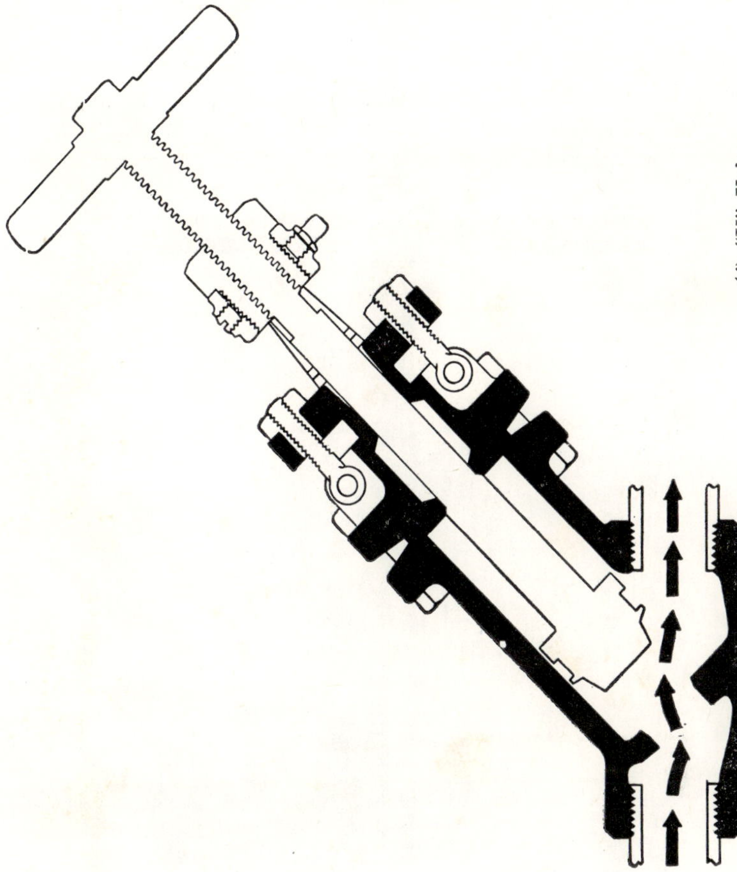

(f) "Y" Valve

Fig. 18-10. (Continued.)

and below, for which sizes it is possible to standardize on a single type for all services.

The so-called angle valve (Fig. 18-10) is a globe-type valve which offers much less resistance than the conventional globe. The angle valve

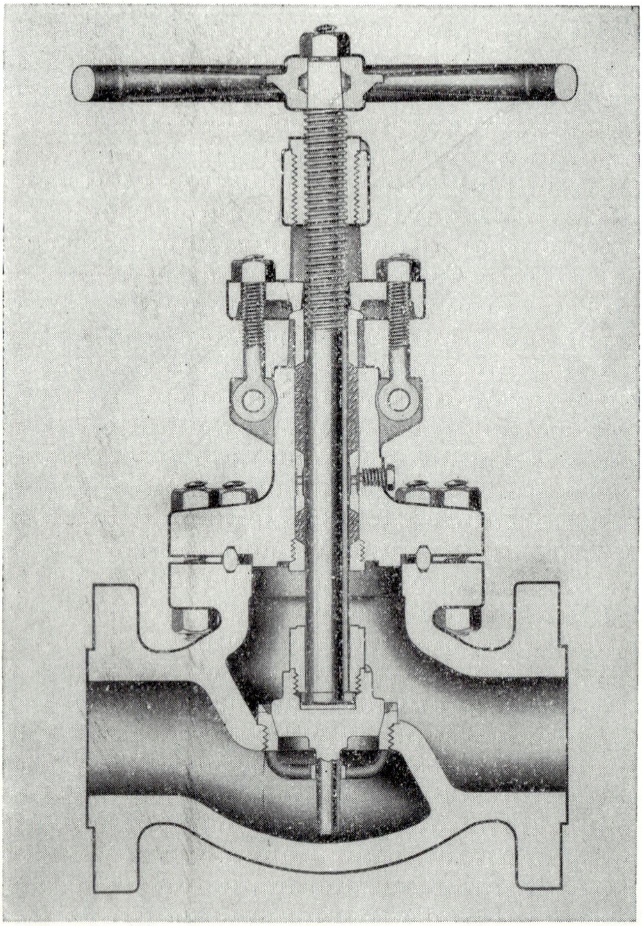

Fig. 18-11. Cast steel globe valve, plug-type disk. (Courtesy of Crane Company.)

functions not only as a valve but also as a 90° piping bend. Flow enters the bottom of the valve, passes upward through the orifice, and turns 90° to leave the valve. Since bends in piping systems are subject to greater strains than straight-runs of pipe, angle valves are not used extensively.

The Y-valve (Fig. 18-10) is similar to a globe valve in throttling characteristics yet provides flow area comparable to a gate valve.

### Plug Valves

The plug valve, or cock (Fig. 18-12) as it is sometimes called, is the ancestor of all valves. It was developed from the wooden spigot. Its outstanding feature is still its simplicity. When the valve is open, fluid flows parallel to the pipe wall through a plug bored the same size as the body orifice. Flow is shut-off by rotating the plug 90° so that the orifice in the plug is turned past the body orifice. Certain plug valves have a venturi-type body orifice so that pressure drop is much lower than for any other type of valve.

Plug valves may be obtained in materials suitable for high temperatures; they have been used in services up to 1200° F. For light hydrocarbons, gas services, and for many corrosive services the plug cock is usually selected, since it may be manufactured in nearly any type of material.

They are not as commonly used as the gate valve. Some arguments against their use for all process work are: (1) The possibility of the plug "freezing" open or closed; (2) no visual evidence of their being open, such as with the *rising-stem* gate valve; (3) they must be lubricated to completely seal in some services;* (4) in the smaller sizes more rigid support must be provided to operate the plug. Many adverse criticisms of these valves, however, can be traced to faulty maintenance.

Modern plug valves have been designed to insure ease of movement and a positive seal. Lubricated plug cocks, for instance, employ a semisolid lubricant which is injected hydraulically. It is possible to lift the valve plug somewhat by rotating the lubricant injection screw, and this action forces the lubricant under the valve plug. Since the plug is tapered it is free as soon as it is moved. Larger sizes of plug valves have a gear and wheel for manual operation. The smaller sizes, 4 in. and below, are operable by a lever wrench.

### Check Valves

Check valves (Fig. 18-10) prevent "backing-up" of fluid in a line, i.e., reversal of flow. Lift check and swing check valves are the two major types. The lift check is provided with a ball or guided plug which is lifted by the passage of fluid in one direction and is seated upon a reversal of flow. The lift check is most often used for sizes 2 in. and smaller. It does not have a great tendency to slam and is, therefore, specified for services having irregular or frequent reversals of flow. It is installed so the direction of lift is vertical.

The swing check valve is more frequently used. It produces less

* Some nonlubricated designs are available.

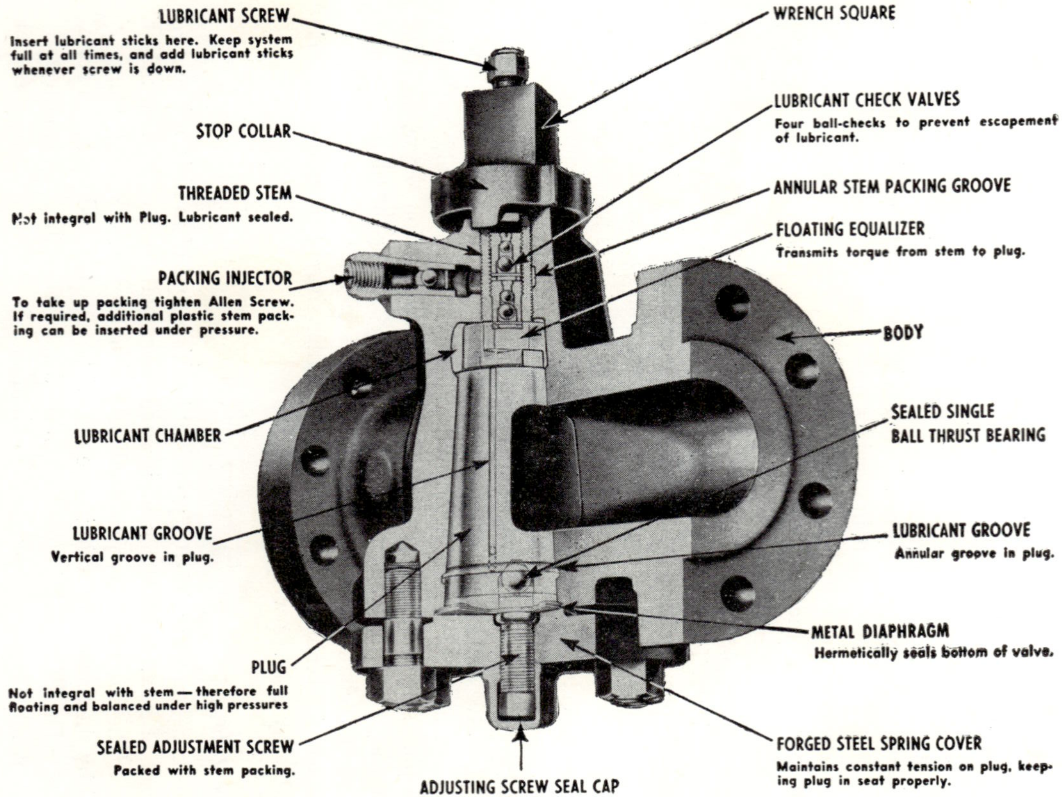


Fig. 18-12. Lubricated plug valve, rectangular opening. (Courtesy of Rockwell Mfg. Company, Nordstrom Division.)

resistance to flow and is specified for sizes larger than 2 in. Frequent flow reversals produce a "chattering" in swing valves which can be alleviated by the use of swing valves equipped with an outside lever and weight. A lever is attached to an external continuation of the hinge. A weight, the position of which can be changed, is attached to the lever so the sensitivity of the valve can be varied. In this manner chatter under low velocity flow conditions and shock on closing can be avoided or reduced. The outside lever can also be used to aid in releasing the disk if it becomes stuck.

## Control Valves

Control valves are fundamentally similar to the globe valve. They are made in all the ranges and materials found in gate or globe valves, but the "trim" and bodies of control valves have higher quality specifications than ordinary valves.

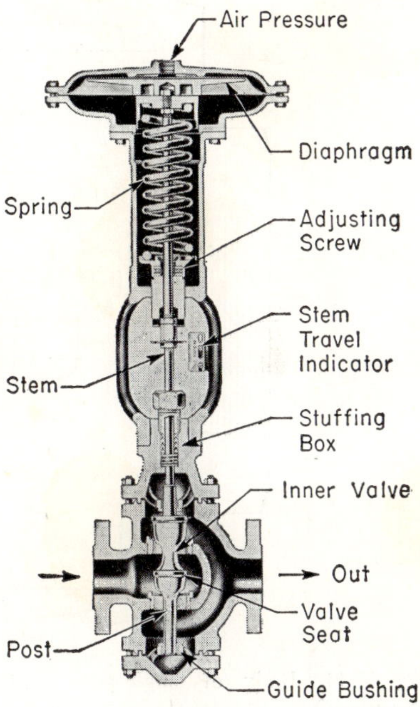


Fig. 18-13. Diaphragm-operated, double-port control valve. (Courtesy of Fisher Governor Company.)

The diaphragm control valve is the most widely used. It is opened or closed by a stem attached to a plate or diaphragm which forms the movable side of a chamber (Fig. 18-13). The force required for movement in one direction is produced by air pressure in the chamber, and in the opposite direction by a spring. Movement of the diaphragm and thus the stem causes the valve plug to open or close the valve orifice. The air pressure on the diaphragm is controlled by the primary sensing instrument.

Control valves may be obtained either with double-ports or with a single-port (see Figs. 18-13 and 18-14). The double-port valve has two orifices; the single-port, a single orifice.

The double-port valve gives a better control range for larger flows. A further advantage is the small force necessary to move the valve stem. The pressure differential across the bottom port is opposite and approximately equal to that across the top which enables movement of the valve stem with the minimum expenditure of power.

The single-ported valve is less costly and is satisfactory for noncritical control services. It provides a tighter seal than the double-ported valve.

There are several inner valve designs. The inner valve illustrated in Fig. 18-14 is designed to reach maximum capacity after only a small

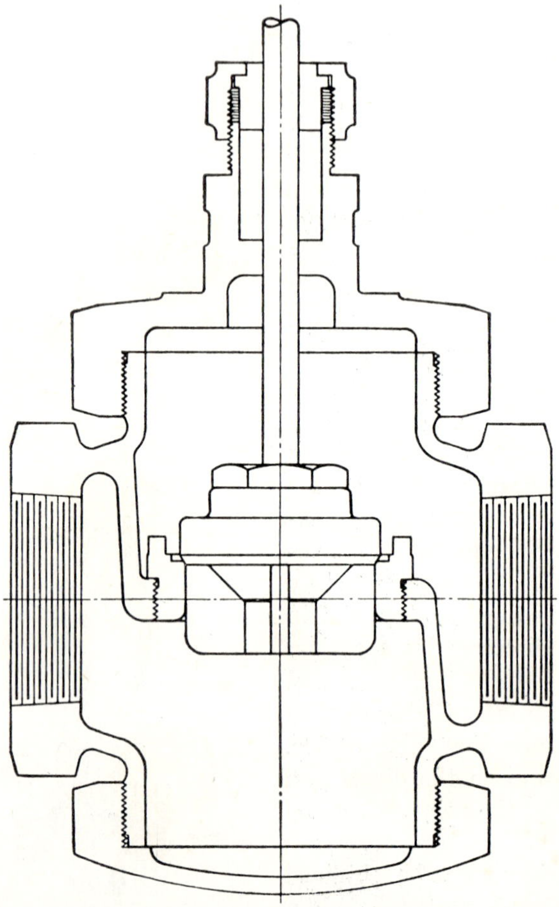

Fig. 18-14. Body of single-port control valve. (Courtesy of Fisher Governor Company.)

movement of the valve stem. Such a characteristic is most desirable for quick open and shut service. The parabolic inner valve in Fig. 18-13 gradually increases its flow area throughout the lift of the stem, producing an excellent throttling action. V-ported inner valves of various designs produce similar desirable characteristics.

Control valves in the larger sizes or in high-pressure or high-temperature services are practically always specified with an auxiliary or booster device called a positioner. This device continues to add or release air from the valve-operating diaphragm, forcing the valve stem to move to the precise position required by the control instrument.

Control valves are used for pressure control as well as flow control. Many pressure controllers employ the fluid being controlled as an actuat-

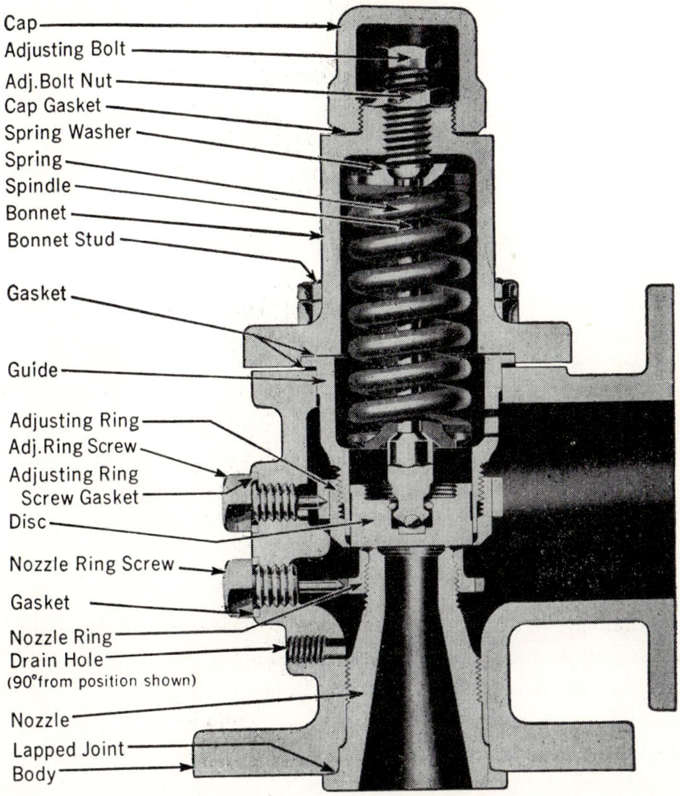


Fig. 18-15. Relief valve. (Courtesy of Crosby Steam Gage and Valve Company.)

ing medium. The sensing element in this type valve is located at the valve and usually consists of a diaphragm or other pressure actuated device. In pressure reducing services the direct connected, differential-pressure valve is often used. This type control valve balances the two fluid pressures, one on each side of a common diaphragm with a spring which can be adjusted to the desired differential.

## Relief Valves

A relief valve is designed for the single purpose of protecting equipment by relieving excess pressure, and is the only valve which is designed for a continuously closed position (see Fig. 18-15). These valves open automatically when the force on the seat exceeds that of the spring. Since they are safety devices which protect life and property, they must be constructed of the finest materials for the service.

The above descriptions include the basic types of valves manufactured. There are, however, countless variations of each of these types manufactured for special services. A listing of miscellaneous process equipment and manufacturers in a general industrial catalog[22] shows over 150 specific types of valves. Each of these types can be obtained in several different materials and each can be obtained in a large number of sizes and pressure-temperature classifications. There are over twenty major manufacturers of the common types of valves exclusive of the manufacturers of specialty items. These manufacturers of common types of valves produce competitive products, the specifications for which are controlled by either manufacturing standards or by the various codes. One publication[10] enables the engineer to compare all manufactured valves on an equivalent basis.

## MISCELLANEOUS PIPING ACCESSORIES

In addition to the necessary pipe, valves, and fittings for transporting process fluids, a number of other items are required, which may be classified as accessories.

### Condensate or Steam Trap

The condensate trap, often erroneously called steam trap, is used in steam piping systems for the removal of condensate which accumulates as heat is lost from steam. Traps are located at low points or pockets in steam piping, at regular intervals in extended runs of steam piping, at steam-driven equipment which could be damaged by condensate, and at steam-heated equipment of all types.

The ideal trap must remove condensate and air from a system with negligible loss of steam and at a rate of removal such that the operating equipment will perform at peak efficiency.

In Table 18-5 seven popular trap types are listed. A brief outline of the operation and distinguishing features is given. Fig. 18-16 illustrates the most widely used trap, the inverted bucket. It is not possible to give exact recommendations for the use of each type. Steam pressures, corrosive conditions, and individual plant experience are important criteria for the selection of a trap type.

The reputable manufacturers are prepared to recommend the use of their traps for services in which they will be successful. These manufacturers should be consulted for assistance in trap selection. Many publish

TABLE 18-5. Steam Trap Characteristics

| | Operation | Features |
|---|---|---|
| 1. Inverted bucket | Steam accumulates in inverted bucket causing it to float by increasing the buoyancy of the bucket. Movement of valve lever attached to bucket closes discharge valve. Entrance of more condensate causes bucket to sink closing valve | Fast, low steam loss, simple, must be protected from freezing |
| 2. Open bucket | Bucket floats with open end up. Condensate fills bucket causing it to sink and open valve. Steam pressure discharges condensate out discharge pipe in bucket which causes bucket again to float in remaining condensate | Operates well under fluctuating loads |
| 3. Piston-operated | Inverted bucket trap which operates a pilot valve which in turn operates a large piston valve | Good for very high capacities (up to 300,000 lb/hr) |
| 4. Thermo-static | Valve attached to bellows containing volatile fluid. When steam contacts bellows, increased vapor pressure of fluid causes expansion of bellows and closing of discharge port | Cannot freeze. Very rapid air venting and high condensate capacity |
| 5. Liquid expansion | Similar in principle to No. 4 except expansion of a liquid (oil) is used to move valve | Cannot freeze. Continuous discharge at any desired temperature |
| 6. Ball float | As condensate collects in trap body it causes a ball float connected to the discharge valve to raise and close the valve | Discharges condensate continuously at steam temperature and without shock. |
| 7. Impulse | One working part, a valve. At or near steam temperature, condensate that has collected in a chamber during discharge flashes and the increase in pressure closes the discharge valve | Cannot freeze. Very small and light weight. High capacity to size ratio |

helpful charts and design procedures which are time saving and are based on considerable operating experience.

In specifying traps for a project a list of all the traps is usually prepared on a drawing or other reproducible form. The list includes a detailed drawing of one or more typical trap installations with a list of the necessary valves and fittings for each type of installation. Numbers for identification and material, pressure and flow specifications are also given. The list is used for (1) material take-off of all the small valves and fittings required for trap installations; (2) procurement operations; and for (3) an installation guide. Trap numbers are used in identification, both on piping drawings and during warehousing and erection.

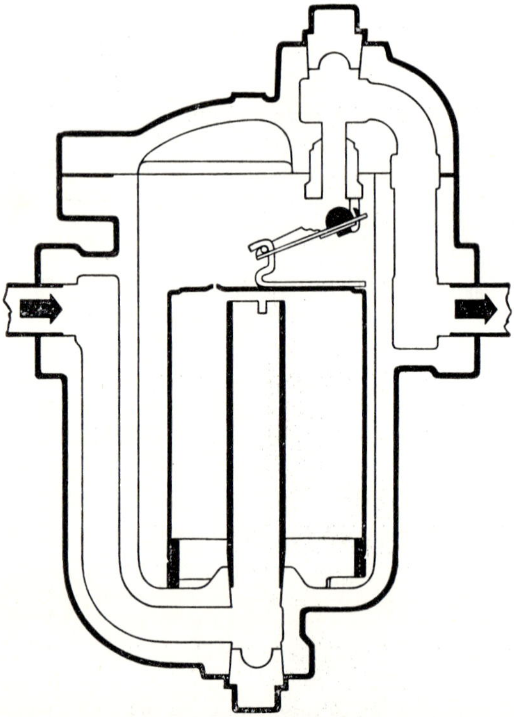

Fig. 18-16. Inverted bucket trap. (Courtesy of Armstrong Machine Works.)

## Expansion Joints

The manufactured expansion joint is used for the reduction or elimination of stresses and for overcoming excessive vibration. Examples of the two most common types, the slip-joint and the packless-joint are shown in Fig. 18-17. Each of these joint types is made in a variety of styles and materials to fit every need. The slip-joint has a stuffing box to prevent leakage; and, therefore, requires more maintenance than the corrugated packless joint. The slip joint simplifies expansion calculations because its length of travel is fixed.

The pressure, temperature, and movement in each expansion joint must be carefully investigated and specified. The specifications must include description of fluid, maximum pressure, maximum temperature, cold or

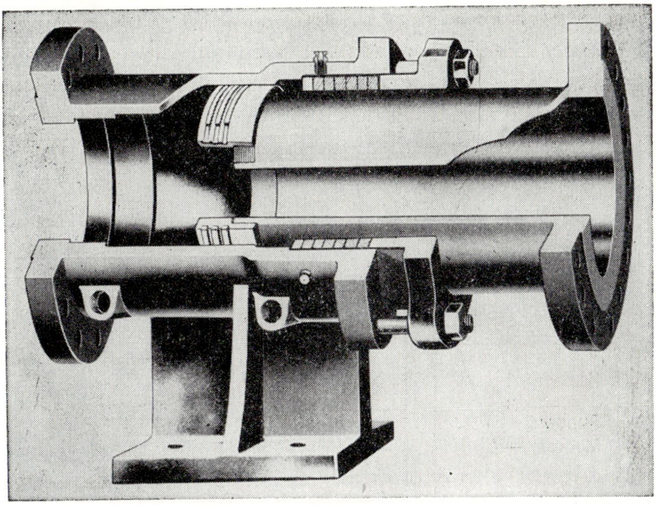

(a)

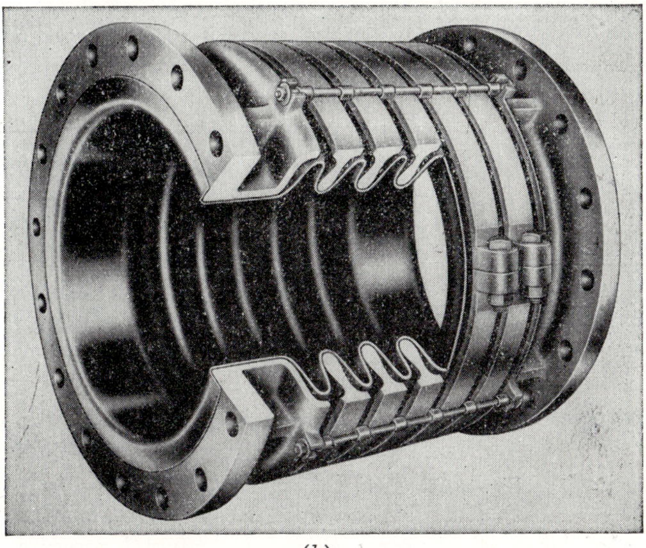

(b)

Fig. 18-17. Expansion joints. (a) Piston-ring expansion joint (internally guided); (b) packless expansion joint. (Courtesy of Adsco Industries, Inc.)

installed position, cold dimension, hot position, hot dimension, axial deflections, and estimated cycle periods from hot to cold.

If a number of expansion joints are to be installed, procurement is simplified by a drawing which shows the various types of joints required in two or more views. A tabulation on the drawing lists the joints by number together with operating data. Using this drawing the manufacturer can prepare his proposal with a minimum of written information.

The data requested from the manufacturer should include recommended joint type and material, magnitude of forces necessary to compress, extend, or cause axial movement in both hot and cold positions, and

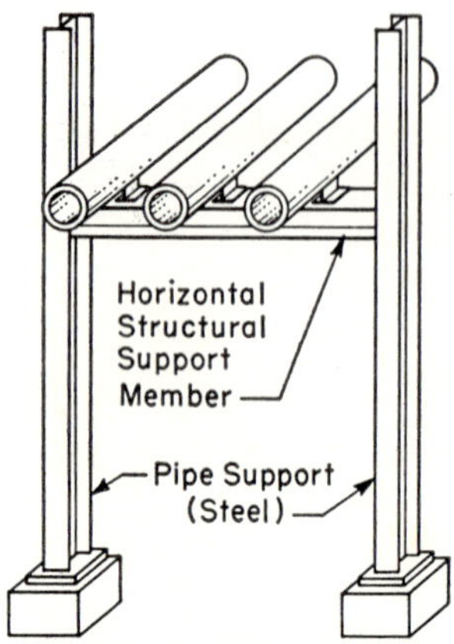


(*a*) Structural Steel Pipe Supports

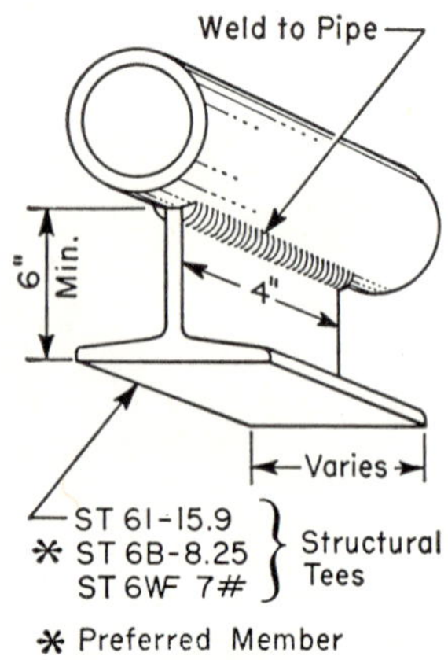


(*b*) Pipe Shoe

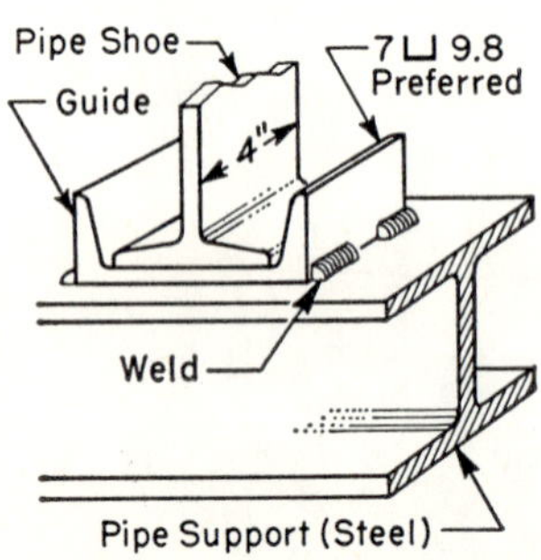


(*c*) Guide for Pipe Shoe

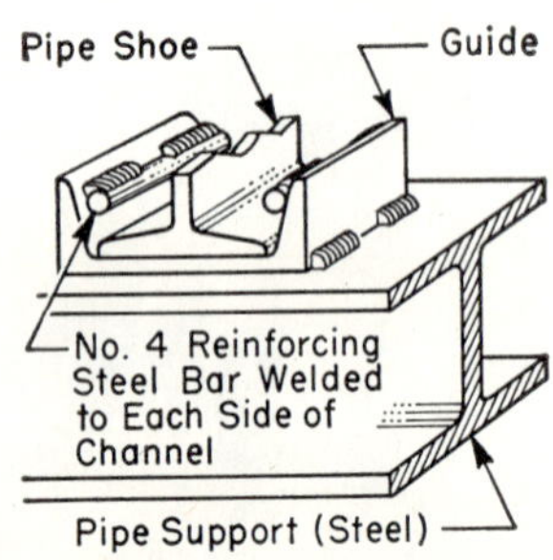


(*d*) Method for Restricting Vertical Movement

Fig. 18-18. Pipe supports and anchors.

magnitude of longitudinal forces caused by internal pressure. These data should be added to the expansion joint list for use in piping stress analysis.

### Pipe Hangers and Supports

In earlier process piping installations nearly all piping was hung by means of clamps attached to rods or chains, to provide greater flexibility or freedom. This method of support is still used for piping installed inside buildings. In the modern process plant, however, most equipment is located outside and piping is run on structural steel supports (Fig. 18-18*a*).

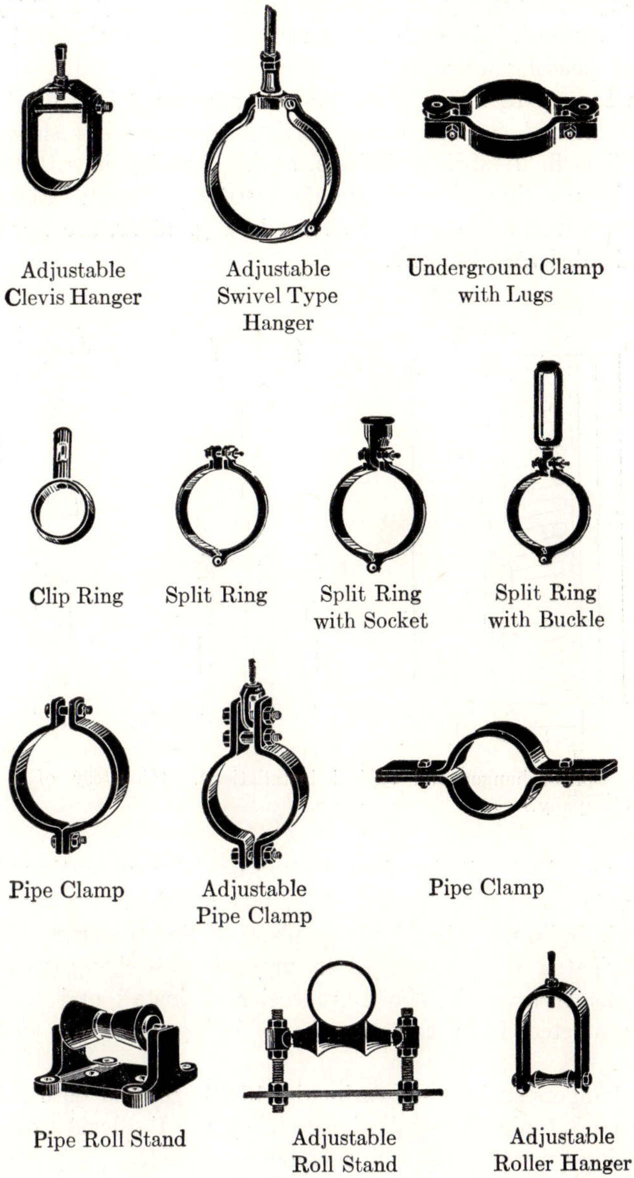


Fig. 18-19. Manufactured hangers and supports. (Courtesy of Elcen Metal Products Company.)

Since the piping is laid across the supports it is simpler to hold the pipe by devices called shoes which rest directly on the structural members.

An inexpensive and simple shoe, such as shown in Fig. 18-8*b*, may be either a tee beam or the more common split structural H or I beam. It is unnecessary and usually undesirable to attach piping securely to each structural support. At each alternate support or each third or fourth support, depending upon the line size, a guide is usually provided as shown in Fig. 18-18*c*. The guide permits movement in a fixed direction. A simple method for restricting vertical movement is shown in Fig. 18-18*d*. The field fabricated shoes, anchors, and guides illustrated are usually cheaper than similar manufactured equipment. Manufactured hangers, however, such as those shown in Fig. 18-19, are less expensive than field-fabricated hangers.

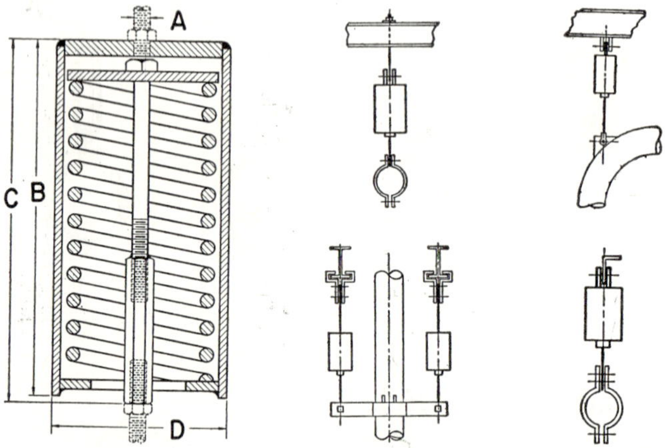


Fig. 18-20. Spring hanger and typical installations. (Courtesy of Elcen Metal Products Company.)

Hangers are best suited and most often used for supporting pipe inside buildings. However, horizontal structural members are used for supports inside buildings when large numbers of lines are run side by side. Field fabricated shoes are used in preference to hangers for outside installations because the shoes are easier to install and can be more readily protected from corrosion by painting since all surfaces are accessible.

Spring hangers such as shown in Fig. 18-20 are used both indoors and outdoors on hot piping installations. They are the least expensive and most reliable type of flexible pipe support available.

In designing pipe supports it is usual and good practice to set a specific bottom-of-pipe (B.O.P.) elevation for all pipe run at the same relative elevation. The word elevation is intentionally used since the dimension is measured from the bottom of the pipe to grade. When shown on drawings as 125, for instance, the number 125 indicates that the piping is 25 ft above nominal area grade set arbitrarily and for convenience at 100 ft.

By setting all pipe in any layer at a fixed B.O.P. elevation, practically all shoes except the guides will be the same height which greatly simplifies their installation. Further, a constant B.O.P. elevation fixes the position of the horizontal structural member of the pipe support which simplifies design and construction.

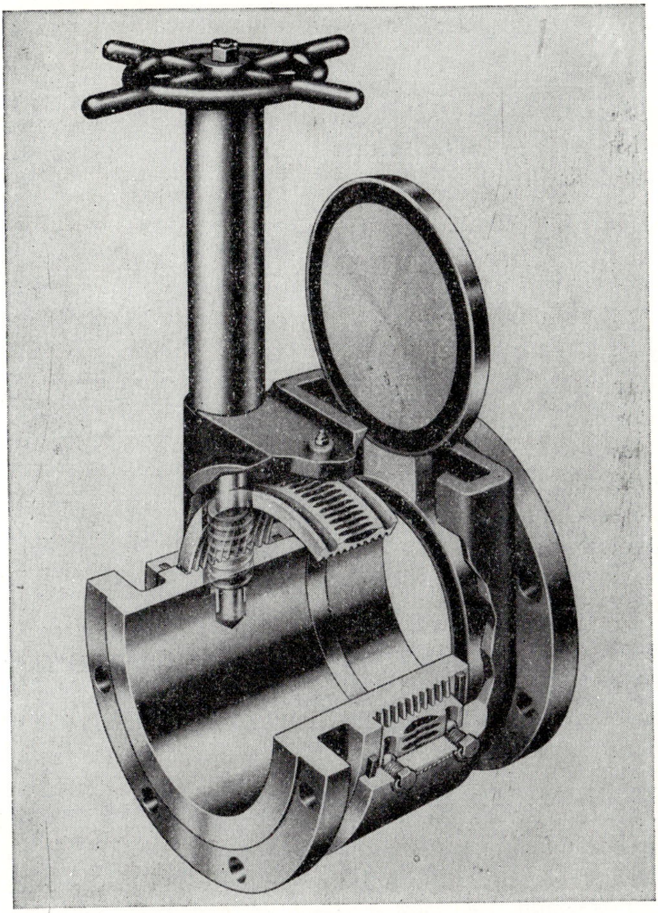

Fig. 18-21. Manufactured line blind. (Courtesy of Hamer Valves, Inc.)

## Line Blinds

The manufactured line-blind (Fig. 18-21) is used to seal off a line or piece of equipment. It provides a more positive and reliable seal than a conventional valve and is valuable when portions of operating systems must be isolated for maintenance purposes.

It is more costly but more convenient than the conventional homemade spectacle or figure 8 blind which is a metal plate in the shape of a figure

8 with a hole the size of the line on one side of the 8. The line is sealed by inserting the solid portion of the plate between flanges. This is accomplished by partially unbolting and springing the flanges apart after first draining the line. With manufactured line blinds the line is sealed merely by manipulation of levers and a handwheel. The spectacle plate is released and clamped as desired by means of a movable hub which is controlled by a hand wheel-operated worm and ring gear.

### Other Accessories

There is a wide variety of additional accessories and specialty equipment used in piping. Manufacturer's catalogs[7,31] should be studied for a complete description.

## PIPING LAYOUT AND ARRANGEMENT

In the layout or initial arrangement of piping for the process plant the designer has six related problems to consider.

1. Process requirements
2. Transmission of excessive stresses from piping to equipment, and the transfer of vibration from equipment into the piping system
3. Economy
4. Operational accessibility
5. Maintenance and replacement
6. Excessive stresses in the piping system

### 1. Process Requirements

Except for some particular "process reason"* the best arrangement is not always accomplished by connecting directly from one piece of equipment to another as shown on a flow diagram. A study of any process plant piping installation will show that essentially all the piping is arranged to run in definite patterns, always changing direction with a 90° turn, whether the change is in the same plane or in a different plane. When hundreds of lines must be routed to various portions of a plant, it is essential that this orderly arrangement be used. Routing each pipe in a straight line would create a confusing and impractical arrangement.

Since all equipment is subject to some kind of mechanical maintenance, and inspection and cleaning, it must be accessible. Safety demands that certain equipment be located at a particular point or as distant as possible from other equipment.

The simplest piping arrangement is made possible when the equipment is located in parallel rows or in some other rectangular pattern. Identical or similar process equipment, such as pumps, vessels, or heat exchangers, is usually grouped in consecutive rows. The location of power-driven

* A peculiarity dictated by the process.

equipment such as pumps in a group simplifies power supply and control systems and thus reduces costs.

Connecting piping, as far as is possible, is always run overhead in racks above the smaller equipment which is located near grade. The primary purpose in elevating piping is to provide clearance around equipment.

It is usual practice to group all parallel piping in a bank at the same elevation changing elevation when direction is changed. In this manner space is provided for connections between successive layers as well as for removal and maintenance. Care must be exercised, however, to avoid elevation changes which create pockets in lines where liquid can accumulate.

### 2. Transmission of Stresses and Vibration

The transmission of stresses from piping to equipment can be eliminated by the installation of manufactured expansion joints in such services as steam or exhaust connections to turbines or other steam-driven equipment. The possibilities of corrosion, erosion, and cyclic stresses in process services, however, often limit the use of mechanical expansion joints. For these reasons the piping designer usually attempts to eliminate piping stresses on equipment by proper piping arrangement and the use of supports and anchors.

There are two reasons for eliminating stresses at piping connections. One is to prevent the overstressing of the body of a machine or valve to which the piping is attached and the other is to prevent misalignment of equipment parts caused by differential movements. Manufacturers of equipment will not recognize any guarantee if a failure can be traced to external stresses.

Vibration in piping is usually encountered when piping is connected to moving machinery. Proper use of supports and hangers or vibration dampeners will eliminate most cyclic vibration. High-frequency vibration caused by high-speed machinery such as centrifugal compressors can be relieved by applying fixed supports. The correct location for these supports may be determined by experimental application of weights while the machine is in operation.

### 3. Economy

Cost considerations are often overshadowed by process requirements. A capable layout designer, however, will avoid wasteful use of fittings and supports. There are no precise rules which can be set down as each system will vary, but the best economies in piping are usually obtained by simplification of the piping specification.

### 4. Accessibility

Valves and other appurtenances which must be used during operation and/or which require periodic maintenance must be located in a convenient position. Such location may be at ground level or immediately adjacent to a platform.

## 5. Maintenance and Replacement

The problem of maintenance and replacement of piping is usually unimportant except for piping conveying extremely corrosive or erosive materials. Carbon-steel or alloy steel piping systems usually have a life of many years and replacement of large portions of a system would be a major problem not ordinarily considered in the overall design. Any particular piece of piping subject to extreme conditions of erosion or corrosion, however, should be arranged so that replacement is simplified.

## 6. Piping Stresses

### *Types of Stresses*

Essentially all piping system stresses or certainly all stresses which require involved stress analysis are the result of temperature changes. There are ordinarily three conditions which cause stresses in piping. These are: first, stress caused by internal or external pressure on the pipe wall; second, stresses remaining in the pipe wall after fabrication or erection; and third, stresses caused by temperature changes produced by the fluid flowing or by external conditions.

By selection of the proper pipe specification and wall thickness according to the rules of the Code, a safe design that will withstand the first condition can be made. The second condition can be eliminated by stress-relieving and proper fabrication. The third condition, however, is more intangible in that, although it can be anticipated, it cannot be precisely evaluated by any presently known simple analysis.

### *Magnitude of "Temperature" Stresses*

The stresses created in a steel pipe rigidly held at both ends when the temperature increases 1° F approximates 195 psi. ($S = ECT$ where $E$ = modulus of elasticity, $C$ = coefficient of expansion, and $T$ = °F temperature change.) The magnitude of possible longitudinal forces caused by one degree change in temperature is illustrated by applying this stress value to several sizes of pipe. Thus for 1 in. Schedule 40 pipe, the force would be $S \times$ metal area or 96.33 lb; for 4 in. Schedule 40, the force would be 618 lb and for 12 in. Schedule 40, 3069 lb.

Temperature changes in process piping usually are in the range of several hundred degrees. With a change of only 100° F the above forces would be roughly 5 tons, 30 tons, and 150 tons, respectively. However, these forces could only exist if movement were restricted so that the pipe was held rigidly in its cold position during heating. Even under these conditions if the stresses reached the yield point of the metal some plasticity would result immediately relieving the system. Some authorities believe that this in-service stress-relieving does occur in most highly stressed carbon-steel piping systems. In other words normal creep or deformation usually relieves the system with a subsequent reduction in stress.

### *Methods for Relieving Expansion Stress*

Some means must be provided for eliminating or reducing the forces that are produced by changes in temperature of piping systems. The piping must be made flexible so that it can relieve itself of the major portion of these forces. This can be accomplished by providing flexibility in the piping layout or by the use of expansion joints or cold springing.

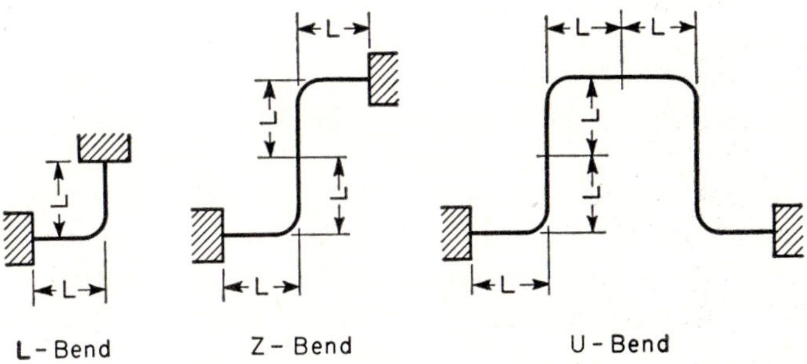


Fig. 18-22. Common expansion bends.

### *Flexibility through Layout*

Piping is rarely run in a straight line from one fixed connection point to another. Each time piping changes direction and is left free to move at the point of change, the system becomes more flexible. Rather than trust chance it is preferable to plan the entire piping layout with the expansion problem in mind. Fig. 18-22 illustrates various offsets and loops which provide flexibility in piping. The sketches are arranged from left to right in increasing order of flexibility, each bend having the same length of leg, *L*. The bends illustrated are made in one plane only. By making bends in more than one plane the flexibility is further enhanced.

### Expansion Joints

The installation of an expansion joint in a straight run of pipe firmly anchored at both ends will alleviate the stresses developed by expansion. The joint connects two sections of pipe in the manner of a coupling but still allows movement.

Manufactured expansion joints are not extensively used in process plant piping. They are used for utility piping and in process applications where piping size or space limitations would render changes in configuration impractical or costly.

#### Cold Springing

Forces generated by expansion are caused by movement of piping as its dimensions change from atmospheric to operating temperatures. One method for decreasing the final intensity of stresses in the hot condition is to cut the pipe short and spring it into place while cold. This springing produces a tensile stress on the system which changes to a compression stress in the expanded hot condition. Both of these stresses will have a lower value than that which would occur if the pipe had been cut to exact dimensions for the cold condition.

This technique of installation is called cold springing. It is discussed in paragraph 620 (*e*) of the Code which defines the limit of stress reduction to be assumed when pipe is so installed. The Code rule applies to actual erected lengths. For this reason when pipe is being shop fabricated, corrections for temperature differentials between shop atmosphere and erection atmosphere must be made. These corrections are required when alloy or stress-relieved pipe is used since these materials cannot ordinarily be reworked at the point of erection. When cold spring is practiced, actual field measurements of the space for any given piece of piping should be made.

### *Design Calculations for Expansion Stresses*

The piping code has set permissible or maximum stress temperature relationships for each acceptable pipe material specification. Chapter 3 of Section 6 of the Code composed of paragraphs 618–620, inclusive, contains various rules, limitations, and requirements for piping stress calculations. Paragraph 620 (*a*) states in part, "Formal calculations shall be required only where reasonable doubt exists as to the adequate flexibility of the system."

Methods for analyzing stress of piping systems are not given in the Code. It functions more as a guide toward good practice and is not a completely specific instrument which covers in detail all possible conditions of piping design. The responsibility for proper interpretation and application of the rules remains with the designer.

The presently available comprehensive methods of piping stress analysis* are extremely rigorous and, therefore, time consuming and

* One excellent method has been developed which is readily adaptable to high speed computing machines.[35] This reference is also a good source of information on all problems of piping flexibility.

costly. Several short methods of piping stress evaluation which utilize charts for the more common factors are offered.[18,34] The simplified methods, however, are not completely acceptable to all designers. Even the more rigorous or formal calculations depend upon assumptions, some of which are admittedly questionable. The designer, therefore, can only expect to err on the safe side regardless of the method used, the degree of error increasing with the more simplified procedures. It would appear that analyses which err on the safe side and yet permit the piping configuration to remain within reasonable economic limits should be acceptable.

In order to illustrate the variation in stress with configuration, the results of a calculation on an expansion U-bend are given in Table 18-6.

**TABLE 18-6. Variation in Stress with Configuration for an Expansion U-Bend**

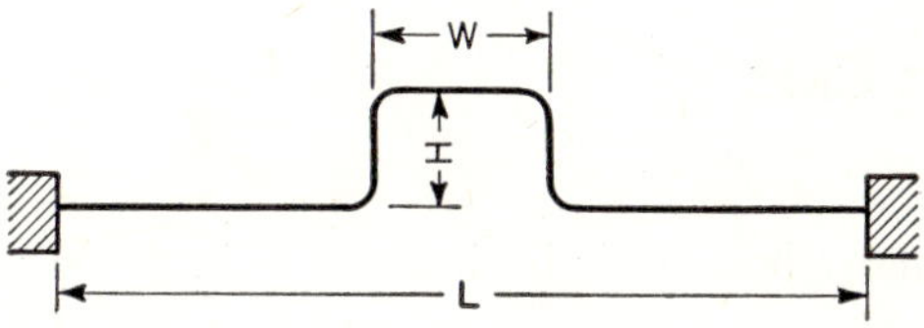


| Case | Length in ft *L* | *H* | *W* | Stress, psi | Total Pipe ($L + 2H$) in ft |
|---|---|---|---|---|---|
| 1. | 200 | 10 | 20 | 20,990 | 220 |
| 2. | 200 | 11 | 20 | 18,544 | 222 |
| 3. | 200 | 11 | 21 | 18,077 | 222 |
| 4. | 200 | 12 | 20 | 16,738 | 224 |

*Basis*

(a) 6 in. Schedule 80 pipe, API-5L, Grade C
(b) 600° F operating temperature, 250 psig operating pressure
(c) Allowable stress = 19,300 psi
(d) Maximum expansion at 70° F is 4.72 in./100 ft
(e) Maximum allowable cold spring permitted by Code is used
(f) Bends are 90° butt-welding, long-radius elbows
(g) For simplification guides were not considered although such a bend would require guides

Case 1 shows an $S$ value of 20,990 psi which exceeds the allowed 19,300 by 1,690 psi. Both Cases 2 and 3 are well below the allowable with an increase of only one ft in $H$ for Case 2 and an increase of one ft in both $H$ and $W$ for Case 3. Case 4 shows a decrease of 4,200 psi by the addition of two ft to $H$ only.

The changes in pipe length cause an increase in piping material cost of approximately $6.00 for Cases 2 and 3 and $12.00 for Case 4. The total cost for installation of the example run of piping would be very close to

\$2,000.00. The possible increase is less than ½ of 1% for 2 and 3 and less than 1% for 4. Since design costs are approaching \$5.00 per manhour of design time (including the costs of engineering, drafting, and overhead), little if any saving can be made by extensive computations for simple problems such as these.

Generally design organizations utilize curves for the calculation of stresses in simple U-bends so that the proper configuration can easily be determined. Reference 18 includes a tabulation giving the relative accuracy of various methods applied to ten configurations. It is recommended that this table be studied in detail.

References 2, 18, 34 are particularly recommended for those interested in the various methods of stress analysis. Reference 18 is incomplete for all configurations at this writing, but, the tables of values repeatedly used in computations which are given eliminate many of the simple or basic computations and are thus very effective time savers.

## SIZING PROCESS LINES

### Basic Equations

The overall energy balance is basic to all fluid flow problems, and it is helpful to recall the principles of the first law of thermodynamics. This law states that the total amount of energy entering a system must be equal to that leaving a system. The expression of the first law, neglecting surface, magnetic, and electrical energy, can be written in differential form as follows for one pound of fluid flowing,

$$dU' + d(Pv) + \frac{du^2}{2g_c} + dZ\,\frac{g}{g_c} = dq - dW \tag{1}$$

where $U'$ = internal energy per lb of fluid
$P$ = pressure, #/sq ft
$v$ = specific volume, cu ft/lb
$u$ = velocity, ft/sec
$Z$ = height above a datum in ft
$g$ = acceleration due to gravity, ft/sec$^2$
$g_c$ = factor to convert poundals to lb of force, lb-ft/#-sec$^2$*
$q$ = heat added to system, Btu/lb
$W$ = work done by system, ft-#/lb*

The expression $Pv$ is often called flow work and represents the amount of energy required to force an element of fluid into or out of a system. Two additional basic concepts of thermodynamics may be stated as follows:

$$dU' = T\,dS - P\,dv \tag{2}$$

* For convenience use # to represent a pound of force and lb for a pound of mass.

$$dq = T\,dS - dF \quad \text{(for an irreversible process such as fluid flow)} \tag{3}$$

where $S$ = entropy
$F$ = irreversible energy losses called friction

Substituting 2 and 3 in 1 the differential equation

$$v\,dP + \frac{du^2}{2g_c} + \frac{g}{g_c}\,dZ + dF = -dW \tag{4}$$

is obtained which can be integrated to give

$$-W = \int_1^2 v\,dP + \frac{u_2^2 - u_1^2}{2g_c} + (Z_2 - Z_1)\frac{g}{g_c} + F \tag{5}$$

Eq. 5 is a basic expression for the work required to move a unit mass of fluid between two points. If the specific volume $v$ is constant as can be assumed for incompressible fluids (liquids), $\int_1^2 v\,dP$ becomes $v(P_2 - P_1)$.

The sizing of lines for a process plant requires the selection of a line size and then the calculation of the expected pressure drop for the proposed line. In such calculations sections of lines before and after pumps and compressors are most conveniently considered separately (suction and discharge lines). The work term then becomes zero for the section of line being considered since the device adding work to the system is not within the boundaries of the energy balance. The equation for pressure drop calculations then becomes

$$-\int_1^2 v\,dP = \frac{u_2^2 - u_1^2}{2g_c} + (Z_2 - Z_1)\frac{g}{g_c} + F \tag{6}$$

The velocity terms (kinetic energy) are usually very nearly the same and the expression $(u_2^2 - u_1^2)/2g_c$ can be neglected. The acceleration due to gravity in most locations is approximately equal to $g_c$(32.17) so that $g/g_c$ is most often called unity. Hence, assuming an incompressible fluid or an average specific volume, eq. 6 becomes

$$v(P_1 - P_2) = Z_2 - Z_1 + F \tag{7}$$

where $v(P_1 - P_2)$ is the "head" change in ft-lb of force/lb of mass or ft-#/lb, between points "1" and "2".
$Z_2 - Z_1$ is the difference in height between points "1" and "2". The units are also ft-#/lb since $g/g_c$ does appear although numerically it is unity. $Z_2 - Z_1$ is often referred to as static pressure.
$F$ = head loss in ft-#/lb due to friction.

It may seem to the reader at this point that undue stress has been placed on the background to the simple equation for pressure drop in a process line. The time is justified, however, if it serves to prevent the confusion that so often results over the term "head." It has been shown above that eq. 7 was derived from an energy balance based on a unit mass (one lb) of fluid flowing and that the energy terms appearing in the final equation have the units of ft-lb of force per lb of mass (ft-#/lb) and not simply ft as is often mistakenly assumed. An awareness of this fact prevents needless confusion especially in converting to other systems of units.

The estimation of the energy lost due to friction $F$ is the major task in calculating pressure drops. Experimental evidence from the studies of the drag exerted by the walls of a conduit on the fluid flowing has yielded the so-called "Fanning" equation:

$$F = \frac{fLu^2}{2\,g_c D} \tag{8}$$

where $F$ = friction loss in ft-#/lb

$f$ = proportionality constant called the coefficient of friction

$L$ = length of circular conduit in ft

$g_c$ = 32.17 ft-lb/(#)(sec)$^2$

$D$ = inside diameter of pipe in ft

By means of dimensional analysis the friction factor $f$ has been shown to be related to the Reynolds' number, $Du\rho/\mu$, and a dimensionless factor called the relative roughness, $\epsilon$, being the depth of irregularities in the pipe. Moody[16] has determined the relationship experimentally and the results of his work are presented graphically in Figs. 18-23 and 18-24. These charts permit the estimation of the friction factor $f$, for any size pipe. It is to be noted, however, that the data are for clean pipe since it is impossible to present in any logical manner the effects of the many types of scale that develop in process lines. Such deposits, however, can increase the ultimate in-service pressure drop over that for clean pipe as much as tenfold.

For some reason that is not completely rational, the Fanning equation has been presented in a number of forms. Most of the variations can be traced to the definition of the friction factor $f$. In some works it is defined such that it is one-fourth the value of $f$ presented here. The final value of friction loss will, of course, be the same if the proper equation is used with the correct friction factors intended for that equation. Because of this confusing situation, it is best to use only friction factor charts that indicate the definition of the particular $f$ value being plotted, otherwise a considerable error in the calculations (fourfold) may result. Since the charts of Moody have been selected as a standard by the Hydraulic Institute, it is good practice to use these charts and thus Moody's form of the Fanning equation.

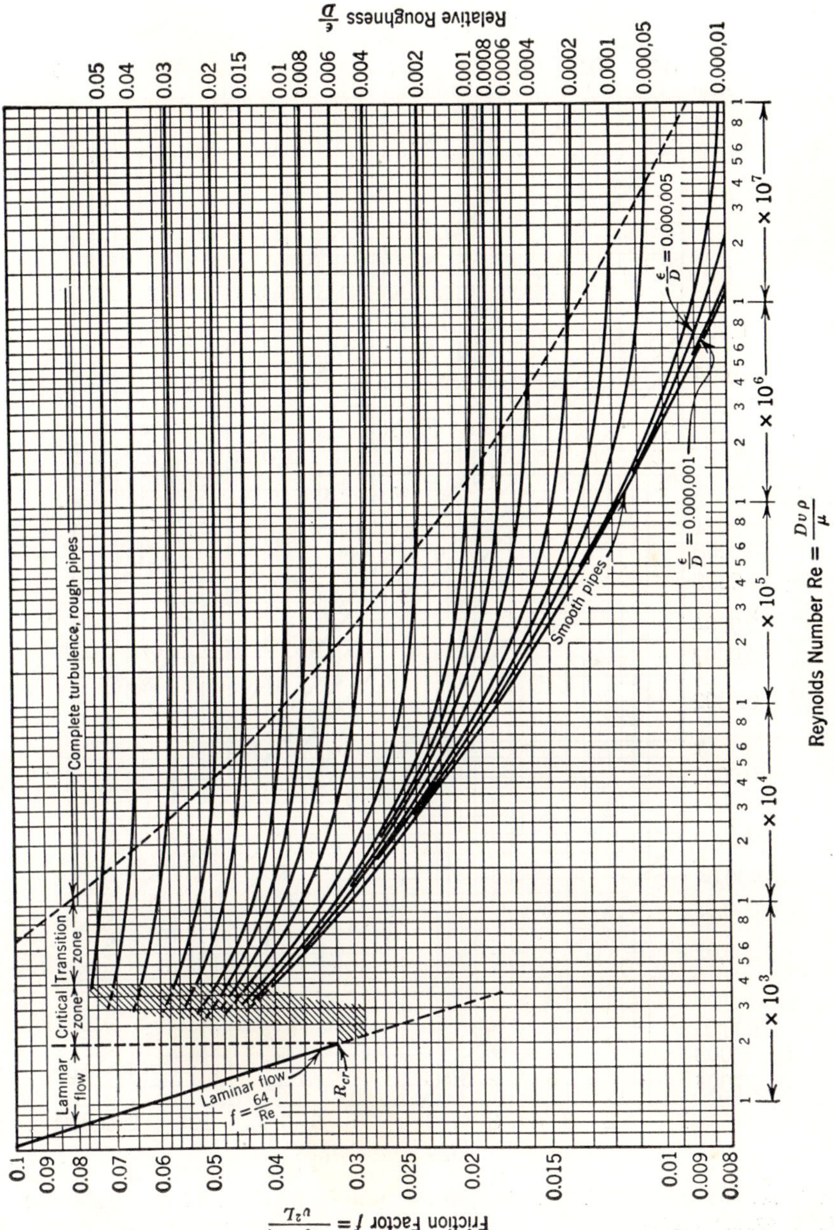


Fig. 18-23. Moody friction factor chart. [Reprinted by permission, L. F. Moody, *Trans. ASME*, *66*, 671 (1944).]

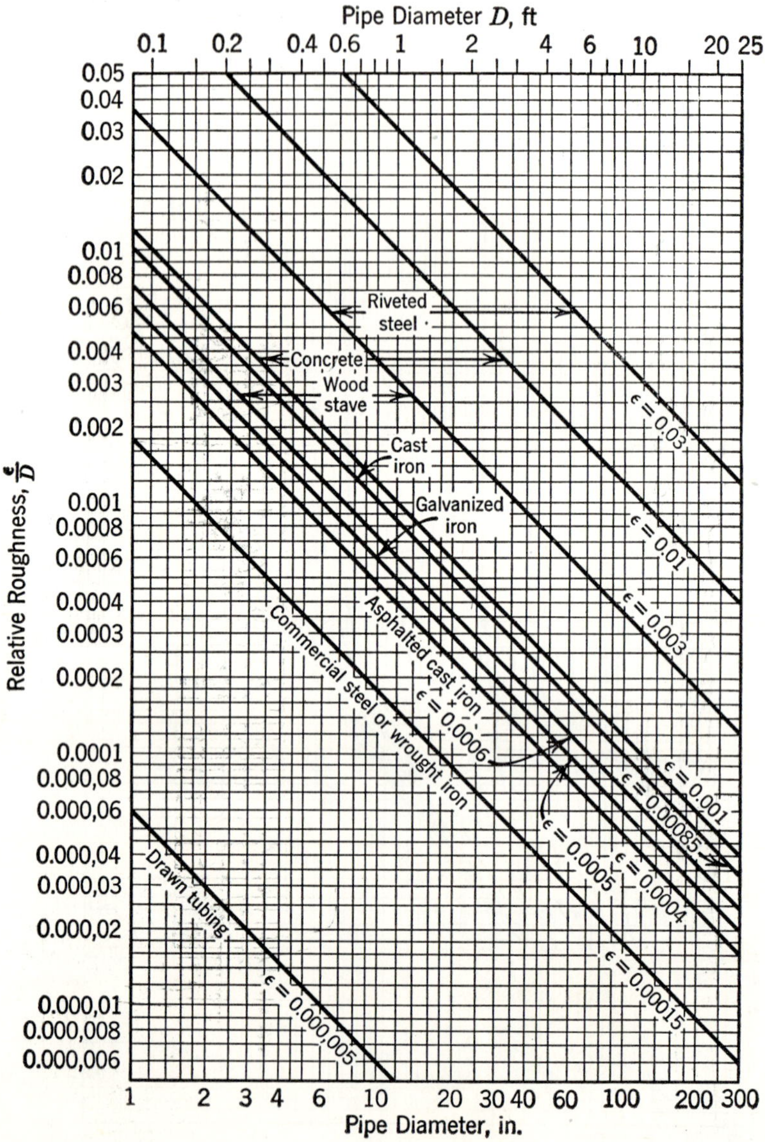


Fig. 18-24. Relative roughness. [Reprinted by permission, L. F. Moody, *Trans. ASME, 66,* 671 (1944).]

### Losses in Fittings

Eq. 8 permits calculation of friction losses in straight lengths of pipe. The various bends, fittings, and valves also require consideration. One of the simplest methods for estimating friction losses for valves and fittings is to add to the actual length of straight pipe an amount of straight pipe that will produce a friction loss equivalent to that which would occur in the fittings and valves. A means of estimating these lengths published by the Crane Company[8] some years ago and now in wide circulation is given in Fig. 18-25. This chart is entirely satisfactory for the usual estimates in process plant calculations. For more accurate information reference should be made to the Standards of the Hydraulic Institute.[23] It is reasoned that the friction loss in fittings consumes a portion of the kinetic energy of the fluid entering the fitting.

$$F = K \frac{u_1^2}{2g_c}$$

where $u_1$ = velocity of fluid upstream of fitting
$K$ = empirically determined constant for the particular fitting: called the resistance coefficient

Values of $K$ for various valves and fittings are given in the Standards of the Hydraulic Institute.

### Line Sizing Techniques and Criteria

The method of sizing process lines involves a technique familiar to engineers. A line size is assumed and the pressure drop is calculated. If the calculated pressure drop is satisfactory, the assumed line size is used.

The problem then becomes one of deciding what is a satisfactory pressure drop. If too high a pressure drop is used, pumping costs can become excessive. If too low a drop is used, costs of pipe, valves, and fittings become too high. This suggests an economic balance between material costs and pumping costs. Having unit costs for these two items, it is a simple matter to develop a general equation for total cost and solve for the minimum by differentiating and equating to zero. Charts for determining the "economic" line size can then be constructed. Many such charts have been presented, but such an approach represents a dangerous over simplification of a much more complex problem. When it is realized that many process lines do not involve pumping, power costs are not always well known or realistic; and most process lines represent a unique situation, it is understandable that generalizations regarding the economic size are very often worthless.

Many operating and design firms have prepared standards which

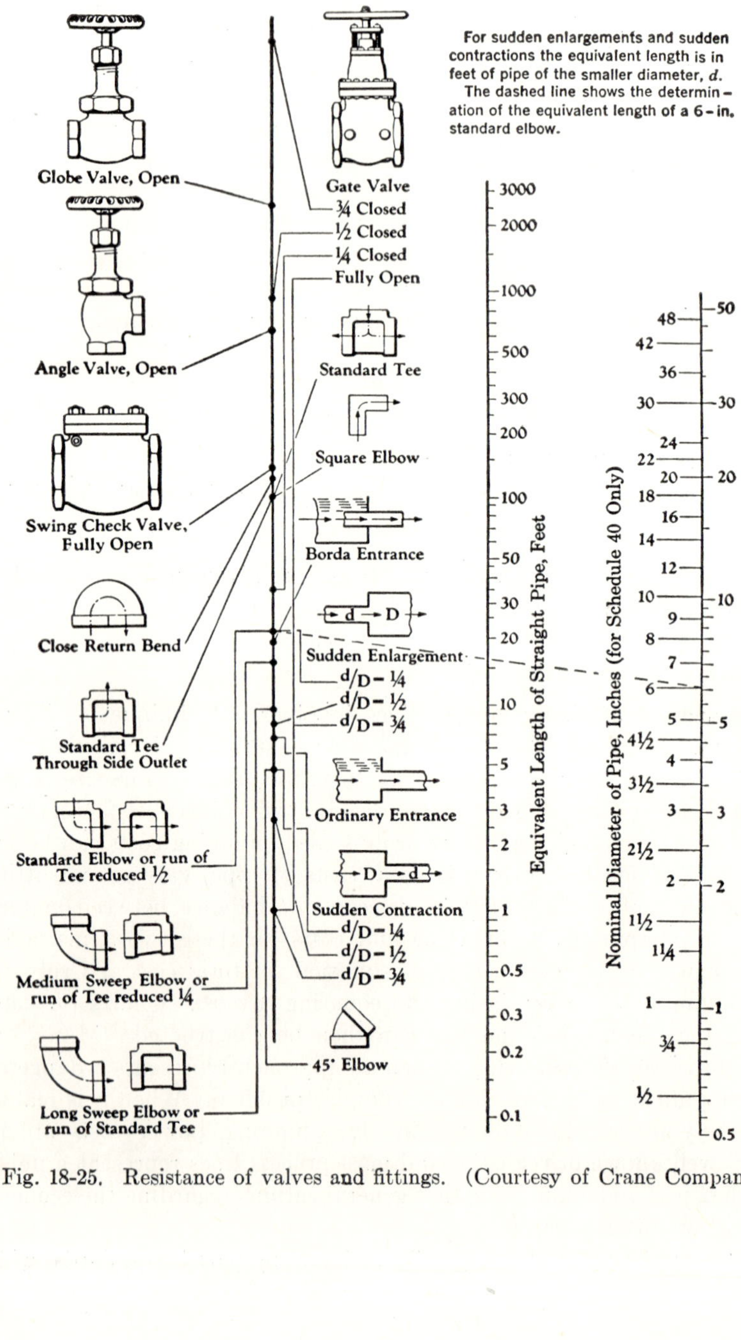


Fig. 18-25. Resistance of valves and fittings. (Courtesy of Crane Company.)

specify allowable pressure drops or reasonable velocities as guides in the sizing of process lines. Some typical standards are:

| | Reasonable velocity (Tube Turns Co.) | Allowable pressure drop due to friction |
|---|---|---|
| Pump discharge | $d/2 + 4$ ft/sec | 2 psi/100 ft |
| Pump suction | One-third the above | 0.5 ft-#/lb per 100 ft |
| Steam or vapor | *$d$ in 1000 ft/min | 0.5 per cent of line pressure per 100 ft |
| Gravity flow of liquids | — | 0.2 ft-#/lb per 100 ft |
| Water lines | 5–7 ft/sec | — |

* $d$ = inside diameter of pipe in in.

These and other similar rules-of-thumb have proved helpful to engineers for many years. They have, however, often been given the status of natural laws and thereby caused more harm than good. Neither the reasonable velocity method nor the allowable pressure drop can yield the exact economical pipe size for all cases. The velocity method does not allow for changes in fluid densities, and the allowable pressure drop method does not recognize that the economical pressure drop should be greater for small than for large pipes. Neither method allows for variations in material and power costs. Despite these objections, line sizing based on such standards is valid in the majority of cases. Although in many cases it may be argued that overdesign results, it should be realized that standards after long use become such a part of an organization that the overdesign that results is unconsciously expected.

It is impossible to make a detailed economic study of each line in a process plant in order to select the proper size nor is such a study usually justified. There are certain cases, however, particularly in the sizing of more expensive alloy lines or large carbon steel lines for which detailed study may be justified to aid in deciding between two possible line sizes.

A short-cut method has been presented[21] for rapidly estimating the costs of process lines so that values for two possible sizes can be compared rapidly. Should a large difference in cost be indicated further detailed study, using costs peculiar to the particular situation, will be justified. If the difference is small, the larger size will automatically be accepted. The data are given in Tables 18-7, 18-8, 18-9, and 18-10 and an illustration of their use is given here.

### Pressure Drop in Gas and Vapor Lines

Most process lines which carry gases and vapors are short and do not vary enough in elevation between the terminuses to justify a consideration of static pressure change or change in density. In such cases an average density is assumed based on an estimated average pressure, and the

**TABLE 18-7. Pipe and Fitting Prices***

Base price: Carbon Steel ASTM 106 Grade A or equal $0.105 per pound
Base weights: (Based on carbon steel pipe)

| Size, in. | Weight (lb/ft) Schedule 10 | Standard weight | Extra strong |
|---|---|---|---|
| 2 | 2.64 | 3.65 | 5.02 |
| 3 | 4.33 | 7.57 | 10.25 |
| 4 | 5.61 | 10.79 | 14.99 |
| 6 | 9.29 | 18.97 | 28.55 |
| 8 | 13.40 | 28.55 | 43.40 |
| 10 | 18.70 | 40.48 | 54.70 |
| 12 | 24.20 | 53.60 | 65.40 |
| 14 | 36.70 | 55.00 | 72.10 |
| 16 | 42.10 | 63.00 | 82.80 |
| 18 | 47.40 | 71.00 | 93.50 |
| 20 | 52.70 | 78.60 | 104.10 |
| 24 | 63.40 | 94.60 | 125.50 |

| Fittings | $L_e/D$, equivalent length to diameter ratio |
|---|---|
| 90° elbow, long and short radius | 1.3 |
| Tees, full size outlet | 2.1 |
| Tees, reducing outlet | 2.6 (12″ and below) |
| | 2.1 (14″ and above) |
| 180° bends, long radius | 3.0 |
| Reducers, concentric | 0.50 |
| Reducers, eccentric | 0.60 |
| Lap-joint stub ends | 0.80 |

| Material factors for pipe and fittings | Multiply by |
|---|---|
| Carbon steel | 1 |
| 4–6 per cent chrome, ½ per cent Moly | 3.25 |
| Stainless type 304 | 10.2 |
| Stainless type 316 | 16.07 |

* Reprinted by permission—Rase, H. F., *Petroleum Refiner, 32,* No. 8, 141–144 (1953).

### TABLE 18-8. Flange Prices*

Base Prices

(Basis: 150 lb ASA RF Welding neck ASTM A-181 (approximately 40¢/lb)

| Size | Cost, $ |
|---|---|
| 2 | 3.5 |
| 3 | 5 |
| 4 | 6.5 |
| 6 | 9 |
| 8 | 14 |
| 10 | 20 |
| 12 | 35 |

| Rating Factors<br>Rating, lb | Multiply by<br>Raised face | <br>Ring type joint |
|---|---|---|
| 150 | 1.0 | 1.5 |
| 300 | 1.22 | 3.5 |
| 600 | 4.3† | 5.0‡ |
| 900 | 5.5† | 6.5‡ |
| 1500 | 9.0† | 9.5‡ |

| Flange Type Factors | Multiply by |
|---|---|
| Welding neck | 1.0 |
| Slip-on | 0.65 |

| Material Factors | Multiply by |
|---|---|
| Carbon steel ASTM A-181 | 1.0 |
| 4–6 per cent chrome, ½ per cent Moly (ASTM 182 F5) | 3.50§ |
| Stainless type 316 | 5.0 § |

* Reprinted by permission—Rase, H. F., *Petroleum Refiner*, *32*, No. 8, 141–144 (1953).

† Use 4.0 for sizes 2, 3, and 4 in. only.

‡ Use 4.5 for sizes 2 in. and 3 in. and 5 for 4 in.

§ For 150 lb and 300 lb welding neck only.

### TABLE 18-9. Fabrication and Erection Costs*

| Fabrication Costs | Multiply carbon steel costs of pipe, fittings, and flanges by |
|---|---|
| Material | |
| Carbon steel | 1.0 |
| All nickel and nickel alloy steels (18-8, Monel) | 2.0 |
| 5 per cent chrome (no nickel) and higher steels | 2.5 |

Erection Costs

Approximately equal to cost of fabricated carbon steel pipe; i.e., material cost of carbon steel pipe, fittings, and flanges plus fabrication cost. This also applies for alloy pipe installations if no field welding is required. If field welding is required, the following multipliers are suggested for application to the cost of an equivalent carbon steel system to allow for special fit-up and field stress relieving.

| | |
|---|---|
| Stainless steels | 1.5 |
| 4–6 per cent chrome | 1.2 |

* Reprinted by permission—Rase, H. F., *Petroleum Refiner, 32,* No. 8, 141–144 (1953).

### TABLE 18-10. Valve Prices*

| Size | Base price (adjusted) cost steel 150 lb O.S. and Y. gate valve |
|---|---|
| 2 | 70 |
| 3 | 85 |
| 4 | 120 |
| 6 | 205 |
| 8 | 290 |
| 10 | 400 |
| 12 | 530 |

| Rating Factors | Multiply by | |
|---|---|---|
| Rating, lb | Raised face | Ring type joint |
| (125) | (See Cast iron below) | |
| 150 | 1.0 | — |
| 300 | 1.6 | 1.7 |
| 400 | 2.4 | 2.5 |
| 600 | 3.4 | 3.6 |
| 900 | 4.3 | 4.6 |
| 1500 | 6.7 | 7.1 |

* Reprinted by permission—Rase, H. F., *Petroleum Refiner, 32,* No. 8, 141–144 (1953).

TABLE 18-10 (continued)

| Material Factors | Multiply by |
|---|---|
| Materials | |
| 5 per cent chrome | |
| ½ per cent moly | 1.25 |
| Monel | 2.1 |
| Stainless steel, type 316 | 1.7 (150 lb only) |
| Cast iron (125 lb) | 0.45 |

| Valve Type Factor | Multiply by |
|---|---|
| Valve type | |
| Gate | 1.0 |
| Globe | 1.2 |
| Check | 1.0 for class 150 |
| | 0.70 for all others |

Illustration—Estimation of a Given Line.* Determine the erected and delivered cost of a 6-in. Schedule 40, 150 lb ASA rating (RF), carbon steel pump discharge line which is 325 ft long and is composed of 12 flanges, 16 weld ells, 1 tee, 1 check valve, and 2 gate valves.

*a.* Pipe and fittings (See Table 18-7.)

$$\frac{L_e}{D}, \text{ equivalent length, ft}$$

| | | |
|---|---|---|
| 325 ft pipe | | 325 |
| 16 ells | (16)(1.3)(6) = | 125 |
| | (1)(2.1)(6) = | 13 |
| | Total: | 463 ft |

$$\text{Total weight: } (463)\left(\frac{\text{lb}}{\text{ft}}\right) = (463)(18.97) = 8780$$

$$\text{Cost: } (8780)(0.105) = \$922$$

*b.* Flanges (See Table 18-8.)
(12)(9) = $108

*c.* Fabrication (See Table 18-9.)
Total cost flanges and pipe = 922 + 108 = $1030
Fabrication cost = (1.0)(1030) = $1030

*d.* Erection costs (See Table 18-9.)
Cost = 1030 + 1030 = $2060

*e.* Valve costs (See Table 18-10.)

| | | |
|---|---|---|
| 2 gate valves | (2)(205) = | $410 |
| 1 check valve | (1)(205)(1.0) = | 205 |
| | | $615 |

* Reprinted by permission—Rase, H. F., *Petroleum Refiner,* 32, No. 8, 141–144 (1953).

Total cost of system delivered and erected

$$(a) + (b) + (c) + (d) + (e) = \$4735$$

If this line were constructed of Schedule 40, 4–6 per cent chrome-moly with 300# RTJ chrome-moly flanges, the cost would be calculated by using the appropriate factors.

| | | | |
|---|---|---|---|
| *a.* | (922)(3.25) | = | \$ 3,000 |
| *b.* | (108)(3.5)(3.5) | = | 1,322 |
| *c.* | (1030)(2.5) | = | 2,580 |
| *d.* | 1030 + 1030 | = | 2,060 |
| *e.* | Gates: (2)(205)(1.7)(1.25) | = | 872 |
| | Check: (1)(205)(1.7)(1.25) | = | 436 |
| | | Total | \$10,270 |

friction pressure drop is calculated as for incompressible fluids.

When the pressure drop is large as in long lines, assumption of an average density is not valid. The pressure drop can be written neglecting static head,

$$v\,dP = \frac{fu^2\,dL}{2g_c\,D} \tag{9}$$

The specific volume can be expressed as $PM/zRT$, assuming an average absolute temperature $T$ and an average compressibility factor $z$ ($P$ is the pressure, $M$ the molecular weight of the gas, and $R$ is the gas constant). The velocity $u$ can be written as $zRTW4/PM\pi D^2$, where $W$ is the flow rate in mass per unit time. Substituting these values in Eq. 9, an expression called Weymouth's equation is obtained which allows for the change in density due to pressure drop.

$$P_1^2 - P_2^2 = \frac{16\,fLW^2zRT}{g_cM\pi^2D^5} \tag{10}$$

where $P$ = psf
$L$ = ft of pipe
$W$ = lb/sec
$D$ = inside diameter in ft
$R$ = 1544
$T$ = °R

When it is not possible to neglect the change in pressure caused by differences in elevation as in deep gas wells more complex equations result. Solutions have been presented,[6,12,19] but the process plant seldom presents such problems.

## PIPING SPECIFICATIONS

Carefully prepared piping specifications are required for the successful economic design and construction of process plants. Specifications describe pipe, valves, fittings, and piping appurtenances for each anticipated type of service and with piping drawings constitute a complete description of the piping system.

The first step in writing piping specifications is the preparation of a list of all the process and utility fluids in the plant. The list should include a description of each fluid, its pressure, temperature, and any service peculiarities. Careful study of this list based on a knowledge of the properties of piping materials as well as the operating conditions will reveal certain categories or ranges of service conditions upon which specifications can be based.

There are two extremes in specification writing that should be avoided. A separate specification can be conceivably written for each fluid and for a number of small operating ranges. The saving in purchase price, however, is spent many times over by the added clerical, engineering, and construction costs caused by too many specifications. The other extreme is preparing only one or two specifications with materials listed that can withstand the most severe operating conditions. All the less severe services will then, of course, be more than adequately served, which would be costly and actually wasteful.

Some optimum number of categories about midway between these two extremes must be found. A careful study must be made for each project to determine this optimum. A few extra weeks spent on the preparation of piping specifications early in the project will produce great savings in time and equipment costs.

### Nomenclature

As explained in Chapter 5 it is convenient to designate each fluid by some letter symbol and each category or specification by a convenient symbol. These symbols with the line number comprise the complete description of each section of piping for use on the mechanical or flow diagram, piping drawings, material lists, and in warehousing operations during construction.

Valves also are described by a numbering system. It is usual practice to include in the piping specifications a typical valve and catalog number with the words "or equal," immediately following the name. This is followed by a valve number. Use of the valve number on piping drawings reduces the size and complexity of notes on the drawing. Furthermore, the same number is used in the lists or bills of material and purchase orders. The manufacturer also places the valve number on a tag attached to each valve, enabling positive identification during storage at the jobsite.

While or after the piping specification is prepared a valve list in numerical order is made. It is common practice to show face-to-face dimensions and other pertinent data such as ASA flange ratings on the valve list. This list aids design-drafting since it provides in a single compilation all pertinent data for use on the piping drawings.

### Specification Preparation

In order to more fully explain the technique of specification preparation an example of a typical specification is given below. It is preceded by the reasoning which usually accompanies the selection of materials and specification categories.

Table 18-11 presents a summary of the fluids to be considered together with their service conditions and letter designations.

**TABLE 18-11. Summary of Fluids for Example Piping Specification**

| Fluid designation | Service and fluid | Operating pressure, psig | Operating temperature, ° F |
|---|---|---|---|
| HC | Hydrocarbon | 250 | 800 Maximum |
| HB | Hydrocarbon | 250 | 800 |
| HA | Hydrocarbon | 250 | 750 Maximum |
| H | Hydrocarbon | 100 | 600 |
| S | Steam | 275 | 410 |
| SE | Steam, exhaust | 40 | 290 |
| SC | Steam, condensate | 15 | 240 |
| A | Air, utility | 100 | Ambient |
| AC | Air, instrument, dried | 50 | Ambient |
| F | Fuel, natural gas, dry | 50 | Ambient |
| C | Caustic | 50 | Ambient |
| IN | Instrument piping | Same as line services | |
| WF | Water, fire protection | 100 | Ambient |
| WC | Water, process cooling | 75 | 85 |
| W | Water, utility | 75 | Ambient |
| WS | Water, sanitary | 40 | Ambient |

The number of specification categories has been restricted to five, M, N, P, Q, and Z. This was made possible by specifying certain items, such as bolting, according to the class of fluid. Further simplification was accomplished by careful planning so that the specifications have been limited to four types of flanges, two types of bolts and five types of pipe.

It is at this point that the procurement section of an organization can be most helpful in keeping the project engineer aware of changes in availability of materials. Before the specification is completed the engineer should determine if the materials specified are available, otherwise time will be wasted later in deciding upon substitutions. It may be desirable in some cases to list alternates for materials of uncertain supply.

The example specification should not necessarily be interpreted as a recommendation of certain types of valves, fittings, or piping materials. The specification is written in a complete form as an illustration of a piping specification and as a guide or check list for those preparing specifications and not as a recommendation of specific items. Indeed one of the most unfortunate practices in engineering design is the copying of specifications used on previous projects. Materials and equipment must be evaluated for each job and a carefully considered decision made on the basis of facts at the time of design.

### *Flange and Bolting and Gasket Selection*

150 lb ASA flanges and flanged fittings, according to ASA B16-5 of the Code, will be satisfactory for all the services in Table 18-11, except S, HA, HB, and HC for which 300 lb ASA flanges are required. F is also within the 300 lb ASA limit, which allows pressures of 340 psi and 300 psi, respectively, for 750 and 800° F. Raised-face flanges can be used for all of these services. For service HC, however, although operating at no higher temperature and pressure than HB (300 psig and 800° F. Maximum), ring-type-joint flanges (RTJ) will be specified since this service is for heater cross-overs, for which at times operating temperatures may be more severe. The RTJ flange facing (Table 16 of the Code) permits a temperature of 850° F and a service pressure of 300 psig. Thus a safety margin of 50 psi and 50° F is provided.

150 lb ASA flanges will be used for services A, AC, F, and C since this rating is the lowest nominal steel flange rating. AC, which is dried air for instrument services is often run in galvanized piping utilizing screwed malleable iron fittings and flanges. In this service galvanized piping will be used; however, all fittings will be forged steel and all flanges will be cast steel because of the possibility of fire. For the same reasons service A, utility air, F, fuel, and C, caustic, will also utilize steel fittings and flanges, but, the piping will not be galvanized.

Service IN, instruments, must have the same specification as the process line when in contact with the process fluid.

### *Gasket Selection*

Compressed asbestos ring type gaskets will be specified for all services except S, HA, HB, and C. For these more severe services corrugated iron, double jacketed, asbestos filled gaskets will be used. For HC, a ring type joint flange facing has been specified, thus oval iron ring gaskets are required.

### *Pipe Selection*

The thickness required for pipe in each service can be determined from the equation presented previously and recommended by the Code $[t = (PD/2S) + C]$. Reference to Table 21 of the Code indicates that ASTM A106 is satisfactory for services above 800° F. The allowable stress for Grade A at 800° F is 9300 psi and Schedule 80 pipe will be satisfactory. Schedule 40, ASTM A106 is adequate for HB and HA. For all other services except WS (and where cast-iron pipe is specified in W services) the less expensive seamless, ASTM A53 Grade A pipe will be specified. For service WS, sanitary water, ASTM A-120, which is a usual galvanized pipe specification, will be specified.

Pipe 2 in. and below, for all services except WS, will be seamless, Schedule 80 ASTM A-83. Schedule 80 pipe is used in the smaller sizes to provide structural strength. At the time of writing ASTM A-83, though more costly (it is made to conform to the more exact tubing specification) was much easier to obtain than ASTM A-106 or A-53.

For small air piping used in instrument air transmission lines 3/8 in. O. D. soft annealed copper tubing and compression type nonferrous fittings will be specified.

## Typical Piping Design Specifications

1.000 *General*

1.100 *Basis of Design*

The design, fabrication, and erection of all piping and accessories shall conform to practices specified in the Code for Pressure Piping ASA B31.1-1951, or latest issue thereof, and to the drawings and the following specifications. All references to the code herein shall be understood to refer to the above ASA Code.

1.200 *Definitions*

The terminology used throughout this specification and on the drawings will be in general that conforming to current trade practices.

*Anchor.* Point where piping is fixed is called the anchor point. The device attached to pipe or the complete structure if it serves the single purpose of fixing the pipe is called an anchor.

*Bleed.* A small valve provided to draw off liquids.

*Blind.* A plate sized to be inserted in a flange to isolate a portion of a system.

*Block Valve.* A valve furnished to shut off a separate system. The term is loosely used and may be understood to mean any valve which is used for shut-off service rather than for throttling.

*B.O.P.* Bottom of outside of pipe. Used for pipe support location.

*Control Valve.* Any one of a number of different types of valves, remotely operated from some type of instrument.

*Directions.* The Plant North direction shown on plot plans and other drawings will serve for orientation of equipment.

*Drip-Leg.* A vertical section of pipe located in horizontal piping to deflect and "catch" condensate.

*F.W.* Field Weld.

*Gradient.* The successive drop in elevation of piping to insure gravity flow and drainage.

*Guide.* Device controlling the direction of piping movement.

*Hanger.* A rod and clamp, a chain, or a spring device used for supporting pipe is called a hanger.

*Line.* A pipe run from one point to another. The designation, Line Number is a number and symbol appearing on the piping drawings which identifies the pipe according to size, process fluid, general location, and specification.

*P.E.* Indicates plain-end. Used to differentiate between threaded-end pipe and mill cut. A procurement term.

*P.S.* The designation P.S. is used to indicate Pipe Support, which support may be one of the above anchors, guides, or shoes or several combined to form the support.

*Random Pipe.* Pipe furnished in random, not cut, lengths.*

*Shoe.* Device welded to or clamped to a pipe which provides a bearing.

*S.P.* Sample point. A small valve with necessary piping to draw samples.

*Spec.* The line specification.

*Spool.* A short piece of pipe provided with flanges. (The term is also used to indicate pieces of fabricated pipe, but a piece of fabricated pipe does not necessarily always have flanges on each end. Lists of spool numbers are called spool-sheets. These are often used as notification to a fabricator that he may proceed with fabrication, i.e., receipt of the spool sheet releases all the listed spools.)

*Spool Number.* A number identifying a piece of fabricated pipe.

*S.T.* Steam trap. Usually schematically indicated, but understood to mean the trap and necessary valving and piping for by-passing in emergencies.

*Strainer.* A perforated metal sheet temporarily placed in a flanged joint to protect equipment from foreign matter in initial operation. Permanent strainers shall also be furnished where indicated.

1.300 *Line Numbering System*

The following numbering system shall be used as line identification on piping appearing on flow sheets and piping drawings. The complete line number shall consist of the following components, as illustrated: 4-HA-101-M—The first numeral is the nominal line size. The HA designates the service; the 100, the area number, and the M designates the piping specification applicable.

Line numbers shall change: (1) When size changes; (2) when specification changes; (3) when branches leave original line; (4) wherever an additional line number would simplify the engineering and drafting.

1.400 *Line Sizes*

Piping shall be sized in accordance with best engineering practice for pressure drops consistent with proper operation of equipment. In general the following limitation shall apply:

1.401 Sizes 3/8 in., 1¼ in., 2½ in., 3½ in., 4½ in., 5 in., and 7 in. shall not be used.

1.402 Drains, sample, and vent lines shall be ¾ in. minimum unless otherwise noted. Drains for hot piping shall be not less than 1½ in. nominal size.

1.403 Except for instrument transmission lines, piping run on support shall

* Carbon steel pipe is never bought in *exact* lengths. The vendor offers his pipe at a price per hundred ft. When pipe is shipped and when received, it is measured. Vendor submits invoices according to the length shipped.

Random lengths of pipe are usually 18 to 20 ft long. Double random is double this length, and is available in certain sizes and is requested when 40 ft sections would be advantageous for straight runs of pipe.

not be smaller than 2 in. Smaller sizes may be used for short runs less than one support space in length in special cases.

1.404 Underground lines beyond process area limits shall be 3 in. or larger. Sanitary water (WS) may be 2 in. minimum. Certain utility lines may be smaller, but must be considered as special cases and approval secured.

1.500 *General Notes on Piping Design*

1.501 *Aboveground Lines*

All process and utility lines within the process area and all yard lines outside from the process area except water lines, sewer lines, and certain designated pump and compressor lines shall be run aboveground on concrete sleepers, or overhead on structural steel supports.

1.502 *Arrangement of Aboveground Lines*

Aboveground lines running in the same plane shall be arranged so that the bottoms of the lines, exclusive of insulation, are at the same elevation. Sufficient height must be allowed to clear the insulation of insulated lines, and this height shall govern the elevation of uninsulated lines as well.

1.503 *Elevation of Aboveground Lines*

When piping is run in parallel groups, one elevation shall be selected for lines running to Plant North and South and another elevation for lines running East and West. If piping is run in "banks" or layers on pipe supports the clear dimension between bottom of pipe in one layer and top-of-pipe in adjacent layer shall be three times the nominal diameter of the largest pipe in either group, plus twice the extension of a flange beyond the outside diameter of the largest pipe in either adjacent layer. Lines shall ordinarily change elevation when they change direction, except at the discretion of the designer in special cases and for aboveground lines not running in a bank of piping. The dimension of clearance shall be the same as that given above. The largest pipe size in the bank shall always govern for clearances. (Plant North is an arbitrary direction set on Plot Plans for orientation purposes in all work.)

1.504 *Elevation of Underground Lines*

Changes in direction of all underground lines, except sewers and drains, shall be accompanied by a change in elevation except in the case of lines 24 in. and over. All underground piping shall be arranged as far as possible for free draining to some low point. Access shall be provided at drainage points.

1.505 *Location of Valves, Use of Chain Operators*

Diaphragm control valves, motor operated valves, special manual control valves and lubricated plug cocks shall be located so that they may be readily accessible from platforms, when elevated, or at grade, when possible.

The same will apply, in general, for block valves, check valves, etc., unless adjustment of the valve is not required during operation in which case extension stems or chain operators shall be used. Seven feet-three inches (7′-3″) is the maximum distance the center of valve wheel may be located above the operating level without the use of a chain operator or extension. Chains shall extend to within three feet (3′) of the platforms, from which the valve will be operated.

Use of chain operated valves, however, shall be kept to a minimum. Valves requiring extensions, stems, or chain operators shall be noted on drawings. Reach-rods or extension stems, field fabricated, shall be provided for valves 1½ in. and below rather than chain wheels.

1.506 *Overhead Clearance*

Overhead lines running from vessels or other elevated equipment to steel pipe supports or to concrete sleepers shall have a minimum clearance of 25′ 0″ above plant roadways, and railroad (rails). 9′-0″ shall be the minimum piping clearance above walkways and platforms and above grade in the immediate process area.

1.507 *Bent Pipe*

Pipe shall not be bent to a mean radius of less than five (5) times the nominal pipe size. Exceptions, when necessary, shall be accomplished by means of short radius welded fittings. Such special fittings shall be noted on arrangement drawings. A minimum straight run equivalent to two or more pipe diameters should be allowed between two adjacent pipe bends wherever possible. Carbon steel pipe 4 in. and below may be cold bent on bending machines. Alloy pipe or other pipe subject to cold working stresses shall be considered as special and shall be bent according to specified procedures.

1.508 *Angle Valves*

The use of angle valves shall be avoided.

1.509 *Drains, Vents and Test Openings*

Drains shall be provided at low points and vents at high points. The drain and/or vent shall consist of a 6000 lb forged steel pipe coupling welded into the line, a short schedule 160 nipple, and a ¾ in. Gate Valve in accordance with the line specification. In hot process lines the drain or vent shall not be less than 1½ in. nominal pipe size. For these connections a hexagon shaped threaded steel plug* shall be made-up into the valve. All such plugs shall extend at least 3 in. beyond threaded portion.

Test connections or threaded connections for future use in piping or vessels shall be fabricated the same as drains and vents except valve may be omitted and connection plugged. Shipping plugs shall be removed from such connections and replaced with hexagon steel plugs as noted above. Plugs for insulated equipment shall be fabricated to extend beyond insulation.

1.510 *Service Connections and Utilities*

¾ in. hose (manufactured, heavy) connections for steam, air, and water shall be provided at convenient points in the process area for general utility purposes. Utility lines at vertical vessels are to be run adjacent to vessels and outside of insulation but not outside of tower platforms.

1.511 *Connections for Pressure Gages and Pressure Instruments*

Connections for pressure instruments to indicate or record pump discharge pressures, shall be located in the discharge piping between the pump flange, near the pump and the first valve. Other points shall be noted on drawings specifically. All such pressure indicating or recording points shall be accessible. Test points may be accessible by ladder. Bleed-off valving for pressure instrumentation on

* The plug consists of a piece of carbon steel hexagon bar stock threaded with ordinary pipe threads on one end. So called threaded bull-plugs often crush during tightening, damaging the pipe coupling or the threads.

hot lines shall be provided with a double block (two valves) and bleed valve located between the two blocks. Instrument bleed sizes need not follow drain and vent sizing, but any bleed piping for hot services shall be not less than ½ in. nominal size and shall be arranged so that a rod may be driven up the valve and fitting. Bleed piping for hot fluids which may ignite on exposure to air should be supplied with a simple pipe coil cooler with water jacket. Bleeds from overhead lines and from gage glasses shall be piped to grade or nearest open drain except that the end of such bleed piping shall be visible from the primary or secondary bleed valve.

Valving arrangements for instrument connections will be shown on instrument connection detail drawings.

1.512 *Thermo-wells, Test Wells, Orifice Flanges, Miscellaneous Instruments*

All temperature instrument points and orifice flange locations shall be noted on piping drawings. The necessary fittings and arrangement shall be shown on instrument schedule and instrument installation drawings. All such installations shall be field fabricated, unless otherwise noted. Orifice flanges shall be located in the exact position shown on drawings. Thermo-wells should be accessible when practical.

1.513 *Steam Traps*

Typical and special steam trap assemblies shall be shown in detail on "Steam-Trap Schedule Drawing" and identified by the trap number. The steam trap location will be shown on piping drawings by symbol and number.

1.514 *Expansion Loops*

If stresses produced by expansion or contraction of piping cannot be reduced to the allowable limits of stress as defined by the Code, by changes in direction or elevation, expansion loops shall be provided. Such loops, for piping 6 in. and above shall be fabricated from welding fittings. Cold spring shall not be used.

All expansion loops shall in general be in a vertical plane, rather than horizontal. Loops may be hung downward from supports, and completely fabricated from welding fittings, i.e., two 90° weld ells—one 180° weld fitting, or four 90° long radius weld ells. Expansion loops shall be designed for a maximum stress of only 75% of that allowed by the code under the service.

1.515 *Piping at Pumps, Compressors, and Turbines*

1.5151 *Pumps*

Check valves shall be provided in discharge lines of all centrifugal pumps upstream of first block valve.

1.5152 Temporary strainers shall be fabricated from machine perforated sheet metal not less than $\frac{1}{16}$ in. thick for piping up to 8 in. nominal size and not less than ⅛ in. thick for all sizes above 8 in.

1.5153 Pumps and piping shall be so arranged that minimum clearance between projections of adjacent pumps or machines shall be 2′ 6″. The minimum operating aisle, which shall be at the driver end, shall be 5′ 0″. Adequate clearance shall be provided for the withdrawal of reciprocating parts or parts which require horizontal withdrawals.

1.5154 Suction and discharge piping shall be arranged in accordance with the "Standards of the Hydraulic Institute." Water pumps shall be installed with flooded suctions wherever possible. Otherwise, foot valves and ejector or injector systems shall be provided.

1.5155 Hot piping connections into equipment such as pumps and turbines shall be arranged and anchored to eliminate all possible piping stress being carried into machine.

1.5156 Piping connections to equipment shall not be run horizontally over or across equipment. Equipment must be left clear for removal by crane or other maintenance equipment. See 1.603.

1.5157 Lubricating system, gland-seal, and small water cooling piping shall be shown in detail on drawings to avoid possible connection errors.

1.5152 *Compressors*

Compressor piping shall be arranged to avoid excessive or cyclic vibration. See 1.5157 for lubricating oil and water systems.

1.5153 *Turbines*

Mechanical expansion joints shall be provided in all exhaust steam and live steam lines connecting to equipment. See 1.5157 for lubricating oil and water systems.

1.516 *Steam, Exhaust and Condensate Lines*

All secondary or saturated steam piping and exhaust steam piping shall be run with a gradient. Drip-legs with steam traps shall be provided in this piping at all junction points and other points considered necessary.

Steam traps shall be provided in all vertical expansion loops. The discharge from all steam traps shall be run to the nearest condensate return header except as noted on drawings.

Main steam distribution headers shall be run with a gradient, and drip-legs shall be provided for condensate removal.

Steam connections from steam headers to equipment drivers shall be provided with a gate valve located in a horizontal run at the header, and a globe valve adjacent to the driver. This valving shall be in addition to any control valves. Exhaust steam connections at the driver shall be provided with a gate valve adjacent to the driver. No block valve shall be provided at exhaust steam headers. ½ in. minimum nominal pipe size condensate blow-off connections shall be provided in both steam and exhaust connections at equipment and run to the nearest open box drain. Such lines shall not be submerged in liquid or run to bottom of the drain box.

1.600 *Fabrication and Fittings*

1.601 *Shop and Field Fabricated Pipe*

All pipe 4 in. and above shall be shop fabricated, utilizing welding fittings. (On foreign projects it is often economical to shop fabricate (weld) pipe down to 2 in. size to save U.S. expatriate labor.) All 3 in. piping shall be field fabricated utilizing welding fittings. All 2 in. piping (2½ in. when used) and smaller shall be field fabricated, utilizing screwed fittings. In certain cases for 2 in. piping only, welded joints may be more practical than screwed. Where welding is to be utilized it shall be clearly noted, (for 2 in. piping only) on drawings. All galvanized piping shall be run screwed. No welding shall be performed on galvanized piping.

1.602 *Changes in Direction*

1.6021 Changes in direction of all screwed piping shall be made with pipe bends where possible (see 1.507); otherwise, use screwed fittings, or as noted above in 1.601.

1.6022 Changes in direction of welded piping shall ordinarily be made with seamless buttwelding elbows. Bent pipe may be used, depending on the specification, the service, or the application. See 1.507. Pipe bends in piping 6 in. and above shall be considered as special cases and must be authorized.

1.603 *Flanged Fittings and Spools*

Flanged fittings shall be kept to an absolute minimum. Short flanged spools (see 1.200) for piping 4 in. and above shall be provided at each piece of equipment for removal purposes. At pumps and turbines the spool may be eliminated if valving or expansion joints provide removable pieces of piping. See 1.5156.

1.604 *Reducers*

Changes in size of screwed lines shall be effected with forged steel (only) screwed reducing fittings, except at screwed control valves where swaged nipples (seamless only) may be used. For flanged or welded lines seamless buttwelding reducers shall be used. The use of reducing screwed tees shall be limited to side outlet reducing only.

1.605 *Branch Connectors*

Branch connections in shop fabricated piping may be made without the use of buttwelding fittings when properly reinforced according to code requirements. Buttwelding fittings shall be used for branch connections in field fabricated welded lines. Flanges shall be provided for the reduced connection near the branch.

Manifold connections may be made by means of 6000 lb forged steel couplings welded into the pipe. All such connections must conform to code practice. Large size (above screwed) piping may be welded directly into headers for manifolds. Thread-end nipples welded into pipe bosses shall not be used.

1.606 *Maximum Length (Shipping)*

No shop fabricated integral piece of pipe shall exceed a maximum length of 40′ 0″, nor a width or height exceeding the limits of a nine (9′) foot rectangle, regardless of whether this limitation is indicated on the drawings. Field welds shall be understood to be required whenever the pieces exceed these limits.

1.607 *Connection Flanges and Blind Flanges*

Flanges shall be supplied in piping outside the process area limits only at selected points and shall be omitted or kept to a minimum. Flanges in steam piping systems particularly shall be kept to a minimum. All piping run as headers shall be terminated by a blind flange.

1.608 *Buttwelding Elbows*

Buttwelding 90° elbows shall be long radius type. Exceptions, where necessary, shall be noted on arrangement drawings.

1.700 *Testing*

1.701 *Field Testing*

After erection but before application of insulation, all lines shall be given a hydrostatic test as follows:

1.7011 Each section of piping shall be tested hydrostatically at a pressure equal to two (2) times the lowest primary service pressure rating of the fittings, valves, and flanges in the line, but in no case less than one and one-half times the actual normal working pressure. (In certain hydrocarbon and chemical processes all water must be removed from any part of the process system before operation. Such processes cannot be tested with water since the complete removal of the water from the system could not be effected without disassembly thus nullifying the test. Some other acceptable process fluid must be used for testing. Air may not be used since it will also leave water or lubrication oil as a contaminant. Such requirements must be developed early so that the fluid will be available.)

1.7012 The test pressure must not be greater than one-half times the maximum allowable working pressure for the pipe or as determined in accordance with testing requirements in the Code.

1.7013 The test pressure must not be greater than the maximum allowable pressure for the weakest piece of equipment installed in the line and included in the hydrostatic test. In general, major equipment such as vessels, exchangers, pumps, and compressors shall be isolated from pipe line hydrostatic test. When necessary for practicability, exchangers and vessels may be included with the connected piping, provided the piping test pressures are within the allowable cold pressure limits of the equipment.

1.7014 The lowest hydrostatic test pressure shall be 100 psig.

1.7015 The following equipment shall be excluded from the general hydrostatic test, but shall be tested with compressed air (and soap suds) at pressure equal to the allowable working pressure of the equipment and lines or equal to the maximum compressed air pressure, whichever is the lower.

1.70151 Instrument Air Lines

1.70152 Airlines to Air Motor Operated Valves

1.70153 Pressure Parts of Instruments in Gas or Vapor Service. All other pressure parts of instruments, however, shall be subjected to the general hydrostatic test, except when test pressure exceeds normal working pressure, in which case the instrument shall be isolated.

1.70154 Plant Air Lines

1.70155 Plant Fuel Gas Lines

1.7016 Relief valves shall be excluded from the general hydrostatic tests. Blinds shall be installed between the relief valve inlet or outlet and the section of pipe being tested.

1.7017 Where the test is to be made using water or other fluid, all air shall be vented from the lines as fluid is admitted.

1.7018 The piping and equipment being subjected to the hydrostatic test shall be maintained under pressure for a sufficient length of time to permit thorough inspection for leaks and defects.

1.7019 All underground pipe anchors for bell and spigot elbows and tees shall be completed previous to testing, to avoid the possibility of blowing out.

1.7020 *Shop Testing*

All flanged, shop-fabricated, carbon steel piping for foreign shipment shall be hydrostatically tested before shipment to twice the pressure corresponding to the ASA flange rating of the piece being tested. All alloy or heat treated shop fabricated piping shall be tested in the same manner whether for foreign or domestic use.

Carbon steel pipe fabricated without flanges will be subject to field test only after erection, and the shop test will be waived.

All shop tests shall be performed in the presence of and be witnessed by the purchaser's inspector or delegated representative.

1.8000 *Pipe Specifications*

1.8010 Specification M

This specification shall apply to the following services:

A 100 lb Air
C 50 lb Caustic
SC 15 lb Condensate
F 50 lb Fuel
H 150 lb Hydrocarbon
SE 40 lb Exhaust

Rating:

230 psi at 100° F
150 psi at 500° F
100 psi at 750° F (Maximum)

1.8011 *Pipe*

Seamless carbon steel, random lengths, P.E. for sizes 2 in. and below, bevelled for welding 3 in. and above, conforming to ASTM Spec. A53 Grade A, latest revision.

2 in. and smaller. Schedule 80. (May be ASTM-A106 or A83 if A53 is unavailable.)

3 in. through 10 in. Schedule 40.

1.8012 *Flanges*

Forged carbon steel, 150 lb ASA welding neck, $\frac{1}{16}$ in. raised face conforming to ASTM Spec. A181, latest revision, Class I, 0.35% maximum carbon. Note: Bore of all weld neck flanges to be the same as the I.D. of pipe with which used.

1.8013 *Fittings*

$1\frac{1}{2}$ in. and smaller. Screwed type, 3000 lb CWP forged carbon steel.

2 in. and larger. Buttwelding type, seamless carbon steel conforming to ASTM Spec. A234 and A106, latest revisions, Grade A. Inside diameter and wall thickness to be same as pipe with which used.

2 in. and larger. Flanged type, 150 lb ASA standard cast carbon steel, $\frac{1}{16}$ in. raised face.

Cast Steel Flanged Fittings. Do not ordinarily use.

1.8014 *Unions*

$1\frac{1}{2}$ in. and smaller. 2000 lb CWP forged carbon steel screwed, ground joint, integral steel seats.

2 in. and larger. Use flanges as specified above.

1.8015 *Bushings and Plugs*

$1\frac{1}{2}$ in. and smaller. Steel hexagon bushings, and steel hexagon barstock plugs. Use reducers where reduction is more than two sizes.

1.8016 *Bolting*

Alloy steel bolt studs, see 1.901.

1.8017 *Gaskets*

Asbestos filled double-jacketed corrugated iron, except for Specs. A and SC sheet asbestos ring gaskets shall be used.

1.80181 *Gate Valves*

1½ in. and smaller. 600 lb SWP forged carbon steel screwed, inside screw stem, union bonnet, 11–13% chrome stainless steel trim (V35). Note: For Spec. A and SC, see under 1.80381.

2 in. and larger. 150 lb ASA standard cast carbon steel, flanged, O.S.&Y., 11–13% chrome stainless steel trim; 1⁄16 in. raised face (V1). Note: For Spec. SC, see 1.80381.

1.80182 *Globe Valves*

1½ in. and smaller. 600 lb SWP forged carbon steel screwed, inside screw stem, union bonnet, 11–13% chrome stainless steel trim (V34). Note: For Spec. A and SC, see 1.80382.

2 in. and larger. 150 lb ASA std. cast carbon steel, flanged, O.S.&Y., 1⁄16 in. raised face, 11–13% chrome stainless steel trim (V2). Note: For Spec. SC, see 1.80382.

Note: Use Gate Valves for sizes above 6 in.

1.80183 *Check Valves*

1½ in. and smaller. 600 lb SWP forged carbon steel screwed, horizontal, piston type, 11–13% chrome stainless steel trim (V19). Note: For Spec. A and SC, see 1.80383.

2 in. and larger. 150 lb ASA Standard cast carbon steel, flanged, 1⁄16 in. raised face, swing-type, 11–13% chrome stainless steel trim (V3). Note: For Spec. SC, see 1.80383.

1.80184 *Plug Valves*

¼ in, Size. 2000 lb WOG forged carbon steel, screwed, lubricated plug cock (V110).

1½ in. and smaller. 300 lb WP cast carbon steel, screwed, wrench operated, lubricated plug cock (V58).

2 in. to 6 in. sizes. 150 lb ASA standard cast carbon steel, flanged, 1⁄16 in. raised face, wrench operated, lubricated plug cock (V54).

1.8020 Specification *N*

*This specification shall apply to services S, HA, HB, HC 300 lb process oil and vapor, 275 lb steam:*

Rating: 500 psi @ 100° F
375 psi @ 500° F
300 psi @ 750° F

1.8021 *Pipe*

Seamless carbon steel, random lengths, P.E. for sizes 2 in. and below, bevelled for welding 3 in. and above, conforming to ASTM Spec. A106 Grade A, latest revision.

10 in. and smaller. Schedule 80.
12 in. and larger. 0.375 in. Wall.

1.8022 *Flanges*

Forged carbon steel, 300 lb ASA welding neck, 1/16 in. raised face, conforming to ASTM Spec. A181, latest revision. Class I, 0.35% maximum carbon, except for HC, use RTJ Flange Facing.

Note: Bore of all weld neck flanges to be same as the I.D. of pipe with which used.

1.8023 *Fittings*

1½ in. and smaller. Screwed type, see 1.8013.

2 in. and larger. Buttwelding type, see 1.8013.

Note: Cast Steel Flanged Fittings; do not ordinarily use.

1.8024 *Unions*

1½ in. and smaller. 2000 lb CWP forged carbon steel screwed, see 1.8014.

2 in. and larger. Use flanges as specified above.

1.8025 *Bushings and Plugs*

1½ in. and smaller. See 1.8015.

1.8026 *Bolting*

Alloy steel bolt studs, see 1.901.

1.8027 *Gaskets*

Asbestos filled corrugated iron, except for Spec. HC which shall be 90 Brinell Soft, carbon steel, octagonal, ring.

1.8028 *Valves*

1.80281 *Gate Valves*

1½ in. and smaller. 600 lb SWP forged carbon steel screwed, inside screw stem, union bonnet, 11–13% chrome stainless steel trim (V35).

2 in. and larger. 300 lb ASA Standard Cast carbon steel, flanged, O.S.&Y., 11–13% chrome stainless steel trim, 1/16 in. raised face (V4).

1.80282 *Globe Valves*

1½ in. and smaller. 600 lb SWP forged carbon steel screwed inside screw stem, union bonnet, 11–13% chrome stainless steel trim (V34).

2 in. and larger. 300 lb ASA Standard cast carbon steel, flanged, O.S.&Y., 1/16 in. raised face, 11–13% chrome stainless steel trim (V5).

Note: Use gate valves for sizes above 6 in.

1.80283 *Check Valves*

1½ in. and smaller. 600 lb SWP forged carbon steel screwed, horizontal, piston type, 11–13% chrome stainless steel trim (V19).

2 in. and larger. 300 lb. ASA Std. cast carbon steel, flanged, 1/16 in. raised face, swing-type, 11–13% chrome stainless steel trim (V6).

1.80284 *Plug Valves*

¼ in. Size. 2000 lb WOG, forged carbon steel, screwed, lubricated plug cock (V110).

1½ in. and smaller. 300 lb WP cast carbon steel, screwed, wrench operated, lubricated plug cock (V58).

2 in. to 4 in. Sizes. 300 lb ASA standard cast carbon steel, flanged, $\frac{1}{16}$ in. raised face, wrench operated, lubricated plug cock (V59).

1.8030 Specification *P*

This specification shall apply to services W, WC, WS, WF:

| | | | |
|---|---|---|---|
| WF | 100 lb Fire Water | WC | 75 lb Cooling Water |
| W | 75 lb Service Water | WS | 45 lb Sanitary Water |

Rating: 125 psi @ 150° F

Note: All pipe and fittings for WS service shall be galvanized.

1.80311 *Pipe Above Grade*

Seamless carbon steel, random lengths, P.E. for sizes 2 in. and below, bevelled for welding 3 in. and above, conforming to ASTM Spec. A53, Grade A, latest revision, for above grade use only. For WS use ASTM A-120 galvanized.

2 in. and smaller. Schedule 80, except for WS which may be Schedule 40.

3 in. thru 10 in. Schedule 40.

12 in. thru 20 in. 0.375 in. wall pipe.

24 in. and above. 0.250 in. wall, seamless or weld.

1.80312 *Pipe Below Grade*

2 in. and smaller. Same as above, except galvanized.

3 in. and above. Cast iron, pipe, mechanical joint in accordance with ASA-A 21.6 or 8, Class 22.

1.8032 *Flanges*

150 lb ASA standard forged carbon steel, $\frac{1}{16}$ in. welding neck, raised face, except to be plain face where connected to plain-faced cast iron valves, fittings, and equipment. Flanges shall conform to ASTM Spec. A-181, latest revision, Class I, 0.35% maximum carbon. For steel pipe only.

Note: Bore of all weld neck flanges to be same as the I.D. of pipe with which used.

1.80331 *Fittings, above grade*

1½ in. and smaller. 300 lb standard malleable iron, screwed. Galvanized for WS up to 2 in.

2 in. and larger. Buttwelding type seamless carbon steel, conforming to ASTM Spec. A234 and A106, latest revision, Grade A, at same I.D. and thickness as pipe with which used.

2 in. and larger. Flanged type. Class 125 lb ASA standard Cast Iron.

1.80332 *Fittings to size required, below grade*

Class 150 mechanical joint fittings having the same laying dimensions as ASA Class 125 standard flanged fittings.

1.8034 *Unions*

1½ in. and smaller. 300 lb standard malleable iron, screwed, ground joint, brass to iron seat.

2 in. and larger. Use flanges as specified above, or mechanical joints. On cast iron pipe in accordance with ASA A 21.11.

1.8035 *Bushings and Plugs*

Hexagon steel bushings and hexagon steel plugs. Use reducers where reduction is more than two sizes. Note: All changes from cast iron to steel shall be effected by a flanged cast iron stub piece connected to a mechanical joint.

1.8036 *Bolting*

Standard carbon steel square head machine bolts with semi-finished steel nuts for steel piping. Bolting shall be supplied with cast iron piping in accordance with ASA A21-11.

1.8037 *Gaskets*

1/16 in. thick compressed asbestos. For raised-faced joints. Use ring gaskets for raised-faced joints and full face gaskets for plain-faced joints. Plain composition rubber gaskets shall be supplied with cast iron pipe.

1.8038 *Valves*

1.80381 *Gate Valves*

1½ in. and smaller. 150 lb SWP brass, screwed, rising stem, inside screw (V47).

2 in. and larger. 125 lb ASA standard cast iron, flanged, O.S.&Y., brass trim; and stem (V26).

2 in. and larger. 125 lb ASA standard cast iron, flanged, O.S.&Y. (fire service double-disk, nonrising stem, brass trim, only) (V117).

1.80382 *Globe Valves*

1½ in. and smaller. 250 SWP brass, screwed, rising stem, union bonnet (V51).
2 in. and larger. Use gate valves.

1.80383 *Check Valves*

1½ in. and smaller. 250 lb SWP brass screwed, horizontal piston type, Ni-Alloy seat and disk (V101).

2 in. and larger. 125 lb ASA standard cast iron, flanged, swing type, brass trim (V28).

1.80384 *Fire Hydrants*

Two hose connection, plant type hydrants (VF).

1.8040 Specification Q

This specification shall apply to service AC only:
50 lb Air for Instruments, dried.
Rating, 125 PSI @ 150° F

1.8041 *Pipe*

Seamless carbon steel conforming to ASTM Spec. A53, Grade A, latest revision, galvanized. ASTM-A83 acceptable for 2 in. and smaller.

2 in. and smaller. Schedule 80.
3 in. thru 10 in. Schedule 40.

1.8042 *Flanges*

Cast iron, 125 lb ASA screwed, flat face, galvanized.

1.8043 *Fittings*

2 in. and smaller. Screwed type. 300 lb standard malleable iron, galvanized screwed.

3 in.–6 in. and larger. Flanged type. 125 lb ASA standard cast iron. Note: Use either screwed or flanged as required.

Note: The use of cast iron flanged fittings shall be kept to a minimum.

1.8044 *Unions*

2 in. and smaller. 2000 lb CWP forged carbon steel screwed, ground, joint, integral steel seats.

3 in. and larger. Flanges are permissible as specified above.

1.8045 *Bushings and Plugs*

1½ in. and smaller. Steel hexagon bushings, and steel hexagon barstock plugs. Use reducers where reduction is more than two sizes.

1.8046 *Bolting.* Carbon steel machine bolts, see 1.901.

1.8047 *Gaskets.* $\frac{1}{16}$ in. compressed asbestos gasket.

1.8048 *Valves.* See 1.8038 for Valve Specifications.

1.8050 Specification *Z*

This specification shall apply to instrument services only:

Instrument Services Rating, see Line Services.

1.8051 *Pipe*

Instrument lead lines, or lines which carry fluids to and from instrument elements shall be the same specification or better, than the line service from which instrument line originates. Schedule 160 nipples shall be used up to first block valve where any danger of physical damage to piping is possible. Extensive runs of small sized lines shall be avoided, but where necessary structural support must be provided. Instrument piping of process fluids exceeding 25 ft must be authorized prior to installation.

1.8052 *Tubing*

Copper tubing shall be used for *air* transmission lines. All tubing shall be ¼ in. O.D. 0.032″ Wall ASTM, B68, nonarsenical cold drawn, vacuum annealed.

1.8053 *Fittings*

Flared-type fittings shall be used for all copper tubing lines.

1.8054 *Valves*

Valving used in instrument lead piping shall be of the same specification as the originating line specification.

1.9000 *Miscellaneous Specifications*

1.9010 *Alloy Steel Bolting*

Alloy steel bolting shall be provided for services S, H, HA, HB and HC and shall conform to ASTM-A193, Grade B7, Class 7 fit for bolt, Class 2B fit for nuts, with two (2) heavy series hexagonal nuts each. Nuts shall conform to ASTM Spec. A194, latest revision, Class 2H, oil quenched, hot forged. All other bolting shall conform to ASTM-A307 with Class 2A fit for bolts and 2B fit for nuts, excepting bolting furnished by pipe vendor for *W* services, which shall meet ASA-A21.11 requirements.

1.9020 *Cleaning and painting*

1.9021 After erection all lines shall be cleaned internally by circulation to remove all dirt and other foreign matter before the plant begins operation. Care shall be taken to thoroughly clean air lines to air actuated diaphragm control valves, before they are connected, and after connection up to last fitting before instrument or valve.

1.9022 All underground (buried) shop fabricated steel pipe 3 in. thru 10 in. size shall have coating on the outside only and wrapping applied as follows:

*a*. Clean by sandblasting.
*b*. Apply one coat of bitumastic primer.
*c*. Apply first coat of bitumastic enamel.
*d*. Apply second coat of bitumastic enamel.
*e*. Spirally wrap one layer of 15 lb asbestos felt.
*f*. Spirally wrap one layer of 60 lb Kraft paper.

1.9023 All underground (buried) steel pipe shall have inside and outside cleaned and wrapping applied as follows:

Specifications Inside (for 12 in. and above)
*a*. Clean by sandblasting.
*b*. Apply one coat bitumastic primer.
*c*. Apply one heavy coat bitumastic enamel.

Specifications for coating and wrapping outside of pipe shall be same as for 3 in. thru 10 in. size as specified above, 1.9022.

1.9030 *Fireproofing*

Pipe supports carrying main steam headers or hydrocarbons through areas subject to fire hazards shall be fireproofed to the point of the cross load bearing members, or as noted on drawings.

1.9040 *Steam Traps*

Inverted bucket-type steam traps shall be provided for all services except those handling small quantities of steam for which case impulse type traps may be used. All bucket type traps shall have forged or cast steel bodies with allowable working pressure of 600 psi minimum. All other traps shall be steel and have service rating at least two times line pressure.

## REFERENCES

Note: Certain manufacturers' catalogs have been included as references to descriptive information. The omission of other manufacturers' printed information should not be construed as a judgment against it. It would be impossible to include a complete list of the many excellent catalogs available.

1. American Gas Association, 420 Lexington Ave., New York, N. Y. Standards.

2. American Petroleum Institute, Standards on Pipe, Flanges and Valves (API-5*A*, *B*, *F*, *G*-1, *G*-2, *G*-3, and *L*, API 600 *A* and *B*). American Petroleum Institute, Division of Production, 1205 Continental Bldg., Dallas 1, Texas.

3. American Society of Mechanical Engineers, 29 West 39th St., New York 18, N. Y.

4. American Standards Association specification, American Standards Association, 70 East 49th St., New York 17, N. Y. (A majority of the standards listed below are approved by the ASA and published by the ASME.)

(*a*) A21.1—1939 Computation of strength and thickness of Cast-Iron Pipe, Manual for (*b*) A21.6—1953 Cast-Iron Pipe Centrifugally Cast in Metal Molds for Water or other Liquids.

(*c*) A21.8—1953 Cast Iron Pipe—Centrifugally Cast in Sand-Lined Molds for Water or other Liquids.

(*d*) A21.10—1952 Short-Body, Cast-Iron Fittings, 3″ to 12″.

(*e*) A21.11—1953 Mechanical Joint for Cast Iron Pressure Pipe and Fittings, Specification for.

(*f*) B16b—1944 Class 250 Cast-Iron Flanges and Flanged Fittings.

(*g*) B16.1—1948 Class 125 Cast-Iron Flanges and Flanged Fittings.

(*h*) B16.3—1951 Malleable-Iron Screwed Fittings, 150 lb.

(*i*) B16.4—1949 Cast-Iron Screwed Fittings, 125 and 250 lb.

(*j*) B16.5—1953 Steel Pipe Flanges and Flanged Fittings.

(*k*) B16.9—1951 Steel Butt-Welding Fittings.

(*l*) B16.10—1939 Face to Face Dimension of Ferrous Flanged and Welding-End Valves.

(*m*) B16.11—1946 Steel Socket-Welding Fittings.

(*n*) B16.15—1947 Brass or Bronze Screwed Fittings, 125 lb.

(*o*) B16.17—1949 Brass or Bronze Screwed Fittings, 250 lb.

(*p*) B16.19—1951 Malleable-Iron Screwed Fittings, 300 lb.

(*q*) B16.20—1952 Ring-Joint Gaskets and Grooves for Steel Pipe Flanges.

(*r*) B16.21—1951 Nonmetallic Gaskets for Pipe Flanges.

(*s*) B31.1—1951 Code for Pressure Piping, including supplement, B31.1—1955 and B31.1.8—1952.

(*t*) B36.1 to 36.26 ASTM Iron and Steel Pipe Specifications accepted as American Standards.

5. American Water Works Association, 521 5th Ave., New York, N. Y. Standards.

6. Clinedinst, W. O., *Oil and Gas Journal, 43,* No. 4, 79 (1945).

7. Crane Company Catalog No. 53, Crane Company, Chicago 5, Illinois.

8. Crane Company Technical Paper No. 409, Crane Company, Chicago 5, Illinois.

9. Crocker, Sabin, *Piping Handbook,* 4th Ed., Rev., McGraw-Hill Book Co., New York, 1955.

10. *Equivalent Valves,* Equivalent Valves Co., 115 West 7th St., Los Angeles 14, California.

11. Federal Specifications, Sup't. of Documents, Washington 25, D.C.

12. Joffe, J., *Chem. Eng., 56,* No. 8, 130 (1949).

13. Ladish Catalog No. 55, Ladish Co., Cudahy, Wisconsin.

14. Lukenheimer, Circular No. 552, 553, 555, Lukenheimer Valve Co., Cincinnati 14, Ohio.

15. Manufacturer's Standardization Society of the Valve and Fitting Industry, 420 Lexington Ave., New York 17, N.Y.

16. Moody, L. F., *Trans. Am. Soc. Mech. Engrs., 66,* 671 (1944).

17. Nordstrom Valve Catalog No. 10, Nordstrom Valve Division, Rockwell Manufacturing Company, 400 Lexington Ave., Pittsburgh 8, Pa.

18. *Piping Engineering,* Tube Turns Research Staff, Tube Turns, Inc., Louisville 1, Kentucky (1947–53). Series as follows:

1.01 Codes and Specifications, Sources of Piping Design Information.

2.01 Pipe Materials and Thicknesses, Pipe Materials and their Allowable Stresses.

3.01 Fluid Flow, Fluid Flow in Pipe.

4.01 Expansion and Flexibility, Introduction to the Piping Flexibility Problem.

4.02 Expansion and Flexibility, *Z, L, U,* and Expansion *U* bends.

5.01 Welded Joints, Joint Preparation for Buttwelding.

5.02 Welded Joints, Welding Rings.

6.01 Mechanical Joints, Standard Steel Pipe Flanges.

6.02 Mechanical Joints, Ratings of American Standard Steel Pipe Flanges.

19. Poettmann, F. H., *J. Petroleum Tech., 3,* 317 (1951).

20. Powell Valve Catalog, Powell Valve Co., Wm. Powell Co., Cincinnati 22, Ohio.

21. Rase, H. F., *Petroleum Refiner, 32,* No. 8, 141 (1953).

22. *Refinery Catalog,* Gulf Publishing Co., 3301 Buffalo Drive, Houston, Texas, 1954.

23. *Standards of the Hydraulic Institute,* Hydraulic Institute, New York (1951).

24. *Standards of the ASTM,* Part I and II, Am. Soc. for Testing Materials, 1916 Race St., Philadelphia, Pa. (1952).

25. *Taylor Forge Catalogs* No. 484 and 423, Taylor Forge and Pipe Works, Chicago 90, Illinois.

26. *Tube Turn Catalog* No. 211, Tube Turns, Inc., Louisville 1, Kentucky.

27. U. S. Dept. of Commerce, U. S. Commercial Standards and Simplified Practices Recommendations, Sup't. of Documents, Washington 25, D. C.

28. U. S. Navy Specifications, U. S. Navy Dept., Bureau of Supplies and Accounts, Washington 25, D. C.

29. *Vogt Company Catalog* No. F9, Henry Vogt Machine Company, Louisville 10, Kentucky.

30. Vogrin, C. M., Frank C. G. Williams, and John S. Worth, *Modern Steel Bolting for Piping and Pressure Vessels* ASME Paper 52-PET-7, 7th Annual Meeting, Kansas City, Mo. (September 22–24, 1952).

31. *Walworth Catalog* No. 52, Walworth Company, 60 E. 42nd St., New York 17, N. Y.

32. *Watson-Stillman Bulletin* A3-50, Watson-Stillman Co., Roselle, N. J.

33. *Welding Handbook,* 3rd Ed., American Welding Society, New York, 1950.

34. Wert, H. A. and S. Smith, *Design of Piping for Flexibility with Flex-Anal Charts,* Blaw Knox Co., Power Piping Div., Pittsburgh, Pa.

35. M. W. Kellogg Staff, *Design of Piping Systems,* New York, John Wiley and Sons (1956).

CHAPTER 19

# THERMAL INSULATION

Thermal insulation, properly applied, assures effective operation of process equipment and conserves valuable heat. There is always a wide variety of insulating materials for a given situation and careful consideration must be given to each type so that an intelligent decision may be made. It is not possible to describe every type of insulating material, but several of the more frequently used types for process plants will be described.

## INSULATION MATERIALS

Insulation materials can be classified into four types: (1) fibrous, (2) granular, (3) cellular, and (4) reflective. In a practical sense, such classifications, however, often have limited value since some commercial products consist of a combination of several of these insulating materials.

### Asbestos

Asbestos is a natural-occurring mineral of the composition $3MgO \cdot 2SiO_2 \cdot 2H_2O$, a hydrated magnesium silicate. It is rock-like in its natural state but can be crushed and changed into soft, silky fibers. In this state it has a texture of wool and has, therefore, amazed people down through the ages, since it does not support combustion. Asbestos is used in many forms to manufacture insulating materials, either by itself or in combination with other substances. By the addition of binders it can be fashioned into paper-like structures, either single flat layers with thin air spaces between layers, or as corrugated asbestos paper built up into layers. The relatively dead air spaces serve to decrease the conductivity of the finished insulation. These asbestos paper insulations are effective up to around 300° F; above this temperature the binding material decomposes. Asbestos can be given a felt-like structure by combining it with small particles of some spongy, cellular material. This felted asbestos resists vibration and withstands rough handling and temperatures up to around 900° F. Asbestos fibers are used to strengthen high-temperature diatomaceous earth insulation and 85% magnesia insulation.

### Diatomaceous Earth

Diatomaceous earth is an unusual form of silica, having a high melting point, 2930° F, and unique properties. It is composed of minute silica skeletons of diatoms, which were microscopic plants that lived millions of years ago. Large deposits occur in California. A mixture of asbestos and diatomaceous earth forms an excellent high-temperature insulation (600 to 1900°F) and is manufactured under various trade names.

### Eighty-five Per Cent Magnesia

For temperatures up to around 600° F 85% magnesia insulation is perhaps the most popular insulating material. It is composed of 85% magnesia and 15% asbestos. The asbestos fiber provides the reinforcing qualities necessary to permit molding and manufacture of pipe sections and other shapes. The so-called magnesia constituent is actually a magnesium carbonate produced from dolomitic limestone by the action of carbon dioxide on slate dolomite. When properly installed, 85% magnesia maintains its excellent insulating properties for many years and is thus an economical material. It has the disadvantage of not being able to support any appreciable load.

Eighty-five per cent magnesia insulation is often used in combination with diatomaceous earth insulation. A layer of diatomaceous earth insulation is placed adjacent to the hot body and is of sufficient thickness to produce a surface temperature of 600° F on its outside surface. The diatomaceous earth insulation is then covered by a layer of 85% magnesia, which has a lower thermal conductivity value and can be used in the lower temperature range below 600° F.

Eighty-five per cent magnesia makes an effective insulation material because 90% of its total volume is dead air space the conductivity of which is low.

### Mineral Wool

Mineral wool is manufactured by blowing molten silica minerals into threads. These threads can be fashioned into insulating blankets or molded into blocks by adding suitable binding materials. When a waterproofing binder is used, an effective low-temperature insulation can be produced, which is highly moisture-resistant and rot-proof. In this form it is generally called mineral or rock cork. Mineral wool is also used for insulating homes and buildings.

### Vegetable Cork

Natural cork contains many thousands of sealed dead-air spaces. It is light, yet strong; it does not absorb vapors nor is it adversely affected by wetting. These properties make cork good for low-temperature insulation.

### Expanded Polystyrene

Styrene, polymerized and molded to occlude many bubbles of air, forms a light and serviceable low-temperature insulation. It has replaced cork in many applications.

### Glass Insulation

The low thermal conductivity, noncombustibility and inertness of glass are desirable features for insulation. Molten glass can be spun into fibers similar to that of mineral wool and these fibers fashioned into blankets, molded pipe covering or blocks. Glass fiber insulation is normally good up to 600°F, although some special designs withstand 1000°F.

Since glass cannot absorb water, glass fibers molded into boards serve as low-temperature insulation when provided with a suitable warm-side vapor seal (see below). Cellular glass insulating material is made of thousands of glass bubbles molded into a rigid structure of extremely light weight. This cellular glass material is suitable for temperatures from − 350 to + 800° F. It is available in blocks and in shapes for pipe coverings.

Although not a glass, another insulating material is made of materials very similar to glass, called hydrous calcium silicate. This material is effective up to 1200°F and is moisture resistant. When blended with small amounts of asbestos fiber, its strength is increased.

Thomas and Turner[11] have prepared a table giving the characteristics of the above insulating materials and certain others of commercial importance. This summary, reproduced in Table 19-1, is notable in its objectivity. The use of trade names has been avoided.

### Reflective Insulation

Heat is transferred across an air space produced by two solid bodies by means of radiation, convection, and conduction. If the width of the air space is properly chosen and the temperature drop across the air space

## TABLE 19-1. Characteristics of Thermal Insulation Materials*

| Materials | General forms | Strength | Resistance to water | Resistance to vapor | Composition |
|---|---|---|---|---|---|
| *Rigid and Semi-Rigid:* | | | | | |
| Asbestos fiber (Standard) | Processed into pipe insulation and block | Good flexural and excellent tensile strength | Excellent | Fair | Long amosite fibers blended with binders |
| Asbestos fiber (Super) | Processed into pipe insulation and block | Good compressive and flexural strength; limited tensile strength | Fair | Moderate | Long amosite fibers blended with binders and diatomaceous earth |
| Diatomaceous earth | Molded and formed pipe insulation, segments and blocks | Good compressive strength; limited flexural and tensile strength | Fair | Moderate | Diatomaceous earth blended with long asbestos fibers |
| Glass fibers | Formed into pipe insulation and block | Soft, flexible fibers; will not break down under flexing | Excellent | Moderate | Glass fibers bonded |
| Glass cellular | Rigid, fabricated pipe insulation, lagging, segments and fitting covers | High compressive strength; fair tensile strength | Excellent | Excellent | Inorganic glass containing microscopic, hermetically sealed cells |
| Hydrous calcium silicate | Molded and formed pipe insulation, segments and blocks | Good compressive and flexural strength; limited tensile strength | Good | Moderate | Hydrous calcium silicate blended with long asbestos fibers |
| 85% magnesia | Molded and formed pipe insulation, segments and blocks | Good compressive strength; limited flexural and tensile strength | Fair | Moderate | 85% hydrated magnesium carbonate |
| Mineral cork for low temp. | Formed pipe insulation and block | Moderate compressive and tensile strength | Good | Moderate | Mineral wool fibers with asphalt binders |
| Mineral wool for low temp. | Semi-rigid blanket, etc. formed into pipe insulation, block and lagging | Moderate compressive and tensile strength | Fair | Moderate | Mineral wool fibers bonded |
| Mineral wool for high temp. | Molded into blocks and lagging | Limited compressive and tensile strength; fair flexural strength | Excellent | Moderate | Mineral fibers blended with binders |
| Polystyrene (expanded) | Block and pipe insulation | Moderate compressive strength; fair tensile strength | Excellent | Excellent | Polystyrene with air cells |
| Vegetable cork | Molded pipe insulation, blocks, lagging and fitting covers | Compressible under light load, but does support heavy loads | Good | Fair | Granulated vegetable cork bonded with natural cork resin |
| *Blankets* | | | | | |
| Asbestos | Blankets | Fair compressive strength | Fair | Poor | Asbestos fiber enclosed by asbestos cloth |
| Glass fibers | Felted glass fiber blanket with various types of facings | Soft, flexible; little compressive resistance after application | Good | Moderate | Felted glass fibers fabricated with metal mesh or other facing |
| Mineral wool | Blankets with various types of facings | Fair compressive strength | Fair | Poor | Felted mineral wool |
| *Felt* | | | | | |
| Hairfelt | Felted into blanket roll | Standard density, very spongy; heavy density, very firm | Little resistance | Little resistance | 100% cattle hair |

* Adapted by permission — R. Thomas and W. C. Turner, *Chem. Eng.* 60, No. 6, 221 (1953) — supplemented by W. C. Turner, Carbide & Carbon Chemical Co.

## TABLE 19-1 (continued)

| Advantages | Disadvantages | Temperature limits, ° F min. | max. | Conductivity Btu/(hr)(sq ft)(° F/in.) 32 | 70 | 212 | 500 | Density, lbs per cu ft |
|---|---|---|---|---|---|---|---|---|
| Excellent for bridging and for use in steam tracing | Dimensions not standard | 33 | 750 | | 0.320 | 0.390 | 0.538 | 14 |
| High shearing resistance; high compressive strength | Not suitable for bridging under heavy load; non-flexible | 212 | 1200 | | | 0.42 | 0.56 | 18 |
| Handles well, used as inner layer under 85% magnesia where operating temperature exceeds 575°F. | Not suitable for bridging under heavy load; non-flexible | 212 | 1900 | | 0.56 | 0.59 | 0.62 | 25 |
| Lightweight, resilient, and flexible | Soft, will not resist mechanical abuse | 32 | 450 | | | 0.25 to 0.30 | | 2 to 10 |
| Easily shaped and fitted; does not depend on vapor barrier for its vapor resistance; incombustible. High temperature limits. | Should be protected against abrasion. Design for expansion and contraction | −300 | 800 | 0.35 | 0.42 | 0.53 | | 9 |
| High shearing resistance; high compressive strength | Not suitable for bridging under heavy load; non-flexible | 70 | 1200 | | 0.400 | 0.450 | 0.540 | 10 to 11 |
| High shearing resistance; high compressive strength | Not suitabler for bridging under heavy load; non-flexible | 212 | 550 | | 0.33 | 0.37 | | 11 |
| Easily fitted at atmosphere temperatures; flexible | Requires careful vapor treatment; depends on vapor barrier for its service in low temperatures | Below −150 | 150 | 0.305 | 0.325 | | | 15 |
| Easily fitted at atmospheric temperatures | Requires careful vapor treatment; depends on vapor barrier for its service in low temperatires | Below −150 | 250 | 0.300 | 0.320 | | | 14 |
| Flexible and withstands considerable expansion and contraction | Will not resist excessive mechanical abuse without indentation | 70 | 1700 | | 0.33 | 0.40 | 0.51 | 18 to 24 |
| Easily fitted, flexible, clean, and light in weight | Depends partly on barrier for effective service | −200 | 200 | 0.23 to 0.40 @ 40°F | | | | 2 |
| High vibration resistance, fair tensile strength; can be pre-fabricated | Should be fire-proofed in critical area; depends on vapor barrier in low temp. service | −200 | 300 | 0.25 | 0.26 | | | 6 to 10 |
| Removable without damage | Each installation must be custom designed | 100 | 950 | | | 0.50 | 0.65 | 10 |
| Flexible, light fibers will not break down under moderate impact; fibers respond to repeated compressing | Soft; little compressive strength | 40 | 1000 | | 0.26 | | | 3 |
| Withstands expansion and contraction | Soft | 50 | 1200 | | | 0.35 | 0.45 | 12 |
| Excellent emergency low temperature insulation; can be formed over any size or shape pipe or vessel | Requires careful vapor treatment; depends on vapor barrier for its service in low temperatures; will burn | Below −150 | 200 | 0.25 | 0.260 | | | 10 (Standard) 15 to 17 (Heavy) |

TABLE 19-1 (continued)

| Materials | General forms | Strength | Resistance to water | Resistance to vapor | Composition |
|---|---|---|---|---|---|
| Glass fibers | Felted into rolls | Soft, flexible; little compressive resistance after application | Good | Moderate | Fine glass fibers lightly bonded |
| Mineral wool | Semi-rigid felt | Fair compressive strength | Good | Poor | Mineral wool felted and bonded |
| *Miscellaneous* | | | | | |
| Air cell | Laminated corrugated felts formed into pipe insulation and sheets | Little compressive strength; moderate tensile strength | Moderate | Moderate | Asbestos felts cemented together |
| Asbestos sheets | Laminated felts formed into pipe insulation, blocks and curved segments | Little compressive strength until compressed approximately 20%; moderate longitudinal strength | Moderate | Moderate | Asbestos sheets separated by spongy cellular nodules or by indentation |
| Wool felt sheets | Laminated sheets formed into pipe insulation | Good compressive and tensile strength | Moderate | Moderate | Bonded sheets of rag felt |
| Cement | Mineral wool-water mix; semi-thermosetting | Resistance to impact normally good | Water will soften the dried cement | Moderate | Nodulated mineral wool and asbestos fibers with inorganic binders |
| Cement | Mineral wool-water mix; hydraulic setting | Impact resistance very good | Will not soften when wetted | Moderate | Nodulated mineral wool and asbestos fibers and inorganic binders |

maintained near some optimum value, convection will be negligible. Transfer of heat by conduction through air is quite low, due to the low thermal conductivity of air. The transfer of heat can be minimized, therefore, by constructing an insulating material with parallel sheets of metal separated by spacers at a distance small enough so that heat transfer by convection is negligible. The metal must have low emissivity, and thus high reflectivity so that the transfer of heat by radiation will also be low. Insulation so constructed is called reflective insulation.

Reflective insulation has become very popular in building and cold storage insulation and is gradually being accepted in the process industries. The desirable air-space width is approximately three-quarters of an inch. An aluminum sheet has proved the most successful material since it maintains a low emissivity even upon oxidation.

Highly reflective corrugated metal jacketing for insulation serves a dual purpose of protecting the insulation and adding its own insulating qualities by providing an air space and a highly reflective surface.

## SELECTION OF INSULATION

There are surprisingly few standardized test procedures for comparing insulating materials. Only one method for determining insulation thermal conductivity has been accepted by the American Society for Testing Materials (ASTM C177-45).[1] The method requires skilled manipulation

**TABLE 19-1 (continued)**

| Advantages | Disadvantages | Temperature limits, °F min. | max. | Conductivity Btu/(hr)(sq ft)(°F/in.) 32 | 70 | 212 | 500 | Density, lbs per cu ft |
|---|---|---|---|---|---|---|---|---|
| Flexible; fibers respond to repeated compressing; good thermal cushioning | Pronounced depression under heavy loading | Not known | 600 | | | | | 1 |
| Resilient; can withstand expansion and contraction | No mechanical abuse can be permitted | Not known | 600 | | 0.18 to 0.25 | 0.30 to 0.40 | 0.60 to 0.80 | 4 to 8 |
| Light weight, low cost | Will not withstand severe abuse | 100 | 300 | | 0.490 | 0.550 | | 9 |
| Fair tensile strength | Sags and becomes soft when wet | 100 | 700 | | 0.390 | 0.450 | 0.560 | 23 |
| Low cost | Not to be used in critical area; will burn | 32 | 225 | 0.510 | | | | 18 |
| Good troweling characteristics; medium smooth finish; good adhesion and cohesion | Slow setting; will not stand beating rain | 100 | 1800 | | 0.690 | 0.820 | 1.058 | 26 |
| Fair troweling characteristics; smooth finish, good adhesion and cohesion; quick setting | Cannot be re-worked | 32 | 1700 | 0.46 | 0.525 | 0.610 | 0.840 | 49 |

in order to obtain reproducible results. The conductivities obtained are for insulation, dried under specific laboratory conditions. The relation between such values and the actual in-service conductivities is not known. Improved techniques are needed for the laboratory evaluation of insulating properties which can be directly correlated with the actual performance of insulation in service.

Because of the lack of correlation between laboratory measurements and service performance characteristics, it is necessary to base selection of insulation largely on previous experience. Such a procedure has a number of limitations the most important of which is that it encourages disregard of newly developed insulating materials. Progressive companies can, however, institute programs for plant testing all promising insulating materials. The following factors must be investigated to properly evaluate an insulting material. The effect of moisture and changes in temperature on the properties listed must be carefully considered.

1. Physical properties
   *a.* thermal conductivity.
   *b.* specific heat.
   *c.* coefficient of expansion.
   *d.* compression, tensile and impact strength.
   *e.* hardness.
   *f.* fatigue resistance.
   *g.* fire resistance.

2. Chemical properties
   *a.* analysis.
   *b.* corrosive nature.
   *c.* effect of chemicals in pipe or vessel being considered.
3. Installation characteristics
   *a.* toxicity.
   *b.* action on skin (does it scratch, etc.).
   *c.* overall workability (cutting, shaping, etc.).

## INSULATION INQUIRIES AND QUOTATIONS

The erection of insulation is usually subcontracted to a firm specializing in such work. The subcontractor must be chosen early in the development of the project so that his plans may be developed in conjunction with the prime contractor.

At the time that inquiries are made to the insulation contractor the actual detailed design of the process plant has not been developed. Usually only the engineering flow diagram, a plot plan, and perhaps several preliminary elevation studies have been made. It is necessary, therefore, that the insulation contractor be supplied with a very complete description of the insulation requirements on which he is to quote. Because of the limited amount of drawings available at the time of inquiry, close cooperation between the insulating contractor and prime contractor is essential.

An insulation inquiry should include a description of the work with reference to appropriate drawings. Complete insulation specifications must also be furnished. These should include not only the insulation to be used, but also the methods of insulation perferred. A clear description of cost figures desired should be given so that all quotations can be fairly judged.

A complete list of all lines to be insulated should be included with the inquiry. This list should show the operating temperatures so that the proper insulation thickness, based on the specifications, can be selected. All other equipment, such as vessels and exchangers, must be adequately described, or suitable drawings furnished so that quantity and type of insulation may be selected. Portions of vessels which are not to be insulated, such as skirts and manhole covers, should be clearly indicated.

In addition to lump sum prices, unit prices* should be requested for all sizes of pipe, fittings, valves, and vessels anticipated. Close cooperation with the vendor will materially reduce the amount of work involved in such a procedure. The unit prices for vessels should be given in price

* Because of the incomplete development of the plant design at the time of inquiry it is necessary to permit the insulation contractor to apply for additional fees if it is agreed by both parties that the scope of the project has changed during the design development. The unit prices are of use in evaluating the additional costs incurred.

per sq ft of shell area and head area. If each vendor quotes on the same basis, the lump-sum prices can then be used with impunity in determining the lowest bid.

## ERECTION OF INSULATION

Despite the wide variety of insulation types, most insulation is manufactured in one or more of three forms; i.e., blocks, blankets, or molded sectional pipe coverings. The methods of installing these three forms of insulation are similar, regardless of the kind of insulation being employed. Two general rules apply to all insulation erection. One is that surfaces should be cleaned thoroughly before insulating begins. The other is that if insulation is installed in several layers, all joints should be staggered. (Fig. 19-3). The staggering of the joints minimizes the possibility of heat leakage at the joints.

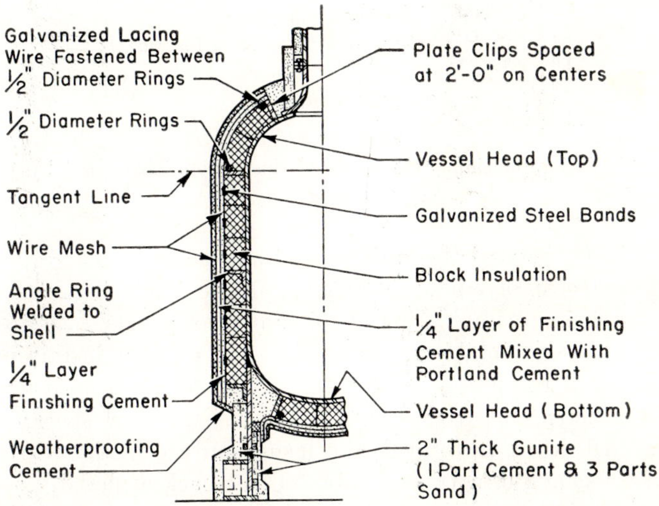


Fig. 19-1. Typical vessel insulation.

### Block and Blanket Insulation

Insulation furnished in the form of blocks is extensively used on vessels, heat exchangers, and other similar equipment (Fig. 19-1). One-half in. by 0.020 in. galvanized bands are used for securing insulation on towers and heat exchangers. Insulation is held on the top and bottom heads with bands which are secured to studs or nuts welded to the heads. To aid in the support of the blocks, steel angles are welded around vertical vessels and spaced at 8- or 12-ft centers. These supports are one-half in. less than the thickness of the insulation. By terminating the block at both ends of the angle and filling the space with loose mineral wool, an expansion joint is fashioned (Fig. 19-2).

The expansion joint is covered with 20-gage galvanized iron or hardware cloth and mesh wire and then finished weather-tight with insulating cement. After the blocks have been wired in place, wire mesh (1-in.) is

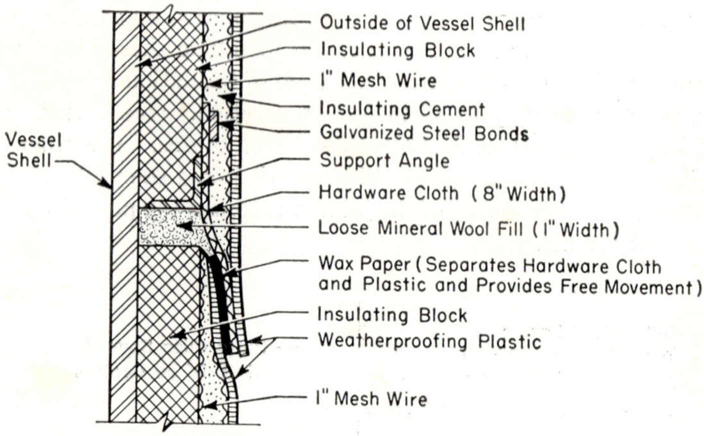


Fig. 19-2. Typical expansion joint for block insulation. (Adapted by permission Aber Company.)

stretched tightly around the entire vessel and laced in place. Over this wire mesh several layers of insulating plastic (finishing cement) are applied, which is a combination of mineral wool pellets, asbestos fibers, clays, and corrosion-inhibiting materials. For outdoor installations another layer of 1-in. mesh is laced in place and followed by a final coating of weather-proofing plastic which consists of a special asphalt emulsion. All openings through the insulation are made weather-tight by use of a flashing cement (e.g., fibrated cut-back asphalt).

Blanket insulation is installed in a similar manner, except that wire instead of galvanized bands can be used to secure the blankets in place. Insulating plastic is applied directly over the blanket area, followed by 1-in. wire mesh to which is applied weather-proofing plastic.

### Pipe Insulation

(In Fig. 19-3 note staggering of joints.) Molded pipe insulation is fitted to the pipe and secured by loops of wire or bands on approximately 12-in. maximum centers. This wire should resist corrosion or high temperatures expected in the area. Irregularities which produce gaps where the joints butt together are filled with insulating plastic or finishing plastic. Indoor piping is finished with 6-ounce canvas jacket, which is pasted on the pipe covering. Outdoor insulation is covered with a weather-proof jacket such as roofing paper or galvanized steel or aluminum sheeting.

Valves and fittings are insulated with segments of blocks or pipe insulation of the same material used on the straight runs of pipe. These segments are wired in place and finished with a coating of insulating

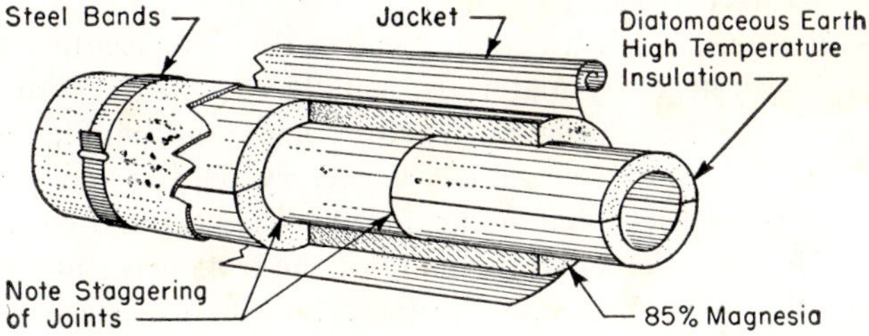


Fig. 19-3. Multiple-layer sectional pipe insulation.

plastic. Fittings smaller than 3 in. are usually insulated with plastic followed by a coating of hard-finish asbestos cement.

It has been found most economical to omit insulation of flanges on all but steam lines. It is so often necessary in process plants to remove sections of lines for inspection or repair that it is less costly to merely omit the insulation around flanges, and thus avoid destroying the entire section of insulation when pipe is removed.

### Low Temperature Insulation

The principal difference in the erection of low-temperature insulation is that some means of preventing water vapor from entering the insulation must be devised. Since the inner surface of the insulation is at a lower temperature than the outer surface, the water vapor pressure at the inner surface is lower than that at the outer surface and the flow of water vapor is toward the inside, where it freezes and causes a marked increase in the conductivity of the insulation and also causes its deterioration. A vapor barrier can be installed to minimize this movement of water vapor. Materials which serve as vapor barriers include vapor-proofing felt, which is finished with a fibrated asphalt sealing compound followed by a weather-proof jacket. Another successful vapor barrier consists of sheet aluminum foil laminated between layers of Kraft paper and finished with a special vapor-seal adhesive. Cork insulation is often finished with several coats of asphalt followed by roofing paper which is applied with additional amounts of hot asphalt. A final layer of weather-proofing cement reinforced with wire mesh is then applied over the paper.

The foregoing represents only a very general and brief description of insulating techniques. Insulating manufacturers and erectors publish detailed specifications and instructions for the installation of their products. These instructions are readily obtainable and should, of course, be followed.

## REFRACTORIES

A temperature of 1900° F is the maximum that any of the insulating materials discussed can withstand. Above this temperature it is necessary to employ refractory materials, usually cast in the form of bricks, to contain the heat in furnaces, reactors, and other similar equipment. The choice of the most economical refractory depends on a number of factors other than price. These include the physical and chemical properties listed below. Fortunately, many of the tests necessary to evaluate such properties have been standardized by the American Society for Testing Materials.

1. Physical properties

*a.* Specific gravity.

*b.* Porosity; indication of resistance to penetration by slags and gases.

*c.* Cold strength; indicates ability to stand up under adverse shipping conditions.

*d.* Softening temperature; determined by use of the pyrometric cone equivalent method which consists of comparing the test sample in the form of a small cone with standard cones of known softening points.

*e.* Load test; measures the subsistance which occurs during the sustained application of a 25 lb per sq in. load under conditions of operating temperature. Bricks, such as silica brick, fail abruptly at a specific temperature, whereas fire clay bricks show a gradual failure.

*f.* Dimension changes at high temperatures; ASTM[5] provides a method of determining the changes in dimensions that occur over prolonged heating periods at operating temperatures. This information is useful in the design of furnaces.

*g.* Thermal expansion; important in the ultimate planning of furnace design.

*h.* Spalling; defined as the breaking off of fragments from the surface of the refractory is an important property which must be considered in the proper design of refractory walls. Three types of spalling occur. Thermal spalling is caused by thermal changes produced when heating or cooling. Magnesite, chrome, and silica bricks (below 1200° F) have the lowest resistance to thermal spalling of the common refractories. Mechanical spalling is caused by strains placed on the brick by mechanical means such as excessive expansion. Structural spalling is caused by changes in the physical properties of one side of the brick compared to the other, which may produce spalling due to unequal expansion. The ASTM[5] has outlined tests for determining spalling characteristics of refractories.

*i.* Thermal conductivity; this is an important property since it determines the thickness of brick required for a given refractory wall.

*j.* Permeability; the permeability of brick to liquids and gases is important since a combination of internal-external attack by some slag or gas during the use of the refractory will produce extremely rapid failure.

2. Chemical properties

*a.* Chemical composition; as a quality check of refractories it is advisable to know the chemical composition of each batch used.

*b.* Chemical reactions with refractories; it is important to be familiar with the chemistry of reactions that can occur between refractories and materials present in furnaces. The term slag is a rather all-inclusive term which serves to designate the molten impurities present in a furnace, kiln, or reactor. The reaction of these impurities with the refractory brick must be well-known and a refractory must be chosen that will withstand such attack. In addition, the resistance of the refractories to the gaseous atmosphere in which they will be exposed is equally as important. Carbon monoxide, halogens, and various acid gases are known to affect certain refractories adversely. It is imperative that manufacturers be consulted on these problems, and it is further important to realize that certain refractory materials, when placed adjacent to each other in forming refractory walls, may react with each other under operating conditions.

Refractories can be divided into four main groups—alumina-silica, silica, basic, and insulating refractories. In Table 19-2 there is reproduced a comparison of some of the more important of these refractories. The so-called basic refractories include magnesite, chrome, and forsterite. In addition, certain special refractories such as silicon carbide and zirconia are listed which prove economical under certain severe conditions.

In Fig. 19-4, thermal conductivities of various refractories are compared. It is interesting to note the striking difference between thermal conductivities of insulating brick and other refractories. Insulating brick is produced from clays (Table 19-3) which are mixed with a combustible material such as saw dust, or powdered cork. Upon firing, the combustible materials are burned leaving a porous structure. Insulating bricks are extremely light in weight and have a low thermal conductivity. Recent discoveries make it possible to produce insulating bricks which withstand temperatures up to 3000°F. It is possible therefore, to econ-

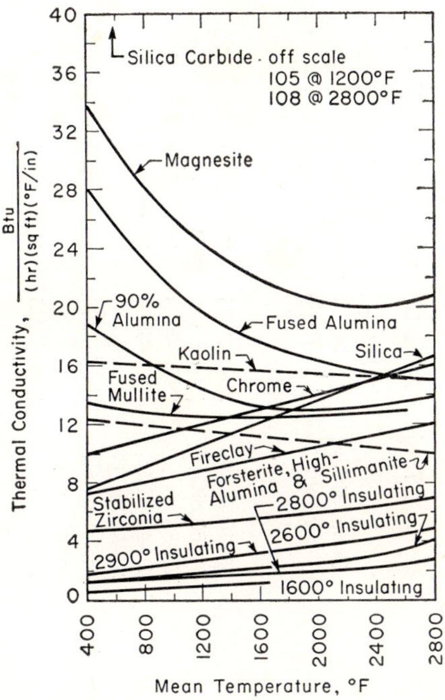


Fig. 19-4. Thermal conductivities of refractory and insulating firebrick. [Adapted by permission C. L. Norton, Jr., *Chem. Eng. 60,* No. 6, 216 (1953).]

### Table 19-2. Characteristics of Refractory Brick*

| Name | Composition, per cent | Melting point, ° F | Normal use limit, ° F ** | Approx. $/M, 9-in. equiv. | True specific gravity | Bulk density, $lb/ft^3$ | Source |
|---|---|---|---|---|---|---|---|
| Alumina-Silica | | | | | | | |
| Fireclay base | | | | | | | |
| First quality | 35–40 $Al_2O_3$<br>54–60 $SiO_2$ | 3090–3175 | 2400–2700 | 95 | 2.60–2.70 | 125–140 | Fireclay |
| Semi-silica | 70–80 $SiO_2$ | 2940–3060 | 2400–2700 | 100 | 2.40–2.45 | 125–140 | Fireclay |
| Superduty | 41–45 $Al_2O_3$<br>51–55 $SiO_2$ | 3175–3200 | 2500–2800 | 115 | 2.65–2.75 | 130–145 | Flint fireclay |
| High burned superduty | 41–45 $Al_2O_3$<br>51–55 $SiO_2$ | 3175–3200 | 2500–2800 | 155 | 2.65–2.75 | 130–145 | Flint fireclay |
| 50% alumina | 50 $Al_2O_3$ | 3200–3245 | 2500–2800 | 160 | 2.75–2.85 | 130–145 | Fireclay, diaspore, bauxite |
| 60% alumina | 60 $Al_2O_3$ | 3245–3310 | 2700–2900 | 200 | 2.90–3.05 | 130–145 | Fireclay, diaspore, bauxite |
| 70% alumina | 70 $Al_2O_3$ | 3290–3335 | 2700–2900 | 235 | 3.15–3.25 | 135–150 | Diaspore, bauxite |
| 80% alumina | 80 $Al_2O_3$ | 3335–3390 | 2800–3000 | 260 | 3.35–3.45 | 140–155 | Diaspore, bauxite |
| 90% alumina | 90 $Al_2O_3$ | 3390–3425 | 2800–3200 | 1175 | 3.55–3.65 | 172 | Calcined alumina |
| Kyanite base | $Al_2O_3 \cdot SiO_2$<br>58–68 $Al_2O_3$ | 3250–3300 | 2800–3000 | 500–680 | 3.00–3.06 | 140–150 | Kyanite |
| Kaolin base (High-fired) | $Al_2O_3 \cdot 2SiO_2$<br>44–45 $Al_2O_3$ | 3190 | 2800–2900 | 170–335 | 2.65–2.75 | 135–145 | Kaolin |
| Fused mullite base | $3Al_2O_3 \cdot 2SiO_2$<br>72–75 $Al_2O_3$ | 3325–3350 | 2900–3200 | 685–1370 | 3.08–3.25 | 150–160 | Bauxite and synthetic |
| Silica | | | | | | | |
| Standard | $SiO_2$ | 3142 (pure) | 3000 | 95–100 | 2.30–2.38 | 100–105 | Quartz sand, ganister |
| Superduty | $SiO_2$ | 3142 (pure) | 3000 plus | 110 | 2.30–2.38 | 100–105 | Quartz sand, ganister |

| | | | | | | | |
|---|---|---|---|---|---|---|---|
| Magnesite | | | | | | | |
| Burned | MgO plus $SiO_2$, $Fe_2O_3$ $Cr_2O_3$ | 5070 (pure) | 3000–4000 | 520 | 3.40–3.60 | 160–165 | Magnesite, sea water, brines, brucite |
| Chemically bonded | MgO plus FeO, $Cr_2O_3$ | | 2900–3100 | 465 | 3.60–3.80 | 170–175 | Magnesite, sea water, brines, brucite |
| Chrome | | | | | | | |
| Burned | $FeO \cdot Cr_2O_3$ pure; plus some $Al_2O_3$, MgO, $Fe_2O_3$ $SiO_2$ | 3540–3990 | 2800–3200 | 430 | 3.60–4.10 | 185–190 | Chromite ores |
| Chemically bonded | $FeO \cdot Cr_2O_3$ plus MgO | 3540–3990 | 2900–3100 | 450–485 | 3.90–4.10 | 170–180 | Chromite ores |
| Forsterite | $2MgO \cdot SiO_2$ | 3461 (pure) | 3000 | 530 | 3.30–3.40 | 150–155 | Olivine or synthetics |
| Silicon carbide | SiC | Dissociates at 4082 | 2800–3200 | 1520 | 3.19 | 155 | Synthetic |
| Fused alumina | $Al_2O_3$ | 3722 (pure) | 3400 | 2000 | 3.70–3.90 | 175–195 | Bauxite |
| Zircon | $ZrO \cdot SiO_2$ | 4532 (pure) | 3400 | 1210 | 4.70 | 205 | Zircon sands |
| Zirconia (Stabilized) | $ZrO_2$ | 4870 (pure) | 4300 | 9430 | 5.75 | 275 | Zircon sands |
| Carbon (Graphite) | C | 6330 | 4000 (reduc. atm.) | 1600 | 2.25 | 137 | Carbon |

* Reprinted by permission, Norton, C. L., Jr. *Chem. Eng. 60*, No. 6, 216, 1953.

** Note "Normal-use-limit" temperatures are approximate only because service conditions sometimes will change allowable temperatures by several hundred degrees. Such conditions may include heavily reducing atmospheres and unusual loading.

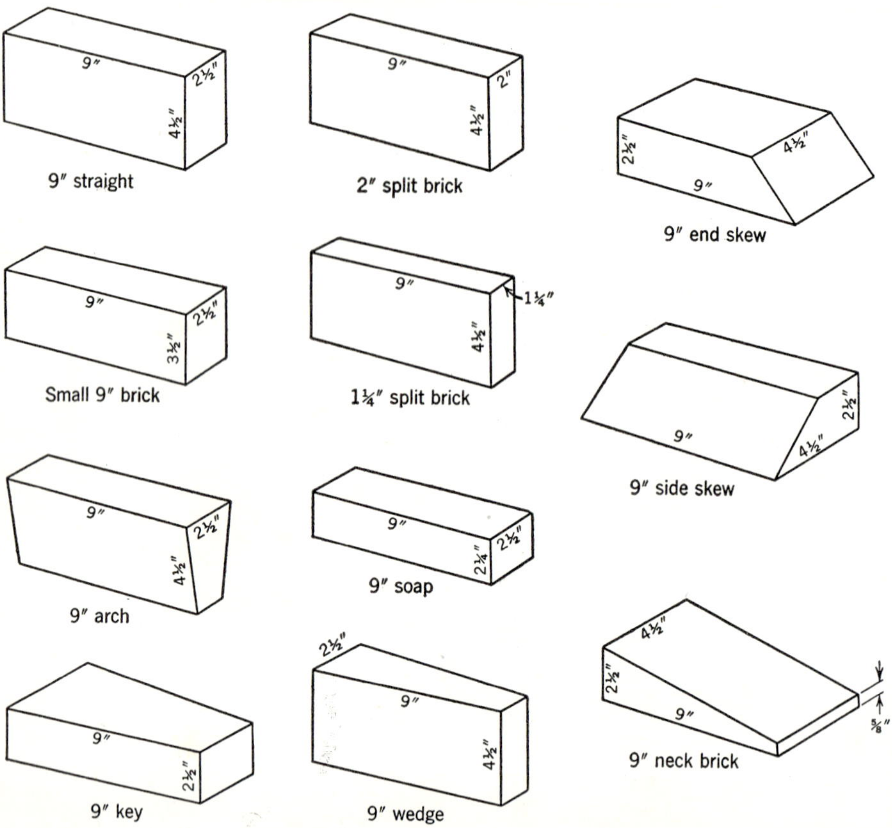


Fig. 19-5. Typical refractory brick shapes. (Courtesy of Harbison-Walker Refractories Company.)

omize under certain conditions by employing insulating firebrick instead of the heavier refractory brick (firebrick) formerly used. Low conductivity makes it possible to use a much thinner wall and the light weight permits much more rapid heating and cooling of a furnace. The use of the insulating firebrick, however, is limited to situations where their low mechanical strength is not a disadvantage and their high permeability can be tolerated.

The basic refractories are most resistant to basic slags. Fire-clay, high-alumina, and silica brick are most advantageously used in contact with siliceous slag. Operating temperature plays an important part in the ultimate selection of a refractory to resist a given atmosphere or slag condition.

Refractory brick is standardized in a number of sizes and shapes Several typical shapes are illustrated in Fig. 19-5.

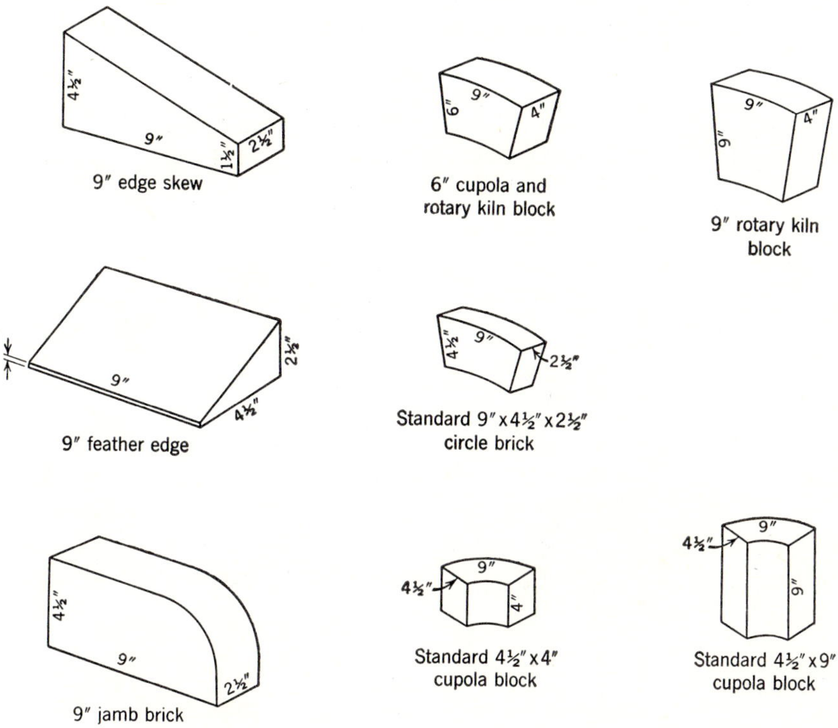


## Installing Brickwork

The laying of refractory brick requires the services of expert brick masons familiar with refractory bricks. Careful job planning suggests ordering 5 to 10% in excess of the estimated requirements because of the inevitable breakage which occurs during construction.

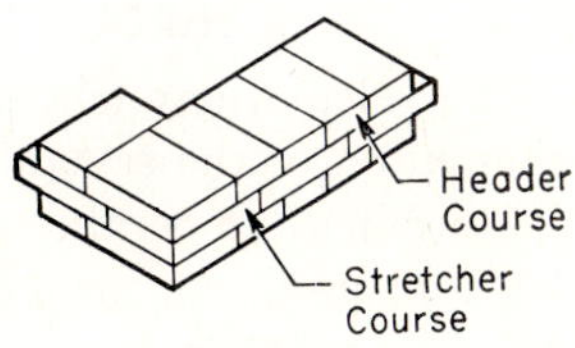


Fig. 19-6. Common methods of laying brick.

The common methods of laying brick are illustrated in Fig. 19-6. The choice of method depends on the required dimensions and stability of the final structure. A typical furnace wall normally requires a combination, the most common of which is alternate headers and stretchers. This arrangement produces a greater structural stability since all vertical joints are thereby offset.

The most complex phase of refractory brickwork is the construction of openings in the walls and suspended arches (the ceiling of an enclosure). Door widths are best kept as small as possible. Suspended arches consist of special refractory units designed to receive metal supports which are in turn attached to an overhead steel framework.

TABLE 19-3. Characteristics of Typical Insulating Firebricks*

| Type† | Composition, per cent | Normal use limit, ° F.‡ | Approx. \$/M, 9-in. equiv. | Bulk density, lb/ft.$^3$ |
|---|---|---|---|---|
| Group 16 | 15–37 $Al_2O_3$<br>30–60 $SiO_2$<br>plus $TiO_2$<br>$Fe_2O_3$, Alk. | 1600 | 115 | 21–37 |
| Group 20 | 26–38 $Al_2O_3$<br>45–61 $SiO_2$<br>plus $TiO_2$,<br>$Fe_2O_3$, Alk. | 2000 | 125 | 26–45 |
| Group 23 | 25–42 $Al_2O_3$<br>45–67 $SiO_2$<br>plus $TiO_2$,<br>$Fe_2O_3$, Alk. | 2300 | 150 | 27–47 |
| Group 26 | 40–46 $Al_2O_3$<br>47–55 $SiO_2$<br>plus $TiO_2$,<br>$Fe_2O_3$, Alk. | 2600 | 195 | 43–64 |
| Group 28 | 45–53 $Al_2O_3$<br>42–52 $SiO_2$<br>plus $TiO_2$,<br>$Fe_2O_3$, Alk. | 2800 | 260 | 45–65 |
| Others | 45 $Al_2O_3$<br>65 $Al_2O_3$<br>90 $Al_2O_3$ | 2900<br>3000<br>3250 | 285<br>480<br>750 | 52<br>69<br>81 |

*Reprinted by permission, Norton, C. L., Jr., *Chem. Eng.*, *60*, No. 6, 216, 1953.
†Group number is std. A.S.T.M. classification, indicating normal use limit (×100).
‡See note under Table 19-2.

## HEAT LOSSES THROUGH INSULATED SURFACES

The calculation of heat losses through insulated surfaces involves the summation of the several resistances to heat flow. In a given situation the terminal temperatures are known. The temperature, for instance, of the fluid inside an insulated pipe and the room temperature are known and the heat flow can be expressed in a fashion analogous to the flow of electricity.

$$q = \frac{\Delta T}{R_T} \tag{1}$$

where $q$ = heat flow in Btu/hr
$\Delta T$ = total temperature drop from inside fluid to outside air
$R_T$ = summation of all resistances

Heat flows through the fluid inside the pipe by convection and by conduction through the pipe wall and insulation. The heat then passes into the surrounding air by a combination of convection and radiation. Each of these processes can be expressed in terms of appropriate resistances. In general the resistance produced by the fluid and the pipe wall is negligible compared to that of the insulation and can be neglected. The equation for heat flow then becomes:

$$q = \frac{\Delta T}{R_i + R_s} \tag{2}$$

where $R_i$ = resistance of the insulation
$R_s$ = "surface" resistance of the insulation

The resistance of any material to the conduction of heat is expressed as $\Delta X/kA$, where $\Delta X$ is the thickness in ft of the material in the direction of heat flow, $k$ is the conductivity [Btu/(hr) (ft)(° F)] and $A$ the area through which the heat is flowing (sq ft). For insulation surrounding a cylinder such as pipe insulation, the surface area varies from inside to outside and some mean area must be determined. It has been shown that this mean area corresponds to the logarithmic mean of the inside ($A_1$) and outside ($A_2$) surface areas, i.e.,

$$\frac{A_2 - A_1}{\ln \dfrac{A_2}{A_1}} \quad \text{or} \quad \pi L \frac{D_2 - D_1}{\ln \dfrac{D_2}{D_1}}$$

where $L$ is the length of insulation in ft and $D$ is the diameter in ft

For spheres, this mean area can be shown to be equivalent to the geometric mean ($\sqrt{A_1 A_2}$) of the inside and outside areas.

The surface resistance is usually much less than that of the insulation and need only be estimated. It is represented as $1/(h_c + h_r)A_2$, where $h_c$ is the convection coefficient and $h_r$ the radiation coefficient. Values of the combined coefficient $h_c + h_r$ for horizontal insulated pipes can be estimated from the empirical equation,*

$$h_c + h_r = \frac{564}{(d_2^{0.19})(273 - \Delta t_i)} \tag{3}$$

where $d_2$ = OD of insulation in in.
$\Delta t_i$ = insulation surface temperature minus air temperature, ° F
$h_c + h_r$ = transfer coefficients, Btu/(hr)(sq ft)(° F)

Values for flat plates can also be calculated. The convection coefficient for air, $h_c$, at vertical plates can be determined from the equation $h_c = 0.27\Delta t^{0.25}$, where $\Delta t$ is the temperature difference between the air

* This equation based on that presented by Heilman, *Trans. Am. Soc. Mech. Engrs.*, *44*, 308 (1922).

and the outside of the plate. The radiation coefficient can be determined from the equation:

$$h_r = \frac{0.173e_1\left[\left(\frac{T_1}{100}\right)^4 - \left(\frac{T_a}{100}\right)^4\right]}{T_1 - T_a} \tag{4}$$

where: $e_1$ = emissivity of the plate at $T_1$
$T_1$ = absolute temperature of plate in ° R
$T_a$ = absolute temperature of air in ° R

To determine conductivities and the values of $h_c$ and $h_r$, it is necessary to assume the surface temperatures of the various layers of insulation since conductivity and $h_c$ and $h_r$ are temperature dependent. These assumptions can be checked after the heat flow is calculated since the flow of heat must be the same through each resistance. The following example will serve to illustrate the above principles.

EXAMPLE. A 2 in. pipe (2.375 in. OD) carrying an oil at 650°F is insulated with an inner layer of 1½ in. of high temperature insulation and an outer layer of 1½ in. of 85% magnesia. The room temperature is 70° F. Calculate the heat loss per ft of pipe length.

Thermal Conductivities Btu/(hr)(ft)(° F)

| Temperature | 85% Magnesia | High Temperature (Diatomaceous earth-asbestos) |
|---|---|---|
| 100 | 0.034 | |
| 200 | 0.038 | 0.054 |
| 300 | 0.040 | 0.056 |
| 400 | 0.043 | 0.058 |
| 500 | 0.046 | 0.060 |
| 600 | | 0.061 |
| 650 | | 0.0615 |

SOLUTION. Assume:

1. Outside pipe wall temperature = 650° F.
2. Average inner layer insulation temperature = 500° F.
3. Average outer layer insulation temperature = 250° F.
4. Outside surface temperature = 150° F.

Basis: 1 ft of length.

Average area of inner layer

$$\frac{\pi}{12}\left(\frac{D_2 - D_1}{\ln \frac{D_2}{D_1}}\right) = \frac{\pi}{12}\left(\frac{3}{\ln \frac{5.375}{2.375}}\right) = 0.96 \text{ sq ft}$$

Average area of outer layer

$$\frac{\pi}{12}\left[\frac{3}{\ln \frac{8.375}{5.375}}\right] = 1.77 \text{ sq ft}$$

Outside area $\quad \frac{\pi}{12} 8.375 = 2.2 \text{ sq ft}$

$$R_i: \quad R_i = \frac{1.5}{(12)(0.06)(0.96)} + \frac{1.5}{(12)(0.039)(1.77)} = 2.17 + 1.81 = 3.98$$

$$R_s: \quad \Delta t_i = 150 - 70 = 80$$

From eq. 3,

$$h_c + h_r = \frac{564}{(8.375)^{0.19}(273 - 80)} = 1.96 \text{ Btu/(hr)(sq ft)(° F)}$$

$$R_s = \frac{1}{(h_c + h_r)(A)} = \frac{1}{(1.96)(2.2)} = 0.23$$

Hence $\quad q = \frac{650 - 70}{3.98 + 0.23} = \frac{580}{4.21} = 138 \text{ Btu/hr}$

Check assumptions:
Inner layer

$$\Delta T = (q)(R)$$

$$\Delta T = (138)(2.17) = 300$$

or, $\quad \text{Average temperature} = \frac{650 + (650 - 300)}{2} = 600° \text{ F, assumed } 500° \text{ F}$

Outer layer

$$\Delta T = (138)(1.81) = 250° \text{ F}$$

$$\text{Surface temperature} = 350 - 250 = 100° \text{ F, assumed } 150° \text{ F}$$

$$\text{Average temperature} = \frac{350 + 100}{2} = 225° \text{ F, assumed } 250° \text{ F}$$

The relatively small change of conductivity with temperature indicates that further calculation is not justified. The value of 138 Btu/hr is satisfactory.

## OPTIMUM INSULATION THICKNESS

Theoretically the determination of the optimum insulation thickness is a straightforward procedure. As the insulation thickness is increased the costs of heat losses go down while the insulation costs go up. The point at which minimum total cost is reached may be described as an optimum value.

McMillan[7] has developed an analytical solution for this particular optimum by equating the first differential of the expression for total cost to zero.

For flat surfaces

$$X = \sqrt{\frac{ak'}{b}} - R_s k \tag{5}$$

where $X$ = optimum insulation thickness in in.
$a = y(t - t_a)M/1{,}000{,}000$
$y$ = hours operation per year
$M$ = value of heat in dollars per million Btu
$b$ = cost of insulation per sq ft per in. of thickness per year
$R_s$ = surface resistance, (hr)(° F)(sq ft)/Btu
$k'$ = conductivity, Btu/(hr)(sq ft/in.) (° F)
$t$ = temperature on hot side of insulation, ° F
$t_a$ = air temperature, ° F

For cylindrical surfaces (one insulating material)

$$\left(r_2 \ln \frac{r_2}{r_1} + R_s k'\right)\sqrt{\frac{2r_2 - r_1}{r_2 - R_s k'}} = \sqrt{\frac{ak'}{b}} \tag{6}$$

where $r_2$ = outside radius of insulation, in.
$r_1$ = inside radius of insulation, in.

A plot of these expressions for various values of $\sqrt{ak'/b}$ is presented in Fig. 19-7. Insulation manufacturers and operating companies have insulation specifications which include recommended thicknesses for various service conditions. Most of these recommendations were based on McMillan's equations and were determined in the late 1920's. Costs have certainly changed since then and a re-appraisal is indicated. One further point about McMillan's equations is that the expressions for insulation costs used in the development do not necessarily apply to all insulation.

[Flat surfaces: cost per sq ft $= bx + C$

Cylindrical surfaces:

$$\text{cost per linear ft} = \frac{2\pi r_2 b}{12}(r_2 - r_1) + C$$

where $C$ is a constant]

Despite these objections, the tables of recommended thicknesses readily available from insulation manufacturers represent a close approximation to a "correct" thickness. In any event, these thicknesses have become familiar through use, and drastic changes must be well substantiated before management will agree. Such evidence is actually difficult to produce.

The effect of wind, rain, and especially sudden changes in weather can be very adverse on the operation of fractionating units. For instance, a sudden change in weather typical of that occurring on the Gulf Coast of the United States has often caused violent upsets in fractionating equipment due to increased heat losses through insulation. In such cases if all factors could be evaluated, it may be possible to justify even thicker insulation though the so-called optimum may be somewhat less.

The effect of wind velocity or heat loss through insulation is a relatively slight 10 to 15%, if the insulation is well sealed. The majority of the

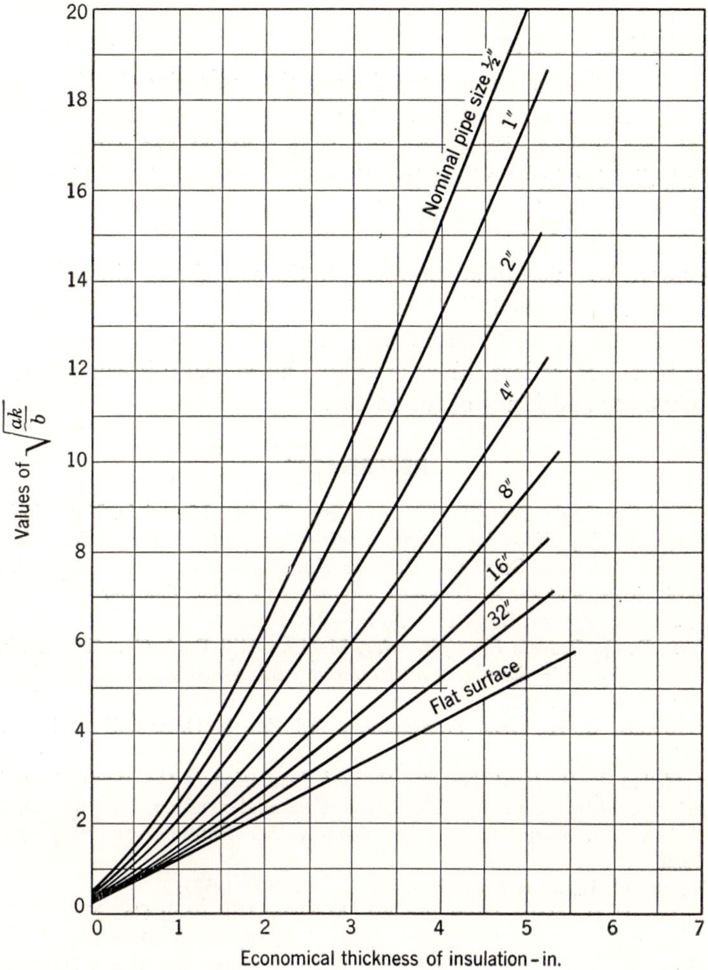


Fig. 19-7. Economical insulation thickness. (Reprinted by permission, L. B. McMillan, *Paper No. 2034,* Annual Meeting of ASME, New York, Dec. 1926; and copyright owner, Johns Manville Corporation.)

resistance is furnished by the insulation and not the surface resistance which is all that is affected by wind. The loss can become large, however, should cracks through the insulation develop. Better methods of installation to prevent losses through cracks and joints must be devised.

Although the situation is somewhat confused, users of insulation should not disregard the processes of objective thinking as new situations arise. New insulating materials, unusual locations (such as extremely cold climates), and other unique features should be examined and evaluated as much as possible on the basis of sound engineering principles.

## INSULATION EFFICIENCY

Insulation efficiency is a term frequently appearing in the literature and in manufacturers' catalogs. It is defined as the difference between the heat loss through bare pipe and that through the same pipe when insulated, divided by the heat lost through the uninsulated pipe. Tables of these efficiencies are reported by manufacturers for various sizes and thicknesses of insulation.

The values are misleading, however, to the casual user since they are all high and what appears to be a small difference in efficiency of only 2% (96 and 94) can represent an increase in heat loss of (96 — 94/100 — 96) (100) = 50%. Heat loss tables are much more useful in comparing various thicknesses and kinds of insulation.

## REFERENCES

1. *ASTM Standards,* Part III, 298, Am. Soc. Testing Materials, New York, 1950.
2. Geller, R. F., *Fire-Clay Brick, Their Manufacture, Properties, Uses, and Specifications,* Bureau of Standards Circular No. 282, Washington, D.C., 1926.
3. Jakob, Max, and G. A. Hawkins, *Elements of Heat Transfer and Insulation,* 2nd Ed., John Wiley and Sons, Inc., New York, 1950.
4. Kern, D. Q., *Process Heat Transfer,* McGraw-Hill Book Co., New York, 1950.
5. *Manual of ASTM Standards on Refractory Materials,* Am. Soc. Testing Materials, New York, 1948.
6. McAdams, W. H., *Heat Transmission,* 3rd Ed., McGraw-Hill Book Co., New York, 1953.
7. McMillan, L. B., *Heat Transfer through Insulation in the Moderate and High Temperature Fields,* Amer. Soc. Mech. Engrs' annual meeting, Dec. 6 to 9, 1926.
8. *Modern Refractory Practice,* Harbison-Walker Refractories Company, Pittsburgh, 1950.
9. Norton, C. L., Jr., *Chem. Eng. 60,* No. 6, 216 (1953).
10. Norton, F. H., *Refractories,* 3rd Ed., McGraw-Hill Book Co., New York, 1950.
11. Thomas, R. and W. C. Turner, *Chem. Eng., 60,* No. 6, 221 (1953).
12. Wilkes, G. B., *Heat Insulation,* John Wiley & Sons, Inc., New York, 1950.

# CHAPTER 20

# PROCESS INSTRUMENTS

Process industries could not exist without instruments which indicate, record, control, and in some cases anticipate the many changes that occur in a process. Even in processes which require manual control, instruments tell the operator when he should turn a valve or press a button.

The advancements in instrument technology have been phenomenal. In the late 1930's instruments and control devices were relatively simple and a few standard types were used. Any engineer could understand and apply them for most conditions. During and since World War II change has been rapid and continues at an accelerated rate. The greatest strides were made in the use of electronic equipment. The complexities of such systems and the rapid changes produced by the freshly stimulated instrument industry, has required the creation of another group of specialists. Some physicists, chemical engineers, mechanical engineers, and electrical engineers have become instrument engineers through self-education and training by the instrument companies. These men are employed not only by instrument companies but also by construction and operating companies. Many schools of chemical engineering offer courses in process instrumentation which will aid in training additional chemical engineers for this type of work and will make others aware of the problems and techniques of instrument engineering.

All instrument selection, application, and purchasing should be directed by an instrument engineer. It is impossible for a process or project engineer to keep up with the latest instrument developments unless his entire attention is devoted to this field.

To provide proper instrumentation the instrument engineer must be thoroughly familiar with all phases of the process, its key characteristics, and the application or function of all the process equipment. Conversely a general knowledge of instrumentation is helpful to the process engineer and project engineer in initial development of the process.

The purpose of the discussion that follows is to supply general knowledge of instrumentation. Complete works on the subject are available[1,5,6,8,9,10] and should be used for specific problems requiring more detailed understanding. Because of the rapid changes taking place, however, the several popular instrument periodicals should also be consulted.

## GENERAL PRINCIPLES

In order to indicate or control a process variable, an instrument must be capable of sensing the change in the variable. The sensing part of the instrument may be called the primary element. The instrument translates this primary impulse by electrical, pneumatic, or mechanical means into a visible indication or record. It may also actuate other devices to change process conditions so that the sensed variable may be returned to some pre-determined point. An instrument, therefore, is composed of some type of primary sensing device in contact with the fluid or substance; an amplification unit, and finally the physical unit which indicates or records and translates the primary sensing impulse into some kind of power or motion.

Automation or instrumentation is, of course, not limited to these basic characteristics, but has been developed to sense and indicate nearly all known physical or chemical characteristics. These include pressure, temperature, density, fluidity, viscosity, color, *p*H, or composition. It is usually unnecessary, however, to control precisely each process variable simultaneously. Ordinarily only several key variables need be controlled to give the desired end-product characteristics.

## INSTRUMENT TYPES

### Classification

The following list of instrumentation or control phases is convenient for discussion purposes.

#### *Primary Sensing Phases*

(1) Flow; (2) temperature; (3) pressure; (4) liquid level; (5) motion.

#### *Primary Elements, Types*

1. Flow; (1.01) mechanical; (1.02) differential; (1.03) area.

2. Temperature; (2.01) solid expansion; (2.02) fluid expansion; (2.03) thermo-electric.

3. Pressure; (3.01) bourdon-type; (3.02) bellows; and (3.03) electrical and diaphragm electronic.

4. Liquid level; (4.01) differential; (4.02) float; (4.03) miscellaneous types; (4.04) visual or manual.

5. Motion; (5.01) tachometric; (5.02) limit; (5.03) counting.

6. Chemical and physical measurements.

### Control Systems

(*a*) Manual, (*b*) mechanical, (*c*) pneumatic, (*d*) hydraulic, (*e*) electrical and electronic.

## Flow Measurement

### Mechanical Flow Measurement

The most elementary form of volumetric measurement consists of filling and emptying a chamber of known volume. This type of measurement is used in the process industries for taking inventory of fluids in storage tanks or vessels. Certain processes require manual or instrument tank gaging at regular intervals, as a check on other flow determinations.

The ordinary displacement meter measures liquids by alternately filling and emptying chambers contained within its body (Fig. 20-1). The velocities and pressure-drops through the device are utilized to move the chambers or diaphragms and the connected valving. Each cycle is counted by a series of numbered gears which are visible. These meters are relatively inexpensive and give long and accurate service when used within their capacity ranges, as illustrated by the residential water and gas meter. The undesirable pulsating effect of the simplest type of displacement meter is eliminated in those designs utilizing a rocking and rotatory motion. The displacement meter is used more frequently for measurement of total production than for recording instantaneous flow rates since it is essentially a totalizing device. Most displacement meters cannot operate at high pressures and temperatures.

Displacement meters are used extensively in blending and filling operations when precise volumetric measurements are required. Mechanically linked displacement meters are used in the blending of costly products such as lubricating oils. Automatic shut-off meters are used in barrel filling or container filling operations.

### Differential Pressure Instruments

Differential pressure flow meters convert a portion of the pressure energy of a fluid to kinetic energy. This is accomplished by placing in

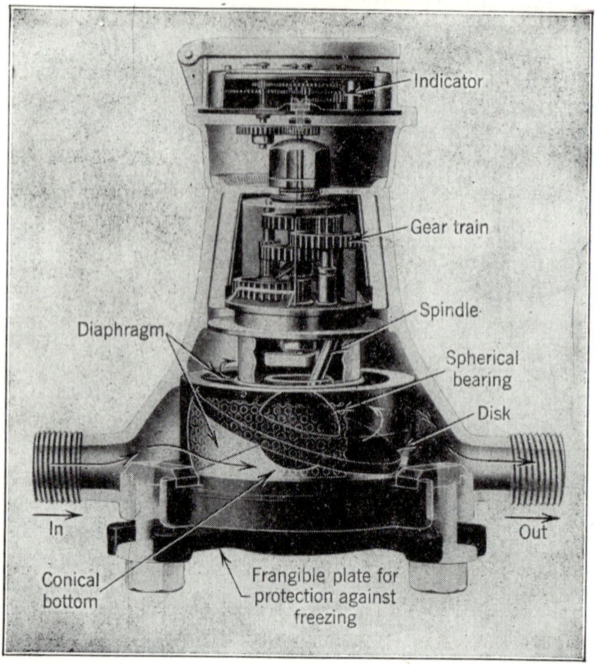


Fig. 20-1. Nutating-disk displacement meter. [Reprinted by permission, G. G. Brown, *Unit Operations,* John Wiley and Sons (1950).]

a pipe a constriction such as an orifice, venturi, or flow nozzle (Fig. 20-2). If friction is neglected the pressure drop across such a restriction is related to the kinetic energy change as follows:

$$P_1 - P_2 = \left(\frac{u_2^2 - u_1^2}{2g_c}\right)\rho \qquad (1)$$

where $P_1$ = pressure upstream from the restriction, psf
$P_2$ = pressure at the restriction or some predetermined point downstream, psf
$u_1$ = velocity, ft/sec, upstream
$u_2$ = velocity, ft/sec, at restriction or downstream
$g_c$ = 32.17
$\rho$ = density of fluid flowing lb/cu ft

Expressing $u_1$ in terms of $u_2$ an equation for velocity at the point 2 can be obtained.

$$u_2 = \frac{1}{\sqrt{1 - \left(\frac{D_2}{D_1}\right)^4}}\sqrt{\frac{2g_c\,\Delta P}{\rho}} \qquad (2)$$

Eq. 2 can be expressed in terms of lb of fluid flowing per second as follows:

$$W' = \frac{\pi D_2^2}{4\sqrt{1 - \left(\frac{D_2}{D_1}\right)^4}} \sqrt{2g_c\rho \, \Delta P} \tag{3}$$

where $D_1$ = diameter of pipe in ft
$D_2$ = diameter of restriction in ft
$\Delta P$ = pressure differential in lb per sq ft

These equations, however, are based on ideal frictionless flow. Actually frictional effects consume a portion of the pressure energy. The most convenient means of allowing for energy loss due to friction has been an empirically determined constant called the coefficient of discharge which is the ratio of actual to theoretical flow. The actual flow then is:

$$W = \frac{\pi D_2^2 C}{4\sqrt{1 - \left(\frac{D_2}{D_1}\right)^4}} \sqrt{2g_c\rho \, \Delta P} \tag{4}$$

where $C$ = coefficient of discharge

$$\frac{1}{\sqrt{1 - \left(\frac{D_2}{D_1}\right)^4}} = \text{velocity of approach factor}$$

Convenient manuals such as the classic ASME publication, *Fluid Meters*,[8] present correlations of $C$ for the various types of orifices, venturi meters, and flow nozzles versus the Reynolds' number through the restriction. The ratio, $C/\sqrt{1 - (D_2/D_1)^4}$ is often called the flow coefficient, $K$. Tables of Reynolds' numbers and corresponding values of $K$ for different pipe sizes and $D_2/D_1$ ratios are published.[8] The trial-and-error solution for flow rate or the required orifice diameter for a given flow rate and pressure differential is made simple by such compilations. A Reynolds' number is assumed and $K$ read from the table for a given $D_2/D_1$ ratio. The values of $K$ do not vary greatly in the range of most industrial applications and hence the solution is rapid.

It is fortunate that standardization of orifice, venturi, and flow nozzle construction was accomplished some years ago. Considerable experimental and operating data have been accumulated and generalized correlations of $C$ with Reynolds' number have resulted. These data make it possible to design and operate differential meters without any necessity for calibration provided the standards of construction and installation outlined[8] are followed.

### Venturi Meter

The venturi meter (Fig. 20-2*a*) has a 21° or less angle of approach which eliminates cavitation. The exit angle is 7° to 9° so that the expansion back to original pipe size is uniform and pressure recovery high. The small loss of energy due to friction which is demonstrated by the low permanent pressure losses ($C = 0.92$ to 0.99) is the unique feature of a venturi meter. It is used, therefore, in systems for which high pressure losses cannot be tolerated such as low pressure gas systems or water distribution systems. Pressure losses in water systems involving large flows increase pumping costs considerably. In such cases the venturi, though more expensive than the orifice or flow nozzle, will be more economical based on overall installation plus operating costs.

### Orifice Meter

The orifice plate is the simplest, cheapest, and most popular restriction for measuring flow. Orifice meters comprise well above 90% of most process plant metering installations. Values of the discharge coefficient range from 0.59 to 0.62 (see published data for tabulations).[8]

Orifice flanges with flange taps are obtainable from numerous manufacturers (Fig. 18-3). The orifice plate is held between these flanges and the pressure connections are tapped into the flange body by the manufacturer. The taps are located so that when the flanges and plate are assembled there is 1 in. between each tap and the plate.

Other locations of pressure taps include *vena contracta* and pipe taps. The *vena contracta,* downstream tap location is at the point of minimum fluid stream cross section (Fig. 20-2*b*). This location varies with the ratio $D_2/D_1$ and must be moved if the orifice size is changed. The upstream tap is placed one pipe diameter from the orifice plate. *Vena contracta* taps are used on special installations for which manufactured orifice flanges are unobtainable or too costly.

Pipe taps are located so that the upstream tap is 2.5 pipe diameters from the plate and the downstream tap, eight diameters. Because of the great downstream distance, a large portion of the pressure drop across the orifice has been recovered. This low pressure differential may be an advantage for high flow rates.

The differential measured by pipe taps represents, in addition to a portion of the original differential across the orifice, frictional losses in the length of pipe between taps. Pipe taps should not be used, therefore, for viscous fluids or fluids with solids in suspension since these produce large frictional losses. Flashing fluids, however, are often successfully measured with pipe taps since the vapors have opportunity to redissolve before reaching the downstream tap.

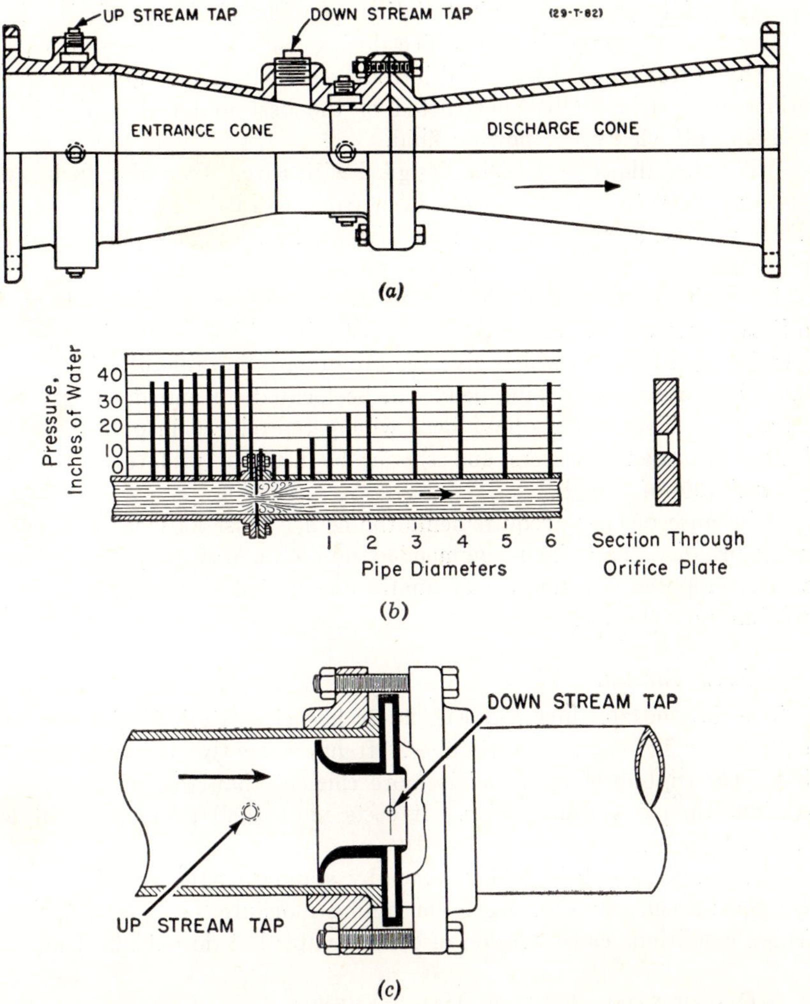


Fig. 20-2. Differential pressure primary elements. (*a*) Venturi meter (courtesy of Minneapolis-Honeywell Regulator Company); (*b*) orifice meter installation showing pressure variation (adapted from *Fundamentals of Instrumentation for the Process Industries,* by permission, Minneapolis-Honeywell Regulator Company); (*c*) flow nozzle (courtesy of Minneapolis-Honeywell Regulator Company).

Flange taps are by far the most popular type of pressure tap. They are easily installed and require less space. They are satisfactory for most types of fluids including viscous materials.

The industrially important orifice plate is the square-edged or sharp-edged orifice as shown in Fig. 20-2*b*. The dimensional characteristics of this design have been very carefully standardized and experimental and operating data on numerous installations provide reliable prediction data.

Any radical change in the orifice during use such as rounding of the edge by corrosion or erosion will introduce serious error. For this reason orifice plates should be constructed of the best material available for resisting attack by the flowing fluid.

Orifice installations must be designed with care. Excessive turbulence at points close to the orifice will cause erroneous readings. The published discharge coefficients are based on flow which is roughly parallel to the pipe axis.

A definite amount of straight unobstructed piping on either side of the orifice is necessary. Sprenkle[11] from studies of various piping systems has developed charts which give the minimum allowable distance that a disturbance such as a pipe bend can be located up and downstream of an orifice. These charts have been widely reproduced in publications of instrument manufacturers and should be used in preference to some general rule-of-thumb.

If because of space requirements the orifice must be located closer to a disturbance than the recommended minimum distance, straightening vanes composed of a number of small tubes nested so as to fill the entire conduit must be utilized.

### Flow Nozzle

The flow nozzle shown in Fig. 20-2*c* has characteristics intermediate between orifices and venturi tubes. It has a gently rounded approach but an abrupt discharge, and is more efficient than the orifice but less efficient than a venturi meter. It costs more than an orifice, but less than a venturi.

The nozzle, which is held in place by standard flanges, is best suited for the measurement of gases containing moisture such as wet steam. These conditions cause erosion of orifice plates but do not affect nozzles.

### Differential Pressure Measurement

Mercury-type (Fig. 20-3*a*) or force-balance type manometers are used for measuring the differential pressure produced by orifices, venturi tubes, and flow nozzles.

The differential pressure across the mercury manometer causes a change in level in both legs. The movement of the float in the larger diameter leg is transmitted by means of mechanical linkage to a shaft through the manometer body to give an external indication. Movement of the shaft may be used to indicate, record, or control (see Fig. 20-3*a*). Certain types of manometers eliminate the shaft extension through the meter body by use of a solenoid or magnet contained within the manometer body.

The longer leg with the smaller diameter is called a range tube. Although the total differential is represented by the movement in both legs, movement in the leg with the float is a function of the total differential. The range of the instrument is increased by decreasing the diameter of the range tube. This causes more movement in the range tube than in the float tube for a given differential.

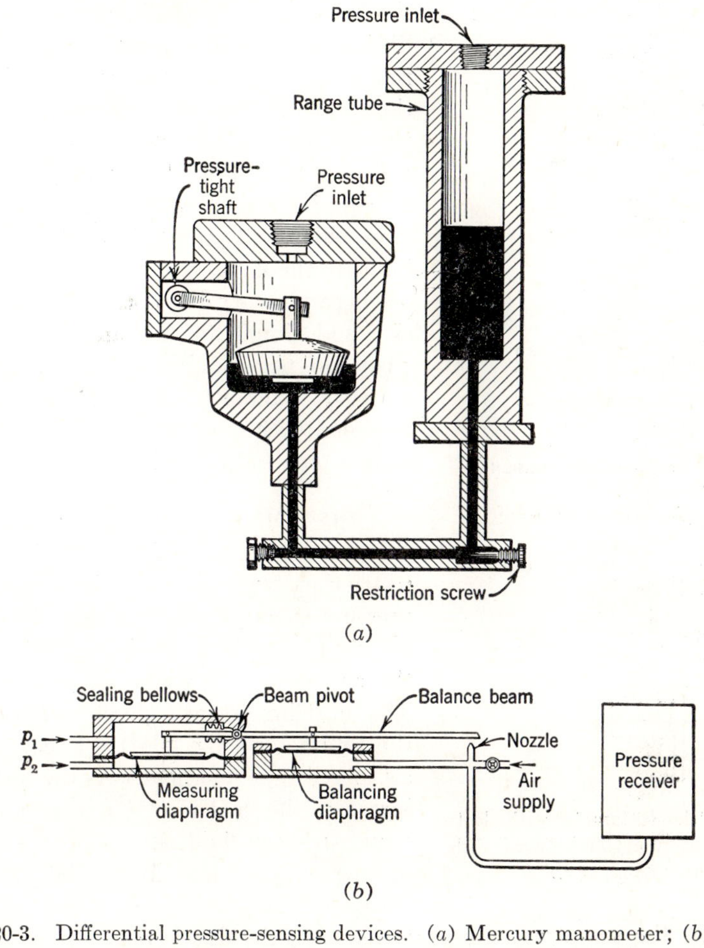


Fig. 20-3. Differential pressure-sensing devices. (*a*) Mercury manometer; (*b*) force-balance meter. [Reprinted by permission, D. P. Eckman, *Industrial Instrumentation,* John Wiley and Sons (1950).]

The mercury manometer has served for many years as the principle measuring device of differential pressure. It is only in recent years that the force-balance or pneumatic-balance type instrument has been introduced and gained popularity (Fig. 20-3*b*). The force-balance manometer operates by developing a force in opposition to the differential pressure.

This opposing force is created by air pressure. As $P_2$ increases (Fig. 20-3*b*) the balance beam covers more of the nozzle through which air is being lost. This movement increases the air pressure in the balancing diaphragm and causes it to force the beam away from the nozzle. An equilibrium of forces is established which is represented by the air pressure attained in the system. This air pressure is directly proportional to the differential pressure and can be readily transmitted to a recording instrument with a pneumatic bellows which will actuate a pen to record or indicate the flow.

These mercury-less meters are small in size and simple to calibrate and install. They do not require seal pots. Unlike mercury manometer assemblies they can be mounted at the orifice supported by a simple pipe saddle and yoke.

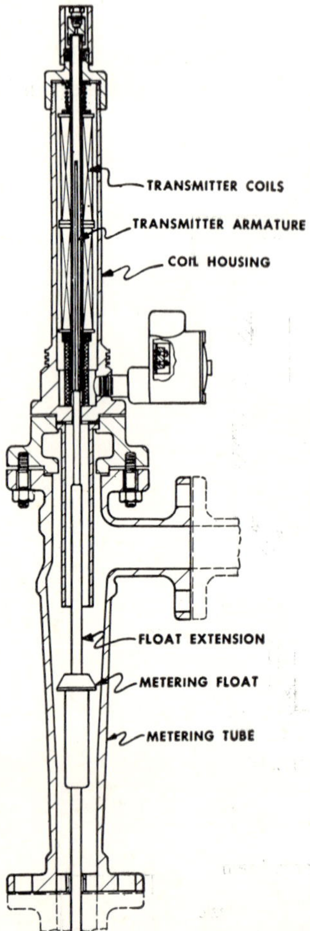


Fig. 20-4. Rotameter with electric transmitter. (Courtesy of Fisher and Porter Company.)

### *Area Meters*

Instead of maintaining a constant area restriction, and allowing the differential pressure to vary with flow rate, the differential pressure is maintained constant and the area allowed to vary in area meters. The most important type of area meter is the rotameter (Fig. 20-4). It consists of a tapered tube and float. As the flow rate increases the float moves upward creating more area for fluid flow and thus maintaining a constant differential pressure across the float.

Using the same reasoning process as in the development of eq. 3 and expressing the constant pressure differential across the float in terms of the conditions at equilibrium $[\Delta P = V_f(\rho_f - \rho)/A_f]$ the following equation is obtained:

$$W = C_r A_2 \sqrt{\frac{2g_c\rho(\rho_f - \rho)V_f}{A_f}} \tag{5}$$

where $W$ = flow, lb/sec

$C_r$ = rotameter coefficient of discharge

$\rho$ = density of fluid
$\rho_f$ = density of float
$V_f$ = volume of float
$A_f$ = effective area of float
$A_2$ = annular area between float and tube

Since $V_f = W_f/\rho_f$ (where $W_f$ = weight of float) the amount of range change produced by reducing the weight of the float is expressed by:

$$\frac{W_2}{W_1} = \sqrt{\frac{W_{f_2}}{W_{f_1}}} \tag{6}$$

Eq. 6 is based on the assumption that $C_r$ does not vary and that the fluid and effective float area remain the same. Rotameter manufacturers offer weight extensions which can be attached to standard float designs to increase the instrument range.

The annular area $A_2$ varies with the flow rate and is directly proportional to the height of the float. Manufacturers' data can be used to calculate height of float versus flow rate curves for rotameters, even for viscous materials.

Rotameters are available with glass metering tubes which permit visual observation of the float or metal metering tubes with indicating extensions. Various electronic and pneumatic transmitting devices are obtainable for transmitting the position of the extension stem to a recorder or controller.

A calibration for viscous fluids can be predicted with greater accuracy for rotameters than for orifices. They are satisfactory and inexpensive for small installations, but are very costly for extremely large flow rates.

## Temperature Measuring Instrument

A wide variety of temperature-measuring devices meet the varied needs of temperature, range, and service. Those in use in process plants will be considered here.

In Table 20-1 a summary of usual temperature ranges for the more common instruments is given. These ranges are subject to change as new developments appear. Many factors other than range, however, are considered in the selection of a temperature device. These include accuracy, speed of response, sensitivity, cost, and corrosion resistance. Manufacturers are anxious to advise customers on the use and limitations of their equipment and should be consulted.

### *Solid Expansion Thermometers*

As is well known, all materials change dimension with temperature. If two strips of metal are welded together, one having a low coefficient of

TABLE 20-1. Usual Industrial Range of Temperature Measuring Instruments

(Based on manufacturer's data)

| Type | Range, ° F |
|---|---|
| 1. Bimetallic | −100 to 1000 |
| 2. Mercury in glass | −38 to 750 |
| 3. Fluid expansion | |
| (*a*) Mercury | −38 to 1000 |
| (*b*) Organic liquid | −125 to 500 |
| (*c*) Organic vapor-liquid | −40 to 600 |
| (*d*) Gas | −400 to 1000 |
| 4. Thermocouple | |
| (*a*) Copper-constantan | −300 to 600 |
| (*b*) Iron-constantan | 0 to 1400 (oxidizing atm) or 1800 (reducing atm) |
| (*c*) Chromel-alumel* | 600 to 2100 |
| (*d*) Platinum-platinum-rhodium | 1300 to 3000 |
| 5. Resistance | |
| (*a*) Copper | −40 to 250 |
| (*b*) Nickel | −300 to 600 |
| (*c*) Platinum | −300 to 1800 |
| 6. Optical pyrometer | 850 to 5200 |
| 7. Radiation pyrometer | 200 to 7000 |

* Registered trademark of the Hoskins Corporation.

expansion (Invar, which is iron-nickel alloy) and the other a high coefficient of expansion (brass or nickel), the combined strip will deflect under the influence of temperature. By making the strips thin a large deflection can be obtained which can be used to move a pointer and indicate the temperature.

A very popular dial thermometer consists of a bimetallic strip wound into a helix. One end is fastened to the casing and the other end to a rod connected to a pointer. As the temperature changes the movement of the helix causes the rod and attached pointer to move. The dials are built in sizes from 1¾ in. to 5 in.

These dial thermometers are rugged and reasonably accurate (within 1% of scale) when properly chosen. The industrial-type mercury-in-glass thermometers are the most economical and convenient means for nonrecording temperature measurement of accessible process points.

The bimetallic strip instrument is used extensively for on-off control of simple devices such as air-conditioning systems.

### *Fluid Expansion Thermometers*

The most familiar temperature measuring instrument, the mercury-in-glass thermometer, is designed for locally mounted applications similar to those of the bimetallic dial thermometer described above. The instrument is made rugged by enclosing the thermometer in a metal case. Fittings make it possible to screw the thermometer into a pipe or vessel nozzle. The glass covered scale can be obtained tilted at convenient angles.

Other types of fluid expansion instruments utilize a closed system. The expanding fluid exerts a pressure on a spiral pressure element (spring) causing it to uncoil. This movement is transmitted to a pen or pointer

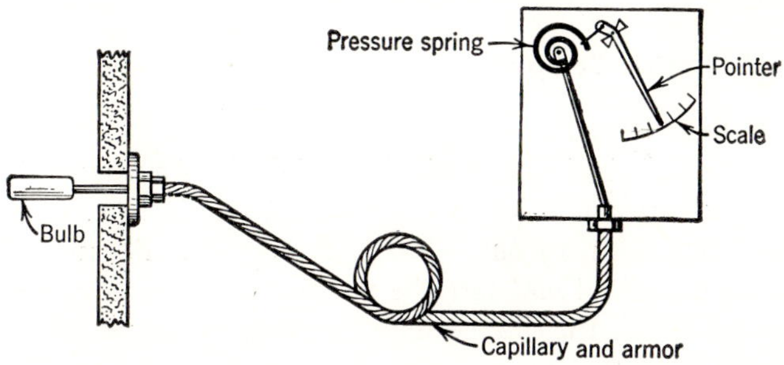


Fig. 20-5. Fluid expansion thermometer. [Reprinted by permission, D. P. Eckman, *Industrial Instrumentation,* John Wiley and Sons (1950).]

which moves across a calibrated scale in direct relationship with temperature (Fig. 20-5). The fluids used are inert gases (usually nitrogen), liquids (mercury or organic liquids), and vapors (propane, methyl ether, ether, or chlorobenzene).

Pressure thermometers filled with liquid or gas must be compensated for ambient temperature changes. If the temperature along the capillary end at the pressure spring is radically different from that at the calibration point, the temperature readings will be in error since volumetric expansion in these areas will be changed. To overcome this effect a bimetallic strip or a second pressure spring with a capillary (without bulb) running adjacent to the measuring capillary can be installed in opposition to the pressure-measuring spring. This arrangement will cancel the ambient temperature effects.

Vapor-filled systems require no such compensation if the liquid level remains in the bulb. The pressure in the system must correspond to the vapor pressure of the liquid at the bulb regardless of ambient conditions.

The use of fluid-expansion, pressure thermometers in process plants is very extensive, particularly in temperature ranges between 0 and 500° F. In order to keep the length of capillary to a minimum, a transmitter-type instrument may be used. This instrument is located near the primary element and converts the expansion pressure impulse into an electrical or pneumatic signal which may be conducted to a central control house. Since the primary sensing device is fundamentally a pressure instrument, pressure-type instrument elements such as bellows and Bourdon elements (helix or spiral) may be used as secondary elements for indicating, recording, and controlling.

The pressure-thermometer instruments are relatively inexpensive compared to thermo-electric instruments. The vapor-filled instrument is the most popular of the three types. It is lowest in cost, requires no compensation, and has rapid response. The scale, however, is nonlinear. Of the three, the mercury-filled thermometer is the most stable and sensitive.

### *Thermo-electric Temperature Measuring Instruments*

#### Thermocouples

The thermocouple is the most widely used device for temperature measurement. Accurate potentiometers make it possible to employ thermocouples for temperatures from $-300°$ F to $3000°$ F. Because of the ease by which the generated emf may be transmitted over large distances to central control rooms, the thermocouple instrument has proved convenient and accurate for modern process control and measurement purposes.

As early as 1821 it was observed that if a circuit of two dissimilar metals was formed and the junction at one end was heated (hot junction) and the other ends (cold junction) were attached to a potentiometer or millivoltmeter, a measurable voltage could be observed on the instrument. This voltage was found to have a definite relationship* for each pair of metals with the temperature difference between the hot and cold junction. If the cold junction, therefore, is maintained at some constant temperature, the hot junction temperature can be obtained over a wide range by measuring the voltage generated.

Experimental evidence and reasoning based on the laws of thermodynamics have produced the following rules of practical significance to industrial thermometry.

1. The emf developed by the difference in temperature of two junctions is independent of temperature gradients along homogeneous wires between junctions. Heat effects alone will not produce a current in a homogeneous circuit.†

* This relationship is not linear but roughly approximated by

$$\text{emf} = m\,\Delta t + n(\Delta t)^2$$

where $\Delta T$ = temperature difference between hot and cold junction

$m$ and $n$ = constants for particular thermocouple

† The well-known Thomson effect is used to describe the emf generated in a homogeneous wire when a temperature difference exists between the terminals of the wire. The temperature distribution along the wires, however, does not affect the Thomson emf. This emf together with that caused by the difference in temperature of the two junctions makes up the total "thermocouple" emf.

2. A third conductor may be introduced into one side of the circuit as in Fig. 20-6*a* without changing the emf produced, provided the temperature along the conductor remains constant.
3. A third conductor may be introduced on both sides of the circuit as in Fig. 20-6*b* without changing the emf produced if the intermediate junctions are at the same temperature. (The algebraic sums of the voltages generated by a number of dissimilar conductors joined together in a circuit is a function of temperature only.)

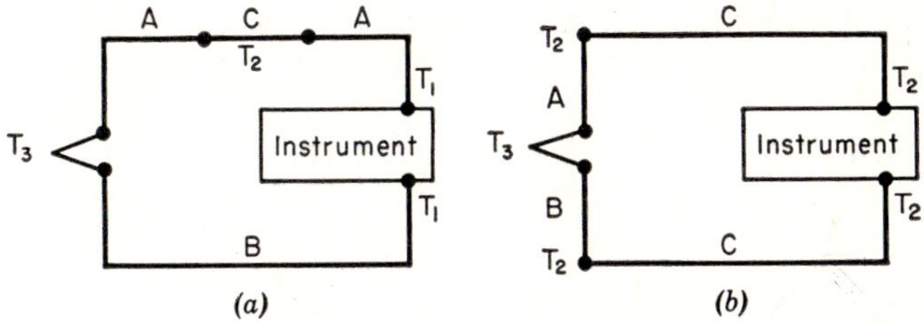


Fig. 20-6. Thermocouple circuits with third conductor.

Thermocouple Combinations. The common thermocouple material combinations are given in Table 20-1. Each pair has particular advantages in degree of accuracy and range. The copper-constantan* thermocouple is superior in the lower temperature range. Above 650° F copper oxidizes rapidly. Below 200° F copper-constantan has greater accuracy than iron-constantan. It produces higher emf's in its range than the platinum couple and is less costly.

Because of their low cost, large range, and reliability when properly applied, iron-constantan thermocouples are the most widely used. The iron, however, is subject to oxidation and precautions must be taken to prevent this.

Chromel [Ni(90%), Cr(10%)]-Alumel [Ni(94%), Mn(3%), Al(2%), Si(1%)] was developed by its manufacturer for use in oxidizing atmospheres where iron-constantan proves unusable. Its most important applications are in the range of 1400 to 2300° F.[4] At high temperatures the Chromel-Alumel couple is unsatisfactory in reducing atmospheres (less than 2% oxygen).

The platinum-platinum rhodium couple is more expensive than the base metal couples and is used in process plants primarily for temperatures above the range of other couples. Platinum is subject to alteration by reducing atmospheres and certain vapors, and is invariably enclosed in gas-tight wells for protection.

Commercial thermocouple wire is manufactured to close tolerances and will, when assembled, match published thermocouple voltages to a degree of accuracy suitable for most purposes. Materials of greater certified

* Constantan is a copper-nickel alloy (50 to 65 Cu and 35 to 50 Ni).

accuracy for precise work such as research can be obtained at a higher cost.

THERMOCOUPLE WELLS AND INSTRUMENTS. Industrial-type thermocouples are made in several ways depending on the type. Chromel-alumel and iron-constantan wires are usually twisted with two turns and welded at the ends with a d-c carbon arc[4] using a flux such as borax. The ends of copper-constantan and platinum are not twisted but usually just fused into a round bead. For platinum a carbon arc is used and for copper-constantan, a soldering procedure. The commonly used wires vary in size from 0.032 to 0.1285 in. in diameter. Platinum is usually 0.02 in. in diameter.

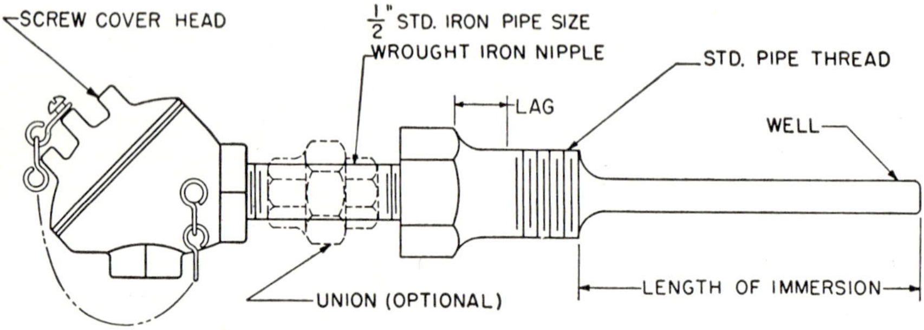


Fig. 20-7. Thermocouple assembly. (Courtesy of Minneapolis-Honeywell Regulator Company.)

So many processes involve corrosive fluids and fluids under pressure that bare thermocouples are seldom used. Instead the thermocouple is inserted into a manufactured well which is provided with fittings for mounting on a pipe or pressure vessel (Fig 20-7). Instrument manufacturers offer these assemblies complete with thermocouple, two-holed ceramic insulators for the wires, well, and terminal head with cover. The most common well materials are carbon steel, stainless steels, and chromium steels.

The speed of response of a thermocouple assembly (well and thermocouple) can be increased and thermocouple errors decreased by using small diameter, thin wells, and couples of low mass. The error for temperature measurement of gases can be further reduced by increasing the flow rate around the assembly and by interposing a radiation screen between the conduit walls and the thermocouple assembly. In all cases the thermocouple well must make good contact with the well body to insure rapid response.

MEASUREMENT OF EMF. The instruments used for recording thermocouple emf's in terms of temperature and controlling processes based on the measured temperature are usually potentiometer-type circuits. The emf of the thermocouple is balanced against a known emf produced by a battery across a calibrated slide wire. The battery is periodically and automatically standardized against a standard cell.

A number of methods are used for balancing the thermocouple emf and thus indicating and recording the temperature by movement of the contact on the slide wire. The simple potentiometer circuit can be made to perform all operations automatically. Unbalance in a galvanometer may be detected by a continuous system which momentarily clamps the galvanometer pointer in position and then moves the slide wire by means of a motor until balance is reached. This type was for years the standard process instrument. More recently, however, high speed instruments based on electronic circuits maintain an almost instantaneous or continuous balance. In several of these the galvanometer has been replaced by a device which converts the d-c thermocouple current to a-c. This is followed by voltage amplification and the voltage is applied to the motor. These instruments are rapid, rugged, and extremely accurate. The length of a 12 in. chart can be traversed in one second. It has, therefore, been possible to offer such instruments for measuring and recording a number of different temperature points on one chart.

The cold junction of the thermocouple circuit is at the instrument. Rather than attempt to maintain it at a constant temperature, an automatic cold junction compensator is provided. This compensator is a resistor in the potentiometer connected in parallel with the slidewire. The resistor is made of a material whose resistance is sensitive to temperature changes. The change in resistance causes a movement of the slidewire balance position so as to compensate for cold junction temperature changes.

LEAD WIRES. Lead wires must be used properly if thermocouple measurements are to be accurate. Usually it is impractical to use the thermocouple wire as a conductor from the junction all the way to a remote recorder. Lead wires which are cheaper and easier to handle are generally connected to the heavier thermocouple wires at the head of the thermocouple well, and conducted to the instrument. If the temperature at the head and the cold junction at the instrument are the same, any suitable conductors can be used. Since these two temperatures are not usually the same in process installations, lead wires of similar thermoelectric properties to the thermocouple must be employed as lead wires. Iron-constantan and copper-constantan of smaller diameter than thermocouple wire are used as lead wires for their respective thermocouples. The lead wire material can often be made of rejected thermocouple material since it must only be similar to the thermocouple metal in the temperature range of the instrument. Iron and copper-nickel alloy are used for connection respectively to Chromel and Alumel; and copper and copper-nickel alloy, to platinum and platinum-rhodium.

Lead wires can be insulated in many ways depending on the atmosphere. Insulation such as enamel, rubber, asbestos, glass fiber, and cotton are on the market.

### Resistance Thermometer

The resistance thermometer, long a standard means for precise laboratory temperature measurement, is also becoming more important as an industrial instrument. It is competitive with the thermocouple in the temperature range of −300° F to 950° F. Although more costly and bulky, and slower to respond, it has greater sensitivity and requires no cold junction compensation.

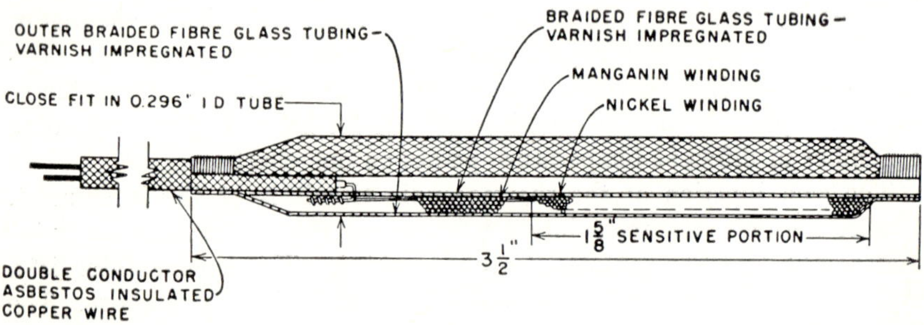


Fig. 20-8. Resistance thermometer assembly. (Courtesy of Minneapolis-Honeywell Regulator Company.)

Resistance thermometry utilizes the increase of resistance of metals with temperature. A typical thermometer bulb as shown in Fig. 20-8 consists of a wire-wound resistor 1⅝ in. in length. The Manganin coil connected in series is used to adjust the temperature-resistance characteristics. These bulbs are installed in wells similar to those for thermocouples. Nickel is the most popular resistance element material since it is inexpensive and has a high change of resistance with temperature. Platinum is more accurate and can be used for higher temperatures.

The resistance element is made a portion of some bridge circuit such as a Wheatstone bridge. A direct current is supplied by a battery. The change in resistance of the bulb unbalances the bridge, and the unbalance is detected and corrected in a similar manner to that described for potentiometer circuits for thermocouples.

### Optical and Radiation Pyrometers

Radiation or optical pyrometers are used for measuring temperature above the ranges of other instruments. Radiation pyrometers measure the total radiant energy coming from a hot object by focusing the energy with a lens on a thermopile (a number of thermocouples connected in series). The thermopile generates an emf proportional to the radiant energy. If an object being measured approximates black body conditions such as a furnace flame, the energy transmitted is proportional to the fourth power of the absolute temperature of the object.

The familiar manually operated optical pyrometer compares the intensity of monochromatic radiation coming from an object to an adjustable standard source. The radiation from the object and from a tungsten lamp are compared through red filters. The current to the lamp can be varied by changing a resistance until the two images match. The meter is calibrated for temperature in terms of the resistance variations.

One type of automatic optical pyrometer utilizes two photoelectric cells to compare the energy from the hot object and a carbon filament lamp. The lamp current is automatically varied to bring the two photocells into equilibrium. This current is then proportional to the temperature and can be used to actuate a recorder. The instrument is instantaneous and can respond to a 200° F change in less than ⅗ of a second. It is thus valuable for measuring temperatures of moving objects.

These high temperature pyrometers are more familiar in steel mills, but they are used in chemical plants particularly for controlling rotary kilns and furnaces.

## Pressure Measuring Instruments

### *Bourdon-Type Gage*

The simple Bourdon tube and its modifications are the most commonly used pressure measuring devices in the process plant. The Bourdon tube as shown in Fig. 20-9 consists of a closed hollow metal tube elliptical in cross section. It is bent in a semi-circular shape, fixed at one end and connected at that end to the pressure source. Application of a pressure higher than atmospheric causes the tube to unwind while a vacuum causes contraction. By means of a connecting linkage and a gear and pinion, a pointer is made to move over a calibrated scale. An approximate linear relationship between pressure and scale movement exists over ranges of moderate size.

The most familiar Bourdon gages are the indicating dial gages which are cheap, rugged, and accurate (½ of 1% of scale range for industrial-type and ½ of 1% of reading for laboratory models). They are common in all parts of process plants, and are available in a number of ranges from 30 in. of mercury vacuum to 10,000 psi pressure. Laboratory test gages capable of measuring up to 100,000 psi are also produced.

Although the simple Bourdon-tube element may be used for moving a pen linkage of a recorder in higher pressure ranges (300 psi to 20,000 psi), spirally or helically wrapped Bourdon elements are more commonly used in recording instruments particularly for the lower pressure ranges (0 psi to 4000 psi). The flat spiral or a helically wrapped element will produce larger deflections at the free end for a given pressure differential than the simple Bourdon tube. The movement is in effect the combined action of

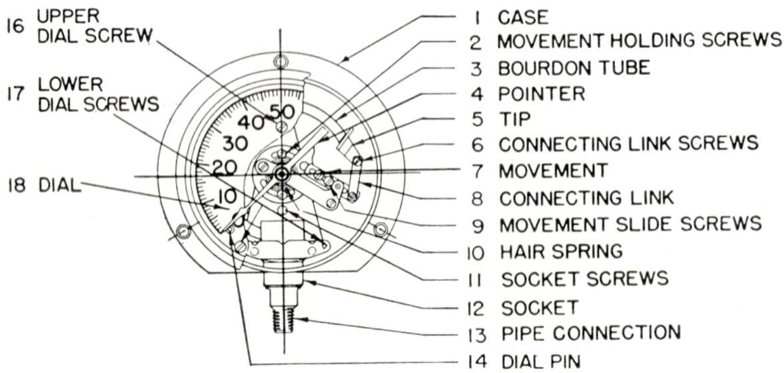


Fig. 20-9. Bourdon tube pressure gage. (Courtesy of Ashcroft Gage Division of Manning Maxwell and Moore, Inc.)

a number of simple Bourdon elements connected in series. This greater movement is necessary for operating through suitable mechanical linkages, electrical contacts, pneumatic systems, and recording pens on controllers and recorders.

Bourdon tube materials used are carbon steel, stainless steel, various chromium steels, bronze, and beryllium copper. The choice of material depends on the pressure range and fluid being measured. Bronze tubes are selected most frequently for pressures below 1000 psig. High pressures and corrosive liquids require alloy steels.

For certain highly corrosive conditions it is preferable to isolate the fluid from the tube. This is accomplished by providing a flexible diaphragm with measured fluid on one side and a sealing or transmission fluid (oil) on the other side in contact with the Bourdon element.

Steam pressure gages can be protected from high temperatures by interposing tubing having one or two coils between the pressure tap and the gage. Steam condenses in this crude atmospheric cooler and is trapped in the coils. The temperature in the gage is thereby kept below that of the steam.

### *Bellows and Diaphragm Instruments*

In the lower pressure ranges, below 30 psig, and for vacuum, more sensitive elements than the Bourdon type are required to produce reliable measurements. Metal bellows and diaphragms operate with much greater sensitivity in these ranges.

Several bellows elements are shown in Fig. 20-10. The higher pressure acting on the outside of the low pressure bellows tends to compress it causing a pointer to move. The range of the instrument is determined by the stiffness of the spring which is located in the sealing bellows on the right. Here as in other instruments the pen movement can be linked

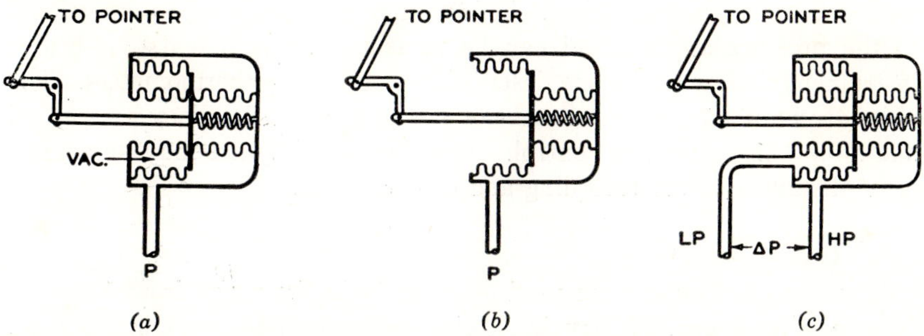


Fig. 20-10. Pressure bellows elements. (*a*) Absolute pressure; (*b*) gage pressure; (*c*) differential pressure or vacuum. (Courtesy of Taylor Instrument Company.)

to pneumatic or electrical systems to affect control of a process. Bellows elements are used in ranges from 100 in. of water to 30 or 40 psig and at 10 to 30 in. of mercury vacuum.

For the lower ranges of positive pressure commonly encountered in measuring furnace drafts (0 to 40 in. of water) a diaphragm element of leather or plastic material is used. It moves with changes in differential pressure across it and the movement can be transferred to a pointer. For higher pressures up to 5 psig metallic diaphragms (brass or bronze) are used.

### *Manometers*

In addition to the familiar laboratory type manometer, differential-pressure mercury manometers and the mercuryless manometers or force-balance instruments discussed under "Flow Meters" provide reliable measurement of pressure differentials up to 600 in. of water. For low pressures one of the several bell differential pressure elements is often used. These elements consist of a bucket or bell inverted in an oil bath. One pressure tap terminates inside the bell and the other terminates outside. The bell then moves up or down depending on the differential. This movement is transferred to a pen. Many other manometer or manometer-like instruments are available, particularly for the low pressure range.

### *Electrical or Electronic Pressure Measurement*

#### Strain Gage

The strain gage which has been used so successfully for stress analysis of structural members under load, has been adapted for the measurement of process pressures and as such is called a pressure transducer. The force acting on the gage changes the tension and thus the electrical resistance of a group of wires. This change of resistance is proportional to the pressure applied and can be measured by a Wheatstone-bridge instrument similar to the one described for resistance thermometers. The units are built for pressure ranges of 100 to 50,000 psi. They are rugged and adaptable to process control. The electrical impulse may be carried great distances to the recording and control point.

#### Vacuum Gages

Mercury manometers are used extensively in the range from 1 mm to 50 mm of mercury. Many industrial gages for high vacuums depend upon electrical or electronic action. In the Pirani and thermocouple gages changes in pressure of the surrounding atmosphere alter the amount of heat loss and thus resistance of a hot wire through which current is flowing. The effect is pronounced below one mm of mercury pressure.

In the Pirani gage the resistance filament is made part of a bridge circuit and the change in resistance with pressure is recorded. The heat effects on the filament produced by pressure change are measured by means of thermocouples. The temperature of the filament in contact with the vacuum being measured is compared to that in a sealed reference or compensating cell. The range of these gages is from 1 micron to about 700 microns of mercury absolute.

For pressures below one micron of mercury, absolute ionization gages are used. One such gage employs a triode tube with constant electron emission. The atmosphere in the tube is the same as that of the vacuum being measured. Current flow from the tube is proportional to the number of ions formed by electron bombardment of gas molecules in the tube. The number of ions formed is, in turn, proportional to the number of gas molecules in the tube which is directly proportional to tube pressure. The filaments of these gages are damaged by pressures over one micron. A similar gage using a permanent radium source for ion production from the gas molecules, requires no filament and can, therefore, operate successfully in the range of normal atmospheric pressures down to 0.1 of a micron of mercury absolute.

## Liquid-Level Instruments

### *Differential-Pressure Type*

Differential-pressure instruments are often used as liquid level instruments. One side of a manometer, for instance, can be connected to a point at the bottom of a vessel and one at the top above the liquid level. Differential pressure between the two points is an indication of the liquid level. Pressure on the lower connection includes liquid static head as well as vessel pressure. Only one pressure connection is needed for open tanks and that at the bottom. This type instrument is often used for level indication of fluidized solids when a displacement type device would not be satisfactory.

In order to avoid mechanical problems with fluids which may settle, solidify, or cause corrosion, other fluids such as inert gas or air may be slowly injected into the vessel* at the two differential points. The difference in pressure of the injected fluid is equal to the static pressure of the liquid. Simple Bourdon tube pressure gages can be used with some fluids for indicating liquid level by the same method of differential pressures.

### *Float Types*

The float or displacement-type liquid level indicator is probably the most common liquid level instrument. This instrument consists of a buoyant float. The device is connected by piping directly to two points in the vessel or container. Changes in liquid level move the float. The simplest type is a ball float connected by an arm to a shaft extending through the chamber or vessel shell (Fig. 20-11). The weight of the float is adjusted by the use of counterweights so that it remains in the half-submerged position throughout its range. In this way the maximum amount of liquid is displaced for all positions of the float and thus the maximum power is generated for both rising and falling levels. This power is transmitted through a lever to a direct-connected control valve and indicator or to a pneumatic relay which transmits air pressure proportional to the level to a recorder or controller. The operating range of these units is narrow (approximately 20-in. maximum level change).

The displacement-type float, unlike the ball-float, varies in displacement depending on the level. A long cylindrical float is designed so that the float hangs free at zero or minimum level and is completely or approximately submerged at maximum level. Fig. 20-12 illustrates the

* The flow of air can be controlled by a rotameter or sight-feed bubbler.

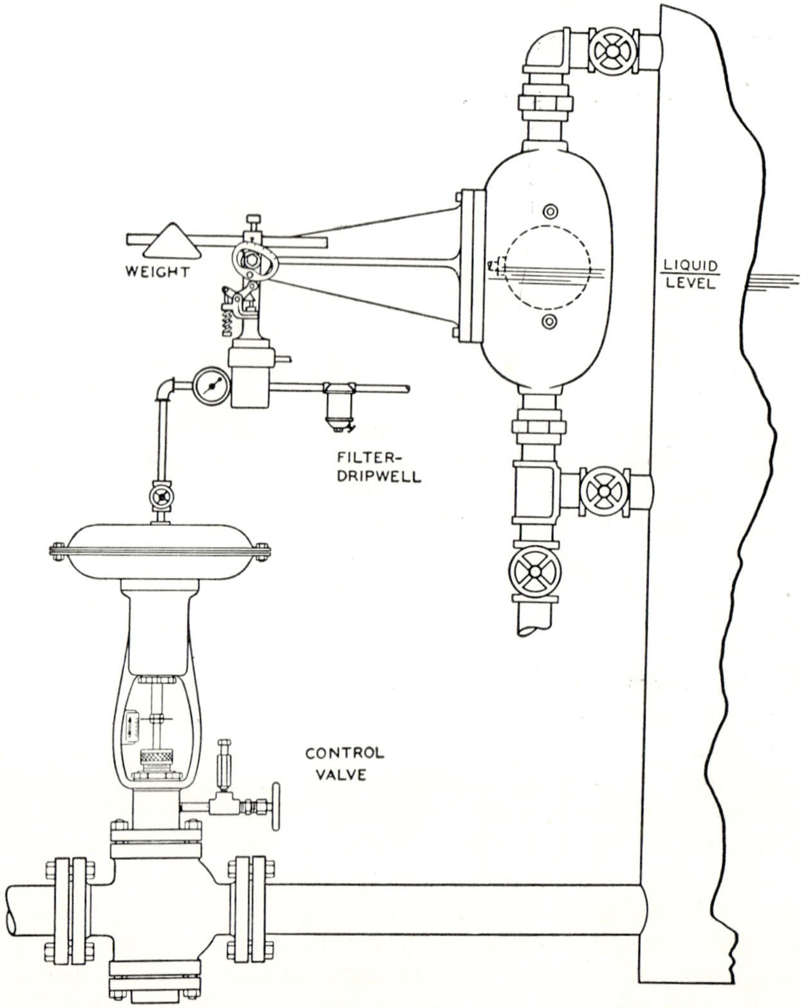


Fig. 20-11. Float-type level controller and valve. (Courtesy of Mason-Neilan Regulator Company.)

basic principles by a simple demonstration. The variation in force on a spring scale and the movement of the float is illustrated for three levels of liquid. If instead of a spring scale a torque tube as in Fig. 20-13 is used, the twisting of the tube is proportional to the level of liquid. This movement is made to operate a pneumatic system which transmits a proportional air pressure to a recorder, alarm, or controller.

The moment acting on the torque tube is equal to the downward force of the float times the length of the float rod. The buoyancy of the float affects the magnitude of the turning moment. Thus by proper selection of float density, dimensions, and length of the float arm, an instrument can be designed for the sensitivity and accuracy required for a given job.

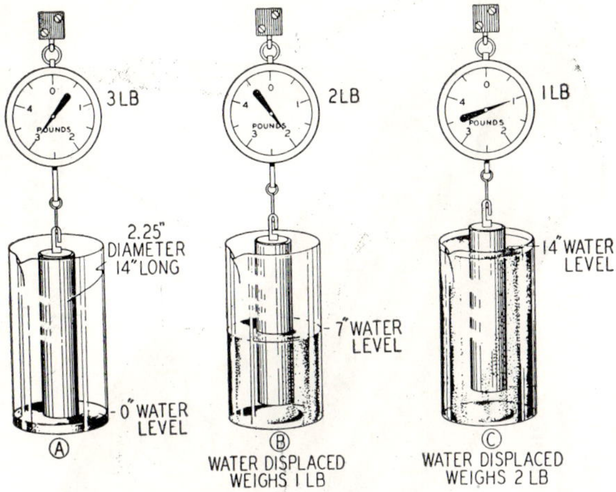


Fig. 20-12. Principles of displacement meters. (Courtesy of Mason-Neilan Regulator Company.)

The range of displacement-type floats is from 14 in. to approximately 120 in. They are very accurate particularly on low density fluids. Their high sensitivity has also made them very popular for interfacial level control and measurement.

### *Miscellaneous Types*

There are many other types of level measuring devices. A diaphragm placed in the bottom of an open tank can be used to indicate level. The movement of the diaphragm can be balanced by a force-balance pneumatic system and the resulting air pressure used as a measure of the liquid height. These units are very sensitive and particularly valuable for viscous liquids which do not function well in the float-type instruments.

Several electronic, electrical, and radiation devices are also available. These employ photoelectric cell units, gamma radiation, and high frequency alternating current. The transfer of photoelectric energy, high frequencies, and gamma radiation are all affected by intervening liquids. Suitable sensing devices can, therefore, be used to detect the position of a level at all times.

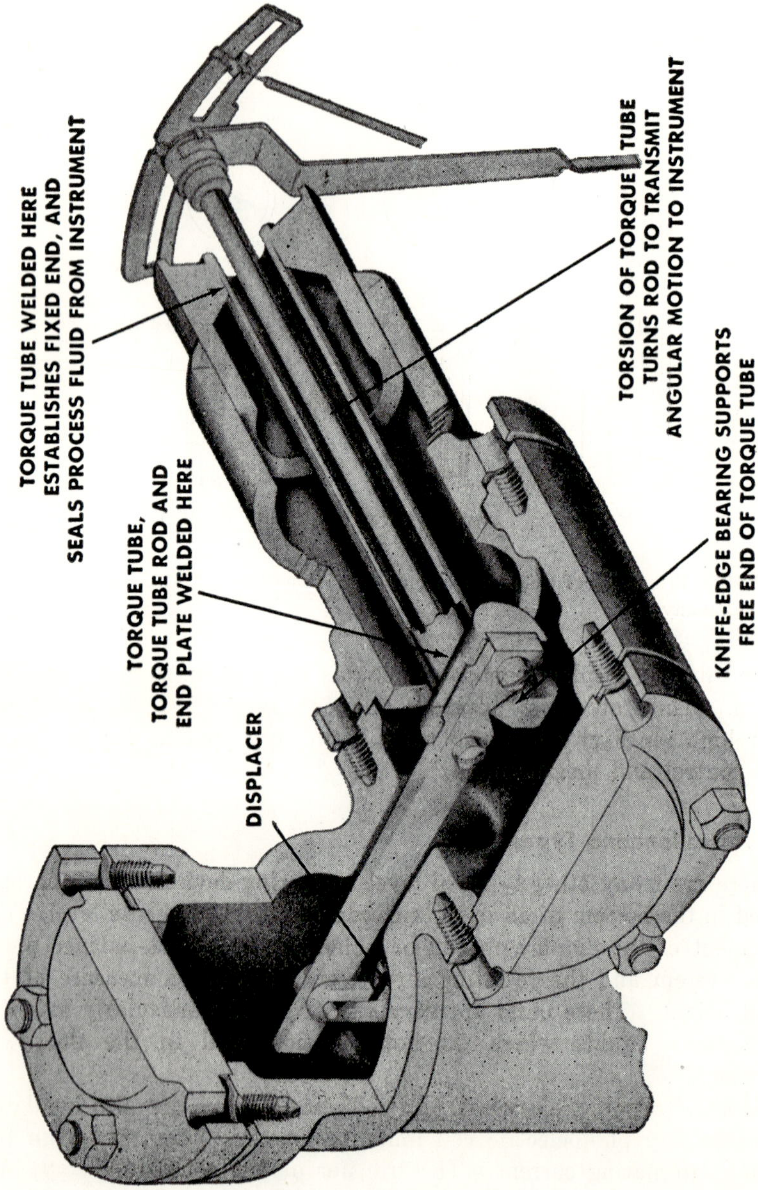


Fig. 20-13. Torque-tube displacement meter. (Courtesy of Mason-Neilan Regulator Company.)

### *Visual or Manual Level Gages*

A number of visual devices are manufactured for indicating liquid level. The most widely used consists of a glass tube connected to a vessel or container at its high and low points. For tall vessels a number of such gage glasses can be connected over the range to be measured. These glasses are provided with protectors to prevent breakage.

Gage glasses for high pressure are constructed of a chamber machined from a solid bar of steel. There are two principal styles. One has thick plain-glass blocks on both back and front (transparent gage) and another has a front glass only. The front glass of the latter is cut so as to produce a number of glass surfaces or prisms which refract the light. The refraction of light is greater for that portion of the gage containing liquid and it appears black while the portion without liquid is clear.

The transparent gage is used for installations requiring knowledge of the color of liquid being measured. In all other installations the other more easily readable refracting gage is used.

Designs are marketed which provide means for heating or cooling the gage fluid. The heating or cooling medium is conducted through a tube attached to the back of the gage or the tube may be run through the liquid chamber.

High-pressure gages have been used with great success. They are rugged and usually exceed the requirements of the several codes for pressure vessels.

Floats connected to a moving tape are sometimes used for measuring levels in large storage tanks. Hand measurement is the oldest method for storage tank level measurement, however. Liquid adhering to a graduated metal tape, which is lowered by hand into the tank, indicates the level.

## Measurement of Motion

### *Tachometric Devices*

Tachometers of various kinds may be obtained for practically any speed. Those for high-speed (7 to 10,000 rpm) machines such as certain turbine generators and centrifugal compressors are usually of the vibrating reed type. This type instrument is used for indicating primarily because its indicating mechanism has little or no force output. Other instruments utilize stroboscopic principles or a driven generating device which in turn has its emf sensed by a translating instrument to indicate, record, or control.

Nearly all high-speed machines such as turbines are equipped with speed-governing devices actuated by centrifugal force. The primary device rotates at some direct speed ratio with the machine. As speed increases weights actuated by centrifugal force operate a valve admitting oil into a hydraulic system; the hydraulic system in turn actuates a diaphragm or piston-type valve controlling steam flow to the machine. On certain units the primary device may be linked directly to the control valve, but the hydraulic system permits greater flexibility. The oil used is taken directly from the turbine lubricating system. Such governors are safety devices primarily since with a sudden load decrease the machine could easily reach critical speeds.

### Limit Controls

The limit-type device is usually an electrical contact which opens or closes at preset limits. The now common electric eye as used to open doors by interruption of a beam of light is a simple illustration of a limit device. The interruption of the light beam changes the out-put of a photoelectric cell, which change can be used to operate micro-switches and in turn relays.

### Counting Instruments

The simplest type of totalizing or counting device is the trip counter. A ratchet rotates a numbered disk one tooth or integer at each cycle. The next consecutive disk is in turn moved one integer for each revolution of the first.

## Physical and Chemical Measurements

In modern process plants many control laboratory functions have been displaced by automatic instruments designed to measure continuously the properties of a flowing process stream. The development of instrumental analytical techniques for the laboratory was followed closely by modifications of laboratory instruments for use as stream analyzers which could be installed in the plant.

Processes requiring a physical measurement or a chemical analysis of the product are burdened with an excessive time lag when operators must await results from a control laboratory before changing conditions. Even though the initial cost may be high the use of stream analyzers such as infrared spectrophotometers have proved profitable in processes requiring rigid quality control. The continuous record eliminates time lag and prevents production of large quantities of product which do not meet specification.

A list of some of the instruments used to measure chemical and physical properties is given in Table 20-2. These instruments have been adapted for continuous stream analysis. Some of these instruments are costly and in such cases the decision to use them must be based on a careful economic study. It would be unwise, for instance, to employ an infra-red analyzer when a process temperature at a critical point could be statistically correlated with product quality.

A list such as that in Table 20-2 is subject to rapid change. Each year existing instruments are improved, and new ones are introduced.

## PROCESS CONTROL

### General

Nearly all process-control applications for the process plant involve an adjustment in rate or change in direction of fluid flow. The elements of control consist of a primary or sensing device, a secondary element or translation device, generally known as the instrument, and a third element which effects the change in stream properties. Most control systems utilize pneumatic systems for operation of the third element.

### Batch Systems

Batch system control usually involves a simple control which will allow a certain desired condition to be reached in a nonflow system and continue to maintain that condition until completion of the process.

A batch system may be illustrated by a steam-jacketed kettle filled with a fluid to be heated. The controller opens a valve admitting steam to the kettle jacket where it is condensed giving up heat to the walls of the kettle. The condensate is continuously withdrawn by means of a steam trap. When the internal fluid reaches the desired temperature the controller maintains the temperature by opening or closing the steam inlet valve.

Such systems are often used in the food industry when a cooking process must continue for an extended length of time. The operator fills the kettle and puts the instrument into operation. By the use of a timing device the system can be made to shut-off at the end of a certain period.

If temperature is not critical the same type heating can be accomplished by a jacket pressure control which controls the steam into the jacket. The condensate trap will continuously withdraw condensate, and the jacket will be held at the same temperature as long as the pressure is constant.

Location of the primary element in a stirred kettle is relatively unimportant since all the fluid must ultimately reach the same temperature, at which time the control will close the steam valve until a temperature drop is sensed.

Fluid-expansion type temperature controllers are most frequently used in such batch-type processes. The instrument may or may not be

## TABLE 20-2. Instruments for Chemical and Physical Property Control and Measurement of Continuous Streams*

NONREVERSIBLE CHEMISORPTION INSTRUMENTS

| Method | Measured effect | Application | Range | Accuracy | Remarks |
|---|---|---|---|---|---|
| Automatic Orsat | Change of volume of gas sample at constant pressure before and after absorption in chemical reagents. | Generally, analysis of acid-forming gases e.g., $SO_2$, $CO_2$, CO, etc. | 0–100% | Basic Orsat principle, 0.2% of theoretical. Industrial automatic instruments, 0.5%–10% of range. | Operates on discontinuous batch sampling at 2 to 3 minute cycle rate. Uses liquid chemical absorption reagent selective to the gas analyzed. Reagent becomes spent and must be replaced. |
| Pressure bridge | Pressure loss in series orifices or pressure bridge as a result of chemical absorption. | Ditto above. | 0–50% | 0.5%–10% of range. | Operates on continuous flow gas sample. Generally uses dry absorbing reagent. Reagent becomes spent and must be replaced. |

REVERSIBLE CHEMISORPTION INSTRUMENTS

Instruments of the types described above employ nonreversible absorbents. The reagent becomes changed in chemical composition by the gas sample and must be replaced. Reversible absorbents behave as catalysts and establish equilibrium with gas concentration and some physical property of the absorbent. They do not become "spent" except by "poisoning."

| Method | Measured effect | Application | Range | Accuracy | Remarks |
|---|---|---|---|---|---|
| Catalytic oxidation — Reduction. | Electrical resistance of sponge silver catalyst planted on glass wool contained in analysis cell. | Oxygen analysis. | Not established. | Not established. | Sensitive to small concentrations of $O_2$. |
| Catalytic combustion | Temperature rise of catalyst, e.g. Hopcalite or silver permanganate contained in gas analysis cell. | Analysis of gases which undergo catalytic combustion e.g., CO, $NH_3$, etc. | Trace quantities and low concentrations below 0.1%. | 0.005% in range of 0.01%:5 ppm (accuracy) in range 10 ppm to 200 ppm CO, or 2% theoretical $O_2$ in range 0.001% to 0.2%. | Applicable to continuous flow analysis. Vulnerable to moisture in sample. |

## TITRIMETERS

Titration implies the measurement of concentration of an unknown material by virtue of its quantitative reversible chemical reaction with a chemical reagent of known concentration; and, the use of a suitable primary element to detect chemical *equivalence* of the reaction. Titrimetric methods are based on a volume-concentration relationship and resolve into methods of metering *volume* of reagent and process liquor. Instruments serve the purpose of detecting end-point equivalence.

| Type | Measured effect | Application | Range | Accuracy | Remarks |
|---|---|---|---|---|---|
| Potentiometric | Potential difference between suitable indicator electrode and standard reference electrode. | Measurement of total *molecular* concentration of electrolytes. | Low concentration range below 10%. | Under favorable conditions well under 0.5%. Poor potential "breaks" may reduce accuracies to 1%–5% of theoretical. | Applicable to continuous process control. Useful in measurement of total acidity. Restricted primarily to aqueous solutions. |

## FILTER PHOTOMETERS

In its original sense, photometry was designated as that phase of radiometry which dealt with the measurement of visual light. Its connotation in analysis instrumentation has been extended to embrace all types of instruments which measure absorption of energy at a specified wavelength throughout the electromagnetic spectrum. The magnitude of absorption (of monochromatic radiation) is a function of concentration in accordance with Beer's law. Only the non-dispersing instruments are treated in this chapter.

| Instrument Type | Measured effect | Application | Range | Accuracy | Remarks |
|---|---|---|---|---|---|
| (*a*) Visible light | Per cent transmission of monochromatic light in the visible range compared to standard. | Ditto colorimeters. | Ditto colorimeters. | Use of photocells to replace human eye increases accuracy tenfold over colorimeters. | Probably most widely used of all laboratory instruments. Double beam instruments are used in continuous flow analysis and control. |
| (*b*) Ultraviolet | Per cent transmission of monochromatic energy in the range 2500A°-3650A°. Usually compared to reference standard. | Absorption of UV energy by conjugate double bonds, ring-structured molecules and some inorganic molecules gives measure of concentration. | Best for trace quantities and concentrations below 10%. | 0.1%–5% of theoretical. | Suitable for analyzing low concentrations $H_2S$, $Cl_2$, $C_6H_5Cl_3$, xylene, acetone, $C_2Cl_2$, $O_3$, aniline $CS_2$. Applicable to continuous process control. |
| (*c*) Infrared | Per cent transmission of monochromatic energy in the range 2.5 to 25 microns, compared to standard. | Usually analysis of gases which absorb IR energy. Has also been applied to liquid in liquid analysis. | Low concentration range. Usually below 20%. | 0.1%–10% of theoretical dependent on specific conditions. Sensitivity can be increased by "pressure broadening." High cone, analysis limited by sensitivity saturation. | Instruments best adapted to gas analysis because the condition of one constituent showing strong absorption in a weak field from interferences is favored more by gases than liquids. |

*Reprinted by permission — M. F. Behar (Editor) *Handbook of Measurement and Control, Instruments and Automation*, *27*, No. 12, 109–115, Part 2 (1954).

## TABLE 20-2. Instruments for Chemical and Physical Property Control and Measurement of Continuous Streams* (continued)

REFRACTOMETERS

Refractive index is an additive function which constitutes a measurement of concentration of liquid mixtures when the total volume of solution is equal to the sum of the volumes of its constituents. Or specific refraction of molecular refraction is characteristic of chemical compounds. Specific dispersion constitutes a measurement of refractive index at two different wavelengths of light and expresses a characteristic function of liquid composition. Transparent or opaque solids may be measured but, for the most part, industrial measurement of refractive index are confined to liquid measurements.

| Type | Description | Application | Range | Accuracy | Remarks |
|---|---|---|---|---|---|
| Minimum deviation method | A prism is made of the material whose refractive index is to be determined. The angle of light deviation is related to its refractive index. | Measurement of transparent solids or liquids. A hollow glass prism through which a liquid is circulated has been applied to continuous process control. | Unlimited. This is the classical method for fundamental determinations. Industrial fluid instrument range 1.1500 to 1.4700. | Very high. See L. W. Tilton, *U. S. Bureau Stds., J. Res.*, Vol. 14 P. 393 (1935) for discussion of factors affecting accuracy. | Widest range of common industrial refractometer methods. See U. S. patent 2,413,208. Industrial type instruments restricted to low pressure transparent fluids. |
| Total reflection method | A transparent cylindrical or trapezoidal prism is surrounded by or contacts the solution to be measured. Ratio of scattered to transmitted light is related to difference of indices of prism and solution. | Measurement of opaque or transparent fluids at atmospheric or elevated pressures. | Limited. Dependent on relative index of prism and fluid measured. Pressure range 0–150 psig. | ±0.00004 to ±0.00005 | Limited rangeability. Principle well adapted to industrial instrumentation. |
| Critical angle dipping type | A prism dipped into or submerged in the fluid measured refracts light and produces a light and dark field divided by a "critical index ray." | Measurement of transparent fluids. | Usually narrow rangeability, i.e., 0.04 (1.32–1.36). | Usually ±0.00003 to ±0.00004. | Only known commercial recording refractometer. |
| Differential refractometer | A beam of light passing through two divided liquids is refracted as a function of the difference in indices of the two fluids. | Difference concentration of two fluids. Or percentage composition of test fluid compared to pure constituent. | Range of coverage usually narrow but sensitivity is high. Pressure range low. | Full scale range 0.0002; sensitivity 0.000002 at 1% accuracy. Full scale at minimum sensitivity 0.01 to 1% accuracy. | Suitable for continuous stream analysis. |

DENSITY AND SPECIFIC GRAVITY

| Instrument type | Measured effect | Application | Range | Accuracy | Remarks |
|---|---|---|---|---|---|
| Tuning fork method | Resonant frequency of vibration of a tuning fork or resonant cavity through which the test fluid flows. | Measurement and control of fluid density at low or high pressures. | Range and accuracy are dependent on relative density of tuning fork and test fluid. Industrial fluids sp. gr. 1.0–1.89, accuracy 100 ppm. | | Applicable to continuous process flow measurement and control. Not flow critical. |
| Hydrometer partially or totally submerged | Magnetic coupling in transformer bridge surrounding hydrometer or intensity of light beam intercepted by hydrometer head. | Liquid specific gravity measurements at atmospheric pressure. | Full range sp. gr. measurements with scale spans 0.005 to .05. | 0.0002 or 3% of range whichever is greater. | Satisfactory for continuous flow measurements but flow critical outside narrow limits. |
| Totally submerged float | Air pressure from pneumatic pilot resulting from buoyant force on totally submerged float. | Liquid sp. gr. or density measurements at atmospheric or very high pressures or vacuum. | Usual range of coverage 1.0–2.9 sp. gr. depending on float dimensions. | Approx. 0.01 to 0.005 sp. gr. | Rugged industrial instruments. Flow critical outside narrow limits. |
| Recording pycnometer | Air pressure which counteracts weight of fixed volume sphere through which test solution flows continuously. | Liquid density or sp. gr. measurements at normal or high pressures. | Full range coverage in scale spans of 0.05 or 0.1 sp. gr. | 0.0005 to 0.001 sp. gr. | Instrument is flow insensitive over wide range. |
| Pneumatic bubble tube or submerged air reaction diaphragms | Difference in back pressure between two air bubble tubes submerged at different elevations. | Measure liquid density or sp. gr. in processing vessel or continuous flow chamber. | Usual range of application 0.8–2.0 sp. gr. | 0.1 to 1 in. W. C. pressure corresponding. | Air must be bubbled at constant rate. Applicable to suspensions and measurements under pressure within limits. |

*Reprinted by permission — M. F. Behar (Editor) *Handbook of Measurement and Control, Instruments and Automation, 27*, No. 12, 109–115, Part 2 (1954).

### TABLE 20-2. Instruments for Chemical and Physical Property Control and Measurement of Continuous Streams* (continued)

MISCELLANEOUS TYPES

| Instrument type | Measured effect | Application | Range | Accuracy | Remarks |
|---|---|---|---|---|---|
| Thermal conductivity | Electrical resistance (nominal 5 to 20 ohms) of a heated wire (2 mil) at constant current or potential in contact with analysis gas. | Analysis of gases having amenable thermal conductivities, e.g., $H_2$, $Cl_2$, $CO_2$, $SO_2$, etc. | 0–10%; 30% or 100% dependent on gas composition. | Usual 1%–15% of theoretical. Error can be 100% or more if instrument not properly applied. | One of the most practical instruments for industrial gas analysis within its limitations. |
| Electrical conductance | A-c. electric resistance of solution contained between appropriate electrodes. | Solution analysis, e.g., $H_2SO_4$ strength, water purity, condenser leakage, salt concentration, etc. | May be limited or 100% coverage depending on solution characteristics. Applicable to dilute solution analysis. | 0.1% to 10% of theoretical dependent on application. | Electrode maintenance limiting factor. D-c. can be used above 1 megohm measured resistance at low concentrations. |
| *p*H | D-c. potential difference between indicator electrode (usually glass) and fixed reference electrode. (Approx. 60 mv per *p*H change.) | Measurement of active acidity in solution, principally aqueous. | 0–14 *p*H | Usually 0.1 *p*H for industrial instruments. | Available as continuous process controllers. |
| Combustion processes | Electrical resistance (temperature) of hot wire which sustains combustion of test gas and reagent fuel. Or thermal conductivity before and after combustion. | Oxygen gas analysis or combustibles gas analysis. | 0–5%; 0–25% . . . 0–30% . . . | 0.25% of range. 5% of theoretical. | Industrial instruments for continuous flow measurement and control. |
| Polarized electrolysis cells | Electrolysis current of cell wherein cathode is depolarized by test gas diffusing through porous walls. | Oxygen analysis | 0–½% $O_2$<br>0–25% $O_2$ | Low range, 0.01% of theoretical. High range 0.25% of theoretical. | Continuous analysis instrument. Limited to gases free of $Cl_2$, $SO_2$ and similar depolarizers. |
| Scrubbing and extraction | Thermal conductivity of gas scrubbed from solution by suitable reagent scrubbing gas. | Dissolved oxygen and dissolved hydrogen in boiler feed water. | 0–0.2; 0–0.5 and 0–1.0 ppm. $O_2$; 0–20 parts per billion $H_2$. | 5% theoretical $O_2$<br>5% theoretical $H_2$ | Continuous stream analysis. Industrial type. |

recording. The controller may be separate from the recording instrument, utilizing fluid expansion to operate the control valve directly.

## Continuous Systems

In the batch system just described there is rarely any necessity for close control since once the kettle reaches its desired temperature the admission of only a small amount of heat should maintain the contents at the desired point. Process control would indeed be a simple matter if the majority of processes were similar to the batch kettle. Process plant control problems, however, usually involve continuous processes in which temperature, pressure, and flow must be considered.

Flow and pressure control problems are usually relatively simple since they only require the adjustment of valves to decrease or increase flow or pressure. The capacity of the equipment and rate of flow or pressure increase can be calculated, and valving can be sized accordingly. If flow or pressure is affected by heat in-put, a more involved control problem results.

Although pressure or flow changes may be sensed almost instantaneously, the primary elements of temperature instruments are usually subject to a considerable lag. The lag produced by both primary element and instrument may be several seconds. Actual fluid temperatures thus may change before the element indicates change. If the instrument engineer has a sound knowledge of the process, he may be able to minimize this lag by careful selection and application of instruments.

## Modes of Process Control

### *Manual Control*

Early processes depended almost completely upon manual or hand control for operation. Sensing devices consisted of the pressure gage and the glass thermometer. Control was based on human experience, sight, sound, and memory.

One of the simplest types of manual control is the operation of the steam locomotive. The boiler is operated by a pressure gage (steam pressure), a water gage glass (liquid level), and the fireman's ability to shovel coal or adjust an oil burner (control). Experience or inspection indicates to the fireman when coal must be added to the fire-box or water to the boiler. When too much fuel is added the safety (relief) valve releases excess steam pressure. The power consumption of the locomotive is erratic depending upon gradient changes. By experience or memory the fireman can anticipate uphill grades and build up pressure to reduce loss of power.

Experience, of course, is the primary requirement of manual control.

A number of processes are manually operated because: (*a*) no instrumentation is manufactured to meet the requirement; (*b*) cost of available instrumentation is excessive, or (*c*) labor costs for maintenance of elaborate instrumentation are high. In general there are relatively few processes which cannot be made completely automatic. A study of almost any process plant will also show that there are few manually operated processes which do not use instrumentation for indicating and recording.

Processes may be started, stopped, or adjusted manually from a central control station by the use of remotely controlled equipment. The pneumatically operated diaphragm control valve is frequently utilized in this service for precise flow control, and hydraulically and electrically operated motor-driven valves are also used. Manual operations can be simplified and operating labor reduced by the utilization of remote control from a centralized control point.

### Mechanical Control

Mechanical control is used in process plant operations for operation of heater damper controls, moving certain valves, counting devices, and limit mechanisms for handling and packaging operations.

### Pneumatic Control

Pneumatic systems are more widely used than any other in process control. Changes in the controlled variable are translated to a specific air pressure which in turn operates a diaphragm control valve (see Chapter 18). The diaphragm moves the valve stem, causing the valve to open or close. A change of pressure of only 5 psi on a diaphragm with an effective diameter of 12 in. yields a force change of 565.5 lb. By varying diaphragm size the necessary power requirements for particular applications can be obtained.

The diaphragm control valve requires a source of air and a method of applying pressure to the diaphragm. Four principle means of accomplishing this action are two-position (on-off), proportional, proportional-reset, and proportional reset-rate control.

#### Two-Position Control

A schematic diagram of a two-position pneumatic control unit is shown in Fig. 20-14. It is an on-off control mechanism. Supply air is bled continuously to the nozzle. A flapper which is actuated by the sensing device either covers or uncovers the nozzle when the measuring element deviates from the control point. An increase in the controlled variable, for instance, causes the flapper to move toward the nozzle and produces an increase in the pressure between the restriction and the nozzle. This pressure is transmitted to a pilot valve (pneumatic relay) causing it to open and place full air pressure on the control valve.

Control valves may be set so that air pressure failure will either open or close the valve. Safety precautions dictate the type of failure action required.

The pilot valve because of its larger ports can transmit full air pressure with negligible lag and is thus preferred to a direct connection between nozzle and control valve.

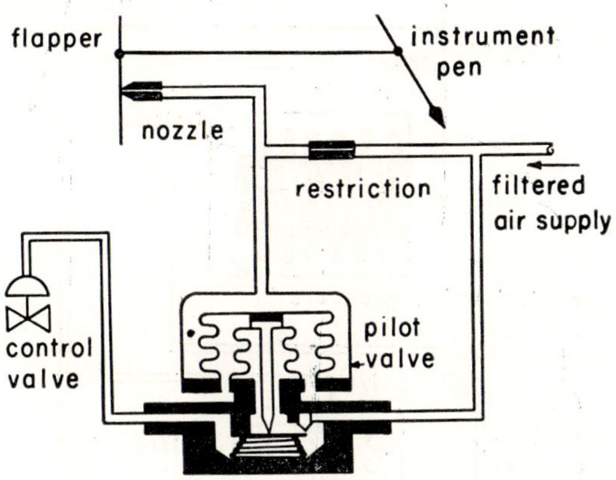


Fig. 20-14. Two-position pneumatic control. (Courtesy of Minneapolis-Honeywell Regulator Company.)

### Proportional Control

Smoother control than cyclical on-and-off can be obtained if in place of two-position control a proportional control is used (Fig. 20-15). These units are designed so that the percentage of full-scale change of controlled variable required to open or shut the control valve can be varied. This percentage is called "proportional band."

The action of a proportional band controller is similiar to that described for the two-position controller except that the control valve air pressure is applied to a throttling bellows, the movement of which counteracts the original movement of the flapper. The final net movement of the flapper is governed by the proportional band setting which may be changed manually. Wide bands produce smaller movements of the flapper and thus smaller changes in the control valve pressure for a given pen movement. Proportional control produces, therefore, a throttling action on the control valve rather than an on- and off-movement of the valve.

The control valve air pressure is proportional to the position of the pen, and there is thus one position of the control valve for each temperature within the range of open and shut conditions. A definite valve opening and air pressure are required to maintain the temperature at the control point. Changes in temperature cause the valve to move toward open or shut, thus returning the system to its original control point. The throttling action produces a minimum of cycling.

If the process load, however, increases appreciably on a steam heater, for instance, the steam control valve opening required to maintain the system at the control temperature must be greater (more steam is

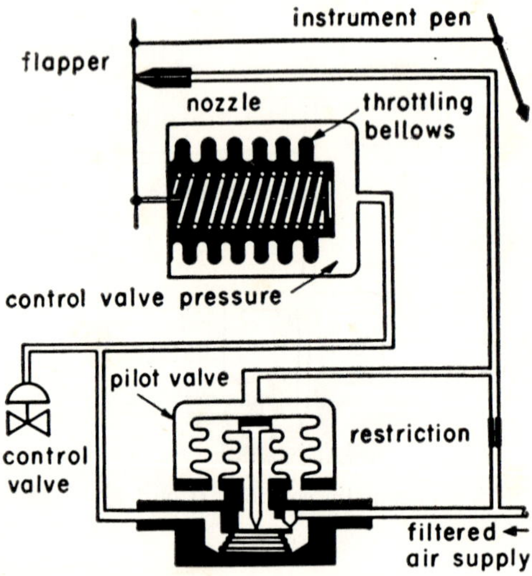


Fig. 20-15. Proportional pneumatic control. (Courtesy of Minneapolis-Honeywell Regulator Company.)

required). The proportional controller cannot correct for this automatically. It, therefore, lines out at a lower equilibrium temperature at which the corresponding valve position is adequate for maintaining constant tank temperature. This deviation or offset depends on the width of the proportional band (larger for wide bands) and the magnitude of the load change. A wide band setting produces the minimum cyclical action, but the greatest offset with large load changes. The narrow band has the opposite effect.

In order to return the instrument to its control point the relationship between pen position and valve position (air pressure) must be changed so that the valve position for the new load will produce the desired temperature. This adjustment or resetting of the pen position-air pressure relationship must be made manually and is impractical for frequent load changes.

PROPORTIONAL-RESET CONTROL

Reset is done automatically in the proportional-reset instrument as illustrated in Fig. 20-16. The throttling action is accomplished in a similar fashion as described. The inner bellows is caused to move by air pressure on the outer bellows which contains a liquid. When the pen moves from the set point in a direction which makes the flapper move from the nozzle, the control valve pressure is lowered. The throttling

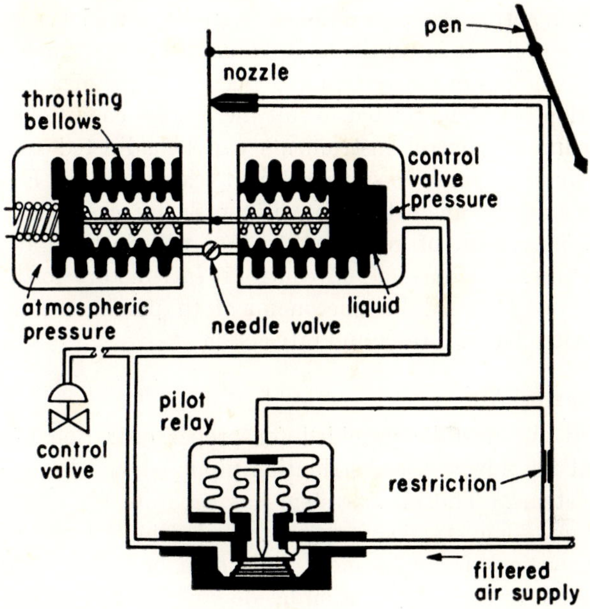


Fig. 20-16. Proportional reset pneumatic control. (Courtesy of Minneapolis-Honeywell Regulator Company.)

bellows expands to the left under the lower air pressure, moving the inner bellows to the left and the flapper back toward the nozzle. The liquid chamber of the two outer bellows (left and right) is connected by means of a restricted passage. The liquid chamber at the right is always at atmospheric pressure. That at the left is momentarily below atmospheric. Liquid from the right, therefore, passes gradually through to the left causing the inner bellows to move back toward the right. This movement moves the flapper away from the nozzle and further lowers the valve air pressure, and the cycle of action just described is repeated.

The movement of the pen away from the flapper will continue in this manner until the control valve reaches the point at which the pen will return to its set or control point. Only at this point will the liquid pressures be equal in each bellows. This reset action, therefore, will maintain the process at the control point regardless of load fluctuation.

### Proportional-Reset-Rate Control

Rate action is accomplished by installing an adjustable restriction between the control valve air pressure line and the throttling bellows. It has the effect of slowing up the throttling action and thus permits the control valve to over-correct temporarily for a change in the controlled variable. Such action produces a faster correction during the early portion of an upset period. It is particularly valuable for systems having large lags between changes in the system and changes in the measured variable. The rate action may be adjusted to suit the process conditions.

Some pneumatic control instruments can be adjusted for throttling, reset, and rate of response to suit most complex processes. Each addition to the complexity of an instrument, of course, increases its price. The least costly instrument of good quality that will do the job properly is all that is required. If on- and off-control is satisfactory for a noncritical process stream, it would be uneconomical to purchase a more costly throttling controller with reset and rate action.

### Instrument Air

Clean, dry air must be provided for proper operation of pneumatic systems, and for this reason distribution piping should be separate from plant air systems. The ideal system has separate air compressors, air receivers, and some type of regenerative dehydrating unit. Plant distribution systems for instrument air should be operated at pressures above 45–50 psi so that distribution headers will not be larger than 3 in. or 4 in. nominal size. A usual pressure is from 75 to 85 psi. Instrument air pressure within a process area need not exceed 35 to 40 psi.

As a precaution against moisture, a drip leg should be provided just before the pressure reducer or area block valve. It is standard practice to provide a separate reducing valve with a filter and drip at each instrument for maintaining constant air supply pressure.

Nearly all maintenance problems other than mechanical wear can be eliminated if the air supply is clean and free of moisture. Distribution piping should be galvanized and instrument air transmission piping should be small diameter copper or aluminum tubing, usually 3/8 in. OD maximum. Fittings may be brass or bronze flare-type. Several manufacturers are now producing multiple tubing assemblies, encased in flexible armored type steel conduit. The assemblies save much time in installation and also protect the small tubing during construction. Various plastic materials are also used for instrument tubing, and some plant operators prefer these materials to copper or aluminum.

The pneumatic type control instrument is perhaps the most fool-proof of all control instruments. The development of such instruments has covered several decades, and practically all the well-known instrument manufacturers offer comparable instruments based on reliability, simplicity, adaptability, and operation. The instrument industry is, of course, by no means stagnant. New developments are constantly being offered, and the instrument engineer must attempt to keep abreast of the whole industry rather than assume that only certain manufacturers are capable of meeting particular applications.

### *Hydraulic Control*

Hydraulically operated controls have long been widely used in some industries. Their use for process control, however, is usually limited to operations requiring considerable power and two-position type control. The application of hydraulics as a self-contained control medium was extended tremendously during World War II. The so-called "servo" systems, which are used extensively in airplane controls utilize hydraulic systems for positive and instantaneous action.

The components of the hydraulic system are a chamber for storage of hydraulic fluid, a gear or other type positive-displacement pump, control valves, and the power unit, which is usually a piston and cylinder. Generally each application of hydraulic control has all these components. It would be possible to have a centralized hydraulic power system in which the fluid would be distributed to each user, similar to pneumatic systems. It can be seen, however, that in each case two sets of piping would be required for pressure service to the user and the return system to the central unit. With the pneumatic system only a single supply or connection is required between instrument and power unit. Furthermore, hydraulic systems are generally operated at higher pressures than pneumatic systems and thus require heavier tubing. While there are a number of uses for hydraulic systems in process control, the characteristics just enumerated generally indicate the selection of pneumatic controls. No doubt as the application of servo-systems becomes greater, the use of hydraulic units will increase.

### *Electrical and Electronic Control*

#### Electrical Controllers

The movement of pressure instruments, potentiometers, and other such instruments may be utilized to operate mercury switches or contacts placed above and below the control point. The electric signal can then open or close an electrically operated valve (such as a solenoid valve) or position a reversible motor-operated valve at the control point.

Some proportional and proportional-reset control instruments have electric and electronic means for positioning an electrically operated valve. These techniques have been described.[6,9] The potentiometer instrument is best adapted to this type of control. The movement of the pen is followed by contact with a slide wire resistor which forms a bridge circuit with a similar resistor in contact with the travel of the motor valve. Suitable relays detect unbalance in the bridge circuit caused by pen movement from the control point. Movement of the motor valve in the proper direction is then initiated in such a way that it balances the bridge circuit.

### Electrically Operated Valves

Several types of electrically operated valves are manufactured. The so-called solenoid valve is perhaps more widely used than any other motor-driven type. The solenoid valve is a two-position unit, operating either wide open or closed.

The solenoid valve is often used in conjunction with pneumatic systems, admitting or shutting off air-pressure to power units. Solenoid valves are very useful in safety applications. If a number of pneumatic systems are fed through a common air supply, a two-way solenoid valve may be installed in the system so that plant electrical failure will cause the solenoid to deenergize and open, releasing air from the system. All control valves on the system will then assume an emergency position.

The air-motor on an air operated valve can be replaced by a reversible electric motor operating through a gear reduction to position the valve in accordance with the signal received. Numerous models are available.

### Signalling

Electrical systems are extensively used in signalling remote process conditions or the operating position of processing equipment by means of colored lights at a control center. Liquid level instruments are frequently provided with contact switches which indicate process liquid level by such means. Critical electrically driven equipment can be provided with on-off indicators.

The annunciator panel, an early development of low-voltage electrical signal systems is used for many of the above purposes. It has the favorable characteristics of small size, low voltage, and centralization, whereas the more conventional signal lights consume much more space and installation labor. Each is usually installed as a separate unit.

The annunciator panel signal system is often used in plants having involved electrical systems. Failure points are connected to the panel and faults are quickly located.

### Electronic Developments

Wherever mechanical movement has sufficient power to operate a microswitch or any other type of electrical switch, countless variations of electric control application are possible. Most of the uses of electrical systems for process control are in on-off services, but the large variety of possible uses of electronics are only beginning to be applied to process control. As development proceeds undoubtedly such control methods will find a wider use.

### System Control

The use of electronic principles in process plant instrumentation is at present limited chiefly to conversion and amplification of primary sensing impulses, but system control by means of electronics is on the horizon.

System control in contrast to point control will permit control of the total process. Anticipatory control must usually be left to manual changes since presently available instruments cannot function until some change occurs at the primary element. The single exception to this is the instrument working in conjunction with another controller. In some respects this may be considered anticipatory control; however, it will be found that the number of variables which can be so controlled are very limited.

In order to provide system control, all considered variables would have to be sensed collectively, their relation weighed by a common instrument, and changes effected according to a pattern, either pre-determined or instrument-determined. It is only in recent years that such a system has proved practicable.

Electronic computers will play a major role in these developments. Such "computer-control," however, requires quantitative knowledge and expression of this knowledge in such a way that it can be used by the machine. It is surprising but true that quantitative knowledge has been developed for only a small number of simple processes. Engineering judgment required in the operation of so many plants cannot, as yet, be reduced to a mathematical expression.

### Pneumatic vs. Electronic Control

Boyd[2] has described the first complete refinery which was controlled by electronic instruments. He has also compared pneumatic control to electronic control. Thermocouple emf signals, for instance, instead of being converted to a pneumatic signal and transmitted in tubing to a distant valve may be transmitted after suitable electronic amplification to the valve as an electrical signal. This arrangement provides much greater sensitivity. The elimination of instrument air piping not only simplifies installations but eliminates the difficulties caused by freezing moisture in air lines during cold weather.

The installation described requires conversion of the electric signal to air pressure at the valve since motor-driven control valves at this time are not as fast as pneumatic valves and gears become excessively worn at a position corresponding to the control point. Boyd[2] suggests the development of hydraulically operated valve with individual pumps and motors actuated by the electrical signal. In this manner the favorable response characteristics of the electronic controller could be used.

### Control Theory

Excellent theories have been developed to describe the numerous types of control systems. At times the project engineer may find limited knowledge of this theory helpful in discussions of difficult control problems with instrument engineers. Several works on control theory are listed in the references and are recommended for engineers with chemical engineering or mechanical engineering background.

### Control Centers

Pneumatic and electronic transmission permits the centralization of all instruments. Multimillion dollar process units producing large quantities of high quality product can now be operated by a few experienced men who seldom need move from the comfortable surroundings of a modern air-conditioned control room. The control house has become not only the nerve center of the process unit but also the show place. Attention is given to the convenience as well as appearance of both inside and outside of the control house as well as the panel board itself. Instrument manufacturers have responded to this trend and are providing shapes and styles of instruments that produce not only accurate control but also an attractive control panel. Typical control panels are shown in Fig. 20-17.

The size and arrangement of control panels has been changed considerably with the advent of graphic panels (see Fig. 20-17). The graphic panel provides a simplified flow diagram of the process. Its utility, however, is more beneficial to uninitiated than to experienced process operators, who must learn to know all process phases regardless of control board arrangements. The most important change caused by graphic panels has been the reduction in instrument size, permitting subsequent reductions in control panel length and thus control rooms. The trend to smaller sizes is also evident in instruments not intended for graphic panel layouts.

In arranging the control center for the process plant there should be adequate space behind the panel for instrument maintenance. A minimum of six ft between building wall and instrument is desirable.

If the process is hazardous or subject to fire, large window openings should be installed on the sheltered side. Small windows may be provided on the side nearest the process unit. All window glass should be shatterproof or wire glass. At least three access openings or exits should be provided in the control room. Door hardware should be anti-panic type or have bar-opening devices toward the outside. Control-room doors should not ordinarily be provided with locks. Exit doors from the area in back of the panel should also be included.

Control rooms should have directional or troffered lighting. Explosion-proof fluorescent lighting fixtures should be used if the control room is adjudged to be within a hazardous area. An intensity of 50 ft-candles

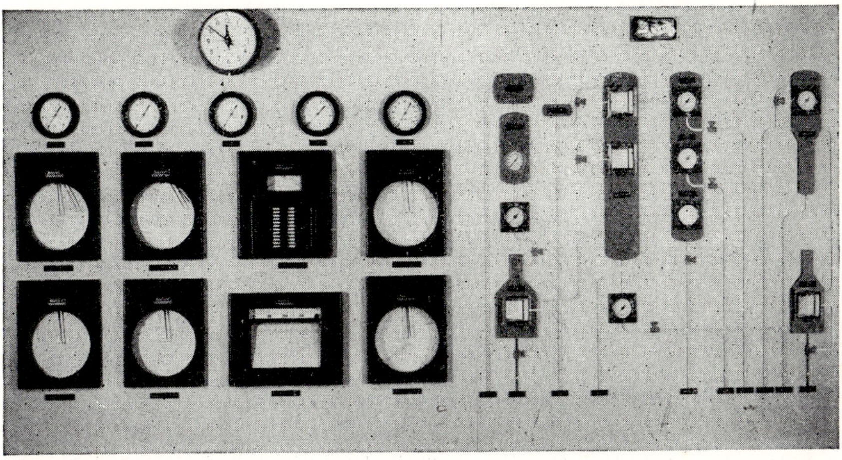

Fig. 20-17. Control panels. Left: conventional panel; right: graphic panel. (Courtesy of Minneapolis-Honeywell Regulator Company.)

or more is sufficient. The same intensity should be provided behind the panel for maintenance. A pleasing control room appearance can be obtained with a dropped ceiling slightly higher than the control panel top. Space is thus provided for troffered lighting, and heating and ventilating ducts.

Color schemes and materials vary with company preference. Ceilings for control centers are often constructed of noninflammable acoustical materials which are usually of a light color but may be painted as desired. Various shades of green or gray have been used for the control panel. In order to provide a clean appearance ceramic tile can be used for exposed walls. Floors are generally concrete slab with terrazzo, asphalt tile, or rubber tile finish. Rubber tile provides the most comfortable walking surface. Both rubber and asphalt tile are much less expensive than terrazzo.

It is common practice for the control building to have an office for the chief or shift operator, toilet facilities, a pantry equipped with an electric refrigerator, a sink, and a hot-plate where the operators can warm their food (shift operators are not usually permitted to leave the unit). An automatically chilled water drinking fountain is usually provided. A locker-room is sometimes included with shower facilities so that operators can change clothing at the process unit.

Summer and winter, air-conditioned control rooms are becoming common, particularly in the warmer climates. If air-conditioning is provided, heating is included in the same system. Ventilation may be accomplished by duct work or by fans located in the building walls. Duct work does not greatly increase costs and eliminates fan noise. For some processes the control room is maintained under a positive pressure. Intake air is received from an elevated stack and noxious fumes near the ground level cannot enter the room.

In general the cost of the control room is a very small part of the cost of process equipment. Attractive control rooms are good morale builders. The slight additional cost of a clean and pleasant working space is indeed well spent.

## INSTRUMENTATION ENGINEERING

Instrumentation developments are continuing at a bewildering rate. Progressive operating companies and instrument manufacturers maintain research divisions concentrating on the development of new techniques, and the instrument engineer who is responsible for the selection and proper operation of instruments in a process plant must keep up with these new developments by attending technical meetings and reading journals. In addition, discussions of new instrument developments with manufacturer's representatives is a necessity.

A process plant design and construction project needs the services of a capable and well-informed instrument engineering group. At present instrument engineers have various backgrounds such as chemical engineering, physics, electrical engineering, mechanical engineering, and chemistry. All have in common an interest in precise machinery and electronic equipment.

The organization of an instrument engineering group for plant design project varies in each company. In general, however, one engineer is assigned responsibility for the instrumentation of a section of a plant or an entire plant. This engineer begins his work early in the development of the plant design. During the process design stage the process engineer will seek his advice on phases of design which would be greatly influenced by instrument selection. Often the design of a chemical reactor, for instance, can be greatly simplified by clever application of control instruments.

When process design is complete and the engineering flow diagram is being developed, all instrumentation including indicating instruments must be determined. This is best accomplished by first beginning with several work conferences between the instrument engineer, project engineer, and process engineer. The project and process engineers must describe the process, explain the customer's thinking on instrumentation, and indicate the budget.

Knowledge of instrumentation is very helpful to the process and project engineer in such discussions. With the information obtained from these talks and subsequent discussions of critical control points, the instrument engineer is able to determine all the variables for each process unit and decide which must be controlled, measured, or indicated. Based on his knowledge of the latest and best instrumentation he can determine all the instruments required for the process unit.

All of these instruments are then shown schematically on the engineering flow diagram by means of symbols (see Table 5-3). Each instrument is assigned a number which is used for identification in design, purchasing, and construction. Instruments are then selected on the basis of competitive bids from manufacturers.

Manufacturers' instrument drawings are used in erection as well as a limited number of installation drawings made by the instrument department. These installation drawings usually include typical installation sketches for orifice meters, thermocouple assemblies and others. Such drawings can be used for a number of different types of jobs. Hence instrument drafting is not a very large task for most projects. Drawings of the panel board layout must also be made since these are designed for each job using standard panels which have been arranged to suit the particular conditions.

Items such as orifice plates are located on piping drawings, but in general few detailed instructions in the form of drawings are required. Instrument installation must be done by highly skilled men supervised by an instrument engineer. The decisions on the various details can be made most economically during construction by the instrument engineer-in-charge. In this way a considerable amount of costly drafting time is avoided.

Successful and economical instrumentation of a process plant demands cooperation between the process, project, and instrument engineers. It also requires a mutual understanding of the other engineers' field so that discussions can proceed beyond the realm of generalities and so that concrete decisions can be reached. The project engineer can maintain a speaking knowledge with new instrument developments by reading the brief accounts designed for such purposes in journals of chemistry and chemical engineering. Following the more detailed accounts in specialized instrumentation journals usually proves too much of a task since the project engineer also must read the current literature on processes, chemical engineering theoretical developments, and process equipment.

## REFERENCES

1. Behar, M. F., "Handbook of Measurement and Control," Part II of *Instruments and Automation, 27,* No. 12 (1954).

2. Boyd, D. M., *Oil and Gas Journal, 53,* No. 27, 139 (1954).

3. Brown, G. S. and D. P. Campbell, *Principles of Servomechanisms,* John Wiley & Sons, Inc., New York, 1948.

4. Dike, Paul H., *Thermoelectric Thermometry,* Leeds and Northrup Co., Philadelphia, Pa., 1954.

5. Eckman, Donald P., *Industrial Instrumentation,* John Wiley & Sons, Inc., New York, 1950.

6. Eckman, Donald P., *Principles of Industrial Process Control,* John Wiley & Sons, Inc., New York, 1945.

7. Farrington, G. H., *Fundamentals of Automatic Control,* Chapman and Hall, London, 1951.

8. *Fluid Meters, Their Theory and Application,* 4th Ed., Am. Soc. Mech. Engrs., New York, 1937.

9. *Fundamentals of Instrumentation for the Process Industries,* Minneapolis-Honeywell Regulator Co., Industrial Division, Philadelphia, Pa.

10. Lajoy, M. H., *Industrial Automatic Controls,* Prentice-Hall, Inc., New York, 1954.

11. Sprenkle, R. E., *Trans. Amer. Soc. Mech. Engrs.,* 67, 345 (1945).

# CHAPTER 21

# PLANT UTILITIES

Utilities are as indispensable to a process plant as they are to a city and as much a part of plant planning as the design of process units.

The following utilities will be discussed with emphasis on design:

*Primary utilities*

(*a*) Water
(*b*) Fuel
(*c*) Power and process steam
(*d*) Storage and movement of raw materials and products

*Secondary utilities*

(*a*) Maintenance facilities
(*b*) Plant buildings (other than process; see Chapter 23)
(*c*) Plant roadways
(*d*) Railroad facilities
(*e*) Fire protection (see Chapter 24)
(*f*) Plant sewer systems and waste disposal
(*g*) Plant air
(*h*) Plant security

## PRIMARY UTILITIES

Utilities which are vital to the operation of a plant are called primary utilities.

### TABLE 21-1. Common Impurities Found in Water*

| Constituent | Chemical Formula | Difficulties Caused | Means of Treatment |
|---|---|---|---|
| Turbidity | None; expressed in analysis as $SiO_2$ | Imparts unsightly appearance to water. Deposits in water lines, process equipment, boilers, etc. Interferes with most process uses | Coagulation, settling, and filtration |
| Color | None; expressed in analysis as "units" of color on arbitrary scale | May cause foaming in boilers. Hinders precipitation methods such as iron removal, hot phosphate softening. Can stain product in process use | Coagulation and filtration. Chlorination. Adsorption by activated carbon |
| Hardness | Calcium and magnesium salts expressed as $CaCO_3$ | Chief source of scale in heat exchange equipment, boilers, pipe lines, etc. Forms curds with soap, interferes with dyeing, etc. | Softening. Distillation. Internal boiler water treatment. Surface active agents |
| Alkalinity | Bicarbonate ($HCO_3$), carbonate ($CO_3$), and hydrate (OH), expressed as $CaCO_3$ | Foaming and carryover of solids with steam. Embrittlement of boiler steel. Bicarbonate and carbonate produce $CO_2$ in steam, a source of corrosion | Lime and lime-soda softening. Acid treatment. Hydrogen zeolite softening. Demineralization. Dealkalization by anion exchange. Distillation |
| Free mineral acid | $H_2SO_4$, HCl, etc. expressed as $CaCO_3$ | Corrosion | Neutralization with alkalies |
| Carbon dioxide | $CO_2$ | Corrosion in water lines and particularly steam and condensate lines | Aeration. Deaeration. Neutralization with alkalies. Filming and neutralizing amines |
| *p*H | Hydrogen ion concentration defined as: $pH = \log \frac{1}{(H^+)}$ | *p*H varies according to acidic or alkaline solids in water. Most natural waters have a *p*H of 6–8 | *p*H can be increased by alkalies and decreased by acids |
| Sulfate | $(SO_4)^-$ | Adds to solids content of water, but, in itself, is not usually significant. Combines with calcium to form calcium sulfate scale | Demineralization. Distillation |
| Chloride | $Cl^-$ | Adds to solids content and increases corrosive character of water | Demineralization. Distillation |
| Nitrate | $(NO_3)^-$ | Adds to solids content, but is not usually significant industrially. High concentrations cause methemoglobinemia in infants. Useful for control of boiler metal embrittlement | Demineralization. Distillation |
| Fluoride | $F^-$ | Cause of mottled enamel in teeth. Also used for control of dental decay. Not usually significant industrially | Adsorption with magnesium hydroxide, calcium phosphate, or bone black. Alum coagulation |

| | | | |
|---|---|---|---|
| **Silica** | $SiO_2$ | Scale in boilers and cooling water systems. Insoluble turbine blade deposits due to silica vaporization | Hot process removal with magnesium salts. Adsorption by highly basic anion exchange resins, in conjunction with demineralization. Distillation |
| Iron | $Fe^{++}$ (ferrous) $Fe^{+++}$ (ferric) | Discolors water on precipitation. Source of deposits in water lines, boilers, etc. Interferes with dyeing, tanning, paper mfr., etc. | Aeration. Coagulation and filtration. Lime softening. Cation exchange. Contact filtration. Surface active agents for iron retention |
| Manganese | $Mn^{++}$ | Same as iron | Same as iron |
| Oil | Expressed as oil or ether extractible matter | Scale, sludge and foaming in boilers. Impedes heat exchange. Undesirable in most processes | Baffle separators. Strainers. Coagulation and filtration. Diatomaceous earth filtration |
| Oxygen | $O_2$ | Corrosion of water lines, heat exchange equipment, boilers, return lines, etc. | Deaeration, Sodium sulfite. Corrosion inhibitors |
| Hydrogen sulfide | $H_2S$ | Cause of "rotten egg" odor. Corrosion | Aeration. Chlorination. Highly basic anion exchange |
| Ammonia | $NH_3$ | Corrosion of copper and zinc alloys by formation of complex soluble ion | Cation exchange with hydrogen zeolite. Chlorination. Deaeration |
| Conductivity | Expressed as micromhos, specific conductance | Conductivity is the result of ionizable solids in solution. High conductivity can increase the corrosive characteristics of a water | Any process which decreases dissolved solids content will decrease conductivity. Examples are demineralization, lime softening |
| Dissolved solids | None | "Dissolved solids" is measure of total amount of dissolved matter, determined by evaporation. High concentrations of dissolved solids are objectionable because of process interference and as a cause of foaming in boilers | Various softening process, such as lime softening and cation exchange by hydrogen zeolite, will reduce dissolved solids. Demineralization, Distillation |
| Suspended solids | None | "Suspended solids" is the measure of undissolved matter, determined gravimetrically. Suspended solids plug lines, cause deposits in heat exchange equipment, boilers, etc. | Subsidence. Filtration, usually preceded by coagulation and settling |
| Total solids | None | "Total solids" is the sum of dissolved and suspended solids, determined gravimetrically | See "Dissolved solids" and "Suspended solids" |

*Reprinted by permission, *Betz Handbook of Industrial Water Conditioning*, 4th Ed., Betz, W. H. and L. O. Betz, Chem. Eng. (1953).

## Water

Regardless of its source, natural water is never pure. As rain descends, gases and dust in the atmosphere are dissolved in it. Water absorbs carbon dioxide both from the atmosphere and from upper layers of soil as it penetrates into the earth. The resulting solution of carbon dioxide in water is a mild acid and dissolves significant amounts of mineral as it flows over soil and rock.

Surface water such as rivers and lakes holds clays and sand in suspension, but underground water is clear because the subsurface strata act as filters. Underground water, unlike surface water, is high in mineral content. Both underground and surface water supplies can become polluted and may contain disease causing (pathogenic) bacteria which must be destroyed before using for drinking purposes.

In Table 21-1 common impurities in natural water, methods of removal, and ill effects caused by these impurities are listed. Waters having high mineral content are called "hard" and those low in minerals, "soft."

The increase in soap consumption caused by "hard" water is well known, but its tendency to form scale causes the most difficulty in process plants. Calcium salts are deposited as scale when water is heated* causing decrease in transfer rates of heat exchangers and cooling systems. Both calcium and magnesium salts deposit as scale upon evaporation of water in steam generating equipment. These deposits not only reduce heat transfer but also restrict fluid flow and if not removed will eventually plug the water side of a heat exchanger.

Removal of calcium and magnesium ions from water is called softening. Softening processes in common use are lime-soda, phosphate, and ion exchange. These are compared in Table 21-2.

### *Types of Water Used in Process Plants*

The required degree of water purity depends on the particular use. If the raw water is impure, as is usually the case, it will prove advantageous to have a number of separate water systems so that complete purification of all incoming water is not necessary. Preliminary treatment of all raw water entering a plant may include screening and sedimentation to remove suspended solids, but subsequent treatment will depend on the ultimate use for each water system. A typical plant water supply might be separated into process, boiler feed, cooling, sanitary, fire water and utility water systems.

#### Process Water

Some processes use water as one of the reactants or as a solvent, and highly purified water is often necessary to prevent product contamination

* $Ca(HCO_3)_2$ decomposes when heated to $CaCO_3$. $CaCO_3$ and $CaSO_4$ solubility decrease with increase in temperature.

or catalyst poisoning. For these purposes water softened by the lime-soda process may be further demineralized by ion exchange or evaporation. Uncontaminated steam condensate is another excellent source of pure water for process use.

### Boiler Feed Water

Modern high-pressure boilers require highly purified water. Ordinarily water treated by the lime-soda process is further purified by ion exchange or by hot phosphate treatment in order to produce boiler feedwater. Further treatment necessary to insure satisfactory operation of boilers and the steam distribution system may include:

(*a*) Acid treatment or treatment by organic surface active agents for prevention of calcium carbonate and calcium phosphate deposits in boiler feed lines.

(*b*) Silica removal to prevent formation of calcium and magnesium silicate scales: silica is removed by magnesium salts which are added during the hot-process lime-soda treatment.

(*c*) Oil removal to prevent carbonaceous deposits and foaming.

(*d*) Deaeration to remove corrosive gases.

In addition to the initial feedwater treatment boiler water may be continuously removed from the system for retreatment. The *p*H is readjusted and the water retreated in a hot process softener. The retreated water is then returned to the boiler along with condensate and make-up water.

### Cooling Water

Water-cooled condensers and coolers use the largest amount of water in a process plant.

Cooling water from a large inexhaustible source can be used once and discarded. Water, however, is a scarce commodity and it is preferable to recirculate the water through a cooling tower and back to the users. In these circulating systems water is supplied at 80 to 90° F and returned to the cooling tower at a maximum of 120°F.*

Some cooling water treatment is necessary even for once-through systems to prevent scale formation, corrosion, and slime and algae formation. The extent of treatment must, of course, be greater for circulating systems since impurities are concentrated as evaporation losses occur.

Calcium carbonate causes the major portion of scale in all cooling water systems although in circulating systems concentrations of calcium sulfate and silicates of calcium and magnesium may also be high enough to produce scale.

* Higher temperatures cause rapid corrosion of piping and heat exchanger parts.

**TABLE 21-2. Comparison of Water-Softening Processes**

| Process | Method | Reactions |
|---|---|---|
| Lime-soda | Calcium and magnesium ions precipitated by lime and soda as $CaCO_3$ and $Mg(OH)_2$ which are then removed by sedimentation. | 1. $Ca(HCO_3)_2 + Ca(OH)_2 = 2CaCO_3 + 2H_2O$<br>2. $Mg(HCO_3)_2 + 2Ca(OH)_2 = Mg(OH)_2 + 2CaCO_3$<br>3. $MgSO_4 + Ca(OH)_2 = Mg(OH)_2 + CaSO_4$<br>4. $MgCl_2 + Ca(OH)_2 = Mg(OH)_2 + CaCl_2$<br>5. $CaSO_4 + Na_2CO_3 = CaCO_3 + Na_2SO_4$<br>6. $CaCl_2 + Na_2CO_3 = CaCO_3 + 2NaCl$ |
| 2. Hot process phosphate | Calcium and magnesium ions precipitated by phosphate and hydroxyl ions as $Ca_3(PO_4)_2$ and $Mg(OH)_2$. $Ca_3(PO_4)_2$ is more insoluble than $CaCO_3$. | 1. $3Ca(HCO_3)_2 + 6NaOH = 3CaCO_3 + 3Na_2CO_3 + 6H_2O$<br>2. $3CaCO_3 + 2Na_3PO_4 = Ca_3(PO_4)_2 + 3Na_2CO_3$<br>3. $Mg(HCO_3)_2 + 4NaOH = Mg(OH)_2 + 2Na_2CO_3 + 2H_2O$ |
| 3. Zeolite softening | Objectionable ions removed by replacement with other ions in the zeolite bed. | 1. $Ca^{++} + \underset{\text{(sodium zeolite)}}{Na_2Ze^*} = CaZe + 2Na^{++}$<br>2. $Ca(HCO_3)_2 + \underset{\text{(hydrogen zeolite)}}{H_2Ze} = CaZe + 2H_2CO_3$ (cation exchange)<br>(Acids produced are neutralized by addition of alkali, raw water, or sodium zeolite-treated water.) |

*$Ze$ = Zeolite.

TABLE 21-2. Comparison of Water-Softening Processes (continued)

| Process | Advantages | Disadvantages | Remarks |
|---|---|---|---|
| 1. Lime-soda | (*a*) Best for high hardness waters.<br>(*b*) Removes iron at same time.<br>(*c*) Removes solids in suspension by occlusion in coagulated mass. | (*a*) Does not reduce hardness as much as zeolite process.<br>(*b*) Control more difficult than zeolite.<br>(*c*) Dissolved solids not decreased in many waters.<br>(*d*) Coagulants required for most efficient sedimentation. | Cold process lime-soda softening is commonly used for municipal systems. The water is treated at its pumping temperature. Hot process softening is used for boiler water treatment. The water is steamheated prior to treatment. The higher temperature (212° F) increases the reaction and sedimentation, permitting use of smaller equipment. |
| 2. Hot process phosphate | (*a*) Greater insolubility of $Ca_3(PO_4)_2$ permits production of water of "zero" hardness.<br>(*b*) Best for water of low hardness. | (*a*) More difficult to control than zeolite.<br>(*b*) Softened water can produce deposits if *p*H is not properly adjusted. | Best for boiler water treatment. If water is very hard, first stage can be a lime-soda process. |
| 3. Zeolite softening | (*a*) Easy to operate and control.<br>(*b*) Produces low hardness waters.<br>(*c*) Anion exchange resins available which in combination with cation resins can produce demineralized water comparable to distilled water.<br>(*d*) Cation exchange resins produce low alkalinity water.<br>(*e*) Cation resins can treat high alkaline water. | (*a*) Turbid water must be filtered.<br>(*b*) Treated water often corrosive.<br>(*c*) May be more expensive than other methods. Each case must be studied.<br>(*d*) High alkaline and acid water should not be used for sodium zeolites. | Many new resins available; cost studies must be made by competent chemical engineer and water specialist to select the proper method of treatment. The sodium zeolites are regenerated periodically by NaCl brine; hydrogen zeolites, by acid. |

The cheapest method for scale prevention is the addition of small quantities of surface active agents to the water. These, in effect, increase the solubility of salts in the water by preventing precipitation even when the salts are supersaturated. Corrosion inhibitors such as chromates or surface active phosphates may be added to prevent corrosion by dissolved oxygen. These methods are usually adequate for once-through cooling water systems.

In circulating cooling systems build-up of salt concentration is inevitable as water is evaporated in the cooling tower. Make-up water required to replace these and other losses (4% including leakage and wind losses) is small. The boiler water is, therefore, often not diluted sufficiently to prevent scaling even when surface active agents are used. When this occurs, a portion of the water may be removed for further softening or for sulfuric acid treatment which converts calcium carbonate to the more soluble calcium sulfate. This treated water is then returned to the system.

Many operators have found it advantageous to use lime-soda or zeolite for circulating cooling water systems. Treated water is the initial charge to the cooling water system and all make-up water is also treated. Corrosion inhibitors are injected and inhibitor concentration checked daily.

Demineralized water or condensate must be used for engine jacket water cooling systems. The jacket water is cooled by air-cooled or water-cooled exchangers and re-circulated to the engine. Corrosion inhibitors such as chromates are used in the closed system.

Cooling tower water pumping equipment should be designed with a 10 to 15 per cent safety factor on capacity. The pump head must be sufficient to circulate water through equipment and return it to cooling tower inlets which are usually from 30 to 40 ft above grade. All piping must be carefully checked for pressure drop in any type of closed circulation system.

Electric motor or turbine-driven centrifugal pumps are used. Alternate power sources and/or spare pumping equipment are installed because cooling water is essential even during emergency shut-down of the plant.

Sea water has been used successfully as cooling water especially in those coastal areas with fresh water shortages. Cathodic protection systems employing magnesium anodes located in the floating head and channel of exchangers prevent excessive corrosion. Deposits are minimized by restricting cooling water temperature increase below the point where the calcium salts begin to precipitate. Deposits which do accumulate can be removed by inhibited acid treatment.

### Sanitary Water

Sanitary water must be potable and free from disease-causing bacteria. In metropolitan areas city water is usually purchased for this purpose. In remote locations or in small towns a portion of the treated water from the plant softening unit may be diverted for sanitary water. The treated water must be chlorinated to destroy bacteria and then pumped in an independent system.

Periodic bacteriological tests establish the residual chlorine that must remain in the treated water after ten minutes to insure destruction of bacteria. The residual chlorine is then checked daily by a simple test.

Sanitary water systems are operated at 20 to 30 psig since higher pressures damage standard plumbing fixtures. An elevated water storage tank is usually installed to insure uninterrupted water supply.

### Fire Water

The requirements for fire water are intermittent, and it may be assumed that other services will be shut down if necessary to provide sufficient water capacity. A cooling tower basin can be used as a reservoir to provide a relatively large capacity surge basin, but the most common storage for emergency water is the standard elevated water tank.

Underwriters-approved centrifugal fire pumps are used to boost the water pressure to 150 psig so that the pressure at the fire hose nozzle will be 100 to 125 psig. Two pumps with at least 1000 gpm capacity each are installed. Each has a different source of power, e.g., one, electric motor and the other, gasoline engine. The pump should be self-priming if a suction lift is required.

Fire water headers must not be used for other purposes. Provisions should be made for emergency connection of the fire water system into the largest reservoir of water. Usually this is the raw water supply since fire water requires no treatment. Sea water or brackish water is often used as fire water by plants located along coastal areas.

### Utility Water

Utility water is used for miscellaneous washing operations such as cleaning an operating area. It should be free from sediment but does not require any other treatment. Notices should be posted to warn personnel not to drink this water.

## Fuel

Fuels are used in process plants for open-fired heaters and reactors, open-fired evaporators, kilns, steam generators, gas turbines, and internal combustion engines. The commonly used fuels are petroleum liquids, natural gas, solid fuels such as coals and coke, and waste products.

### Natural Gas

Natural gas is the most ideal fuel for the process plant. It requires no handling or storage facilities other than piping and pressure controls. Gas flames are clean and propagate rapidly, burning satisfactorily at very low fuel rates.

Petroleum refineries and petrochemical plants have large quantities of natural gases available for use as a fuel. Ethane and heavier hydrocarbons are stripped from raw natural gas or from gases evolved in cracking operations. Gas, in addition to its use as a primary fuel, is commonly used for initiating combustion in industrial furnaces that burn other types of fuel. Liquid fuel sprays are ignited by gas burning pilots, and combustion chambers for pulverized coal are heated to the ignition temperature by gas burners.

### Liquid Petroleum Fuels

A wide variety of fuel oils are marketed for industrial use. The U. S. Bureau of Standards and other governmental organizations have issued standards for many of the common types. In Table 21-3 the properties of some typical fuel oils are given.

The heavy industrial fuels, because of their high viscosity, are delivered in steam-heated tank cars. The fuel is stored in heated tanks and in large installations circulated continuously between heated storage and furnace. By this means the fuel is kept at the correct temperature for burning and constant pressure is maintained at the burners so that control of firing will be simple. Isolated furnaces can have combined pumping and heating units located near the point of use.

The lighter fuels such as kerosene and diesel fuel are used for internal combustion engines. Gasoline engines are rarely used in process plants except for certain stand-by services, but diesel engines are used extensively. Fuel storage for internal combustion engines is simple since the fuels used require no heating other than in the engine. A gravity head tank is often located near the engine and is filled intermittently from larger storage facilities.

Liquefied petroleum gas is used extensively as a stand-by fuel in areas subject to interruption of natural gas supplies.

### Waste Products

Combustible by-products of process units may be valuable as fuels. In some cases, however, they prove more costly than conventional fuels. If, for instance, the waste material is a highly viscous liquid, special pumping systems will be required. In other cases, the by-product may contain water and corrosive materials. Finally, extensive alterations to

# Table 21-3. Detailed Requirements for Fuel Oils [a]

| Grade of Fuel Oil [b] | | Flash point, ° F | Pour point, ° F | Water and sediment, % | Carbon residue on 10% residuum, % | Ash, % | Distillation temperatures, ° F | | | Viscosity | | | | | | | | Gravity, ° API |
|---|---|---|---|---|---|---|---|---|---|---|---|---|---|---|---|---|---|---|
| | | | | | | | | | | Saybolt | | | | Kinematic centistokes at— | | | | |
| Number | Description | | | | | | 10% point | 90% point | End point | Universal at 100° F | | Furol at 122° F | | 100° F | | 122° F | | |
| | | *min* | *max* | *max* | *max* | *max* | *max* | *max* | *max* | *max* | *min* | *max* | *min* | *max* | *min* | *max* | *min* | *min* |
| 1 | Distillate oil intended for vaporizing pot-type burners and other burners requiring this grade. [c] | 100 or legal | 0 | Trace | 0.15 | — | 420 | — | 625 | — | — | — | — | 2.2 | 1.4 | — | — | 35 |
| 2 | Distillate oil for general purpose domestic heating for use in burners not requiring No. 1. | 100 or legal | [d] 20 | 0.10 | 0.35 | — | [e] | 675 | — | 40 | — | — | — | (4.3) | — | — | — | 26 |
| 4 | Oil for burner installations not equipped with preheating facilities. | 130 or legal | 20 | 0.50 | — | 0.10 | — | — | — | 125 | 45 | — | — | (26.4) | (5.8) | — | — | — |
| 5 | Residual-type oil for burner installations equipped with preheating facilities. | 130 or legal | — | 1.00 | — | 0.10 | — | — | — | — | 150 | 40 | — | — | (32.1) | (81) | — | — |
| 6 | Oil for use in burners equipped with preheaters permitting a high-viscosity fuel. | 150 or legal | — | [f] 2.00 | — | — | — | — | — | — | — | 300 | 45 | — | — | (638) | (92) | — |

[a] Recognizing the necessity for low-sulfur fuel oils used in connection with heat treatment, nonferrous metal, glass and ceramic furnaces, and other special uses, a sulfur requirement may be specified in accordance with the following table:

| Grade of fuel oil | Sulfur, max, percent |
|---|---|
| No. 1 | 0.5 |
| No. 2 | 1.0 |
| Nos. 4, 5, and 6 | No limit |

Other sulfur limits may be specified only by mutual agreement between the buyer and seller.

[b] It is the intent of these classifications that failure to meet any requirement of a given grade does not automatically place an oil in the next lower grade unless in fact it meets all requirements of the lower grade.

[c] No. 1 oil shall be tested for corrosion in accordance with par. 15 for 3 hours at 122° F. The exposed copper strip shall show no gray or black deposit.

[d] Lower or higher pour points may be specified whenever required by conditions of storage or use. However, these specifications shall not require a pour point lower than 0° F under any conditions.

[e] The 10-percent point may be specified at 440° F. Maximum for use in other than atomizing burners.

[f] The amount of water by distillation plus the sediment by extraction shall not exceed 2.00 percent. The amount of sediment by extraction shall not exceed 0.50 percent. A deduction in quantity shall be made for all water and sediment in excess of 1.0 percent.

Reprinted from *Fuel Oils*, Commercial Standard CS12–48, U. S. Dept. of Commerce the burner system itself may be necessary to supply the correct air-fuel ratio for proper combustion of the waste fuel. If changes such as these are necessary, it may be cheaper to waste the by-product in a flare or burning pit.

If quantities of waste products suitable for use become available periodically, it is best to use them by mixing with the regular fuel so that the fuel to the furnace will be constant in quality.

The most common waste or by-product fuels are gases from cracking furnaces, sugar cane waste (bagesse) and wood waste. A number of interesting uses of waste products as fuels have been made. One refiner, for instance, uses the products of combustion from a regenerator of a catalytic cracking unit. These gases are high in carbon monoxide and are burned in a special furnace.

### Coal and Coke

Coal and coke were standard fuels for the early process industries, but as oil and natural gas became cheaper and pipe lines were constructed, the use of these solid fuels has declined.

The use of coal or lignite continues to prove advantageous in plants located near large deposits. It is expected that coal consumption will increase as oil and gas deposits are depleted.

## Power and Process Steam

### Electric Power

Electricity is the primary source of power in the process plant. It is used for driving pumps, compressors, and other mechanical equipment, for electrolytic processes, heating, instrument power, and lighting.

Electric power may be purchased from a utility company or generated at the plant by steam-turbine driven or internal-combustion-engine driven generators. Unless a process plant is located in an isolated area, purchased power is usually cheaper than that generated by the process plant. Progressive utility companies serving industrial communities provide excellent service at low rates. In order to maintain constant load they also offer off-peak* power at attractive rates. Under such conditions it is often difficult to justify the construction of a complete power plant particularly when the same capital can be invested in additional process facilities that yield higher returns.

One interesting case of mutual benefit is an agreement between a large west coast public utility and several oil refiners.[14] The utility has constructed power plants adjacent to the refineries. The refineries sell fuel

* Power used during periods of low residential consumption.

to the utility company and purchase a large portion of the power and process steam generated by the utility.

In most cases, however, process steam is generated by the process plant. If these process steam requirements are high (75,000 to 100,000 lb of steam per hr), an economic study may indicate that power should also be generated using high-pressure turbines. Exhaust steam removed from various stages of the turbines is then used as process steam.

If electric power needs are very great, such as in electrochemical plants, power must be generated at the plant or purchased from some large hydroelectric project. Electrochemical plants may consume more electric power than entire municipalities, and utility companies are often unwilling to increase their power output so greatly for just one user. Furthermore, a plant-owned power generating facility when designed for large power requirements can be operated as efficiently and cheaply as a public utility. In areas having low cost natural gas, generators driven by internal combustion engines operating on natural gas may prove less costly than operation of steam-driven generators. This is especially true when direct current is required since low-speed, gas-driven engines can be directly connected to d-c generators. As gas turbines are improved even greater economies may be possible.

An economic study is necessary to select electric generating equipment—internal-combustion engine, steam-turbine, or gas-turbine.* Each must be studied and compared with the cost of purchased power.

### Electric Power Distribution Systems

Any electric system must be designed by technically qualified and experienced persons. In addition to the fundamental principles involved, a working knowledge of the Natural Electric Code[1] and complete familiarity with manufactured equipment is imperative.

The basic electric system is composed of the power source, which may be generated or purchased power, transformation equipment, switching and protective devices, distribution wiring, and the ultimate users. Fig. 21-1 is an abbreviated single-line drawing of an electrical system illustrating the diagrammatic type of representation used on electrical drawings. References are also given to the article in the National Electric Code[1] applicable to each piece of equipment shown.

Distribution Voltages. Purchased power is distributed at very high voltages and must be reduced to some lower voltage for use in the plant. Power generated by the plant must also be reduced. In modern plants

* Internal combustion plants require 24 manhours per 24-hr day. Steam plants require 72 manhours per 24-hr day. The rapid development of electric power derived from heat generated in atomic piles may change the entire economic picture within 15 or 20 yr.

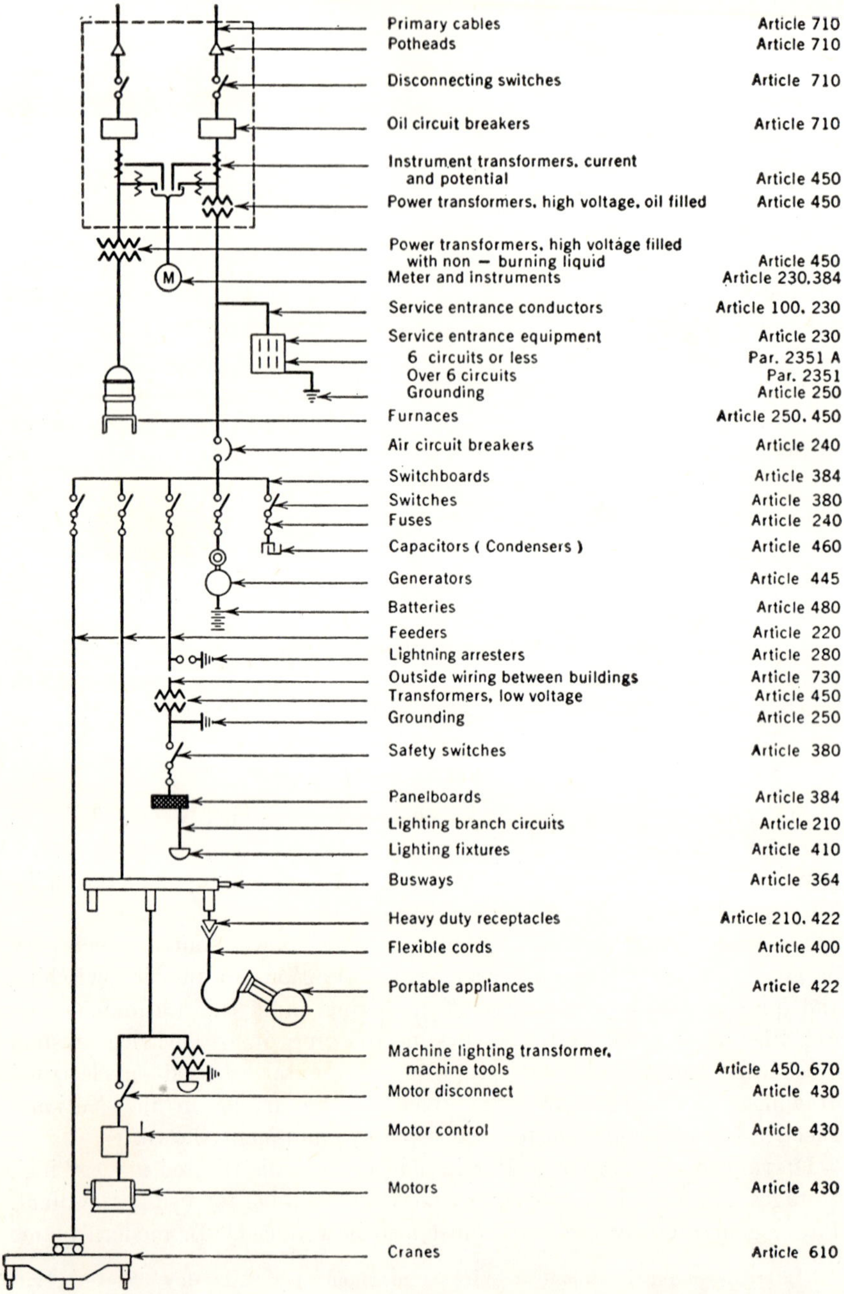


Fig. 21-1. Electrical system schematic with references to 1953 National Electric Code. (Courtesy of Square-D-Company.)

the present practice is to distribute power at high voltages (4,160 to 13,800 v), and then transform it to lower voltages (120, 120/208Y, 240, 480, 600, and 2400)* near the area in which it is to be used. High voltage systems have lower power losses, smaller conduit and circuit breakers, and lower installation costs,

Heavy-duty switchgear best suited for plant distribution systems has been standardized at 4160 v. It is, therefore, more economical to select 4160 v for primary feeders than 2400 v. Systems having capacities up to 20,000 kva[9] may use 4160 v. Above this capacity 13,800 v is used.

Regardless of voltage levels all distribution systems should have a grounded neutral to insure safe operation and ease of maintenance.

EQUIPMENT OPERATING VOLTAGES. The operating voltage recommended by manufacturers of motors and other electrical equipment is given on the nameplate or in operating instructions. The ratings are slightly below the nominal distribution voltages in order to allow for line losses.

Typical motor ratings are given in Table 16-2 of Chapter 16. Typical voltages for lamps are 120 for incandescent and 118, 208, 236, and 265† for fluorescent.

CONDUIT. Power may be distributed through the plant by (*a*) open wiring attached to poles or structures, (*b*) cable run in metal conduit above ground, (*c*) cable run in metal or other types of conduit buried under ground, (*d*) cable buried directly in the soil. Methods (*a*) and (*d*) are avoided in modern plants since they are unsafe and pose a constant hazard during maintenance operations. Method (*b*) is used for cable run inside buildings and structures.

Cable run underground is laid in light-wall galvanized steel or concrete-asbestos conduits. The conduit is encased in a red-colored concrete envelope (red pigment added during mixing) to insure ready identification of the conduit during any future excavations.

Conduit runs are located below the frost line and spaced far enough apart to allow dissipation of heat. Manholes are located at regular intervals (200 ft maximum for straight runs and 100 ft for runs with bends) so that cable may be pulled through the conduit with ease. The shorter the run between pulling points the simpler the pulling operation becomes.

Bends in conduit should be long radius bends and no more than one 90° bend should be permitted between manholes.

MANHOLES. Manholes must be large enough to enable two men with the necessary tools to make connections and pull cable. They are con-

* 600 volts is less frequently used than the other voltages given. 120/208Y is a three-phase four-wire system with the fourth wire grounded neutral.

† Line to neutral on 480 v systems. A separate lighting system is not needed when lamps of this voltage are specified.

structed of reinforced concrete, and are rectangular or polygonal in shape with cast-iron pre-cast concrete slab covers. The bottom may be soil or concrete, but if the water table is near ground surface, the manhole bottom should be made waterproof.

Ladder rungs are set in the concrete sides of the manhole for access and pull irons are located opposite conduit entrances for use in cable pulling. Cable racks are also provided.

SYSTEM PROTECTION. The plant circuits and equipment must be protected by devices which open the circuits under over-load or faulty conditions so that current will not flow. The power feeders or high voltage services entering a plant are protected by manually or power-operated switchgear and disconnecting (isolating) switches (see Fig. 21-1). Reduced voltages from transformers feed plant circuits each of which has its separate disconnect (isolating) switch and circuit-breaker.

The disconnect switch is used to open a circuit so that it can be completely isolated for repairs. They are not operating switches designed for opening or closing under load.

Circuit breakers are designed to open a circuit when it becomes overloaded due to faulty operation or short circuit. In this way all equipment in the circuit is constantly protected.

ELECTRICAL EQUIPMENT FOR HAZARDOUS AREAS. Electrical equipment, transformers, and switchgear must be explosion-proof when located in hazardous areas. Large switching equipment and transformers, however, may often be located outside hazardous areas or in specially ventilated enclosures so that the expense of explosion-proof construction may be avoided.

Explosion-proof equipment is designed to withstand an internal explosion and prevent propagation of flame through its openings. Any flame generated internally must travel through restrictions which cause extinguishing of the flame. All types of electrical equipment including telephones and drinking fountains may be obtained in explosion-proof construction. Vapor-tight equipment which is merely sealed to prevent vapor penetration should never be used in place of explosion-proof equipment in hazardous areas.

### Steam Power and Process Steam

Steam is not only a source of direct power for operation of pumps and turbines but is also the most useful heating medium in a process plant. Although some plants may purchase electric power, most process plants usually produce process steam.

A central boiler house can supply steam for the entire plant or smaller (packaged) boilers may be located near the area of use. It may prove economical to provide several boiler houses in large plants to reduce distribution piping costs. A careful economic study should be made to determine the type and number of boilers required and the most advantageous arrangement of the distribution system.

In the initial stages of steam generation, water is evaporated and is in equilibrium with the boiling water at the pressure of generation. This water vapor is called saturated steam and any lowering of temperature or increase of pressure will cause condensation to begin. Moisture in steam is undesirable for most uses and is harmful to equipment such as turbines, the blades of which are eroded severely by wet steam.

In most steam boilers, therefore, the saturated steam passes through a heat exchanger called a superheater which is located in a hot zone of the furnace. The steam is heated above its saturation temperature so that it will be sufficiently dry to provide moisture-free steam at all users.

### Steam Generators

A steam generator is composed of a furnace section which supplies the heat and a boiler section in which the heat is transferred to circulating water to form steam.

*Furnace.* Fuel is burned under or adjacent to the boiler and the heat of combustion is transferred to the drums and tubes by radiation and convection. The design of the furnace depends on the fuel to be used.

Solid fuels such as coal are all fed to the furnace by mechanical means. These include the following:

(*a*) Retort stoker; introduces fuel beneath burning bed.

(*b*) Spreader stoker; distributes coal above burning bed.

(*c*) Traveling grate.

(*d*) Pulverizer; pulverized coal is mixed with air and distributed vertically, horizontally, or tangentially into the combustion zone.

(*e*) Cyclone; crushed coal in high velocity stream of hot air issues tangentially from burner and flows in spiral path. Particles form molten slag on wall which burns readily.

Walls of large coal furnaces are usually water cooled to minimize undesirable clinker formation and decrease radiation from the walls.

Liquid fuels must be atomized so that the drops of fuel will burn as completely as possible. A number of different types of burners can be obtained using either steam or mechanical atomization. In the first type, steam flows through the nozzle with the oil, causing atomization. In the second type, atomization is produced by forcing oil at high pressure through a spray type nozzle.

Gas burners operate at 2 to 25 psi and are made in a number of styles. They are usually designed for alternate fuels (pulverized coal or oil).

Air- or water-cooled furnace walls are also installed in gas- and oil-fired

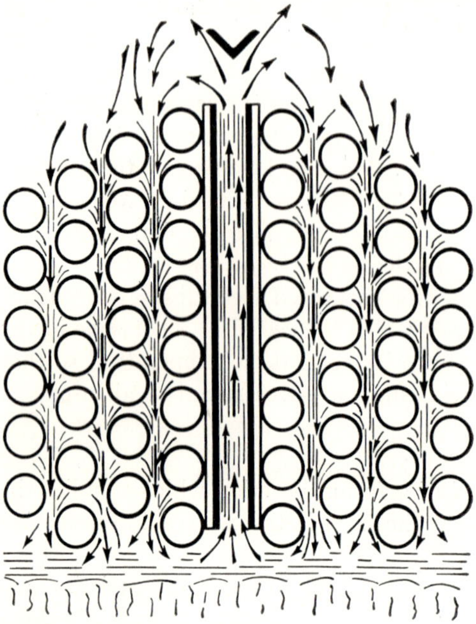

Cross section of boiler drum showing water circulation.

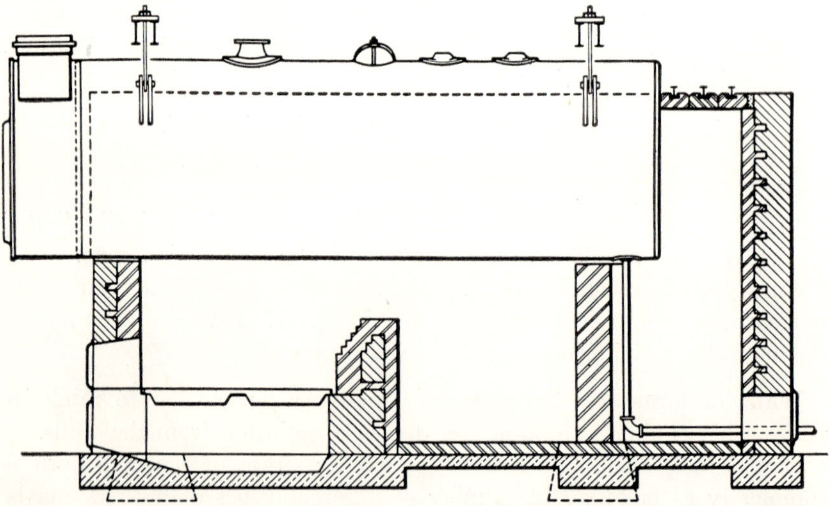

Fig. 21-2. Fire-tube boiler. (Courtesy of Combustion Engineering-Superheater, Inc.)

furnaces. In this type direct flame contact with refractories is avoided by locating burners several feet above the hearth.

Types of Boilers. Boilers consist of surfaces through which heat is transferred and are designed for circulation and separation of water and steam. They may be classified generally as fire-tube and water-tube types.

*Fire-tube Boilers.* In fire-tube boilers (Fig. 21-2) water circulation, steam formation, and separation all take place in a large horizontal drum (3 to 7 ft in diameter) situated above the furnace floors so that it will receive both radiant and convective heat. Hot gases pass over the drum and then through a number of tubes (2 to 4 in. in diameter) extending through the drum. Feedwater is maintained above the top row of tubes.

Fire-tube boilers are low in first cost and because of the large capacity

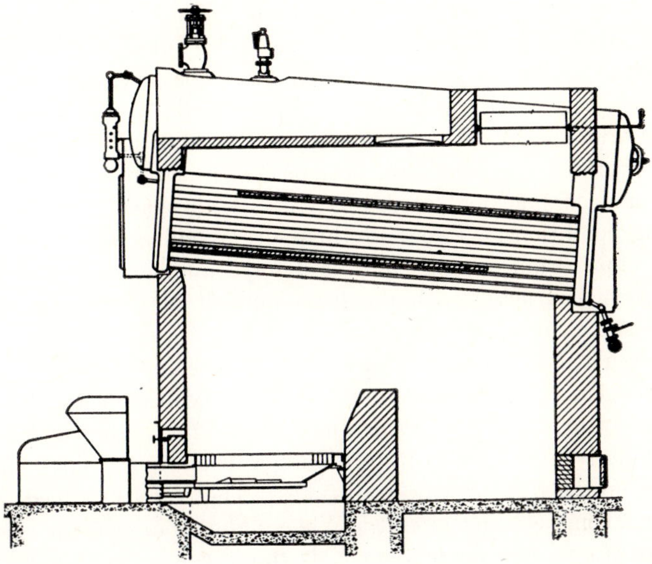

Fig. 21-3. Straight-tube, box-header type, water-tube boiler. (Courtesy of Combustion Engineering-Superheater, Inc.)

of the drum operate well under fluctuating loads common to process plants. They are, however, limited to low capacity (1000 to 15,000 lb steam/hour) and low pressures (15 to 250 psig).

*Water-tube Boilers.* In water-tube boilers water and steam flow inside the tubes and hot gases are directed across the tubes by baffles. They may be classified as straight-tube and bent-tube boilers.

The straight-tube design (Fig. 21-3) has a drum parallel or transverse to inclined tubes (10°). Feedwater enters the bottom of the drum and passes down the rear header and up the inclined tubes (3½ to 4 in. OD) where partial vaporization occurs. Steam and water then rise through the front header and back to the steam drum. These boilers are simple in design, require a minimum of space, and are easier to maintain than bent-tube designs.

Because of the box-like design of headers, straight-tube boilers are limited to pressures of 300 to 500 psig. A sectional-header boiler, however, can be designed for pressures up to 2000 psig. Sectional-headers are made of forged seamless tubing capable of withstanding high pressures.

The tubes in straight-tube boilers are expanded into the headers and a handhole must be provided at each tube. With the advent of efficient welding techniques for large drums the bent-tube design became popular. It eliminated the necessity for handholes and increased the flexibility possible in design. Drums are used in these designs to act as headers as well as separators for steam and water.

The drums range from 3 to 6 ft in diameter and have manholes which permit access for tube rolling and cleaning.* Tubes (2 to 2¼ in OD) must be bent so that they may enter the drums perpendicular to the drum surface (see Fig. 21-4).

Upper drums are used for steam separation and the lower drum (mud drum) serves as an accumulator of deposits formed by evaporation. The number and arrangement of drums varies with space requirements and operating conditions. Two and three drum boilers are most common although four drum designs were used for a number of years. Modern boilers are constructed with the least number of drums because large high-pressure drums are expensive.

The circulation system becomes complex with fewer drums and many tube arrangements have been developed. A typical circulatory system for a two-drum boiler with a side and front header is shown in Fig. 21-4. Water flows from the lower drum, or upper drum, to the side-wall header and then up the front or side-wall tubes. The rear tubes feed the lower drum and complete the circuit. When more than one upper drum is used, separate tube banks serve each drum but connections between drums equalize water level and pressure.

The outlet drum is equipped with baffles to divert incoming cold water to the circulating system and centrifugal separators to remove water from the steam.

All large high-pressure boilers are water-tube types. Since the pressure is on the inside of the tubes instead of the outside, a given thickness and diameter tube can withstand a higher pressure.

*Forced Circulation Boilers.* Forced circulation of water and steam through water-tube boilers may be used instead of natural circulation under the influence of gravity and density changes. In once-through designs water is pumped through a continuous coil and is evaporated and superheated before reaching the outlet. No drums are required and large savings in first cost are possible, but tubes are subject to overheating and scaling in the section where steam is formed and water storage capacity is low.

* Cleaning of boilers by chemical means is gaining wide acceptance and eliminates the need to enter the drums for cleaning.

Another type of forced-circulation design overcomes these disadvantages by recirculation. Water passes through a coil in which it partially vaporizes. The mixture of steam and water discharges in a drum in which steam is separated from the water before passing through a superheater. Water returns to the pump suction for recirculation.

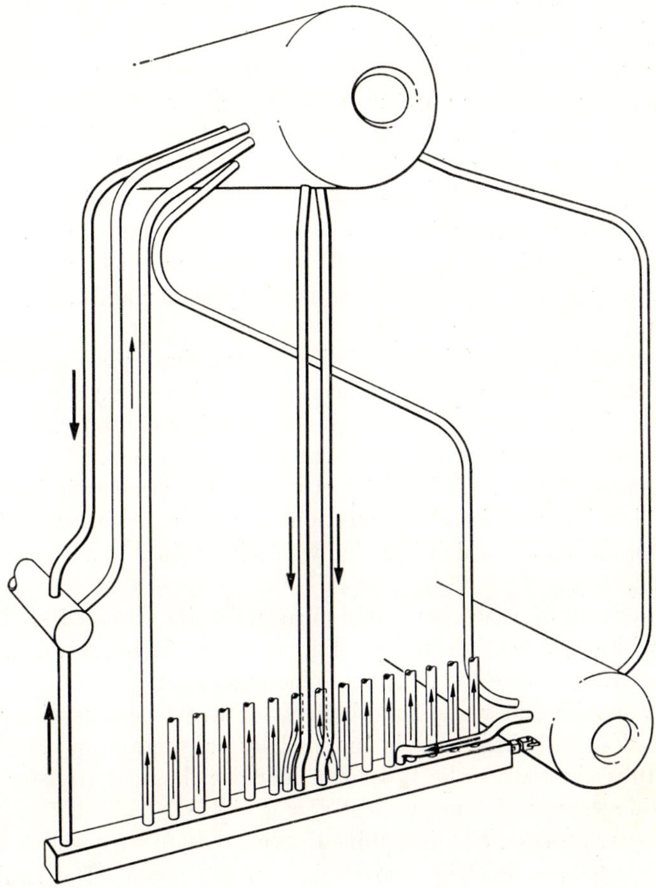

Fig. 21-4. Circulation pattern in a typical bent-tube boiler. [Courtesy of Combustion Engineering-Superheater, Inc., and reprinted by permission, C. E. Miller, *Mech. Eng.*, *73*, No. 1, 10 (1951).]

All forced circulation boilers have smaller tubes and less heat transfer surface for a given capacity than natural circulation. Higher flow rates and pressure drops can be tolerated, and the higher liquid and vapor velocities produce better heat transfer.

*Standardized and Packaged Steam Generators.* The boiler industry has standardized construction of many boiler styles in the past 15 years. Complete boilers have been designed in a number of sizes and standard parts are maintained in stock. If a standard boiler is ordered, a minimum of engineering and drafting work is required and savings are substantial.[19] Sizes having capacities below 100,000 lb of steam per hr are common.

Smaller standardized boilers are obtainable as packaged units which are completely assembled units that are shipped ready for installation. Fire-tube units are built for capacities from 500 to 17,250 lb of saturated steam per hr and pressures from 10 to 600 psig.[4] Water-tube units produce twice as much steam and up to 900 psig.

Packaged units are automatically controlled and unlike large power plant steam generators are designed to operate efficiently both at low and high capacity. They can be installed out of doors and are ideal for remote locations. These characteristics as well as low cost have made packaged units very popular in process plants requiring steam at or near a process unit. They are especially suitable for the varying demands of many batch operations.

The locomotive-type (oil-field) boiler is sometimes used for temporary installations in process plants. Since its drum is not exposed to radiant heat from the burners, the locomotive-type boiler can operate at higher pressures (400 psig) than conventional fire-tube boilers.

*Waste-heat Boilers.* A by-product of many processes is heat. Exothermic chemical reactions, regeneration of cracking catalyst, and atomic piles all produce heat which is called waste-heat since if not used it must be wasted.

In a number of cases generated heat must be removed so that the process will operate satisfactorily. For such situations waste-heat boilers can be used to remove the heat and generate steam.

Both fire-tube and water-tube boilers for waste-heat service must be specially designed since heat is supplied by convection only. Extended surface tubes in water-tube types can increase the heat transfer area on the gas side where resistance to heat flow is high.[7]

Studies are necessary to determine the economy of waste-heat recovery. In some cases heat must be recovered at such a low level of availability or at such great cost that it is cheaper to waste it.

### Steam Generator Selection

Design of modern steam generators is a complex problem requiring the efforts of experts. Many different designs are produced and each is applicable to a particular situation. The range of operating conditions for some popular types of steam generators is given in Table 21-4. These data may serve as a guide, but consultation with manufacturers and study of many proposals is necessary before intelligent selection can be made.

**Table 21-4. Steam Generator Operating Ranges***

| Steam-generating equipment | Continuous capacity range lb of steam per hour: 1000 to 15,000 | 15,000 to 35,000 | 35,000 to 150,000 | 150,000 to 1,000,000 | Design pressure range | Steam temperature |
|---|---|---|---|---|---|---|
| Firetube boiler | √ | — | — | — | Up to 250 psi | Up to 150° F superheat |
| 3-drum low-head boiler | √ | √ | — | — | Up to 400 psi | Up to 150° F superheat |
| 2-drum vertical unit-type boiler | √ | √ | √ | √† | Up to 1000 psi | Up to 900° F |
| 3- or 4-drum vertical boiler | — | √ | √ | √ | Up to 1500 psi | Up to 925° F |
| Sectional-header boiler | — | √ | √ | — | Up to 2000 psi | Up to 900° F |
| Special utility-type boiler | — | — | — | √ | Up to 2500 psi | Up to 1050° F§ |
| Controlled forced-circulation boiler‡ | √ | √ | √ | √ | Up to approx. 3000 psi | Up to 1050° F§ |

*Reprinted by permission, *Combustion Engineering*, Combustion Engineering-Superheater, Inc. (1947).
†Up to approximately 225,000 lb per hr on coal, and 300,000 lb per hr on oil and gas.
‡Controlled forced-circulation boilers are designed in various types and sizes for the full range in capacities.
§Present maximum used. Design and materials available for higher temperature.

Operating pressures and temperatures for steam generators vary considerably. Efficiency generally increases with an increase in pressure and temperature, but equipment costs also increase accordingly. Ordinarily generating pressures and temperatures in the range of 400 psig and 150° F superheat are satisfactory for a process plant. If steam requirements for power generation are large, higher pressure generation is usually more economical. Lower pressure process steam is then obtained from the exhaust of turbines operating the power generators.

The selection of steam generator and generating pressure depends on a study of plant requirements and operating costs. Although steam generation efficiency is of prime importance in the operation of a public utility power plant, first cost or flexibility may be the most important consideration for a process plant.

### Steam Generation Auxiliaries

The modern steam generating plant requires a large amount of auxiliary equipment. The more important items are discussed below.

Feedwater Treating Facilities. The equipment required for feedwater treating includes chemical storage and handling facilities, filters, water pumps, tanks for settling and storage, mixing equipment, proportioning pumps, and sludge disposal facilities.

A large portion of process steam in some plants is returned to the boilers after having been condensed in heat exchangers and other users. If the make-up required is small, evaporators can be used to supply distilled make-up water to the boilers. Large power plants using condensing turbogenerators invariably use evaporators to replace small water losses from the system. Partial treatment of the raw water is necessary to avoid frequent descaling of the evaporator.

Feedwater Heating (Economizers and Feedwater Heaters). Steam generating plants exhaust some heat at low levels of availability such as exhaust steam and flue gases. These must be used for maximum fuel economy. One user of this heat supply is the feedwater heater. In addition to conserving heat, feedwater heating reduces thermal shock on boiler parts and in some cases increases boiler capacity.

Feedwater heaters use exhaust or live steam either on the shell side* of shell and tube heat exchangers or by direct contact of steam and water in a deaerating heater.

In deaerating heaters water is sprayed across baffles or trays countercurrent to exhaust steam. The water is heated and some steam condensed. All uncondensed steam and noncondensable gases pass to a vent condenser which condenses the steam and vents the gases.

Deaeration of feedwater is necessary to reduce boiler corrosion. Some

* Some heaters have steam on the tube side.

power plants use shell and tube feedwater heaters followed by separate deaerators operating at subatmospheric pressure. These vacuum deaerators reduce dissolved oxygen in the water to a minimum.

Feedwater can be further heated by exchangers called economizers located in the path of flue gases from the boiler. The most common of these consists of a number of U-tube sections spaced closely together. Fin-tubes are used on many designs to increase heat transfer. If water scales badly, the removable return bends can be installed on one end.

FEEDWATER SURGE DRUM. Heated feedwater and returned condensate are usually accumulated in a vessel which provides a surge chamber for the feedwater. The deaerating feedwater heater can be made a part of the vessel. Such a reservoir is more necessary for the process plant power facility with varying loads than for the steady loads of public utility power plants.

BOILER FEEDWATER PUMPS. The treated and deaerated feedwater is taken from a feedwater heater or surge chamber and pumped into the steam generator. Feedwater pumps may be driven by electric motor or steam turbine. Generally turbine drives are more easily controlled. Multistage centrifugal pumps are used in nearly all installations and at least one spare pump is necessary.

In larger installations several pumps may be used in parallel or series to provide flexibility. Sufficient net positive suction head must be provided to prevent vaporization regardless of arrangement.

FUEL SUPPLY. Storage facilities must be provided for all types of fuels except natural gas. Manufactured gas, however, sometimes requires a gas-holder or pressure storage system for load surges.

FORCED AND INDUCED DRAFT EQUIPMENT. The function of the draft system is to provide sufficient air for complete combustion of the fuel. Steam generators may operate on natural draft, induced draft, forced draft, or combinations of the three. Natural draft is used for small furnaces, and the stack must be designed to develop the necessary pressure differentials.*

Practically all large steam generators, however, use forced and induced draft equipment. The forced draft system injects primary air with the fuel and secondary air into the furnace at or near the burner openings. The induced draft system imposes a reduced pressure on the furnace by removing the hot flue gases through a fan, thereby inducing primary and secondary air.

Since the air-fuel ratio is critical, fire-box conditions must be closely

*Head available $= H(\rho_a - \rho_c)$

where $H$ = chimney height
$\rho_a$ = density of outside air, and
$\rho_c$ = average stack-gas density

controlled. Both the forced draft and the induced draft fan and suitable dampers can be used for control. Natural draft units must be controlled by dampers alone.

Air Preheaters. The incoming air is frequently forced through air preheaters to exchange heat with the hot flue gases. Heat transfer surface may be thin plates or tubes. If preheaters are not used, part of the incoming air is often conducted through channels in the furnace refractory, sometimes the floor, so that the entering air is heated.

Soot Blowers. All fuels deposit noncombustible dust on furnace surfaces. The deposits are heaviest for solid fuels and are negligible for most natural gases. These deposits insulate heat transfer surfaces and reduce efficiency and must be removed. Soot blowers are located so that parts of the heating surface on which soot accumulates may be cleaned. High pressure steam, air, or a mixture of steam and air passes through tubes having a series of nozzles. As the cleaner is rotated axially, soot is forced from the heating surfaces.

Superheaters. Equipment motivated by steam requires dry, superheated steam in order to operate efficiently and without erosion. The efficiency of steam generating units with capacities above 15,000 lb per hour are also improved if the steam produced is superheated. Superheaters, therefore, are installed in most modern boilers. They may be heat exchangers exposed to hot stack gases (convection-type) or radiant energy from the flame (radiant-type).

As the load on a boiler is increased, flue gas flow and temperature increase faster than steam flow, and convection-type superheater outlet temperature increases. The reverse temperature effect occurs in a radiant-type superheater, but a combination of the two gives a reasonably constant steam outlet temperature regardless of load.

Superheated steam at a particular temperature for operations such as steam stripping is often supplied near the point of use by heating plant steam in the convection section of a direct-fired reboiler.

Desuperheaters. Some evaporators for chemicals which boil at low temperatures require saturated steam. If the central steam plant produces superheated steam, saturated steam can be supplied where necessary by means of a desuperheater.

Desuperheaters may be spray-type or shell and tube exchangers. In the spray-type, pure water such as condensate or feedwater is sprayed into the flowing steam. Regular cooling water is used in the shell and tube design since there is not direct contact between water and steam.

Instrumentation of Steam Plants. Instrumentation for steam generation plants is relatively simple compared to process plant instrumentation. Controlled conditions are:

(*a*) Steam pressure. (In certain high-pressure, high-temperature generators temperature may be as critical or more critical than pressure. Temperature control is not ordinarily important except for large electric power generation plants.)

(*b*) Air-fuel ratio.

(*c*) Feedwater control.

The action briefly is as follows:

Increased demand on the steam distribution system causes a drop in pressure. Pressure instruments sense the drop and increase the fuel flow. Air flow control is correspondingly increased by a ratio controller which increases the speed of air fans or other controls which operate dampers. As boiler pressure drops, steam generation increases, lowering boiler water level which causes the feedwater level control to admit more feedwater. The same action takes place in reverse when steam pressure increases.

Feedwater in-put cannot be controlled satisfactorily by level alone. A drop in steam pressure will increase feedwater flow because of the reduction in pressure. If water level alone controls, the feedwater input will be increased both by a steam pressure drop and a drop in feedwater level. For this reason differential pressure controls utilizing steam generator pressure and pump discharge pressure are used for feedwater pumps. An increase or drop in steam pressure causes a controlled change in feedwater pump discharge pressure to maintain the same differential pressure across the feedwater level-control valve.

Steam generator size and the type fuel influences the selection of control systems. Solid fuels require much more complicated control systems, because of the mechanical firing equipment. Liquid and gaseous fuel systems can be controlled by changing the pressure on the fuel system.

In addition to the above instrumentation, many recording and indicating instruments are necessary. Draft gages, temperature indicators or recorders, flow and pressure instruments are used extensively. Continuous analyzing equipment record the components of flue gas for the larger generators.

Many power plants are being constructed out of doors similar to process units. All instruments are centralized in a control room and controls are completely automatic so that it is rarely necessary for operators to leave the control room. These units are operated by a small crew and contrast sharply with early "hand-operated" boilers.

### Steam Distribution Systems

The piping for steam distribution systems must be carefully designed to insure long trouble-free operation. The ideal steam system should supply steam continuously with the minimum of maintenance.

Line Sizing. Line sizes for steam lines should not be governed by pressure drop alone. Heat losses can be reduced by using high velocities, as suggested in Table 21-5.

TABLE 21-5. Conventional Velocities for Steam Distribution Piping*

| Steam condition | Pressure, psig | Service | Reasonable velocity ft/min |
|---|---|---|---|
| Saturated | 0–15 | Heating (short lines) | 4000 to 6000 |
| Saturated | 50 and up | Miscellaneous† | 6000 to 10,000 |
| Superheated | 200 and up | Miscellaneous | 7000 to 20,000 |

* Reprinted by permission, Crane Technical Paper No. 409 (1942), Crane Co., Chicago, Ill.

† The velocity of steam, in the case of boiler leads, should be lower than large turbine leads because of the stop-check valves which are necessarily installed in these lines. A high velocity through the stop-check valve would cause an excessive pressure drop which may be detrimental to efficient operation.

Layout of Distribution System. Steam distribution piping may be run at grade on concrete pipe "sleepers," or preferably above grade on structural-steel or reinforced-concrete supports. Piping run at grade is an obstacle and must be run over or under roadways, which creates a pocket in the line where condensate can be trapped.

Condensate must be accumulated at predetermined points in distribution piping and at users, and then discharged continuously by steam traps into a condensate system for return to the boiler house. The return piping is most conveniently run near the main distribution piping so that steam traps on the main line may discharge into the condensate line. If several steam systems at different pressures are used, steam traps can be arranged to discharge into each succeeding lower pressure service.

Condensate accumulation at points other than those serviced by steam traps must be avoided by careful arrangement of valves and piping.‡ Branch lines from main headers should rise vertically from the top of the header and then run horizontally as required. Block valves can be installed in the horizontal run of piping and condensate will then drain freely on either side of the valve. By-passes around valves should be located above the valve and a small bleed installed between the by-pass valves.

Drip-legs are installed in long horizontal runs of piping to accumulate condensate. They are constructed of pipe and pipe caps in two sizes larger than line size and are welded vertically in the run of pipe with the major portion extending below the horizontal line. A vertical plate is installed in the drip-leg to deflect steam and moisture into the leg. As

‡ In cold climates freezing of condensate trapped in lines causes great difficulties.

the steam makes a 180° turn, the water drops out. A steam trap is connected to the bottom of the drip-leg for condensate removal.

BLOCK VALVES. Block valves are located in all branch lines as they leave the main header. Small (¾ in. or 1 in.) by-passes are installed for warm-up purposes so that a cold line may be warmed up gradually before admitting full steam flow. A block valve is usually located at each user as an added precaution.

Some companies, however, install only one block valve at the header on exhaust steam lines since exhaust steam is less dangerous. Failure to close the header valve, however, will cause the discharge of large amounts of steam around the area.

INSULATION. Insulation for steam piping, especially main headers, should be selected and designed for long life. Economic studies of optimum insulation thickness for steam lines are usually straightforward since costs are more clearly defined. Such studies should be made for the large diameter lines.

### *Internal Combustion Power*

Internal combustion engines are used in the process plant as low-speed, heavy-duty drivers of reciprocating pumps and compressors, and electrical power generators. Large reciprocating compressors are generally built integral with the engine driver—the same crank shaft is used for both power and reciprocating machine (see Chapter 15). Reciprocating pumps are furnished as separate units with power transmitted through reduction gears and shafting.

Fuel systems for internal combustion engines should be designed to prevent foreign material from entering the engine. All fuel should be filtered regardless of origin, and piping following the filter should be nonferrous or stainless steel so that scale will not be formed. A centralized lubricant supply with piped distribution for single or multiple machines also reduces the quantity of foreign materials which may be introduced.

Particular care should be taken with air-intake piping. Generally these runs are below grade from air-intake filter to engine manifold piping. They are not supplied by the engine manufacturer. Since this piping is not readily accessible for inspection, treatment and inspection of the inside must be thorough prior to installation. All underground air-intake piping should be pickled or sand-blasted (or both), inspected on the inside (most intake piping is of sufficient size to permit entrance), immediately treated with some reliable type of rust inhibitor, and then painted or enameled. Aluminum metallizing, when carefully applied, or light stainless steel or nonferrous liners are also used.

Instrumentation for internal combustion engines usually consists of:

(*a*) Fuel control.

(*b*) Automatic shut-down on cooling water or lubricating oil failure, operable from pressure elements and temperature elements; audible and visible signals are required but not necessarily supplied by engine manufacturer.

## Storage and Movement of Raw Materials and Products

Adequate storage facilities for both raw materials and products are essential to the operation of a process plant. A supply of raw materials permits operation of the plant regardless of temporary procurement or delivery difficulties. Storage space for finished products makes it possible to fill customer's orders even in a time of plant difficulty.

### *Liquid Storage*

Vertical, cylindrical tanks are used for liquids which can be stored at low pressures. Most such tankage is constructed in accordance with American Petroleum Institute Standard 12c. Although this specification was developed primarily for the petroleum industry, it consists of an excellent set of rules for almost any type of tank. The standardization of tank construction techniques and tank appurtenances by this code aids in engineering and procurement. The National Board of Fire Underwriters and the National Fire Protection Association also publish rules for storage tank construction.[27] (Also see Chapter 24.)

The most common type of tank for storage at atmospheric pressure is a cone-roofed tank provided with gage hatch, inspection openings, manholes, a ladder or stairway, and a vent for maintaining atmospheric pressure. For materials with low vapor pressures the vent may be an open-type sized in accordance with tank area and pumping rate.* The open vent is constructed of pipe with a large opening around the top circumference. This opening is protected from wind and rain by a cap which directs flow of vapor downward.

Tanks for storing liquids with significant vapor pressures at atmospheric temperature (above 1 psia) require pressure-vacuum (conservation) vents. These vents open only when the tank pressure becomes excessively low or high and thus conserve valuable vapors.

Floating-roof tanks are used to conserve valuable products with vapor pressures at operating temperature below atmospheric pressure (up to 8 psia approximately). The roof floats on the surface of the liquid and thus eliminates vapor space. The space between the roof and tank wall is sealed by a fabric curtain coated with synthetic rubber.

If flammable liquids are stored in vented tanks, flame arrestors must be installed in all openings except connections made below the liquid

* Air capacity in standard cu ft should be at least twice the pumping rate. Detailed venting specifications have been published.[15]

level. As a flame arrestor a 40-mesh screen or a purchased flame arrestor may be used. The latter consists of a number of thin plates installed in a small chamber. Propagation of flame into the tank is prevented in both types through rapid dissipation of heat by the metal parts of the arrestor.

Fire foam connections are installed on all vented storage tanks containing flammable liquids. Large storage tank areas can be served by a central system. Since floating-roof tanks have no vapor space, they need not be equipped with a permanent extinguishing system.

Liquids with vapor pressures above atmospheric must be stored in vapor-tight tanks capable of withstanding internal pressure. In the pressure range between 2–215 psig thin steel similar to that for vertical storage tanks can be used if tanks are built in the shape of spheres (30–215 psig working pressure) or spheroids (2–30 psig working pressure).

For higher pressures, it is more economical to use a number of standard cylindrical pressure vessels of small diameter (3 to 6 ft). These are installed horizontally and side by side. Relief valves are located on each drum and manifolded into a common header.

Tankage for water storage is usually elevated, and, if part of the fire water system, is designed in accordance with fire insurance codes (see Chapter 24).

In addition to flammable and volatile products and raw materials, many corrosive liquids must be stored in process plants. Steel tanks lined with plastic, rubber, lead, and special brick and other ceramic materials have been used successfully, particularly at lower temperatures. At higher temperatures alloy steels, or nonferrous clad steels are necessary.

### *Gas Storage*

Gases are stored at atmospheric pressure in wet or dry-seal gas holders. The wet-seal holder is the familiar type used for municipal storage. Gas is admitted into a tank (holder) which is inverted in a water-filled tank. As gas enters the holder, water is displaced and the holder rises.

The dry-seal holder, a more recent development, is similar to a floating roof. The seal between the roof and the tank wall is made by a flexible synthetic rubber or plastic curtain, and the roof rises or falls with pressure. The dry seal holder requires no heating coils to prevent freezing of water as does the wet-seal and is particularly useful for storage of gases that must remain dry.

Natural gas and even gaseous products may be stored underground by pumping the gas into those underground strata known to be practically impermeable to gas.

High-pressure gas is stored in horizontal cylindrical pressure vessels. Gases such as $N_2$, $H_2$, and $O_2$ are shipped in gas cylinders designed in accordance with Interstate Commerce Commission regulations.

### *Storage of Solids*

Solid products and raw materials are stored in weather tight tanks and bins with sloping bottoms or in outdoor piles. Solid products discharging from flakers or dryers are often packed directly in bags and metal or fiber drums. These bags and drums are stored on pallets in warehouses to await shipment.

### *Movement of Raw Materials and Products to Shipping Point*

Movement of liquid and gaseous raw materials and products within the plant to and from the point of shipment is done by pipeline. Transfer pumps are grouped together to simplify power distribution and reduce labor. A shelter is provided for transfer pump personnel, but is not necessary for pumps.

Suction and discharge piping is manifolded for greatest flexibility. This is done by running all piping entering the transfer area parallel and at one level (12 in. to 18 in. above ground). Transfer pump suction and discharge lines are run above these transfer lines and are connected to them by risers. Valves are located in the connecting risers. In this way a single discharge or suction line may be connected to a number of transfer lines and valves can be easily reached by the pump operators.

Some plants transport raw materials and products by pipe line. When this is done local and federal regulations must be strictly followed in designing and laying the pipe. Land must be purchased or leased, but this land must be selected with care since the pipe line should not pass near residential or future residential areas.

Lines passing under highways or railroads must be provided with an external pipe jacket welded to the pipe and vented to the atmosphere at specified distances from the highway or track. Every other possible safety precaution should be used.

Solid materials are transported within the plant by conveyors, bucket elevators, chutes, lift trucks, and pneumatic systems. New materials handling equipment continues to be introduced, and the task of moving solids is no longer a difficult hand operation.

## SECONDARY UTILITIES

### Maintenance Facilities

Every process plant must have a machine shop equipped with standard tools such as lathes, grinders, and drill presses. Large plants usually have complete shops for machine and boiler work, pipe fabricating, instrument repair, electrical equipment repair, and carpentry. Since a large mechanical labor force must be maintained for regular shut-down and repair of operating units, the same personnel may be used during operating periods for regular maintenance and for building special equipment.

Small repair jobs can be done at the process unit. Specially outfitted trucks and motor scooters carrying tools and small parts are used to answer emergency calls and for making quick repairs. More extensive repairs can sometimes be handled better in the shop and mobile cranes and smaller lift trucks are used to transport the equipment to the shop.

Planning of maintenance facilities is greatly facilitated by complete spare parts lists and maintenance instructions from the manufacturer for each piece of equipment. These can be used as a guide to the scope of maintenance work to be expected in a plant as well as an aid in planning such widely varying needs as:

*a.* personnel requirements.
*b.* quantity and type of machine and hand tools.
*c.* storage space for spare parts and new equipment.
*d.* salvage area for old equipment and parts.
*e.* quantity and types of handling equipment (trucks, hoists, etc.).

### Plant Roadways

Roadways for process plants are designed to permit easy access to all portions of the plant for mobile servicing equipment, trucks, and fire fighting apparatus. Roads are located so that each process area is accessible from all sides.

Plant roadways are designed in accordance with usual highway practice, and entrances from public roads must conform to local design methods. Reinforced concrete is used for main plant roads since maintenance work on concrete is negligible. Secondary roads may be macadam or gravel but must be designed to handle loads imposed by such equipment as travelling cranes.

Main two-way roads should not be less than 20 ft wide. Short entrance roads into process areas may be narrower but must be adequate for trucks and cranes to pass into the area for servicing.

### Railroad Facilities

Railroad facilities are necessary for shipping and receiving materials. Generally local railroads will provide without cost all the required specifications and a certain amount of the necessary equipment, such as turnouts from their system, and in some cases, they may install sidings or spur tracks. Before preparing specifications or designing railroad trackage local railroad operators should be consulted. There are limitations set by local statutes or railroad operators on rail size (weight per yard), gradients, curve radii, switch design and type, and clearances.

Spur tracks (sidings) and switching equipment must conform to the requirements of the serving carriers. If a spur track must cross a public highway either the railroad operator or the plant owner must obtain a formal permit, and follow the requirements for such crossings.

Railroad trackage is usually laid on wooden ties. Tie plates are located under each rail at each tie, and splice bars are used to connect rail ends. Standard gage for U.S. railroads is 4 ft 8½ in. between inner edge of railheads. Rails, splice-bars and tie-plates for plant spur tracks can be obtained second-hand and should be used since no high speed traffic is anticipated.

The rail weight for plant trackage may be 60–100 lb/yd. Rail lengths vary from 25 to 30 ft. Tie sizes are approximately 8 in. × 8 in. × 8 ft 0 in. long.

## Plant Sewer Systems and Waste Disposal

The primary sources of sewage and waste in a process plant are sanitary waste, process drains, and surface drainage. The plant sewer system must be designed to conduct these wastes to the disposal system without becoming clogged with solids or filled with dangerous concentrations of explosive gases.

### *General Considerations*

Drainage systems for process plants may consist of one or more main sewer lines with branch lines from each area to be drained. The elevation of the main sewer must be of sufficient depth to receive drainage from any area or from contemplated extensions to the plant.

Sewer lines are sloped to provide sufficient head to produce velocities of 2½ to 3 ft per sec (5 ft/sec maximum). These velocities are necessary to prevent settling of solids. Contour maps of the plant area are helpful in planning the location of the main sewer so that slopes can be obtained for branch lines from all areas.

To prevent clogging and to facilitate cleaning, sewer lines are run without change in direction between manholes and between a catch basin and manhole. Clean-out connections should be provided for branch lines not originating at a catch basin.

Manholes (Fig. 21-5) are located in main sewer lines at each change of direction and at points about every 200 ft apart in straight lengths of pipe. Branch lines (laterals) should connect to mains at a manhole when possible. Manholes in hazardous areas have sealed covers with vents discharging in a nonhazardous area. Manholes through which flammable materials pass should be trapped by means of a water seal to prevent spreading of fires or explosions (Fig. 21-5).

Manholes are constructed of common brick, acid-resisting brick, concrete with brick lining, or plain concrete. Manhole covers, gratings, and other appurtenances are usually cast iron.

Catch basins are used to collect surface or process drainage. Water makes a 180° turn to leave the catch basin and to enter the branch sewer line. In so doing sediments are trapped in the bottom of the basin. Catch basins are constructed of concrete and have cast iron slotted covers.

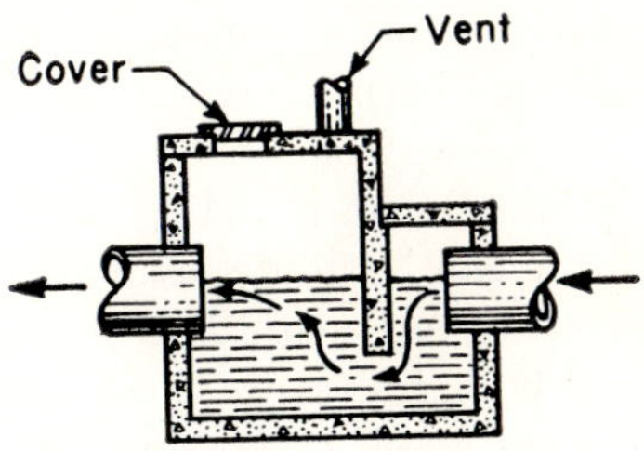


Fig. 21-5. Sewer manhole with seal for isolating explosions and fires.

Reinforced concrete pipe is the most widely used sewer pipe. Cast iron is preferred under paved roadways and near foundations or other structures. Vitrified tile is used extensively for areas in which corrosive waste materials may enter the drainage system and lead pipe is used for strong acid wastes. Soil pipe is used for sanitary sewers.

### Surface Drains

Rainfall intensity and run-off are the basic data required for designing surface drainage systems. Intensity is measured in in. of water per hr, and may be obtained from local weather bureaus and the U.S. Department of Agriculture. Rainfall intensity varies over the continental U. S. from 0.25 in. in some parts of the northwest, to 1.50 in. in the mid-continent areas, 1.00 in. in some parts of the east, and 2.25 in. along the Gulf coast. In the same areas in the northwest annual rainfall may be 40 in., while some of the arid areas of the southwest have annual rainfalls of 8 in. with intensities of 0.75 in. Since maximum intensities are more important than averages, local weather data should be searched for the infrequent but extremely heavy rainfalls.

Run-off is the per cent of rainfall which is not absorbed by the soil and thus is the quantity which determines conduit size. The value may vary from 100% for paved areas or roofs to zero for dry pervious soils. Most of the published data is based on city drainage systems and ranges from 10 to 75%. These values are used for extremely large acreages and are not of much use as criteria for process plants where high percentages of total area may be paved or roofed structures. It is preferable, therefore, to assume 100% run-off for the comparatively small areas served by a single sewer since often these areas are paved and are adjacent to roofed buildings.

The run-off in paved process areas enclosed by firewalls may be governed by fire hose capacity in that area rather than rainfall. Both should be checked.

### Process Wastes

Disposal of process wastes is often a difficult problem. State and local air and stream pollution regulations are becoming more strict as population grows in industrial areas. Plant operators are becoming more aware of community responsibilities and usually institute modern waste disposal methods even if regulations do not apply to them.

Some wastes can be burned provided products of combustion do not pollute the surrounding atmosphere. Others such as acid and alkaline waste can be diluted and neutralized, and then discharged into a flowing stream. All such wastes must be treated and discharged in a manner that will be harmless to stream life. Progressive plant operators maintain a careful check of stream health.

Inert wastes or wastes insoluble or immiscible in water can be discharged into the surface drainage system and removed in settling basins along with other similar contaminants present in the drainage water. Strong acid streams or extremely dangerous materials are preferably discharged into special sewer systems for disposal in a separate treating plant.

### Sanitary Wastes

Sanitary wastes must be discharged into a separate sewer system and treated in a sewage disposal plant or in septic tanks. Ordinarily, septic tanks are satisfactory for the moderate sized plant.

The septic tank is a retention chamber in which solid wastes are accumulated and then disintegrated by bacteria. The effluent is a clear liquid which may be discharged in a separate sewer or in the plant surface drainage system. Solids are removed once a year.

Table 21-6 summarizes water usage for various plumbing fixtures. This information is necessary in sizing sanitary sewer lines.

**TABLE 21-6. Water Usage by Plumbing Fixtures**

Based on 15 to 20 psig water pressure

| Fixture | Gallons per use |
|---|---|
| Toilet | 2.0–3.0 |
| Urinal | 0.5–0.75 |
| Lavatory basin/faucet | 2.0–2.5 |
| Shower | 2.0–3.0 (gpm when used) |
| Fountain | 0.5–0.75 (gal/day/person) |

### Disposal of Solid Wastes

Waste solids which cannot be burned must be hauled to dump areas for burial or be towed to sea in dump-bottom barges and discharged many miles from shore.

In some cases solids can be disintegrated by chemical treatment, dissolved in water, and, after careful neutralization discharged in the plant sewer.

## Plant Air

The process plant's use of compressed air as a power source for portable equipment continues to increase. Air-driven tools are designed for all types of work. They are simple to control and may be used in hazardous areas.

A service air system, therefore, is usually installed in the modern plant. Piping is run above grade along with process piping, and outlets are located in all process areas. Compressed air may be distributed over the entire plant in a single system or may be supplied by separate compressors in each area.

Plant service air and instrument air (Chapter 20) are run in separate systems. Service air pressure is maintained at 100–125 psig, while instrument air pressure need not exceed 35–40 psig. Plant service air is sometimes used in emergencies as a source for instrument air, but it must first pass through the instrument air dryer.

The usual auxiliaries for air compressor installations are:

*a.* Air compressor: electric motor driven or steam-driven, two-stage, reciprocating.

*b.* Air intake filter.

*c.* Cooling water for compressor jackets, inter-cooler and after-cooler.

*d.* Air receiver.

*e.* Pressure control.

Provision must be made at various points in the system to remove condensed moisture. Air receivers located at each point of heavy use will increase system volume and provide a separation or condensation point for moisture. Moisture may be removed manually or by trap. Traps are preferable for systems in humid climates.

The air compressors for a central plant service air system are most conveniently located adjacent to the power and steam generating plant. Power-house operators can then check compressor operation periodically.

## Plant Security

Process plants are surrounded by fence topped with barbed wire to prevent entrance by unauthorized visitors and prowlers. Legal decisions which have become part of common law make plant owners responsible for injury to persons within the boundaries of a plant even though their presence is unauthorized.

Fences are constructed of galvanized steel wire and are 7 to 10 ft high. Fence posts are galvanized steel pipe set in concrete.

Guard houses are located at each gate and all plant personnel and visitors must pass through a gate house. Employees wear badges containing an identifying photograph. Visitors are issued temporary badges and must be accompanied by an authorized employee.

If time clocks are used for employee records, they are located in the gate house.

Uniformed guards are stationed at each gate house to inspect persons and trucks entering and leaving the plant. Other guards are assigned to night patrol duty.

It is preferable to locate office buildings outside the plant security area, particularly buildings for sales, engineering, and personnel divisions. Visitors can then be received in a proper and friendly manner and in pleasant surroundings, conditions which seem to be difficult to create at plant gate houses.

## REFERENCES

1. Abbott, A. L., *National Electrical Code Handbook,* 7th Ed., McGraw-Hill Book Co., New York, 1952.

2. Babbitt, H. E., *Sewerage and Sewage Treatment,* 7th Ed., John Wiley & Sons, New York, 1952.

3. Betz, W. H., and L. D. Betz, *Betz Handbook of Industrial Water Conditioning,* Philadelphia, Pa., 1950.

4. *Chem. Eng., 61,* No. 6, 122 (1954).

5. Cook, A. L., and C. C. Carr, *Elements of Electrical Engineering,* John Wiley & Sons, New York, 1954.

6. Croft, Terrell, *Steam Power Plant Auxiliaries and Accessories,* McGraw-Hill Book Co., New York, 1946.

7. Cubberly, R., *Chem. Eng., 57,* No. 5, 140 (1950).

8. de Lorenzi, Otto, *Combustion Engineering,* Combustion Engineering-Superheater Co., Inc., New York, 1948.

9. Eliason, J. R., *Chem. Eng., 57,* No. 5, 129 (1950).

10. Erickson, William H., and N. H. Bryant, *Electrical Engineering—Theory and Practice,* John Wiley & Sons, New York, 1952.

11. Fair, G. M., and J. C. Geyer, *Water Supply and Waste-Water Disposal,* John Wiley & Sons, New York, 1954.

12. *Flow of Fluids,* Tech. Paper No. 409, Crane Co., Chicago, Ill., 1942.

13. *Fuel Oils,* 5th Ed., Commercial Standard CS12-40, National Bureau of Standards, Supt. of Documents, Washington, D. C., 1940.

14. Gordon, D., *Chem. Eng., 57,* No. 5, 119 (1950).

15. *Guide for Tank Venting,* R. P. 2000 American Petroleum Institute, 50 West 50th St., New York, 1952.

16. Hart, W. B., *Industrial Waste Disposal for Petroleum Refineries and Allied Plants,* National Petroleum News Publishing Co., Cleveland, Ohio, 1947.

17. Heuman, G. W., *Magnetic Control of Industrial Motors,* John Wiley & Sons, New York, 1954.

18. *Manual on Disposal of Refinery Wastes,* Vols. I, II, and III (Oil, Gas and Chemicals), American Petroleum Institute, New York, 1951–53.

19. Miller, C. E., *Mech. Eng., 73,* No. 1, 9 (1951).

20. Pender, H., *Electrical Engineer's Handbook—Electric Power Volume,* John Wiley & Sons, New York, 1949.

21. *Permutit Water Conditioning Data Book,* The Permutit Co., New York, 1953.

22. Powell, S. T., *Water Conditioning for Industry,* McGraw-Hill Book Co., New York, 1954.

23. Ryan, W. J., *Water Treatment and Purification,* 2nd Ed., McGraw-Hill Book Co., New York, 1946.

24. Salisbury, J. K., *Kent's Mechanical Engineer's Handbook—Power Volume,* 12th Ed., John Wiley & Sons, New York, 1950.

25. Severns, W. H., H. E. Degler, and J. C. Miles, *Steam, Air and Gas Power,* 5th Ed., John Wiley & Sons, New York, 1954.

26. Skrotzki, G. A. B., and W. A. Vopat, *Applied Energy Conversion,* McGraw-Hill Book Co., New York, 1945.

27. *Storage, Handling and Use of Flammable Liquids,* NFPA No. 30-L, pp. 30–15, National Fire Protection Association, Boston, Mass., 1954.

# CHAPTER 22

# FOUNDATIONS

Foundations for equipment, buildings, and steel structures not only must transmit the loads to the soil or piling but also fix the supported equipment in position. For practical purposes it may be said that all foundations are constructed of concrete or reinforced concrete. This material is relatively inexpensive and may be formed to nearly any desired shape while plastic.

In order to attach the equipment directly to a foundation, anchor-bolts are cast in the concrete when it is poured (Fig. 22-9). After a curing period, the equipment is bolted in position. Most equipment is firmly bolted down whether it is located on a structural steel support which is in turn bolted to a foundation, or whether the equipment is bolted directly to the concrete foundation. Large-diameter, flat-bottomed tanks, however, are usually not bolted since they may be set directly on concrete slabs or on a prepared soil surface with no attachment.

In some installations small pumps or machines and light conveyor systems are supported by placing the equipment base directly on the finished concrete floor, and by providing expansion bolts which are drilled and set into the concrete.

## FOUNDATION TYPES

The most common types of foundations used in process plants are spread footings, mats (or rafts), massive, and pile foundations. This listing does not represent a system of classification, but serves more as convenient subdivisions for the discussion below.

### Spread Footings

The footing, which is the oldest type of foundation, is the least expensive type if soil conditions are favorable. In simple terms, a spread

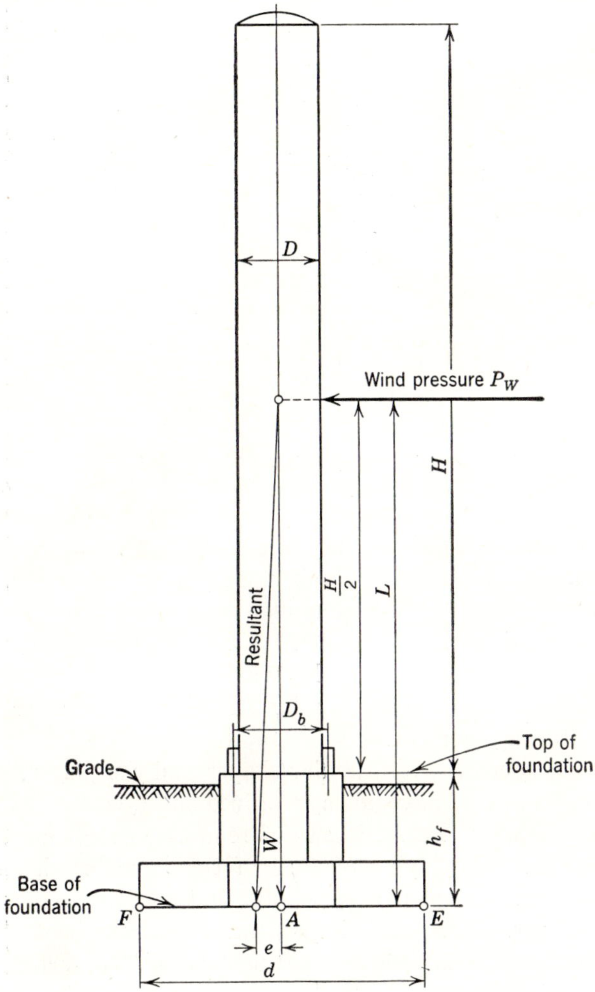


Fig. 22-1. Spread-footing for a self-supporting tower. [Reprinted by permission, V. O. Marshall, *Petroleum Refiner, 22,* No. 8, 101 (1943).]

footing as used in the process industries consists of a column or pedestal cast monolithically with an enlarged base (the footing) as shown in Fig. 22-1. The enlarged portion, if properly designed, transmits the imposed load to the soil. The more area covered by the base of such a foundation, the less will be the unit pressure on the soil.

Spread footings are usually constructed of reinforced concrete. Their bases are ideally located at a depth not less than the normal frost line or the maximum depth of appreciable soil shrinkage and expansion, whichever is the greater. The shrinkage area of many clay soils, however, may be so deep that excavation for footings proves too costly. In such cases other types of foundations may be indicated.

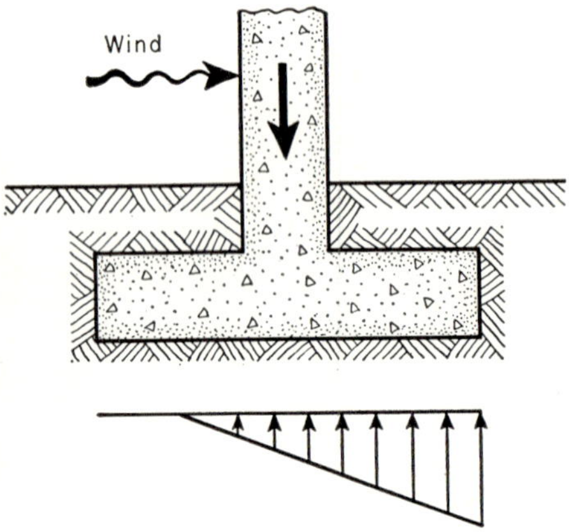


Fig. 22-2. Distribution of soil pressure beneath a footing subject to wind load.

The spread footing is used most extensively for supporting towers (Fig. 22-1), stacks, and tall structural columns. For such applications the circular footing is the idealized shape. Tall structures subject the footing to simultaneous forces due to weight and overturning moment produced by earthquake and wind. The unbalanced soil pressures produced can in some cases cause areas of zero upward pressure between soil and footing (see Fig. 22-2). There is a strong tendency for the footing to crack under such conditions. Because of their corners, square and rectangular shapes crack more easily than circular shapes. Since it is possible that the overturning moment can be in any direction depending on the wind, the circular shape has the further advantage of providing similar stress distribution for any direction of overturning moment.

The forming operation for circular shapes and the placement of reinforcing steel, however, have always proved more costly than that for straight-sided shapes. Parallel rows of reinforcing steel must be cut to different lengths. The laying of steel of the same length through the center (radial reinforcement) is possible but requires too much space toward the center. Octagonal shapes, therefore, have been widely used for such structures as tall self-supporting towers. The octagon requires little additional concrete and equal length reinforcing rods may be placed perpendicular to the sides.

As new forming techniques are perfected, costs may change. It may, for instance, prove more economical to have standard sizes of circular metal forms for self-supporting towers of various sizes.

Because of the simple forming operation required for square footings, they remain the most generally used type of spread footing. If the overturning moment is small or negligible, the design is simple. Square footings are used for supporting equipment and structures, especially those with high cross-sectional area to height ratios.

## Mats

Due to soil conditions some equipment and buildings may require a number of footings that have a combined area approximating that of the supported structure. In such cases it will prove economical to use a flat reinforced concrete slab covering the entire area. This slab is called a mat.

Mat foundations have been successfully used with compressible soil and when so used are called "floating" foundations. A weight of soil is removed for the mat equal to the weight of the mat plus the structure to be supported. The entire structure will, therefore, "float" on the undisturbed subsoil.

Mat foundations are often used for supporting storage tanks. The simplest mat foundation for large flat-bottomed storage tanks that do not require anchoring is a sand pad about four in. deep. The portion of the sand pad extending beyond the tank known as the berm should be about 5 ft 0 in. It is usually protected from weathering by an asphaltic coating. The surrounding soil must be graded and prepared for good drainage.

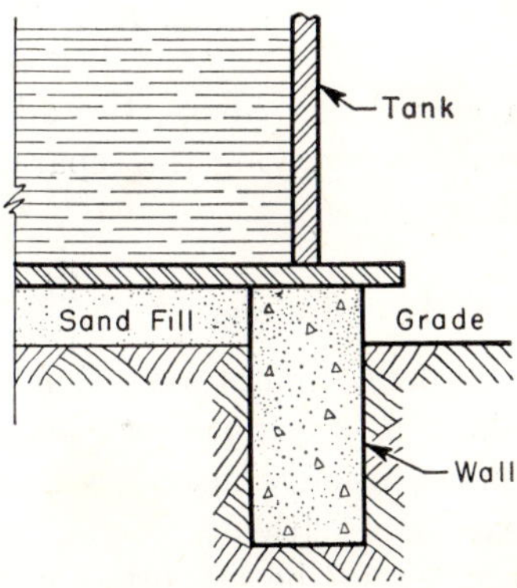

Fig. 22-3. Ring-wall construction for storage tank.

Perhaps the most popular mat-type foundation for large storage tanks employs a concrete wall with a minimum width of 8 in. The tank shell rests on this reinforced concrete wall which extends below the frost line. Four in. of sand contained within the wall provides additional support (Fig. 22-3).

## Massive Foundations

Machinery such as reciprocating compressors, centrifuges and other similar moving equipment must be supported so that vibration and unbalanced inertia forces are dampened. Many clever methods have been developed for accomplishing this dampening. Various vibration isolators or dampeners employing absorbent materials or springs can be purchased. However, wherever possible, in the process plant heavy or massive foundations are used to which the machinery is firmly anchored. The massive foundation balances the periodic inertia forces set up by reciprocating machines as the acceleration reaches a maximum at the ends of the stroke.

The vibrations produced by high speed rotating machinery are also absorbed by foundations of sufficiently high mass.

A typical mass foundation for a compressor is shown in Fig. 22-4. If a number of machines such as compressors are located together as in

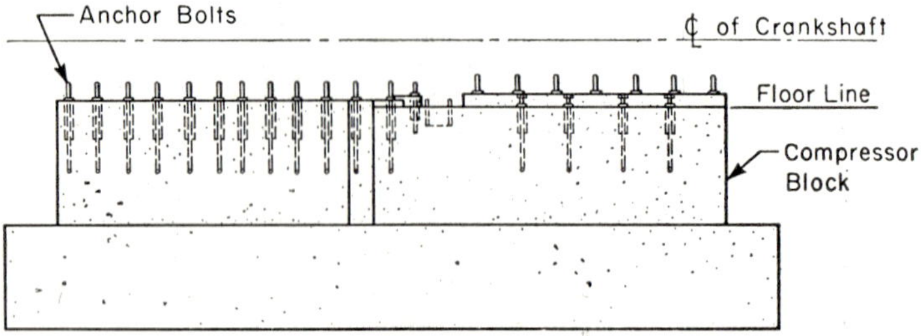


Fig. 22-4. Mass-type compressor foundation.

a compressor house, the advantages of a larger effective mass can be realized by tying the separate compressor blocks into a common monolithic mass.

## Pile Foundations

Process plants are often advantageously located in coastal areas or along rivers. Such locations, though convenient for transportation economies, usually have low bearing capacity soils and/or soils subject to excessive settlement under load. In such cases the foundations discussed above must be supported by piles as shown in Fig. 22-5.

Piles support the loads on foundations by direct bearing on stable soil strata many feet below the surface or by means of the shearing resistance created by friction between the soil and pile surface. The first type is called a point-bearing pile since most of the support is developed at the bottom or point of the pile. The second is called a friction pile. Under most conditions, however, such a simple classification is inadequate. Often the support is obtained by a combination of both effects.

### *Pile Types*

The most common type of piles are pre-cast reinforced concrete, cast-in-place concrete, creosoted wood, structural steel members such as H-beams, pipe, and composite wood-concrete.

Wood piles are the lowest in first cost and for this reason are widely used. The wood must be thoroughly creosoted to prevent fungus growth. To further prevent decay the top of the wood pile must be below permanent ground water level.

Because of strength limitations wood piles are best used as friction rather than end-bearing piles.

Raw material for wood piles as long as 90 ft can be found in the forests of Washington, Oregon, and Northern California, but is becoming less

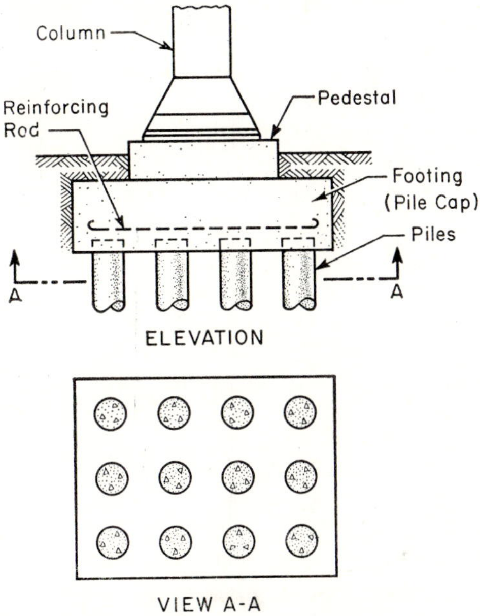


Fig. 22-5. Piles supporting a structural steel column.

available. Concrete piles are, therefore, being used more frequently in process plants.

Concrete is not subject to decay or corrosion* and when reinforced is capable of carrying large loads. Concrete piles are either pre-cast or cast-in-place. They are produced by companies which both manufacture and drive the pile.

### Cast-in-Place Concrete Piles

Cast-in-place piles are made by driving a corrugated or fluted sheet-steel shell or pipe into the earth to the required distance. To strengthen the thin shell a core or mandrel is placed inside before it is driven into the ground. Then the core is removed and the opening filled with concrete. A number of patented cast-in-place pile designs are available, some of which are tapered (Fig. 22-6). Tapered piles are preferred for soils which

* Sea water attacks concrete at or near the surface.

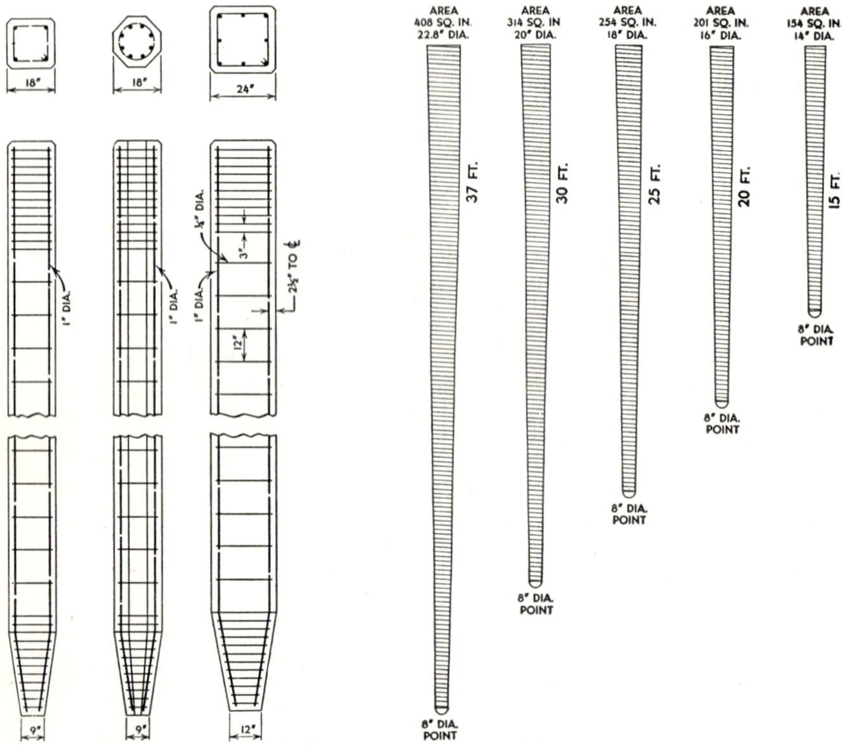


*(a)* *(b)*

Fig. 22-6. Typical piles. (*a*) Precast concrete piles; (*b*) cast-in-place concrete piles; (*c*) steel-taper piles; (*d*) wood-composite pile. (Courtesy of Raymond Concrete Pile Company.)

have high frictional resistance, as sand, since in this type of soil they produce larger shearing resistances than piles having parallel sides.

### Pipe-Concrete Piles

Pipe piles serve primarily as end-bearing piles. In the casting of certain types the pipe is withdrawn as the concrete is being rammed down the pipe. A bulb or pedestal may be formed at the bottom of the pile to increase the load bearing area. This is accomplished by applying a large force on the first few feet of concrete poured.

### Precast-Concrete Piles

Precast-concrete piles (Fig. 22-6) are made of high-strength concrete and reinforcing steel. The shorter piles can be obtained with a taper. They are used in the installations for which casting in place would be

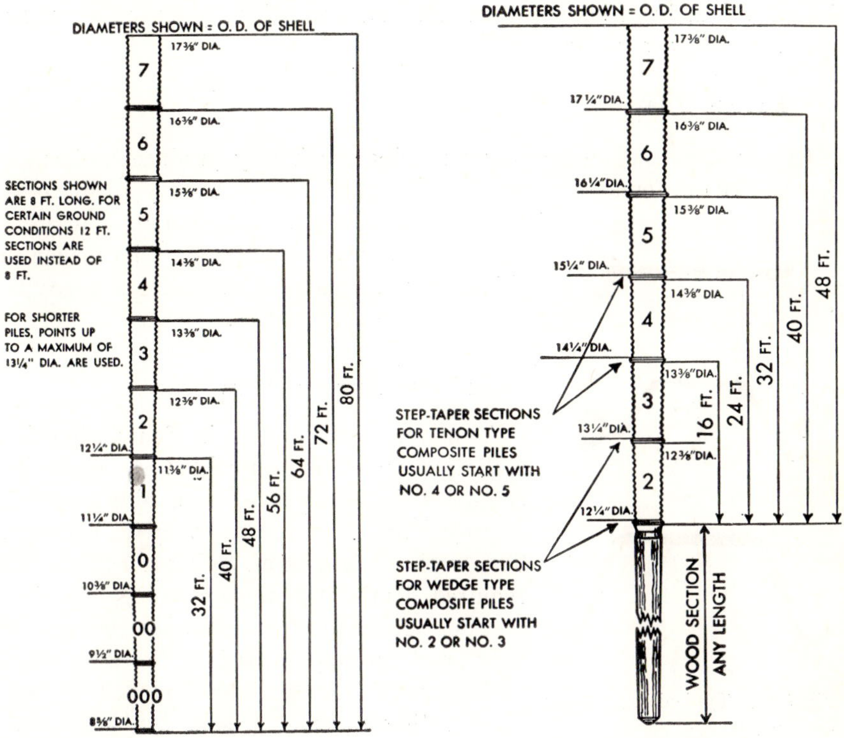


(c) (d)

difficult such as locations under water. In addition they are of great value for installations requiring immediate use of the pile as a load bearing member. Precast and prestressed concrete piles are being used for heavy industrial plants particularly if the above-grade structure is also concrete.

### Steel Structural Shapes as Piles

Sometimes it is necessary to use a pile that is strong and can easily penetrate the most dense soil. Steel structural shapes, although expensive, meet those requirements and are used for difficult piling operations requiring long lengths and large forces in driving which would break weaker materials.

### Combination Piles

In certain installations requiring long piles it is possible to combine the low cost advantages of wood with the corrosion decay free qualities of concrete. The wood pile is driven first followed by a tapered concrete pile section which is cast in place. Several types of joints are used to insure a firm bond between concrete and wood thereby excluding water and preventing bending.

### Batter Piles

Compressors and other heavy moving machinery produce horizontal forces. If the foundation must rest on piles, it is necessary to drive at least a portion of the piles at an angle in order to resist the horizontal forces. Piles driven in this manner are called batter piles and may be of any of the above types.

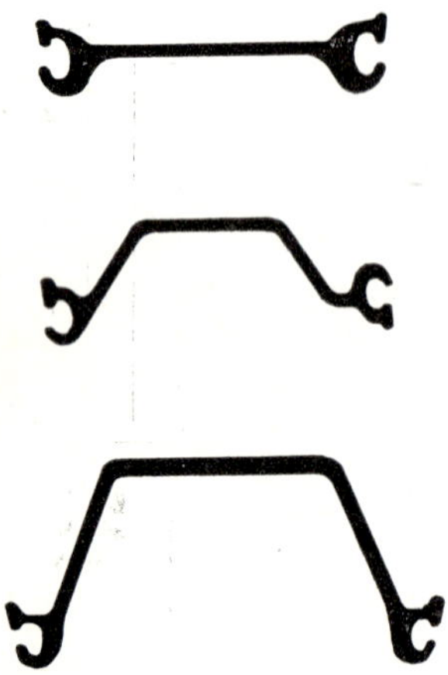

Fig. 22-7. Typical sheet piling sections.

### Miscellaneous Piles

A newer development in foundation installation which is being used satisfactorily combines the characteristics of piling, spread footings, and reinforced concrete columns. The method consists of boring a hole to the required depth, then underreaming a conical section. After boring and underreaming is completed the apparatus is withdrawn, reinforcing steel is lowered into place and the hole filled with concrete. If soils are sufficiently stable to prevent caving, the method is ideal. The upper ends of the cylindrical sections may be located below grade, with steel exposed for dowelling into any required surface structure.

Economically, this method has advantages over both spread footings and piling, if the soil is stable. Metal casings can be used in unstable soils but at greater expense.

Another type of pile valuable in construction, although not used as a structural member, is the sheet pile. As illustrated in Fig. 22-7, these steel shapes are designed to slide vertically and to interlock. They are used as strong walls to hold back water and shifting soils during construction.

## SELECTING THE FOUNDATION TYPE

It is not possible to state conclusively the particular type of foundation which should be used for a each type of project. This decision can be made only after careful analysis of data accumulated by exploration of subsurface soil. If piling is required, economic studies must be made to determine the most desirable type.

Arrrangement of equipment over a given area to meet process conditions is often a determining factor in selecting the type of foundation. Rough estimates of the required spread of the footings, if the soil load bearing characteristic is known, can be determined early in the project, by approximating the loads over each area. Soil loading can be estimated to aid placement of equipment within an area in order to spread the loads as much as possible.

The economics of equipment location involve the costs of interconnecting piping, access platforms, and supporting structures. Such studies may indicate that the use of piling will be less costly than spreading the equipment to obtain a reduction in load concentration.

## FOUNDATION CONSTRUCTION

Foundation construction operations primarily are the concern of the civil engineer. The project engineer, however, must be familiar with the terminology, procedures, and sequence of operations.

Briefly, the installation of spread or mass-type foundations consists of excavating to a firm stratum, setting the reinforcing steel in place, and pouring the initial concrete mat or spread. Form work which encloses and contains the concrete while it is still plastic, usually is not necessary for the initial pours below grade level. In stable soils the sides of the excavation are sufficiently firm to form the peripheral shape required to contain the concrete. The completed pour is left rough with reinforcing rod extending above the concrete so that additional pours required to complete the structure will be securely bonded.

The superimposed load on piled foundations is distributed to each pile by means of a reinforced concrete pile cap, which is cast around the tops of the piles. This type of foundation is installed by excavating to the required pile-cap depth, driving the piles, cutting off the tops of wood piles to make them even, setting the reinforcing steel for the pile cap, and pouring a concrete mass which is the pile cap.

### Scheduling of Foundation Work

In the ideal construction program, all underground operations should be completed before any above-grade work is started. The continuation of underground work after the erection of above-grade equipment has begun is extremely costly. Opening a trench, for instance, across a construction area produces a safety hazard and seriously handicaps operations. Late excavations in a congested area nearly always must be done by hand and the surface work is slowed by the interference.

The release of foundation drawings to the erection forces should be coordinated with the release of all other underground work such as piping, electrical grounding systems, and electrical conduit work.

### Subsurface Soil Investigations

The load bearing characteristics of the soil on which a plant is to be erected should be explored and determined at some early stage of the project. In fact, such characteristics should have been known before purchase of the site. It is impossible to judge subsurface conditions by visual inspection, and competent investigation is usually most economically accomplished by firms specializing in soil analysis or subsurface explorations. Reliable data permit foundation design and selection to proceed on a realistic basis.

In addition to the general characteristics of the underground stratum more extensive studies should be made at the specific locations for each footing.

### Borings

The most reliable method for determining the characteristics of the subsurface soil is from samples of each stratum obtained from borings. Soil borings can be divided into major or deep borings, which locate the governing strata, and secondary or shallow borings, which should be made under each footing. It is sometimes possible to utilize drilling logs from deep water wells to determine the practical depth for the major borings. Water well logs usually yield detailed data since all water bearing sands must be identified.

Major borings are made by portable truck-mounted machines. They may be driven one to two hundred feet deep or deeper if necessary. One method in wide use employs a fish-tail bit which moves in long vertical strokes. Wash water removes the loosened material as in most drilling operations. As each change of soil is encountered, the bit is removed and a sampler driven into the soil not yet penetrated. The dry samples thus obtained are preserved in hermetically sealed jars carefully identified for laboratory investigation. By noting the number of blows at uniform force required to drive the sampler into each section of strata, the compactness of the soil is ascertained. Fig. 22-8 shows a typical boring report.

The boring operation must be discontinued when rock is reached (noted as "refusal" in Fig. 22-8). The extent and character of this rock or bedrock can be determined from core borings. These are obtained using rotating coring devices similar to those employed in the oil industry.

Extensive subsurface borings are usually mandatory in areas which have been subject to water encroachment, such as coastal areas, stream valleys, and tidal flats. If the area is located in a known stable region, not coastal or stream-bed, and one or two deep borings show that the section is composed of stable sedimentary formations or formations overlying rock, only a minimum amount of deep boring is necessary. Shallow cores to locate uppermost firm stratum should be taken, however, under each major footing. These shallow cores which can be hand-driven indicate the depth of topsoil or humus, and the depth of excavation necessary to locate the bottom of the footing on the uppermost frim

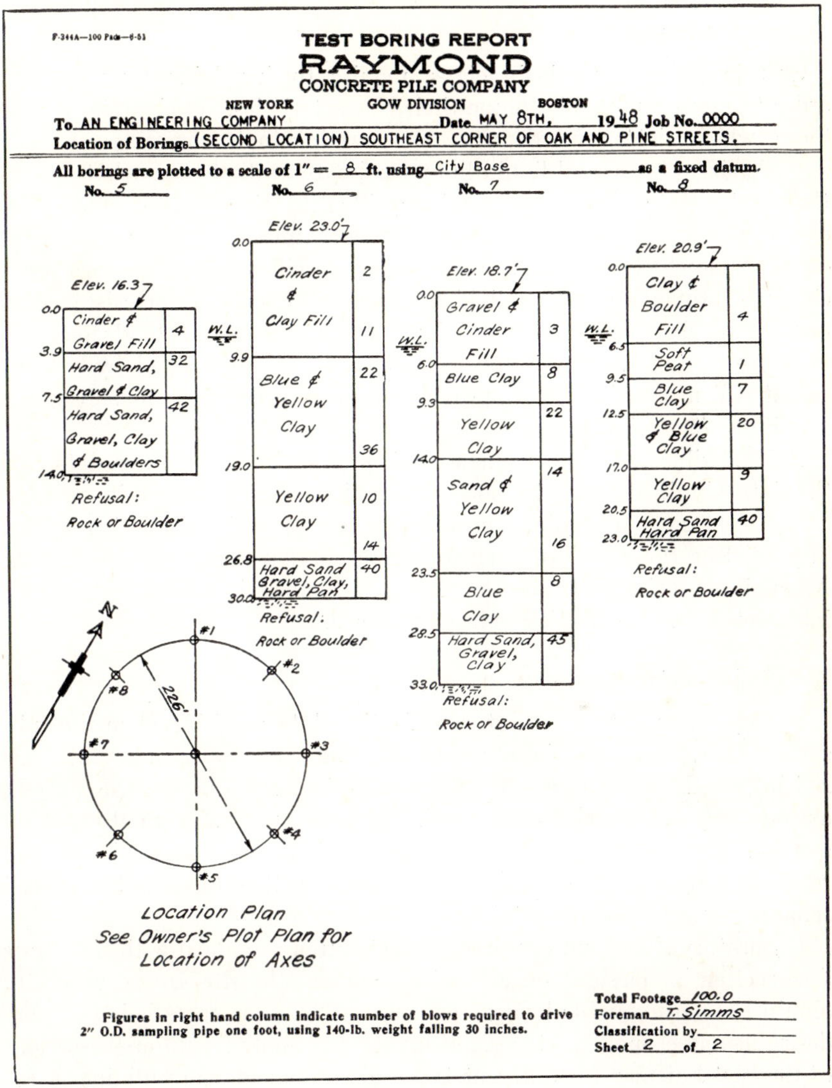


Fig. 22-8. Test boring report. (Courtesy of Raymond Concrete Pile Company.)

stratum. The topsoil, or humus layer is never satisfactory for load bearing and must all be completely removed from any area which is to be used for foundations or is to be covered with concrete slabs.

Borings also indicate the location of subsurface water. This information is of value throughout construction in enabling the erector to foresee and to make provisions for removing water from excavations.

### Soil Analysis

Certain soils, particularly clays, require further laboratory tests in order to more completely define their characteristics. Such tests include those which determine density, moisture content, shearing strength, particle size, plastic and liquid limit (moisture content range that defines the plastic state of the soil permitting identification of clays and silts), coefficient of internal friction, consolidation (measure of the settlement of an "undisturbed" soil sample under specified load) permeability, chemical properties (*p*H, colloidal nature, etc), cohesion (resistance to particle displacement), and shrinkage (volume loss due to decrease in moisture content). These and other tests when interpreted by a competent soil mechanics expert provide accurate predictions of subsurface conditions for design purposes.

It is possible to draw subsurface soil profiles based on these data and the data obtained from the boring operation. Since an infinite number of borings cannot be made a considerable amount of judgment is required in predicting the extent of each formation and stratum between borings. All engineers using these profiles should remember the limitations of such interpretation. No profile can be guaranteed.

### Soil Bearing Capacity

Perhaps the most important quantitative data resulting from the subsurface investigation is the allowable bearing capacity or pressure of the soil in pounds per square foot. This value is used in design calculations as the carrying capacity of the soil. Although the value finally recommended by soil experts is based partially on speculation, it is used religiously by the designers. It is imperative, therefore, that the most reliable value possible be obtained.

Laboratory shear and compression tests often combined with field tests (see section on Design) on critical portions of the site are necessary to obtain the most reliable bearing capacities. In some cases it is far more desirable and economical to establish values not only for different sections of the plot but also for different types of equipment and buildings. Installations subject to occasional heavy loads can be designed for a lower bearing value than those upon which a continuous heavy load is imposed. Flexibility in establishing load requirements is more advantageous than establishing a single safe value for all conditions.

### Data for Piling

Soil analysis data may indicate the necessity for piling. When piling is necessary decisions based on soil analysis data can be used to indicate the type of pile, maximum load per pile, minimum piling spacing, and pile length. The exact type of piling to be used depends upon the depth to firm strata, whether or not the pile will support its load by skin friction or point bearing, the type of soil through which pile is to be driven, and the possibility of failure caused by soil or ground water conditions.

If the soil characteristics are extremely poor necessitating sheet piling, load bearing piling, and water removal equipment, it is usually more economical to subcontract all foundation work, including subsurface investigation. These operations require specialized skill as well as special equipment which is costly to rent.

Foundation problems are usually different for each locality, varying with the type of plant, geology of the area, and load concentrations in the plant. Complete studies and analysis of the subsurface stratigraphy must be made by competent specialists in the field. For this work experienced local firms are preferred since local experience is an important factor in developing reliable data and conclusions.

## Location of Boundary Lines

The first field operation in plant layout is the exact location of property or boundary lines, references, and elevations. These may be located from monuments on adjacent properties, local highways, railroads, and other permanent measurable landmarks, identifiable in the field, and legally recorded. The prime contractor may be requested to perform this work for the owner or the owner may retain a local civil engineering firm to check records, and identify the boundaries. At the same time that this work is being done, ground elevations should be determined and contour maps prepared.

Construction work being performed in an existing plant will not require such extensive checking, but reference points for elevations and definite boundaries for the area in which the work is to be performed should be shown on drawings (usually a plot plan) and recorded in field books. The boundaries of work, or area limits in an operating plant must be clearly defined, since under certain conditions labor jurisdictional disputes may arise between the union representing the owner's personnel and that representing the contractor's personnel.

In order to locate the various facilities in the plant, the master plot plan (see Chapter 6, Fig. 6-1) showing all the areas is usually cross-sectioned by imaginary horizontal and vertical lines, numbered from some arbitrary zero point and located at regular intervals such as 100-, 200-, or 500-ft centers depending upon the size of the project. These lines may be identified as Range and Station or North and East. Major points such as a building corner or a corner of a particular area are thus easily identified by showing the point by coordinates, for instance, 5 plus 58.33

East and 5 plus 25.00 North would indicate a point 558.33′ east of, and 525.00′ north of the Zero corner. Some consistent system such as this should be used for a project since it specifically ties certain points to the original datum.

At least two permanent markers must be installed so that dimensioning may be checked easily. Such markers or monuments should be located out of the work area and should be constructed in such a way that they cannot be easily dislocated. One simple marker consists of a section of six in. steel pipe four to six ft in length, set vertically in a concrete envelope, 18 in. to 24 in. square, with the upper portion projecting 6 in. to 12 in. above grade. The pipe marker is located so that the reference point will fall within the diameter. A piece of 2-in. diameter brass shafting approximately 18 in. long is then located as near as possible to center of the point, with the top projecting several in. above the top of the pipe. The pipe is then filled with concrete. After the concrete has set the exact point can be punch marked on the brass rod with the use of a plumb bob and transit. To make the marker even more permanent, the upper portion of the bar can be threaded and a brass cap provided to cover the top. The cap is very desirable when the marker is also to be used as an elevation reference.

The arbitrary setting of a plant north, is equally as important as elevation and dimensional references. Once the plant north has been set, right-angled arrangements can be used on drawings much more simply than if a true north location is used.

### *Elevations*

After dimensional references are located, plant elevation planes must be set. The civil engineering survey report should include a reference to the nearest U. S. Geodetic (or other reliable) survey datum. This datum can then be used in establishing the elevation of the permanent markers with reference to sea level. For convenience, the plant datum, however, should be established with reference to a low point or to the lowest point on the project, arbitrarily called 100.00. Thus all elevations used in engineering the project will be 100.00 or above. The markers set for dimensional reference can then have an arbitrary plant elevation, which will be plus 100.00 and at some later time may be checked back to the local datum.

Large projects are usually separated into areas. A normal area grade or ground elevation above the plant datum must be set for each area as early as possible so that foundation design and field work such as rough grading and other preliminary construction operations can proceed.

Area elevations, or area grades as they are more commonly known, are best developed from project contour maps. The area grade generally refers to a highpoint in the area since a certain amount of sloping is necessary for surface drainage. After the nominal area grade has been determined, height of equipment above grade can be set and foundation design can be completed.

The elevation of equipment above grade depends on process requirements, maintenance considerations, and possible operating hazards. As a typical illustration, vessels carrying fluids at temperatures and pressures near vaporization must be elevated above pumping equipment to give sufficient liquid head to maintain the fluid in a liquid state. Thus the type fluid, its process state and the characteristics of the pump, all must be considered in determining the proper elevation. All process pumps or other mechanical equipment should be located sufficiently above grade to permit operation, normal inspection, and maintenance from a standing position.

Area grades or equipment elevations cannot be determined arbitrarily, but must be planned with care. While the elevation of foundations above grade is usually set for process or maintenance reasons, the minimum depth below grade of a foundation depends upon the depth of frost penetration or the depth to firm stratum whichever is the larger.

### *Need for Accuracy in Dimensioning*

The large amount of interconnecting prefabricated piping requires more precise field engineering and equipment location for process plant projects than for other types of construction. The draftsman's accuracy in showing correct dimensions on drawings is wasted if the field engineer misaligns or fails to use the same dimensions when locating foundations. Shop-fabricated pipe is usually held to close tolerances ($1/16$ in. for some work). Prefabricated piping may require heat treatment, or may consist of materials which cannot be reworked in the field, thus an error in equipment location of as little as $1/2$ in. can be serious. Errors in anchor-bolt locations on integral foundations for major equipment can be even more serious or costly. Usually anchor-bolts for large equipment are firmly fixed and held in place by metal or wooden templates, which reduce the possibilitiy of error during the pouring of concrete.

Manufacturers sometimes make errors in their preliminary dimensional drawing or have to make changes as fabrication proceeds. For this reason the manufacturer must furnish his final prints with dimensions certified to be correct. Experienced erectors will not release foundation drawings for construction before the manufacturer's final certified prints are received on all equipment involved.

## Excavation

After the site has been cleared and graded, and locations of equipment foundations established, excavation can begin.

Excavation may be done by machine or by hand. Machine excavation is performed by a number of types of equipment the choice of which depends upon the soil and depth of the excavation. Trenches for extensive straight runs of underground piping or conduit are often excavated with trenching machines. The drag-line, which is a self-propelled crane equipped with a bucket, is the most popular machine method of excavation for heavy construction since the crane may also be used for lifting equipment during erection. Bulldozers, which are simply tractors with a blade attached to the front, are extensively used for backfilling and for compacting by means of other attachments. Sizes and types of excavating and backfilling equipment vary with each project.

Machine excavation, being less costly than hand excavation, is used wherever possible. The process plant usually consists of a number of separated but congested areas. It is ordinarily much more economical to machine excavate a total area to the firm stratum, when a number of separate small footings are to be poured, than to hand excavate each footing. Although this may require certain additional forming, the advantage of a better working space for craftsmen reduces form building time and cost of reinforcing steel placement costs. Other underground facilities such as piping and piping thrust anchors, electrical conduit and grounding systems can be installed simultaneously, thus eliminating practically all hand excavation.

Water seepage caused by penetration of the water-table or by opening of water-bearing strata in excavation operations is a common problem in subsurface operations. If the general area has sufficient slope, water may sometimes be sealed off by sheet piling on the upstream side of the excavation.

In any major excavation provisions should be made to pump out seepage water or rain water. One or two low points with pumps are provided.

A thin (2 in. or 3 in.) seal coat of weak concrete over the bottom of the excavated area is often effective in stopping water seepage. This method is regularly used in extensive excavations to improve working conditions and to avoid contamination of the concrete. Unstable water-bearing soils or sand soils that are subject to caving require extra safety precautions for the protection of both personnel and equipment.

Errors made in excavation depths or the accidental removal of soil around existing foundations require careful refilling. This operation is termed backfilling. Heavy gravel or rubble concrete is sometimes used for this purpose if the depth of backfill is not excessive. If soil conditions are poor and the bearing value is low, excessive weights of gravel and concrete must be avoided. Ordinarily backfilling with weak concrete is the simplest and most reliable practice. Backfilling an area where a floor slab or work slab is to be poured should always be carefully done and each successive layer of soil should be machine tamped. Specifications for backfilling, including the type of permissible backfill material, should be prepared at the beginning of the project and specifically followed.

Backfilling also refers to the filling in of excavations with soil after the foundation has been poured. Low cost grades of sand and gravel are often used for this levelling of excavations. If backfilling materials are subject to movement by ground water, precautions must be taken to see that they are contained, either by the concrete work itself, by sheet piling, or by some impervious soil.

It is poor practice to attempt a saving in concrete by placing a footing on backfilled material or on disturbed soil. Drawings prepared for the installation of foundations should note that "All footings shall be carried to firm and undisturbed soil," and no deviation should be permitted without some important justification.

Great care must be exercised in making excavations near major footings, though proper scheduling of the work will eliminate the need for such excavations.

## Form Building

The various shapes which retain the concrete while it is in the plastic state are called forms. The building of these forms is the most costly operation in concrete work.

Forms may be constructed of wood or metal. Metal forms require less labor than wood forms, but foundation shapes must be designed in accordance with metal form lengths and shapes. Metal forms are sturdy and last for many years with reasonable care. Wooden forms may be used several times, but cleaning concrete from wood forms is costly.* Finishing cement to a smooth appearance is a costly hand treatment and may be almost completely eliminated with metal forms since the form surfaces are flat and smooth.

## Reinforced Concrete

Since concrete is the primary material in reinforced foundation construction, a brief description of its composition and characteristics must be included in any discussion on foundations.

### *Cement and Concrete*

Concrete is composed of cement, water, and aggregates, such as sand and crushed stone or gravel. Cement, as its name indicates, is the agent which solidifies the aggregates into a firm mass. Portland cement, which is used for most construction purposes, is composed of essentially four basic compounds: tri-calcium silicate ($3\ CaO{\cdot}SiO_2$), dicalcium silicate

* Forms can be oiled before using to facilitate concrete removal. Various silicone greases have proved highly successful for this purpose.

($2\,CaO{\cdot}SiO_2$), tri-calcium aluminate ($3\,CaO{\cdot}Al_2O_3$), and ferrites similar to the aluminates.

These materials are produced by calcining the proper combinations of raw materials containing $CaO$, $SiO_2$, $Al_2O_3$, and Fe. Lumps or "clinkers" from the cement kiln are pulverized to form a dry powdery material.

When cement is mixed with the proper fine and coarse aggregates and water, in correct proportions, it hardens into the solid mass known as concrete. The strength and other characteristics of concrete vary, depending upon the type of cement used, the proportions of the mixture, the kind and size of aggregates, the quantity of water used in its preparation, and the curing process after pouring.

While the ultimate strength of concrete may approach 8,000 psi in compression, its strength in tension is approximately one-tenth of this compressive strength. In design, its tensile strength in tension is considered zero. It is for this reason that another material must be added to concrete, to provide strength in tension or bending. Steel wires or rods are embedded in the concrete in definite positions to absorb stresses or reinforce the concrete mass. The design of reinforced concrete, therefore, is primarily determination of the required amounts and locations of reinforcing steel.[4,9]

### *Strength of Concrete*

The physico-chemical reaction between the cement and water requires a certain time to initially set, and further time to harden to its design strength. In order to standardize test procedures, concrete compressive strength used for design is that stress which it will withstand after 28 days. The use of the 28-day time interval is the result of extensive research. Although concrete continues to harden the compressive strength curve flattens appreciably after 28 days for concrete cured in air without extraordinary curing procedures.[5]

Many factors influence concrete strength; water is perhaps the most important. Results from many controlled tests show that the compressive strength varies from a maximum of 7500 psi to a minimum of 2700 psi at 28 days for water-cement ratios of 4 to 8 gal per sack, respectively.

Curing methods also affect strength. Concrete continuously moist-cured for one year shows an increase up to 160% of 28-day strength. The increase is 140% up to 100 days. Air curing (without moisture) produces a strength of about 75% of design at 30 days. Apparently moist curing at any time after a dry-curing period will give a sharp increase in strength.

Curing temperature also influences strength. Seven-day strengths of 44% and 65% of 28-day strength for curing at 33°F and 72°F, respectively, are reported. The Portland Cement Association[5] recommends that tests be run to determine compressive strength curves for 7 days and

28 days with materials to be used. Such curves will indicate the best water-cement ratio and possibly the most effective curing procedures.

The water-cement ratio, as indicated above, should be closely controlled. This ratio is expressed in gal per 94 lb sack of cement and varies according to the required strength, the aggregate size and type, mixing conditions, delivery conditions, and placement conditions. The range is about 4 to 8 gal per sack. Openly stored aggregates must be checked for water content and this water deducted from the water to be used. This value is easily checked by drying a measured quantity of aggregate at 200°F and weighing it. Equal quantities of undried aggregate measured and weighed thereafter will show the approximate water volume per cu ft.

There are a number of high early-strength cements which advance the hardening process. The design or 28-day strength with certain mixes, may be reached 12 and 24 hours after pouring. Since these materials are more expensive than portland cement and since the increased rate of reaction often generates large amounts of heat which may require a more costly curing procedure, they are not commonly used in ordinary construction.

Certain additives are available for increasing the density, water resistance, surface characteristics, and resistance to wear of concrete. The curing procedure of freshly poured or exposed concrete surfaces is also very important to the wear resistance and weathering characteristics of the finished concrete.

### Specifications

Concrete strengths commonly used are 2000, 2500, 3000, and 3750 psi after 28 days. Specification of strength in effect sets the water-cement ratio which must be used since this ratio determines the 28-day strength. Ordinarily suppliers of ready-mix concrete are familiar with this method of specification and will supply concrete to meet or better the specifications.

In order to check the concrete strength, samples are taken from each delivered mix by a testing laboratory in prepared cartons 6 in. diameter by 12 in. These cylinders are preserved under controlled conditions and tested in compression, until failure, after 7-, 14-, and 28-day intervals. For certain types of work it is desirable to specify minimum sacks of cement per cubic yard of concrete, and maximum gallons of water per sack of cement. The slump test which is described in detail in ASTM Designation C 143-39 is often used as a consistency check test while concrete is being poured.

The American Society for Testing Materials has prepared a number of tests for evaluating cement, concrete, and aggregates. These tests are the basis for all carefully prepared concrete specifications. The Portland

Cement Association[5] has issued suggested concrete specifications as well as literature on the proper use of both plain and reinforced concrete.

### *Reinforcing Steel*

The basic strength material used in concrete is steel in the form of rods, bars, and steel wire fabrics.[4] The most common reinforcing material is the round steel rod or bar. Common bar specifications are: for billet bars, ASTM A15-50T; rail bars, ASTM A16-50T; cold drawn steel wire, ASTM A82-34; and welded steel wire fabric, ASTM Al 85-37. The so-called deformed bar has forged or rolled surface extrusions. For dimensions of deformations, see ASTM A305-50T. At present only deformed bars are used and their sizes are specified by number (not size) according to U. S. Department of Commerce Simplified Practice Recommendation R-26. The bar number refers approximately to the number of ⅛ in. diameter increments up to No. 8 bar which is 1 in. Only three square bar sizes are listed in the recommended standard, starting at 1 in., which is No. 9; 1 ⅛ in., No. 10; and 1 ¼ in., No. 11. The smallest bar is a No. 2, ¼ in. round. This bar is furnished plain or undeformed and is sometimes called pencil rod. It is used for miscellaneous functions such as form-ties but is ordinarily not used for reinforcement.

Strength or stress limits for reinforcing steel vary from 14,000 psi to 30,000 psi depending upon the design requirements and the steel used. The usual specification is 18,000 or 20,000 psi. The various strengths may be found in the above named ASTM specifications.

Another common reinforcing material is known as welded wire fabric. This material consists of steel wire welded together to form a screen-like fabric. The spacing and gage must be specified according to commercial practice. A 6 × 6–10/10 mesh indicates gage wire on 6 in. squares.

## Pouring Concrete

After the forms are built and the reinforcing steel is in place, the concrete can be poured to form the foundation. Concrete may be mixed at the point of pour or elsewhere. For a large project a central mixing plant may be provided, where all concrete is mixed, and then conveyed to the various points of use by hand "buggies" or by trucks. Ready-mix concrete is also extensively used when it is available. This type concrete is sold ready-mixed by mixing companies and is delivered by specially designed trucks to the point of pour on the project.

The pouring of concrete is a relatively simple procedure. The plastic mass must be poured or shovelled into the forms. For large pours (several hundred cubic yards) or pours covering a large area metal chutes are usually provided to convey the wet mass to the proper place. Large buckets are sometimes used, holding from one-half to two cubic yards of concrete. The concrete is poured in the bucket which, in turn, is carried by a crane to the necessary point. Chutes must be high enough to permit the concrete to flow with ease. In certain types of work a system of conveying chutes or pipes may be provided.

Once the pouring of concrete for any massive slab is started, such as for a large foundation, it must not be stopped for any length of time. A stoppage would permit partial setting (solidifying). The successive pour would then not bond properly and a cleavage plane might be produced. When very large pours are contemplated arrangements must be made to insure continuity of all operations, mixing, conveying, and placing.

As the concrete is poured into the forms it is continuously prodded by hand with wooden paddles or poles so that it flows into all the contours of the forms. Pneumatic vibrators are also used continuously during the pour.

Water must never be added after the concrete is poured. The degree of fluidity cannot be arbitrarily changed by adding water since this will change the specified water-cement ratio necessary for the required concrete strength. There are, however, several additives sold which will lower the viscosity of the concrete without requiring an increase in the water-cement ratio.

### Curing of Concrete Foundations

In order to develop maximum strength fresh concrete must be protected from loss of water by evaporation and from excessively low or high temperatures. This is especially true in the first 7 to 10 days.

Concrete can be kept moist by sprinkling, covering with water, applying moisture resistant seal coats, and by using moisture-retaining covers such as burlap or cotton cloth.

Air around concrete should be maintained at 70°F or above for at least 3 days. In cold weather it is sometimes necessary to heat the water and aggregates before mixing in order to provide an initial temperature sufficient to promote the hydration reaction at a favorable rate. The fresh concrete as poured, however, should not be above 80°F since the heat of hydration further increases the temperature of the concrete. Excessively high temperatures cause undesirably fast stiffening which may produce cracks in the concrete.

In order to maintain an atmosphere of 70°F during cold weather it is necessary to cover the concrete with tarpaulin. Straw or other material beneath the tarpaulin will increase the insulating barrier. During extremely cold weather heat must be added in the form of live steam within a tarpaulin enclosure. This procedure is costly.

Although the strength of concrete increases with curing temperature up to a certain point (the maximum is somewhere between 100 and

200°F), the cracking tendency also increases. It is, therefore, necessary to guard against excessively high temperatures during hot weather by shading the concrete from the direct rays of the sun and using cooled mixing water.

## The Finished Foundation

After the concrete is cured, the forms are carefully removed to avoid injuring the surfaces. Projections are removed by chipping. Holes may be repaired by chipping out to produce at least a 1 in. deep rectangular area. The chipped area is then filled with successive layers of ½ in.

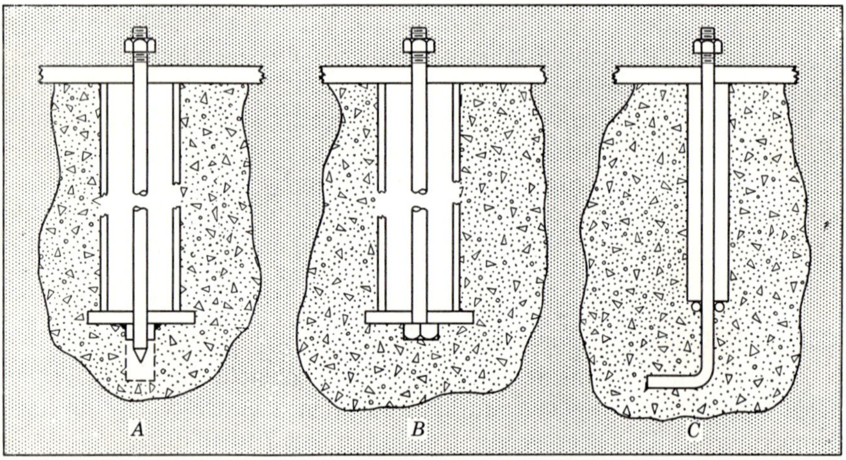


Fig. 22-9. Typical anchor bolts. [Reprinted by permission, Geo. Stevens, *Power, 91,* No. 2, 111 (1947).]

thick concrete. Compressor and pump foundations must be left rough at the top after pouring so that grout will adhere to the foundation when the machine is installed.

During and after the completion of the foundation all anchor bolts should be carefully protected from damage and bending. For large machinery such as compressors, removable bolts (Fig. 22-9*A*) are preferred. Machines can then be skidded into position while the bolts are removed, and broken bolts can be replaced easily.

## Piling

When piles are required they are driven by a hammer supported on vertical leads hung on an ordinary crane or a machine made specifically for pile driving. The hammer is driven by a single-acting or double-acting steam or air reciprocating engine. The single-acting engine lifts the ram (3000 to 5000 lb) and allows it to fall by gravity onto the pile from a distance of about 3 ft. The double-acting hammer aids the force of gravity by supplying energy on the downward stroke. Drop hammers are often used for driving light piles. These are simple devices employing winches and hoists to lift a ram and then permit it to fall again.

The driving of piles is judged by skilled operators on the basis of the inches of penetration for each blow. They are usually driven to a predetermined resistance measured as the number of blows required for the final inch of penetration. The last blow in turn is related empirically to the safe load that the pile may bear.

The penetration of piles through dense strata can be aided greatly by the loosening action of a strong stream of water near the tip of the pile. This process is called jetting.

## FOUNDATION PLANNING AND ARRANGEMENT

### General Practices

The layout and preliminary design of foundations starts with the arrangement of the major process equipment. Pumps, compressors, vessels, heat exchangers, or other equipment located essentially at grade, are set on separate raised concrete sections. The base of the equipment or top of concrete is always elevated at least six inches above nominal area grade. In general, all major equipment whether housed in an enclosure or located in the open, has a separate or independent foundation. An independent foundation is one that is unattached to the floor or work area slab and extends to the depth required to transmit the load to a firm stratum when using spread foundations, or to a pile-cap in the case of piled foundations. Steel column bases of structures or structural steel supports for equipment (Figs. 22-1, 4 and 5) are set on reinforced concrete pedestals, extending above area grade and provided with anchor bolts cast in place. The steel structure like other equipment must be securely fixed in position. The separate concrete pedestals supporting the columns extend to independent spread footings or to a pile cap.

### Equipment Drawings

Since equipment varies in dimensions, and foundations cannot be designed until dimensional data are known, the various manufacturers prepare drawings showing the dimensions of their equipment necessary for installation. The manufacturer also utilizes these drawings for securing approvals from the purchaser. Thus a series of transmittals takes place as described in Chapter 9. Depending upon the type of equipment, several months may elapse from the time of order placement and receipt of the manufacturer's data or vendor prints.

Vendor prints are the only reliable source of dimensional data and accurate dimensions are necessary for the preparation of all installation drawings including foundation drawings. Manufacturers must, therefore, certify the correctness of the dimensions on the drawings they furnish. These prints are known as Certified Vendor Prints and guarantee in effect that the equipment will be furnished in accordance with the dimension shown on the drawing. Although the lower portion of foundations constituting massive pads may be poured, the upper or final portions, which contain the anchor bolts and conform to the shape of the equipment cannot be poured until this certified dimensional data is received. The lack of vendor prints can delay construction as seriously as the delay in delivery of the equipment itself.

## Economics

As in most design and planning for plant construction, economics is important. The following cost factors must be considered.

### *Concrete Costs*

Concrete in its mixed state, ready for pouring or placement in a form, is a relatively cheap material costing from $5.00 to $15.00 per cu yd. These costs, up to the point of pouring, depend on: (1) quantity; (2) method of mixing; and (3) mix specification, in that order. These are the costs only through the mixing phase. Final cost of concrete work, however, is considerably higher, ranging from $40.00 for simple pours to $200.00 per cu yd for complex pours. This increase is due to: (4) material and labor for form building; (5) material and labor for bending and placement of reinforcing steel; (6) labor for placement in the forms, including vibration; (7) labor and material for curing procedure; (8) labor for form stripping; (9) labor and material for cement finishing. In items 4 through 9, labor is the governing factor varying from 50% to 80% of the total costs. The labor costs of form building will usually exceed all others.

Cost of concrete foundations seldom exceeds $65 per cu yd based on 1954 labor costs. Approximately 60% is for labor. This figure should not be used indiscriminately for all foundation work. The cost per cu yd for large mats consisting of many hundreds of cu yd of concrete with little or no forming will be much below $65. Other factors affecting concrete labor costs are labor productivity in the area, climatic conditions, working conditions, and type of work.

### *Ready-Mixed versus Site-Mixed Concrete*

Certain other basic costs, such as the source of concrete materials, sometimes seriously affect foundation and other concrete material costs. Concrete may be mixed at the jobsite or may be purchased ready-mixed. In nearly all domestic industrial areas ready-mixed concrete is delivered to the point of pour at a fixed price. Generally the delivered price of ready-mixed concrete will be fairly close to the cost of plant-mixed concrete when the requirements do not exceed 4000 or 5000 cu yd. In comparing the two the cost of plant-mixed concrete should include costs of the mixing equipment on a rental basis, material handling at the mixer, and concrete transportation over the jobsite. If concrete is to be mixed at the site, provisions must be made for dry storage of cement, aggregate storage, and a water supply. In cold climates provisions must be made for heating water and usually aggregate. Thus if the amount of concrete required is not excessive, it is usually more economical to purchase concrete mixed and have it delivered to the forms.

#### *Forms and Dimensioning*

Since the major costs in foundation construction are labor costs for form building, simplification of concrete shapes and thus form work should reduce costs. Thus all possible changes in surface which might be made to effect small economies in concrete volume should be avoided. Flat panels are less expensive to form than protrusions, offsets, or any change in form surface.

The use of metal forms for all flat surfaces conserves both form material and labor since these forms may be erected in about ⅓ the time required for wooden forms and may be re-used as soon as removed. Furthermore, the concrete surface will require little or no finishing since metal forms leave a smooth and clean surface. Modular metal forms can be used for practically all types of concrete work, but the dimensions of the foundation must conform to the form sizes available.

Costs will be reduced if dimensions of foundations are based on standard lumber sizes or standard metal forms. Fractions of inches should be avoided unless required for anchor bolt dimensions.

#### *Standardization*

Further economies can be obtained by standardizing certain types of foundations. Most projects require a number of square pedestals for supporting structural steel columns used for pipe supports. By setting one standard size for these pedestals and standard anchor bolt sizes based on the severest service, costs of design, drafting, field engineering, and form building are all greatly reduced.

## FOUNDATION DESIGN

### Design Basis and Soil Mechanics

A foundation consists of the concrete mass, the reinforcing rod, and the surrounding soil. Since the man-made portion of the foundation must be supported and surrounded by the subsurface soil, it is impossible to select a foundation type and to design it for a given application without careful consideration of soil mechanics.

Because of the large variety of subsurface conditions peculiar to each area, a designer must rely on judgment based on experience in that area. For this reason it has become common practice to engage local foundation engineering consultants to aid in the soil analysis and testing program as well as in decisions necessary for foundation selection and design.

Foundation design is empirical in many respects. This is particularly true if the foundation must rest on clay, sand, or silt. Load tests made at the jobsite are necessary for any such soil with questionable supporting capacity. These tests when properly made subject the subsurface strata to conditions similar to those anticipated for the plant equipment.

Load tests are made by applying weights (sand, rock, steel) to a platform resting on test piles or bearing plates, if piling is not contemplated. The bearing plates should rest on the soil at the same level planned for the footings. As the load is applied, settlement observed for each load increment is recorded. These tests may at times prove costly, but nevertheless should be made when recommended by competent soil experts. Load test information together with soil analysis data provide the only sound basis for design in questionable soils.

### Settlement and Foundation Failure

Two important factors must be considered in the design of each foundation. First a foundation must be designed so that the soil or rock upon which it is resting supports both the load imposed by the foundation and the equipment being carried by it. Wind loads and other external forces must be considered also. The single most important factor in this phase of design is the bearing capacity of the soil or rock. If the bearing capacity is exceeded the soil or rock will give way and the foundation will collapse into the ground.

It is possible, however, to construct a foundation that will not exceed the bearing capacity of a certain soil, but that will settle into the ground. If this settlement becomes excessive it can cause cracking of the foundation, particularly if the settlement is uneven, and produces undue strain on structures which are supported by the foundation. Differential settlement between multiple footings supporting a single structure can cause difficulty since concrete columns begin to crack when the differential settlement exceeds ¾ in. Steel and brick can withstand greater amounts, but all are adversely affected.

### Design Procedures

No attempt will be made here to outline or recommend design techniques because foundations, due to the empiricism involved, must be designed and selected by experts in that field.

The following brief description of design considerations for various types of process plant foundations will provide a superficial knowledge which, when properly used, will aid the project engineer in understanding the problems of the foundation engineer.

### Soils and Design

As already emphasized, the character of the underlying soil or rock must be an important factor in foundation design. Each type requires a different basis for design. It will usually be found, for instance, that the soil pressure permitted for maximum allowable settlement for sand is less than required for bearing capacity failure.* Hence, the allowable settlement usually governs the design of foundations in sand. The width of a spread footing or a mat located in sand influences the allowable bearing capacity of the soil (the wider the footings the greater the capacity).

Clays, on the other hand, do not exhibit such increases in allowable or safe capacity with width. Further, it cannot be stated definitely whether settlement or bearing capacity will control foundation design for clays. Both must be checked. After the foundation is designed for a clay soil, a prediction of the settlement based on laboratory consolidation tests must be made.

In clays and silts, the only reliable method for determining the safe load on piles is by means of pile load tests. Empirical safe-load pile formulas should be used only when previous experience with the same soil conditions has indicated reliability of the formulas.

### Design Calculations

Once the allowable safe soil pressure or allowable bearing capacity is determined by a combination of soil tests and judgment, actual design of the concrete portion of the foundation can proceed.

In industrial practice standard computation procedure forms are usually prepared for specific designs. The arrangement of these forms follows the logical steps in the computation. Since foundations are concerned primarily with weight, all weights are tabulated. The kip or 1000 lb unit is employed by most designers.

Although standard forms are used by many organizations the procedures vary widely. Many of these techniques have been published and those common to process plant practice are included in the references.[2,4,6,7,8,9,10]

### Spread Footings

A simple calculation yields the area required for a concentrically loaded spread footing, when the depth of placement of the footing and the safe soil pressure for settlement or the allowable bearing capacity are given.

* The usual safety factor for bearing capacity is 3.0.

The area is related to the allowable soil pressure or bearing capacity, whichever controls, as follows

$$S_1 = \frac{W}{a} \tag{1}$$

where $S_1$ = soil loading psf
$W$ = weight of structure and foundation
$a$ = area of base of foundation

Although the area can be calculated or the soil loading checked for a given area, a number of judgments and assumptions must be made to perform even this simple calculation. First, the allowable bearing capacity or permissible soil pressure for settlement must be determined on the basis of soil studies. Second, the depth of the foundation must be based largely on judgment. Of course, it must be below the frost line. Finally, some correction to the assumption that the foundation is concentrically loaded must usually be made since most process plant foundations are not so loaded.

Self-supporting towers (Fig. 22-1) such as fractionating towers, and other tall equipment produce lateral forces due to wind or earthquake which cause eccentric loads on foundations. An accurate analysis of these situations is not possible because both experimental and theoretical information is lacking. They must, however, be analyzed as well as possible since such foundations are common.

V. O. Marshall*,[8] has written a widely accepted procedure for the design of these footings. The design is necessarily trial and error. A foundation must be assumed. Then the soil loading and stability are checked. The maximum soil loading imposed by the tower is a combination of the dead load (weight of tower, appurtenances, and foundation divided by the area of the base of foundation) and the overturning load produced by wind or other lateral forces. This combination will be highest on the leeward side (side farthest from the wind) of the foundation. The overturning load can be calculated for a cylindrical tower, as follows:

$$S_2 = \frac{M_f}{Z} \tag{2}$$

where $S_2$ = unit soil loading due to overturning psf
$Z$ = section modulus for the base of the foundation
$0.1016d^3$ for octagon
$0.118d^3$ for square
$0.098d^3$ for circle
$d$ = shortest distance between edges, ft

* The material in this section is based in part on the article by V. O. Marshall.

$M_f$ = overturning moment about the base of the foundation. (ft-lb) = $P_w L$
$P_w = P_c D_0 H$*
$P_c$ = wind pressure psf for round towers
$P_c = 0.0025V^2$
$V$ = wind velocity in mph
$D_0$ = outside diameter of tower including insulation, ft
$H$ = height of tower, ft
$L$ = lever arm in ft = height of foundation in ft plus $H/2$

The total soil loading is

$$S = S_1 + S_2 \tag{3}$$

In addition to this loading the stability of the tower must be checked. The sum of the stresses at any point on the foundation must be positive or the structure will be subject to overturning. It should be realized that $S_2$ is positive at point $F$ in Fig. 22-1 and numerically the same but negative at point $E$ (positive being considered in the downward direction). The minimum soil load at point $E$, therefore, is the difference between $S_1$ and $S_2$. The conditions for greatest instability always occur at time of erection of the tower before the additional weight of internals, ladders, and piping increase the dead load enough to counteract the negative load on the windward side. In determining the degree of stability $S_1$, based on the weight of tower without internals plus weight of foundation, should be used.

The following example will serve to illustrate the calculation procedure discussed thus far.

EXAMPLE (Reproduced by permission; Marshall, V. O., *Petroleum Refiner, 22*, No. 8, 101 (1943).

Design the concrete foundation for a tower 4 ft diam by 54 ft high, including a 4 ft skirt, and weighing 30,000 lb empty. The insulation, platforms, and piping weigh 9000 lb, the maximum wind velocity is 100 mph, and the frost line at the location of the proposed installation is 4 ft below grade. The maximum safe soil loading is 2000 psf.

SOLUTION: Since the frost line is 4 ft below grade, the foundation will be 6 ft deep, with the top 1 ft above grade, making the bottom of the foundation 5 ft below grade, or 1 ft below the frost line.

The foundation will be octagon shaped, which is recommended for such cases, as it combines the features of stability, ease of construction, and minimum material better than other shapes. The top course will have a short diam of 6 ft since the tower is 4 ft diam and allowance must be made for foundation bolts, etc. The short diam of the base will be assumed to be 13.5 ft. The thickness of the base will depend on the bending and shearing forces; however

* Some designers include an additional multiplier to allow for wind forces on ladders, platforms, etc.

that phase of the design will not be discussed, and the thickness will be assumed to be 2 ft.

The weight of the foundation and tower will be calculated as follows:

| | |
|---|---|
| Area of 6 ft octagon | $= 0.828d^2 = 0.828 \times 6^2 = 29.8$ sq ft |
| Volume of top course | $= 4$ ft $\times 29.8 = 119.2$ cu ft |
| Area of base (octagon) $(a)$ | $= 0.828 \times 13.5^2 = 151$ sq ft |
| Volume of base | $= 2$ ft $\times 151 = 302$ cu ft |
| Total volume | $= 119.2 + 302 = 421.2$ cu ft |
| Weight of concrete | $= 421.2$ cu ft $\times 150$ lb/cu ft $= 63{,}000$ lb |

Volume of earth fill

$$(4 \text{ ft} - 1 \text{ ft}) \times (151 \text{ sq ft} - 29.8 \text{ sq ft}) = 363 \text{ cu ft}$$

Weight of earth fill $= 363$ cu ft $\times 90$ lb/cu ft $= 32{,}700$ lb
Weight of empty tower $= 30{,}000$ lb

$$^*W_t = 30{,}000 + 63{,}000 + 32{,}700 = 125{,}700 \text{ lb}$$

$^*W_a$ will be as follows:

| | |
|---|---|
| Insulation, platforms, piping, etc. | = 9,000 lb |
| Water required to fill the tower (4 ft diam)(50 ft high) | = 39,500 lb |
| Total $(W_a)$ | = 48,500 lb |

$$W = 125{,}700 + 48{,}500 = 174{,}200 \text{ lb}$$

$a = 151$ sq ft, and from eq. 1

$$S_1 = \frac{174{,}200 \text{ lb}}{151 \text{ sq ft}} = 1155 \text{ psf} = \text{maximum dead load on soil}$$

Allowing 3 in. for the thickness of the insulation, the effective diameter of the tower exposed to the action of the wind is 4.5 ft. A wind velocity of 100 mph is equal to 25 psf of projected area $[P_c = 0.0025(100)^2]$.

Therefore

$$P_c = 25 \text{ psf}$$
$$D_0 = 4.5 \text{ ft}$$
$$H = 54 \text{ ft}$$
$$P_w = 25 \times 4.5 \times 54 = 6080 \#$$
$$\text{Height of foundation} = 6 \text{ ft}$$

$$L = 6 + \frac{54}{2} = 33 \text{ ft}$$

$$M_f = 6080 \times 33 = 200{,}000 \text{ ft-\#}$$
$$Z = 0.1016d^3 = 0.1016 \times 13.5^3 = 248.5$$

From eq. 2

$$S_2 = \frac{200{,}000 \text{ ft-\#}}{248.5} = 803 \text{ psf}$$

= maximum soil load due to overturning moment.

The total maximum soil load (toe pressure) can be calculated from eq. 3, as follows:

$$S = 1155 + 803 = 1958 \text{ psf}$$

* $W_t$ = weight of tower and foundation.
$W_a$ = weight of appurtenances and water necessary to fill tower for pressure test.

This loading is satisfactory, as the soil will safely support 2000 psf.

The minimum soil loading due to dead load before internals, ladders, piping, and liquid are added is:

$$\frac{125{,}700 \text{ lb}}{151} = 830 \text{ psf}$$

This is the dead load under the worst stability condition, and since it is greater than the overturning stress ($S_2 = 803$), the soil below the foundation will always be under compression at all points, thus indicating that the foundation is stable.*

Usually it is found that the first assumption of foundation size is not correct, in which case another assumption is made, and the calculations repeated.

The calculations shown in the above example do not complete the design of the foundation. Next, the various stresses in the foundation must be calculated to be sure that they are below the allowable. Permissible stresses for design are recorded.[4,8,9] These stresses include diagonal tension produced by the upward action of the soil. This is an eccentric reaction which tends to break the foundation along a diagonal starting at the foot of the pedestal and on the leeward side. The foundation must also resist the tendency to shear along the bottom leeward edge of the pedestal. The shearing load is produced by the dead load and the wind load and in its calculation allowance must be made for the load-carrying capacity of the soil.

Other stresses include the bending produced by the upward reaction of the soil which must be resisted by the reinforcing steel, bond stress between concrete and steel (must not be excessive or steel will not act with full strength), and bearing stresses at the base of tower. These and many other considerations must be carefully analyzed before a foundation can be completely designed. Much judgment and many assumptions are necessary. Only experienced designers should attempt the design of the more complex foundations.

### *Mats*

The design of mats is influenced by the type of soil. Because of the large area of a mat the soil loading in sands is usually considerably below the allowable bearing capacity. Soft materials such as clays, however, have low bearing values. In such cases excavation to a greater depth will uncover strata capable of greater load bearing capacity.

The eccentric loadings produced by wind on the equipment ordinarily supported by large mats are never as great as for unsupported towers. But these loadings and the stresses produced thereby should be checked. More commonly, however, the eccentric loadings caused by differential settlement of large mats which support tanks can produce stresses sufficient to crack the foundation.

* $S_1 - S_2 = +27$ #psf minimum soil loading. Positive sign indicates compression.

Tanks being supported by ring walls filled with sand must be carefully designed. The ring wall must be so proportioned that the soil pressures under the tank bottom are equal at the base of the wall. Because of the tremendous weights carried in storage tanks it is essential that the characteristics of the soil under the entire tank area be known accurately. Only then can supports be designed which will not undergo differential settling.

### Massive Foundations

No satisfactory method has been developed for designing massive foundations to resist kinetic reaction of machinery. Equipment manufacturers are prepared to provide values for the unbalanced forces and vibration produced by their machinery. Experience[10] has shown that the total mass of equipment plus concrete foundation should be equal to 15 to 20 times the disturbing force for reciprocating equipment. Specific recommendations for each type of equipment should be obtained.

Heavy foundations must, of course, be checked carefully for soil loading. Piling may be necessary for massive foundations more often than for other types.

### Piles

The stresses in the pile cap must be determined in the same manner as for spread footings. The allowable load on each pile can be determined by an empirical formula, as mentioned above, but load tests are preferable if doubt exists.

### Smaller Equipment

Foundations for smaller equipment such as pumps, filters, and dryers are often poured in the form of rectangular solids with straight sides and no offsets. The periphery of the support must conform to the periphery of the equipment and the foundation base must be extended to firm stratum. Small process equipment usually has a low mass–area ratio and thus unit soil loading is invariably below the bearing capacity because of the large foundation area. Reinforcing steel is rarely required in these installations unless it is needed to insure secure attachment of consecutive concrete pours.

It would, of course, be possible to undercut the base of the foundation since the large area is not needed for support. In such cases, however, tension is created in the foundation and reinforcing steel is required. The added cost of the steel and additional form work usually offsets any saving in concrete.

Numerous excellent articles and handbooks have been published[1,2,4,6,7,8,9,10] in addition to standard texts on foundation design which outline design procedures. The clarity and logical presentations in these references, however, may cause the novice to consider foundation design simple. As in any specialty field, handbooks, codes, and papers have been written to reduce the routine of design. Implicit in all of these presentations, however, is the assumption that an experienced designer or a beginner under the supervision of an experienced man will use them with the judgment necessary for safe design.

A plant is certainly no better than its physical foundation. Ill-considered "savings" on foundations can cause much needless expense and unsatisfactory plant operation.

## REFERENCES

1. *A.C.I. Building Code* (ACI 318–51), American Concrete Institute, Detroit, 1951.

2. Boberg, Irving, *Petroleum Refiner, 30,* No. 4, 125 (1951).

3. *Concrete Piles,* Portland Cement Association, Chicago, 1951.

4. *C.R.S.I. Design Handbook,* 1st Ed., Concrete Reinforcing Steel Institute, Chicago, 1952.

5. *Design and Control of Concrete Mixtures,* 10th Ed., Portland Cement Association, Chicago, 1952.

6. Fork, C. H., *Petroleum Refiner, 30,* No. 3, 81 (1951).

7. Fork, C. H., *Petroleum Refiner, 31,* No. 11, 145 (1952).

8. Marshall, V. O., *Petroleum Refiner, 22,* No. 8, 101 (1943).

9. *Reinforced Concrete Design Handbook,* American Concrete Institute, Detroit, 1948.

10. Steven, George, *Power, 91,* No. 2, 111 (1947).

11. *Subsoil Investigations for Foundations,* Cat. B4, Raymond Concrete Pile Co., New York, 1953.

# CHAPTER 23

# STRUCTURES AND BUILDINGS

The modern process plant is located almost entirely out of doors and contrasts sharply with the totally enclosed plant of a decade ago. Now only instrument control centers, critical machinery, and operations requiring controlled atmospheres are enclosed by buildings. Automatic control instruments, and weather-proof and explosion-proof motors have enabled this striking change to take place. Today the most imposing structures in a process plant are the large unenclosed structural steel skeletons which are used to support equipment that is operated from the comfort of an air-conditioned control room by means of automatic instrumentation.

There are three types of structures for process plants: (*a*) the totally enclosed structure, the conventional industrial or commercial building; (*b*) the partially enclosed structure which is a simple type of roofed shelter; and (*c*) the unenclosed structure which is an unsheltered structural support.

## TOTALLY ENCLOSED STRUCTURES — BUILDINGS

The types of buildings usually found in process plants are relatively simple ones, similar to the mill buildings used in heavy industrial plants, with the exception that process plant buildings are generally of much lighter construction.

Single-story structures in the colder climates of the United States are usually masonry structures with flat roofs which are supported by the walls and partitions. When several floors are required, the structure framing may be either steel or reinforced concrete with the roof and exterior walls above the first floor supported from this framing. The lightest possible materials are used for exterior walls in order to reduce the dead load, and insulation is used to provide weather protection instead of many layers of brick and masonry.

In milder climates, simple steel-framed structures are used for housing mechanical equipment. Metallic or corrugated asbestos-cement siding and gabled roofs covered with the same materials are generally satisfactory for single-story structures. Flat roofs are more common for multistory buildings.

Some of the more common buildings in process plants include buildings for housing process equipment, office buildings, change houses, cafeteria and recreation center, medical facilities, laboratories, and shops and warehouses.

### Buildings for Process Equipment

In cold climates some companies have found it advantageous to locate process units in buildings, but the necessity for enclosures in any climate is rapidly diminishing. Some pieces of equipment, however, must be enclosed regardless of climatic conditions. These include: (1) certain heavy mechanical apparatus such as electrical power generating machinery, (2) small batch or kettle processes, (3) processes or mechanical equipment handling solid materials which must be protected from the weather, (4) packaging equipment, (5) certain types of mechanical filters, (6) major or master control panels; and (7) any equipment or process which requires constant operating attention.

In warm climates buildings for some of these items will be open-sided at points where wind will not interfere with the process. Those buildings which are completely enclosed such as control rooms will be air-conditioned.

### Office Buildings

In designing office buildings the most effective studies can be made by making cut-outs of desks, chairs, file cabinets, and office machines. Various arrangements of these cut-outs can be made on a plan view of the proposed office space. Heat and ventilation, lighting, traffic, and office equipment requirements can also be studied in this manner.

Good office equipment arranged in such a way that it causes inefficiency or constant annoyance to the staff is of little value. If, for example, several persons must work together in a room, the seating arrangement should provide as much privacy as possible. Two persons should never be seated so they must face each other. Regardless of personality, such arrangements lead to problems. In general, separation of each individual even if only by glass semi-partitions is better than group arrangements.

The proper floor upon which to locate a given group must be given careful consideration. Assume that an engineering department and a purchasing department are to be housed in a two-storied structure. Location of these two departments in the same building is functional since in any organization where engineering and purchasing is performed a considerable amount of collaboration will be necessary. The next question is which department should be located on the ground floor. Purchasing department personnel are usually more active in dealing with outside firms than any other department except sales, and a steady flow of representatives from other firms can always be anticipated. It follows then that the purchasing department should be on the ground floor to facilitate the reception of salesmen and visitors.

### Change Houses

The design of employee change houses is relatively simple. A count must be made of employees under various work classifications, and personnel in clean types of jobs separated from heavy maintenance workers. Foremen and supervisors are also often provided with separate space.

Generally each employee is provided with two lockers, one for work clothes and one for ordinary clothing; however, many different arrangements are made depending upon company policies.

The problem of ventilation is best solved by using single rather than back-to-back arrangement of lockers. This permits the lockers to air from both back and front although requiring more floor space.

Ceiling heights for change houses should not be less than 12 ft to permit better natural air circulation. Shower rooms should be provided with doors and completely partitioned from other parts of the building to reduce humidity as much as possible.

Lighting may be kept to 25 foot-candles for extreme economy. Careful design of circuits, however, can provide better lighting at little additional cost.

In larger plants there are free laundry facilities for outer clothing (coveralls etc.), and these are operated by the locker room attendant. The existence of such facilities removes any excuse for continued uncleanliness.

Heating may be by forced hot-air circulation in the colder climates or by ceiling and wall radiators or unit heaters. Generally it is desirable to provide some means of forced air circulation for year-round ventilation in all climates.

### Cafeteria and Recreation Center

Cafeterias for industrial plants may be equipped with complete kitchen facilities or may be designed for short-order cooking with complete dinners furnished by catering services. Hotel equipment supply houses are helpful in the planning of complete kitchen facilities and will usually prepare a list of recommended equipment.

Adequate storage space should be provided for canned goods, bulk goods, and vegetables, and refrigeration (walk-in) facilities must be built for meat storage. The cafeteria counter is usually designed as a unit to suit the specific location and service. Traffic must be studied to avoid congestion. The total space required for tables can be estimated by providing from 10 to 15 sq ft per person; however, the best procedure is to lay out four-person tables. If larger tables are ever substituted, more persons may be accommodated.

Costs for recreation facilities may be reduced by combining the recreation area with the plant cafeteria. This arrangement offers many possibilities. It provides more space if large groups of personnel or visitors are to be entertained and it enables better service if food is to be provided for these occasions. Movable partitions which fold into the wall can be used for isolation of certain areas. One part of the combination room can become a library or reading room equipped with shelves and library tables. Wall space of another part may be arranged for moving picture projection. Folding chairs can be provided and stored away when not in use.

Lavatory facilities should be provided adjacent to the kitchen for cafeteria help and larger facilities (5 to 10 persons each male and female) located conveniently for personnel being served.

The design of heating, ventilating, and air-conditioning services will depend upon the company's policies. If an air-conditioning system is properly designed and the structure has been adequately insulated the operating costs will be relatively low. With proper zone control whole areas may be isolated, thereby reducing the load. In warm and humid climates air-conditioning is a necessity.

The combination cafeteria and recreation center represents the minimum in facilities for worker convenience in the process industries. Many plants provide country-club style recreation buildings, swimming pools, libraries, and sports equipment. Each company must decide the extent of recreation facilities necessary for its particular situation. Plants located in large metropolitan areas do not require as much as plants located in remote areas.

### Medical Facilities

Plant size and location are the primary considerations in determining the type of medical facilities needed. Plants located near cities with hospital facilities can use ambulance service, and provision need only be made for temporary bed patients. Waiting rooms, examination rooms, and offices for the physician and nurse are required, however. A combination examination and surgery room is sometimes provided if the nearest hospital is one or two hours away.

Operating rooms require anesthesia facilities, special lighting, hospital-type lavatories, sterile containers, and sterilizers. All equipment must be grounded to prevent explosions. Grounded types of floor tile are available for this purpose.

An analytical medical laboratory is necessary if extensive physical examinations are contemplated. Medical supply firms will recommend the equipment necessary for routine examinations.

## Process Plant Laboratories

Laboratories for the process plant may be designed as Control Laboratories, General Plant Laboratories, and Research Laboratories.

### *Control Laboratories*

In many plants simple control tests may be performed in the operating area. Usually a sink with water, air, and other liquid connections will suffice. The extent of apparatus provided varies with the function.

### *General Plant Laboratories*

In addition to the actual laboratory work rooms required for General Plant Laboratories, space may be provided for a chief chemist, assistant chemists, shift chemists, library, file room, supply and store rooms, and locker and shower rooms. Isolated and air-conditioned space is sometimes necessary for analytical equipment such as spectrophotometers. In some cases lead shielding may be needed for floors, walls, doors, and ceilings of rooms housing radioactive materials.

Arrangements for the general analytical laboratory equipment follow conventional patterns. A number of firms specialize in custom-built laboratory benches and related equipment, and it is more economical to purchase equipment from them. The furniture, though standard in design, is unitized and can be arranged to suit almost any situation.

Proper ventilation for the laboratory is usually a problem which requires considerable study. All office, technical, and clerical personnel and facilities not related to actual analytical work should be located together in one end of the building or on a separate floor. If a choice of floors is possible all such personnel and facilities should be located on the ground floor.

If the laboratory is air-conditioned the two general areas, office and analytical, should be on separate recirculation systems. An alternate arrangement could be recirculation in the office area and venting or 100% make-up in the analytical area. This, of course, wastes cooled air. The amount of air circulation can be reduced if operations evolving noxious or disagreeable fumes are conducted in hoods.

Lighting for the analytical areas of the laboratory should be of high foot-candle intensity (75 to 100). Fixture selection depends on the types of laboratory work. In more hazardous areas all lighting may have to be vapor- or explosion-proof.

### *Research Laboratories*

Process plant research laboratories must be flexible in design and layout since the operations performed in any one area may change. Research is often conveniently divided into bench scale and pilot plant research. The bench scale operations are best conducted in two-man laboratories complete with hoods and all utilities. Special racks for assembling tall glassware such as laboratory distillation columns are required, and convenient desks and files in the laboratory are necessary for each researcher.

Pilot plant facilities may be located outdoors or inside light prefabricated buildings. Buildings and structures should be constructed with the knowledge that change in such facilities is inevitable. Safety precautions should not be neglected, however, just because the existing equipment is known to be temporary.

In designing research facilities, every attempt should be made to provide working conditions conducive to creative thinking. This implies comfortable, quiet, and pleasing surroundings. Conference rooms and adequate library facilities are essential.

## Shops and Warehouses

Shops and warehouses are no longer ugly, dark, and dirty places in which to work. Modern mechanical shops and warehouses in a process plant are designed to provide the maximum in worker comfort and thus worker efficiency.

Machines are usually painted pastel shades which are easy on the eyes and enable more rapid location of knobs and push-buttons. Better illumination together with well-designed painting schemes assure a minimum of eyestrain.

New shops in warm climates are being furnished with complete air-conditioning even in the areas used for welding. The increased costs are apparently offset by greater worker efficiency.

Anyone involved in the design of shops or warehouses should thoroughly study the needs for the particular plant. If possible, the shop or warehouse foreman should aid in all the planning stages.

### Economics in Building Construction

In planning buildings for process plants, the designer is often confronted with attempts by management to reduce costs. Process equipment costs are usually fixed, and when total estimates must be reduced, it is a common procedure to attempt savings in buildings or building facilities. In general savings effected by reduction of building size, or by limitation of facilities such as lighting, heating, ventilation, and lavatories only result in poor employee morale.

Building costs, however, can be reduced by the use of new and better materials. While the characteristics of the basic or fundamental materials of construction such as concrete, steel, and masonry have not changed appreciably in the last several decades many hundreds of new materials or new uses for existing materials have been developed which reduce labor costs. The cost of exterior walls, for instance, may be greatly reduced by using light metallic panels instead of masonry. Not only is there saving in labor cost but also in time. Light panels with fixed insulation may be erected in days while it takes weeks or months for a masonry wall. Many similar economies can be effected by alert engineering organizations.

## PARTIALLY ENCLOSED STRUCTURES — SHELTERS

Some process equipment, such as large process pumps and compressors, need not be enclosed in a building but do require protection from rainfall and snow so that maintenance of the equipment may be done in a sheltered area.

The various parts of compressors must be handled by mechanical means when disassembled because of their great weight. The sheltering structure, therefore, must provide support for overhead cranes or monorails. Compressor piping is also large and must be supported. Although compressor shelters are not necessarily designed to withstand horizontal forces from piping, it is common practice to design the major structural members to carry the dead loads of connecting piping.

Other shelters that are common in process plants include those for loading racks, barrel storage, and salvaged equipment storage.

## UNENCLOSED STRUCTURES — OUTDOOR PROCESS UNITS

In most areas of the U.S., particularly south of latitude 40°, it is common practice to install the major portion of the process or operating equipment without shelter (Fig. 23-1). Shelters of some kind must be provided for operating personnel, but improvements in instrumentation permit all major instruments to be located in a control center. This not only provides a shelter for operators but also enables them to observe the process from a single location.

The location of equipment outdoors eliminates the problem of heating and ventilating, reduces the initial costs by elimination of walls and roofs, and permits simpler and less costly maintenance operations. Outdoor plants, in general, are also safer plants. Accumulations of noxious or

Fig. 23-1. Typical process plant; fluid catalytic cracking unit. (Courtesy of Humble Oil and Refining Company.)

flammable and explosive vapors are rare in outdoor plants. For these reasons, nearly all process plants, particularly where climatic conditions are not severe, are designed for outdoor installation of most process equipment.

Pumps, motors, and other equipment have been designed to operate satisfactorily when exposed to all types of weather. The additional costs for such weatherproofing is negligible in comparison to building costs.

## Types of Equipment Supports

Unenclosed equipment which must be elevated above grade may be supported by either structural steel or reinforced concrete. Structural steel is less expensive particularly for multi-story structures, but reinforced concrete columns have the advantage of requiring little or no maintenance since they are unaffected by many of the corrosive vapors common to process plants. In addition, concrete will not fail when subjected to fire while steel must be protected by the application of a concrete coating. When "fireproofing" is necessary, therefore, reinforced concrete structures may prove the more desirable, especially for single level structures (10 ft to 15 ft above grade).

## General Considerations in Planning Equipment Structures

### *Arrangement*

In addition to supporting elevated equipment, space and access must be provided at the elevated level for operating inspection and for maintenance. Heat-exchangers and other equipment subject to dismantling will require more space than simple chambers or vessels. Any equipment which must be opened or disconnected periodically must be arranged for disassembly in place or for removal and lowering to grade, and equipment at lower levels must be arranged to avoid any interference during this operation. The structure must also be designed for the forces encountered when lowering of equipment is in progress.

Wherever heavy equipment or its parts are dismantled or lifted, hoisting apparatus is provided. Gantry-type cranes are sometimes installed if many pieces of heavy equipment subject to dismantling are grouped together. A monorail with trolley is provided for isolated pieces of equipment. A portable air, electric, or manual hoist can then be attached to the trolley when required for lifting.

Modular or fixed spacing of columns, as used for office or other multistory building structures, is seldom practical for structures designed for the process plant. Instead, the weight and the desired arrangement of equipment should govern the column arrangement. Main structure columns, of course, must be located on common centerlines to simplify design and improve stability. When a number of floors are planned, floor spacing should be made equal if possible to simplify general design and stair arrangements, and to improve stability.

### *Floors*

Concrete slab floors are installed when maintenance operations for elevated equipment are extensive, or when operating personnel must spend considerable time on the operating floor. Steel plate or grating floors are used for areas requiring only occasional attention. Grating is preferred to steel plate in areas in which snow occurs, since snow accumulates on steel plate and creates a hazard. Grating also requires a minimum of cleaning and reduces the weight of the structure, an important factor for tall structures subject to high wind loads.

### *Attachments*

All equipment must be securely bolted to the supporting members. Horizontal cylindrical vessels, drums, and heat-exchangers are usually supported at two points only. These supports are two transverse shoes welded to the equipment. Equipment which changes dimension under operating conditions is arranged so that one end is free to move in a guided direction. Large vertical towers located in a structure or above grade are usually supported at the lower end. Lugs welded to the periphery are used for smaller equipment. Since heat exchangers have removable heads at both ends, they must always be supported along the length of the equipment so that the heads may be removed.

Interconnecting piping for hot equipment must be restrained and/or guided so that stresses caused by changes in temperature, either in the piping or the equipment, are controlled or absorbed by the supporting structure rather than the equipment. Since many of these stresses cannot be calculated before all piping design is completed, it is common practice to overdesign structures for process equipment rather than delay the final design of structural steel. And further, since nearly all processes are subject to modification, it is usually more economical to design major members slightly heavier than required rather than design for the absolute minimum as in commercial building practice.

## BASIS FOR DESIGN OF STEEL STRUCTURES

The detailed design of steel structures must follow the practices outlined by the American Institute of Steel Construction,[8] and reinforced concrete structures those of the American Concrete Institute.[1,7] Either of these codes may be amended according to the requirements for the particular type of plant.

Specifications which must be developed as a basis for design include the following:

I. Structures thirty ft in height or above

1. Weight of equipment.

(*a*) Empty; (*b*) operating or filled with process fluid; (*c*) testing (filled with water or other testing fluid).

2. Wind loads; use maximum anticipated wind and calculate load for each of the three conditions above under item 1.

3. Live loads; loads caused by personnel or by operating equipment; cyclic changes of liquid level of fluids in vessels should be checked if large tonnages are involved. Maintenance operations also must be considered. Structures designed to resist overturning with equipment in place may be inadequate under disassembled or maintenance conditions. In brief, the structure must be self-supporting under all conditions.

II. Structures less than thirty ft in height

Conditions above for I.1 and I.3 apply. Wind loads are usually very small for these lower structures.

III. Allowable or design loads and stresses

| | | Notes |
|---|---|---|
| 1. Wind | 30 psf | Reduce 60% for cylindrical surface |
| 2. Live loads | | |
| *a*. Work areas | 150 psf | |
| *b*. Walkways and platforms | 100 psf | |
| *c*. Cat-walks, foot-bridges | 50 psf | |
| 3. Exchanger bundle pull | 150% of weight of bundle | |
| 4. Crane trolley | 150% of weight of equipment | |
| 5. Roof (bare) | 40 psf minimum | |
| *a*. Roof with equipment | 150 psf minimum | |
| *b*. Snow | 20 psf | Add to snow and wind |
| *c*. Wind | 30 psf for Type I | Reduced by Duchemin's formula |
| | 20 psf for Type II | |
| 6. Structural steel | | According to AISC |
| 7. Reinforced steel (concrete) | | According to ACI |

In addition to these, specifications for the particular installation must include a number of specific requirements similar to those illustrated in the following specifications. All paragraphs should be numbered for later identification or revision. Involved general statements in specifications often lead to misunderstandings which usually result in additional costs. Any statement should be examined to be sure the exact meaning is understood by both fabricator and customer.

IV. Miscellaneous requirements

1. If local building codes require increased design requirements such increased values shall form the basis for design. When local code requirements are less than the requirements herein or would permit lighter designs, the requirements herein shall govern.

2. Clearances.

*a*. Railroad clearances. No structure, nor equipment of any kind shall be permanently located within a rectangle seventeen (17) ft wide by twenty-two (22) ft high, the centerline of which is the centerline of the railroad track, and the height of which rectangle is measured from the top of the railroad rails. (Note: Many states have statutes which fix railroad clearances. These dimensions conform to most requirements.)

*b*. Electrical power line clearances. The minimum overhead clearance for open wiring carried on structural supports or entering buildings, or run adjacent to structures above roadways or railway track rails shall be twenty-eight (28) ft. The minimum horizontal clearance from structures shall be ten (10) ft. Wiring entering buildings or run inside the above limits shall be enclosed in conduit.

*c*. Piping clearances. Piping run overhead on structural supports or otherwise shall clear roadways or railroad rails by not less than twenty-five (25) ft from highest point of road crown or top of railroad rail. Horizontal clearance for railroads shall be a minimum of 8½ ft from the centerline of the railroad tracks. Horizontal clearance for roadways shall be at least equal to the clearances for railroads; however, piping support structures shall not be located closer than four ft to the edge of paving. If roadways are run with graded shoulders or ditches, supporting structures shall be located at least four ft beyond the far slope of such a graded area.

The minimum vertical clearance above grade for piping run in general process areas, except for headers adjacent to equipment shall be 12 ft. The minimum vertical clearance in operating areas shall be 7½ ft above grade.

*d*. Structural clearances. Secondary or bracing structural members shall be arranged to clear process equipment by at least 18 in.

3. Fabrication.

*a*. All shop connections shall be riveted.

*b*. All field connections shall be riveted.

*c*. All erection bolts and rivets shall be furnished to 150% of requirements.

*d*. All steel shall be cleaned free of rust and scale and given one coat of red-lead and oil primer before shipment.

These are only partial specifications for the design and fabrication of structural steel. The Manual of the American Institute of Steel Construction[8] includes the standard specifications used throughout the steel fabricating industry. Standard practices are also included which describe the usual policies followed by the fabricators if specific instructions are not given in the specifications.

## STRUCTURAL SHAPES

The use of steel as a supporting framework for elevated structures is probably familiar to everyone. Many decades ago certain shapes were developed which would give maximum strength with minimum weight. Design formulas were proposed based on these shapes, and tabulations of properties were made so that the structural steel designers' task is much simplified. In general, the common "rolled" structural steel shapes, such as beams, columns, channels, and angles, suffice for nearly all types of heavy structural design.

With the advent of electrical arc welding and its general acceptance in industry, other special and light shapes are built, such as the well-known bar joist (Fig. 23-2) used as a supporting member. Many other types of welded plate or builtup shapes are used in the manufacture of prefabricated, rigid frame steel structures.

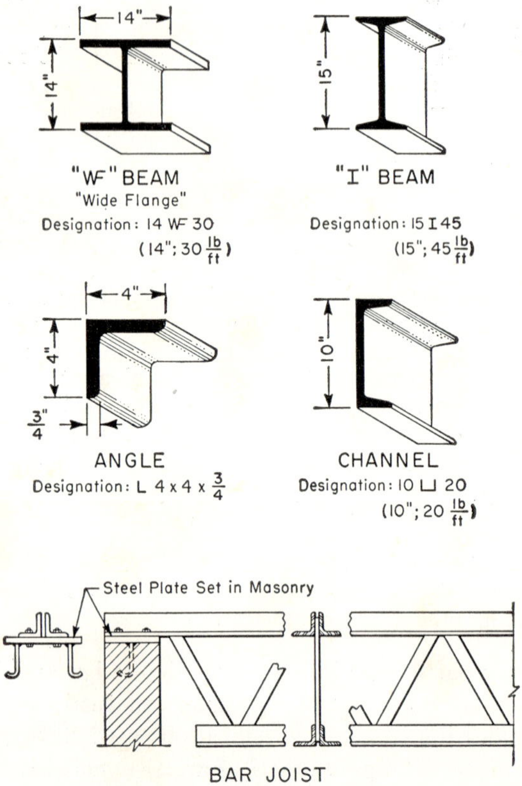


**Fig. 23-2.** Commonly used structural shapes.

Common structural members are shown in Fig. 23-2. Each section is manufactured in a number of different sizes. For instance, "equal-leg" angles may be obtained from 1 in. by 1 in. up to 8 in. by 8 in. in a number of thicknesses totaling 80 different sizes and weights. About 120 different sizes and weights are available in unequal-leg angles. Standard manufacturing practices and the properties of various carbon-steel structural sections are given by the AISC[8] and manufacturer's catalogs.

In addition to carbon-steel as a structural framing material many other materials are available, such as the various stainless steels, aluminum, aluminum alloys, magnesium, magnesium alloys, and the many copper bearing alloys. In most cases either these materials are more costly than carbon steel or the required fabrication processes are more costly.

Alloy and nonferrous materials have been used economically, however, when corrosion or weight problems exist. Aluminum and its alloys have been used to good advantage for exterior panels and window frames, and stainless steel structures are often used in particularly corrosive atmospheres because of the rapid disintegration of other less expensive materials. Stainless steels also have greater strength than carbon steel, and some reduction in weight for a given design is possible.

## REFERENCES


1. *Building Code Requirements for Reinforced Concrete* (ACI-51), American Concrete Institute, Detroit, Mich., 1951.
2. Carrier, W. H., R. E. Cherre, and W. A. Grant, *Modern Air Conditioning, Heat and Ventilating,* 2nd Ed., Pitman Publishing Corp., New York, 1950.
3. Dunham, C. W., *Planning Industrial Structures,* 1st Ed., McGraw-Hill Book Co., New York, 1948.
4. Gay, C. M., and H. Parker, *Materials and Methods of Architectural Construction,* 2nd Ed., John Wiley & Sons, New York, 1943.
5. Ketchum, M. S., *The Design of Steel Mill Buildings,* 5th Ed., McGraw-Hill Book Co., New York, 1932.
6. *Procedure Handbook of Arc Welding, Design and Practice,* 9th Ed., Lincoln Electric Co., Cleveland, Ohio, 1950.
7. Reese, R. C., *CRSI Handbook,* Concrete Reinforcing Steel Institute, 1st Ed., Chicago, Ill., 1952.
8. *Steel Construction Manual,* 5th Ed., American Institute of Steel Construction, New York, 1952.
9. Stoddard, R. P., *Brick Structures,* 11th Ed., McGraw-Hill Book Co., New York, 1946.
10. Sutherland, H., and H. L. Bowman, *Structural Theory,* 4th Ed., John Wiley & Sons, New York, 1950.

CHAPTER 24

# SAFETY IN PLANT DESIGN

Safety must be a paramount consideration in the design of a process plant. The mass of steel that is fashioned into a plant will ultimately be operated by human beings whose health and well-being must be protected.

Disregard of safety in the design of a process plant is comparable to criminal negligence. Each design engineer should feel a responsibility for providing a safe design. He should ask himself: "Would I be willing to operate this equipment as it is now designed?" If the answer is "no" or if the "yes" is weak, the reasons for the uncertainty should be found and corrected. The designer should, whenever possible, discuss his design with operating men and the plant safety engineers. These men are keenly aware of safety and will have many good suggestions.

## HAZARDS IN PROCESS PLANTS

The hazards to be considered in the design of any process plant can be grouped roughly into three major categories: flammable and explosive, hygienic, and mechanical. A majority of chemical processes are potentially dangerous because flammable and explosive materials are produced. The design of equipment for such plants must be dictated by characteristics of the materials being processed. One's memory on the combustion characteristics or explosive limits of materials should never be trusted. Careful investigation of these characteristics of the materials to be handled in the plant should precede all design calculations.

A number of chemical products constitute definite health hazards. Prolonged breathing and/or skin contact over a long period of time of such materials as benzene and carbon tetrachloride can cause permanent impairment of health or even death. It is surprising that many scientifically trained people are unaware of the hazards to health presented by some common chemicals. Although there are excellent references on industrial toxicology which can be used as guides, all but the most common chemicals should be referred to a trained industrial toxicologist for study before handling techniques are devised.

Mechanical hazards include all the hazards created by moving equipment and arrangement of equipment and structures.

These classifications are somewhat arbitrary and overlap, but they serve to emphasize the primary hazards in process plants that create or are capable of creating unsafe working conditions.

Careful design combined with well-planned accident prevention programs have produced amazing safety records. Manufacturers of tetraethyl lead, for instance, were faced with an unusual health problem due to the poisonous nature of the raw materials and product. Realizing this problem even before the design of the plant was begun created an awareness that influenced all design decisions. The result has been an enviable record of safe operation.

## Sources of Information

Many safe practices are no more than application of common sense. The problem is to recognize unsafe situations before they are incorporated in plant design. Basically, this involves determination of the hazards that exist. In addition to many books on safety and industrial hygiene,[1,2,6,11,12,13] there are a number of handbooks, codes, and similar publications by numerous agencies interested in safety and fire prevention which are helpful in defining potential hazards. A selection of these is given below:

I. Fire and explosions

1. National Fire Protection Association; 60 Batterymarch Street, Boston 10, Mass.; a nonprofit technical and educational organization; publishes:

*a*. Standards: widely adopted throughout U.S.A. as basis for construction. Following summaries of standards are available.

(i) National Fire Codes for Flammable Liquids, Gases, Chemicals and Explosives.

(ii) National Fire Codes for the Prevention of Dust Explosions.

(iii) National Fire Codes for Building Construction and Equipment.

(iv) National Fire Codes for Extinguishing and Alarm Equipment.

(v) National Electrical Code.

*b*. Quarterly of the National Fire Protection Association: up-to-date information on fires and fire prevention.

*c*. NFPA Handbook of Fire Protection: a handy reference work.

2. National Board of Fire Underwriters, 85 John Street, New York 7, New York; an educational, factual and engineering organization supported by the capital stock fire insurance companies publishes:

*a.* Standards for fire protection many of which are based on NFPA standards.

*b.* Research reports.

*c.* Special reports on unusual disasters. (Most of this information is sent free of charge upon request on company stationery. Advice and counsel on certain phases of fire protection problems not now covered by codes or standards can be obtained.)

3. Other insurance organizations: The Factory Insurance Associaton, 555 Asylum Street, Hartford 2, Conn.; Associated Factory Mutual Fire Insurance Company, 184 High Street, Boston 10, Mass. These also operate research laboratories and publish standards.

4. Federal Agencies, Washington, D.C.

*a.* 1. U.S. Bureau of Mines—research reports and recommendations, minerals, dusts, and oils.

*b.* U.S. Dept. of Agriculture; dust explosions and fires.

*c.* U.S. Dept of Labor, Division of Labor Standards; safe practices recommended for various industries.

*d.* National Bureau of Standards; fire hazards and fire resistance research.

*e.* U.S. Public Health Service; industrial hygiene publications.

5. National Safety Council, 20 N. Wacker Drive, Chicago, Ill.; mainly safety education. Research work conducted-advice available (nonprofit).

6. Underwriters Laboratories, Inc., 207 E. Ohio Street, Chicago 11, Ill.; (nonprofit). Tests equipment (electrical, fire protection, etc.) to insure safety. Lists of tested and approved equipment issued annually with bi-monthly supplements.

7. Laws; federal, state, and local laws relating to fire protection and buildings. Many refer to various provisions of the NFPA codes.

8. Insurance companies and local and state fire prevention bureaus can provide valuable advice.

9. American Petroleum Institute, 50 W. 50th Street, New York, N.Y.; publishes bulletins on *Fire Protection in Refineries.*

II. Industrial hygiene

1. Manufacturing Chemists' Association; publishes chemical safety data sheet for the safe handling and use of many chemicals.

2. National Safety Council; publishes health and safe practices pamphlets relative to the handling of various chemicals.

3. American Industrial Hygiene Association; publishes technical quarterly on industrial hygiene.

4. U.S. Bureau of Mines, Washington, D.C.; dusts and toxic vapors.

5. Atomic Energy Commission, Washington, D.C.; radioactive materials.

6. Mellon Institute, Industrial Hygiene Foundation, Pittsburgh, Pa.; research.

7. Division of Industrial Hygiene, U.S. Public Health Service, Washington, D.C.; publications of various types and advice.

8. American Medical Association, Committee on Physical Agents of a Hazardous Nature.

9. Chemical companies; publish information on the safe handling of their own products. Producers of tetraethyl-lead, for instance, recommend and furnish complete design for equipment handling the ethyl fluid.

10. Sax, N. I., *Handbook of Dangerous Materials,* Reinhold Publishing

Corp., N.Y. (1951); gives hazardous properties, treatment, and antidotes; also suggestions for safe handling of numerous toxic materials of commerce.

III. Mechanical

1. Equipment manufacturers; will recommend safe installation procedures.
2. U.S. Department of Labor, Division of Labor Standards; publications on safe practices and recommendations for industry.
3. National Safety Council; safe practices, recommendations, presented in pamphlets.
4. Codes for construction of vessels, tanks, piping, electrical equipment:
   *a.* ASME and API-ASME Codes for Unfired Pressure.
   *b.* ASME Boiler Construction Code.
   *c.* American Standards Association Code for Pressure Piping.
   *d.* API Standard No. 12C and 12E for Storage Tanks.
   *e.* National Electrical Code.
5. Liability insurance companies; publish helpful data which aid in the design of safe plants.

## PLANT LAYOUT

The safety record of a plant depends partially upon the arrangement of the various units and the location of equipment within those units. For this reason alone safety considerations should permeate even the early planning stages of plant design.

### Topography and Weather

In selecting a site topography and meteorological phenomena should be carefully considered. Hilly areas present certain inherent hazards. Overflowing tanks or vessel failures can send thousands of gallons of flammable materials into surrounding areas. If the area is hilly, these materials can reach a wider area and if ignited cause a rapid spread of fire.

Prevailing winds should be considered in the layout of a plant so that flammable and toxic gases can be properly dissipated without endangering the lives or comfort of plant personnel or people in the surrounding areas.

The occurrence of such weather phenomena as inversions should be analyzed. The atmospheric temperature normally decreases with altitude. During periods of inversion the reverse is true. In such situations the lower atmosphere remains stagnant, and smoke and noxious vapors hang over the earth's surface and create extreme health hazards. The DeNora, Pennsylvania disaster in which a number of people died was caused by such a condition. Any area that has frequent inversions lasting more than three to four days should be rejected as a possible site for a process plant.

### Division of Plant into Units

The operating units of a process plant should be separated not only for more efficient operation and maintenance but also for safety reasons. By separating the process units it is possible to prevent the spread of fire and explosion.

Dividing a plant into city-type blocks has many advantages. The blocks are usually separated by roadways making access to all parts of the plant convenient. These roadways may vary from 50 to 100 ft in width. Dead-end roads should be avoided and convenient access from any part of the plant should be possible in at least two different directions.

Although roads provide a definite service to the plant, they also can constitute a safety hazard if they are not properly arranged. Many serious fires and explosions have been caused by vehicles striking tanks or vessels containing flammable materials. The possibility of such accidents should be considered in planning roadways.

In Table 24-1 some suggestions are made for spacing between process units and other facilities in a plant. Any such list to be complete would be extremely voluminous since each type of process unit should be considered individually.

**TABLE 24-1. Suggested Minimum Spacings for Hazardous Process Units**

| Description of hazard | Distance to adjacent operating unit, ft |
|---|---|
| 1. Ordinary flammability and low to medium pressures | 25–75 |
| 2. High flammability and high pressure | 100–150 |
| 3. Direct fired boilers and furnaces | 100–150 |
| 4. Blowdown stacks with flare | 100–200 |
| 5. Loading facilities | 100 |
| 6. Public roads and railroads | 100 |
| 7. Cooling tower | 100 |
| 8. Storage tanks* | 75–150 |

* Minimum distances between storage tanks are given in the National Fire Codes.[10] Wherever possible a distance equivalent to 1 or 1½ times the diameter of the larger tank should be used.

## Utilities

The generation and distribution of utilities in a process plant is such an important function that these facilities should be located as remotely as practicable from the operating units. During emergencies caused by fire or explosions, the continued operation of the power plant, the water pumping station, and the steam generating unit is imperative. Open flames and sparking equipment are common in power plants and boiler houses, and location near a unit processing flammable products constitutes a real hazard.

## Buildings

Plant offices, mechanical shops, and laboratories should be located as far as possible from the operating units. There is no real need for proximity of offices to process units, and since open flames and sparks are common to both shops and laboratories it is also advisable to separate these buildings from the process units.

Warehouse and loading facilities should be located at the plant property line so that they are easily accessible from the public roads and are as far as possible from the areas of possible danger.

### Equipment Layout

A primary rule for equipment layout is that an operator should have at least two different routes of escape from any point in a unit. Extremely hazardous areas should be provided with escape chutes for rapid exit from the area.

Since a clean well-maintained area is often a safe area, equipment must be arranged to enable ease of maintenance and housekeeping.

Equipment should be located well above the average height of a man or considerably below since any location near usual height can result in frequent head injuries. As a rule the simpler the arrangement, the safer the unit. A simple arrangement facilitates prompt action in emergencies.

### Firewalls

When it is necessary to space units closer than is considered safe, shielding walls can be constructed of concrete. High walls of this type are also used to protect personnel from operating equipment which could explode or cause a severe fire.

Dikes or firewalls should be provided around all storage tanks containing flammable material. These walls are especially necessary for flammable materials which tend to boil over. In these cases, the firewall should be high enough to contain the entire contents of the tank and thus prevent the spread of fire to other tanks or other portions of the plant. Rules and suggestions for firewall construction have been outlined in detail.[5,8,9]

Units processing large quantities of flammable material are often paved and enclosed by low walls 12 to 24 in. high. Should a vessel rupture, these walls will contain the flammable material and prevent the spread of fire. In some process areas the sewage system can be designed to drain off such spillages rapidly in which case firewalls are not needed.

## MECHANICAL DESIGN AND SAFETY

To insure safe construction all mechanical design should be in accordance with existing codes. All practices important to safe design, however, are not specified by the codes. Some of the more common practices are listed below in the form of a safety check list for designers.

A. Vessels and tanks

1. Careful selection of materials to withstand corrosion. Use adequate corrosion allowances on materials that are expected to corrode.

2. Provide a sufficient number of manholes and handholes for inspection and maintenance, and orient to permit easy access.

3. Avoid screwed nipples for small nozzles. Use 6000 lb couplings.

4. Nozzles above 1½ in. should be flanged. For high pressure, use flanged nozzles for all sizes.

5. Careful inspection during and after fabrication; a competent inspector can do much to prevent vessel failures.

*a.* Inspection of materials.

*b.* Inspection of workmanship (especially quality of welds).

*c.* Witnessing of radiographing and stress relieving.

6. Atmospheric storage tanks should be provided with adequate vents for breathing (the taking in or discharging of vapor as the pressure in the tank increases or decreases due to change in temperature or liquid level).

*a.* Use vacuum-pressure breather valves (conservation vents).

*b.* Use floating-roof tanks for volatile liquids.

*c.* Connect vapor spaces of gas-tight tanks to a single tank serving as a gas holder. Provide relief valves for all tanks, however.

7. Automatic totally enclosed tank level gages preferred; manual gage hatches subject operator to needless hazards.

B. Heat exchangers

1. If both inlet and outlet of exchanger require block valves, provide relief valve on cold side to prevent excessive thermal expansion. Many explosions have occurred because of such expansion.

2. Sufficient drains should be provided for maintenance.

3. Each fluid space of an exchanger should be given an individual test.

4. If units are to be stacked, they must be stacked during the testing process.

5. If mixture of shell and tube fluids constitutes a hazard, extreme care in design and workmanship must be exercised.

C. Moving equipment (pumps, etc.)

1. Provide guards at all couplings, belts, and chain drives.

2. Provide adequate space between units for safe and convenient servicing.

3. Automatic overload, overspeed, or other devices are good investments on machinery such as compressors.

4. Protective devices such as strainers on pumps and liquid knockout drums for compressors are essential to safe operation.

D. Piping

1. Above-ground process lines are preferred. These may be elevated or placed at ground level on sleepers. Buried lines constitute a hazard since leaks are not readily detected.

2. If lines are installed in trenches, fire stops and drains must be provided at regular intervals.

3. Arrangement of piping and valves is important to safe operation.

*a.* Large valves in particular must be readily accessible from platforms. Approximately 7 ft 0 in. is the maximum distance from centerline to platform level. Above this chain operators are used.

*b.* Use double-block valves for hazardous equipment which must be removed while remainder of unit continues to operate. A small tell-tale valve in between double blocks indicates leakage.

*c.* Arrange long yard lines to avoid congestion so that in case of fire any portion can be isolated from the remainder by closing a block valve.

*d.* Headroom for overhead lines should be approximately 15 ft 0 in. above roads; 6 ft 6 in. above platform; 9 ft 0 in. above grade. If a line must be located below 6 ft 6 in., locate considerably below and provide a walkway over it.

*e*. Fuel supply valves for furnaces should be in a remote location.

*f*. Emergency dump valves should be located at a safe distance from the equipment to be dumped and preferably behind a heavy wall or building.

*g*. Install valves discharging into open tanks so that operator can operate without undue exposure to vapors or splashing.

4. Check valves leak; and if such leaking is intolerable, provide a block valve in addition to the check valve.

5. Avoid angle valves.

6. For structural reasons avoid pipe smaller than 1 in. for overhead lines.

7. Drains for low points and vents for high points should be provided. Use ¾ in. or larger sizes.

8. Water lines, especially fire lines, should be looped.

9. Loop steam lines and provide block valves so steam can be provided from several directions in case of failure of one portion of the system.

10. Separate sewer systems may be necessary for certain hazardous materials. Venting of such systems must be carefully planned.

11. Fire stops are necessary in all surface drainage systems to prevent spread of fire.

12. Lines should be pressure tested prior to operation. Care must be taken to avoid overstressing certain pieces of equipment or foundations during hydrostatic tests.

E. Electrical equipment and grounding

The safe installation and design of electrical equipment is carefully outlined by the *National Electrical Code*. The problem requires the attention of competent electrical engineers. They must be supplied with information on the hazards in each part of the plant.

1. Means for remote shut-down of operating equipment in addition to locally mounted starters is desirable.

2. Critical items such as switchgear and transformers should be located as remotely as practicable from operating units.

3. Identify underground conduit by including a dye in the concrete for the casing.

4. Protect overhead conduit from exposure to excessive heat.

5. Provide emergency power supplies.

6. All portions of the plant must be adequately lighted so that operators can avoid bodily injuries from bumps or falls.

7. Static electricity and lightning constitute real hazards to safe process plant operation. Neither can be prevented but protection is possible. Watch for static electricity accumulation and discharges in the following situations:

*a*. Moving machinery (belts, conveyors, etc.).

*b*. Flowing fluids or dust.

*c*. Human beings.

*d*. Lightning.

Protection is accomplished by proper grounding of all equipment in areas where static electricity is anticipated. When all metal equipment rests on concrete foundations, for instance, grounding must be installed. Instructions for proper grounding of equipment are given in Handbook of Fire Protection,[8] the National Electrical Codes, and various government publications[9] (See also sources of information).

F. Structural steel

1. Protect from failure due to fire by covering with concrete to the main support level or by providing emergency water sprays for cooling steel in event of fire.

2. Avoid interferences of structural steel with piping and equipment. Interferences are sometimes remedied in the field by removing or "cutting-out" portions of steel, thereby reducing the strength of the structure.

3. Platforms should be provided at all operating or inspection areas with a 2 ft 6 in. minimum width. 3 ft to 3 ft 6 in. is preferred in maintenance areas around manholes.

4. Stairways are preferable to ladders and at least two are required for larger units. Use gentle slope, 45° or less, and widths of 3 ft 0 in. or more.

5. Safety cages are required for ladders over 15 to 20 ft.

6. Handrails are required around all platforms and stairways.

7. Toe plates (6 in. high) around the edges of platforms are necessary to protect personnel from falling objects.

G. Buildings

1. Fireproof construction is essential in hazardous areas.

2. Unless absolutely necessary process equipment should not be housed. Operation of outdoor plants is much safer.

3. Provide explosion doors to handle any possible explosions that may occur.

4. Provide automatic sprinkler systems when practical.

5. At least two exits are required.

H. Insulation

1. Insulate all hot lines with which personnel may come in contact. Approximately 1 in. to 1½ in. of 85% magnesia is adequate for protection of personnel. More, of course, may be required for heat conservation. Detailed specifications are available from insulation manufacturers.

2. Insulate portions of critical instrument leads and electrical conduit in areas of excessive heat or areas where fire is probable.

## VENTILATION

When process equipment must be located indoors, ventilation becomes a major problem. Both toxicity and flammability of a material must be considered in design of a ventilation system.

### Toxic Concentrations

Materials such as aniline, the halogens, hydrogen fluoride, and phosgene are toxic in concentrations as low as 5 parts per million, whereas toxic concentrations of such familiar materials as ammonia, benzene, chloroform, hydrogen sulfide and carbon tetrachloride range from 20 to 100 parts per million. Maximum allowable concentrations of various chemicals have been published as tables.[2] The user of such tables should be cautioned, however, that research on toxic properties of some materials is incomplete, and, as new information becomes available, the values of allowable concentrations may be revised either upward or downward.

### Limits of Flammability

Information on flammability limits of gases and vapors and methods of estimation have been published.[3,11,15] The limits of flammability are expressed as upper and lower limits. The lower limit represents the least amount of the material which will sustain flame propagation when mixed with air, and the upper limit represents the greatest amount. These limits of flammability as determined experimentally are affected by the type of testing apparatus used. Extremely small apparatus in particular produces narrow limits of flammability, and results from such equipment should be used with caution. The flammability limits for mixtures of several combustible gases and air can be estimated from an equation which has been verified experimentally:

$$L_m = \frac{100}{\dfrac{C_1}{L_1} + \dfrac{C_2}{L_2} + \dfrac{C_3}{L_3} + \cdots}$$

where $L_m$ = limit of flammability of mixture. Mole % (upper or lower)

$L_1, L_2, L_3$ = limits of flammability (upper or lower) for gases 1, 2, and 3, mole % (e.g., moles of gas/mole of gas plus air)

$C_1, C_2, C_3$ = mole % on air and inert free basis of gases 1, 2, and 3

The units of this expression are consistent since the summation of $C/L$ terms has units of moles of air plus moles of combustible per mole of combustible.

## Required Ventilation Rate

Knowing the rate of release of hazardous vapor and the allowable concentration of contaminate in an enclosure, the minimum fresh air rate can be calculated from a simple material balance.

$$\frac{r_V}{r_A + r_V} = \frac{L_A}{100}$$

where $r_V$ = moles of contaminate entering enclosure per unit time

$r_A$ = moles of fresh air per unit time

$L_A$ = lower flammability limit or toxic limit whichever is smaller, mole % contaminate per mole of air plus contaminate

The practice of expressing ventilating capacity in air changes per hour should be avoided. If 15 air changes per hour proves the correct air rate for one installation, it does not follow that this same rate will produce satisfactory results in an enclosure of smaller size.

$$\left(\text{Air changes} = \frac{\text{Air rate, cu ft/hour}}{\text{Enclosure volume cu ft/change}}\right)$$

## Local Ventilation

General room ventilation can be supplemented by localized ventilation over the equipment from which offending vapors are being emitted. Such local units are arranged as exhaust systems, and consist of hoods designed to catch the maximum amount of vapor or dust without interfering with the operation or servicing of the equipment. Hoods can be placed over filters, at pump packing glands, around crushing and grinding operations, and at other similar installations.

### Positive Air Pressure

It is often desirable to maintain a higher air pressure in a hazardous area than in the surrounding rooms or the outdoors. The electrical equipment, for instance, which cannot be designed for explosion-proof operation can be installed in a hazardous area by placing it in a room under positive air pressure.

### Inert Gas Purge

The use of inert gas is another valuable aid to ventilation. Inert gas is manufactured by burning fuel gases in conventional furnaces or, preferably, specially designed inert gas generators. The gas can be used to purge tanks and vessels of toxic and flammable vapors before purging with air so that men may enter the vessel.

Men have died after entering inefficiently purged equipment. Every process plant which handles toxic or flammable materials must be equipped with inert gas generators or air or steam eductors for removing the toxic vapors and flammable vapors from tanks and vessels which must be entered by personnel.

### Ventilation for Comfort

In addition to removing flammable or toxic materials generated in enclosed areas, ventilation and conditioning of the air in operating buildings is also important to workers' comfort and morale. It may be considered safe for men to work under hot and humid or extremely cold conditions, but the worker's efficiency and dexterity are reduced under such conditions and accidents may develop merely because of fatigue, discomfort, or low morale.

### Design of Ventilating Systems

In process plants extreme care should be exercised in locating inlet and exhaust ducts to avoid short-circuiting of fresh air (entrance of fresh air into exhaust ducts). The degree of mixing of the toxic or flammable gas with the room air must be known in order to decide upon location of exhaust openings.

If possible, vapors removed from rooms should be treated before releasing to the surrounding atmosphere. Centrifugal or cyclone separators, spray or packed columns, air filters, and electrostatic precipitators can render exhaust gases less objectionable. In any event, the gases should be discharged from the building in the same direction as the prevailing wind. Revolving-type roof ventilators permit such a discharge even with variable winds.

## PRESSURE-RELIEVING DEVICES

It is impracticable to design process equipment for every possible condition that may arise through abnormal situations. Fortunately, however, abnormal situations can be handled by the use of pressure-relieving

**TABLE 24-2. Conditions Requiring Relieving Devices**

<table>
<tr><th>Condition</th><th>Design Basis</th></tr>
<tr><td>Failure of cooling water supply to fractionating tower</td><td>Vapor equivalent to maximum total overhead rate</td></tr>
<tr><td>Failure of fractionating tower reflux</td><td>Maximum possible flow equivalent to total overhead rate</td></tr>
<tr><td>Entrance of foreign highly volatile material</td><td>Calculate vapor flow based on instantaneous vaporization</td></tr>
<tr><td>Exposure to fire</td><td>1. Calculate heat input on basis of PAW data which in equation form becomes approximately:<br>$$\log q = (0.796) \log S - 1.7$$<br>where $q$ = heat input to vessel in millions of Btu per hour<br>$S$ = surface area exposed to flame in sq ft<br><br>2. Use for exposed surface area
<table>
<tr><td>Horizontal vessels less than 20 ft in diam</td><td>One-half cylindrical surface plus area of heads</td></tr>
<tr><td>Horizontal vessels more than 20 ft in diam</td><td>All area up to 20 ft above ground</td></tr>
<tr><td>Vertical vessels less than 20 ft high</td><td>All circumferential area plus bottom head</td></tr>
<tr><td>Vertical vessels greater than 20 ft high</td><td>Lower 20 ft of area</td></tr>
</table>
For insulated vessels (2 in. fireproof insulation or above) reduce calculated input by 20%</td></tr>
</table>

TABLE 24-2. Conditions Requiring Relieving Devices (continued)

| Condition | Design Basis |
|---|---|
| Exposure to fire (*continued*) | 3. Calculate vapors generated $$W = \frac{q}{H}$$ where $W$ = lb/hr vapors, $H$ = latent heat (Alternate method) $$W = \frac{\left(\frac{q}{S}\right)(S)}{H}$$ Use $q/S = 22{,}000$. This alternate method is far more conservative than the PAW method. Other methods give intermediate values. |
| Closed valves | Amount of fluid leaving vessel under normal operating conditions |
| Thermal expansion of confined liquids | $$W = \frac{Bq}{C_p}$$ where $B$ = coefficient of thermal expansion, $1/T$; $q$ = heat supplied to fluid; $C_p$ = heat capacity; $W$ = lb/hr |
| Thermal expansion of confined gases | Calculate increase in volume using Gas Laws |
| Excessive heat loads during abnormal operation | Vapor generated by additional heat input |

devices such as relief valves or rupture disks. The design of pressure-relieving systems requires careful study. In a badly planned system, too many or too few are installed or those installed are not well placed.

Sylvander and Katz[14] have prepared an excellent discussion on the design and construction of pressure-relieving systems. Some typical applications for relieving devices are given in Table 24-2, with suggested design basis for each application. This table is based in part on the work of Sylvander and Katz.

### Discharge Rates of Relief Valves and Rupture Disks

The rate of discharge for a relief valve or rupture disk is calculated by the following formulas for vapors and liquids. The vapor formula is derived assuming an ideal gas, adiabatic expansion, a value of 1.001 for the ratio of specific heats, and vapor flowing at the acoustic velocity. Higher values of specific heat ratio will give a lower discharge rate; and, therefore, the simplified formula given below is conservative. The formula for liquids is based on the simple orifice equation. Manufacturers furnish handy charts and tables which facilitate the calculations of relief valve and rupture disk sizes.

Discharge rates of relief valves and rupture disks

Vapor:

$$W = 315APC\sqrt{\frac{M}{T}}$$

$A$ = area of orifice, sq in.
$P$ = upstream pressure, psia
$C$ = nozzle coefficient, 0.97 for relief valves (10% accumulation), 0.81 for rupture disks (consult manufacturer)
$M$ = molecular weight
$T$ = absolute upstream temperature, °F

Liquid:

$$Q = 38AC\sqrt{\frac{\Delta P}{S.G.}}$$

$Q$ = liquid capacity in gpm
$A$ = orifice area, sq. in.
$C$ = coefficient of discharge, 0.40 for relief valves (10% accumulation), 0.61 for rupture disks
$S.G.$ = specific gravity

### Relief and Blow-down Discharge Systems

The proper disposal of materials discharged by automatic relieving devices or manually operated blow-down valves is an important phase of plant safety. Relief valves which are located on top of high fractionating towers can often be vented to the atmosphere. Such atmospheric venting, however, is not always possible when large quantities of flammable vapors or liquids must be discharged at a safe distance from the plant for disposal. The discharges of relief valves and blow-down valves can be manifolded into a single discharge line connected to the disposal section (a burning pit, flare stack, or a quench tower).

If there is a large area far enough from the process section and smoke is not objectionable to the surrounding community, burning pits with pilot flames can be employed for the disposal of both flammable liquids and gases. As industrial areas grow and plants expand, however, the burning pit becomes more and more objectionable.

### *Flare Stacks*

Flare stacks are constructed of pipe or boiler plate of sufficient diameter to satisfy the overall pressure-drop requirements for the system and high enough that the burned vapors will not constitute a nuisance or hazard and will be diffused readily by the prevailing winds. The stacks usually range in height from 100 to 200 ft.

A knock-out or entrainment separator is installed for the separation of liquids contained in the vapor stream. The gases then pass through a seal drum, the inlet of which extends below a water or oil level in the drum. The gas bubbles through this liquid and then enters the flare stack. The sealing liquid prevents flame flash-back from the flare. The upper 8 to 10 ft of the flare stack is best constructed of an alloy such as Type 304 stainless steel since this upper portion becomes extremely hot and ordinary steel will not withstand the high temperature.

A pilot light is provided at the top of the flare which burns continuously and is thus able to ignite gases as they escape. To prevent extinguishment of the pilot flame, a mixture of air and gas is piped to the pilot so that a strong, blue flame will result. This flame is difficult to observe from the ground, however, and an additional small luminous burner is also used so that a ground observer can easily detect failure of the pilot. The pilot is lighted by a sparking device inside of the pipe through which the mixture of gas and air travels to the top of the flare. The flame is, thereby, propagated to the pilot. Other methods are used. One employs a series of gas jets located inside an open pipe spaced at regular intervals starting at the location of ignition and terminating at the pilot.

## Quench Towers

Large amounts of hot heavy liquids can be discharged to a quench tower where condensation of a majority of the liquids will take place. The small amount of vapor yet remaining can be vented to the atmosphere or to a flare stack depending upon its quantity. The heavy liquids can be recovered in an oil separator.

### *Relief Valve Piping*

Relief valves should be installed so that they are easily accessible for servicing. Dual relief valves can be employed so that one may be serviced while the other is in place for operation. In such installations plug cock valves* are installed on the inlet side of the relief valve and arranged with an interlocking mechanism so that one valve must be opened when the other is closed.

Piping upstream from the relief valve should be as short as possible to avoid excessive pressure drops. The pressure drop in the suction line should not exceed one per cent of the allowable pressure for full capacity relief. Discharge piping should also be designed to prevent excessive pressure drop, and for most installations the pressure drop should not exceed 10 per cent of the set pressure of the relief valve. The line should never be smaller than the outlet size of the relief valve. The discharge piping must be well supported at all points of strain, especially that produced by the impact of the discharging gases.

Common relief valve and blown-down headers should slope gradually toward the knock-out drums so that liquids may readily drain from the system. Pockets in discharge lines cannot be tolerated. Discharge lines venting to the atmosphere must be provided with weather protection to prevent accumulation of water in the vertical pipe. Steam snuffing lines venting to the atmosphere are also needed so that discharging gases might be diluted to prevent ignition. The steam can also be used in case of fire in the discharge line.

Though relief and blown-down systems are vital, an overdose of caution in their design may result in a costly and cumbersome system. Each section and item of equipment must be considered separately and the most probable abnormal condition or conditions selected for the design basis. Careful consideration will usually show that all possible abnormalities will not occur simultaneously on one item of equipment or one section of the plant. Experience plays an important part in such decisions and the engineer should seek the advice and counsel of the safety department of his organization.

## FIRE EXTINGUISHING EQUIPMENT

Extinguishing process plant fires is a highly specialized type of fire fighting, the know-how for which has evolved from long experience. Fortunately most of the useful regulations that aid in the selection and application of fire fighting equipment have been published by such agencies as American Petroleum Institute, National Fire Protection Association, National Board of Fire Underwriters, and others (see section on sources of information). As well as taking advantage of such information, the project engineer should obtain the advice of men skilled in the art before attempting the design of fire extinguishing facilities for a process plant.

* Gate valves are also used.

### Water as a Fire Extinguishing Agent

Water continues to be the major fire fighting agent, and its uses expanded after the introduction of water fog nozzles and sprinkler systems. A steady stream of water on a flammable liquid often does more harm than good, but the use of a fog or mist of water which tends to smother the fire in addition to producing a cooling effect has proved extremely successful.

The great dependence upon water as a fire fighting agent suggests the careful planning of fire water distribution systems. Such a system invariably requires fire pumps to boost the normal water pressure to the higher pressures required for the fire hose and sprinkler systems. Fire pumps must be carefully selected and have an alternate drive to insure operation under the most adverse conditions. A typical dual drive fire pump is equipped with an electric motor and a gasoline engine drive. The gasoline engine is used during power failure.

### Other Extinguishing Agents

Another common method of extinguishing fires quickly in confined areas is by smothering the fire. This can be done by the use of foam or inert gases such as carbon dioxide. Steam smothering is also valuable and is particularly useful in extinguishing small fires around furnaces.

### Emergency Valving

Main shut-off valves for lines carrying flammable materials should be located so they can be operated in case of fire. It may be preferable in some fires to simply stop the flow of flammable material and allow that already escaped to burn. This avoids large concentrations of explosive gas and vapor.

### Portable Extinguishers

In addition to the permanent installations of fire hydrants, automatic sprinkler systems, and automatic carbon dioxide extinguishing systems, portable fire extinguishers should be supplied throughout a process plant since a small fire can often be extinguished with ease by the rapid use of an extinguisher. Most large fires begin as very minor conflagrations, and immediate attention to a small fire will often prevent a disaster.

### Selecting Fire Extinguishing Equipment

The project engineer should be cautious in using handy fire extinguishing tables designed to indicate the exact extinguishing method for various types of flammable material. They give general suggestions and are not intended to be specific for all situations. These tables can be a helpful guide, but advice of authorized agencies should be obtained.

When a new plant is being designed all materials which constitute a fire hazard in the operation of the plant should be carefully studied. Case histories of previous fires for similar plants should be obtained so that fire extinguishing equipment can be intelligently selected and placed. The National Board of Fire Underwriters, The National Fire Protection Association, The Associated Factory Mutual Insurance Agency, and other such groups will supply reliable information of this type.

### Plant Alarm System

An alarm system should be installed in all process plants so that personnel can be alerted in case of fire. In most plants the alarm system can be manual, although in some process areas normally requiring little attention, automatic alarm is preferable.

## REFERENCES

1. Armistead, George, Jr., *Safety in Petroleum Refining and Related Industries,* John G. Simmonds and Co., Inc., New York, 1950.

2. Brandt, A. D.. *Industrial Health Engineering,* John Wiley & Sons, Inc., New York, 1947.

3. Burgess, M. J., and R. V. Wheeler, *J. Chem. Soc., 99,* 2013 (1911).

4. Duggan, J. J., *Chem. Eng., 58,* No. 6, 125 (1951).

5. *Fire Protection in Refineries,* 3rd Ed., American Petroleum Institute, New York, 1941.

6. Gafafer, W. M., *Manual of Industrial Hygiene,* U.S. Public Health Service, W. B. Saunders Co., Philadelphia, 1943.

7. *Heating, Ventilating and Air Conditioning Guide,* Vol. 31, Amer. Soc. Heating Ventilating Engrs., New York, 1953.

8. Moulton, R. S., *NFPA Handbook of Fire Protection,* Crosby-Fiske-Forster, 10th Ed., National Fire Protection Assoc., Boston, 1948.

9. *National Bureau of Standards Handbook H40,* Supt. of Documents, Washington, D.C., 1945.

10. *National Fire Codes,* National Fire Protection Association, Boston (1945).

11. Patty, Frank A., *Industrial Hygiene and Toxicology,* Interscience Publishers, Inc., New York, 1948.

12. Reigeluth, R. J., *Safety and Economy in Heavy Construction,* McGraw-Hill Book Co., New York, 1933.

13. Robinson, C. S., *Explosions, Their Anatomy and Destructiveness,* McGraw-Hill Book Co., New York, 1944.

14. Sylvander, N. E., and D. L. Katz, *The Design and Construction of Pressure Relieving Systems,* Engr. Research Inst., Bulletin No. 31, Univ. of Michigan Press, Ann Arbor, Mich., 1948.

15. Thornton, W. M., *Phil. Mag., 33,* 140 (1917).

# PART IV

# CONSTRUCTION OF THE PLANT

AFTER design work is complete, construction may begin. At this point most of the responsibility shifts to the construction superintendent. The project engineer, however, must maintain active interest in the construction operations in order to interpret wherever necessary the intent of the design engineers and act as liaison between the construction forces and the engineering design groups. Such work requires some knowledge of construction methods.

# CHAPTER 25

# CONSTRUCTION

When construction begins the work of the project engineer is essentially finished and completion of the project depends on the construction department. The project engineer, however, cannot be complaisant about the techniques and problems of construction. On the contrary he must be so well schooled in the methods of construction that he can foresee and avert possible difficulties in construction while the plant is still in design stages.

To understand the problems of construction the project engineer should have some understanding of construction planning, labor, and labor relations, and actual construction operations.

## CONSTRUCTION PLANNING

Design drawings and specifications give the construction department detailed information about what is to be built, but they do not prescribe the methods of erection. Long before actual erection begins, therefore, the construction superintendent and his staff study preliminary engineering data and scale models of the plant to be erected. In this way the most advantageous location for derricks and hoisting equipment can be chosen; major construction equipment can be selected, and the most practical construction methods determined for each major piece of equipment. Such preliminary planning in the office always results in lower construction costs.

## CONSTRUCTION PERSONNEL

The field or construction organization has two general classifications of personnel, clerical personnel for the field office and the craftsmen who perform the actual construction work.

The routine office operations for a construction job consist of local purchasing, receiving, and accounting operations, preparation of a weekly payroll, and maintenance of daily labor cost records for each portion of the project.

The executive personnel for a construction project consists of a field project manager or superintendent, or both for large projects, assistant superintendents, general and subforemen, and an office manager.

The key field organization usually consists of from three to five general supervisors including the superintendent. Office operations require from eight to fifteen clerical workers. The following organization is typical of one which could be used on a large project costing ten million dollars or more.

1. Field project manager
2. Superintendent
3. Assistant superintendent

3.1. Field engineer (civil)

4.0 Field

4.01 General foreman
4.02 Area foremen
4.03 Piping superintendent
4.04 Pipefitter foremen
4.05 Carpenter foreman
4.06 Iron worker foreman
4.07 Boilermaker foreman
4.08 Bricklayer foreman
4.09 Cement masonry foreman
4.10 Electrician foreman
4.11 Machinist foreman
4.12 Painter foreman
4.13 Asbestos worker foreman
4.14 Sheet metal worker, foreman
4.15 Rodman, reinforcing steel workers, foreman.

5.0 Office

5.01 Office manager
5.02 Chief accountant
5.03 Chief payroll clerk
5.04 Buyer
5.05 Warehouseman
5.06 Receiving clerk
5.07 Time clerks
5.08 Secretaries and stenographers
5.09 Typist, switchboard operators

Engineering groups sometimes work at the jobsite, but they are usually separated from the construction operations and are not considered here.

The titles under 4.0 are also classifications of union labor. Organized labor requires a particular craft for each type of work. When the work of a given craft is completed those workers are discharged, but most crafts on process plant projects are used during the entire project.

The greater portion of construction work is performed by organized labor. The construction worker is essentially an itinerant worker and is employed permanently only when he becomes a supervisor.

Construction employee labor organizations make agreements with the contractor which fix the construction worker's wage, hours, overtime rates, and working conditions. In general the union controls the employee, his employment, and his specific work as well as the area labor practices. Unions constantly seek improvement in working conditions, wage rates and length of employment. The length of employment is increased by strictly limiting the functions of each craft so that more types of workers must be retained throughout the project. Most of the problems with construction labor arise from this practice which often causes one craft whose work is slack to arbitrarily take over the work of another craft. A dispute results which is called a jurisdictional dispute.

### Jurisdictional Disputes and Labor Relations

The jurisdictional dispute is a disagreement between two or more crafts over which craft shall perform certain functions or parts of the work. The reasonable solution to such questions would seem to be mutual agreements between the several labor organizations without employer participation. The employer is not necessarily interested in whether one craft or another performs an operation unless erection costs are increased or quality of work impaired.

Regardless of his neutrality, the contractor may find himself in a dilemma when trying to choose between two crafts for a certain job. The craft not favored may "walk-off" or the dispute may induce a "slow-down" of other work. Some of these difficulties can be averted by agreements on division of work with the national offices of the various craft unions.

In any discussion of labor problems it must be remembered that the construction laborer performs the most difficult types of work. Furthermore, he has no job security, and must attempt to earn higher wages than regularly employed personnel. As understanding increases between labor, management, and labor leaders many of the problems such as jurisdictional disputes will be more satisfactorily resolved.

### Construction Labor Distribution

The scope of work and type of project governs the types of craftsmen required on a project. The following work distribution is typical of a process involving fluids. The manhour distributions shown vary considerably with each project.

The project illustrated is one which is to be erected in 12 months. All foundations are spread-type and no piling is used. The total cost of the project, material, labor, engineering, procurement, fee and profit amounts to $4,200,000 of which $3,150,000 includes material, equipment, fee, and profit. The remainder, $1,050,000 is for construction costs, erection labor primarily.

**Craftsman Distribution**

Maximum Craftsmen at job peak = 350

| Craft designation | Craftsmen | Maximum used | Manhours |
|---|---|---|---|
| *P* | Pipefitter | 135 | 115,000 |
| *L* | Laborer | 55 | 68,000 |
| *B* | Boilermaker | 30 | 51,000 |
| *C* | Carpenter | 45 | 44,000 |
| *E* | Electrician | 35 | 37,000 |
| *I* | Ironworker | 45 | 35,000 |
| | | Subtotal | 350,000 |
| *M* | Miscellaneous crafts, clerical, and supervision | 40 | 50,000 |
| | | Total | 400,000 |

**Craftsman Distribution by Month**

| Month | 1000 manhours per month | Number of Men Used | | | | | | | | |
|---|---|---|---|---|---|---|---|---|---|---|
| | | *P* | *L* | *B* | *C* | *E* | *I* | Subtotal | *M* | Total |
| 1 | 13 | 8 | 25 | 2 | 15 | 8 | 6 | 64 | 20 | 84 |
| 2 | 14 | 10 | 35 | 2 | 25 | 10 | 8 | 90 | 20 | 110 |
| 3 | 22 | 22 | 50 | 2 | 35 | 15 | 10 | 134 | 22 | 156 |
| 4 | 28 | 42 | 45 | 8 | 38 | 10 | 20 | 163 | 28 | 191 |
| 5 | 35 | 54 | 55 | 8 | 38 | 16 | 30 | 201 | 30 | 231 |
| 6 | 50 | 120 | 50 | 15 | 40 | 26 | 35 | 286 | 32 | 318 |
| 7 | 62 | 135 | 55 | 35 | 45 | 35 | 45 | 350 | 38 | 388 |
| 8 | 48 | 135 | 35 | 30 | 20 | 20 | 25 | 265 | 40 | 305 |
| 9 | 42 | 130 | 25 | 30 | 10 | 29 | 20 | 244 | 35 | 279 |
| 10 | 35 | 110 | 25 | 15 | 10 | 29 | 10 | 199 | 30 | 229 |
| 11 | 31 | 90 | 30 | 10 | 5 | 25 | 10 | 170 | 25 | 195 |
| 12 | 20 | 45 | 35 | 5 | 5 | 15 | 5 | 110 | 20 | 130 |
| | 400 | | | | | | | | | |

The columns indicating subtotal and total are peak requirements rather than averages. The total of craftsmen per month constantly fluctuates because of the discharging of some personnel, changes in the work, adverse weather conditions, and off-schedule equipment delivery. The totals shown, if constantly at work, would exceed the manhour level by 10%.

## Labor Rates

The typical rates listed in Table 25-1 are so-called "Journeyman" rates. Foremen receive from 12½¢ to 25¢ per hour over journeymen and general foremen receive 25¢ per hour over top foreman's rate. There are approximately 300 different classifications of organized labor used in construction operations. Many of the classifications slightly overlap. Those shown are most common for process plant installations.

**TABLE 25-1. Construction Labor Rates 1954–1955 Southwest Area U. S.**

| Craftsman | Rate/hr | |
|---|---|---|
| 1. Painters | 2.50 –2.75 | |
| 2. Teamsters | 1.615–1.70 | |
| Warehousemen | 1.70 | |
| 3. Laborers | 1.60 –1.80 | |
| 4. Cement finishers | 2.625 | |
| These craftsmen (nos. 1–4) receive time-and-one-half for overtime except for Sunday and any authorized holiday which are paid at double time. | | |
| 5. Electricians | 3.00 | 1% Benefit fund, $0.075/hr health, and welfare |
| These craftsmen (No. 5) receive time-and-one-half for hours from 4:30 PM to 10:00 PM Monday through Friday. All hours past 10:00 PM, Saturday, Sunday, and holidays are paid at double time. If personnel are called from home after working hours they receive double time from home to home, i.e., they are paid for travel time as well as work. | | |
| 6. Ironworkers | 2.90 | |
| 7. (Reinforcing) rodmen | 2.65 | |
| 8. Carpenters | 2.625 | |
| 9. Millwrights | 2.875 | |
| 10. Operating engineers | 1.975–2.75 | |
| Heavy cranes, tractors, derricks | 2.75 | |
| Oilers | 1.975 | |
| 11. Boilermakers | 2.975 | Health and welfare $0.075 |
| 12. Asbestos workers | 3.0625 | Health and welfare $0.075 |
| 13. Pipefitters | 3.10 | |
| 14. Bricklayers | 3.45 | |
| These craftsmen (nos. 6–14) receive double time for all overtime. | | |
| 15. Sheetmetal workers | 2.85 | |
| These craftsmen (no. 15) receive time-and-one-half for the first three hours over any regular working day and for the first four hours on Saturday. All other overtime is paid at double rate. | | |

## CONSTRUCTION OPERATIONS

Every project engineer should accept the first opportunity to observe the erection of a process plant from start to finish. An understanding of the many complex operations can be developed only by actual observation.

The erection of a plant must proceed with the minimum of delay and a visitor to a project is invariably impressed by the spirit of action that pervades the jobsite.

### Site Preparation

A site which may be filled with trees, undergrowth, and boulders is "attacked" by a construction crew and the appearance changes at an unbelievable rate. As fast as surveyors lay out the plant boundaries and reference points (see Chapter 22) large earth moving equipment such as bulldozers and power shovels move in to uproot trees and level hills and valleys. Within one day the appearance of the site will change completely. Each successive day activity increases and more workers join the construction forces.

### Driving of Pile

Pile driving (see Chapter 22) should be started as soon as possible so that it may be completed before foundation work begins. In this way damage to new concrete work can be avoided. If piling is to be driven over a large area, topsoil is removed during rough grading below the planned pile cap elevation. Pile driving can then proceed immediately and no additional excavation around the piles will be required.

### Temporary Buildings

During the first few days of construction the number of workers is small and facilities required are correspondingly small. However as the project proceeds, the working forces increase; clerical and office personnel arrive, and delivery of equipment begins. Buildings are required to handle this influx of people and equipment.

These buildings are temporary and must be moved or torn down at the close of the job, but they must be adequate for the needs of the project. They are constructed either of wood or of prefabricated steel frames with aluminum or steel roofing and siding.

Although the number of buildings varies with the type of project, the following is typical:

*a.* Construction office building: 1500 to 2000 sq ft, for housing general superintendent, assistant superintendent, office manager, payroll clerk, field buyer, registered nurse, and 6 to 8 clerks and stenographers.

*b.* Shelters for pipe fabrication, carpentry, and limited electrical work.

*c.* Change houses: lavatory and washing facilities. Separate space must be allocated for each craft; may be located on ground floor of two-story building, the upper floor of which is used for office space.

*d.* Tool room: 800 to 1000 sq ft, used for storage and checking of hand tools, supplies, and portable equipment.

*e.* Warehouse: used for storing plant equipment which must be protected from the weather; valves, small pipe and pipe fittings, electric conduit fittings, cement, and insulation.

Several large buildings providing space for many functions are usually cheaper than a number of separate buildings. Many contractors have found portable steel buildings economical especially when space requirements are small.

### Temporary Water Supply

Long before permanent water service is constructed for the plant a temporary supply is installed early in the project so that water will be available for drinking, concrete mixing, supplying steam-operated equipment and fire protection. It often proves convenient to install the plant fire protection system early so that it may be used as a distribution system for construction. If a city is nearby, drinking water is hauled to the jobsite thus avoiding the added expense of chlorination.

### Roadways and Railroad Spurs

As temporary buildings are being constructed other crews of men are constructing railroad spurs for delivering large equipment into the site and roadways for access to all parts of the site by trucks and construction equipment.

The plant roads are laid out, topsoil is removed, compacting and backfilling with sand and gravel is completed. Paving, however, especially with soft material like macadam, is done toward the end of the job to avoid damage to the pavement by construction equipment.

If a permanent railroad spur is not required for the plant, a temporary spur is often constructed when large equipment is to be shipped by rail. By this method equipment can be lifted directly from the freight car and installed. The temporary spur need not be ballasted, but may be welded together at the joints and braced by structural members so that entire sections can be moved without misalignment.

### Excavation Operations

Once temporary facilities are built construction can proceed at top speed. The next logical step is to complete as much of the excavation work as possible since underground facilities and foundations are more conveniently installed before other work is started.

The areas to be excavated are laid out by surveyors, and most of the effort on the project is centered on the actual excavations. Equipment of all types is brought in at this time: drag-lines, power shovels, tractor-operated carry-alls, bulldozers, ditching machines, and various scarifying and compacting apparatus. These machines when operated skillfully can complete excavations in days that would require weeks if done by hand. Some hand work, however, may be required in congested areas but even this is made simpler by portable pneumatic tools, and at times hand work can be avoided by use of portable self-powered hydraulic shovels.

During and after excavation, soil removed that cannot be used for backfill is hauled from the area so that the site will be as clear as possible and ready for the installation of underground facilities.

### Installation of Underground Facilities

As ditches are completed, pipe fitters begin the task of installing underground piping, and electricians, underground conduit and cable.

The method of pipe installation varies with the type. Bell and spigot cast-iron pipe as used for water, process sewers, and drains is laid on a sand cushion to insure continuous support. The joints are packed with plain or tar-saturated jute and carefully caulked with molten lead. After the joints are made-up, firm soil is packed around the lower half of the pipe and as the ditch is backfilled the soil is tamped.

Mechanical joint cast-iron pipe, most often used for water, is laid in a similar manner except making up of joints requires only the setting in place of a gasket and tightening of bolts.

Cement-asbestos combination pipe can be used instead of cast-iron for some services. A collar and heavy rubber ring compose the joint which can be made up in one-tenth the time required for leaded joints.

Vitrified tile is usually used for sanitary sewers and corrosive process fluid drains. The joints are sealed with neat* cement or corrosion resistant mastic. Ungalvanized carbon steel pipe to be installed underground is first coated and wrapped. After the welders complete their work it is cleaned and coated with rust-resisting paint and then bitumastic, and finally it is wrapped with tarred felt or tarred glass fiber blanket. If the quantity of pipe to be wrapped is large, it is done by a machine which moves along the ditch and wraps and lowers the pipe into the ditch.

All underground joints not threaded or bolted are protected against leakage and water hammer by means of concrete anchors. The tamped soil alone around a joint is not sufficient to overcome strains introduced by sudden pressure surges.

### Electrical Conduit

Steel conduit is cut, threaded, and bent with tools similar to those used by pipe fitters. Because of the thin walls and light weight, however, handling and fabricating is very simple. After cutting and threading, sharp edges are completely removed since they can damage the cable insulation during cable pulling. The conduit is then lowered into the ditch and joined together by couplings.

Cable is pulled through the conduit between manholes by air-driven winches or by hand, using a lubricant to reduce friction. To avoid injury to the cable, pulling tension is carefully checked by spring scales when pulling tension is believed high.

* Cement without aggregate.

As cable is installed the insulation is tested by an instrument which applies high voltage across the insulation and indicates the resistance of the insulation. This testing is supervised by a competent electrical engineer, as is all electrical equipment testing and installation.

Red-colored concrete is poured around the conduit and after it sets, the ditch is backfilled and tamped. The red color is plainly visible and acts as a warning to workers when excavations are made around it.

The only other underground electrical work proceeding at this time is the installation of grounding systems and cathodic protection. Bare grounding cable, either solid or stranded copper, is laid directly in the ground and brought to the surface at predetermined points for connection to equipment. Insulated wire used for cathodic protection is connected electrically to the underground pipe to be protected and brought to the surface for attachment to the cathodic protection system.

### Foundation Construction

Foundations are poured as soon as underground piping and conduit work in the immediate area has been completed. Larger foundations require several pours and work on a single foundation may last a week or more.

The details of these operations are described in Chapter 22. Once a concrete pour has begun it must be continued uninterrupted. Hence the major effort of the construction forces and supervision is directed toward the pouring operations.

### Erection of Guyed Derrick

A foundation is also constructed for the guyed derrick which must be installed next so that equipment may be lifted in place. The erection of a guyed derrick is a major operation often requiring some 3000 man-hours of construction labor. The derrick (Fig. 25-1) consists of a vertical structural steel mast, a boom, and a hoisting engine. The mast is often 200 ft or more in height and is held in position by 6 or 8 guy cables extending radially from its top to ground where each guy is securely attached to a "deadman." A "deadman" is a heavy length of pipe buried perpendicular to the direction of pull and encased in concrete.

The boom is pin connected to the mast near its bottom. At the lower end of the mast is a ball-and-socket joint, the base of which is supported by and securely bolted to a concrete foundation poured for the derrick. The upper end of the boom is raised or lowered by heavy cables running through a set of blocks on the boom and mast, and on to the hoisting engine. Another cable and set of blocks is extended from the boom to the load for vertical lifting. A wheel-like contrivance is provided at the lower end of the mast so that mast and boom may be rotated. The load may thus be hoisted with the boom extended, then horizontally moved by hoisting the boom or by rotating the mast. The hoisting engine may be a steam engine, a diesel engine, or an electric motor.

Loads of several hundred tons may be lifted, within the boom radius, up to heights of 175 or more ft. Derricks are constructed in various sizes depending upon the work area, equipment heights, and the loads to be lifted.

Experienced erection planners must select the proper location of the guyed derrick so that it can be used to the greatest possible extent. Since its foundation, the various hoisting cables running to the hoisting engine, and the engine itself require a large space, the installation of some process equipment may have to await the removal of the derrick. The location thus may have some effect on final construction.

### Erection of Elevated Reinforced Concrete Structures and Structural Steel

Before most equipment can be installed major structures must be completed. These include reinforced concrete structures, constructed in a manner similar to foundations using a number of successive pours, and steel structures.

Fabricated steel for steel structures is delivered with each piece marked in accordance with a numbering system common to the industry. Steel fabricators furnish erection drawings showing by means of these numbers the correct location for each piece. These erection drawings must arrive with or before the arrival of the steel since they are often used to aid in unloading the steel members in the most convenient order.

Structural steel is often further identified by a color code which designates the area in which the steel is to be used.

Structural steel members are lifted into place by a guyed derrick or a large crane. The members are bolted together with temporary erection bolts until permanent connections are made. As soon as there are enough main members to provide stability the structure is aligned and leveled with the aid of a transit and level and held in the proper position by cables until permanent connections can be completed.

Riveting, bolting, or welding* are used to make permanent connections between the flat surfaces of two structural steel shapes.

The rivet driving crew consists of an operator for the pneumatic rivet gun, a person to "buck-up" or hold a heavy steel billet against the opposite end of the rivet, a person to catch the rivet and place it in the punched or drilled rivet hole, and a person to heat the rivets. The "rivet-heater" may work for two or more crews. The heater tosses white hot rivets to a worker (catcher) who catches them with a conical metal cup (sometimes

* The American Institute of Steel Construction outlines the requirements for connecting structural steel [5]

a water bucket), removes each rivet with a pair of tongs and places it in the hole. Immediately the pneumatic gun is placed against one end, the heavy billet against the other and driving proceeds. The rivet gun and buck-up billet are each provided with die ends which form the rivet heads. Generally button-head or hemispherical heads are used in preference to other shapes. The operation is essentially a hot forging process which forces the rivet to fill the rivet hole. When it cools the rivet contracts and draws the steel members together tightly.

If rivets are overheated, they may rupture internally when driven. These faulty rivets, however, may be found by striking each rivet with a hammer. The ruptured rivet makes a dull sound which is easily detected. Faulty rivets so discovered are chipped out with pneumatic tools and replaced.

The usual four-man crew required for riveting may be reduced to two for permanent bolting of structural steel. The structural members are aligned by reaming and a bolt slightly larger than the bolt hole is driven into the hole. Good craftsmanship is required to insure a strong connection.

Many new types of bolts are on the market. One has a ribbed shank with its outside diameter slightly larger than the bolt hole. When the bolt is driven, the ribs are forged into the hole filling all the space. These bolts force alignment of the structural steel. In general only secondary structures and galvanized steel members are bolted, but as new bolting developments continue to appear, bolting of larger structures may become more popular.

After the bolt is driven the nut is fastened and the bolt is tightened-up. Faulty bolts and faulty rivets are detected in a similar way.

Structural steel is welded in the field with standard welding equipment. Two flat plates are placed together and fillet-welded along the plate edges or plug-welded through a hole in one of the plates. Butt welds are also used. If made in a shop special jigs or clamps are used to hold the pieces in position. Field-welded structural members are held in position by the erection bolting.

Welded joints are chipped away at spots to check weld penetration and soundness. Faulty welds are chipped out and rewelded.

## Erection of Major Equipment

It is simpler to install heavy equipment before smaller associated equipment and so after supporting structures are erected the installation of major equipment begins.

The job of lifting and setting equipment in place is called rigging. Rigging techniques have been handed down through the centuries and improved with new machinery and materials. Equipment to be lifted is held by steel wire slings. A crawler crane, a guyed derrick (Fig. 25-1), gin poles (Fig. 25-2), various portable truck mounted cranes, and even chain hoists may be used for the lifting, depending on the load.

Large equipment which is to be located at or near ground level can often be skidded over wood or pipe rollers to the desired position. In

Fig. 25-1. Guy derrick (right) and crawler crane (left) at construction site. (Courtesy of M. W. Kellogg Company.) Note guy wires.

such cases the equipment is protected at its base by wooden beams called skids. Short lifts of large equipment can be made, when necessary, by the use of heavy duty jacks.

The lifting of heavy equipment is probably the most interesting phase of the construction operation. It requires highly skilled operating engineers and helpers. The lifting and precise placing of a 200-ton piece of equipment, for instance, requires strong nerves and a great deal of confidence. The hoist operator controls his derrick with the aid of hand signals from a helper standing near the load. These signals, as well as all phases of crane operation, have been standardized by the American Standards Association[1] since misinterpretation may prove costly and dangerous. During big lifts even seasoned construction workers stop to watch and admire the work of the derrick operator.

Any large prefabricated vessel which is to be installed vertically is unloaded from truck or freight car so that lifting equipment can operate from directly above the point of attachment. The lift must be as nearly

Fig. 25-2. Gin-poles being used to erect a 360-ton distillation tower. (Courtesy of M. W. Kellogg Company.)

*Note:* Bottom of tower rests on sled which was pulled on rollers starting with the tower in a horizontal position. The speed of pull was such that the lifting cables remained vertical. The illustration shows the tower hanging in an almost vertical position just before the sled was removed and the tower placed on its foundation.

vertical as possible since hoisting equipment cannot withstand high horizontal forces. Vessels fabricated on the jobsite are usually built within the effective radius of the fixed lifting apparatus.

Equipment may be provided with lifting lugs for attachment of hoisting apparatus or special slings may be wrapped around the equipment.

Lifting lugs, which are welded on the vessel, simplify the attachment of hoisting cables. They add to the cost of the vessel, however, and make insulation of the vessel more difficult.

In general nearly all types of cylindrical equipment are lifted with braided steel wire slings. Vertical vessels are lifted directly above their foundations and the slight degree of slope in the vessel caused by the sling aids in starting the anchor bolts into the vessel base. Anchor bolts for large vessels are usually provided with special threaded cone-shaped caps which protect the threads as well as guide the bolt into position. Steel plate shims are placed at each space between anchor bolts so that when the vessel is set down on its bolts it will be essentially level. Bolt nuts are tightened after the vessel is checked with an instrument (transit) for perpendicularity. After the vessel is satisfactorily located, nuts are tightened and the anchor bolt sleeves and base ring are grouted with concrete.

Other large equipment such as machines are handled in much the same manner. Even though machines may weigh several hundred tons they are not usually so unwieldy as extremely long and heavy vessels.

Machines are carefully aligned and leveled during the bolting operation, and grouting-in of large machines is usually done only after a manufacturer's representative has checked the machine for proper alignment and level. When bolting and grouting is completely supervised by the manufacturer's representative, the supervision is a condition of the manufacturer's guarantee.

### Installation of Piping

The erection of above-ground piping begins after major equipment is installed and continues throughout the construction period.

The majority of process plant piping is connected by flanges and welds. The various pieces are fabricated in a shop and delivered to the site or fabricated in a field shop at the jobsite. In either case the methods are similar. The pipe is cut to the proper lengths by means of a cutting machine or an oxyacetylene cutting torch. A piping piece or assembly which may, for instance, consist of a straight length with a 90° bend attached and two nozzles in different planes is set up on a work table or on trestles. The setup is carefully aligned and then tack welded.

The tack-welded assembly is checked dimensionally and final welding is done usually by the electric-arc process. A full-penetration weld free from slag and cracks must be made. It is economical to do as much of the welding as possible on the ground since the pipe may be rotated on rollers and welding may be done with little or no movement of the welding electrode.

After welding, each piece is checked for alignment and weld quality.

Slight misalignment can be corrected by pulling into place by suitable machinery while the assembly is held stationary.

Screwed piping is fabricated at the jobsite and as much as possible at ground level. Threading and cutting tools, many of them power operated, are part of a pipefitter's standard equipment. The fitter cuts, threads, and partially fits the pipe on a bench about 3 ft wide by 12 ft long.

Cutting oil is used during the threading operation to prevent excessive heat and help in the penetration of the metal. Threads are made in accordance with standard lengths recommended by the codes. They are thoroughly cleaned before making up joints. A jointing compound which serves as a lubricant and filler is placed on the threads. Various compounds are used for different operating temperature ranges and fluid conditions. Screwing of threaded joints is done slowly to prevent binding caused by excessive heat.

Pipe assemblies and sections of straight runs of pipe are lifted into position by means of chain hoists or slings attached to cranes. The several pieces are then aligned. Flange faces are thoroughly cleaned,* gaskets applied and the bolts tightened. Tightening of bolts is done in such a way that the tension is equal on opposite bolts. If field welds are to be made, aligning clamps are used.

Expensive items such as valves must be handled with care. They are well supported at all times during erection and their faces protected from injury.

A safe adequate work area is necessary for the pipefitters who install overhead piping. Scaffolding must be strong and roomy since it is not possible for pipefitters to align heavy piping with great accuracy when their own bodily safety is in jeopardy or when forced to assume extremely uncomfortable positions. As piping is completed, it is washed or pickled, pressure-tested, cleaned, painted, and insulated. The painting of insulated pipe depends on process conditions. In general all carbon-steel piping, whether left bare or insulated, is treated with a rust inhibitor after thorough cleaning.

### *Pipe Identification*

Each separate piece of shop-fabricated piping is given a piece number or spool number on the original piping design drawings. These numbers are used as identifying numbers and are painted and stamped with a steel die on the fabricated pipe. The original design drawings are then used as erection drawings.

A color code is often used for identification on extensive projects having a number of process units which utilize larger sizes of piping. Mark, piece, or spool numbers must still be assigned to identify the pieces. If

* Ring-type joints must be cleaned exceptionally well to insure a tight seal.

several separate units are being fabricated, color can be selected for each unit so that piping in the storage yard can be easily separated without finding the mark number. The color coding is applied near each end of the piece and completely circles the pipe in a band sufficiently wide to be seen from a reasonable distance.

Random pipe delivered to a jobsite is also color coded when received. This code should be designed to indicate the size and type of pipe (seamless, alloy). Much of the random pipe purchased must often be stored for several months. Only certain mills carefully identify pipe, but the original identification may be lost or defaced if it existed. Color marking can be quickly done as pipe is being unloaded and will save much time as the job proceeds.

### Insulation

After cleaning, pipe is pressure-tested and then insulated. Insulation is done by a subcontractor specializing in this work. These workers begin the insulation of vessels at about the same time as pipe insulation. (See Chapter 19.)

### Buildings

Office buildings, change houses, warehouses, and machine shop buildings are erected at any time during the project as men become available. Process buildings and shelters are erected after major equipment and large piping are installed.

### Final Stages of Construction

After all the major equipment is in place, small equipment which must be supported by the larger apparatus is installed. Fragile materials such as instruments, electrical apparatus, and small piping are also installed in the final stages in order to minimize possibility of damage.

The last weeks of a construction project are busy ones. Equipment is pressure-tested. Operating tests of equipment are made. The site is cleaned and put in order. Temporary buildings are removed, and many official visitors arrive to inspect the plant.

The contractor's and the customer's engineers who began working on the design of the plant a year or two before can at last see the tangible proof of their efforts. Many portions of the design could be improved and criticism of some portions of the plant can easily be made. But as the construction equipment is moved from the site and the contractor's operators conduct the acceptance test of the equipment and demonstrate that the quantity and quality of the product can be produced with ease, adverse criticism seems inappropriate.

The plant now ready to operate is a tribute to a tremendous group effort, involving chemical, mechanical, electrical, and civil engineers, draftsmen, stenographers, accountants, buyers, craftsmen, and office boys.

As one leaves the plant site on this final day two people will probably be busily conferring: the project engineer and the construction superintendent. These two men have done more than any to assure a successful operating plant. The project engineer has directed the project since the early process design. He has been a good organizer and a competent engineer capable of dealing with people of widely different backgrounds.

The slightly harassed looking gentleman whom the project engineer is congratulating is the construction superintendent, who will shortly move to a new assignment. Because of his intelligence, high mechanical aptitudes and good judgment, he has not only been able to construct the plant as outlined by the engineering drawings, but also has corrected, with the help of the project engineer, the inevitable errors that occur in drawings and procurement. He has improvised where necessary, but most important, he has completed the job on time and built a plant that pleases the customer.

## REFERENCES

1. *Cranes, Derricks and Hoists,* ASA-B30.2, American Standards Association, New York, 1943.

2. Kellogg, F. H., *Construction Methods and Machinery,* Prentice-Hall, Inc., New York, 1954.

3. Rossnagel, W. E., *Handbook of Rigging,* McGraw-Hill Book Co., New York, 1950.

4. *Standard Manual on Pipe Welding,* Heating, Piping, and Air Conditioning Contractors National Association, New York, 1951.

5. *Steel Construction Manual,* 5th Ed., American Institute of Steel Construction, New York, 1950.

6. Underwood, G., *Standard Construction Methods,* 2nd Ed., McGraw-Hill Book Co., New York, 1931.

# INDEX


Above ground lines, 458
Acceptance order, 131
Accessories for piping, 426–434
Accounting procedure of contractor, 164
Acetylene gas welding, 189
Actual gases, compressor calculations for, 339
Additives for concrete, 603
Affinity laws, as applied to centrifugal pumps, 280
Air, design of instrument air system, 536
  design of plant service air system, 581
Air chambers for reciprocating pumps, 256
Air changes, calculation of, 641
Air conditioning, costs of, 16–17
  design of facilities, 621–623
  in control rooms, 542
Air-cooled exchangers, 215, 223
Air line service and plant location, 12
Air-motor valve, 538
Air preheaters, 570
Alarm systems, 649
Alignment of pumps, 294
Alloy pipe, effect of various metals on physical properties, 400
Alternating-current motors, 352
Alumina-silica refractories, 485
American Bureau of Shipping Rules for Classification and Construction of Steel Vessels, 198
American Concrete Institute, design practices of, 627
American Institute of Steel Construction, design practices of, 627
American Petroleum Institute; Specifications for Oil Storage Tanks, 198
American Standards Association, Code for Pressure Piping, 396
  flange classification, 403
  flange facings, 408
  specifications for centrifugally cast-iron pipe, 412
Anchor bolts, 606, 666
Anchors (and supports), marking on drawings, 111
  definition of, 456
  for pipe, 430
Angle-type reciprocating compressor, 301–304
Angle valves, 418, 420
Annual return, 54
API-ASME Code for Unfired Pressure Vessels, 195, 198
Arc welding, 189
Area elevation, 598

Area grades, 598
Area meters, 506
Areas, separation of project, 598
Armature, 350
Arrangement, of equipment in structures, 626
of pumps in-line, 293
ASA, *see* American Standards Association
Asbestos-impregnated graphite packing, 276
Asbestos insulation, 473
ASME, Boiler Code, Section VIII, 198
Code for Unfired Pressure Vessels, 195
ASTM, specification numbers for pipe, 399, 401
specifications, for pipe, 399
Atmospheric temperature, maximum and minimum, for design, 27
Atomization of liquid fuels, 561
Attachments for equipment, 627
Autoclaves, 373–374
design of, 374
Automatic sprinkler systems, 640
Automatic steam valve cut-off type for reciprocating compressors, 307
Auxiliary piping for pumps, 296
Average temperature, for heat transfer calculations, 237
AWWA specification for cast-iron pipe, 412
Axial compressors, 317, 318
regulation of, 318
Axial flow impeller, 265
Axial force on centrifugal pumps, 270
Axial thrust, balancing of on multistage pump, 274

Back filling, 600–601
Back pressure, recommended for steam cylinder of reciprocating pumps, 261
Bacteriological tests on water, 553
Baffle spacing, minimum for exchangers, 224
Balancing drum for centrifugal compressor, 317
Ball bearings, on centrifugal pumps, 275
on motors, 358
Ball-float level indicator, 519
Ball valves, 254–255
Base plates for centrifugal pumps, 277
Basic refractories, 485
Basis of design, 55
Batch system control, 525
Batter piles, 592
Bearing capacity, allowable, 596
for soil, 596, 612
Bearing housings on centrifugal pumps, 275
Bearings, for centrifugal compressors, 317
for centrifugal pumps, 274–275
for turbines, 364
Bed plates for centrifugal pumps, 277
Bell and spigot joint, 409
Bellows elements for pressure measurement, 517
Belt-line railroads, 11
Bench-scale laboratory study, 45
Bending rolls, 187–188
Bent-tube boiler, circulation pattern in, 565
Best efficiency point (BEP), 279, 284
Bid comparisons, 127
Bill of lading, 128
Billing procedure of contractor, 164
Bimetallic strip, 508
Birmingham Wire Gage (BWG), 225
Blanket insulation, erection of, 481
Blanket orders, 123
Bleed, definition of, 456
Blind, definition of, 456
line, 433–434
Block diagram, 65
Block insulation, erection of, 481
Block valve, definition of, 456
Block valves for steam, 573
Blow-down systems, 645–646
Blow-down valves, 645
Blowers, comparison with compressors, 315
drives for, 317–318
regulation of, 318
Blue-printing, 116
Body of valve, 414
Boiler feedwater, 549
Boiler feedwater pumps, 569
Boilers, drums for, 564
types of, 562–571
Boiling coefficients, 235
Bolt stud, for flanges, 407
Bolted-flanged bonnet valve, 414–415
Bolting, for flanges, 407, 410
of equipment, 627
of structural steel, 662, 663
Bolts and nuts, recommended types for various services, 409–410
Bonnet of valve, 414
Borings for soil tests, 594–595
Bottom of pipe (B.O.P.), definition of, 456
setting of elevation of, 432–433
Boundary line location, 597
Bourdon tube, 516
Bourdon-type gage, 515–516

Brake horsepower (BHP), 259, 278, 282–283, 331–335
Branch pipe connections, 411
Brick, laying of, 489
 *see also* Refractory brick and Insulating brick
Bubble plates, 192, 193
Bubble trays, 192, 193
Buckling of pressure vessels, 203, 204
Building codes, 628
Building drawings, 104
Buildings, design of, 325, 618–624
 design of for process equipment, 619
 erection of, 658, 668
 location of to minimize hazards, 636–637
Bulldozers, 600, 659
Burning pits, 646
BWG numbers, *see* Birmingham Wire Gage

Cable pulling, 660
Cables, 148, 150
Cafeteria, design of, 621
C. & F., 126
Caloric temperature for heat transfer calculations, 237
Cancellation clauses, 134, 135
Capacity, actual of reciprocating pumps, 257
 theoretical for reciprocating pumps, 256
 variation in, for centrifugal pumps, 281, 283
Carbon steel pipe, specification for, 401
Car-loading company, 11
Carry-alls, 659
Casings (centrifugal pump), circular, 269
 diffusion, 269
 mounting of, 269
 volute, 269
Cast-in-place concrete piles, 589, 590
Cast-iron fittings, 412
Cast-iron flange, 406
Cast-iron joints, types of, 413
Cast-iron pipe, description and use of, 402, 412
 installation of, 660
Catch basins, 579
Cathodic protection, installation of, 661
Cavitation in centrifugal pumps, 288
Cement, 601
Cement-asbestos pipe, installation of, 660
Center-packed plunger pump, 251
Centrifugal compressors, and axial blowers, 310–320
 bearings for, 317
Centrifugal compressors, design calculation for, 343
 diaphragms for, 315
 drives for, 317–318
 efficiency of, 333
 impeller types, 312–314
 in-line arrangement of, 315
 lubrication of, 317
 multistage, 311
 performance curves for, 336–337
 regulation of, 318
 selection of impeller size and speed, 337–338
 shaft seals for, 316–317
 terminology for, 310–311
Centrifugal filter, 375
Centrifugal governor, 365
Centrifugal pumps, affinity laws for, 280
 auxiliary piping for, 296
 base plates for, 277
 bearings, 274, 275
 calculation of NPSH for, 288
 casings of, 268
 chemical, 266–268
 classification of, 263
 corrections to performance curves of, 285–287
 coupling for, 277
 double-suction, 270
 estimating effect of speed and impeller diameter on performance of, 280–289
 general service of, 266
 gland oil piping for, 296
 horsepower and efficiency of, 278
 hot oil, 266
 impeller types, 265
 installation of, 293
 mechanical seals for, 276, 277
 multistage, 272
 NPSH, 287
 operating points for, 289
 operation in parallel of, 289
 operation in series of, 289
 packing for, 276
 performance curves for, 278–280
 process type, 266
 radial impeller types, 271
 range of service for, 263
 rotation of, 268
 shaft and shaft sleeve for, 275
 single-stage, 272
 single-suction, 270
 specific speed of, 284–285
 stuffing box for, 275
 wearing rings for, 271–272
Centrifuges, description of, 375

Centrifuges, performance of, 375
tests of, 375
Certified prints, description of, 109
for centrifugal pumps, 277
use in foundation design, 608
Chain operators, use of, 458
Chambers of Commerce, aid in site selection of, 17
information available from, 24
Change houses, construction for, 658
design of, 620
Change orders, 128, 134
Changing impeller diameters, 283
Characteristic curve for centrifugal pump, 279
Check valves, 295, 421, 423
Checking drawings, 103–105, 211
Chemical feeders, 262
Chemical pumps, 266–268
Chemisorption instruments, 526
C.I.F., 126
Chilton's method, 51
Chilton and Genereaux, pressure drop equation for exchangers, 239
Chromel-alumel thermocouple, 511
Circuit and conduit schedules, 114
Classifiers, 381
Clearance control for reciprocating compressors, 308
Clearance pockets, effect on compressor efficiency of, 308
Clearances, typical values for pipe, structural steel, and power lines, 629
Closed impeller, 265, 271
Coal, as a fuel, 31, 556
Code for pressure piping, 396
Codes, *see* name of sponsoring body
state and city, 210
Coil-in-box cooler, 221
Coke, as a fuel, 556
Cold junction of thermocouple circuit, 513
Cold spring of pipe, 438
Color code, for pipe, 667
for structural steel, 662
Columns, for structures, spacing of, 626
Combination piles, 591
Commodity buying, 122
Commutator, 350
Compacting, 600
Completion date for process design, 55
Compound-wound motor, 351
Compressed Air and Gas Institute, reciprocating compressor lubrication suggestions by, 307
reciprocating compressor nomenclature by, 297–298
Compressed Air and Gas Institute, terminology for centrifugal compressor by, 310–311
Compressibility factor, 340–341
chart for, 340
Compression ratios, typical, 339
Compression tests for soil, 596
Compression theory, 326–330
Compression-type fitting, 409
Compressive strength of concrete, 602–603
Compressor efficiencies, 331, 333
Compressor horsepower, calculation of, 331
Compressors, *see* type
Compressors, accessories for, 322
cost of maintenance for, 322
cylinders for, 304–305
for plant air, 581
foundations for, 587–588
installation of, 325
piping at, 461
pistons for, 304–305
valves for, 304–306
selecting type of, 320–322
Concrete, costs of, 608
curing of, 602–603, 605, 606
pouring of, 604–605
reinforced, 601
repair of, 606
strength of, 602–603
use in piles of, 589
Concrete foundations, curing of, 605, 606
Condensate, piping for, 572
Condensate lines, typical rules for, 461
Condensate traps, *see* Steam traps
Condensation curve, 235
Condensers, 220
Condensing coefficients, 232
Conduit, 559
Cone-roofed tank, 574
Conference leader, 137
Conference notes, 137
Conferences, 136, 137
Confidential information on flow diagrams, 79
Construction, final stages of, 668
organization for, 654
personnel for, 654
planning of, 653
preliminary data for construction planning, 32–42
steps in, 88
Construction department, planning by, 653
Construction interferences, anticipation of, 34

Construction labor, distribution of, 655
man hour requirements, 94
on foreign jobs, 41
rates, 657
Construction office building, 658
Contingencies for estimates, 47
Continuous system control, 531
Contracts, bases for, 159
cancellation of, 166, 179
definition of terms in, 173
form of, 162
types of, 160, 161, 162
Contract buying, 122
Contractor and customer, division of work, 2
Contractor, definition of, 2
function of, 156
selection of, 157
Control centers, 540
Control Laboratories, design of, 622
Control of processes, 525, 531–542
Control of turbines, 364
Control panels, 105, 540–541
Control point of an instrument, 534
Control room, design of, 540–542
Control theory, 540
Control valves, air pressure of, 533
definition of, 456
description of, 423–425
double-port, 423
single-port, 424
Controlled volume pumps, 261, 262
Convection coefficient, for horizontal pipes, 491
Conveyors, 389–390
Cooling between stages for centrifugal compressors, 315
Cooling systems, circulating, 552
Cooling towers, design of, 377–378
Cooling water, 549
Copper-constantan thermocouple, 511
Cork insulation, 483
Corrected mean temperature difference, 237
Correspondence, and drawings, transmitting for foreign jobs, 40
design instruction, 143
forms, 141, 148
instruction letters, 141
interoffice, 140
letters of intent, 146–147
memo forms, 141
owner-contractor, 144
restrictions on, 144
style, 151
transmittal letters, 142, 148
vendor, 146–148
Correspondence, volume of, 139
Corrosion allowance for vessels, 210
Corrosion data, 58
Corrosive fluids, pumping of, 277
Corrosion inhibitors, 552
Cost estimates, costs of, 47
precision of, 47
rapid methods for, 47
types of, 46
Cost-plus contract, typical example of, 163
Cost-plus with fixed-fee contracts, 161
Costs, estimation of for piping, 447–452
labor, 54, 170, 657
slip-on versus welding neck flanges, 404
Counterflow 1-1 exchanger, 216
Counting instruments, 524
Couplings, for centrifugal pump, 277
making joints by, 402
use of on vessels, 192
Cracking furnaces, design of, 387–388
Craftsman, manhour distribution, 656
Crane, crawler type, 664
gantry-type, 626
self-propelled, 600
Cross-fins, 216
Crossflow, in heat exchangers, 232
pressure drop in, 239
Crushers and grinders, description of, 380–381
selection of, 379
test of, 380
Crystallizers, description of, 381
design and selection of, 381
tests of, 381
Cultural facilities, effect on plant location, 17
Customer, definition of, 2
Cutouts, use of in plot planning, 82
Cycle for gas turbines, 369
Cylinders, for reciprocating compressors, 304
Cylinder liners, 254
Cylinder sizes, designation of for reciprocating pumps, 254
Cylindrical vessel, stresses on, 196

Deadweight loads on vessels, 203
Deaerating heaters, 568
Deaeration of boiler water, 549
Delivery date, 91, 127
Delivery times for equipment, 91
Den Hartog formula, 196
Density measurement, 529
Design data, 59
Design drafting, 103
manhour requirements, 94

Design formulas for pressure vessels, 199, 201–203
Design information, for autoclaves, 374
for centrifuges, 376
for cooling tower, 378
for crushers and grinders, 380
for dryers, 383
for evaporators, 385
for filters, 386
for furnaces, 388
for mixers, 390–391
for thickeners, 393
Design instruction letters, 142
Design methods for heat exchangers, 242
Design resistances for shell-and-tube heat exchangers, 244
Design specifications, 4
Desuperheaters, 570
Desuperheating in condensers, 235
Dial thermometer, 508
Diaphragm-actuated tank level indicator, 521
Diaphragm element for pressure measurement, 517
Diaphragm-operated control valves, 423–425
Diaphragm pump, 263, 264
Diaphragms, on centrifugal compressors, 315
Diatomaceous earth insulation, 474
Diesel drive, for reciprocating compressor, 301–304
Diesel fuel, 554
Differential-pressure instruments, 499
Differential pressure, measurement of, 504
Differential-pressure type of liquid-level gage, 519
Diffuser, for multistage centrifugal compressor, 314
Diffuser-type multistage pump, 272, 274
Dikes, 637
Dimension drawing and data sheet, 107
Dimensional analysis applied to centrifugal pumps, 280
Dimensioning, need for accuracy, 599
Direct-current motors, advantages of, 351
principles of, 350, 351
starting of, 351
types of, 351
Directions, definition of, 456
Discharge piping for pumps, 294–295
Discounts, 125
Disk valves, 254–255
Displacement for reciprocating compressor, calculation of, 345–346
Displacement meters, description of, 499
Displacement meters, principles of, 521
Displacement-type level indicator, 519
Distribution systems for electric power, 557
Disturbances at orifices, 504
Ditching machines, 659
Dittus and Boelter equation, 228
Dock charges, 37
Donohue equation for outside coefficient of heat transfer, 232
Double-action centrifugal pump, 270
Double-pipe heat exchanger, 215
Double-suction centrifugal pumps, 270
Double, volute on multistage pumps, 274
Draft equipment for boilers, forced and induced, 569
Draft gages, 571
Drafting labor, reduction of, 116
Drafting manhour requirement, calculation of, 96
Drag-lines, 600, 659
Drains, 192
Drains, vents and test openings, typical rules for, 459
Drainage systems for process plants, 578
Drip-legs, definition of, 456
in steam lines, 572
Drip-proof motor, 355
Driver-speed, control for reciprocating compressors, 307
Drives, for centrifugal compressors and blowers, 317–318
for centrifugal pumps, 278
selection of for compressors, 324
Drawings, for equipment, 607–608
Drawing scales, 115
Dryers, description of, 382–383
design and selection of, 382–383
tests of, 382
Dry-seal gas holder, 575
Duplex pump, 251
Dust-explosion-proof motor, 356

Early-strength cements, 603
Earnings, 54
Earthquake, overturning moment produced by, 586
Economic evaluation, 47
Economic studies, 45
Economical thickness of insulation, 495
Economics, in building construction, 624
of foundation design, 608–609
Economizers, 568
Effective fin area, 230
Effective mass velocity, 231
Efficiency, of compressors, 332
of gas turbines, 371

Efficiency, of insulation, 496
of multistage turbines, 367
of pumps, 259–260, 279–282
of single-stage turbines, 366
Eighty-five per cent magnesia insulation, 474
Electric power, 556–560
data on for design, 559
distribution systems, 557
drawings of, 104
Electrical conduit, installation of, 660
Electrical control, 537
Electrical controllers, 537
Electrical design, 104
Electrical drawings, 104, 114
Electrical equipment, safety rules in design, 639
specifications for, 104
Electrical power line, clearances for, 629
Electrical system, schematic of, 558
Electrically operated valves, 538
Electronic computers, use in process control, 539
Electronic control, 537–540
compared to pneumatic, 539
Elevation of equipment, selection of, 539
Elevations, location of, 597–598
Ellipsoidal heads, 190
Ells, 409, 411
Emergency governors for steam turbines, 365
Emergency valving, 648
EMF, measurement for thermocouple, 512
Empirical correction to volumetric efficiency, 333
End-bearing piles, 590
End-packed plunger pump, 251
Energy balance, 326
Engineering and construction, example of contract, 172
firm, 2
manhours for, 92
Engineering and News Record, 52
Engineering flow diagram, arrangement of, 71
examples, between 67 and 68
information on, 69
item numbers for, 68
representation of equipment on, 68
standard symbols for, 68, 70, 71
Engineering research, 43
Enthalpy, 328
Entropy, 328
Equipment, erection of heavy, 663–666
Equipment, selection of, 57
specifications of, 58
Equipment costs, approximation of, 52
Equipment drawings, use of in foundation design, 607–608
Equipment specialty engineering, 101–102
Equipment supports, types of, 626
Equivalent diameter, 229
Erection, of insulation, 481–484
of major equipment, 663
Erection drawings, 107
Escape chutes, 637
Estimate, 3
example of contractor's, 182
Estimating departments, 47
Estimating group, 3
Estimation, of construction labor, 95
of drafting manhour requirements, 96
of total plant costs, 51
Ethics, 79, 134, 183
Evaporators, 384–385
comparing quotations on, 384–385
design of, 384
test of, 384
heat transfer coefficients of, 384
types of, 384
Excavation operations, 599–601, 659
Ex-Factory, 126
Exhaust steam, from turbines, properties of, 368
Ex-Mill, 126
Expanding of exchanger tubes, 225
Expansion bends, common type of, 437
Expansion joints, 428–430, 437
for block insulation, 482
tabulation of, 429–430
Expansion stresses, design calculation for, 438–440
Expansion U-bend, 439
Expediter, duties of, 133
Expediting, 124, 132
Explosion doors, 640
Explosions, information on prevention, 633
Explosion-proof electrical equipment, 560
Explosion-proof motor, 356
Export license, 35
Export packing, 37, 38
Export permits, 35
Export shipping, 36
Extended surface tubes, 216
Extended surface exchangers, 215
External gear pump, 290
Extraction turbines, 363
Ex-Warehouse, 126

Fabrication-in-transit, 11

Fabricator's drawings for vessels, 211
Fanning equation, 442
Fans, 318
F.A.S., 126
Feedwater, heaters for, 568
  surge drums for, 569
  treating facilities for, 568
Field coil, 350
Field-fabricated pipe, typical rules for, 461
Field fabrication of vessels, 195
Figure 8 blind, 433–434
Filing and distribution of vendor-prints, 108
Filing, systems of, 151
Filter photometers, 527
Filters, tests of, 386
Fin efficiency, 230
Fin tubes, 215
Fire extinguishing equipment, 648–649
  selection of, 648
Fire foam connections for storage tanks, 575
Fireproofing of steel, 626
Fire pumps, 553
Fires, information on prevention, 633
Fire-tube boilers, 563
Firewalls, 637
Fire water, 553
Fittings, cast-iron, 412
  energy losses in, 445
  flanged, 411
  pressure drop in, 445
  ratings for, 403
  screwed, 412
  welding type of, 409, 411
Five-step clearance control, 308–309
Fixed capital costs, 50
Fixed price contracts, 161
Fixed-tube sheets, 221
Flame arrestors, for storage tanks, 574
Flammability limits, 640–664
  calculation of for mixtures, 641
Flange and bolting, selection of, 455
Flange facings, 407
  American standard, 408
  flat-faced, 407
  male-and-female, 407
Flange fittings, 411
Flange prices, estimation of, 449
Flange taps for orifices, 502
Flange types, 405
Flanged joints, 403–409
Flanges, 403–409
  companion for pumps, 295
  ratings for, 404
  special types of, 407
Flapper for pneumatic control, 533
Flare stacks, design of, 646
Flashing cement, 482
Flat heads, 190
Flexible couplings, 277
Floating head, pull-through type of, 221
  split-ring type of, 221
Floating-roof tanks, 574
Float-type level controller, 519, 520
Flood conditions, design for, 28
Flow diagrams, block, 65
  engineering, 67
  graphic, 67
  process, 65
  schematic, 64
Flow measurement, mechanical, 499
Flow nozzle, 503, 504
Fluid expansion thermometer, 509
F.O.B., 126
Force-balance type of manometers, 504–506
Forced circulation boilers, 564
Foreign projects, 35–42
Form work for foundations, 593
Formal conferences, 137
Forming operations for spread footings, 586
Forms, building of, 601
  economics of, 609
Forwarding agents, 36, 37
Fouling factors, 236
Foundations, construction of, 593, 661
  design of, 609–617
  example design of, 613–615
  for compressors, 325
  for pumps, 294
  for small equipment, 616
  layout of, 607
  scheduling of work for, 593
  selection of type of, 592–593
  types of, 584
Four-cycle engines for compressor drives, 302
"Free Air" unloaders, 308
Freight forwarders, 36
Freight planning, movement within foreign country, 38
Freight rates, 9
  and plant location, 10
Friction factor, 442–443
Frost line, design for, 28
  location of conduit runs with respect to, 559
  location of foundation with respect to, 586
Fuel oils, 31, 554, 555
Fuel sources, 30

Fuels and plant location, 14–15
Fuels for process plants, 553–556
Fundamental research, 43
Furnaces, 387–388
design of, 387
types of for boilers, 561
F. W., definition of, 456

Gage glasses, 523
Galvanized bands for insulation, 482
Galvanizing on structural steel, 114
Gantry-type cranes, 626
Gas burners, 561
Gas diesels for reciprocating compressor, 301–304
Gas engines for reciprocating compressor, 301–304
Gas storage tanks, 575
Gas turbines, 368–371
applications of, 370, 371
cycle for, 369–370
efficiencies of, 371
use of on centrifugal compressors and blowers, 317
Gaskets, for flanges, 407
selection of, 455
Gate valves, 417, 418
installation of in pump piping, 295
Gear pumps, 290
General files, 152
General service pumps, 266
Geometrically similar pumps, 281
Gin poles, 664–665
Glands for valves, 414
Gland oil piping for pumps, 296
Glass insulation, 475
Glass pipe, 399
Globe valves, 417–420
Governors, for steam turbines, 365
throttling type of for reciprocating compressors, 307
Gradient, definition of, 456
Graphic diagram, 67
Graphic panels, 540–541
Grating, use of, in outdoor structures, 627
Grid plates, 193
Grinders, *see* Crushers and grinders
Grounding, 639
Grounding systems, installation of, 661
Ground joint unions, 403
Grout, 606
Grouting, 666
Guarantee clause, 129
Guaranteed maximum contracts, 162
Guarantees, 181
Guards and guardhouses for process plants, 582
Guide for pipe, 430, 456
Guide vanes for centrifugal compressors, 314
Guyed derrick, 661–664

Hairpins, 215
Handrails, 640
Hangers, definition of, 456
for pipe, 430–433
Hazardous areas, electrical equipment for, 560
Hazardous equipment, minimum spacing for, 636
Hazardous locations, for reciprocating compressors, 310
National Electric Code, definition of, 357–358
Hazards in process plants, 632–635
Head, calculation of horsepower and psi from, 288
definition of, 272
of centrifugal pump, 278, 288
variation in, for centrifugal pumps, 281, 283
Head room, recommended values, 638
Headers for fire water, 553
Heads for vessels, selection of, 190
spinning of, 190
types of, 190, 191
Heat exchangers, design of, 226, 241, 242
fabrication of, 224
nomenclature, and parts for, 221, 222
safety rules in design of, 638
shell size of, 224
specification sheet for, 240
tube lengths for, 224
types of, 214–220
Heat losses, through insulated surfaces, 490–493
through insulation, example calculation of, 492–493
Heat transfer research, University of Delaware, 231
Heaters, *see* Furnaces
Heating and ventilating facilities, design of, 621
Heating, costs of, 16–17
Heavy equipment, erection of, 663–666
Heavy machinery, planning for, 39
Height of cooling tower, determination of, 377
Helical flow stage of a turbine, 361
Helical pumps, 291
Horizontal lift check valve, 419
Horizontally split casings for centrifugal pump, 269

Horsepower, calculation of for compressors, 331 (*see also* Work of compression)
  for centrifugal pumps, 278
  for motors, standard ratings of, 358
  for reciprocating pumps, 259
Hose connections, 459
Hot oil pumps, 266, 267
Hot process phosphate, softening of, 551
Hotel facilities, effect on plant location, 17
Houdry process, use of gas turbines in, 369
Housing, for labor on foreign jobs, 42
Human relations, need for understanding of, 154
Hurricane frequency, design for, 27
Hydraulic control, 537
"Hydraulic" speed governor, 365
Hydrostatic test, of heat exchangers, 226
  of vessels, 195

Impellers, cutting of, 283
Impeller diameter, effect on centrifugal pump performance, 283–284
Impeller size, selection of, for centrifugal compressors, 337–338
Impeller types, for centrifugal compressors, 312–314
    backward curved, 313
    forward curved, 313
  for centrifugal pumps, 265
Impulse turbine, 361, 362
Induction motors, calculation of actual speed, 353
  description of, 352
  use of, on reciprocating compressor, 301
Indicator horsepower, 259
Industrial Commissions, aid in site selection by, 17
Industrial hygiene, information on, 634
Inert gas purges, 642
Infrared photometer, 525, 527
In-line arrangement of centrifugal compressors, 315, 316
Inquiries, for insulation, 480
  forms for, 124
  minimum number of, 124
  parts of, 124, 240–241
Inside film coefficient, 227
Inspection, of equipment, 124, 131
  of vessels, 195
Inspector, duties of, 131
  reports by, 131
Instability, conditions for in foundations, 613
Installation, drawings for, 106
  of compressors, 325
  of pumps, 293
Instruction letters, 141
Instructions to vendors, 134
Instrument air, design of system, 536
Instrument drafting, 105, 115
Instrumentation engineering, 542–543
Instrumentation, for centrifugal compressors, 324
  for internal combustion engines, 573
  for reciprocating compressors, 324
  for steam plants, 570–571
  on flow diagrams, 71
Instruments, chemical and physical property measurement by, 526–530
  general classification of, 498–499
  types of, 498–499
Insulating brick, 485
Insulating cement, 482
Insulating firebricks, characteristics of, 490
Insulating refractories, 485
Insulation, characteristics of various materials for, 476–478
  economical thickness of, 495
  efficiency of, 496
  erection of, 481–484, 668
  factors in evaluation of, 479–480
  for personnel protection, 640
  for steam piping, 573
  heat loss through, 490–493
  inquiries and quotations for, 480
  jackets for, 483
  optimum thickness of, 494–496
  safety rules in design of, 640
  selection of, 479
  types of, 473–478
Insurance for construction labor, 42
Intercoastal canal, 12
Internal combustion power, 573
Internal gear pump, 290
Interoffice correspondence, 140
Interoffice letter, 140
Interstate Commerce Commission, 11
Inversions, effect of on plant location and layout, 635
Inverted bucket trap, 428
Invoice, 128
Iron-constantan thermocouple, 511
Isentropic compression, reversible, 328
Isentropic efficiency, 333
  for centrifugal compressor, chart for, 335
Isentropic head for centrifugal compressors, 336
Isentropic horsepower, from enthalpy diagrams, 332

Isothermal compression, reversible, 328
Isothermal efficiency, 333
Item numbers, on flow diagrams, 68

Jacks, heavy duty, 664
Joints, for pipe, 402
  staggering of for insulation, 482–483
  types of cast-iron pipe, 413
Jurisdictional disputes, 655

Kerosene as a fuel, 554
Kettle-type reboiler, 219
Kingsbury-type bearing, for centrifugal compressor, 317
  for centrifugal pumps, 275
Knockout drums, 323, 647
Kraft paper, 483

Labor, and plant location, 15
  check list for construction planning of, 33
  costs of, 54, 170, 657
  distribution of, on construction jobs, 655
  on foreign jobs, 41
  unrest of, 16
  U.S. expatriates, 41
Labor market studies, 15
Labor organizations, 655
Labor practices, 655
Labor rates, 656
  on construction jobs, 657
Labor relations, 655
Laboratories, design of, 622
Labyrinths, for centrifugal compressors, 316
Ladders, 194
Lantern ring, 275
Lap-joint flanges, 404,
Large equipment, planning for, 391
Large pressure vessels, maximum sizes, 40
Lavatory facilities, design of, 621
Layout, for steam distribution system, 572
  of equipment, 637
  of foundation, 607
  of office equipment, 619–620
  of piping, 103, 434–440
  of transfer pump system, 576
LCL shipments, 10
Lead wires for thermocouples, 513
Leased wire services, 150
Lens flange, 406
Letter of intent, 146
Letters, *see* Correspondence
Level gages, 523
Lift check valve, 421
Lifting lugs, 665–666
Light hydrocarbons, pumping of, 277
Lighterage charge, 37
Lighting, 620, 623
Lime-soda softening process, 551
Limit controls, 524
Limits of flammability, 640–641
Line blinds, 433–434
Line numbers, 71, 457
Line sizes, determination of, 440–452
  usual sizes, 457
Line sizing, for steam distribution, 572
  techniques and criteria for, 445
Linen drawing sheets, 116
Ljungstrom turbine, 362
Liquefied petroleum gas as a fuel, 31, 554
Liquid cylinders for reciprocating pumps, 253
Liquid horsepower, calculations of, 259
Liquid-level instruments, 519–523
Liquid storage, 574
Load tests on soil, 610
Loads, allowable for structures, 628
Local flood or tide conditions, design considerations for, 28
Location of subsurface water, 595
Locomotive boiler, 566
Log mean temperature difference, 236
Longitudinal baffle, 218
Longitudinal fins, 215, 216
Longitudinal stresses, 196
Low point, establishment of, 598
Low temperature insulation, erection of, 483
LTL shipments, 11
Lubrication, of centrifugal compressor, 317
  of plug valves, 421
  of reciprocating compressor, 306–307
Lump-sum or fixed-price contracts, 161

Machine bolts for flanges, 407
Machine shops, 576
Machinery, foundations for, 587–588
Maintenance costs for compressors, 322
Maintenance facilities in process plants, 576
Manholes, electrical, 559–560
  for sewers, 578, 579
Manhours, distribution on construction job of, 656
  requirements for engineering, drafting, and construction of, 92
Manometers, bell type, 517
  force balance, 504–506
  mercury, 504–506

Manual control, 531
Manufacturer's drawings and data, 106
Manufacturing costs, estimation of, 52
Marine freight carriers, 37
Markers, permanent, for construction site, 598
Marshall and Steven index, 52
Mass ratios, for massive foundations, 616
  for reciprocal compressor foundations, 325
Massive foundations, 587–588
  design of, 616
Master plot plan, 80–82
  coordinates on, 597
  example of, 81
  factors influencing, 82
Mat foundations, description of, 587
  design of, 615
Material and energy balance, 55
Materials handling equipment, 389–390
  selection of, 389
Materials of construction, for centrifugal pumps, 278
  for exchanger tubes, 225
  for pipe, 399–400
  for pump casings, 268
  for reciprocating pumps, 250
  for structural shapes, 630–631
  for vessels, 209
  selection of, 57
Mean effective pressure (MEP), 260
Mean temperature difference, 236
  corrected, 237
Mechanical control, 532
Mechanical design, and safety, 637–640
  of vessel, 210
Mechanical efficiency, definition of, for compressors, 333
  of reciprocating pumps, 259–260
Mechanical expansion joints, schedules of, 111
Mechanical flow diagram, 67
Mechanical joint, 409
Mechanical seals, 275–277
  for centrifugal compressors, 317
Mechanical specifications, 101
Medical facilities, design of, 621–622
Memo forms, 141
Mercury-in-glass thermometer, 508
Mercury-type manometers, 504–506
Meridional stress, 196
Mineral wool insulation, 474
Mixed flow impeller, 265
Mixers, selection of, 390
  tests of, 390
Mixing equipment, 390
Mixing plants, for concrete, 604
Monolithic mass, 588
Monthly time chart, 100
Moody friction factor chart, 443
Motor-driven reciprocating compressor, 299
Motor ratings, 358
Motors, 348–361
  alternating-current, 352–355
  as drives for centrifugal compressors, 317
  classification by N.E.M.A. of, 355–356
  cost comparison of, 356–357
  direct-current, 350–351
  elementary principles of, 349, 350
  selection of, 359
  standards for, 349
  typical applications of, 360
Movement of raw materials and products, 576
Multicylinder compressor, 300
Multistage blowers, 314–316
Multistage centrifugal compressor, 311
Multistage centrifugal pumps, 272
Multistage compressors, 314–316, 338
Multistage steam turbines, 362–363

National Electric Code, 349
National Electric Manufacturers Association (N.E.M.A.), 349
Natural gas, design of pipeline connections, 30
  plant pressure of, 31
  use of in process plants, 30, 554
Net positive suction head (NPSH), 287, 288
Nipples, use of on vessels, 192
Noise abatement, 14
Nominal pipe sizes, 396
Nozzles, arrangement on centrifugal pumps of, 269
  control of steam turbines by, 364, 365
  details of, 211
  for process vessels, 192
  orientation of, 210, 211
  reinforcement of standard-length, 192
Nusselt equation, 227, 232–233
Nutating-disk displacement meter, 500
Nuts, *see* Bolts and nuts

Ocean freight rates, 36
Octagonal footings, design of, 613–615
Office buildings, design of, 619–620
Off-peak power, 556
Off-the-shelf materials, procurement of, 123
Oil-field boiler, 566
Oil malleable union, 402
Oil seal for centrifugal compressors, 316

One-two exchanger, 218
Open impellers, 265, 271
Open motor, 355
Operating company, 2
Operating labor, 15, 18
Operating points of centrifugal pumps, 289
Operating tests of equipment, 668
Optical pyrometers, 514–515
Orientation of nozzles, 210, 211
Orifice meters, 502–503
  equation for flow rate for, 501
Orifice-type baffles, 221
Orsat, automatic, 526
O.S. and Y. (outside stem and yoke), definition of, 416
Outdoor equipment location, advantages of, 625
Outdoor process units, 624–629
Outline drawing for vessels, 112
Outside coefficients for shell-and-tube exchangers, 231
Outside film coefficient, for double-pipe exchanger, 229
  for extended surface, 229
Outside stem and yoke valve, 415
Overall heat transfer coefficients, calculation of, 227
  estimating of, 244–246
Overhead costs, 54
Overhead clearance for piping, typical, 459
Overtime payments to labor, 171
Overturning load for cylindrical tower, 612
Overturning moment, 586

Package-type steam generators, 29, 565
Packed stuffing box, 275
Packing, for centrifugal pump, 276
  for export, 37, 38
  for reciprocating compressor, 304
  for reciprocating pumps, 254–255
  for turbines, 364
  *see also* Shaft seals
Packing gland on 1-1 exchanger, 217
Packing list, 128
Packless expansion joint, 429
Packless valves, 416
Painting of pipe, 470
Parallel centrifugal pump operation, 289
Patent protection, 130
Pedestals for centrifugal pumps, water-cooled, 270
Perforated plates, 193
Performance curves, estimation of, 282
  for centrifugal compressors, 336–337
Performance curves, for centrifugal pumps, 278–280
  for viscous liquids, correction to, 285–287
Peripheral pumps, 291
Peripheral velocity, design valve for centrifugal compressor, 338
  of centrifugal pump, 283
Permits, 35
Personnel distribution, typical, 95, 96
Personnel morale, 154
*p*H meter, 530
Photoelectric cell, 524
Piece marks on structural steel, 662
Piece numbers for pipe, 667
Pile drivers, 606–607
Piles, cast-in-place concrete, 589
  driving of, 606–607, 658
  penetration of, 607
  types of, 588–592
Pilot light for flare stack, 646
Pilot plant facilities, design of, 623
Pilot plant, use of, 45
Pin-fins, 216
Pipe, ASTM specifications for, 399, 401
  coating of, 470
  design codes for, 396
  fittings for, 409–413
  history of, 394
  identification of, 667
  joints for, 402
  manufacture of, 395
  plastic, 399
  required thickness of, 398
  schedule number for, 398
  selection of, 455
  sizes of, 396
  steel, 399
  types of, 395
Pipe and fitting prices, estimation of, 448
Pipe bending, 459
Pipe-cap, 409, 411
Pipe coils, 223
Pipe-concrete piles, 590
Pipe fittings, screwed, 403
Pipe foundations, 588
Pipe hangers, 430–433
Pipe insulation, erection of, 482
Pipe shoe, 430
Pipe supports, 430–433
  design of, 432–433
  fire proofing of, 470
  indication on drawings, 111
Pipe taps for orifices, 502
Pipe wall resistance to heat transfer, 235
Pipe wall temperature, calculation of, 238
Pipelines and plant location, 13

Piping, accessories for, 426–434
changes in direction of, 434, 462
clearances required for, 629
estimating installed costs of, 447–452
expansion of, 436–440
fabrication of screwed piping, 667
for compressor installations, 325
for pumps, 294–295
for relief valves, 646–647
for steam distribution, 571–573
installation of, 666
layout and arrangement of, 434–440
lifting of, 667
safety rules in design for, 638
shop-fabricated or field-fabricated, 110
sizing of, 440–452
specifications for, 452–470
testing of, 464–465
Piping costs, use of in line sizing, 447
Piping design, 103
Piping drawings, checking of, 103
methods for, 110
scales for, 110
Piping material listing or take-off, 112
Piping representation on flow diagrams, 70
Piping specifications, typical example of, 456–470
Piping stresses, 436
Pirani gage, 518
Piston pump, 251–253
Piston-ring expansion joint, 429
Piston rods, 254
Piston speeds recommended, for reciprocating pumps, 257–258
for reciprocating compressor, 304–305
Pistons, for reciprocating pumps, 251–253
Pitch, minimum for exchanger tubes, 224
Plant air, 581
Plant datum, 598
Plant design, steps in, 88
Plant laboratories, design of, 622
Plant layout, *see* Master plot plan
and safety, 635
Plant location, factors in, community, 17
fuel and power, 14
labor, 15
noise abatement, 14
process water, 13
raw materials, 9
transportation
air freight, 12
railroad freight, 10
trucking, 11
waste disposal, 14
water, 12
weather, 16
Plant location studies, information for, 19–24
Plant north, setting of, 598
Plant service air system, design of, 581
Plant site, *see also* Plant location
procedure for selection of, 8
Plant terrain studies, 28
Plant wastes, 31
Plastic, use in chemical pumps, 266–268
Plastic pipe, 399
Plates for shells, usual sizes of, 189
Platforms, dimensions of, 640
for vessels, 194
Platinum–platinum rhodium thermocouple, 511
Plot plan, *see* Master and unit plot plans
Plug cock, 421
Plug valves, 421, 422
Plumbing fixtures, water usage of, 580
Plunger pump, 251–253
Plungers for reciprocating pumps, 251–253
Pneumatic control, 532
proportional, 534
proportional reset, 536
proportional-reset-rate, 536
Polyphase induction motors, 354
Polystyrene insulation, 475
Polytropic compression coefficient, 331
Polytropic efficiency, 333–335
to isentropic efficiency for centrifugal compressors, conversion of, 334
Portable extinguishers, 648
Potentiometer circuit for thermocouple emf, 513
Pouring of concrete, 604, 605
Power costs, study of, 15
Power data for design, 29
Power distribution, 559
Power shovels, 659
Prandtl number, 227
Precast-concrete piles, 590
Preliminary design, 158
Pressure drop, allowable for process lines, 447
in gas and vapor lines, 447
in heat exchangers, 238
in process lines, equations for, 441, 442
Pressure gages, 515–516
connections for, 459
Pressure-measuring instruments, 515–518
Pressure regulator for steam turbine control, 365
Pressure-relieving devices, 643–646
Pressure-vacuum vents, 574
Prestressed concrete piles, 591

Prevailing wind direction for design purposes, 27
Prevailing winds, effect on plant layout of, 635
Prices, discussion of, 134
re-negotiation of, 134
Process control, 525, 531–542
Process design, data for, 59
procedures for, 55–58
Process development, 44
Process engineer, 56, 57
and project engineer, relation between, 58
Process engineering, 43–63
manhour requirements for, 93
Process evaluation, 45
Process flow diagrams, 58, 65
Process information on engineering flow diagrams, 69
Process lines, sizing of, 440–452
Process pumps, 266, 267
Process specifications, for pumps, 292
use of, 101, 157
Process steam, 557, 560
Process wastes, disposal of, 580
Process water, 548
Procurement, engineering participation in, 122
manhour requirements for, 94
preliminary data for, 32
purchaser-vendor practices for, 134
Procurement Department, organization and operation of, 121
Procurement methods, 122
Production Department, 109
Project engineer, and procurement, 135
cooperation of with instrument engineer, 543
definition of, 1
duties of, 2–4
Project engineering, manhour requirements for, 93
Project estimate, 3
Project files, 153
Project manager, definition of, 1
Project organization, 3
Project schedule, 85
Proportional control, 533
Proposal, by contractors, 158, 169
Proposals, *see* Quotations
Public transportation, effect on plant location of, 17
Pumps, *see* type of pump
Pump casings, 268
Pump house, 293
Pump selection chart, 249
Pumps, alignment of, 294
Pumps, arrangement of, 293
boiler feedwater, 569
for cooling water, 552
for fire, 553
foundations for, 294
installation of, 293
piping for, 294–295, 460
safety factors for, 291
tank transfer, 576
Purchase order, guarantee clause in, 129
inspection clause in, 129
labor clauses in, 130
patent protection in, 130
quality clause in, 129
shipping instructions in, 130
standard notes on, 130
subcontractor clauses in, 130
Purchasing, *see also* Procurement
Purchasing agent, 122
Purchasing methods, 122

Quantity discounts, 125
Quench towers, 646
"Quickie" estimates, 51
Quotations, 125–128
comparison of, 127
contents of, 125
evolution of for heat exchangers, 246
for insulation, 480

Radial bearing on centrifugal pump, 274
Radial impellers, closed, 271
for centrifugal compressor, 311
open, 271
semi-closed, 271
semi-open, 271
Radiation coefficient for horizontal pipes, 491–492
Radiation pyrometers, 514–515
Radiographing of vessels, 194
Railroad clearances, 628
Railroad facilities in process plants, 577
Railroad shipment and plant location, 10
Railroad spurs in process plants, design of, 577–578, 659
Railroad-type union, 402
Rainfall, for design, 27
Rainfall intensity, for design of surface drains, 579
Raised-face flange facing, 407
Random pipe, definition of, 457
Rankine cycle, 365
Raw materials and plant location, 9
Reaction turbine, 361, 362
Ready-mixed concrete, costs of compared with site-mixed concrete, 608

Ready-mixed concrete, use of, 604
Reboilers, 219
Reciprocating compressors, 297–310
  angle-type, 301–304
  clearance pockets of, 308
  design calculations for, 343–346
  efficiencies of, 334–335
  engine-driven, 301
  five-step clearance control for, 308
  lubrication of, 306–307
  motor-driven, 299
  parts of, 304–306
  regulation of, 307
  types of, 298–304
  selection of drives for, 308–310
  steam-driven duplex, 298
  suction valve unloaders for, 308
  turbine-driven, 301
  two-stage, 299
Reciprocating pumps, actual capacity of, 257
  air chambers for, 256
  application of, 249
  classification of, 250
  cylinders for, 253
  flanges for, 256
  materials of construction for, 250
  mechanical efficiency of, 259–260
  packing for, 254–255
  parts of, 253
  piston and plunger types of, 251–253
  piston rods for, 254
  recommended piston speeds for, 257–258
  sizing of, 256
  theoretical capacity of, 256
  valves for, 254
  volumetric efficiency for, 257
Recreation center, design of, 621
Recreation facilities, effect on plant location of, 18
Reducer, 409, 411
Re-entry nozzle, 361
Reference location, 597
Reflective insulation, 475, 479
Refractometers, 528
Refractories, 484–490
  selection of, 484–485
  types of, 485
Refractory brick, characteristics of, 486–487
  installation of, 489
  typical shapes of, 488–489
Refrigeration, principles of, 347
Refusal, 594
Regulation, of centrifugal and axial compressors and blowers, 318
Regulation, of reciprocating compressor, 307
Reinforced concrete, 601
Reinforced concrete structures, erection of, 662
Reinforcing steel, for concrete, 604
  specifications for, 604
Relative humidity, for design, 27
Relative roughness of pipe, 444
Release for fabrication, 113
Relief valves, description of, 425
  design basis, 643–644
  discharge rates of, 645
  discharge systems for, 645–646
  in pump piping, 295
  piping for, 646–647
Relieving devices, conditions requiring, 643–644
Research laboratories, design of, 623
Reset, 535
Resistance of valves and fittings, 446
Resistance thermometer, 514
  assembly of, 514
  range of, 514
Return on investment, 55
Reynolds number, 227, 280, 285, 442, 501
  for centrifugal pumps, 285
  for orifice meters, 501
Ridgway's approximation of compressor work, 342
Rigging techniques, 663
Right-angle compressor, 302, 303
Ring-joint flange facing, 407
Ring-wall foundation, 587
Riveting of structural steel, 662, 663
Rivets, detecting faulty, 663
Roads, safety hazards created by, 636
Roadways, in process plants, 577, 659
  suggested width of, 577
Roller expander, 225
Rpm effect of on centrifugal compressor performance, 337
Rotameters, description of, 506
  effect of changing float weight on, 507
  flow equation for, 506
Rotary compressors and blowers, 318
Rotary-liquid piston compressor, 319, 320
Rotary pumps, 290
Rotation, of centrifugal pump, 268
  of induction motors, 354
Routing fluids in shell-and-tube exchangers, 242
Run-off, for design of surface drains, 579
Rupture disks, design basis for, 643–644
  discharge rates of, 645

Saddles, 211

Safety cages, 640
Safety factors, for pumps, 291
for pressure vessels, 208
for reboiler design, 220
for soil-bearing capacity, 611
use of, 55
Safety hazards, in plant design, 632–635
information on mechanical, 635
Sanitary wastes, disposal of, 580
Sanitary water, 553
Scaffolding, 667
Scale, in boilers, 549
in cooling-water systems, 549
prevention of, 552
Scales for drawings, 100, 115
Schedule numbers of pipe, 398
Schedule preparation, 90
Schedule sheet, 86, 87
Scheduling departments, 100
Scheduling drafting manhours, 99
Scheduling foundation work, 593
Schematic drawing of an electrical system, 558
Schematic flow diagrams, 55, 64
Scope of work, changes in, 4
Screwed bonnet valve, 414, 415
Screwed fittings, 412
Sea water, use of for cooling, 552
Seal oil pump for centrifugal compressors, 324
Seal oil systems, 276
Sealing liquid, 275
Sealing oil, 276
Seamless pipe, 395
Seat of valve, 417
Secretarial help, 151
Sectional pipe insulation, 483
Security for process plants, 581
Sedimentation equipment, 391–392
Sedimentation tests, 392
Segmental baffles, 221
Selection, of insulation, 479
of pipe materials, 402
of pump types, chart for, 249
of pumps, 291
Self-supporting towers, design of footings for, 612–615
Semi-closed impellers, 271
Semi-open impellers, 265, 271
Series-wound motor, 351
Service connections for hose, 459
"Servo" system, 537
Set pressure, of relief valves, 647
Settlement of foundations, 610
Sewer systems, 578
Sewers, design of, 578
manhole for, 579
Sewers, slope of lines for, 578
Shaft for centrifugal pump, 275
Shaft seals for centrifugal compressors, 316
Shaft sleeves for centrifugal pump, 275
Shaft work, 327
Sharp-edged orifice, 503
Shear tests for soil, 596
Shearing load in foundations, 615
Sheet piling, 592
Shell-and-tube heat exchangers, 216-220
Shell plates, 189
Shell-side pressure drop, 239
Shelters, for construction, 658
for process equipment, 624
Shielded arc welding, 188
Shipment of products, 576
Shipper, 37
Shipping, costs of, 36
instructions for, 130
Shipping lengths, for piping, 462
for vessels, 40
Shoes, definition of, 457
for supporting equipment, 627
Shop drawings, description of, 107
for structural steel, 113
for vessels, 211
Shop-fabricated pipe, practices for, 110, 111
typical rules for, 461
Shop fabrication, of pipe, 666, 667
of vessels, 195
Shop testing of pipe, 464
Shopping facilities, effect on plant location, 17
Shops, design of, 623
Short-cut methods of heat exchanger design, 243
Shunt-wound motor, 351
Sieder and Tate equation, 228
Signals, standard for derrick operations, 664
Signaling, electrical systems for, 538
Silica refractories, 485
Similarity of flow, 280
Simplex pump, 251
Single-phase induction motor, 354
Single-stage blowers, 314
Single-stage centrifugal pumps, 272
Single-suction centrifugal pumps, 270
Site, *see* Plant site and plant location
preparation of, 658
selection of, 18
Sizing steam cylinders for pumps, 261
Skidding large equipment, 664
Skirts, 194
Slab floors, 626

Sliding-vane type compressor, 318
Slings for lifting cylindrical equipment, 666
Slip, of electric motors, 353
of reciprocating pumps, 257
of rotary pump, 290
Slip-on flanges, 404
Slump test for concrete, 603
Smothering gland on centrifugal pump, 276
Snow conditions for design, 27
Snow load, 628
Socket-welding flange, 406
Soil, analysis of, 596–597
bearing capacity of, 596
effect of on foundation design basis, 611
experts on, 597
investigations of, 593
profiles of, 596
sample of, 596
Soil pressure, allowable, 596, 612
beneath a footing, 586
Solid wastes, disposal of, 580
Soot blowers, 570
Spacings between hazardous process units, 636
Spare pumps, selection of, 293
Specialty engineering, 101
Specific speed, for centrifugal pumps, 263–265, 284–285
Specification sheet for pumps, 292
Specifications, for concrete, 603
for equipment, 58
for structural steel, 627–629
nomenclature for piping, 453
preparation of for piping, 453
Specifying pumps, 291
Speed, calculation of, for centrifugal compressor, 343
control of, on steam turbines, 364
effect of, on centrifugal compressor performance, 337
effect of, on centrifugal pump performance, 281
selection of, for centrifugal compressors, 337–338
Spherical tanks, 575
Spherical vessels, stresses on, 197
Spheroidal tanks, 575
Splash-proof motor, 356
Split-phase motor, 354
Spool, definition of, 457
Spool numbers, definition of, 457
for pipe, 667
Spread footings, description of, 585–586
design of, 611–615
example design of, 613–615
Spring hanger, 432
Spur tracks in process plants, 577–578
Square-edged orifice, 503
Square footings, 586
Square pitch, 225
Squirrel-cage motor, 352–353
Stages, calculation of number for centrifugal compressor, 343
Stairways, slope of, 640
Stalling pressure of reciprocating pumps, 295
Standard instrument symbols, 71
Standard notes on purchase order, 128
Standard pipe sizes, 397
Standard welding symbols, 211, 212
Standards, for motors, National Electrical Manufacturers Association, 349, 355, 356
for shell-and-tube exchangers, 223–224
Tubular Exchanger Manufacturer's Association, recommended fouling factors, 236
Start-and-stop control for reciprocating compressors, 307
Starters for alternating-current motors, 355
Starting, of alternating-current motors, 355
of induction motors, 354
of synchronous motors, 354
Static electricity, protection from, 639
Statistical information for plant location studies, 19–24
Steam cylinder diameter, calculation of, for reciprocating pumps, 261
Steam cylinders, 254
Steam distribution systems, 571–573
Steam-driven reciprocating compressor, 298
Steam generators, auxiliaries for, 568
operating ranges of, 567
packaged, 565–566
selection of, 556–568
standardized, 565–566
types of, 561–571
Steam lines, typical rules for, 461
Steam power, 560
Steam smothering, 648
Steam traps, characteristics of, 427
description of, 426–428
listing of, 428
piping specification for, 470
specifying of, 427
types of, 426
Steam turbines, applications of, 363
calculation for, 365
control of, 364

Steam turbines, description and theory of, 361–368
efficiency of, 366, 367
example calculations for, 368
rate of steam for, 366
steam consumption of, 366–368
use on centrifugal compressors and blowers of, 317
use on reciprocating compressor of, 301
Steel taper piles, 590
Steel structures, basis for design of, 627–629
Stem of valve, 414
Step-control for reciprocating compressors, 308
Storage, location of, 31
of raw materials and products, 574
of solids, 576
Storage tanks, foundations for, 587
Straight-tube, box-header-type water tube boiler, 563
Strain gage, 518
Strainer for pumps, 295
Stream analysis, 524
Streamline flow, 228
Stress, allowable, for structures, 628
working, 199, 201, 208
design calculations for, 438–440
due to dead weight wind loads, 203
in foundations, 615
in piping, 435, 436
limits of for reinforcing steel, 604
on cylinder under internal pressure, 196, 197
on sphere under internal pressure, 197
variation with configuration of piping, 439
Stress relieving of vessels, 194
Structural clearances, 529
Structural design, 103
Structural details, 113
Structural drawings, 113
Structural shapes, commonly used types of, 630
design formulas for, 629
materials used for, 630
Structural steel, detail drawings of, 104
erection of, 662
marking of, 662
piles of, 591
safety rules in design for, 690
Structures, basis for design of, 627–629
partially enclosed, 624
totally enclosed, 618–624
unenclosed, 624–629
Stub end for lap joint, 404
Stud bolt for flanges, 407
Stuffing box, circulating type of, 275
for centrifugal pump, 275
water-jacketed, 256, 276
Stuffing box packing, for reciprocating pumps, 254
Subcontractor, in purchase orders, 130
Subcooling in condensers, 235
Subsurface soil investigations, 593
Subsurface water, location of, 595
Suction headers for pumps, 295
Suction piping for pumps, 294–295
Superheaters, convection type of, 570
radiant type of, 570
use and description of, 570
Superintendent of construction, 654
Support details, 211
Supports, for equipment, 626
for pipe, 430, 431
Surface drains, design of, 579
run-off to, 579
Surge drums for reciprocating compressors, 323
Surplus materials, handling of, 181
Swing check valve, 419, 421
Switch gear for plant distribution systems, 559
Symbols, for equipment, 66
for instruments, 71
for valves and pipe, 70
Synchronous motors, advantages of, 355
description of, 354
use of for reciprocating compressors, 301
System control, 539

Tabulation of bids, 127, 128
Tachometers, 523
Tangential stresses, 196
Tank gaging, 499
Tanks, design of foundations for, 616
storage of, 574–575
Tax ordinances, effect on plant location, 181
Technical writing, 138
Tee, 409, 411
Telegrams, 149
Telephone, use of, 151
Teletype, 148, 150
Temperature measuring instruments, 507–515
usual industrial range of, 508
Temperature rise, allowable for motors, 358
calculation of for compression, 335, 344
efficiency for centrifugal compressors, 333
of cooling water, 549

Temporary buildings, 658
Terminal companies, 11
Terms of payment, 125
Test curve for centrifugal pump, 279
Testing of pipe, field, 463
  shop, 464
Tests of equipment, *see* type of equipment
Theoretical work of compression, 327
Thermal conductivity meter, 530
Thermal conductivities, of refractory and insulating fire brick, 485
  of thermal insulators, 479
Thermocouple circuits, 511
Thermocouple combinations, 511
Thermocouple errors, minimizing of, 512
Thermocouple wells, 512
Thermocouples, range of, 510
Thermo-electric temperature measuring instruments, 510–515
Thermo-syphon reboilers, 219, 220
Thermometers, solid expansion type of, 507
  fluid expansion in, 508–510
Thickeners, estimation of required area and volume for, 392
  tests of, 392
Thickness calculation, for pipe, 398
  for vessels, 199, 201, 205
Thick-walled vessels, 201
Thin-walled vessels, definition of, 196
Threaded flange, 406
Threaded pipe joints, 402
Throttling bellows, 533
Throttling, control of steam turbines, 364, 365
Thrust bearings on centrifugal pumps, 270–271, 274
Tide conditions, for design, 28
Titrimeters, 527
Toe plates, 640
Ton of refrigeration, 347
Tongue-and-groove flange facing, 407
Tool room, for construction, 658
Topography, effect on plant safety, 635
Top-suction, top-discharge centrifugal pumps, 269
Tornado frequency, for design, 27
Totally enclosed motors, 356
Tower base plates, 211
Tower, self-supporting, 585
Toxic chemicals, information on, 634
Toxic concentrations of chemicals, typical, 640
Toxic fluids, pumping of, 277
Trade terms, 126
Transfer pumps, 576
Transmittal letters, 142, 148
Transverse fins, 215
Triangular pitch, 225
Trickle coolers, 223
Trombone coolers, 223
Truck shipments and plant location, 11
Tube bundle, assembly of, 224
Tubing, 398
Tubular exchanger, Manufacturing Association Standards for, 223–224
Turbine drives, for centrifugal compressors and blowers, 317
Turbine pumps, 291
Turbines, 361–371
  piping at, 461
  *see also* types
  steam classification of, 361
Turbulence at orifice, 504
Turbulent flow, 228
Two-cycle engines for compressor drives, 302
Two-four exchangers, 218
Two-impeller blowers, 319
Two-position control, 532, 533
Two-wheel turbine, 361
Typing, on drawings, 117

Ultimate strength of concrete, 602
Ultraviolet photometer, 527
Unbalanced forces, effect of, on foundation design, 616
Underground facilities, installation of, 660
Underground lines, 458
Union bonnet valve, 415
Unions, for labor, 655
  for pipe, 402
Unit plot plans, description of, 82–84
  steps in preparation of, 84
U.S. Geodetic survey datum, 598
Utility companies, purchase of power from, 15
Utility water, 553
Utilities, *see* type
  location of generating plants to minimize hazards for, 636
  primary, 545
U-tubes, 221

V-angle compressor, 302, 305
Vacuum equipment selection chart, 323
Vacuum gages, 518
Vacuum pumps, 322
Valves, 413–426
  accessibility of, 435
  classification of, 414

Valves, electrically operated, 538
emergency, 648
estimation of prices for, 450
flow characteristics for, 418–419
lists for, 453
location of, 458, 638
nomenclature for, 414
parts of, 414
reciprocating pump types of, 254
seat for, 417
symbols for pipes and, 70
trim for, 417
Vapor barriers for all insulation, 483
Vapor-filled thermometer, 509
Vegetable cork insulation, 475
Velocity-compounded turbine, 362
Velocity diagrams of centrifugal pumps, 280
Velocities, for steam distribution piping, 572
reasonable for process lines, 447
*Vena contracta,* 502
Vendor-prints, 106
filing and distribution of, 108
identification of, 109
submission of, 109
use in foundation design of, 607
Vendors, relations with, 123
Ventilating systems, design of, 261–262, 642–643
Ventilation, for comfort, 642
for safety reasons, 640
localized, 641–642
of compressor buildings, 325
of pump houses, 293
requirements for, 641
Vents, fittings for, 192
for tanks, 574
Venturi meter, 502–503
Vertical exchangers, 220
Vertically-split casings for centrifugal pump, 269
Vessel design, development of, 211
detailed mechanical, 210
example of, 207
selecting materials of construction for, 209
setting basic dimensions for, 209
theory of, 195–197
Vessel drawings, 112, 209–211
standard details for, 211
Vessel insulation, details of, 112
procedure for, 481
Vessels, fabrication of, 187–195
heads of, 190
installing of large vertical, 665
internals of, 192
Vessels, safety rules in design for tanks and, 637, 638
sketch of, 206, 210
Vibration, in piping, 435
isolation of, 587
Viscosity correction curves for centrifugal pump performance, 286
Viscous fluids, pumping of, by rotary pumps, 291
Vitrified tile, installation of, 660
Voltage ratings for motors, suggested, 359
Voltages, typical, for equipment, 559
for power distribution, 557
Volumetric efficiency, for reciprocating compressors, 332
with actual gases, 341
for reciprocating pumps, 257
Volute-type multistage centrifugal pump, 272, 273

Warehouses, design of, 623
for construction, 658
Waste disposal, 14, 580–581
Waste-heat boilers, 566
Waste products, as fuels, 554, 556
Water, as a fire extinguishing agent, 648
common inpurities in, 546–548
information on, for design, 28
sources of, 548
temporary supply of, for construction, 659
Water-bearing strata, 600
Water-cement ratio, 602, 605
Water-cooled bearing housings, 275
Water horsepower, 259
Water jacket, 276
Water-proof motor, 356
Water seepage, 600
Water-softening processes, comparison of, 548, 550–551
Water sources, and plant location, 13
investigation of, 14
Water-tube boilers, 563
Water transportation and plant location, 12
Water usage of plumbing fixtures, 580
Waterways, 12
Wearing rings, advantages of, 272
for casing, 272
for impeller, 272
on centrifugal pumps, 271–272
Weather, data for construction projects, 26
effect on plant location, 16
effect on plant safety, 635

Weather-proofing cement, 483
Weather-proofing plastic, 482
Wedge gate valve, O.S. and Y., 416
Weight, of foundations, 614
 of towers, 614
Weighted overall heat transfer coefficient, 234
Weighted temperature difference, 234, 237
Welding, acetylene, 189
 arc, 188
 faulty, 189
 of vessels, 188
 of structural steel, 663
Welding efficiencies, allowable, 200
Welding fittings, 409
Welding neck flanges, 404
Welds, detecting faults in, 663
Wet-seal gas holder, 575
Weymouth's equation, 452
Williams plot for steam turbines, 367–368, 369
Wind, for design purposes, direction of, 27
 overturning moment produced by, 586
 pressure of, 613
Wind loads, 203, 628
Wind velocity, effect of, on heat loss through insulation, 495
 for design, 27
Wing values, 254–255
Wiring diagrams, 114
Wood-composite pile, 590
Wood, use of, in piles, 589
Work, calculation of, for compressors, 327, 346
  for pumps, 259, 278
  for steam turbines, 366
 for multistage compressor, 339
 per stage for centrifugal compressor, 338
Work of compression, based on ideal gas law, 329–330
 example calculation of, 345–346
 for actual gases, 342
Working capital, 54
Wound-rotor induction motors, 353

"Y" valves, 419, 420

Zeolite softening, 551